COMPRESSIBLE FLUID DYNAMICS with Personal Computer Applications

B. K. HODGE

Thermal and Fluid Dynamics Laboratory
Department of Mechanical Engineering
Mississippi State University

KEITH KOENIG

Department of Aerospace Engineering
Mississippi State University

Prentice Hall
Englewood Cliffs, New Jersey, 07632

Library of Congress Cataloging-in-Publication Data

Hodge, B. K.
Compressible fluid dynamics with personal computer applications /
B.K. Hodge, Keith Koenig.
p. cm.
Includes index.
ISBN 0-13-308552-X (cased)
1. Fluid dynamics—Data processing. 2. Microcomputers.
I. Koenig, Keith, . II. Title.
TA357.5.D37H63 1995
533'.28—dc20 94-46332
CIP

DEDICATION

To the wind beneath our wings: Gayle and Brenda
And to the lights of our eyes: Benjamin, Lauren, and Bryan

Acquisitions editor: Bill Stenquist
Production editor: Judy Winthrop
Cover designer: Wendy Alling Judy
Buyer: Lori Bulwin
Copy editor: Kristen Cassereau
Editorial assistant: Meg Weist

A Simon & Schuster Company
Englewood Cliffs, New Jersey 07632

Printed in the United States of America

10 9 8 7 6 5 4 3 2 1

ISBN 0-13-308552-X

Prentice-Hall International (UK) Limited, *London*
Prentice-Hall of Australia Pty. Limited, *Sydney*
Prentice-Hall Canada Inc., *Toronto*
Prentice-Hall Hispanoamericana, S.A., *Mexico*
Prentice-Hall of India Private Limited, *New Delhi*
Prentice-Hall of Japan, Inc., *Tokyo*
Simon & Schuster Asia Pte. Ltd., *Singapore*
Editora Prentice-Hall do Brasil, Ltda., *Rio de Janeiro*

Contents

Part IV
MULTIDIMENSIONAL COMPRESSIBLE FLOW

APPENDICES 569

Preface

This compressible flow textbook differs from other compressible flow textbooks in at least three significant ways: (1) the order of presentation of some compressible flow topics, (2) the extensive use of generalized flow concepts for one-dimensional flows, and (3) the integration of software elements into the narrative and problem solutions. We have used the approach developed in this textbook in both aerospace and mechanical engineering compressible flow courses and have been quite pleased with the results. All of the topics usually presented in an introductory compressible flow textbook are contained in this book, albeit in different order and perhaps in a different manner. In the order of topical appearance this book is divided into the following five parts: Part I Fundamental Principles, Part II Waves in Compressible Flow, Part III One-Dimensional Generalized and Simple Flows, Part IV Multi-Dimensional Flows, and Part V High Temperature Gas Dynamics.

The purpose of Chapter 1 and Part I is to establish the physical and thermodynamic bases for the study of compressible flow and to present some fundamental concepts of the flow of a compressible fluid. Chapter 1 builds from thermodynamic principles and conservation principles to compressible flow concepts.

Chapters 2, 3, and 4 and Part II represent the first order of topics different from most compressible flow books since wave processes are not usually the first topics considered. Shock waves and expansion waves are not only among the most interesting of compressible flow phenomena, they also unambiguously demonstrate the wave nature and the importance of waves in compressible flows. Moreover, with normal shock waves, oblique shock waves, and the Prandtl-Meyer process developed, subsequent developments of internal compressible flows, such as those associated with various nozzle and duct systems, are greatly simplified. Software that replaces the familiar tables and charts used for the wave processes is integrated into the narrative and example problems.

Part III, which encompasses Chapters 5, 6, 7, and 8, deals with the traditional one dimensional compressible flows. However, the mode of development is different from that of the usual compressible flow textbook. Most compressible flow textbooks develop sequentially and in detail each of the so-called simple, one-dimensional compressible flows [isentropic (area change), frictional (Fanno), heat addition or rejection (Rayleigh), and mass addition]. Generally, each of the simple flows are examined, the equations derived, and some applications considered. This traditional approach, while effective in presenting details of each simple flow, does not foster consideration of the relationship between the simple flows and does not lead naturally to consideration of combinations such as area change with friction or heat transfer.

Part III starts with Chapter 5 which contains a control volume analysis for generalized one-dimensional compressible flow that includes area change, friction, heat addition or rejection, and mass addition. From this analysis, a table of influence coefficients is derived, and from the table of influence coefficients, the usual closed-form expressions for the simple, one-dimensional compressible flows are obtained. This sequence relates the simple flows to each other and provides the necessary basis for the consideration of more generalized flows. Chapters 6 and 7 then use the simple, one-dimensional compressible flow equations derived in Chapter 5 to develop an understanding of and to present applications for each of the simple flows. Chapter 6 is concerned with isentropic flows and included nozzle flows as well as supersonic diffusers and inlets. Chapter 7 treats Fanno flow, Rayleigh flow, and simple mass addition. As with Part II, software that replaces the familiar tables and charts is integrated into the narrative and example problems. Part III closes with Chapter 8 which develops techniques for analyzing one-dimensional flows with combinations of area change, friction, heat transfer, and/or mass addition. Much of Chapter 8 is devoted to numerical approaches and details of the solutions for these more complex flows. Software elements are provided.

Part IV, Chapters 9 and 10, are rather traditional presentations of the derivations of the equations of inviscid, multi-dimensional compressible flow and the development of the method of characteristics (MOC). MOC unit processes are fully addressed. A MOC software element, suitable for use by students, is included and is discussed.

Part V, comprised of Chapter 11, is the last chapter and contains an introductory treatment of high temperature gas dynamics. Particular attention is paid to distinctions in the hierarchy of compressible flow descriptions beyond calorically perfect. A useful applications software element is provided for thermally perfect, but calorically imperfect gases. This software permits the routine assessment of the caloric imperfection assumption and can be utilized independently without development to assess the effects of caloric imperfection. The chapter and textbook closes with a general examination of chemically reacting flows and presents examples involving such flows.

Software elements are provided that can replace all of the usual tables and charts associated with calorically perfect compressible flows. None-the-less, tables and charts for a ratio of specific heats of 1.4 are provided in the appendices. A significant advantage offered by the software is that the value of the ratio of specific heats is input, so that gases with specific heats ratios different from 1.4 can be considered. The appendices also contain a discussion on friction factors and the standard atmosphere.

We started this project with the idea that a non-traditional approach to the development of the subject might prove advantageous and that the availability of the personal computer could be used to eliminate arithmetical drudgery and the table and charts lookups that have traditionaly been used in compressible flow. Flexibility is provided for a wider selection of gases since the ratio of specific heats can be input. Moreover, we have endeavored to keep an applications flavor throughout the textbook. We would be interested in receiving feedback on our approach and implementation.

We would like to thank the following reviewers: E. M. Sparrow, University of Minnesota; Frank Lu, University of Texas at Arlington; George S. Campbell, University of Connecticut; J. C. McDaniel, University of Virginia; W. E. Castro, Clemson University; Eugene Brown, Virginia Polytechnic Institute and State University.

B. K. Hodge
Keith Koenig

Part I

FUNDAMENTAL PRINCIPLES

Fundamental Concepts of Compressible Fluid Dynamics

1

1.1 INTRODUCTION

Concern for the behavior of fluids pervades nearly every scientific and engineering discipline. Processes ranging from the diffusion of ions across a biological membrane to the formation of galaxies involve fluid dynamics. The nature of the fluids involved, the physical scale of the phenomena, and the energy of the motions vary widely among these different fluid dynamic situations.

This text is devoted to one general category of fluid dynamics: *compressible fluid flow*. Often called *gas dynamics*, compressible fluid flow is concerned with relatively low-pressure gases that exhibit significant changes in density as a result of the flow. Within this category is a vast array of interesting and highly useful phenomena. We shall explore these phenomena with three goals in mind:

1. to understand the physics involved
2. to develop the analytical tools necessary to describe and predict the phenomena
3. to apply our understanding and tools to problems of practical interest.

1.2 THERMODYNAMIC CONSIDERATIONS

The Continuum Approach versus the Molecular Approach

Processes involving fluids can be described either from the macroscopic or phenomenological standpoint or from the microscopic or molecular standpoint. As Arpaci (1966) points out, error-free, exact solutions from both approaches must ultimately yield consistent results. The macroscopic approach treats the fluid as an infinitely divisible substance—a continuum. The microscopic approach models the molecular structure of the fluid and provides details of the fluid molecules' interactions.

The continuum assumption is the basis of both classical fluid mechanics and classical thermodynamics. Since each fluid property in a continuum has a unique value at each point in space, all fluid properties are continuous functions of position and time within the framework of the continuum assumption. The microscopic approach considers a fluid as a collection of particles (molecules, atoms, ions, electrons) which are in random motion and which may be acted on by intermolecular forces. The microscopic approach, when applied to the physics of gases, leads to the well-developed kinetic theory of gases and to statistical thermodynamics. In general, the large number of particles (for example, 2×10^{19} molecules/cm^3 for air at room temperature) that must be considered for most cases of practical interest precludes the routine use of the microscopic approach for engineering calculations. The number density of particles in a dilute gas is given by Anderson (1989) as

$$n = P/kT \tag{1-1}$$

where n is the number of particles per unit volume, P is the pressure, T is the temperature, and k is the Boltzmann constant, 1.38×10^{-23} J/K-molecule.

The kinetic theory of gases and the subject of statistical thermodynamics provide information that can be used in the macroscopic formulation. Statistical thermodynamics is especially useful since it permits the evaluation of thermodynamic properties that can be used in classical thermodynamics. In Chapter 11, which deals with high-temperature flows, the properties of air at elevated temperatures are based on the results of statistical thermodynamics. The kinetic theory of gases is concerned with the behavior of gases that are dominated by translational motion. Textbooks by Kennard (1938), Jeans (1940), and Present (1958) are generally recognized as classic presentations of the kinetic theory of gases. Statistical thermodynamics considers the more general situation of a gas with translational, rotational, vibrational, and electronic energies. Since these effects, which are examined in more detail in Chapter 11, become increasingly important as the temperature is elevated, we shall find the results from statistical thermodynamics to be very useful for classical thermodynamic applications. Lee et al. (1973), Zemansky (1957), and Vincenti and Kruger (1965) are well-respected references on the subject.

Two techniques are available for fluid dynamic description: the Eulerian and the Lagrangian. The Eulerian description is often associated with the macroscopic formulation and the Lagrangian description with the microscopic formulation, but either descriptive technique can be used with either formulation. The Eulerian approach seeks to make quantitative statements about what occurs in a selected volume in space—a control volume—by considering the transport of mass, momentum, and energy across the boundaries of the control volume. One of the purposes of this chapter is to develop such quantitative statements for the Eulerian description. The Lagrangian approach seeks to make quantitative statements about what occurs in a flow by considering the trajectories of the fluid material. The Lagrangian description is thus associated with tracking identifiable fluid material through the flow field. The distinction between the two approaches is illustrated schematically in Figure 1-1.

In this text we shall follow the Eulerian description as applied to the macroscopic approach. Nevertheless, an initial examination of the microscopic approach is useful because it will permit us to define a limit on the continuum assumption and it will also provide insights and alternate interpretations of some concepts from the macroscopic approach.

An Examination of the Microscopic Approach

The microscopic approach is an appealing technique for fluid mechanics since it is based on conservation of mass, momentum, and energy as applied to a system of discrete particles: an approach with which most engineers feel comfortable. The kinetic theory of gases models a gas as being composed of a large number of particles whose volumes are negligible. These particles are in motion and undergo elastic collisions with each other and with the walls of any confining container. The particles are assumed not to attract each other; thus, intermolecular binding forces are ignored. The average distance between collisions, the *mean free path*, for air at standard temperature and pressure is on the order of 10^{-7} m (Liepmann and Roshko, 1957). The collision frequency for a given molecule is of the order of 10^{10} per second. Quantitative descriptions of the flow are interpretations of the detailed microscopic

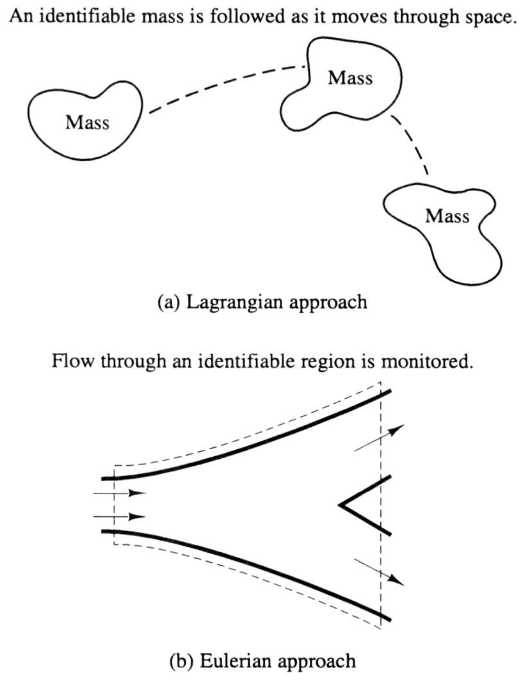

Figure 1-1 Lagrangian and Eulerian Descriptions.

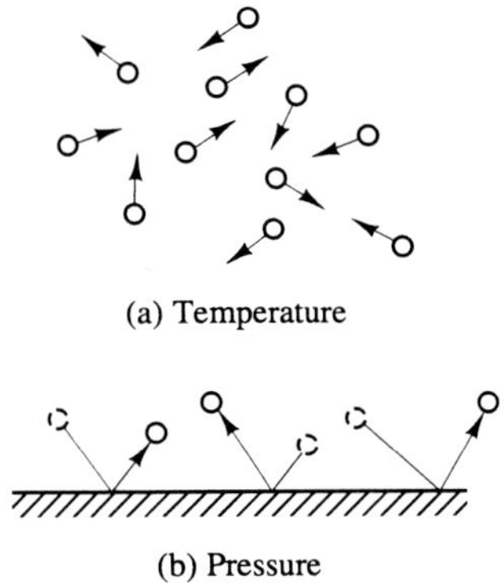

Figure 1-2 Random Molecular Motion for Microscopic Description Interpretation

solutions. The molecular motion, as in Figure 1-2, gives rise to physical properties such as temperature and pressure.

The temperature, T, in the kinetic theory of gases interpretation is a measure of the average gas velocity due to random molecular motions (as illustrated in Figure 1-2a); the more energetic the motions, the higher the temperature. For air at room temperature, the mean velocity is of the order of 450 m/s (Liepmann and Roshko, 1957). The mean translational kinetic energy, mc^2, where c is the mean molecular velocity and m the molecular mass, is proportional to the absolute temperature. The usual expression is

$$mc^2/2 = 3kT/2, \tag{1-2a}$$

where k is the Boltzmann constant.

The pressure, P, is the force per unit area exerted on a surface, solid or fluid, from impacts on that surface of the random moving particles. The force per unit area represents the average rate of transfer of momentum across any unit area in the fluid. Pressure is thus a normal stress that always acts perpendicular to and toward the surface in question. For a particle number density of n, the usual expression is

$$P = nmc^2/3. \tag{1-2b}$$

The continuum hypothesis is valid as long as any length scale (i.e., any characteristic linear dimension pertinent to the flow field) of practical interest is large compared with the mean free path. The ratio of the mean free path, l, to any characteristic linear dimension, L, is a dimensionless parameter called the *Knudsen number*, Kn, and is defined as

$$Kn = l/L. \tag{1-3}$$

The continuum assumption is valid for flows in which $Kn < 0.01$. The Knudsen number is often used to delineate various flow regimes. The usual flow regime ranges are given in Table 1-1. Since the mean free path for air at standard conditions is on the order of 10^{-7} m, the continuum assumption is satisfied for air at standard conditions for virtually any practical-sized system or component. However, the decrease in density (and mean free path) with altitude means that the continuum assumption for a given length scale system will eventually break down. As an example, consider a system with a characteristic length of 10 m. The Knudsen number would be 10^{-8} at sea level—well into the continuum regime. But at an altitude of 131,000 m (U.S. Standard Atmosphere, 1976) the mean free path is 1.0 m. The Knudsen number would greatly exceed 10^{-2} m, and the continuum assumption would no longer be valid. For our purposes in this book we shall always be interested in the continuum regime. The *free molecular regime* possesses such large mean free paths in comparison with any characteristic length of the body that particle interactions are unimportant since intermolecular collisions become infrequent. The *slip flow regime* is so named because the flow at the surface does not assume the velocity of the surface but is said to "slip" over the surface with a finite velocity with respect to the surface. Transition flow bridges the gap between the free molecular and slip flow regimes. For any regime other than the continuum, the discrete particular nature of the gas must be considered.

TABLE 1-1 KNUDSEN LIMITS FOR VARIOUS FLOW REGIMES

Knudsen number	Flow regime
>3.0	Free molecular
0.1–3.0	Transition
0.01–0.1	Slip
<0.01	Continuum

Thermodynamic Properties of a Gas

Classical thermodynamics gives no considerations to the molecular structure of a fluid and focuses on a *system*, any three-dimensional region of space bounded by an arbitrary geometric surface. A *closed system* is always concerned with the same identifiable quantity of matter. Frequently in fluid mechanics a system is understood to be a closed system. An *open system* is a system in which mass crosses the system boundaries. In fluid mechanics an open system is often called a *control volume*, an identifiable or designated region in space. A control volume is bounded by *control surfaces*. All matter external to the system is called the *surroundings*. Control volume terminology is illustrated in Figure 1-3. Thermodynamics treats the behavior of the system and the interaction of the system with the surroundings in terms of readily observed or measured characteristics called *properties*.

The *state* of a system is characterized by the values of its properties. According to the *principle of state*, two independent properties are required to describe an equilibrium state of a pure, simple compressible substance. Substances composed of a single chemical species are called *pure*. Air, which is a nonreacting mixture of N_2 and O_2 plus traces of other molecules, can be treated as a pure substance for virtually all cases of interest in this textbook. "Simple" and "compressible" imply that the only work form is Pdv. The specification of two independent properties fixes

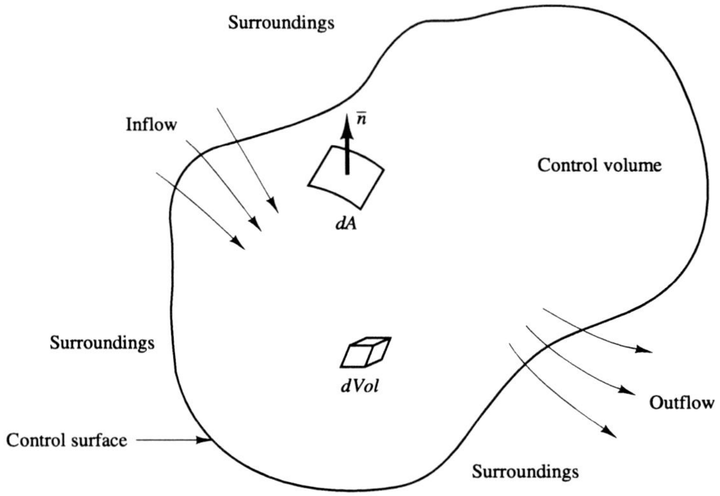

Figure 1-3 Control Volume Terminology

the values of all other properties. Thermodynamics is concerned only with the behavior of a system as it passes from one equilibrium state to another. Wark (1983) defines a system in thermodynamic equilibrium as one not capable of a finite, spontaneous change to another state without a finite change in the state of the surroundings.

Thermodynamic properties are used to describe quantitatively the state of a gas and are usually labeled as extensive or intensive. Properties are based on the state or condition of the system, not on how the system got to a particular state. *Extensive* properties depend on the mass of the system, while *intensive* properties are independent of the system's mass. In this text, except for pressure, P, an extensive property divided by the mass of a system will be denoted by a lowercase letter. For example, density, ρ, is the mass per unit volume and is an intensive property. "Compressible" means that the volume of a given mass of material can be altered significantly by compression or expansion processes. The inverse of density, the specific volume, v, will also be used.

1.3 THE LAWS OF THERMODYNAMICS

In addition to states, thermodynamics is also concerned with the processes involved with going from one state to another. Heat addition or rejection and work are the two principal processes. The relationships among the state variables and processes are described by the three fundamental laws of thermodynamics. These laws essentially serve to define the properties of temperature, energy, and entropy and form part of the foundation on which compressible fluid dynamics is based.

The zeroth law of thermodynamics postulates the existence of temperature by saying that two systems in contact such that heat can flow from one to the other are in equilibrium only if their temperatures are the same (Fowler and Guggenheim, 1952).

The first law of thermodynamics postulates the existence of energy as the state variable that changes if a system undergoes heat addition (or rejection) and/or work. This has a more obvious mathematical form:

$$de = dq - dw, \tag{1-4a}$$

where de is the total specific energy change of a system in response to an amount of heat, dq, transferred to the system and an amount of work, dw, done by the system, all quantities being per unit mass. In Eq. 1-4a energy is a property that depends only on the beginning and ending states, while both heat addition/rejection and work are processes that depend on the path traversed from the beginning to the ending states. We shall be concerned with only three types of energy:

1. *kinetic energy*, the directed energy of motion
2. *potential energy*, the energy associated with elevation above a datum
3. *internal energy*, the energy associated with the random motions of the molecules of a system.

In many situations of interest in gas dynamics, the kinetic and internal energies are the only important energies in a flow. In the absence of either potential or kinetic energy, only *internal energy*, u, remains, and Eq. 1-4a becomes

$$du = dq - dw. \tag{1-4b}$$

The differential work is usually regarded as the result of a differential change in the volume per unit mass of a system due to pressure. That is,

$$dw = Pdv, \tag{1-5}$$

where an increase in the volume occurs when work is done by the system. The *enthalpy* is defined as

$$h = u + Pv = u + P/\rho. \tag{1-6}$$

An alternate form of the first law of thermodynamics can be obtained by combining Eqs. 1-5 and 1-6 with Eq. 1-4b to produce

$$dh = dq + vdP. \tag{1-7}$$

Consideration of the second law of thermodynamics, which defines the property of entropy, usually starts with the concept of a *reversible* process. At any time a reversible process can be completely undone or reversed such that the original states of the system and the surroundings are obtained. Thus during a reversible process, the system remains continually in equilibrium. This means that the process must occur very slowly so that the entire system can adjust to each incremental change. Because the adjustment time in an actual gas is typically very short, this requirement is approximately met by many real processes. During an irreversible process, on the other hand, the system does not remain in equilibrium, and the process cannot be undone without affecting the surroundings of the system.

If 1 and 2 represent the initial and final states, respectively, of a system, we can imagine a reversible reference process that produces the same final pressure, temperature, and density for the given initial conditions. In general this reference process will involve heat transfer. The second law of thermodynamics then defines the state variable entropy, s, on a per-unit mass basis as

$$s_2 - s_1 = \int_1^2 \frac{dq_r}{T}, \tag{1-8}$$

where dq_r is the heat transferred during the reversible reference process and $s_2 - s_1$ is the entropy difference between the two states. For differential changes Eq. 1-8 becomes

$$Tds = dq_r. \tag{1-9}$$

These statements become particularly significant when it is realized that the heat transfer in the reference process is never less than the actual heat transfer dq_{act} in the real process. That is,

$$dq_r \geq dq_{act}, \tag{1-10}$$

where the equality holds if the actual process is reversible. Using Eq. 1-10 in Eq. 1-9 then gives

$$ds = dq_{act}/T + (dq_r - dq_{act})/T, \tag{1-11a}$$

or

$$ds = dq_{act}/T + ds_{irr}, \tag{1-11b}$$

where ds_{irr} is a positive quantity reflecting an increase in entropy due to irreversibility. Equation 1-10 says that a process can proceed only in a direction such that the actual entropy change is at least as great as the contribution due to the actual heat transfer. Any irreversibilities, spontaneous or dissipative, will increase entropy above the heat transfer contribution.

1.4 EQUATIONS OF STATE FOR A PERFECT GAS

The principle of state infers that there are specific relations, known as *equations of state*, which exist among the state variables. As Table 1.2 shows, equations of state can take many functional forms. Equations 1-12a–e each express one state variable in terms of two other state variables. A gas is thermally perfect if it obeys the *thermal equation of state*

$$P = RT/v = \rho RT \tag{1-13}$$

TABLE 1-2 EQUATIONS OF STATE FORMS AND NAMES

Functional form		Name
$P = P(v, T)$ or $P(\rho, T)$	(1-12a)	Thermal equation of state
$u = u(v, T)$ or $u(\rho, T)$	(1-12b)	Caloric equation of state
$s = s(v, T)$ or $s(\rho, T)$	(1-12c)	
$u = u(s, v)$ or $u(s, \rho)$	(1-12d)	Canonical equation of state
$h = h(s, P)$	(1-12e)	Canonical equation of state

where R is the *gas constant* and has a particular value for each gas. The value of R is determined by the molecular weight, MW, of the gas through

$$R = \overline{R}/MW, \tag{1-14}$$

where $\overline{R}$ is the universal gas constant, a number that is the same for all perfect gases. In several common sets of units, $\overline{R}$ is 8.3135 kJ/kg-mol K, or 1545 ft-lbf/lbm-mol R or 1.9842 Btu/lbm-mol R. The thermal equation of state, Eq. 1-13, is an accurate representation for gases and mixtures of gases except for conditions at high pressure and/or low temperature. For example, a gas at 15,000 psia and 100 R would not be accurately described by the thermal equation of state, and a more involved expression would be required. Additional discussion of the limitations of the thermal equation of state are presented in Chapter 11. Table 1.3 contains gas constants, molecular weights, and other thermodynamic constants for several common gases, including air. Despite its simplicity and because of its relative accuracy for engineering calculations, Eq. 1-13 will be the equation of state for virtually our entire study.

The *specific internal energy* for a normally perfect gas is a function of temperature only. That is,

$$u = u(T). \tag{1-15}$$

Enthalpy obeys a similar relation:

$$h = u + Pv = u(T) + RT = h(T). \tag{1-16}$$

Specific heat, c, is the quantity of heat per unit mass necessary to raise the temperature of a system by one degree. Thus,

$$c = dq/dT. \tag{1-17}$$

The amount of heat added depends on the actual process; two processes—constant volume and constant pressure—are of particular interest. For these cases Eq. 1-17 becomes

$$c_v = (\partial q/\partial T)_v \tag{1-18a}$$

and

$$c_p = (\partial q/\partial T)_p, \tag{1-18b}$$

TABLE 1-3 THERMODYNAMICS CONSTANTS FOR SEVERAL COMMON GASES

Gas	Chemical formula	Molecular mass	Gas constant R (kJ/kg K)	c_p (kJ/kg K)	$\gamma = c_p/c_v$
Acetylene	C_2H_2	26.038	0.3193	1.711	1.23
Air		28.967	0.2870	1.004	1.40
Ammonia	NH_3	17.030	0.4882	2.092	1.30
Argon	Ar	39.948	0.2081	0.519	1.67
Butane	C_4H_{10}	58.124	0.1430	1.674	1.09
Carbon dioxide	CO_2	44.009	0.1889	0.845	1.30
Carbon monoxide	CO	28.010	0.2968	1.042	1.40
Ethane	C_2H_6	30.070	0.2765	1.766	1.18
Ethylene	C_2H_4	28.054	0.2964	1.720	1.21
Helium	He	4.003	2.0770	5.200	1.67
Hydrogen	H_2	2.016	4.1240	14.350	1.40
Methane	CH_4	16.043	0.5182	2.223	1.32
Nitrogen	N_2	28.014	0.2968	1.038	1.40
Oxygen	O_2	31.998	0.2598	0.916	1.40
Propane	C_3H_8	44.097	0.1885	1.690	1.12
Water vapor	H_2O	18.015	0.4615	1.866	1.33

Adapted from John (1984).

which apply to any substance and where the subscripts denote the constant property. Specializing to a thermally perfect gas, for which internal energy is only a function of temperature, then gives

$$c_v = (\partial u/\partial T)_v, \tag{1-19a}$$

and in a similar fashion

$$c_p = (\partial h/\partial T)_p. \tag{1-19b}$$

A consequence of Eq. 1-16 is

$$R = c_p - c_v. \tag{1-19c}$$

It is also convenient to define the ratio of specific heats, $c_p/c_v = \gamma$. We then may write $c_v = R/(\gamma - 1)$ and $c_p = \gamma R/(\gamma - 1)$.

Equations 1-19 can be integrated to give

$$u(T) = \int c_v\,dT \tag{1-20a}$$

and

$$h(T) = \int c_p\,dT. \tag{1-20b}$$

A special case is the *calorically perfect gas* for which c_v and c_p are constant. A gas that is both thermally and calorically perfect is sometimes called a *perfect gas*. Air approximates a thermally and calorically perfect gas over a broad temperature and pressure range. For calorically perfect gases, Eqs. 1-20 reduce to

$$u(T) = c_v T + \text{constant} \tag{1-21a}$$

and

$$h(T) = c_p T + \text{constant}. \tag{1-21b}$$

The constants in these last expressions depend on the definitions of u and h at absolute zero temperature. It is common engineering practice to treat the constants as zero. Although this is not strictly correct, we are usually concerned just with differences in u and h between different states of a nonreacting system. In these situations the constants subtract out regardless of their values, and thus the assumption of zero constants is acceptable. If the gas is undergoing chemical change, however, these constants must be included.

Some idea of the validity of the calorically perfect gas assumption can be obtained from the data in Table 1-4, which presents for air the temperature variation, at constant pressure of 1 atmosphere (atm), of c_p/R and γ. Over much of the temperature range of usual interest, the variations in c_p/R and γ are relatively small. These thermodynamics constants are also nearly invariant with pressure. Hence, in terms of engineering accuracy, the assumption of a calorically perfect gas is acceptable over wide temperature and pressure ranges. Chapter 11 deals extensively with thermally perfect gases with nonconstant specific heats. Table 1-5 provides the thermodynamic constants for air in several different sets of units, including the International System of Units (SI) and English Engineering (EE).

TABLE 1-4 THERMODYNAMIC PROPERTIES OF AIR[a] at 1 ATM PRESSURE AND VARIOUS TEMPERATURES

$T(K)$	c_p/R	γ	$(h - h_{ref})$ kJ/kg	$\frac{s - s_{ref}}{R}$
200	3.506	1.406	199.7	22.497
300	3.506	1.402	300.2	23.917
400	3.533	1.396	401.2	24.929
500	3.588	1.387	503.3	25.723
600	3.663	1.376	607.4	26.383
700	3.746	1.365	713.7	26.954
800	3.828	1.354	822.4	27.460
900	3.906	1.345	933.4	27.915
1000	3.979	1.336	1046.6	28.330
1200	4.109	1.322	1278.8	29.068
1400	4.230	1.310	1518.2	29.711
1600	4.352	1.299	1764.5	30.284
1800	4.487	1.288	2018.2	30.804
2000	4.662	1.274	2280.6	31.284

[a] R for air = 0.2870 kJ/kg K

Adapted from John (1984).

TABLE 1-5 PERFECT GAS THERMODYNAMIC CONSTANTS FOR AIR

Symbol	$\frac{\text{ft}^2}{\text{sec}^2\, R}$	$\frac{\text{ft-lb}}{\text{slug}\, R}$	$\frac{\text{ft-lbf}}{\text{lbm}\, R}$	$\frac{\text{Btu}}{\text{lbm}\, R}$	$\frac{\text{J}}{\text{kg K}}$	$\frac{\text{kJ}}{\text{kg K}}$
R	1716.5	1716.5	53.35	0.0685	287.04	0.28704
c_p	6007.8	6007.8	186.70	0.2400	1003.50	1.00350
c_v	4291.3	4291.3	133.40	0.1714	716.80	0.71680

Ratio of specific heats: $\gamma = 1.4$

1.5 CHANGES IN ENTROPY AND ISENTROPIC RELATIONS

The first law of thermodynamics for a reversible process becomes $dq_r = dh - vdP$. Assuming a perfect gas and using Eqs. 1-13 and 1-19b, we have

$$dq_r = c_p\,dT - RT\,dP/P \tag{1-23a}$$

and for a reversible process

$$Tds = c_p\,dT - RT\,dP/P. \tag{1-23b}$$

Although derived for a reversible process, Eq. 1-23b is also valid for irreversible processes since only state variables are involved. Integrating yields

$$s_2 - s_1 = \int_1^2 \frac{c_p\,dT}{T} - R\ln(P_2/P_1), \tag{1-24}$$

which applies to any thermally perfect gas. For a calorically perfect gas c_p is constant, and Eq. 1-24 thus reduces to

$$s_2 - s_1 = c_p \ln(T_2/T_1) - R\ln(P_2/P_1). \tag{1-25}$$

An important class of processes involves no change in entropy. Such processes are termed *isentropic*. By Eq. 1-11 isentropic processes involve no heat transfer; that is, they are *adiabatic*, as well as reversible. Although real processes are never truly isentropic, irreversible effects are often quite small. With the isentropic assumption, Eq. 1-25 leads to the following very important result:

$$T_2/T_1 = (P_2/P_1)^{(\gamma-1)/\gamma} = (\rho_2/\rho_1)^{(\gamma-1)}. \tag{1-26}$$

Equation 1-26 relates the changes in T, P, and ρ from one state of a system to another due to an isentropic process in a calorically perfect gas.

An alternative form of Eq. 1-26 is

$$Pv^\gamma = \text{constant}. \tag{1-27}$$

1.6 THE INTEGRAL CONSERVATION EQUATIONS

The laws of thermodynamics establish the existence of the properties of temperature, energy, and entropy and describe what external influences cause these properties to change. Additional relationships are required to define completely the system and the interactions between the system and its surroundings. These relationships arise from the fundamental notions that certain properties of a system are *conserved*: Either they do not change, or they change only in response to particular external influences. Engineers are familiar with conservation concepts that are applied to systems of fixed mass—a rigid body, for example. The purpose of this section is to develop conservation expressions for mass, momentum, and energy for a control volume using our understanding of conservation concepts for fixed-mass systems.

Conservation Concepts

The definition of a fixed-mass system—a closed system—actually constitutes the first conservation relation. Since such a system contains an identifiable mass or quantity of matter, by definition the system's mass cannot change. Thus the conservation of mass, or continuity equation, says

$$dm/dt = 0, \tag{1-28a}$$

where m is the system mass and d/dt is the time derivative. For consistency with the relations to follow, the system mass can be thought of as the integral over all the elemental masses, dm, within the system. We then have for continuity

$$d/dt \int_{\text{system}} dm = 0. \tag{1-28b}$$

The second conservation equation is Newton's second law:

$$d(\text{system } \overline{\text{momentum}})/dt = \sum \text{applied } \overline{\text{forces}}, \tag{1-29a}$$

which states that the momentum of a system is conserved unless external forces act upon the system. Newton's second law is a vector equation since both momentum and force are vector quantities. Vector quantities will be indicated with an overbar. For a rigid body, momentum is simply $m\overline{V}$, mass times the vector velocity. For a deformable system, an integration over the system is necessary, that is:

$$\text{system } \overline{\text{momentum}} = \int_{\text{system}} \overline{V}\, dm, \tag{1-29b}$$

where $\overline{V}\, dm$ is the momentum of each elemental mass within the system. The momentum equation then becomes

$$d/dt \int_{\text{system}} \overline{V}\, dm = \sum \overline{\text{forces}}. \tag{1-29c}$$

We will need one further conservation law, and this is a generalization of the first law of thermodynamics, Eq. 1-4. Here we will find it more convenient to deal with rates of change. Considering internal energy and kinetic energy (such as $mV^2/2$ for a rigid body), the first law for a deformable system is

$$d/dt \int_{\text{system}} (u + \frac{1}{2}V^2)\, dm = \text{rate of heat addition to system} - \text{rate of work done by system.} \tag{1-30}$$

Generally in the study of fluids we are concerned with open systems such as those associated with a pipe, an engine, or the vicinity of a wing, through which a gas moves. Such a region is known as a *control volume.* We must convert the conservation expressions for a fixed-mass system, Eqs. 1-28, 1-29, and 1-30, to forms that are applicable to control volumes. Thompson (1972) provides a rigorous derivation of the control volume conservation equations.

If the conditions inside the control volume change with time, the process is said to be unsteady. A conserved quantity (mass, momentum, energy) within the control volume can change if that quantity is added to or discharged from the control volume by fluid entering or leaving the volume or if that quantity is accumulated in or removed from the control volume by other processes. Fluid entering the volume brings in new contributions to mass, momentum, and energy; departing fluid, in turn, removes these conserved quantities. The process of adding or removing conserved quantities by the motion of fluid across the boundaries of the control volume is known as *convection*, *flux*, or *transport*. The net amounts of mass, momentum, and energy can be changed by unequal inflow and outflow fluxes or by nonconvective processes. Consider, for example, the process of heating a closed container of some fluid. As heat is added, the internal energy of the fluid within the closed container increases.

We need to quantify these mechanisms of change. Let β represent a conserved quantity (mass, momentum, or energy) per unit volume, and let B represent the total amount of β in the control volume at any time. B is obtained by summing the contributions of each volume element, $\beta dVol$, over the entire control volume,

$$B = \int_{\text{cv}} \beta dVol. \tag{1-31a}$$

The time rate of change of B for a system (fixed-mass) that coincides with the control volume is

$$dB/dt = d/dt \int_{\text{cv}} \beta dVol \tag{1-31b}$$

and is the result of the unsteady and convective mechanisms acting within the control volume. That is,

$$d/dt \int_{\text{cv}} \beta dVol = \partial/\partial t \int_{\text{cv}} \beta dVol + \int_{\text{cs}} \beta \overline{V} \cdot \overline{dA}. \tag{1-31c}$$

The first term on the right-hand side of Eq. 1-31c is the unsteady rate of change of B within the control volume. The partial derivative signifies that this term depends explicitly on time. The second term on the right-hand side of Eq. 1-31c expresses the convective changes. An element of the control surface can be regarded as a vector having a magnitude of dA and a direction given by the unit outward normal to the surface, $\bar{n}$. Fluid crosses the control surface only if there is a component of fluid velocity perpendicular to the surface. The rate at which fluid volume crosses an element of the control surface is $\bar{V} \cdot \bar{n}dA$, and the rate at which the conserved quantity crosses is $\beta\bar{V} \cdot \bar{n}dA$. This rate is positive when fluid leaves the volume and is negative when fluid enters. The net flux is the result of the integral over the entire surface of the control volume.

The changes within the control volume are in response to the external influences acting on the control volume. B changes because of the external influences which act upon the control volume. This can be expressed as

$$\sum \text{external influences} = \partial/\partial t \int_{cv} \beta dVol + \int_{cs} \beta\bar{V} \cdot \overline{dA}, \tag{1-32}$$

where the left-hand side is from the appropriate conservation statement, Eqs. 1-28, 1-29, or 1-30 for mass, momentum, or energy, respectively. Equation 1-32 is the schematic form of the conservation equations as applied to a control volume and is often referred to as the *Reynolds transport equation.* It relates the conservation expressions available for particles to what is happening inside a control volume. We will now consider each particular equation individually.

Conservation of Mass

The schematic variable β is replaced by mass per unit volume, ρ. No external influences act to change mass, so the right-hand side is zero. Equation 1-32 becomes

$$0 = \partial/\partial t \int_{cv} \rho dVol + \int_{cs} \rho\bar{V} \cdot \overline{dA}. \tag{1-33}$$

Mass within a control volume can change only due to a net inflow to or outflow from the volume.

Example 1-1

An upright cylindrical tank with a base area of 50 m^2 is being filled at the rate of 10 m^3/s. Fluid is flowing out of the tank through a hole near the tank's bottom. The cross-sectional area of this exit is 0.25 m^2, and the velocity of the fluid at the exit is $V = \sqrt{19.6h}$, where h is the elevation in meters of the free surface of the liquid above the exit. Find h for steady-state conditions and develop an expression for $h(t)$.

Solution. The situation is illustrated schematically in Figure 1-4, which shows the control volume selected for the problem solution. Conservation of mass for the control volume is

$$\frac{\partial}{\partial t}\int_{cv} \rho dVol + \int_{cs} \rho\bar{V} \cdot \overline{dA} = 0.$$

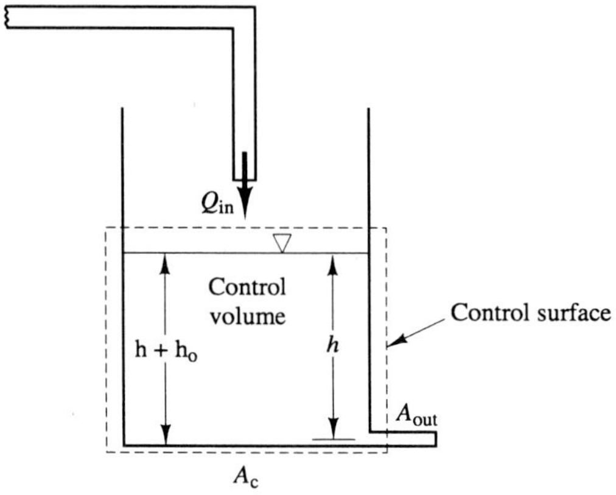

Figure 1-4 Unsteady Tank Problem Schematic

Time is the only independent variable, and so

$$\frac{\partial}{\partial t}\int_{cv} \rho dVol = \frac{d}{dt}\rho(h + h_o)A_c = \rho A_c\frac{dh}{dt}.$$

Fluid crosses the control volume at the inlet and the exit; hence

$$\int_{cv} \rho\overline{V}\cdot\overline{dA} = -\int_{A_{in}} \rho VdA + \int_{A_{out}} \rho VdA = -\rho Q_{in} + \rho A_{out}\sqrt{19.6h}.$$

Then

$$\rho A_c\frac{dh}{dt} = \rho Q_{in} - \rho A_{out}\sqrt{19.6h},$$

or

$$A_c\frac{dh}{dt} = Q_{in} - A_{out}\sqrt{19.6h}. \tag{1-34}$$

Steady-state conditions will be reached when $dh/dt = 0$ or when

$$0 = Q_{in} - A_{out}\sqrt{19.6h_{ss}},$$

for which

$$h_{ss} = \left(\frac{Q_{in}}{A_{out}}\right)^2\frac{1}{19.6} = 81.63 \text{ m}.$$

The relationship between h and t can be obtained by integrating Eq. 1-34. The initial condition is that $h = 0$ at $t = 0$. Then

$$\int_0^{h(t)} \frac{dh}{Q_{in} - A_{out}\sqrt{19.6h}} = \frac{1}{A_c}\int_0^t dt.$$

The solution to the preceding equation is

$$t = \frac{2A_c}{A_{out}\sqrt{19.6}}\left(\frac{Q_{in}}{A_{out}\sqrt{19.6}}\ln\frac{Q_{in}}{Q_{in} - A_{out}\sqrt{19.6h}} - \sqrt{h}\right).$$

For $Q_{in} = 10\ m^3/s$, $A_c = 50\ m^2$, and $A_{out} = 0.25\ m^2$, the solution becomes

$$t = 816.3 \ln \frac{10}{10 - 1.1068h^{1/2}} - 90.35h^{1/2}.$$

This expression is explicit in time, implicit for $h(t)$, and valid for $0 < h(t) < 81.63$ m. The behavior of $h(t)$ can be observed by evaluating t as a function of h and plotting the results. Figure 1-5 presents such a plot, where the exponential nature of $h(t)$ is evident.

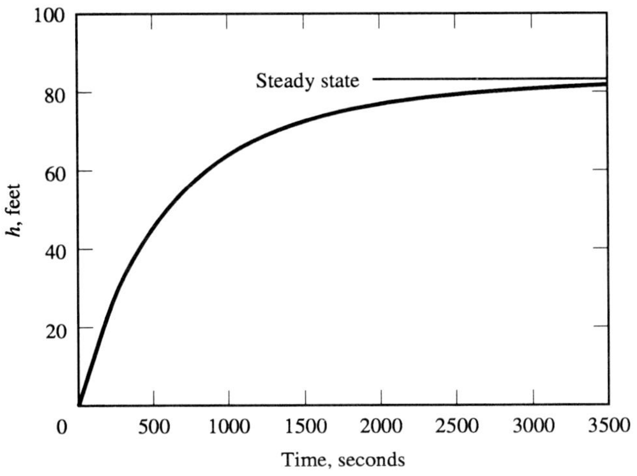

Figure 1-5 Liquid Level versus Time for Example 1-1

The sequence of the solution procedure is important:

1. develop a schematic of the system
2. choose a control volume including a sign convention
3. consider all stream entering and exiting the selected control volume
4. apply each term of Eq. 1-33 to the control volume
5. reduce the results of step 4 to the desired solution.

Conservation of Momentum

Here we replace the variable β by momentum per unit mass, which is $\rho\overline{V}$. The forces are the external influences that act on the volume. Equation 1-32 cast in terms of momentum becomes

$$\partial/\partial t \int_{cv} \rho\overline{V} dVol + \int_{cs} \rho\overline{V}\,\overline{V} \cdot \overline{dA} = \sum \overline{F}, \qquad (1\text{-}35)$$

where $\sum \overline{F}$ is usually identified as the sum of all forces acting on the control volume. An alternate statement is that $\sum \overline{F}$ represents the sum of all forces acting on the mass within the control volume. Equation 1-35 is restricted to control volumes that are fixed in inertial space; a fixed coordinate system as well as one that moves in a straight line with constant velocity are inertial. Fluid mechanics textbooks such as Shames

(1992) and Fox and McDonald (1992) treat conservation of momentum as applied to accelerating control volumes.

Two classes of forces can act on a control volume: surface and body. *Surface forces*, as the name implies, act on the surface of the volume and can be further subdivided into tangential and normal forces. Friction is the only tangential force generally important in gas flows.

Pressure is the only normal stress that is generally important in gas flows. Consider the control volume shown in Figure 1-6a. For the phenomena of interest in this textbook, the viscous normal stress contribution will be taken as negligible in comparison to the pressure; see White (1991) and Warsi (1993) for discussions on this assumption. The surroundings of this volume will impose pressures on the volume surface. Absolute pressure is always a compressive stress and always acts into a control volume. The force pressure exerts on an element of the surface is perpendicular to and toward the element and has a magnitude equal to the pressure times the element area. Expressing this gives

$$d\bar{F} = -P\bar{n}\,dA. \tag{1-36a}$$

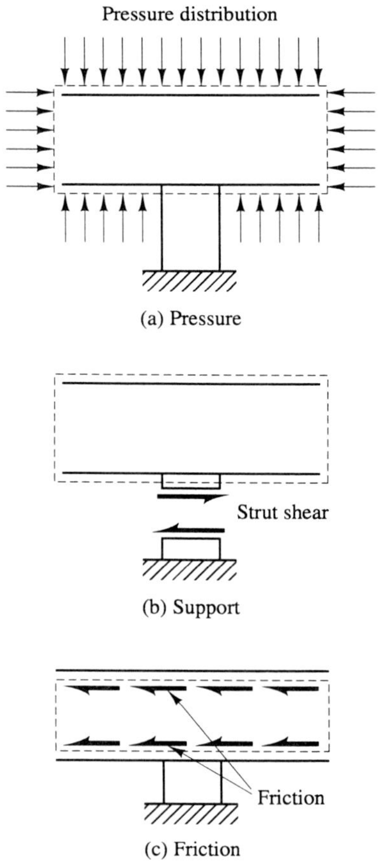

Figure 1-6 Common External Surface Forces

The negative sign in Eq. 1-36a indicates that the direction of action of pressure is opposite to the direction of the unit outward normal vector of the area element. The net pressure force acting on the control volume is the integral of the elemental forces over the entire control surface. That is,

$$\bar{F}_p = -\int_{cs} P\,\overline{dA}. \tag{1-36b}$$

The integral of a constant pressure acting over the entire control surface must always be zero, because an irregularly shaped object placed in a constant pressure environment does not accelerate. Equation 1-36b can therefore be written as

$$-\int_{cs} P\,\overline{dA} = -\int_{cs} (P - P_{ref})\overline{dA}, \tag{1-36c}$$

where P_{ref} is a constant reference pressure. An appropriate choice for P_{ref} will often simplify the computation of the net pressure force acting on a control volume.

Body forces act throughout a volume and depend on the mass of the material involved, and they are generally expressed in terms of force per unit mass. If $\bar{f}$ is the body force per unit mass, then the net body force acting on the fluid in the control volume is

$$\int_{cv} \bar{f}\rho\, dVol. \tag{1-37}$$

For neutrally charged gases, body forces are usually quite small and will typically be neglected in our discussions.

It often occurs that the surface of the control volume actually cuts through solid objects, such as the engine support stand in Figure 1-6b, or that the control surface lies against a solid surface as in Figure 1-6c. These solid surfaces may exert both tangential and normal stresses on the control surface, and these should be included in the summation of viscous or pressure forces.

Example 1-2

The rocket booster shown in Figure 1-7 has four nozzles, each canted at 10° with respect to the longitudinal axis. In the nozzle exit planes the exhaust gas velocity is 2000 m/s, the density is 0.30 kg/m^3, and the pressure is 50 kPa. If the total exit area of all four nozzles is 1 m^2, what is the thrust when the atmospheric pressure is 10 kPa?

Solution. First we make the assumption that the vehicle is not accelerating; hence, a reference frame attached to the vehicle, which is moving at constant velocity in a straight line, is inertial. Several control volume choices are possible; perhaps the simplest one to consider is a control volume consisting of the rocket engines, the fluid within the engines, and the vehicle structure that bears the thrust load. The control volume is indicated in Figure 1-8 as the rear portion of the vehicle. The force acting on the mass within the control volume is denoted as F. The reaction of the control volume on the remainder of the vehicle is equal but opposite in sense to F and is denoted as T. Since T is the reaction of the engines on the vehicle, it is usually called the thrust. Because the engines are exposed to atmospheric pressure, the pressure at every point on the

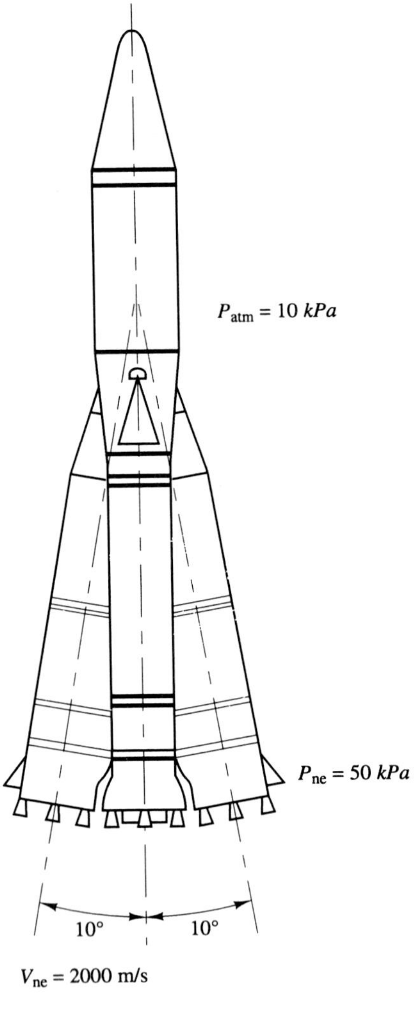

Figure 1-7 Rocket Booster with Canted Nozzles. *Adapted from Bertin.*

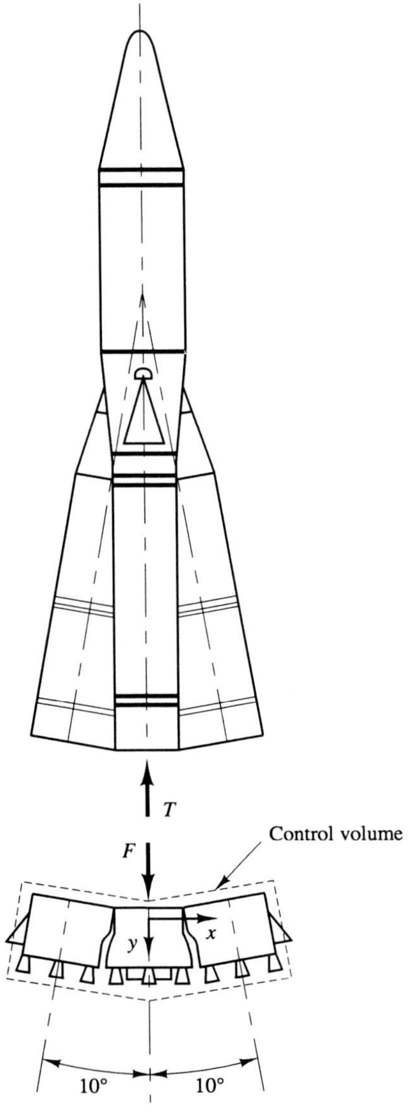

Figure 1-8 Control Volume for Example 1-2. *Adapted from Bertin.*

control surface, except the nozzle exit planes, will be atmospheric. The control surfaces of the control volume are coincident with the external surface of the rocket booster and the exit plane of the nozzles. The y-component of the momentum of the fuel and oxidizer as they are injected into the combustion chamber of the rocket engines is usually quite small in comparison to the momentum of the exhaust gases and will therefore be neglected. Within the stated assumptions, conservation of linear momentum in the y-direction as applied to the control volume becomes

$$\sum F_y = \frac{\partial}{\partial t}\int_{cv} \rho V_y \, dVol + \int_{cs} \rho V_y \overline{V} \cdot \overline{dA},$$

where V_y is the y-component of the exit velocity. The first term on the right-hand side is negligibly small since $V_y \approx 0$ for virtually all of the mass contained within the control volume.

The summation of the external applied forces becomes

$$\sum F_y = F - \int_{cs} P\overline{dA}\,|_y,$$

where the last term represents the y-component of the pressure forces exerted on the control surfaces of the control volume. And using Eq. 1-36a we derive

$$\sum F_y = F - \int_{cs} (P - P_\infty)\overline{dA}\,|_y.$$

Since $P = P_\infty$ except for the nozzle exit planes,

$$\begin{aligned}\sum F_y &= F - (P_{ne} - P_\infty)A_{ne}|_y \\ &= F - (P_{ne} - P_\infty)A_{ne}\cos 10^\circ,\end{aligned}$$

where the subscript "ne" represents the nozzle exit plane. The nozzles are canted 10° from the longitudinal axis so that the y-component of the resultant force is simply cos(10°) times the product of the exit pressure and the exit area for each nozzle.

Since the momentum of the entering propellants is taken as negligible, the term

$$\int_{cs} \rho V_y \overline{V} \cdot \overline{dA}$$

represents the flux of the y-momentum out of the control volume. Uniform flow in the nozzle exit planes gives

$$\begin{aligned}\int_{cs} \rho V_y \overline{V} \cdot \overline{dA} &= V_y \int_{cs} \rho \overline{V} \cdot \overline{dA} = V_y \dot{m} \\ &= V_{ne}\cos 10^\circ\,(\rho_{ne} A_{ne} V_{ne}),\end{aligned}$$

where, as with the pressure, the cos 10° term yields the y-component. Finally

$$F - A_{ne}(P_{ne} - P_\infty)\cos 10^\circ = V_{ne}\cos 10^\circ\,\rho_{ne} V_{ne} A_{ne}.$$

F becomes

$$\begin{aligned}F &= A_{ne}(P_{ne} - P_\infty)\cos 10^\circ + \rho_{ne} V_{ne}^2 A_{ne}\cos 10^\circ \\ &= 1\ \text{m}^2(50{,}000 - 10{,}000)\text{Pa}\cos 10^\circ \\ &\quad + 1\ \text{m}^2(2000\ \text{m/s})^2\ 0.3\text{kg/m}^3\cos 10^\circ \\ &= 39{,}392\text{N} + 1{,}181{,}700\text{N} \\ &= 1{,}221{,}162\text{N}.\end{aligned}$$

The thrust of the exhaust gases on the rocket booster is equal in magnitude, but opposite in sense, to F. The thrust is thus 1,221,162N, is directed in the negative y-direction, and is shown as T in Figure 1-8. Because the system is symmetrical, there is no resultant side force, the force in the x-direction.

Conservation of Energy

The variable β is now energy per unit volume and includes both the internal and kinetic energies of the gas, $\rho(u + V^2/2)$. Equation 1-32 for energy becomes

$$\partial/\partial t \int_{cv} \rho(u + \frac{1}{2}V^2)dVol + \int_{cs} \rho(u + \frac{1}{2}V^2)\bar{V} \cdot \overline{dA} = dq/dt - dw/dt,$$

where dq/dt represents the rate of heat addition and dw/dt the rate at which work is done.

The external forces that act on a control volume also, in general, do work on the volume. A force does work only if it is applied through a distance in the same direction as the force. The rate of doing work is the force times the fluid velocity at which the force moves, again in the same direction as the force. Mathematically we have

$$\text{rate of doing work} = \bar{F} \cdot \bar{V},$$

where the dot product ensures that the force and velocity are in the same direction for a nonzero contribution. Pressure acts normal to each surface, and at each control surface elemental area the work rate contribution is $-P\bar{V} \cdot \bar{n}dA$. The total rate of work associated with the control surface pressures is called the *flow work* and is the integral of the above expression taken over the control surface of the control volume:

$$-\int_{cs} P\bar{V} \cdot \overline{dA} = -\int_{cs} \frac{P}{\rho} \rho \bar{V} \cdot \overline{dA}. \tag{1-39}$$

The work rate term, dw/dt, is composed of two contributions: the flow work and the shaft work. The flow work is associated with the movement of mass into and out of the control volume. *Shaft work*, w_s, is equivalent to work done by a propeller immersed in the fluid and driven by a shaft that passes through the control volume.

The complete energy equation can now be assembled. The schematic form of the conservation equation, Eq. 1-32, thus becomes

$$\partial/\partial t \int_{cv} \rho(u + \frac{1}{2}V^2)dVol + \int_{cs} \rho(u + 1/2V^2)\bar{V} \cdot \overline{dA}$$
$$= dq/dt - \int_{cv} \frac{P}{\rho} \rho \bar{V} \cdot \overline{dA} - dw_s/dt. \tag{1-40a}$$

The last integral term in Eq. 1-40a is the flow work. That term can be combined with the convection term in Eq. 1-40a to yield

$$\partial/\partial t \int_{cv} \rho(u + \frac{1}{2}V^2)dVol + \int_{cs} \rho(u + P/\rho + 1/2V^2)\bar{V} \cdot \overline{dA}$$
$$= dq/dt - dw_s/dt. \tag{1-40b}$$

But we have already identified $u + P/\rho = u + Pv$ as the enthalpy, and so

$$\partial/\partial t \int_{cv} \rho(u + \frac{1}{2}V^2)dVol + \int_{cs} \rho(h + 1/2V^2)\bar{V} \cdot \overline{dA} = dq/dt - dw_s/dt. \tag{1-40c}$$

Example 1-3

Air enters a turbine at 1400 R and 150 ft/s. Measurements at the turbine exit show an exit velocity of 300 ft/s and an exit temperature of 800 R. If 2630 Btu/hr are lost from the turbine because of heat transfer, find the turbine output under steady conditions for flow rate of 1300 lbm/hr.

Solution. Consider the control volume, shown in Figure 1-9, as containing the turbine.

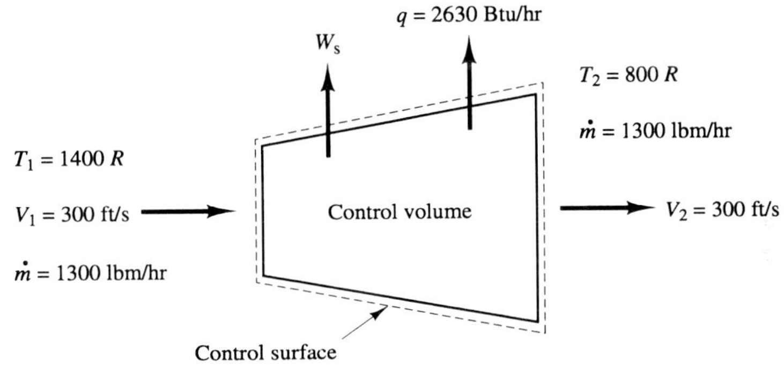

Figure 1-9 Example 1-3 Schematic and Control Volume

The energy equation is

$$\frac{dq}{dt} - \frac{dw_s}{dt} = \frac{\partial}{\partial t}\int_{cv} \rho\left(u + \frac{1}{2}V^2\right)dVol + \int_{cs} \rho\left(u + \frac{P}{\rho} + \frac{V^2}{2}\right)\overline{V} \cdot \overline{dA}.$$

For steady-state flow,

$$\frac{\partial}{\partial t}\int_{cv} \rho\left(u + \frac{1}{2}V^2\right)dVol = 0.$$

Uniform flow at the inlet and exit of the turbine gives

$$\frac{dq}{dt} - \frac{dw_s}{dt} = -\dot{m}_1\left(h_1 + \frac{V_1^2}{2}\right) + \dot{m}_2\left(h_2 + \frac{V_2^2}{2}\right).$$

Negligible leakage corresponds to $\dot{m}_1 = \dot{m}_2 = \dot{m}$, and

$$-\frac{dw_s}{dt} = -\frac{dq}{dt} + \dot{m}\left(h_2 + \frac{V_2^2}{2} - h_1 - \frac{V_1^2}{2}\right)$$

$$-\frac{dw_s}{dt} = -\frac{dq}{dt} + \dot{m}\left(c_p(T_2 - T_1) + \frac{V_2^2 - V_1^2}{2}\right)$$

$$-\frac{dw_s}{dt} = 2630\text{ Btu/hr} + 1300\text{ lbm/hr}\left(0.24\text{ Btu/lbm }R(800 - 1400)R\right.$$

$$\left. + \frac{1}{2}(300^2 - 150^2)\text{ ft}^2/\text{s}^2\,\frac{\text{lbf} - \text{s}^2}{32.174\text{ ft-lbm}}\,\frac{\text{Btu}}{778\text{ ft-lbf}}\right)$$

$$= 2630\text{ Btu/hr} - 185{,}447\text{ Btu/hr}$$

$$= -182{,}817\text{ Btu/hr} = -71.83\text{ hp}.$$

The power extracted from the turbine shaft is 71.83 hp.

Example 1-4

Air passes through a constant area duct as shown in Figure 1-10. The flow is steady and properties are uniform at the entrance and exit. Determine the rate at which energy is being added to or removed from the flow by the duct system.

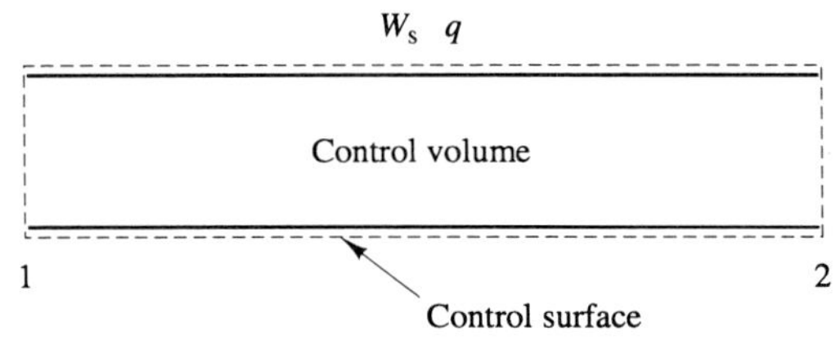

Figure 1-10 Example 1-4 Control Volume

Solution. The energy equation, Eq. 1.40a, will be applied to the control volume in Figure 1-10. Notice that no information is given concerning details of the flow inside the duct; only external flow details are known. Then

$$\text{rate of energy transfer} = \partial/\partial t \int_{cv} \rho(u + \frac{1}{2}V^2) dVol + \int_{cv} \rho(u + \frac{1}{2}V^2)\overline{V} \cdot \overline{dA}. \tag{1-41}$$

The steady condition eliminates the time derivative while uniform flow properties reduce the integrals to simple products. Flow crosses the control surface only at stations 1 and 2. The energy equation thus becomes

$$\text{rate of energy transfer} = [\rho(u + \frac{1}{2}V^2)(-VA)]_1 + [\rho(u + \frac{1}{2}V^2)(VA)]_2. \tag{1-42a}$$

Again, we use the perfect gas law to express the density in terms of pressure and temperature. Equation 1.42a is used to express the internal energy, u, in terms of temperature. With these substitutions we have

rate of energy transfer

$$= [P/(RT)(c_v T + \frac{1}{2}V^2)(-VA)]_1 + [P/(RT)(c_v T + \frac{1}{2}V^2)(VA)]_2. \tag{1-42b}$$

This last expression is the desired result and when it is evaluated with specific numerical values, its sign will indicate whether energy is being added or removed. A positive answer indicates that the duct is supplying energy to the flow.

1.7 THE SPEED OF SOUND

The creation and motion of sound are important concerns in compressible fluid dynamics since many of the phenomena associated with compressible flows involve sound or sound-like processes. Sound is a wave or signaling phenomenon in which the consequences of a small disturbance occurring within some region of a fluid are transmitted or propagated to other regions. A microscopic interpretation of the mechanism for transmission is the collision of gas molecules. A disturbance in a gas perturbs the gas molecules in the immediate neighborhood of the disturbance. These perturbed molecules then disturb their neighbors, and so on. In a macroscopic sense the disturbances involve extremely small—essentially infinitesimal—variations or fluctuations in all the thermodynamic properties of the gas. Sound waves can be moving or stationary with respect to any particular observer, but with respect to the fluid sound waves always move at a particular speed defined by the local thermodynamic state. This speed, the speed of sound, we shall now determine.

Consider a long tube as shown in Figure 1-11a in which the gas motion is constrained to be essentially in the x-direction. A movable piston is at one end of the tube. Initially the piston and gas are stationary. At some time the piston is impulsively accelerated to the right at a very small velocity, dV. The sudden acceleration of the piston acts as a disturbance and, in the absence of viscosity and wall effects, creates a planar sound wave. This sound wave propagates ahead of the piston at the speed a, "warning" the fluid of the approaching piston. The speed a is called the *speed of sound* since it represents the rate at which information about a disturbance, in this case the motion of the piston, is propagated into a medium at rest. Ahead of the wave the gas is at rest with conditions P, ρ, and T, while downstream of the wave the gas is moving at the piston speed, dV, and has slightly disturbed conditions $P + dP$, $\rho + d\rho$, and $T + dT$. This situation is depicted in Figure 1-11b. This wave is called *vanishingly weak* since only infinitesimal changes in fluid properties occur across the wave. Since only small temperature gradients are associated with the wave propagation, the process is effectively adiabatic. Moreover, if the flow is taken as inviscid, the fluid shearing stresses are zero and the process can be considered reversible.

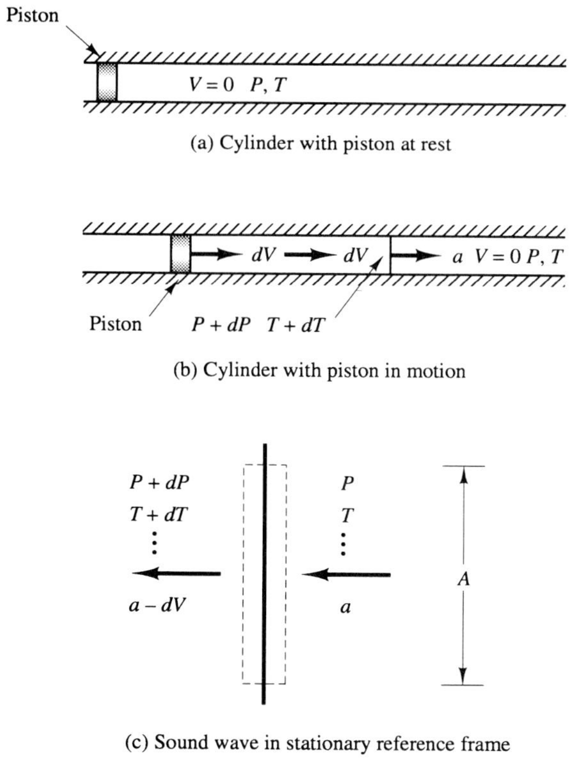

Figure 1-11 Speed of Sound Derivation Schematic

We will apply the integral conservation equations to a control volume drawn about a small segment, away from the walls, of the moving wave. As shown in Figure 1-11b, the process will be unsteady because of the motion of the sound wave. Generally, applications of the integral expressions are simpler if the process is steady. A transformation can be applied to the situation in Figure 1-11b to make the process steady. This is accomplished through a Galilean transformation in which a velocity of $-a$ is added to all velocities in the problem; see Figure 1-11c. The wave and the mass within the control volume are now in a steady state and gas flows into the volume from the left at speed a and leaves the volume at speed $a - dV$.

The flow is uniform and steady on both sides of the wave. As is indicated in Figure 1-11c, the selected control volume encloses the sound wave. For the purposes of this analysis we will consider the sound wave to be thin enough that the vertical sides of the control surface can be taken arbitrarily close together. The top and bottom sides of the control surface are parallel to the flow and no flow crosses them. The continuity equation applied to this situation is

$$-\rho aA + (\rho + d\rho)(a - dV)A = 0. \tag{1-43a}$$

Dividing through by the area and multiplying out gives

$$-\rho a + \rho a + d\rho a - \rho dV - d\rho dV = 0. \tag{1-43b}$$

The first two terms in this equation cancel. The last term, $d\rho dV$, is the product of two very small quantities and is negligibly small compared to the terms containing only one small quantity. This negligibly small, higher-order (here, second-order) term is thus dropped, leaving

$$d\rho a - \rho dV = 0. \tag{1-44}$$

The momentum equation for this situation (with no shear forces) is

$$(\rho a^2 A) - (\rho + d\rho)(a - dV)^2 A = -[PA - (P + dP)A], \tag{1-45}$$

which reduces to

$$\rho a dV = dP. \tag{1-46}$$

Using Eq. 1-44 to eliminate dV in Eq. 1-46 and solving for a^2 gives

$$a^2 = dP/d\rho. \tag{1-47}$$

With no heat transfer or external work, the energy equation, Eq. 1-40c, becomes

$$0 = -\rho A\left(h + \frac{a^2}{2}\right) + (\rho + d\rho)(a - dV)\left[h + dh + \frac{(a - dV)^2}{2}\right]A. \tag{1-48}$$

The mass flow rate entering the control volume is ρaA and is equal to the mass flow rate leaving, $(\rho + d\rho)(a - dV)A$. Thus

$$h + \frac{a^2}{2} = h + dh + \frac{a^2}{2} - adV + \frac{dVdV}{2}. \tag{1-49}$$

Equation 1-49 reduces to

$$dh = adV. \tag{1-50}$$

The result from the momentum equation, Eq. 1-47, when substituted into Eq. 1-50, yields

$$0 = dh - dP/\rho = dh - vdP. \tag{1-51}$$

Equations 1-7 and 1-9 can be combined as

$$Tds = dh - vdP. \tag{1-52}$$

A comparison of Eq. 1-51, valid for the propagation of a sound wave, and Eq. 1-52, a form of the first law of thermodynamics, reveals that the sound wave propagates in such fashion that $ds = 0$, that is, the process is isentropic. This is a formal confirmation of our previous assertion that a sound wave can be considered to be reversible in the thermodynamic sense. And so, the speed of sound is

$$a = \sqrt{(\partial P/\partial \rho)_s}. \tag{1-53}$$

For a calorically perfect gas the derivative in Eq. 1-53 can be evaluated with the aid of Eq. 1-26 and the thermal equation of state. The speed of sound can then be cast as

$$a = \sqrt{\gamma RT}. \tag{1-54a}$$

For air ($\gamma = 1.4$ and MW = 38.966 kg/kg-mol), Eq. 1-54a reduces to

$$a = 49.021\sqrt{T} \text{ ft/s } (T \text{ in } R); \tag{1-54b}$$

$$a = 20.046\sqrt{T} \text{ m/s } (T \text{ in } K). \tag{1-54c}$$

The speed of sound is a thermodynamic variable as are pressure and temperature, and, in fact, by Eqs. 1-54 it can be regarded as a measure of temperature.

The speed of sound is the speed at which information about an infinitesimal disturbance or pressure pulse is transmitted through a medium. Waves involving finite disturbances also exist and can likewise move or propagate through a medium. Such finite waves are called *shock waves* and are an important part of compressible flow. Shock waves will be treated extensively in the following chapters. As we will see, the speed of a flow relative to the speed of signal propagation is an extremely important aspect of that flow. The ratio of flow speed to sound speed is defined as M, the Mach number (named after Ernst Mach, an Austrian scientist-philosopher who briefly studied high-speed flows in the 1870s):

$$M = V/a. \tag{1-55}$$

One interpretation of this ratio is as a measure of the relative importance of directed and thermal energies in a flow. The kinetic energy per unit mass of a flow is V^2, while the thermal or random kinetic energy per unit mass, c^2, is proportional to T. Thus, the ratio of bulk kinetic energy to thermal energy $\sim V^2/c^2 \sim V/\sqrt{T} \sim V/a = M$.

Many other important consequences of the Mach number will be encountered during our study.

The Mach number is a very important dimensionless parameter in the flow of a calorically perfect gas. A glance through the remainder of this textbook will show that many expressions are cast in terms of the Mach number. The extensive tabular and graphical information normally utilized in compressible flow will also be presented with Mach number as the independent variable; see the appendices for examples.

If the Mach number is less than unity, the flow is called *subsonic*; if the Mach number is greater than unity, the flow is termed *supersonic*. The condition of the local velocity being equal to the local speed of sound is designated as *sonic*. The classification of a flow as subsonic, sonic, or supersonic is fundamental, and our understanding of compressible flow processes will be enhanced if we develop an appreciation of what each of these terms represents.

Example 1-5

Find the Mach number and classify as subsonic, sonic, or supersonic an aircraft flying at 100 ft/s, 1000 ft/s, and 3000 ft/s at an altitude of 20,000 ft.

Solution. The Mach number is

$$M_\infty = V_\infty / a_\infty.$$

The speed of sound, a_∞, depends on the local temperature. An abbreviated version of the U.S. Standard Atmosphere is given in Appendix J. At 20,000 ft, the temperature is 447 R. From Eq. 1-54a or 1-54b,

$$a_\infty = \sqrt{\gamma R T} = 49.021\sqrt{447} = 1036 \text{ ft/s},$$

and

$$M(100 \text{ ft/s}) = 100/1036 = 0.0965;$$

$$M(1000 \text{ ft/s}) = 1000/1036 = 0.965;$$

$$M(3000 \text{ ft/s}) = 3000/1036 = 2.896.$$

The aircraft is subsonic for $V_\infty = 100$ ft/s and $V_\infty = 1000$ ft/s, although the latter velocity is nearly sonic. An aircraft velocity of 3000 ft/s is supersonic.

The Mach number, by definition, is intimately associated with sound waves and the speed of sound. Consider a point source of sound moving subsonically at a velocity of V. Hence, V must be less than the speed of sound, a. Sound waves are continuously emitted as the source moves. With the waves traveling at the speed of sound, the wave fronts are spheres of radii na, with n the number of seconds since the sound was emitted and a the speed of sound. In two dimensions, the wave fronts are circles of radii na. The situation in two dimensions is schematically indicated in Figure 1-12. The figure illustrates the current location of the sound source and the sound waves emitted the past one, two, and three seconds from the source. Additionally, the locations of the sound source one, two, and three seconds in the past are indicated as "e," "f," and "g," respectively. Sound waves would be emitted

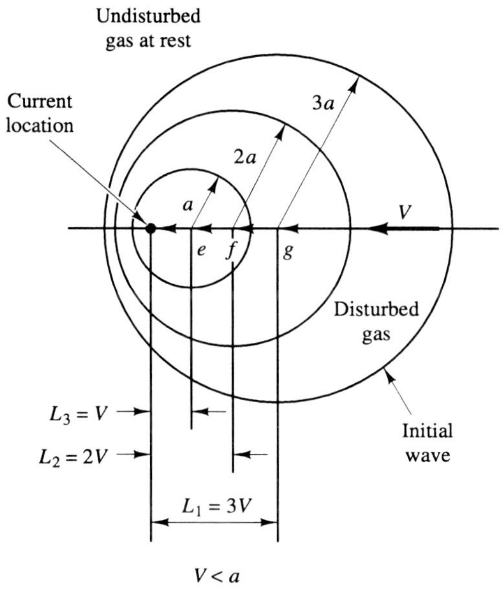

Figure 1-12 Sound Source Moving Subsonically

continuously, but for the sake of clarity, only the three wave fronts corresponding to the three discrete times are shown. Since the waves are traveling faster than the source, the sound waves move ahead of the sound source. The gas in front of the source moving at a subsonic speed thus has warning that the source is approaching.

Now consider the same situation, but with the sound source moving at exactly the speed of sound. Thus, V is equal to a. The sound wave front schematic corresponding to sonic motion of the source is depicted in Figure 1-13. This figure shows the wave fronts from sound waves emitted one, two, and three seconds previously, as well as the current and previous locations of the source. Because the source and the sound waves are moving at the same velocity, the wave fronts never outrun the sound source. The gas ahead of the source does not receive any warning of the source. A zone of silence exists ahead of the moving sound source, and a zone of action exists behind it. The demarcation between the two zones is perpendicular to the motion of the sound source.

Finally, examine the situation when the speed of the sound source is greater than the speed of sound. Figure 1-14 portrays the two-dimensional wave fronts for this, the supersonic case. As in the previous two figures, the wave fronts originating one, two, and three seconds before are illustrated along with the sound source locations. The zone of silence is now markedly increased from the sonic case; the zone of action is markedly decreased and is wedge-shaped. Until reaching the zone of action, an observer would be unaware that the moving sound source was present. The wave fronts form a wedge whose half-angle, μ, is given by

$$\mu = \arcsin\left(\frac{na}{nV}\right) = \arcsin\left(\frac{1}{V/a}\right) = \sin^{-1}\left(\frac{1}{M}\right). \tag{1-56}$$

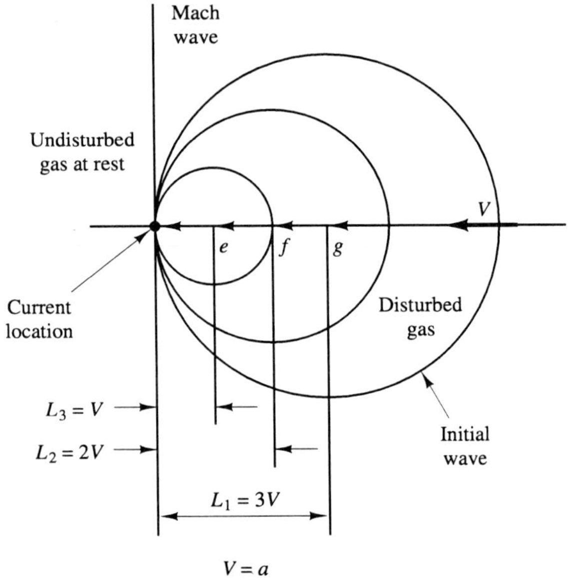

Figure 1-13 Sound Source Moving at Speed of Sound

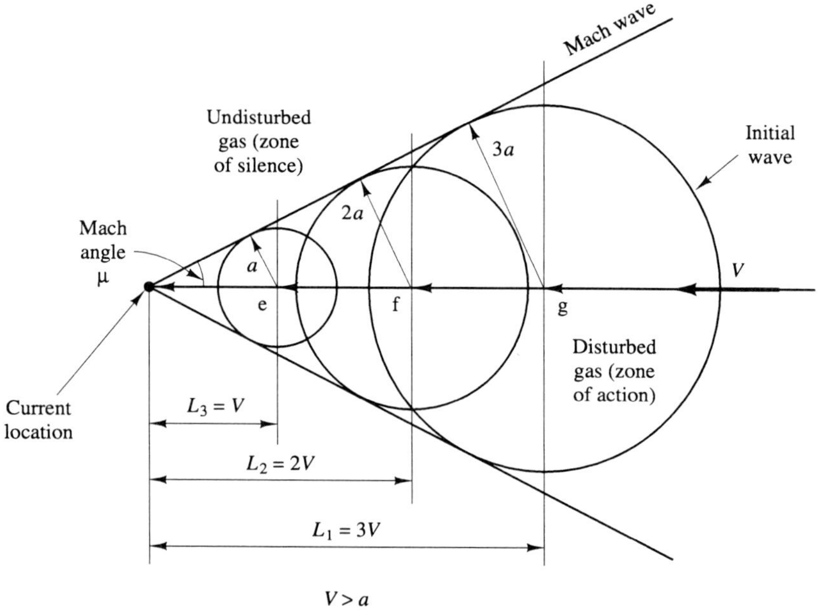

Figure 1-14 Sound Source Moving Supersonically

The angle μ is called the *Mach angle* and is defined only for supersonic flow (the arcsin of a quantity larger than unity is in the complex plane). The Mach angle as a function of the Mach number is tabulated in Appendix D and varies from 90 degrees for $M = 1$ to 0 degrees as M approaches ∞. In a supersonic stream, Mach angles define Mach waves. Mach waves are often viewed as sound waves across which only infinitesimal changes in fluid properties occur.

Contrast the behavior of the three cases. Actually, the sonic case is a limiting case of the supersonic case. The Mach angle is 90 degrees for the sonic case. *In supersonic flow, a disturbance cannot propagate upstream* since upstream lies in the zone of silence. This is a very important conclusion and is vital for the analysis and understanding of supersonic flow. Subsonic flow possesses no such limitation. The preferential, downstream-only, propagation of sound waves in a supersonic flow permits some aspects of supersonic flow analysis to be simpler than the analysis of those same aspects for subsonic flow with no such preferential propagation.

1.8 THE STEADY-STATE ENERGY EQUATION

Consider steady, adiabatic flow with negligible viscous effects and negligible body forces. Under these conditions the energy equation becomes

$$\int_{cs} \rho\left(u + \frac{1}{2}V^2\right)\overline{V} \cdot \overline{dA} = -\int_{cs} P\overline{V} \cdot \overline{dA}. \tag{1-57}$$

Moving the pressure term to the left-hand side of Eq. 1-57 and introducing enthalpy gives

$$\int_{cs} \rho\left(h + \frac{1}{2}V^2\right)\overline{V} \cdot \overline{dA} = 0. \tag{1-58}$$

Recall from fluid mechanics that streamlines are lines drawn in a flow field so that at any given time they are tangent to the velocity at every point in the flow field. No flow can cross a streamline since it is always tangent to the velocity. A streamtube is formed by all of the streamlines passing through a closed curve. Now consider a streamtube formed by all the streamlines passing through the closed curve that defines dA, as shown in Figure 1-15, and apply Eq. 1-58 to it. By definition, no flow crosses the long sides of the control volume. Furthermore, because of the differential cross-sectional area, the flow may be regarded as uniform over the inflow and outflow faces. The energy equation then becomes

$$\left[\rho\left(h + \frac{1}{2}V^2\right)(VdA)\right]_1 = \left[\rho\left(h + \frac{1}{2}V^2\right)(VdA)\right]_2. \tag{1-59}$$

The continuity equation applied to the streamtube yields

$$(\rho V dA)_1 = (\rho V dA)_2, \tag{1-60}$$

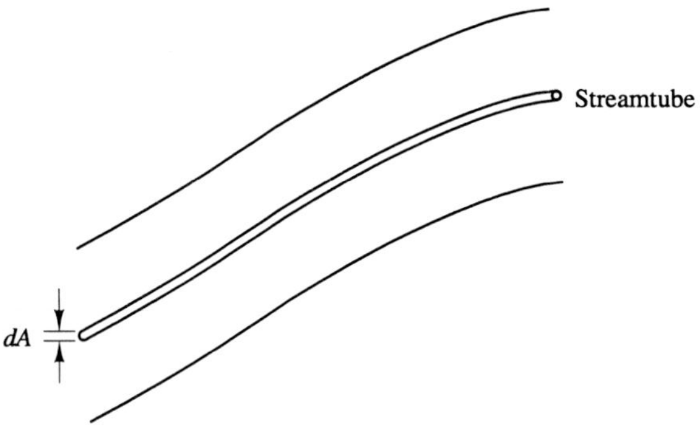

Figure 1-15 Streamtube Schematic

which, when substituted into the energy equation, gives

$$(h + \frac{1}{2}V^2)_1 = (h + \frac{1}{2}V^2)_2. \qquad (1\text{-}61)$$

This expression for the energy equation applies to any streamtube in a flow which satisfies the assumptions of steady, adiabatic flow with negligible viscous and body forces. It applies even in the limit as the streamtube reduces to a streamline. Since stations 1 and 2 are arbitrary, the energy equation can be generalized to

$$h + V^2/2 = \text{constant along a streamline}. \qquad (1\text{-}62)$$

The constant value is called the *stagnation enthalpy*, h_o. To see this, imagine a steady flow streamline that is adiabatically brought to rest at the stagnation point. Since $V = 0$ at the stagnation point, the value of enthalpy at that point must be the constant. That is,

$$h + V^2/2 = h_o. \qquad (1\text{-}63)$$

It is not necessary that the flow be physically brought to rest for the notion of stagnation enthalpy to be of use. At every point along a streamline there are particular values of h and V. Their combination in the form of $h + V^2/2$ will always be the same value, h_o. We can imagine that somewhere each streamline is brought to rest adiabatically so that $h = h_o$ occurs, but it is not necessary for this to actually happen.

For any steady, adiabatic flow h_o is a constant along a streamline. However, the value of h_o may change between streamlines. If the flow originates in a uniform region, h_o will have the same value on every streamline. This is most commonly the case, and unless specified we will take h_o as the same value everywhere.

For a calorically perfect gas we can use Eq. 1-21b to replace h with T in Eq. 1-63:

$$c_p T + V^2/2 = c_p T_o. \qquad (1\text{-}64)$$

Rearranging and using Eq. 1-20 to replace c_p gives

$$T_o = T[1 + (\gamma - 1)V^2/(2\gamma RT)]. \qquad (1\text{-}65)$$

Finally, recognizing that $a^2 = \gamma RT$ and $M = V/a$, we have

$$T_o/T = 1 + (\gamma - 1)M^2/2. \tag{1-66}$$

T_o is the *stagnation temperature*, the temperature that occurs when a flow is adiabatically brought to rest at a point. From Eq. 1-62 this temperature is a constant in a steady, adiabatic flow originating in a uniform region and is determined from the local values of V and T (Eq. 1-65) or M and T (Eq. 1-66). The stagnation temperature is an important reference temperature whether or not there is actually a stagnation point in the flow.

Traditionally, values of the thermodynamic properties of fluid measured by an instrument moving with the fluid are called *static properties*. If the instrument is moving with the fluid, then the relative velocity between the instrument and the fluid is zero—hence the term "static." A thermometer moving with the fluid would measure the local static temperature of the fluid. Although we have not indicated so, all references to thermodynamic properties up to this point have been to static properties, with the exception of the stagnation enthalpy and stagnation temperature. We adopt the convention that any property without the subscript o is a static property. Static properties are independent of the motion of any observers; the measurements must be made by instruments moving with the flow. A point in a flow field in which the flow is brought to rest is called a stagnation point, and properties at that point are called stagnation properties. Thus, in contrast to static properties, measurements of stagnation properties require that the flow be brought to rest with respect to the observer's instruments. Since the fluid must be brought to rest with respect to an observer, the motion of the fluid relative to the observer influences the stagnation values. This is reflected in the appearance of V in Eqs. 1-63 to 1-65 and M in Eq. 1-66.

Equation 1-64 can be developed into a useful expression for relating the speed of sound and velocity for a constant enthalpy, calorically perfect gas flow. The maximum kinetic energy, and thus the maximum velocity, associated with a constant enthalpy flow would occur if we envisioned that all the thermal energy ($c_p T$) of the flow is converted into kinetic energy. That is,

$$c_p T_o = \frac{V_{max}^2}{2}, \tag{1-67a}$$

from which

$$V_{max} = \sqrt{2 c_p T_o}. \tag{1-67b}$$

This represents the highest velocity that the gas could possess for a given stagnation temperature. V_{max} is a fictitious quantity in the sense that we would never measure it in an actual flow since the calorically perfect gas assumptions break down well before the temperature vanishes; nonetheless, it is a useful reference quantity, and it does indicate that the fluid velocity must be bounded for a given stagnation temperature or enthalpy. At a location associated with V_{max}, the local static temperature vanishes, and the local speed of sound becomes zero. Thus, the Mach number is unbounded for V_{max}, even though the velocity is finite. The unbounded Mach number occurs because the speed of sound in the denominator approaches zero, not because the velocity becomes unbounded.

Example 1-6

For air at a stagnation temperature of 1000 K, find

a. V_{max}
b. The static temperature, T, and velocity for a Mach number of 0.8
c. The Mach number and velocity for a static temperature of 800 K
d. The Mach number and static temperature for a velocity of 1000 m/s

Solution.

a. The thermodynamic constants for air are taken from Table 1-5:

$$\gamma = 1.4$$
$$R = 287 \text{ J/kg K}$$
$$c_p = 1004 \text{ J/kg K}.$$

Then using Eq. 1-67b,

$$V_{max} = \sqrt{2c_p T_o}$$
$$= \left(2\ 1004 \frac{\text{J}}{\text{kg K}}\ 1000 \text{ K} \frac{\text{kg-m}}{\text{N-s}^2} \frac{\text{N-m}}{\text{J}}\right)^{1/2}$$
$$= 1417 \text{ m/s}.$$

b. Equation 1-66 is a convenient form for dealing with temperatures and Mach numbers. Thus

$$\frac{T_o}{T} = 1 + \frac{\gamma - 1}{2} M^2 = 1 + \frac{1.4 - 1}{2}(0.8)^2,$$

for which

$$\frac{T_o}{T} = 1.128$$

$$T = \frac{T_o}{1.128} = \frac{1000}{1.128} = 886.5 \text{ K}.$$

The velocity is

$$V = Ma$$
$$= 20.046\sqrt{T}M.$$

where Eq. 1-54c is used for the speed of sound:

$$V = (0.8)(20.046)\sqrt{886.5} = 477.5 \text{ m/s}.$$

c. As in part (b) of this example, Eq. 1-66 is convenient and can be solved for M

$$M = \sqrt{\left(\frac{T_o}{T} - 1\right)\frac{2}{\gamma - 1}}$$

$$= \left[\left(\frac{1000}{800} - 1\right)\frac{2}{1.4 - 1}\right]^{1/2} = 1.118;$$

$$V = 20.046\sqrt{T}M$$

$$= 20.046\sqrt{800}\ 1.118 = 633.9 \text{ m/s}.$$

d. For situations involving velocity and temperature, Eq. 1-64 is useful:

$$c_p T + \frac{V^2}{2} = c_p T_o,$$

from which

$$T = \frac{1}{c_p}\left(c_p T_o - \frac{V^2}{2}\right)$$

$$= T_o - \frac{V^2}{2c_p}$$

$$= 1000 - \frac{1}{2}(1000 \text{ m/s})^2 \frac{\text{kg K}}{1004 \text{ J}} \frac{\text{N-s}^2}{\text{kg-m}} \frac{\text{J}}{\text{N-m}}$$

$$= 1000 - 498 = 502 \text{ K}.$$

The Mach number is

$$M = \frac{V}{a} = \frac{V}{20.046\sqrt{T}}$$

$$= \frac{1000}{20.046\sqrt{502}} = 2.227.$$

Although we shall eventually use tables or the computer software that comes with this book to solve the problem statements just presented, full dexterity with the steady-state energy equation is an absolute necessity. One important aspect of the various problem statements in this example is the proper definition of speed of sound for use in Mach number related calculations. The speed of sound must always be calculated using the value of the static temperature associated with velocity. Thus, in parts (a), (b), and (c) the speeds of sound are different since the static temperatures are different.

Using the definition of V_{max} from Example 1-6, we can rewrite Eq. 1-64 as

$$2c_p T_o = V_{max}^2 = 2c_p T + \frac{V^2}{2}. \tag{1-68}$$

Multiplying $2c_p T$ by $\gamma R/\gamma R$ and reducing changes Eq. 1-68 as follows:

$$V_{max}^2 = \frac{2}{\gamma - 1}a^2 + \frac{V^2}{2}. \tag{1-69}$$

However, this is the equation of an ellipse in terms of a and V. This ellipse, illustrated for $\gamma = 1.4$ in Figure 1-16, is called the *Prandtl velocity ellipse* and is a useful device for examining the relationship between velocity and speed of sound. The various regimes are indicated on the figure, and their usual Mach number ranges are delineated in Table 1-6.

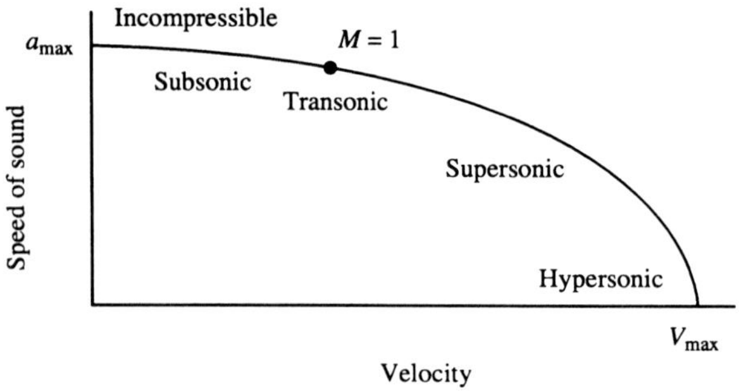

Figure 1-16 Prandtl Velocity Ellipse

TABLE 1-6 COMPRESSIBLE FLOW REGIMES

Regime	Mach number range
Incompressible (effectively)	0–0.3
Subsonic	0.3–0.8
Transonic	0.8–1.2
Supersonic	1.0–5.0
Hypersonic	>5.0

Unique configurations may exhibit characteristics of these various regimes at either higher or lower Mach numbers.

If the velocity is small (in the incompressible and subsonic regimes), the change in speed of sound with velocity is small and changes in Mach number occur primarily because of changes in velocity. Transonic flow is important because configurations in transonic flow usually possess regions of both subsonic and supersonic flows. The loci of the sonic locations, the *sonic line*, define the boundary between the subsonic and supersonic regions. As the Mach number moves from the transonic to the supersonic regime, the speed of sound decreases because of the decreasing static temperature (or enthalpy), and changes in Mach number result from both increases in velocity and decreases in speed of sound. Finally, in the hypersonic regime, the speed of sound exhibits large changes while the velocity exhibits small changes. Changes in hypersonic Mach number for a constant enthalpy flow then arise because of large decreases in the speed of sound for small increases in velocity.

1.9 THE ISENTROPIC FLOW EQUATIONS

Two important relations can be formed if we assume that the stagnation state is obtained through a process that is reversible as well as adiabatic. In this case the process is isentropic and Eq. 1-26 can be used to give

$$P_o/P = [1 + (\gamma - 1)M^2/2]^{\gamma/(\gamma-1)}; \tag{1-70}$$

$$\rho_o/\rho = [1 + (\gamma - 1)M^2/2]^{1/(\gamma-1)}. \tag{1-71}$$

Equations 1-66, 1-70, and 1-71 are referred to as the isentropic flow relations, although Eq. 1-66 is a form of the steady-state energy equation. The isentropic relations are tabulated for $\gamma = 1.4$ in Appendix A. The computer software available with this book also contains an isentropic flow option for use with arbitrary values of the ratio of specific heats.

Stagnation properties, denoted with the subscript o, are defined as the properties that would be attained at a point in the flow if the flow were isentropically brought to rest, or if we *imagined* that the flow were isentropically brought to rest, with respect to an observer. The isentropic flow relations then relate, at every point in the flow, the local static conditions and the local Mach number to the local stagnation conditions. Even in a flow that is not steady nor isentropic, Eqs. 1-66, 1-70, and 1-71 are valid at every point at any instant of time and serve to define the stagnation conditions. Equation 1-26 is valid at different points in a flow field only in regions where the flow is isentropic. The isentropic relations are used in two distinct situations: (1) to define the stagnation conditions at a point, and (2) to relate the properties at any two arbitrary points in an isentropic flow field. We shall make extensive use of the stagnation condition concept in the remainder of this book.

The change in entropy for a process involving a calorically perfect gas was derived earlier and presented as Eq. 1-25, which is repeated here for convenience:

$$s_2 - s_1 = c_p \ln(T_2/T_1) - R \ln(P_2/P_1).$$

The entropy change in this expression is given in terms of the static temperature and pressure ratios. If Eqs. 1-70 and 1-71 are used to form the quotients of T_2/T_1 and P_2/P_1, then Eq. 1-25 reduces to

$$s_2 - s_1 = c_p \ln[(T_{o_2}/T_{o_1})(P_{o_1}/P_{o_2})]. \tag{1-72}$$

This expression casts the entropy in terms of the stagnation temperature and stagnation pressure ratios. In a steady, isentropic flow originating in a uniform region, the stagnation properties are constant throughout the flow.

Example 1-7

A large tank, shown in Figure 1-17, supplies air to a supersonic wind tunnel. In the wind tunnel test section the flow conditions are to simulate flight at a Mach number of 2 and an altitude of 53,000 ft. Assuming steady, isentropic flow, what are the conditions in the tank?

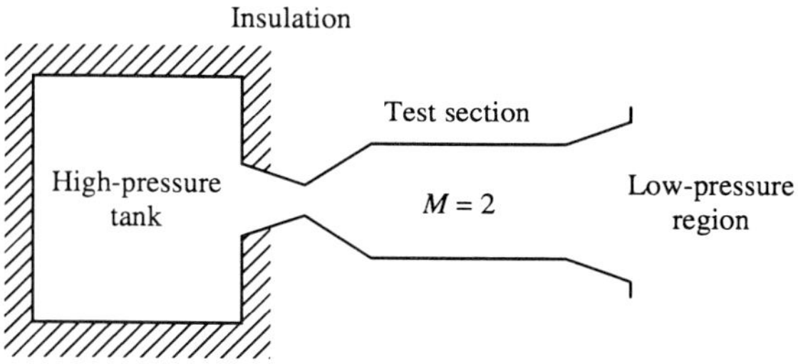

Figure 1-17 Schematic for Example 1-7

Solution. The large tank and the steady flow mean that the air is nearly at rest within the tank; thus, the tank conditions are essentially $M = 0$ with constant temperature and pressure. Since $M = 0$, the static and stagnation values are the same. As the air flows out of the tank and through the tunnel, however, the static properties will change although the stagnation values will not. The requirement that the flow simulate 53,000-ft altitude means that the local static properties are the same as the conditions at that altitude; these conditions are from Appendix J: $T = 390°R$ and $P = 0.1$ atm. We use these values and $M = 2$ in Eqs. 1-66 and 1-70 to find

$$T_{\text{tank}} = T_o = 1.8*T = 702 \text{ R} = 242 \text{ F};$$

$$P_{\text{tank}} = P_o = 0.7824 \text{ atm}.$$

When compared to sea-level air at standard temperature and pressure, the contents of the tank must be heated and depressurized to achieve the desired conditions. The low test section temperature (−70°F) requires that the tank air be thoroughly dried to avoid the formation of ice as the flow proceeds through the tunnel. The tank conditions are also nearly what would be experienced by the fuselage nose and wing leading edges of a flight vehicle where the flow stagnates. The high temperature may require special materials and insulation to prevent damage to the vehicle.

1.10 CONCLUSION

In this chapter we have examined concepts from thermodynamics and have developed the fundamentals needed for our study of compressible flow. Some of the nomenclature that will be used in our study of compressible flows has also emerged. Everything we do in this textbook will be structured on the foundations presented in this chapter. For example, the integral conservation equations will be the starting point for many of the developments that we shall pursue. We will find that an understanding of topics from both thermodynamics and fluid mechanics will greatly aid our study. The importance of sound waves and the local Mach number in compressible flows has also been introduced.

CHAPTER 1 REVIEW QUESTIONS

1. What is a continuum?
2. Contrast the Eulerian and Lagrangian descriptions.
3. Define and cite an example of an intensive property.
4. What is the principle of state?
5. An isentropic process is always
   - **a.** irreversible and adiabatic
   - **b.** reversible and adiabatic
   - **c.** frictionless and adiabatic
   - **d.** frictionless and irreversible
   - **e.** none of the above
6. What is flow work?
7. How do systems and control volumes differ?

8. Does $M \to \infty$ imply $V \to \infty$?
9. If $R = 62$ ft-lbf/lbm R and $c_p = 217$ ft-lbf/lbm R, find the ratio of specific heats.
10. Find V_{max} for air if the stagnation temperature is 100°C. What is the Mach number corresponding to V_{max}?
11. At a given temperature, how does the molecular weight of a gas affect its speed of sound?
12. What is the difference between static and stagnation properties?
13. How do the possible signal (information) propagation directions differ in subsonic and supersonic flow?
14. At the inlet to a duct the stagnation temperature is measured to be 1000 K. Further down the duct the stagnation temperature is measured to be 1200 K. Is the process between the two measuring points adiabatic? Is heat added or rejected? Justify your answer.
15. State in your own words the zeroth, first, and second laws of thermodynamics.
16. What is a Mach angle? What is a Mach wave?
17. How are stagnation properties defined? How are they determined?
18. What is a thermally perfect gas? What is a calorically perfect gas?

CHAPTER QUESTIONS

1. Steam enters a turbine with a velocity of 100 ft/s and an enthalpy of 1000 Btu/lbm. At the outlet, the velocity is 300 ft/s and the enthalpy is 800 Btu/lbm. The heat transfer from the turbine to the surroundings is 2000 Btu/hr. Determine the output of the turbine in horsepower if the flow rate is 1000 lbm/hr. How does this compare with the energy lost to the surroundings?
2. Air is flowing in a duct and at stations 1 and 2 the following are measured:

Station 1:	Station 2:
$V = 629$ ft/s	$T_o = 233$°F
$P = 10.0$ psia	$P_o = 10.5$ psia
$T = 200$°F	$V = 1300$ ft/s

   a. Find M, T_o, and P_o at station 1.
   b. Is the flow adiabatic? Why?
   c. Is the flow isentropic? Why?
   d. Find T and M at station 2.
3. Air flow in a duct has $P = 10$ psia, $T = 200$°F, and $V = 1000$ ft/s at a point where the area is 1 ft^2. If the flow is adiabatic, what are the fluid velocity, the duct area, and the speed of sound at a location where the static temperature is 100°F and the static pressure is 30 psia? What is V_{max} for this flow?
4. Air at 50°C and $M = 0.75$ is accelerated to $M = 2$. If the static temperature at $M = 2$ is 100°C, how much energy must be added to or rejected from the flow?
5. Nitrogen is flowing adiabatically with a stagnation enthalpy of 500 kJ/kg. Find T_o, V_{max}, and the static temperature for $M = 2.5$.

6. Conditions for developing flow in a circular pipe are given in Figure 1-18. At a distance L down the pipe, the velocity profile is fully developed and is parabolic as indicated. At station 2 the pressure is P_2 and is uniform across the pipe. What force does the fluid exert on the pipe?

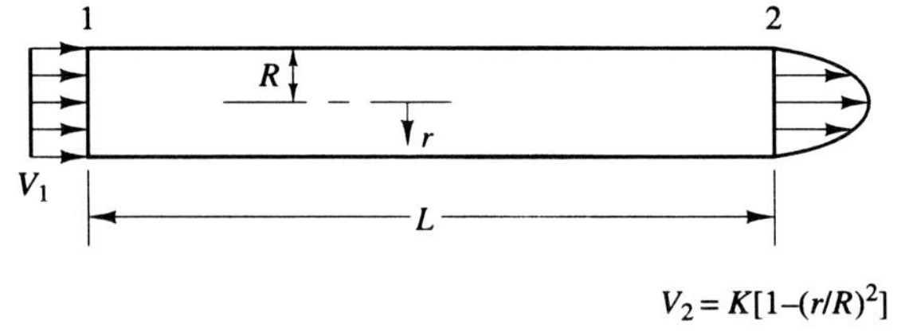

Figure 1-18

7. The dam shown in Figure 1-19 is 125-ft long and 27-ft high. During a flood, water flows over the top of the dam. Far upstream of the dam the water is 33-ft deep and flows at a uniform velocity of 8 ft/s. Far downstream of the dam the flow is 7-ft deep. Estimate the force on the dam. State all assumptions.

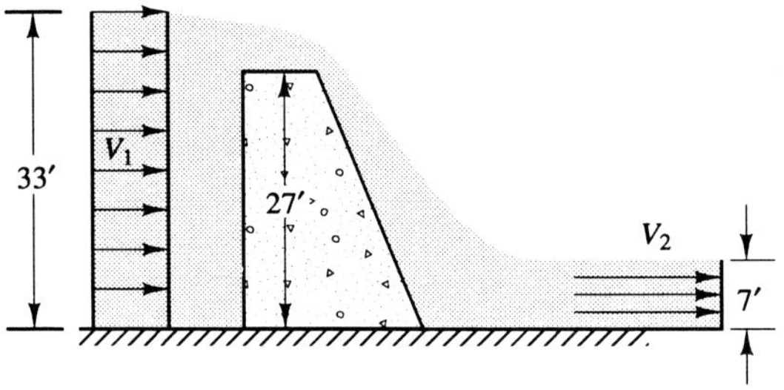

Figure 1-19

8. A sluice gate is used to regulate the flow of water from a reservoir. The reservoir is 10-m deep. With the sluice gate as shown in Figure 1-20, estimate the force per meter needed to hold the gate in place. Make sure all assumptions are stated. Sketch the pressure distribution of the water on the gate.

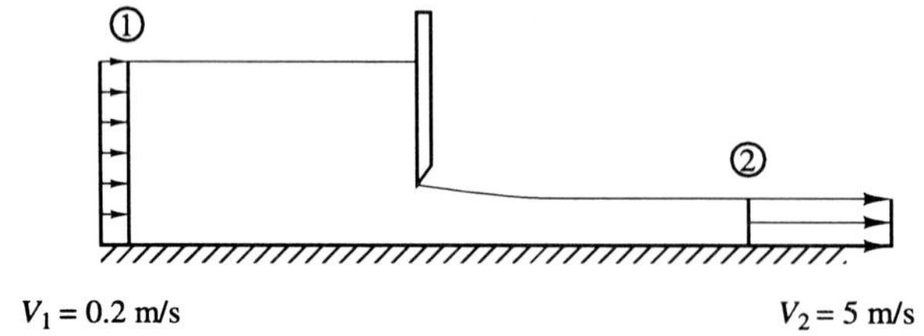

Figure 1-20

9. A wing section placed in a channel produces the velocity profile shown in Figure 1-21. If the pressures upstream and downstream of the object are equal, find the drag per unit length of the object. Is this situation possible? State all assumptions.

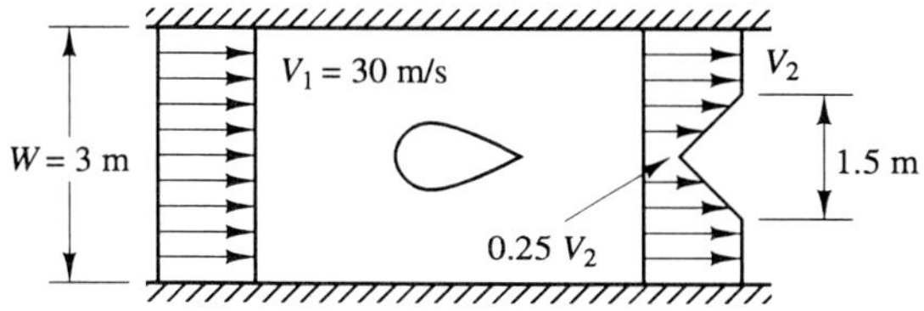

Figure 1-21

10. A two-dimensional object (a wing) is tested in a wind tunnel with an air density of 1 kg/m^3. The wind tunnel is 1-m high, with upstream and downstream pressures of 1.5 kPa gage and 1 kPa gage, respectively. If the mean velocity of the entering air is 30 m/s and the downstream velocity profile is as shown in Figure 1-22, what is the drag per unit length of the wing? Neglect the wind tunnel wall shear stress.

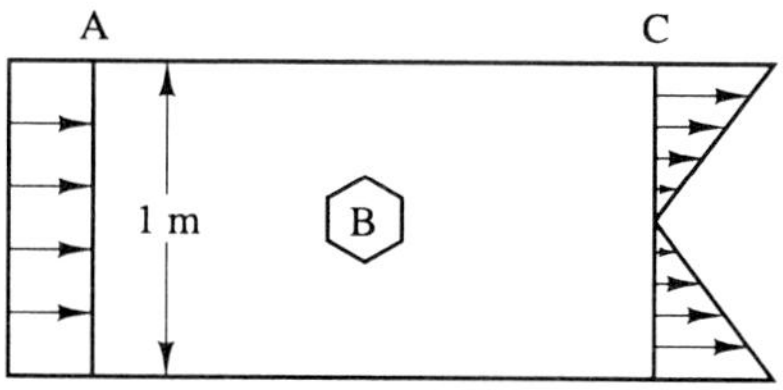

Figure 1-22

11. A jet of water from the upper tank in Figure 1-23 is being discharged at 1.5 cfs. At station 1 the jet diameter is measured to be 5 inches. The empty tank weighs 150 lbf, has a cross-sectional area of 5 ft^2, and has a water depth of 6 ft. The upper tank is located 6 ft from the surface of the water in the lower tank. What forces will be measured by scales at A and B? State all assumptions.

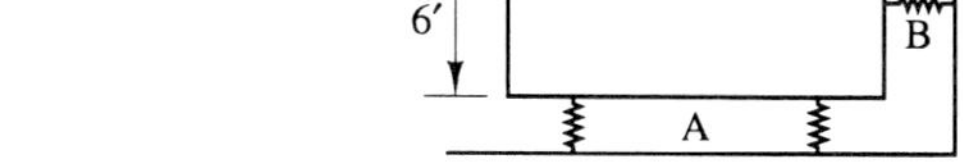

Figure 1-23

12. The wake behind a long, elliptically shaped pole placed in an unbounded stream is shown in Figure 1-24. Estimate the drag of the pole. Make sure that all assumptions are stated.

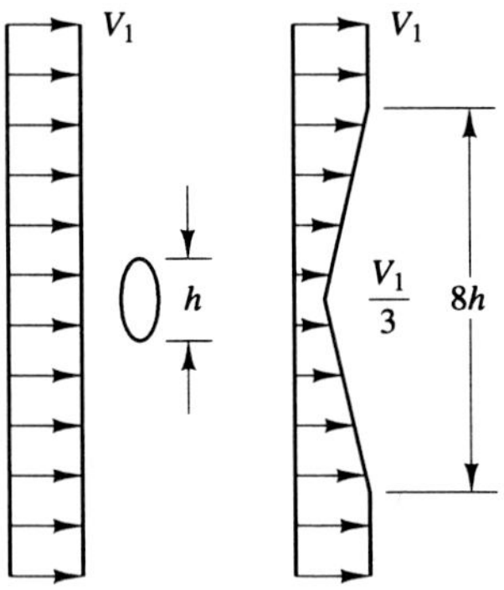

Figure 1-24

13. Two streams, both initially at 30 psia but with different velocities, mix together. For the system in Figure 1-25, find the pressure difference between station 3 and stations 1 and 2. Neglect frictional effects between the stream and the walls of the device.

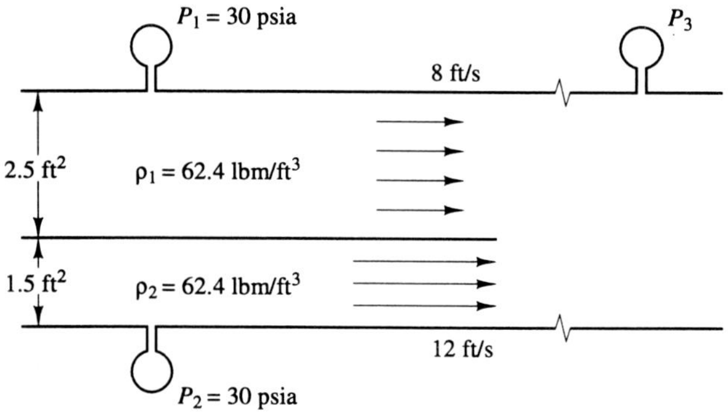

Figure 1-25

14. A rocket nozzle has the velocities and pressures shown in Figure 1-26. Find the force required to hold the nozzle in place. If the combustion chamber of the rocket has a pressure of 1500 kPa, the same as the nozzle inlet pressure, find the thrust of the rocket.

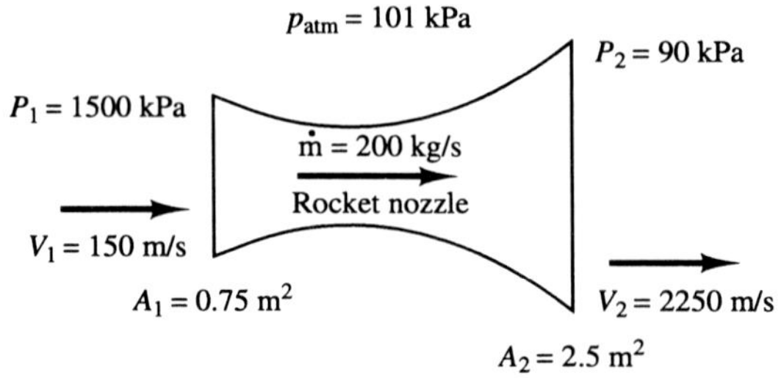

Figure 1-26

15. The jet engine in Figure 1-27 is anchored to a test stand. If the pressures and velocities are as shown, compute the thrust.

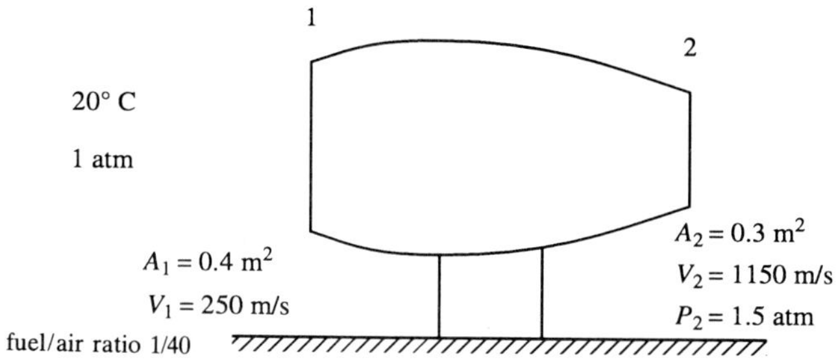

Figure 1-27

16. A jet engine on a test stand produces a thrust of 180,000 N. For the circumstances of Figure 1-28 find the air and fuel mass flow rates.

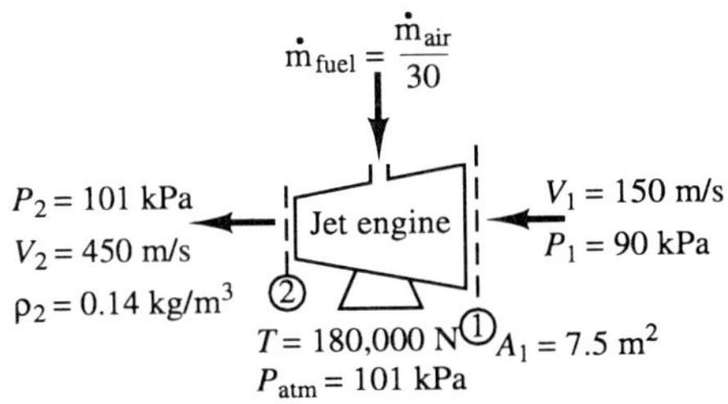

Figure 1-28

17. Figure 1-29 illustrates the configuration of the center (tail-mounted) jet engine on an airliner. For the situation shown, compute the thrust of the engine.

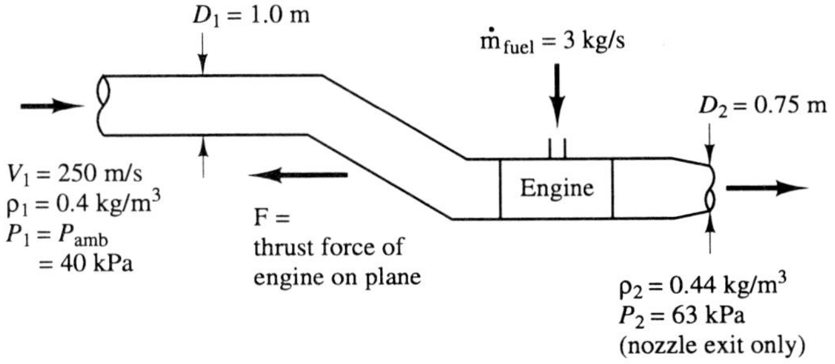

Figure 1-29

18. Figure 1-30 presents a schematic of a rocket engine mounted on a test stand in standard atmospheric conditions. The area of the nozzle exit plane is 0.025 m^2, the velocity of the exhaust gases is 1600 m/s, and the mass flow rate is 1.2 kg/s. If the pressure in the nozzle exit plane is 200 kPa, find the thrust of the rocket engine.

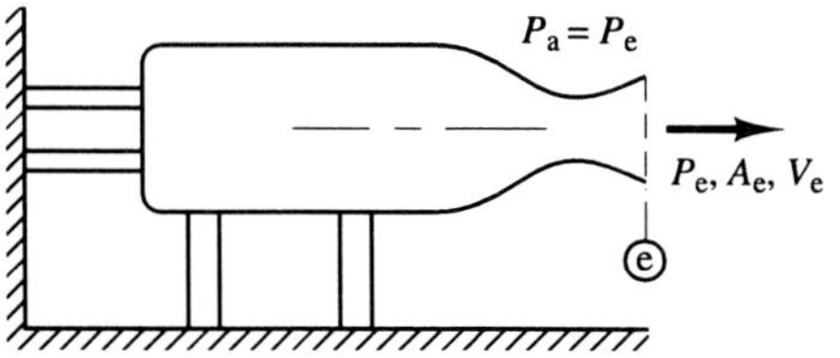

Figure 1-30

19. The rocket engine illustrated in Figure 1-31 operates with a chamber pressure of 2000 kPa and an exit pressure of 160 kPa. The atmospheric pressure is 101 kPa. If the exit area of the rocket is 1 m^2, the exit velocity (V_e) is 2000 m/s, and the mass flow rate is 250 kg/s, find the magnitude and direction of the horizontal force that the rocket exerts on the test stand.

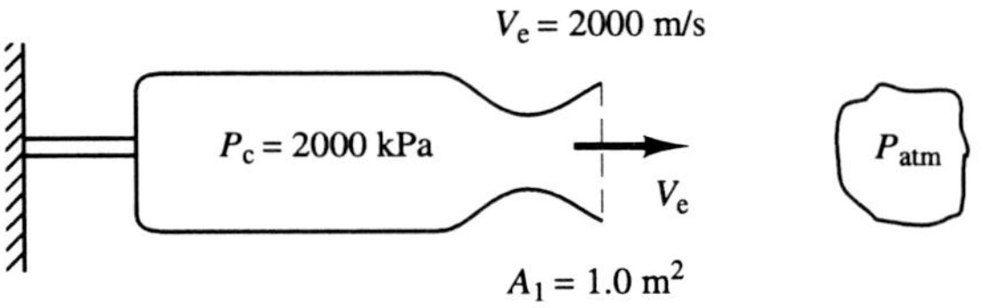

Figure 1-31

20. Air is flowing one-dimensionally and adiabatically in a varying-area duct. At a location in the duct where the area is 1 ft^2, the static conditions are $P = 10$ psia and $T = 200$ F, and the velocity is 1000 ft/s. Find the fluid velocity and the duct area at a location where the static conditions are $P = 30$ psia and $T = 100$ F. Is the process isentropic?

21. Air initially at 300 K, 80 kPa, and a velocity of 1000 m/s undergoes a process to the end state of 1000 K stagnation temperature, 200 m/s velocity, and 240 kPa static pressure. Is heat transferred to or from the fluid? What is the change in entropy? What is the percentage change in cross-sectional area?

22. An initially evacuated chamber is fed by air with a stagnation temperature of 300 K. A system schematic is shown in Figure 1-32. What is the final temperature of the fluid inside the chamber, assuming no heat transfer or shaft work? Explain in physical terms what is happening.

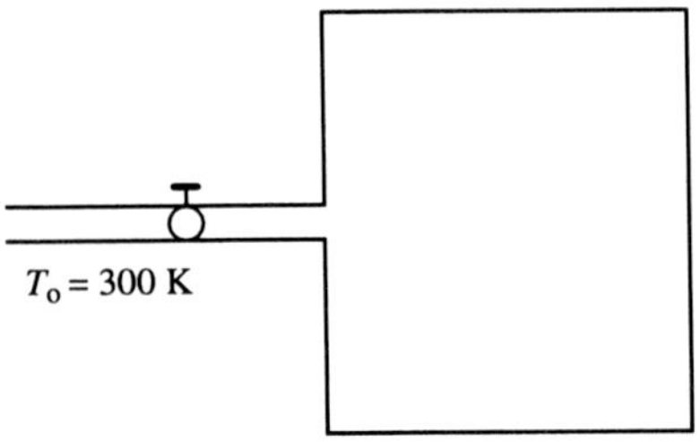

Figure 1-32

23. The aircraft is flying at Mach 2 at 70,000 ft. Construct Mach lines showing the zone of silence and the zone of action for these conditions (see Figure 1-33). If the aircraft flies directly over an observer's head, how far will it be from the observer when the observer hears its sound? When the observer hears the sound, at what angle should the aircraft be from the observer?

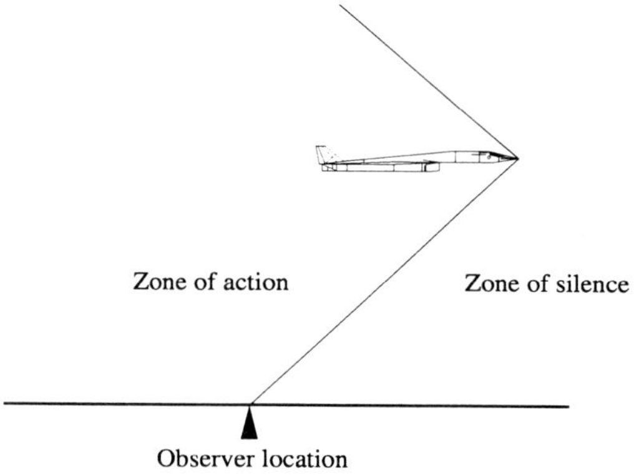

Figure 1-33

Part II

WAVES IN COMPRESSIBLE FLOW

Normal Shock Waves

2

2.1 INTRODUCTION

The fundamental concepts needed for the development of one-dimensional compressible flow were introduced in Chapter 1. Some of these concepts, such as conservation of mass or momentum, were familiar to us from incompressible fluid mechanics; others, like an equation of state, arose because the density of the fluid is no longer a constant in compressible flow. One of the ramifications of a variable density fluid is a finite speed of sound. In contrast, an absolutely incompressible fluid (no real fluid, however, can be absolutely incompressible) possesses an unbounded speed of sound, implying that information or events at one point are communicated instantaneously to all other points in the flow field. The finite speed of sound associated with compressible fluids leads to a richer environment with more interesting and diverse fluid phenomena possible than for an incompressible fluid.

Perhaps the most interesting of all compressible flow phenomena are shock and expansion waves. These are the subjects of the next three chapters, in which we shall consider normal shock waves, moving normal shock waves, oblique shock waves, conical flows, and expansion waves. After the fundamentals are introduced and discussed, applications using the various wave concepts are examined. As with the other chapters and topics covered in this text, the software element COMPQ is integrated into the discussion and used in example problems.

2.2 WAVE MOTION IN A COMPRESSIBLE MEDIUM

We previously suggested that in a compressible fluid the finite speed of sound means that some time must elapse before information concerning events at one point is communicated to other points in the flow field. The means of communication are infinitesimal or finite compression waves and infinitesimal or finite expansion waves.

Consider first what might happen if an infinitesimal disturbance or change occurs at one point. Since the disturbance is vanishingly small, its effects are transmitted from the point of occurrence to other points in the flow field by a wave of vanishing strength, a sound wave. This wave travels at the speed of sound; remember in Chapter 1 that we developed expressions for the speed of sound in gases.

Examine the piston–cylinder arrangement in Figure 2-1a. If the piston is impulsively displaced to the right with a velocity dV, then the fluid adjacent to the piston surface at the time of the displacement must have the velocity of the piston. But in order for the fluid adjacent to the piston to possess a velocity dV, it must displace some of the fluid to the right of it, and so forth. The information that the piston is moving is transmitted by a sound wave. At the start of the piston displace-

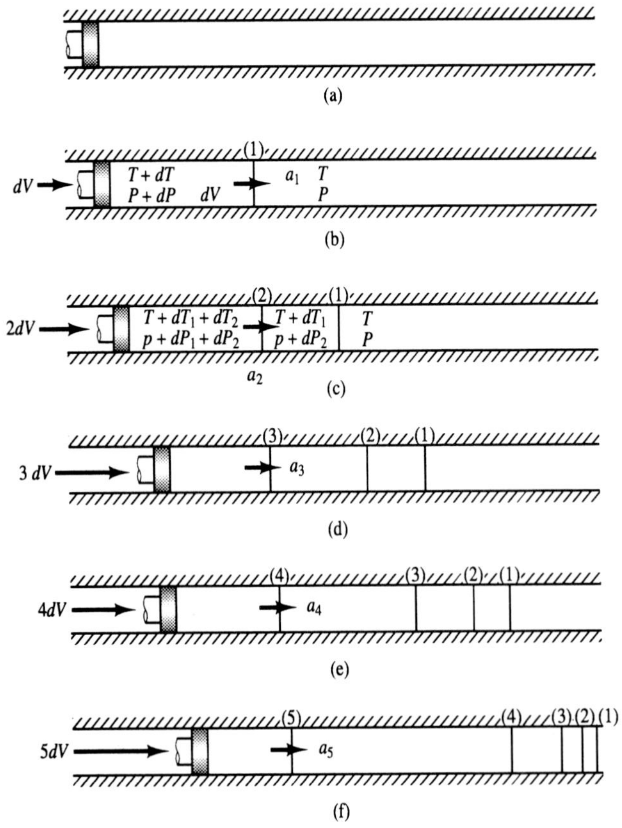

Figure 2-1 Sound Wave Train in a Compression Situation ($a_1 < a_2 < a_3 \ldots$)

ment, a sound wave, originating at the face of the piston and moving with the speed of sound, a, to the right, induces a velocity dV to the fluid as it passes. If wall effects are neglected, the situation is as represented in Figure 2-1b. The flow field at any time t is divided into two parts: the undisturbed fluid to the right of the sound wave and the disturbed fluid to the left of the sound wave.

The states of the fluid to the left and right of the sound wave are different. Ahead, or to the right, of the wave, the fluid velocity is 0, the pressure is P, the temperature is T, and the entropy is s. Behind, or to the left of, the wave, the velocity of the fluid is dV, the pressure is $P + dP$, the temperature is $T + dT$, and, since the sound wave is isentropic, the entropy is still s. Because the piston motion reduces the volume available, the sound wave acts to compress the fluid; since the process is isentropic, the change in temperature dT and the change in pressure dP are related isentropically.

Now, imagine that the piston is impulsively accelerated a second time to the velocity $2dV$. The sequence of events described for the initial acceleration occurs for the second increment of velocity. Now, however, the temperature of the fluid into which the second sound wave is moving is $T + dT$, and the second sound wave is traveling faster than the first sound wave, by a factor of $(1 + dT/T)^{1/2}$. The second wave is thus gradually overtaking the first. This is illustrated in Figure 2-1c.

If a large number of impulsive velocity increments dV is given to the piston, a train of sound waves results, each one traveling faster than the preceding one. We could envision the situation as in Figure 2-1d. Some finite times later the situation would be as in Figure 2-1e and Figure 2-1f. It is clear that as time passes, the individual sound waves are coming closer together; eventually for a finite piston velocity, the train of sound waves would merge together—coalesce—into something stronger than a sound wave, a *shock wave*. A shock wave would form with finite changes in velocity, pressure, temperature, and perhaps entropy across the shock wave. We shall pursue more about shock waves in Section 2.3.

Consider the same situation, but let the piston displacement be to the left. The situation is similar to that portrayed in Figure 2-1, except that the sound waves act as expansion waves. The pressure is decreased by dP, and the temperature is decreased by dT by the passage of an individual sound wave. The sequence of events is illustrated in Figure 2-2. Since the temperature of each subsequent sound wave is lower than that of the previous one, the speed of sound decreases for each subsequent sound wave. The net result for the sequence of sound waves for the expansion process is that the waves spread farther apart as time passes; they do not tend to coalesce as in the compressive piston motion example.

These two possible processes with the piston–cylinder arrangement provide some useful information about waves in compressible flow. The compressive piston process leads to a moving shock wave when a finite piston velocity is considered. The compressive process results in the coalescence of the sound waves to naturally form a moving normal shock wave—one that is perpendicular to its direction of motion. Since the sound waves are coming together, we must suspect that the resulting shock wave is thin; that is, the changes occur over a small distance. Experimental data confirm that shock waves are indeed very thin, with a thickness on the order of

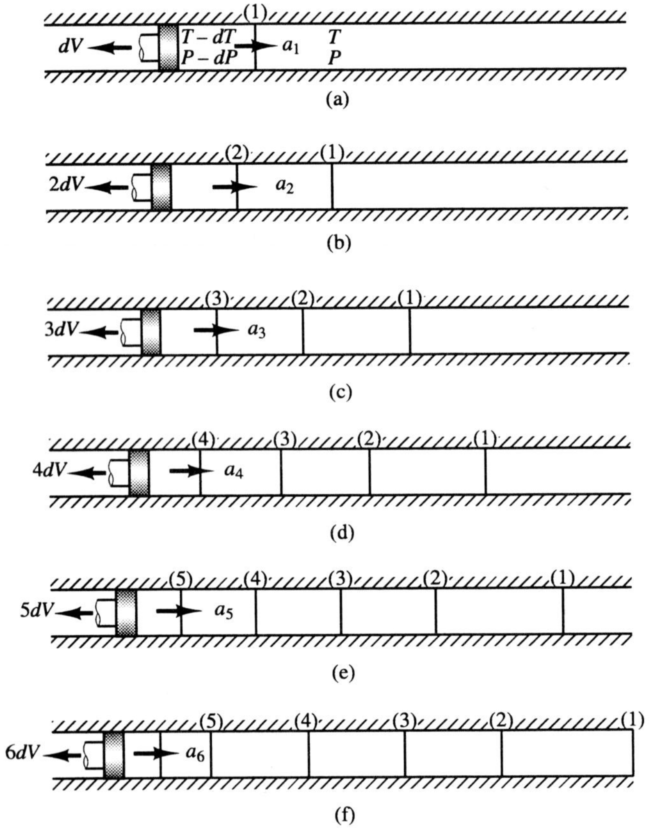

Figure 2-2 Wave Train in an Expansion Situation ($a_1 > a_2 > a_3 \ldots$)

10^{-7} m. The expansion piston process does not lead to the formation of a shock wave. The expansion process results in a dispersion, not a coalescence, of the sound waves. We shall use these ideas in the next sections to increase our understanding of wave behavior.

2.3 NORMAL SHOCK WAVE ANALYSIS

Shock waves are one of the more unique phenomenon associated with compressible fluid flow. The discussion in Section 2.2 indicated that a shock wave is a wave across which the velocity, pressure, and temperature change by finite amounts. In order to put these changes to use in engineering terms, we must quantify those changes. Moreover, we need to establish guidelines about the thickness of a shock and to understand the thermodynamics of shock waves.

Accepting for the moment that a shock wave is thin, we represent the control volume about a steady, stationary, one-dimensional shock wave represented by two parallel lines drawn closely together (see Figure 2-3). This representation will be used in the remainder of the text to identify shock waves. The shock wave is assumed to have zero thickness but finite changes in the physical properties. Section 2.2 proposed the idea that a shock wave can be formed as a result of a compressive process being used. Steady, uniform flow, which is moving from the left to the right, enters the control volume at point (1) and exits at point (2) in Figure 2-3. The shock wave is normal to the flow, and the control volume is rectangular and oriented as shown. The control surface area A_1 is equal to the control surface area A_2. The assumptions of uniform inflow and a one-dimensional wave imply all velocities and properties are invariant with x. The locations of control surfaces A_3 and A_4 are therefore arbitrary. Since the flow velocities and fluid properties are the same for corresponding locations on surfaces A_3 and A_4, no net mass, momentum, or energy crosses these surfaces.

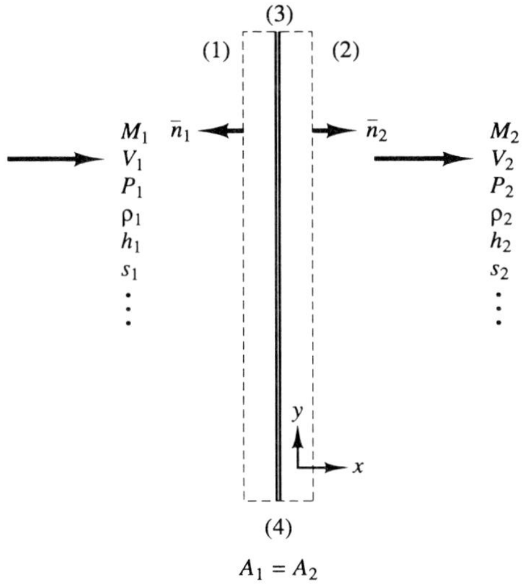

Figure 2-3 Control Volume for Steady, Stationary Normal Shock Wave Derivation

In addition to the conservation statements (mass, momentum, and energy), the analysis of the normal shock wave requires an equation of state,

$$P = \rho RT, \tag{1-8}$$

as well as the definition of Mach number and the expression for the speed of sound. Conservation of mass,

$$0 = \frac{\partial}{\partial t}\int_{cv} \rho\, dVol + \int_{cs} \rho\bar{V} \cdot \overline{dA}, \tag{1-33}$$

as applied to the control volume in Figure 2-3 yields

$$0 = -\rho_1 V_1 A_1 + \rho_2 V_2 A_2 \tag{2-1a}$$

or

$$\rho_1 V_1 = \rho_2 V_2. \tag{2-1b}$$

Conservation of momentum in the x direction,

$$\sum F_x = \frac{\partial}{\partial t}\int_{cv} \rho V_x \, dVol + \int_{cs} \rho V_x \overline{V} \cdot \overline{dA}, \tag{1-35}$$

as applied to the control volume produces (neglecting friction)

$$P_1 A_1 - P_2 A_2 = -\rho_1 V_1^2 A_1 + \rho_2 V_2^2 A_2, \tag{2-2}$$

which can be written as

$$P_1 + \rho_1 V_1^2 = P_2 + \rho_2 V_2^2. \tag{2-3}$$

However, since

$$\rho V^2 = \frac{P}{RT} M^2 \gamma RT = \gamma P M^2, \tag{2-4}$$

the final form of the momentum equation becomes

$$P_1(1 + \gamma M_1^2) = P_2(1 + \gamma M_2^2). \tag{2-5}$$

Conservation of energy,

$$\frac{\delta q}{dt} - \frac{\delta w_s}{dt} = \frac{\partial}{\partial t}\int_{cv} \rho u \, dVol + \int_{cs} \rho\left(h + \frac{V^2}{2}\right)\overline{V} \cdot \overline{dA}, \tag{1-40c}$$

as applied to Figure 2-3 in the absence of shaft work and heat addition/rejection to the control volume yields

$$0 = -\rho_1 V_1 A_1\left(h_1 + \frac{V_1^2}{2}\right) + \rho_2 V_2 A_2\left(h_2 + \frac{V_2^2}{2}\right). \tag{2-6}$$

With Eq. 2-1a the above equation takes the form

$$h_1 + \frac{V_1^2}{2} = h_2 + \frac{V_2^2}{2}, \tag{2-7}$$

which we recognize as

$$h_{o_1} = h_{o_2}. \tag{2-8}$$

For calorically perfect gases this reduces to

$$T_{o_1} = T_{o_2}. \tag{2-9}$$

Thus, the normal shock process possesses a constant stagnation temperature. An alternate form of the energy equation expressed in terms of the static temperature is

$$T_1\left(1 + \frac{\gamma - 1}{2} M_1^2\right) = T_2\left(1 + \frac{\gamma - 1}{2} M_2^2\right). \tag{2-10}$$

The density ratio across a normal shock wave can be expressed as a function of the pressure and temperature ratios:

$$\frac{\rho_2}{\rho_1} = \frac{P_2}{RT_2}\frac{RT_1}{P_1} = \frac{P_2/P_1}{T_2/T_1}. \tag{2-11}$$

The analytical solution to the normal shock wave is contained within Eqs. 2-1 through 2-11; the problem at hand is to uncover that solution. The sequence is to start with Eq. 2-1b and eliminate all variables associated with station 2 except M_2 by selectively using Eqs. 2-2 through 2-11. To start, conservation of mass can be written as

$$\rho_1 V_1 = \frac{P_1}{RT_1} M_1 \sqrt{\gamma R T_1} = \rho_2 V_2 = \frac{P_2}{RT_2} M_2 \sqrt{\gamma R T_2} \tag{2-12}$$

using the equation of state, the Mach number, and the speed of sound relations. If Eq. 2-12 is divided by the final results from momentum, Eq. 2-5, then

$$\frac{M_1}{\sqrt{T_1}(1 + \gamma M_1^2)} = \frac{M_2}{\sqrt{T_2}(1 + \gamma M_2^2)}. \tag{2-13}$$

We use Eq. 2-10 to eliminate the temperature ratio and then we square the results to obtain

$$\frac{M_1^2}{(1 + \gamma M_1^2)^2}\left(1 + \frac{\gamma - 1}{2} M_1^2\right) = \frac{M_2^2}{(1 + \gamma M_2^2)^2}\left(1 + \frac{\gamma - 1}{2} M_2^2\right). \tag{2-14}$$

This is a quartic equation in M_2 (or a quadratic in M_2^2) if the ratio of specific heats and M_1 are known. Solving for M_2^2, we obtain (with the help of some algebra)

$$M_2^2 = \pm \frac{2 + (\gamma - 1)M_1^2}{2\gamma M_1^2 - (\gamma - 1)}, \qquad \pm M_1^2. \tag{2-15}$$

The root $M_1^2 = M_2^2$ simply says that a shock wave is not present; the negative roots are imaginary. Only the first positive solution is of interest in conjunction with the normal shock wave. Symbolic manipulation languages such as MACSYMA, MATHEMATICA, and DERIVE can accomplish such algebra rapidly and error-free and have been used extensively in equation development in this textbook.

Equation 2-15 is an important relation since now given γ and M_1, we can compute M_2. Once M_1 and M_2 are known, the temperature ratio T_2/T_1 can be found from Eq. 2-10; the pressure ratio P_2/P_1 can be found from Eq. 2-5; the density ratio can then be computed from Eq. 2-11, and the velocity ratio V_2/V_1 can be found from Eq. 2-1b. The stagnation pressure ratio across the normal shock can be related to the static pressure ratio, P_2/P_1, and the stagnation-to-static pressure ratios at the upstream and downstream states using

$$\frac{P_{o_2}}{P_{o_1}} = \frac{P_{o_2}}{P_2}\frac{P_2}{P_1}\frac{P_1}{P_{o_1}} = \frac{P_2}{P_1}\frac{P_1/P_{o_1}}{P_2/P_{o_2}}. \tag{2-16}$$

Cast in the form of Mach numbers, this becomes via Eqs. 2-5 and 1-70:

$$\frac{P_{o_2}}{P_{o_1}} = \frac{1 + \gamma M_1^2}{1 + \gamma M_2^2}\left(\frac{1 + \frac{1}{2}(\gamma - 1)M_2^2}{1 + \frac{1}{2}(\gamma - 1)M_1^2}\right)^{\gamma/(\gamma-1)}. \tag{2-17}$$

We had previously concluded, in Eq. 2-9, that for a calorically perfect gas the stagnation temperature remains constant across a stationary normal shock wave.

The most useful forms of the calorically perfect gas expressions for the various property ratios across a stationary normal shock wave can be obtained by substituting Eq. 2-15 for M_2 in the previously obtained equations for the property ratios. The ensuing forms contain only the ratio of specific heats and the upstream Mach number and are particularly useful for computational procedures. The usual versions appear as follows (Eqs. 2-18a through f):

$$\frac{P_2}{P_1} = \frac{2\gamma M_1^2}{\gamma+1} - \frac{\gamma-1}{\gamma+1} \tag{2-18a}$$

$$\frac{T_2}{T_1} = \frac{\left(1 + \frac{\gamma-1}{2}M_1^2\right)\left(\frac{2\gamma}{\gamma-1}M_1^2 - 1\right)}{M_1^2\left(\frac{2\gamma}{\gamma-1} + \frac{\gamma-1}{2}\right)} \tag{2-18b}$$

$$\frac{\rho_2}{\rho_1} = \frac{V_1}{V_2} = \frac{(\gamma+1)M_1^2}{(\gamma-1)M_1^2 + 2} \tag{2-18c}$$

$$M_2^2 = \frac{M_1^2(\gamma-1) + 2}{2\gamma \mathrm{M}_1^2 - (\gamma-1)} \tag{2-18d}$$

$$\frac{T_{o_2}}{T_{o_1}} = 1 \tag{2-18e}$$

$$\frac{P_{o_2}}{P_{o_1}} = \left(\frac{\frac{1}{2}(\gamma+1)M_1^2}{1 + \frac{1}{2}(\gamma-1)M_1^2}\right)^{\gamma/\gamma-1}\left(\frac{1}{\frac{2\gamma}{\gamma+1}M_1^2 - \frac{\gamma-1}{\gamma+1}}\right)^{1/\gamma-1}. \tag{2-18f}$$

The change in entropy between two states for a perfect gas is

$$s_2 - s_1 = c_p \ln\frac{T_2}{T_1} - R\ln\frac{P_2}{P_1}. \tag{1-20}$$

Using the previous results, for the normal shock wave this becomes

$$\frac{s_2 - s_1}{R} = \frac{\gamma}{\gamma-1}\ln\frac{1 + \frac{1}{2}(\gamma-1)M_1^2}{1 + \frac{1}{2}(\gamma-1)M_2^2} - \ln\frac{1 + \gamma M_1^2}{1 + \gamma M_2^2}; \tag{2-19}$$

or expressed in terms of only M_1 and γ:

$$\frac{s_2 - s_1}{R} = \frac{1}{\gamma-1}\ln\left(\frac{2\gamma M_1^2}{\gamma+1} - \frac{\gamma-1}{\gamma+1}\right) + \frac{\gamma}{\gamma-1}\ln\left[\frac{2 + (\gamma-1)M_1^2}{(\gamma+1)M_1^2}\right]. \tag{2-20}$$

Recall from Chapter 1 that for an adiabatic process in terms of the stagnation pressure ratio,

$$\frac{s_2 - s_1}{R} = -\ln P_{o_2}/P_{o_1}. \tag{1-72}$$

Equation 2-20 is a very interesting and useful result as it can be used to determine what restrictions the second law of thermodynamics imposes on normal shock behavior. Figure 2-4 presents the plot of Eq. 2-20 as a function of upstream Mach number for a γ of 1.4. A *supersonic upstream* Mach number (M_1) produces a positive change in entropy, while a *subsonic upstream* Mach number (M_1) produces a negative change in entropy. Thus, the second law prohibits normal shock waves with subsonic upstream Mach numbers. In order for a normal shock wave to be possible, M_1 must be greater than unity. Equation 1-72 contains some of the bad news about normal shock waves. The required increase in entropy demands that $P_{o_2} < P_{o_1}$; hence, stagnation pressure is always lost across a stationary normal shock wave. Since stagnation pressure is often viewed as an indicator of the capability of a flow to do work, loss in stagnation pressure represents a degradation in the ability of the flow to be used. The fact that stagnation pressure is lost across a shock wave should come as no surprise since the shock is a near-discontinuous process involving large gradients in properties.

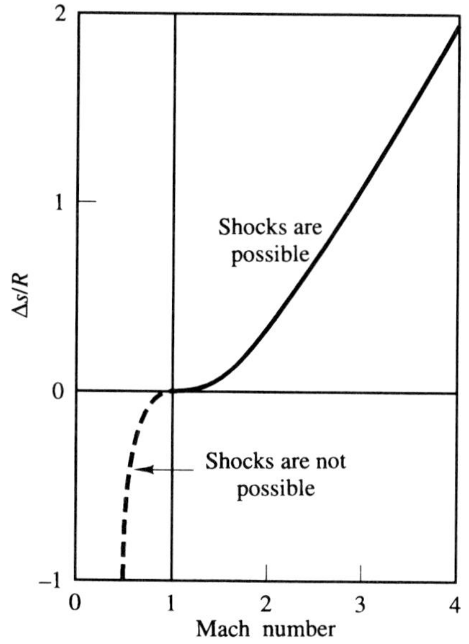

Figure 2-4 Entropy Change across a Normal Shock (γ = 1.4)

An examination of Eq. 2-15 reveals that for a supersonic value of M_1, M_2 is subsonic and a compression process occurs. For subsonic values of M_1, M_2 is supersonic and an expansion process takes place. Since the second law prohibits the latter, expansion shock waves cannot occur. We might have suspected this from the discussion in Section 2.1 where we saw that compression piston motion produced a coalescence of sound waves into a shock wave. Piston motion corresponding to an expansion process resulted not in coalescence, but in dispersion of the sound waves. Piston motion, therefore, cannot lead to expansion shock waves, those across which the pressure decreases.

One of the important points to notice is that for a given value of γ, M_2 is only a function of M_1. Equations 2-18 present the property ratios as functions of M_1 for a specified ratio of specific heats. Thus, the downstream Mach number M_2 and all the property ratios can be tabulated as functions of the upstream Mach number M_1 and γ. The normal shock tables, which contain tabulated values of downstream quantities of interest as a function of upstream Mach number are presented in Appendix B-1 ($\gamma = 1.4$). The normal shock **NS** option of the software element COMPQ uses the relations discussed here and provides several input options for use with normal shock waves. Figure 2-5a reproduces the menu for option **NS** of COMPQ. A variety of cases of **NS** are available: the single input can be the upstream Mach number, the downstream Mach number, or the static or stagnation pressure ratio. A sample output of the **NS** option is given in Figure 2-5b. The upstream and downstream Mach numbers as well as various property ratios and the change in entropy are provided in the output. COMPQ possesses several advantages over the normal shock wave table:

1. Since the ratio of specific heats is an input variable, COMPQ may be used for any realistic value of γ.
2. No interpolation is necessary as the appropriate closed-form equations are used in the calculations.
3. Hard copies of COMPQ output are available.

How does the normal shock wave behave as a function of upstream Mach number? The simplest way to answer this question is to use Eqs. 2-18 or COMPQ

```
NORMAL SHOCK WAVE CALLED FOR
INPUT PARAMETERS ARE:
     UPSTREAM MACH NUMBER        U
     DOWNSTREAM MACH NUMBER      D
     STATIC PRESSURE RATIO       P
     STAGNATION PRESSURE RATIO   S

ENTER THE DESIRED LETTER?
```

(a) NS Menu

```
ENTER THE UPSTREAM MACH NUMBER? 1.43

    M1         M2        P2/P1      T2/T1      RHO2/1     PO2/PO1
  1.4300     0.7274     2.2190     1.2741     1.7416     0.9504

STAGNATION TEMPERATURE IS CONSTANT ACROSS A NORMAL SHOCK.

THE CHANGE IN ENTROPY IS DEL S/CP =    0.0145
```

(b) NS Output

Figure 2-5 Sample COMPQ Output

or Appendix B-1 and generate a plot of important downstream quantities (M_2 and property ratios) as a function of upstream Mach number. Such a plot for $\gamma = 1.4$ is presented in Figure 2-6. This figure is not very useful for finding numerical values, but it does illustrate the behavior of downstream quantities with increasing upstream Mach number. Although it is certainly possible to apply COMPQ or Appendix B-1 for large Mach numbers (such as $M_1 > 5$), the validity of the assumption of a calorically perfect gas starts to break down as large Mach numbers are reached. Large upstream Mach numbers result in large downstream temperatures. At elevated temperatures the assumption of constant specific heats is no longer valid. For completeness, the limiting downstream values are

$$\lim_{M_1 \to \infty} M_2 = \sqrt{(\gamma - 1)/(2\gamma)} \tag{2-22a}$$

$$\lim_{M_1 \to \infty} \frac{\rho_2}{\rho_1} = \frac{\gamma + 1}{\gamma - 1} \tag{2-22b}$$

$$\lim_{M_1 \to \infty} \frac{P_2}{P_1} = \infty \tag{2-22c}$$

$$\lim_{M_1 \to \infty} \frac{T_2}{T_1} = \infty \tag{2-22d}$$

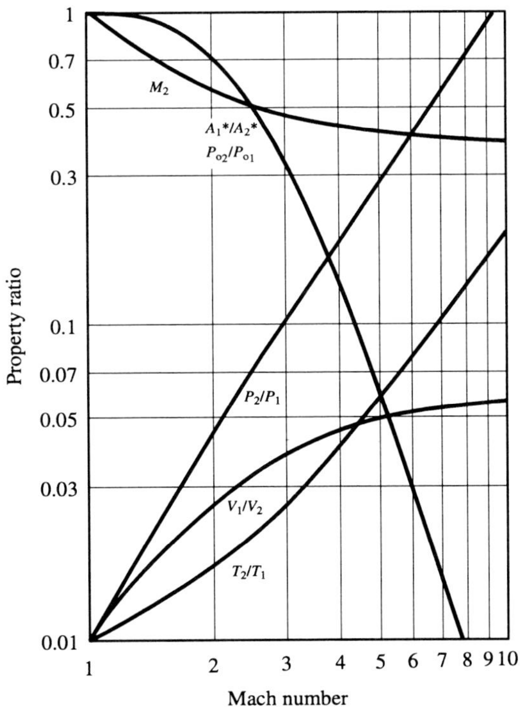

Figure 2-6 Normal Shock Relations ($\gamma = 1.4$)

The beginning of the approach to asymptotic limits for M_2 and V_1/V_2 (or P_2/P_1) can be observed in Figure 2-6.

Much of this analysis is based on the supposition that shock waves are thin. Experimental results, analytical results using continuum fluid mechanics, and analytical results using kinetic theory of gases all point to shock waves as being, at most, a few mean free paths thick for usual conditions. Typical thicknesses are of the order of 10^{-7} m for air. Thus, the supposition that the typical shock wave is very thin is quite acceptable.

In determining the downstream Mach number and various property ratios for normal shock waves, the conservation equations were applied to a control volume containing the shock wave. We made no attempt to determine what happens inside the shock wave; obviously something dissipative is occurring since the entropy increases across the shock wave. Because the changes in velocity, pressure, and temperature are finite—and possibly significant for many cases—large temperature, velocity, and pressure gradients exist within the shock wave. Within the structure of the shock wave, viscosity plays an important role. Viscosity acting over the region of high-velocity gradient is responsible for much of the dissipation encountered in the shock wave. The viscous mechanism converts part of the kinetic energy of the flowing gas into heat; this dissipative mechanism results in the conversion of the energy of the motion (ordered) of the gas into random motion, that is, increased temperature in the interpretation of the kinetic theory of gases.

2.4 THE RANKINE-HUGONIOT RELATIONS

A traditional way of looking at a process is to compare the pressure and density ratios of the beginning and ending states. For example, for an isentropic process that relationship is

$$\frac{P_2}{P_1} = \left(\frac{\rho_2}{\rho_1}\right)^{\gamma}. \tag{1-26}$$

The pressure–density relationship for the normal shock can be obtained by solving for M_1^2 from the pressure ratio expression given by Eq. 2-18a and substituting that into the density ratio expression given by Eq. 2-18c. The resulting expressions, cast as $P_2/P_1 = f(\gamma, \rho_2/\rho_1)$ or $\rho_2/\rho_1 = f(\gamma, P_2/P_1)$, are called the *Rankine-Hugoniot relations* and appear as

$$\frac{P_2}{P_1} = \frac{\dfrac{\gamma+1}{\gamma-1}\dfrac{\rho_2}{\rho_1} - 1}{\dfrac{\gamma+1}{\gamma-1} - \dfrac{\rho_2}{\rho_1}}, \tag{2-23a}$$

or

$$\frac{\rho_2}{\rho_1} = \frac{\dfrac{\gamma+1}{\gamma-1}\dfrac{P_2}{P_1} + 1}{\dfrac{\gamma+1}{\gamma-1} + \dfrac{P_2}{P_1}}. \tag{2-23b}$$

The temperature ratio expressed as a function of the pressure ratio and γ can be recovered by using the equation of state in conjunction with the Rankine-Hugoniot relation for ρ_2/ρ_1. The usual form is

$$\frac{T_2}{T_1} = \frac{P_2}{P_1} \frac{\dfrac{\gamma + 1}{\gamma - 1} + \dfrac{P_2}{P_1}}{\dfrac{\gamma + 1}{\gamma - 1}\dfrac{P_2}{P_1} + 1}. \tag{2-23c}$$

The Rankine-Hugoniot relations express the pressure and density ratios across a normal shock wave.

The isentropic relations and the Rankine-Hugoniot relations both relate the pressure, density, and temperature ratios: the former for a reversible, adiabatic process and the latter for an irreversible adiabatic process. The pressure ratio versus the density ratio for both the isentropic and Rankine-Hugoniot relations are plotted in Figure 2-7. For pressure and density ratios less than approximately 2, the two processes are essentially indistinguishable from each other. This corresponds, for a γ of 1.4, to an upstream Mach number of about 1.6. Thus, for situations in which the local upstream Mach number does not exceed 1.6, the shock wave is very close to an isentropic process. An important ramification of this observation is that applications involving low supersonic Mach numbers can, with acceptable accuracy, be considered isentropic.

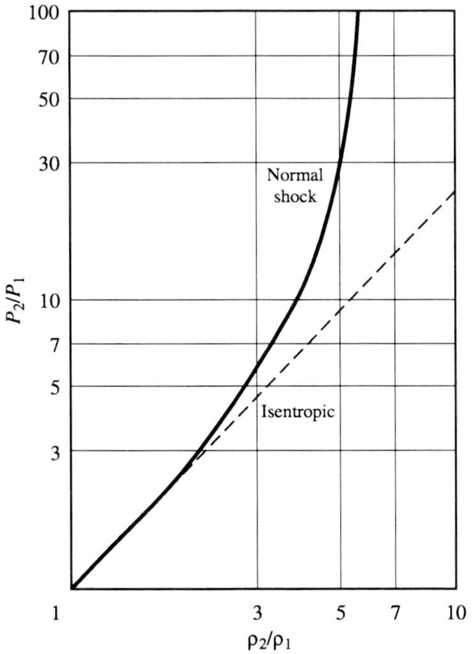

Figure 2-7 The Normal Shock (Rankine-Hugoniot) and Isentropic Curves for $y = 1.4$

2.5 NORMAL SHOCK WAVE APPLICATIONS

The previous sections have given us an analysis of stationary normal shock waves for an ideal gas. What is needed now are examples of how to apply the analysis to meaningful situations. Table B-1 (in Appendix B) (for $\gamma = 1.4$) and COMPQ are available to use for normal shock wave applications.

Consider the following examples.

Example 2-1

Upstream of a stationary normal shock wave the static pressure is 10 psia, the static temperature is 0 F, and the velocity is 1500 ft/s. If the gas is air, find the remaining static and stagnation pressures and temperatures upstream and downstream of the shock wave, the Mach numbers, and the loss in stagnation pressure. Discuss the significance of the results.

Solution. COMPQ will be used. Since air is specified as the fluid, the ratio of specific heats, γ, is 1.4, and the gas constant, R, is 1716 ft^2/s^2 R. Figure 2-8a represents a schematic of the shock wave with important quantities labeled. The upstream Mach number is first calculated:

$$M_1 = \frac{V_1}{a_1} = \frac{1500 \text{ ft/s}}{[(1.4)1716 \text{ ft}^2/\text{s}^2\ R(460\ R)]^{1/2}} = 1.43.$$

If we use the normal shock option **NS** of COMPQ and select the upstream Mach number as the input option, case **U**, the normal shock solution for $M = 1.43$ is as presented in Figure 2-5b. Using the results of conservation of mass, Eq. 2-1b, yields

$$V_2 = \frac{V_1}{\rho_2/\rho_1} = \frac{1500}{1.7416} = 861.3 \text{ ft/s},$$

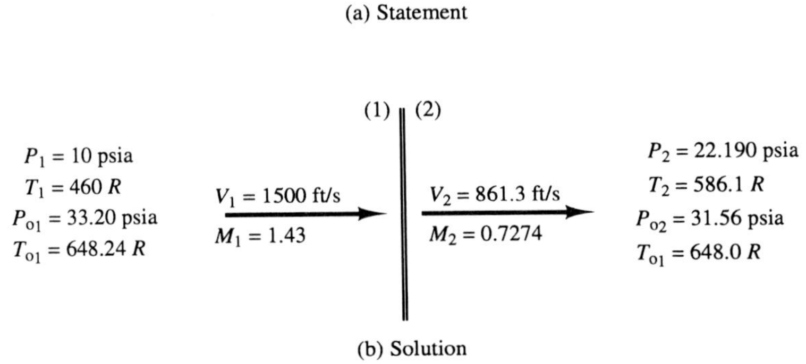

Figure 2-8 Example Problem 2.1

and from Figure 2-5b:

$$P_2 = 10(2.2190) = 22.190 \text{ psia};$$

$$T_2 = 460(1.2741) = 586.1\ R.$$

From the isentropic option of COMPQ with Mach number selected as the input, the following data are produced:

$M_1 = 1.43$	$T_1/T_{o_1} = 0.7097$	$T_{o_1} = 648.2\ R$
	$P_1/P_{o_1} = 0.3012$	$P_{o_1} = 33.20$ psia
$M_2 = 0.7274$	$T_2/T_{o_2} = 0.9043$	$T_{o_2} = 648.0\ R$
	$P_2/P_{o_2} = 0.7032$	$P_{o_2} = 31.56$ psia

Thus, the change in stagnation pressure is

$$\Delta P_o = P_{o_2} - P_{o_1} = 31.56 - 33.20 = -1.64 \text{ psia}.$$

The complete solution with all quantities identified is presented in Figure 2-8b. This problem could be worked using the normal shock tables in Appendix B-1 and the isentropic tables in Appendix A-1, instead of COMPQ. Within the range of accuracy of reading the tables, the results would agree with the quantities above. Although the local upstream Mach number is less than 1.6, some loss, about 5 percent, in stagnation pressure occurred. Considering what can happen in stronger shock waves, this is a relatively small loss, and it reinforces the idea that the shock process for this low supersonic Mach number is nearly isentropic.

Example 2-2

The pressures on either side of a normal shock wave are measured as 100 kPa and 400 kPa. Find the upstream and downstream Mach numbers and the loss in stagnation pressure if air is the fluid.

Solution. COMPQ will be used. The ratio of specific heats for air is 1.4. Since a shock wave is compressive, the upstream pressure is 100 kPa, and the downstream pressure is 400 kPa. The pressure ratio is

$$\frac{P_2}{P_1} = \frac{400}{100} = 4.0.$$

Using case **P** of option **NS** of COMPQ, which requires the pressure ratio as the input quantity, the following data are recovered:

$$M_1 = 1.8898$$

$$M_2 = 0.5976$$

$$P_{o_2}/P_{o_1} = 0.7720$$

$$T_2/T_1 = 1.6000.$$

For an upstream Mach number of 1.8898, the isentropic option for COMPQ yields

$$\frac{P_1}{P_{o_1}} = 0.1516,$$

from which

$$P_{o_1} = 659.63 \text{ kPa}$$

and

$$\Delta P_o = P_{o_2} - P_{o_1} = -P_{o_1}(1 - P_{o_2}/P_1)$$
$$= -659.63(1 - 0.7720) = -150.40 \text{ kPa}.$$

As with Example 2-1, the normal shock tables and the isentropic tables could have been used with similar results. This example makes the point that M_1 need not always be specified for normal shock problems.

2.6 MOVING NORMAL SHOCK WAVES

The analysis presented in Section 2.3 covered stationary normal shock waves; however, many useful applications involve *moving* normal shock waves. This section explores how to use what we know about stationary shock waves to solve problems involving moving normal shock waves.

A schematic representation of a moving normal shock wave is given in Figure 2.9a. The shock wave is moving at a constant velocity V_s into a gas with velocity V_a. After passage of the shock wave, the velocity behind the shock wave is V_b. A compressive shock wave, the only kind permitted by the second law of thermodynamics, will result in V_b greater than V_a. The moving shock wave must then induce motion in the direction of the shock's motion. Static properties ahead of and behind

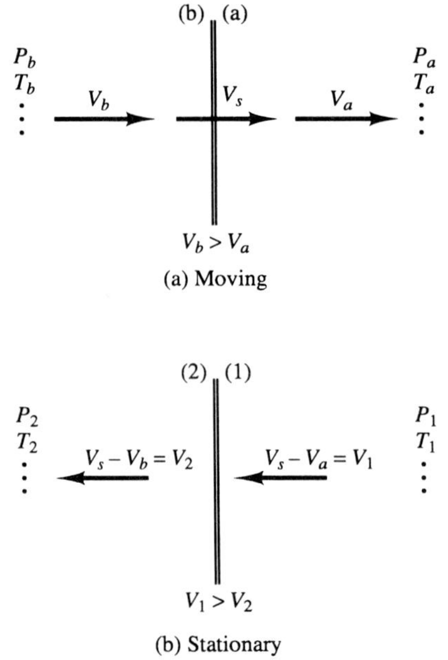

Figure 2-9 Moving and Stationary Normal Shock Wave Nomenclature

the shock wave are as indicated. Because the situation portrayed in Figure 2-3 (which is in steady flow with respect to a fixed control volume) is different from that shown in Figure 2-9a (which is in unsteady flow with respect to a fixed control volume), the analysis of Section 2-3 cannot be directly applied. The observer in Figure 2-9a is viewed as stationary. Consider the situation, as presented in Figure 2-9b, that an observer on the moving shock wave would see. Since static properties would be measured by an instrument at rest with respect to the gas velocity, it follows that static properties do not depend on the reference frame and are the same for Figures 2-9a and b. Stagnation properties would be measured by an instrument at rest with respect to a reference frame. Thus, while static properties are independent of the observer, stagnation properties depend on the observer and are different for Figures 2-9a and b.

TABLE 2-1 MOVING AND STATIONARY NORMAL SHOCK WAVE RELATIONS

Moving	Stationary
$P_a = P_1$	$P_a = P_1$
$P_b = P_2$	$P_b = P_2$
$T_a = T_1$	$T_a = T_2$
$T_b = T_2$	$T_b = T_2$
$M_a = \dfrac{V_a}{a_a}$	$M_1 = \dfrac{V_s - V_a}{a_1} = \dfrac{V_1}{a_1}$
$M_b = \dfrac{V_b}{a_b}$	$M_2 = \dfrac{V_s - V_b}{a_2} = \dfrac{V_2}{a_2}$
$T_{o_a} = T_a(1 + \frac{1}{2}(\gamma - 1)M_a^2)$	$T_{o_1} = T_1(1 + \frac{1}{2}(\gamma - 1)M_1^2)$
$T_{o_b} = T_b(1 + \frac{1}{2}(\gamma - 1)M_b^2)$	$T_{o_2} = T_1(1 + \frac{1}{2}(\gamma - 1)M_2^2)$
$P_{o_a} = P_a(1 + \frac{1}{2}(\gamma - 1)M_a^2)^{\gamma/\gamma-1}$	$P_{o_1} = P_1(1 + \frac{1}{2}(\gamma - 1)M_1^2)^{\gamma/\gamma-1}$
$P_{o_b} = P_b(1 + \frac{1}{2}(\gamma - 1)M_b^2)^{\gamma/\gamma-1}$	$P_{o_2} = P_2(1 + \frac{1}{2}(\gamma - 1)M_2^2)^{\gamma/\gamma-1}$
$T_{o_a} \neq T_{o_b}$	$T_{o_1} = T_{o_2}$

Table 2-1 provides a summary of the relationships for the moving and the stationary normal shock waves as depicted in Figure 2-9. Moving normal shock problems can be solved by a three-stage procedure:

1. Transform the moving normal shock wave problem into the equivalent stationary normal shock wave problem.
2. Solve the stationary normal shock problem using the analysis of Section 2-3.
3. Transform the equivalent stationary normal shock wave solution back to the moving normal shock problem.

An example is appropriate at this point.

Example 2.3

A normal shock wave, with a velocity of 600 m/s, moves into ambient air. The ambient air is at 300 K and 101 kPa. Find the velocity after the passage of the shock and find the stagnation temperatures and pressures both ahead of and behind the moving normal shock wave.

Solution. A sketch of the problem with the salient features labeled is shown in Figure 2-10a. Step 1 is to transform the moving normal shock problem into the equivalent stationary shock problem. The transformed problem is presented in Figure 2-10b, where upstream is identified as the side of the shock with the higher velocity—the right-hand side in this case. The upstream Mach number is

$$M_1 = \frac{600 \text{ m/s}}{[(1.4)(287.0 \text{ J/kg K})(300 \text{ K})]^{1/2}} = 1.728,$$

and from COMPQ using option **NS** with case **U**,

$$P_2/P_1 = 3.3170$$

$$T_2/T_1 = 1.4785$$

$$\rho_2/\rho_1 = V_1/V_2 = 2.2434.$$

Conservation of mass, Eq. 2-1b, can be used to calculate the downstream velocity:

$$V_2 = \frac{600}{2.243} = 267.45 \text{ m/s}.$$

The solution to the equivalent stationary normal shock problem is given in Figure 2-10c. Transforming the solution back to the original moving normal shock problem requires that

$$V_2 = 267.48 = 600 - V_b,$$

from which

$$V_b = 332.55 \text{ m/s}.$$

Cast in the form of the original moving normal shock problem, the solution appears in Figure 2-10d. The stagnation conditions ahead of the shock wave are 300 K and 101 kPa. The Mach number after passage of the shock wave is

$$M_b = \frac{V_b}{a_2} = \frac{332.52}{[(1.4)(287.0)(443.55)]^{1/2}} = 0.7877.$$

The stagnation conditions after passage of the shock then become

$$P_{o_b} = \frac{P_b}{P_b/P_{o_b}} = \frac{335.02}{0.6640} = 504.55 \text{ kPa}$$

and

$$T_{o_b} = \frac{T_b}{T_b/T_{o_b}} = \frac{443.55}{0.8896} = 498.59 \text{ K}$$

An important point in this example is that the solution, which was obtained from the equivalent stationary normal shock problem, must be transformed back into the original

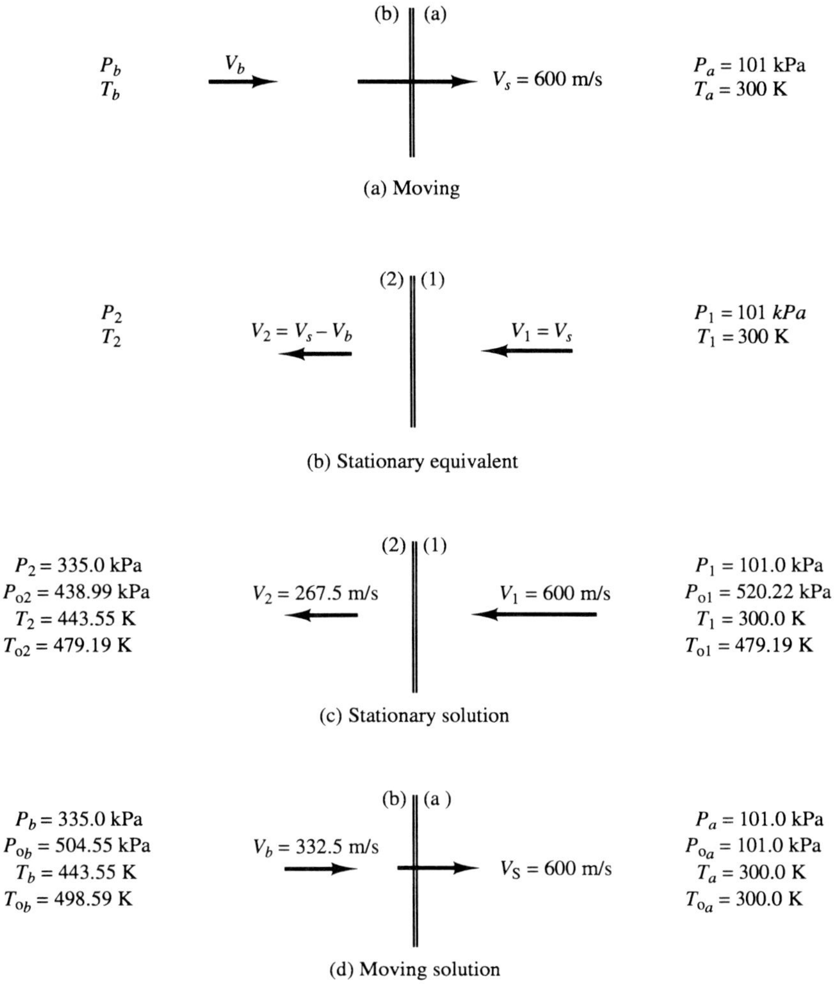

Figure 2-10 Moving and Stationary Normal Shock Wave Representations for Example 2.3

moving normal shock problem for the correct interpretation. In the stationary shock solution space, the stagnation temperature is constant across the shock, and stagnation pressure is lost across the shock. The stagnation property values for this condition are indicated in Figure 2-10c. When the problem involves the moving normal shock, both the stagnation temperature and pressure increase after passage of the shock; see Figure 2-10d for the stagnation property values. This problem is an excellent illustration of the dependence of stagnation properties on the observer reference frame.

Example 2-3 was relatively straightforward since sufficient information was available to evaluate the upstream Mach number of the stationary equivalent problem. Once the upstream Mach number was known, the normal shock relations were applied to solve the problem. However, in many problems of interest, the shock wave velocity is the unknown. When the shock wave velocity (or upstream Mach number) is not known or if a property ratio is not known, then either an iterative process must be employed or further analytical work must be done. A typical iterative process will be delineated in Example 2-4.

For situations in which the given information is insufficient to determine the upstream Mach number, Eq. 2-18c can be utilized to solve for the shock wave velocity. Consider, as in Figure 2-9a, an alternative to an iterative procedure if V_a and V_b are specified, and V_s is not. The moving normal shock wave problem is first transformed to the equivalent stationary one, as illustrated in Figure 2-9b. In order to be meaningful, the resulting shock wave must be a compression process. The sign convention takes V_a and V_b as positive in the direction of the shock wave velocity, as in Figure 2-9a. Equation 2-18c applied to the conditions of Figure 2-9b becomes

$$\frac{V_s - V_a}{V_s - V_b} = \frac{(\gamma + 1)(V_s - V_a)^2}{(\gamma - 1)(V_s - V_a)^2 + 2a_1^2}, \tag{2-24}$$

where

$$V_1 = V_s - V_a$$
$$V_2 = V_s - V_b$$
$$a_1 = \sqrt{\gamma R T_1}.$$

Equation 2-24 is quadratic in the unknown V_s. Solving for V_s yields

$$V_s = \frac{(\gamma + 1)V_b - (\gamma - 3)V_a}{4} \pm \left[a_1^2 + \left(\frac{\gamma + 1}{4}\right)^2 (V_b - V_a)^2\right]^{1/2}. \tag{2-25}$$

Many problems of interest concern a shock wave moving into a region of quiescent flow. For this condition V_a is zero, and Eq. 2-25 reduces to

$$V_s = \frac{(\gamma + 1)V_b}{4} \pm \left[a_1^2 + \left(\frac{\gamma + 1}{4}\right)^2 V_b^2\right]^{1/2}. \tag{2-26}$$

Example 2.4

A normal shock wave is moving into 443.55 K air already moving at 332.52 m/s in the same direction as the shock wave. The velocity behind the shock wave is 400 m/s. Find the shock wave velocity.

Solution. The conditions for the problem are sketched and labeled in Figure 2-11a, and the equivalent stationary normal shock problem is presented in Figure 2-11b. Since the shock velocity and property ratios are unknown, the problem will require either iteration or the use of Eq. 2-25.

The iterative solution, which we will examine first, is devised for the stationary normal shock representation. The unknown variable is V_s, the shock wave velocity. Several different iterative procedures can be formulated; however, virtually all involve

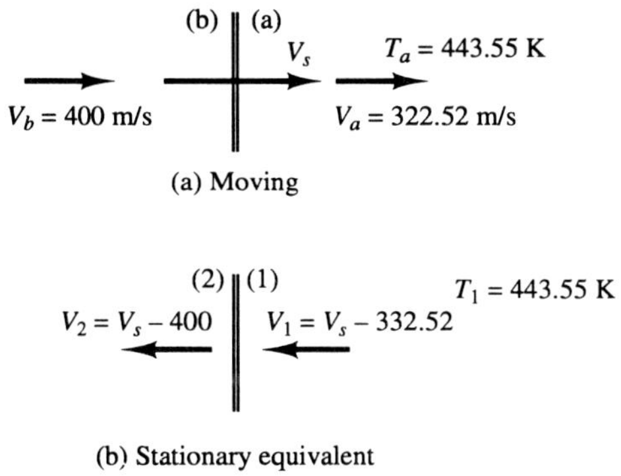

Figure 2-11 Moving and Stationary Normal Shock Representations for Example 2.4

using the equality of the density (ρ_2/ρ_1) and the inverse velocity ratio (V_1/V_2) to check for convergence. The procedure adopted for the solution of this problem is to guess the shock wave velocity (V_s), then compute V_1, V_2, and M_1. From M_1 the density ratio across a normal shock is obtained, and that number is compared with the ratio of velocities for the assumed value of V_s. For state 1 the static temperature is known, so that the speed of sound for state 1 is known (and constant). The order of the iteration is presented in Table 2-2. Other procedures are possible, but most are structured about $\rho_1 V_1 = \rho_2 V_2$ from conservation of mass for a normal shock wave. The iterative sequence is presented in Table 2-2; the converged solution for the shock wave velocity is $V_s = 797.0$ m/s. Once V_s is known, other quantities can be obtained.

TABLE 2-2 ITERATION SEQUENCE FOR MOVING SHOCK WAVE

Guess V_s	$V_1 = V_s - 332.52$	$V_2 = V_s - 400.0$	V_1/V_2	$M_1 = V_1/a_1$	ρ_2/ρ_1
754.68	422.16	354.68	1.190	1.000	1.000
900.00	567.48	500.00	1.135	1.344	1.592
800.00	467.48	400.00	1.169	1.107	1.181
790.00	457.48	390.00	1.173	1.084	1.142
795.00	462.48	395.00	1.171	1.096	1.162
798.00	465.48	398.00	1.170	1.1026	1.174
797.00	464.48	397.00	1.170	1.1002	1.169

An alternate approach not involving iteration is to use Eq. 2-25. Comparing the schematic representation in Figure 2-11a with that shown in Figure 2-9a for the derivation of Eq. 2-25, a_1, V_a, and V_b are as follows:

$$a_1 = [(1.4)(287.0)(443.55)]^{1/2} = 422.16 \text{ m/s}$$

$$V_b = 400.0 \text{ m/s}$$

$$V_a = 332.52 \text{ m/s}$$

Hence,

$$V_s = \frac{(\gamma + 1)V_b - (\gamma - 3)V_a}{4} \pm \left[a_1^2 + \left(\frac{\gamma + 1}{4}\right)^2 (V_b - V_a)^2\right]^{1/2}$$

$$= \frac{(1.4 + 1)400 - (1.4 - 3)332.52}{4}$$

$$\pm \left[422.16^2 + \left(\frac{1.4 + 1}{4}\right)^2 (400 - 332.52)^2\right]^{1/2},$$

and

$$V_s = 373.01 \pm 424.10 = 797.11 \text{ m/s}.$$

Since the negative root for a shock wave velocity is not meaningful, the solution is $V_s = 797.11$ m/s. This compares well with the solution from the iterative procedure. The sign convention takes V_a and V_b as positive in the direction of V_s.

This example was included to illustrate the analytical and iterative procedures available for solution of moving normal shock problems if M_1 is not available from the initial formulation. A variation on the use of Eq. 2-25 is to apply Eq. 2-18c for the particular circumstances involved. This was actually done in order to derive Eq. 2-25, but in some cases starting with Eq. 2-18c may be less ambiguous than using Eq. 2-25.

So far in this section, we have extensively discussed shock waves moving at constant velocities. Also of interest is the reflection of a moving shock wave either from a closed conduit or from a constant pressure boundary. Both correspond to physically realistic conditions: the former to shock wave motion in a conduit with an end closure and the latter to shock wave motion in a conduit open to an ambient pressure. The situation illustrated in Figure 2-12a shows a normal shock wave moving toward the closed end of a duct. The velocity of the fluid ahead of the shock is taken as zero; a velocity V_b is induced behind the shock wave. Since the end closure is impervious, when the shock wave reaches the end, the velocity of the fluid in contact

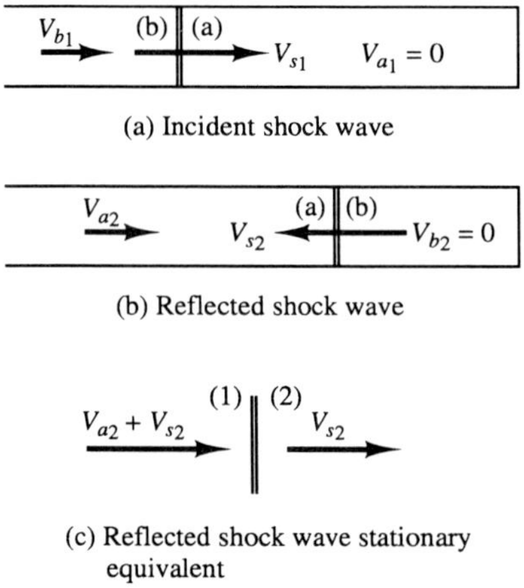

Figure 2-12 Shock Wave Reflection from a Closed Conduit

with the end must be zero. This is accomplished by a *reflection* of the original shock in the upstream direction. (The reflection is illustrated in Figure 2-12b.) The velocity of the reflected shock wave must be such that the velocity of the gas between the end closure and the shock wave is zero. We already know from previous discussion that the first shock wave, the one moving to the right, is a compression wave. Consider, as illustrated in Figure 2-12c, the stationary shock equivalent to the reflected wave. Since the velocity is reduced across this system, the reflected shock wave is also compressive. Thus, a compression wave reflects from a solid boundary as another compression wave.

The case of a reflection from a constant pressure boundary is somewhat different. Initially the shock wave moves down the pipe as shown in Figure 2-13a. As the shock wave moves out the exit of the conduit, the "reflected" wave that moves into the conduit must reduce the pressure inside the conduit to atmospheric. Thus the reflection of the compression wave from the constant pressure boundary must be an expansion process since the pressure is reduced. The reflection is illustrated in Figure 2-13b. But if the original compression wave reflects from the constant pressure boundary as an expansion shock wave, the second law of thermodynamics is violated. What happens is that the incident compression wave reflects as a series of expansion waves, which spread as they travel into the conduit.

Further consideration would reveal that an expansion process reflects both from an end closure as an expansion process and from a constant pressure boundary as a compression process that coalesces into a compression shock wave. In general, the wave that reflects from an end closure is always of the same kind as the incident on the closure; in contrast, the wave that reflects from a constant pressure boundary is of the opposite kind as the incident. The word *process* is used here since an expansion shock wave is not possible. As we saw in Section 2.2, expansion processes tend to spread with time and distance while compression processes tend to coalesce with time and distance. The same ideas must be followed in the reflection process. The name *reflection* is something of a misnomer since the characteristics of the reflected wave differ from those of the incident wave. The reflected and incident waves are related, but in general their strengths are not equal. The characteristics of the reflected wave are such that the proper boundary condition—zero gas velocity for the end closure and ambient pressure for the open duct—is satisfied.

This section closes with an example.

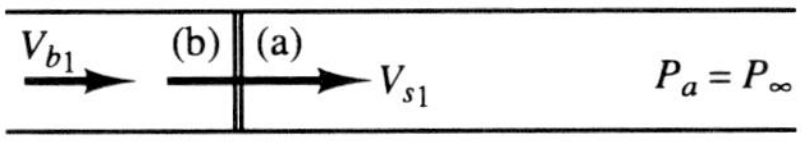

(a) Incident shock wave

(b) Reflected expansion process

Figure 2-13 Shock Wave Reflection from an Open Conduit

Example 2.5

A shock wave moves down a circular tube that is closed at the end, containing undisturbed air at 14.7 psia and 70 F. The velocity behind the moving shock wave is 100 ft/s. Find the pressure to which the end of the tube is subjected.

Solution. The solution to this problem requires analysis of two different moving shock systems: the incident and the reflected. The first system is illustrated in Figure 2-14a with the equivalent stationary system shown in Figure 2-14b. Since the shock wave velocity is not known, either an iterative process or Eq. 2-25 must be used. Examining Figure 2-14b, we obtain

$$a_1 = [(1.4)(1716.0)(530.0)]^{1/2} = 1128.4 \text{ ft/s}$$

$$V_a = 0$$

$$V_b = 100 \text{ ft/s}$$

Using Eq. 2-26 produces

$$V_{s_1} = \frac{\gamma + 1}{4} V_b \pm \left[a_1^2 + \left(\frac{\gamma + 1}{4} \right)^2 V_b^2 \right]^{1/2}$$

$$= \frac{1.4 + 1}{4}(100) \pm \left[1128.4^2 + \left(\frac{1.4 + 1}{4} \right)^2 (100)^2 \right]^{1/2}$$

$$= 1190 \text{ ft/s.}$$

This gives

$$M_1 = \frac{V_1}{a_1} = \frac{1190.0}{1128.4} = 1.0546,$$

and using the normal shock option of COMPQ,

$$\frac{T_2}{T_1} = 1.0358 \qquad T_2 = 549.0 \text{ K}$$

$$\frac{P_2}{P_1} = 1.1309 \qquad P_2 = 16.6 \text{ psia.}$$

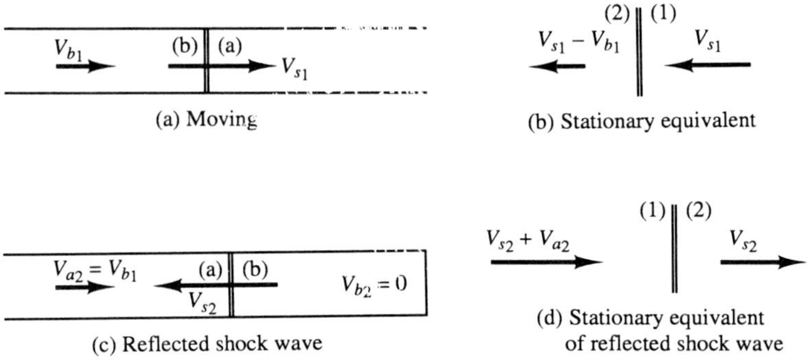

Figure 2-14 Moving and Stationary Equivalent Shock Waves for Example 2.5

The second analysis involves the reflection of the incident wave from the end closure. The reflected shock must be strong enough to reduce the velocity between it and the end closure to zero. Figure 2-14c shows the reflected shock. Again, the shock wave velocity is not known, so either an iterative process or Eq. 2-25 must be used. From the situation in Figure 2-14c, the following values are obtained:

$$V_b = 0$$

$$V_a = -100 \text{ ft/s}.$$

V_a is negative since it is directed in the opposite direction of the V_s.

Using Eq. 2-25 gives

$$V_{s_1} = \frac{-(1.4 - 3)(-100)}{4} \pm \left[1148.4^2 + \left(\frac{1.4 + 1}{4}\right)^2 (0 + 100)^2\right]^{1/2}$$

$$= 1110.0 \text{ ft/s},$$

from which

$$M_1 = \frac{1210.0}{1148.14} = 1.0536,$$

since the Mach number must be computed in the stationary shock equivalent problem as illustrated in Figure 2-14d. Using COMPQ option **NS** yields

$$\frac{P_2}{P_1} = 1.1284 \qquad P_{b_2} = 18.76 \text{ psia}$$

$$\frac{T_2}{T_1} = 1.0352 \qquad T_{b_2} = 568.30\ R$$

The second (reflected) shock wave has a velocity of 1110 ft/s and results in a pressure of 18.76 psia on the end closure. Because the gas velocity in the region between the end closure and the reflected shock wave is zero, the static and stagnation conditions in the region are equal.

An alternative to using Eq. 2-25 is to use an iterative procedure to compute the velocity of the reflected shock wave. As in Example 2-4, a value of V_s will be assumed, V_1 and V_2 calculated, M_1 calculated, and finally ρ_2/ρ_1 compared with V_1/V_2 to test convergence. Table 2-3 presents the iteration sequence to find the velocity of the reflected shock wave. Once the shock velocity is known, the solution proceeds as previously described.

TABLE 2-3 ITERATION SEQUENCE FOR REFLECTED SHOCK WAVE

Guess $V_s = V_2$	$V_1 = V_s + 100.0$	V_1/V_2	$M_1 = V_1/a_1$	ρ_2/ρ_1
1150	1250	1.087	1.088	1.149
1100	1200	1.091	1.045	1.076
1125	1225	1.089	1.067	1.113
1112	1212	1.090	1.055	1.092
1110	1210	1.090	1.054	1.091

This section has developed procedures for treating moving normal shock waves using the results of the stationary normal shock wave analysis. Chapter 3 will deal with another type of shock wave: the oblique shock wave.

CHAPTER 2 REVIEW QUESTIONS

1. Why do successive infinitesimal compression waves tend to reinforce and successive infinitesimal expansion waves tend to attenuate?
2. Is stagnation pressure always lost across a stationary normal shock wave?
3. In a normal shock wave:
   a. Velocity, pressure, and density increase.
   b. Pressure, temperature, and density increase.
   c. Velocity, temperature, and density increase.
   d. Pressure, density, and stagnation pressure increase.
   e. Entropy remains constant.
4. Can an expansion shock wave exist? Justify your response.
5. For a normal shock wave as $M_1 \rightarrow \infty$, does $M_2 \rightarrow 0$? Justify your response.
6. Explain why stagnation temperature is constant across a stationary normal shock wave but is not constant across a moving normal shock wave.
7. Why does a moving normal shock wave reflect from the end of a closed duct as a compression wave?
8. For the situation in Figure 12-15, are P_a and T_a >, <, or = P_b and T_b? Set up, but do not complete, the algebra for an expression for the shock wave velocity, V_s.

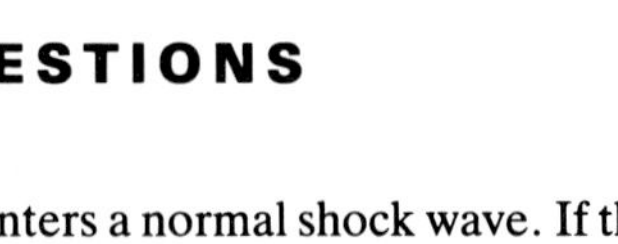

Figure 2-15

9. What happens to stagnation temperature across a stationary normal shock wave?

CHAPTER QUESTIONS

1. Air at 1000 K encounters a normal shock wave. If the pressure ratio across the shock wave is 5, determine the upstream and downstream Mach numbers, the static temperature downstream of the shock wave, the stagnation pressure ratio across the shock wave, the density ratio across the shock wave, and the dimensionless entropy change. Determine the same parameters if the fluid is helium.
2. A helium flow undergoes a normal shock wave. For an upstream Mach number of 3 and an upstream stagnation pressure of 1 atmosphere, find the downstream Mach number, the static and stagnation pressure, and the dimensionless entropy change if the fluid were helium.
3. A shock wave is moving at a Mach number of 3 into undisturbed air (T = 20 C and

$P = 101.3$ kPa). What are the velocity and stagnation temperature (with respect to a stationary observer) in the air after passage of the shock wave?

4. A shock wave is moving in air at 1500 ft/s. The velocity of the air ahead of the shock wave is 150 ft/s directed toward the shock wave. If the static properties ahead of the moving normal shock wave are 2 psia and 600 R, find the static pressure and the Mach number after the shock wave has passed.
5. Air at 90 kPa and 280 K is flowing in a constant area duct with a velocity of 200 m/s. Suddenly, the end of the tube is closed, causing a normal shock to propagate back through the air stream. Find the velocity of the shock wave and the stagnation pressures and temperatures ahead of and behind the moving shock wave.
6. Air is moving in a constant area duct at 100 m/s with a static pressure of 10 kPa and a static temperature of 300 K. If the duct end is suddenly encountered, determine the pressure and temperature behind the resulting shock wave.
7. A shock wave moves into undisturbed air (101.3 kPa and 50 C). If the pressure downstream of the shock wave is 250 kPa, find the speed of the shock wave and the absolute velocity of the gas downstream of the shock wave.
8. The Mach number downstream of a stationary normal shock wave in argon is 0.7. If $T_2 = 1000$ R and the stagnation pressure downstream of the shock wave is 50 psia, find M_1 and the upstream and downstream densities and velocities.
9. A normal shock wave moves through a constant-area tube into air at rest at 25 C. The velocity of the air behind the wave is measured to be 175 m/s. Find the shock wave velocity and the stagnation temperature in the pipe after passage of the shock wave. If the end of the duct is suddenly closed, find the velocity of the reflected shock wave and the properties after passage of the reflected shock wave.
10. A steady, normal shock wave in methane has $V_2 = 500$ m/s and $T_2 = 1000$ K. Determine the upstream Mach number, static temperature, and velocity. Repeat the computations for argon and air. Which parameters are affected by γ and which by both γ and R?
11. If the stagnation pressure across a normal shock wave decreases by 66 percent, find the upstream and downstream Mach numbers if the fluid is air. What happens if the fluid is argon?
12. Helium flows with a velocity of 7500 ft/s and a static temperature of 400 R. After helium passes through a normal shock wave, find the velocity, the static temperature, and the stagnation temperature.
13. A blunt reentry vehicle enters the atmosphere at a Mach number of 15. As shown in Figure 2-16, in the region near the blunt nose the shock wave can be approximated as a normal shock wave. If the vehicle is at 30,000 m altitude, determine the stagnation pressure to which the nose is subjected assuming the air acts as a calorically perfect gas. What are the stagnation pressure and temperature if the vehicle is at sea level?

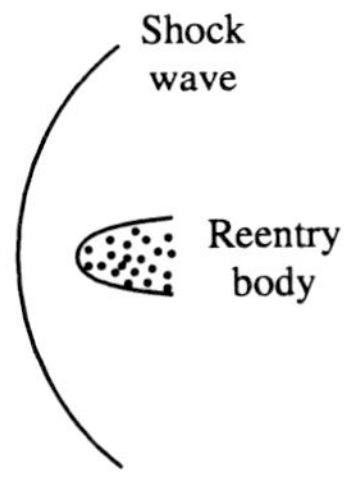

Figure 2-16

14. A projectile moves down a gun barrel with a velocity of 1850 ft/s. Calculate the velocity of the normal shock wave that would move ahead of the projectile. The pressure and density of the undisturbed air ahead of the shock wave are 70 F and 14.7 psia, respectively. How fast would the projectile have to move in order for the speed of the moving shock wave to be two times that of the projectile? What differences would there be if the gun were fired at 40,000 ft?

15. A normal shock wave moves into still air at 800 m/s. The air ahead of the shock wave is at 101.3 kPa and 30 C. Find the velocity of the air after passage of the shock wave, the static pressure behind the shock wave, and the stagnation temperature and pressure behind the shock wave.

16. A shock wave moves through a constant-area tube into air at rest at a temperature of 520 R. The velocity of the air behind the wave is measured to be 400 ft/s. Calculate the shock wave velocity and the stagnation temperature of air after passage of the shock wave.

17. A piston in a tube filled with helium is suddenly accelerated to a velocity of 25 m/s. A normal shock wave moves into the tube ahead of the piston. Half a second after the piston is accelerated to 25 m/s, it is suddenly accelerated to 50 m/s. How much time will be required for the two moving normal shock waves to intersect? Repeat the computations for a tube filled with air.

18. As illustrated in Figure 2-17, air at 14.7 psia and 60 F is flowing in a constant-area tube with a velocity of 300 ft/s. The end of the tube is suddenly closed. Find the absolute velocity of the shock wave reflected from the closed end of the tube. What is the stagnation temperature of the air after passage of the reflected shock wave?

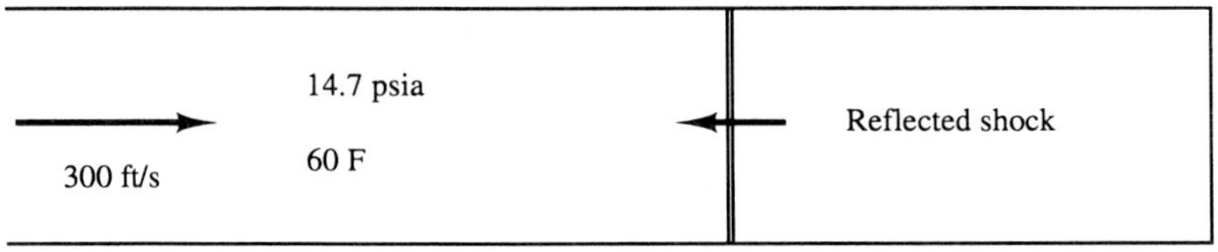

Figure 2-17

19. A normal shock wave traveling at 1250 m/s moves into still air at sea level conditions and reflects from a plane wall. Compare the pressure ratio across the reflected shock with that across the incident shock. Explain why the pressure ratios are different. What are the stagnation conditions after passage of the reflected shock wave?

20. Engine knock can be modeled as a moving normal shock wave in the cylinder of an automobile engine. Consider, as in Figure 2-18, a piston moving upward at 30 m/s and a normal shock wave moving downward at 2500 ft/s into a mixture of unburned air and gas at 70 psia and 400 R. Assuming that the properties of the unburned mixture are similar to those of air, find the pressure acting on the piston face after the reflection of the shock wave.

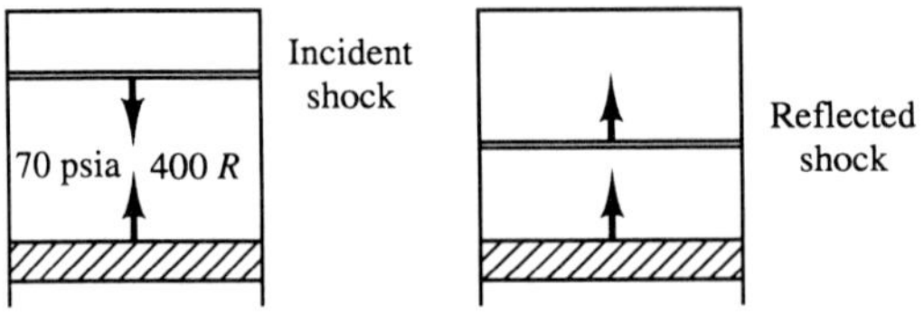

Figure 2-18

21. A normal shock wave moves at 600 m/s down a tube into a gas with static conditions of 50 kPa and 300 K. At the end of the tube, a piston is moving with a velocity of 50 m/s, as illustrated in Figure 2-19. Find the velocity of the reflected shock wave and the static and stagnation conditions behind the reflected shock wave. Choose any gas for which R and the ratio of specific heats are known.

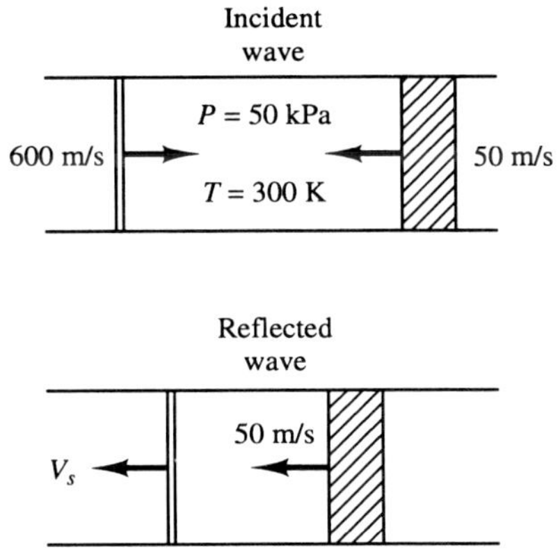

Figure 2-19

22. A blast wave from an explosion passes through still air (60 F and 14.7 psia). The pressure increase is 0.1 psia, which is a very loud noise. Find the velocity of the blast wave and the velocity of the air after passage of the wave. If the shock strength of the wave is defined as

$$\xi = \frac{P_2}{P_1} - 1,$$

what is the shock strength of this wave?

23. How fast must a shock wave move into air initially at 25 C and 101 kPa, if the velocity of the air behind the shock wave is 300 m/s? What are the static and stagnation pressures behind the moving shock wave?

24. Generate a set of normal shock wave tables for a perfect gas with $\gamma = 1.67$. Use Mach number increments of 0.05 over the range $M = 0$ to $M = 4$. Check the results using COMPQ.

25. A normal shock wave is moving at 900 m/s into undisturbed air at 101 kPa and 30 C. Find the velocity of the air behind the shock wave, the static pressure behind the shock wave, and the stagnation temperature behind the shock wave.

26. A normal shock wave with a velocity of 700 m/s moves down a tube into a gas with static properties of 40 kPa and 320 K. At the end of the tube, a piston is moving at 60 m/s. The arrangement is as sketched in Figure 2-19. Find the velocity of the reflected wave and the static and stagnation pressures behind the reflected wave.

Oblique and Conical Shock Waves

3

3.1 INTRODUCTION

An *oblique* shock wave forms an angle of less than 90° with the upstream flow. A plane, oblique shock wave is straight (not curved). In this text, unless otherwise indicated, a plane, oblique shock wave will be called an oblique shock wave. Although the normal shock wave can be treated as a special case of the oblique shock wave, we shall distinguish between the two. As with the normal shock wave, the oblique shock wave represents a very thin region across which near discontinuous changes in properties occur.

One obvious distinction between normal and oblique shock waves is that oblique shock waves are inclined with respect to the incident flow direction, while normal shock waves are always perpendicular to the incident flow. A second important distinction is that while the upstream and downstream flow directions are unchanged across the normal shock, they are different for flow across the oblique shock wave. For example, the change in flow direction caused by a wedge in a supersonic flow is illustrated in Figure 3-1a, which also presents the nomenclature used in the analysis of oblique shock waves. The oblique shock wave makes an angle θ with respect to the upstream flow. A compressive oblique shock results in the flow turning an angle δ into the shock wave. Downstream of an oblique shock wave, the flow is displaced by the wedge; hence, the nature of the compression. In a similar manner to the normal shock wave, we can conclude that an oblique expansion wave is a violation of the second law of thermodynamics.

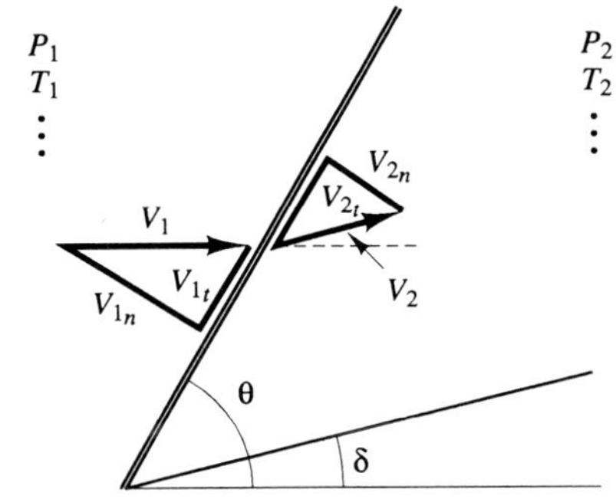

(a) Oblique shock wave nomenclature

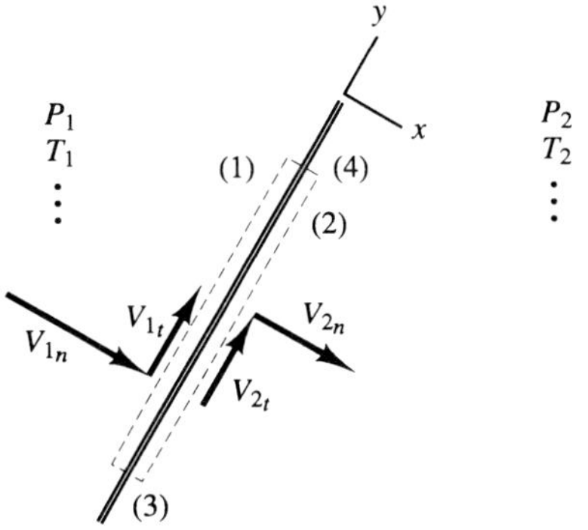

(b) Control volume for oblique shock wave analysis

Figure 3-1 Oblique Shock Wave Schematics

3.2 OBLIQUE SHOCK WAVE ANALYSIS

As Figure 3-1a suggests, one of the most common ways of producing an oblique shock wave is to place a wedge in a compressible flow. The flow is divided into two regions by the plane, oblique shock wave. The region upstream of the oblique shock is uniform with the incident flow parallel. Downstream of the oblique shock wave the flow is also uniform and parallel, but with properties and the flow direction different from those upstream of the oblique shock. Because flow through an oblique shock is dissipative, the entropy increases and the stagnation pressure decreases across the shock wave.

Figure 3-1a shows a two-dimensional "wedge" as producing a plane, oblique shock wave. Some texts used the term *wedge flow* to refer to supersonic flows with plane, oblique shock waves because of the frequency with which such waves are associated with wedges. A more generalized interpretation can be used. The plane, oblique shock wave illustrated in Figure 3-1a divides the flow field into two regions and results in the downstream flow being deflected in the direction of the oblique shock. Downstream of the oblique shock wave, the uniform flow must be parallel to the bounding surface since the surface that conforms to a uniform change in flow direction is inclined at an angle with respect to the incident flow.

The analysis for an oblique shock wave proceeds in the same fashion as for the normal shock wave. Consider, as illustrated in Figure 3-1b, the control volume placed around a segment of a plane, oblique shock wave. The rectangular control volume is very thin and is parallel to the shock wave. The coordinate system is as shown. Conservation of mass as applied to the control volume results in

$$\rho_1 V_{1_n} = \rho_2 V_{2_n}, \tag{3-1}$$

since the areas of control surfaces 3 and 4 are vanishingly small.

Steady-state conservation of momentum in the x- and y-directions becomes

$$\sum F_x = \int_{cs} \rho V_x \overline{V} \cdot \overline{dA}; \tag{3-2a}$$

$$\sum F_y = \int_{cs} \rho V_y \overline{V} \cdot \overline{dA}. \tag{3-2b}$$

In the x-direction,

$$P_1 A_1 - P_2 A_2 = -\rho_1 V_{1_n}^2 A_1 + \rho_2 V_{2_n}^2 A_2, \tag{3-3}$$

where there are no contributions from surfaces A_3 and A_4. After some rearrangement, conservation of momentum in the x-direction becomes

$$P_1 + \rho_1 V_{1_n}^2 = P_2 + \rho_2 V_{2_n}^2. \tag{3-4}$$

In the y-direction,

$$0 = -\rho_1 V_{1_n} V_{1_t} A_1 + \rho_2 V_{2_n} V_{2_t} A_2, \tag{3-5}$$

where again the contributions from surfaces A_3 and A_4 cancel. Conservation of momentum in the y-direction leads to

$$V_{1_t} = V_{2_t}. \tag{3-6}$$

Hence, the tangential component of velocity across an oblique shock wave remains constant.

Conservation of energy becomes

$$0 = -\rho_1 V_{1_n} A_1[(h_1 + \tfrac{1}{2}(V_{1_n}^2 + V_{1_t}^2)] + \rho_2 V_{2_n} A_2[(h_2 + \tfrac{1}{2}(V_{2_n}^2 + V_{2_t}^2)], \tag{3-7}$$

since the oblique shock process is adiabatic and involves no shaft work. With Eq. 3-1, this reduces to

$$h_1 + \frac{V_{1_n}^2 + V_{1_t}^2}{2} = h_2 + \frac{V_{2_n}^2 + V_{2_t}^2}{2}. \tag{3-8}$$

However, since the tangential component of velocity is constant across an oblique shock wave,

$$h_1 + \frac{V_{1_n}^2}{2} = h_2 + \frac{V_{2_n}^2}{2}. \tag{3-9}$$

The equation of state for a perfect gas remains unchanged from the normal shock development.

Compare Eqs. 2-1b and 3-1, Eqs. 2-3 and 3-4, and Eqs. 2-7 and 3-9. The only differences between the pairs of equations are that the velocity components V_{1_n} and V_{2_n} appear in the latter, whereas V_1 and V_2 appear in the former. This leads to a very important conclusion: if the normal components of velocity are considered, an oblique shock wave behaves just as a normal shock wave, while the tangential velocity components remain unchanged. Since the tangential component of velocity is invariant across an oblique shock wave and since the conservation expressions reduce to those equivalent for the normal shock except that only the normal components of velocity are involved, the oblique shock wave can be viewed as a normal shock wave with a uniform velocity superimposed in the tangential direction. The results of this imposition are to incline the shock and deflect the downstream flow. Some textbooks exploit this concept in the development and discussion of oblique shock waves.

All downstream properties, such as pressure and temperature, and the downstream normal Mach number component can be determined using the normal shock wave results and the upstream normal Mach number component, M_{1_n}. Hence in order for an oblique shock wave to exist, M_{1_n} must be greater than unity. M_{1_n} greater than unity will always yield a compressive oblique shock and a supersonic upstream Mach number.

The downstream conditions for oblique shock waves are derived from the downstream conditions for normal shock waves (Eqs. (2-18)) by simply replacing M_1 in the normal shock versions with $M_1 \sin\theta$ for the oblique shock versions. The conditions across an oblique shock wave appear as follows:

$$M_{2_n}^2 = \frac{(\gamma-1)M_1^2\sin^2\theta + 2}{2\gamma M_1^2\sin^2\theta - (\gamma-1)} \tag{3-10a}$$

$$\frac{P_2}{P_1} = \frac{2\gamma M_1^2\sin^2\theta}{\gamma+1} - \frac{\gamma-1}{\gamma+1} \tag{3-10b}$$

$$\frac{\rho_2}{\rho_1} = \frac{V_{1_n}}{V_{2_n}} = \frac{(\gamma+1)M_1^2\sin^2\theta}{(\gamma-1)M_1^2\sin^2\theta + 2} \tag{3-10c}$$

$$\frac{T_2}{T_1} = \frac{\left(1 + \dfrac{\gamma-1}{2}M_1^2\sin^2\theta\right)\left(\dfrac{2\gamma}{\gamma-1}M_1^2\sin^2\theta - 1\right)}{M_1^2\sin^2\theta\left[\dfrac{2\gamma}{\gamma-1} + \dfrac{\gamma-1}{2}\right]} \tag{3-10d}$$

$$T_{o_1} = T_{o_2} \tag{3-10e}$$

$$\frac{P_{o_2}}{P_{o_1}} = \left[\frac{\dfrac{\gamma+1}{2}M_1^2\sin^2\theta}{1 + \dfrac{\gamma-1}{2}M_1^2\sin^2\theta}\right]^{\gamma/(\gamma-1)}\left[\frac{1}{\dfrac{2\gamma}{\gamma+1}M_1^2\sin^2\theta - \dfrac{\gamma-1}{\gamma+1}}\right]^{1/(\gamma-1)}. \tag{3-10f}$$

For the normal shock case $\theta = 90$, and Eqs. 3-10 reduce to Eqs. 2-18. The Rankine-Hugoniot relations, Eqs. (2-23) are also valid for oblique shock waves since they contain only static property ratios.

To illustrate the use of the normal shock relations to solve an oblique shock wave problem, consider the solution of the oblique shock wave problems given in Figure 3-2a. The shock wave angle is 40°, and the incident Mach number is 3. The normal component of the incident Mach number is

$$M_{1_n} = M_1 \sin\theta = 3.0 \sin 40° = 1.928. \tag{3-11}$$

Using the normal shock tables or case **U** of option **NS** of COMPQ leads to

$$\frac{P_2}{P_1} = 4.170$$

$$M_{2_n} = 0.5903 \tag{3-12}$$

$$\frac{T_2}{T_1} = 1.630.$$

Since the oblique shock process is adiabatic, $T_{o_1} = T_{o_2} = T_o$, and

$$\frac{T_2}{T_{o_2}} = \frac{T_2}{T_o} = \frac{T_2}{T_1}\frac{T_1}{T_o} = (1.630)(0.3571) = 0.5824, \tag{3-13}$$

and the downstream Mach number for $T_2/T_{o_2} = 0.5821$ becomes

$$M_2 = 1.895. \tag{3-14}$$

With M_{2_n} and M_2 as well known, the downstream Mach number components appear as in Figure 3-2b. Then

$$\sin(\theta - \delta) = \frac{M_{2_n}}{M_2} = \frac{0.5903}{1.89}, \tag{3-15}$$

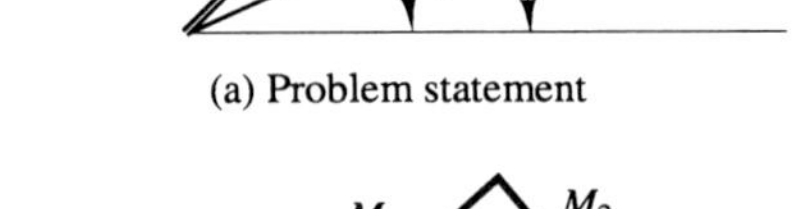

(a) Problem statement

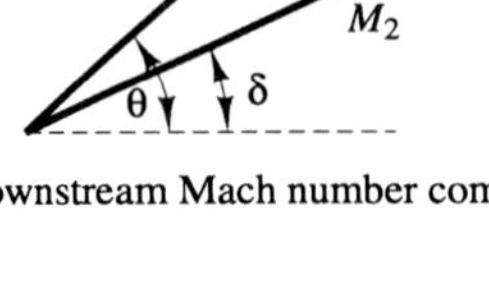

(b) Downstream Mach number component

Figure 3-2 Oblique Shock Wave Example

and

$$\theta - \delta = 18.15°, \tag{3-16}$$

so that

$$\delta = 40 - 18.15 = 21.85°. \tag{3-17}$$

The solution to the oblique shock wave problem has been obtained using the normal shock analysis. In summary:

$$\begin{aligned} M_1 &= 3.0 & M_2 &= 1.895 \\ \theta &= 40.0° & \delta &= 21.85° \\ \gamma &= 1.4 & P_2 &= 4.170P_1 \\ & & T_2 &= 1.630T_1. \end{aligned} \tag{3-18}$$

Had the shock wave angle been changed to 81.19°, the following values would have been obtained:

$$\begin{aligned} M_1 &= 3.0 & M_2 &= 0.5549 \\ \theta &= 81.19° & \delta &= 21.85° \\ \gamma &= 1.4 & P_2 &= 10.087P_1 \\ & & T_2 &= 2.638T_1. \end{aligned} \tag{3-19}$$

Thus, for the same incident Mach number, but for shock wave angles of 40.0° and 81.19°, the same turning angle, $\delta = 21.85°$, was obtained. Hence, for a given upstream Mach number and turning angle, the shock wave angle is double-valued!

The double-valued shock wave angle for a given upstream Mach number and turning angle is intriguing. Are both possible? If both are possible, under what circumstances will each occur? What is the behavior of shock wave angle as a function of upstream Mach number and turning angle? Both the 40.0° and the 81.19° shock wave angles result in upstream normal Mach number components (1.928 and 2.965, respectively) that are greater than unity. Neither shock wave angle then produces a violation of the second law of thermodynamics, so both are possible. Moreover, experimental data indicate that both occur in reality.

The shock wave with the larger shock wave angle is commonly called the *strong shock wave solution*. Because it possesses the larger upstream normal Mach number component, and hence, the larger increase in pressure, it is a stronger shock wave than the oblique shock wave for the smaller shock wave angle. The larger the upstream normal Mach number component, the larger the increases in downstream pressure and temperature, the greater the loss in stagnation pressure, and the smaller the downstream Mach number. The oblique shock wave corresponding to the smaller shock wave angle is called the *weak shock wave solution* since it corresponds to a weaker shock wave than the oblique shock wave for the larger shock wave angle. Which one occurs depends on the downstream conditions, in particular the downstream pressure—the back pressure. A high back pressure will be associated with the strong shock wave and a lower back pressure with the weak shock wave. In actuality, the weak shock wave is usually the one observed since the high back pressure required for the strong shock wave is difficult to establish.

Two ways exist to explore the shock angle behavior with respect to upstream Mach number and turning angle. The first is to continue to use the same procedure as for the previous example and to do parametric studies involving upstream Mach number and shock wave angle. An alternative is to continue the analysis and to develop expressions that can guide us. We shall pursue the latter.

Refer to Figure 3-2a; the quotients of the upstream and downstream normal and tangential components of the velocity yield

$$\tan\theta = V_{1_n}/V_{1_t}$$
$$\tan(\theta - \delta) = V_{2_n}/V_{2_t}, \tag{3-20}$$

from which

$$\frac{\tan(\theta - \delta)}{\tan\theta} = \frac{V_{2_n}}{V_{1_n}}. \tag{3-21}$$

Previously, conservation of mass across an oblique shock wave had been expressed as

$$\rho_1 V_{1_n} = \rho_2 V_{2_n}. \tag{3-22}$$

Combining Eqs. 3-1, 3-21, and the oblique shock wave expression for the density ratio, Eq. 3-10, gives

$$\frac{V_{2_n}}{V_{1_n}} = \frac{\rho_1}{\rho_2} = \frac{(\gamma - 1)M_1^2 \sin^2\theta + 2}{(\gamma + 1)M_1^2 \sin^2\theta} = \frac{\tan(\theta - \delta)}{\tan\theta}. \tag{3-23}$$

A more useful form of Eq. 3-23 is obtained by using the trigometric identity for the tangent of the difference of two angles and simplifying to yield

$$\tan\delta = 2\cot\theta \frac{M_1^2 \sin^2\theta - 1}{M_1^2(\gamma + \cos 2\theta) + 2}. \tag{3-24}$$

Equation 3-24 relates the incident Mach number, the shock wave angle, and the turning angle δ (or deflection angle) for a given ratio of specific heats. It is explicit in the deflection angle, δ, and implicit in the incident Mach number and shock wave angle. Generally, the behavior of oblique shock waves can be explored more easily with Eq. 3-24 than with parametric studies. For an upstream Mach number of 3.0 and a shock wave angle of 40.0°, Eq. 3-24 yields a deflection angle of 21.85°. The equation yields the same deflection as the earlier procedure, but with much less arithmetic.

Equation 3-24 can be used to infer a number of interesting properties of plane, oblique shock waves. If the deflection angle vanishes, Eq. 3-24 must reduce to

$$M_1^2 \sin^2\theta - 1 = 0$$

or

$$\cot\theta = 0. \tag{3-25}$$

For the former,

$$\sin\theta = \frac{1}{M_1}, \tag{3-26}$$

which is the definition of the Mach wave. And for the latter, the cotangent of the shock wave angle must approach zero, which corresponds to a normal shock wave. Thus, as the angle through which the flow is turned approaches zero, the oblique shock wave approaches either a Mach wave or a normal shock. A shock wave angle less than the Mach wave angle results in an upstream normal Mach number component less than unity, precluding the existence of a compressive shock wave. For a given upstream Mach number, the oblique shock wave is bounded by the Mach wave and the normal shock wave.

Consider a parametric study of the oblique shock wave angle as a function of Mach number and flow deflection angle. Figure 3-3 presents the results of such a parametric study with upstream Mach number as the independent variable and flow deflection angle as the parameter. Appendix C-1 presents the parametric study results with flow deflection as the independent variable and upstream Mach number as the parameter. Both figures contain the same information; only the choice of independent variable is different. As expected from the earlier discussion, Figure 3-3 confirms that the oblique shock wave envelope is contained between the Mach line and the normal shock for all supersonic Mach numbers. For a given Mach number only three outcomes are possible:

1. No solution
2. A multivalued solution with two distinct shock wave angles
3. A special case of outcome (2) with only one distinct shock wave angle.

All three possible outcomes can be seen in Figure 3-3. Each possibility will be discussed in turn.

No oblique shock wave solution exists for a specified M_1 and δ if there is no intersection of the deflection angle curve with the specified M_1. For example, in Figure 3-3, with $M_1 = 3$ and $\delta = 35°$, no oblique shock wave solution exists. Certainly we can insert a 35° wedge in a Mach = 3 flow; the indication of no solution from Figure 3-3 or Eq. 3-24 simply means that no oblique shock wave solution exists for that M_1 and δ. What happens in this situation is discussed ahead.

Generally the intersection of a specified M_1 and δ yields two values of the shock wave angle: the larger value being the strong shock solution and the smaller value being the weak shock solution. For example, for $M_1 = 3$ and $\delta = 30°$, the two oblique shock wave angles are approximately 52° and 75°. An examination of Figure 3-3 also reveals that a maximum deflection angle exists for a given incident Mach number. For deflection angles greater than this maximum, no oblique shock wave solutions are found.

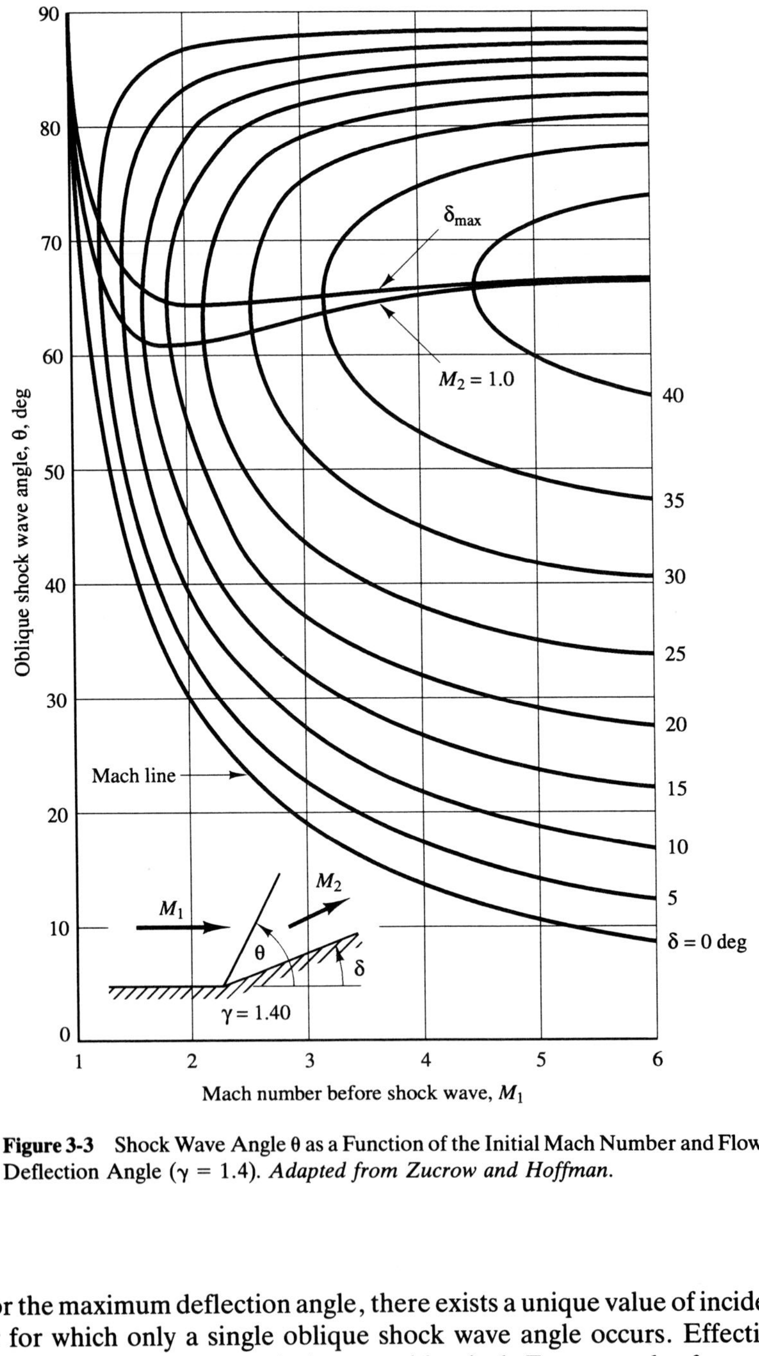

Figure 3-3 Shock Wave Angle θ as a Function of the Initial Mach Number and Flow Deflection Angle (γ = 1.4). *Adapted from Zucrow and Hoffman.*

For the maximum deflection angle, there exists a unique value of incident Mach number for which only a single oblique shock wave angle occurs. Effectively, the strong and weak shock wave solutions are identical. For example, for an incident Mach number of 3, the maximum deflection angle for an attached oblique shock wave to exist is approximately 33.5°. Flow deflections larger than about 33.5° for an incident Mach number of 3 do not possess an attached oblique shock wave. Figure 3-3 demonstrates that only for the maximum deflection angle for a given incident Mach number does a single shock wave angle occur. As the Mach number is increased, the maximum deflection angle for an attached oblique shock wave increases.

The maximum deflection for an oblique shock for a given incident Mach number is obviously an important delimiter in determining what is happening in supersonic flow over a wedge. The NACA TR 1135 (1953) presents the following expression for the shock wave angle corresponding to the maximum deflection angle:

$$\sin^2\theta_{max} = \frac{1}{\gamma M_1^2}\left[\frac{\gamma+1}{4}M_1^2 - 1 + \sqrt{(\gamma+1)\left(1+\frac{\gamma-1}{2}M_1^2+\frac{\gamma+1}{16}M_1^4\right)}\right]. \quad (3\text{-}27)$$

Given the shock wave angle corresponding to the maximum deflection angle, Eq. 3-24 can be used to calculate the maximum deflection angle.

Figure 3-3 contains additional useful information. The downstream Mach number, M_2, is always subsonic for the strong solution and is usually supersonic for the weak solution. The $M_2 = 1.0$ line is shown in the figure. Between the $M_2 = 1.0$ line and the maximum deflection line, the weak oblique shock wave solutions possess a region of subsonic downstream Mach numbers. The region where subsonic Mach numbers exist for the weak shock solutions is relatively small and the Mach numbers are only slightly subsonic. An increase in Mach number for a constant deflection angle results in a decrease in the shock wave angle for the weak solutions and an increase in the shock wave angle for the strong solutions.

Earlier the question was posed of what happens in a uniform supersonic flow over a wedge if no oblique shock wave solution exists. For such a situation, a *detached, curved shock wave* is formed. A detached, curved shock wave that might form ahead of a wedge apex in supersonic flow is illustrated in Figure 3-4. Important features of the curved shock are indicated on the figure. At the apex of the wedge, the shock is normal with a subsonic downstream Mach number. Starting from the apex and moving along the shock, the shock wave is curved with decreasing local shock wave angle. As the local shock wave angle decreases, the downstream Mach number increases since the shock wave is becoming progressively weaker. Eventually the indicated sonic line is reached, and the flow field between the shock wave and the surface of the body becomes supersonic. Near the apex, an embedded subsonic region is present.

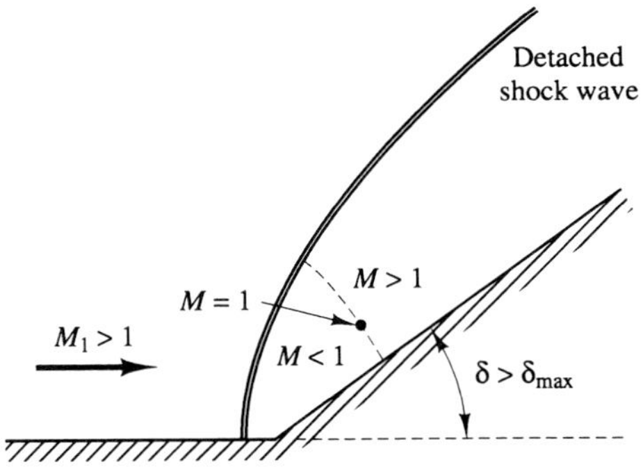

Figure 3-4 Detached Shock Wave Illustration

A detached shock wave represents a much more complex phenomenon than an attached oblique shock wave. Since the detached shock wave is curved, with the local shock wave angle changing, the strength of the shock wave is not constant. This gives rise to a flow field behind a curved shock wave in which each streamline has a different value of entropy, and, hence, stagnation pressure. Such a flow field is two-dimensional and is not amenable to treatment with the techniques discussed in this chapter.

3.3 OBLIQUE SHOCK WAVE APPLICATIONS

The preceding section has developed and discussed many useful concepts associated with oblique shock waves. We shall now examine some computational aids and techniques for oblique shock waves. Aids for use in oblique shock computations are available in the form of graphs, and personal computer software. Appendix C contains both graphical for oblique shock waves. Entries C-1 through C-3 contain graphs that present shock wave angle, downstream Mach number, and surface pressure coefficient as a function of the deflection angle with the incident Mach number as the parameter. The pressure coefficient is defined as

$$C_p = \frac{P - P_\infty}{\frac{1}{2}\rho_\infty V_\infty^2}. \tag{3-28}$$

Solving for P/P_∞ and using the definitions of Mach number and speed of sound as well as the equation of state, yields

$$\frac{P}{P_\infty} = 1 + \frac{\gamma M_\infty^2}{2} C_p. \tag{3-29}$$

For the particular case of Figure C-3, this becomes

$$\frac{P_2}{P_1} = 1 + \frac{1}{2}\gamma M_1^2 C_p. \tag{3-30}$$

These graphs are traditional computational aids used universally in the engineering workplace. The strong and weak shock wave solutions are indicated on each of the graphs. Usually the weak shock solution is what is observed.

COMPQ contains an oblique shock wave option, **OS**. As a glance at Entries C-1 through C-3 will confirm, the graphical presentations are cluttered and difficult to read accurately. The **OS** option of COMPQ combines the ease of use of tables with the flexibility of graphs and the accuracy of closed-form expressions. The closed-form solution of Mascitti (1969) was used for case **D** of option **OS**. Figure 3-5 presents the menu for the oblique shock option of COMPQ. Whereas the input for the normal shock option, **NS**, was a single parameter, the input for the **OS** option is two parameters. This should not be surprising since the plane, oblique shock wave has an additional dimension associated with it. Consider the use of COMPQ in the following example.

```
OBLIQUE SHOCK WAVE OPTION SELECTED
INPUT PARAMETERS ARE:
      M1 AND TURNING ANGLE    D
      M1 AND SHOCK ANGLE      S
      SHOCK AND TURNING ANG   B
      M1 AND STAG P RATIO     P
      M1 AND STATIC P RATIO   PR

ENTER THE DESIRED LETTER?
```

Figure 3-5 Option OS Menu

Example 3-1

At a particular location in a uniform flow, the local Mach number is 3, the local static pressure is 12 psia, and the local static temperature is 300 R. If the ratio of specific heats is 1.34, determine the pressure on the surface of a 20° half-angle wedge.

Solution. COMPQ will be used. Figure 3-6a sketches the physical situation. For this example, the incident Mach number and the deflection angle are known; thus, case **D** is chosen. The output is presented in Figure 3-7. Case **D** of COMPQ will compute the shock wave angle for the weak solution and then query to see if the strong shock solution is also desired. In this example, the answer to the query was yes (**Y**). Complete solution parameters for both the strong and weak shock solutions are then computed. Table 3-1 summarizes the results for the strong and weak oblique shock wave solutions.

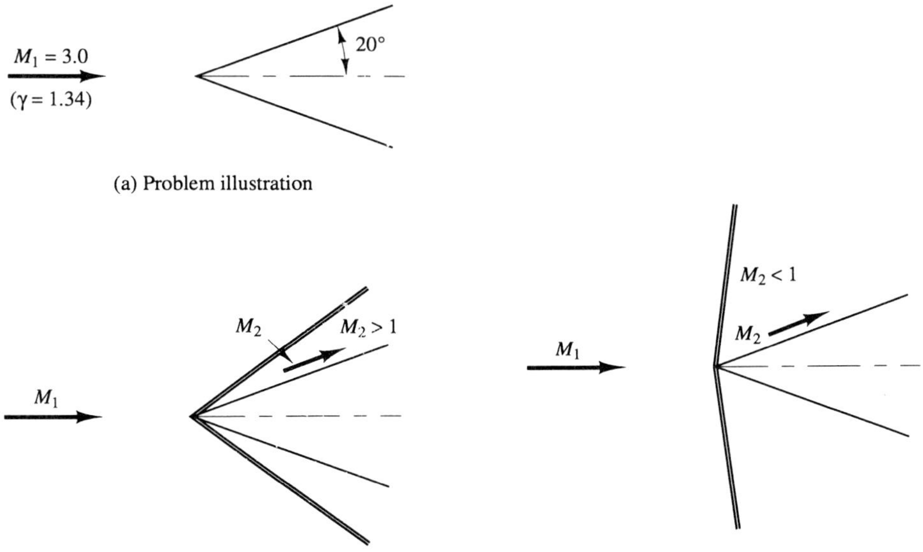

(a) Problem illustration

(b) Weak solution

(c) Strong solution

Figure 3-6 Example Problem 3-1

```
ENTER THE DESIRED LETTER? d
ENTER THE UPSTREAM MACH NUMBER? 3
INPUT THE TURNING ANGLE IN DEGREES? 20

WEAK SOLUTION:    MACH =  3  DELTA =  20 THETA =  37.07892  DEGREES
IS THE STRONG SHOCK SOLUTION DESIRED? y
STRONG SOLUTION: MACH =  3  DELTA =  20 THETA =  82.97202  DEGREES

WEAK SHOCK WAVE SOLUTION
     M1         DELTA        THETA          M2          P2/P1
   3.0000      20.0000      37.0789       2.0692       3.6016
   PO2/PO1      T2/T1      RHO2/RHO1      DELS/CP
   0.8013       1.4642       2.4597       0.0562

STRONG SHOCK WAVE SOLUTION
     M1         DELTA        THETA          M2          P2/P1
   3.0000      20.0000      82.9720       0.5194      10.0081
   PO2/PO1      T2/T1      RHO2/RHO1      DELS/CP
   0.3079       2.4190       4.1372       0.2989
```

Figure 3-7 COMPQ Output for Example 3-1

Table 3-1 affirms that the strong oblique shock wave solution is associated with much more change than the weak solution. Figures 3-6b and c illustrate the inclinations of the weak and strong oblique shock waves, respectively. With the information given, either the weak or the strong shock wave could exist. Usually, we would expect the weak solution to be the appropriate one.

TABLE 3-1 WEAK AND STRONG OBLIQUE SHOCK WAVE SOLUTIONS

Solution	M_2	θ (deg)	P_2 (psia)	T_2 (R)
Weak	2.0692	37.08	43.22	439.3
Strong	0.5194	82.97	120.10	725.7

The preceding example is typical of using COMPQ to solve an oblique shock wave problem. The remaining cases of the **OS** option provide the capability to solve directly many problems of interest. Consider a problem that cannot be solved using the oblique shock wave analysis.

Example 3-2

Sketch the flow field if a 45° half-angle wedge were placed in the uniform flow of Example 3-1.

Solution. The physical situation is sketched in Figure 3-8a. The case **D** of option **OS** in COMPQ is utilized. For a ratio of specific heats of 1.34, an incident Mach number of 3, and a deflection angle of 45°, COMPQ shows no attached oblique shock wave solution. The COMPQ output for this situation is shown in Figure 3-9. Hence, a detached, curved shock system is present. This is sketched in Figure 3-8b, but it is a qualitative representation only. The output shows that the maximum deflection angle for an attached shock at this Mach number and specific heat ratio is 35.804°. For a given M_1, case **D** with M_1 and a large (near 90°) turning angle input will output the maximum turning angle; alternately, the figures in Appendix C can be used.

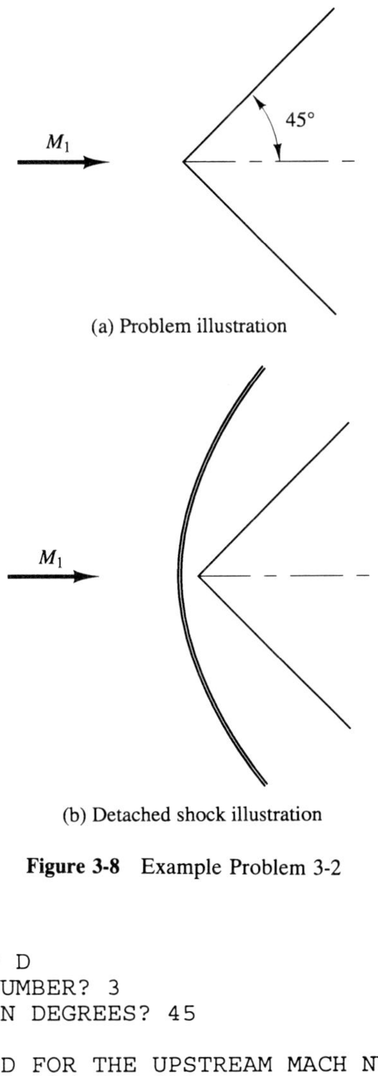

Figure 3-8 Example Problem 3-2

```
ENTER THE DESIRED LETTER? D
ENTER THE UPSTREAM MACH NUMBER? 3
INPUT THE TURNING ANGLE IN DEGREES? 45

THE TURNING ANGLE SELECTED FOR THE UPSTREAM MACH NUMBER RESULTS
IN A DETACHED SHOCK WAVE; NO OBLIQUE SHOCK SOLUTION EXIST.

FOR M1 =  3  THE MAXIMUM TURNING ANGLE FOR AN ATTACHED SHOCK IS
 35.80367 DEGREES
```

Figure 3-9 COMPQ Output for Example 3-2

Oblique shock wave applications are not limited just to wedges, but also can involve reflections, intersections, and cancellations, as well as numerous external aerodynamic examples. Some external aerodynamic applications concerning oblique shock waves will be covered in Chapter 4. The central idea involved with the reflection, intersection, or cancellation of oblique shock waves is the concept of flow deflection across the wave.

Consider, for example, the reflection of an oblique shock wave from a plane surface. The situation is portrayed in Figure 3-10a, where the required flow directions downstream of both the incident and reflected shock waves are shown. Ahead of the incident oblique shock wave the flow is parallel to the plane surface. After passage through the incident shock, the flow is deflected toward the oblique shock wave; the deflection depends on the incident Mach number and the shock wave angle. However, the flow direction in region 2 is into the plane surface. Clearly, some mechanism must exist to turn the flow parallel to the plane surface downstream of the incident shock wave. This mechanism is the reflection of the incident oblique shock wave. The reflected shock wave has the strength required to deflect the flow in region 2 back parallel to the plane surface in region 3. The net results of the incident–reflected shock system are to decrease the stagnation pressure and reduce the Mach number of the incident flow. Because of losses associated with the shock wave, the strength and the surface orientation of the reflected shock wave are different from those of the incident shock wave.

Computations involving oblique shock wave reflections are better visualized if imaginary wedges are drawn. In Figure 3-10b the incident shock is visualized as being caused by a wedge of angle δ_1 and resulting in a shock wave angle θ_1. Likewise, the reflected oblique shock wave is visualized as being caused by another wedge of angle $\delta_1 = \delta_2$. This second wedge must be offset from the plane surface by the angle δ_2 since the upstream flow direction for the reflected shock wave is into the plane surface. The shock wave angle for the reflected shock must also be measured from this offset. With uniform flow in region 1, the flow in regions 2 and 3 will also be uniform.

An oblique shock reflection such as in Figure 3-10b is called a *regular reflection* and involves only plane, weak oblique shock waves. A regular reflection can occur only if the Mach number in region 2 and the deflection angle required for the reflected shock admit an oblique shock solution. Since this requires M_2 to be supersonic, the incident shock wave cannot be a strong shock if a regular reflection is to exist. But, even if M_2 is supersonic, the required deflection angle for the reflected shock can exceed the maximum for an attached oblique shock wave. A Mach reflection occurs if a regular reflection is not possible. In a Mach reflection, a normal shock is formed at the plane surface. The flow is parallel to the plane surface ahead of and behind the normal shock. Away from the plane surface the normal shock wave becomes curved and intersects the incident oblique shock wave. At the point of intersection of the curved shock from the surface and the incident oblique shock, another curved shock wave is formed as the reflected shock. Qualitative aspects of the Mach reflection are presented in Figure 3-10c. The dashed line demarcates the flow that passes through two shock waves from the flow that passes through only one. A slip line across which some of the properties are discontinuous is formed at the junction of the three shock waves. The shock wave from the plane wall to the junction is called a Mach shock wave. Subsonic flow exists downstream of the Mach shock wave. Because of the curvature of the shock waves involved in the system, the flows in regions 2, 3, and 4 are not uniform. The Mach reflection is much more complex than the regular reflection and is not amenable to analysis using techniques discussed in this chapter.

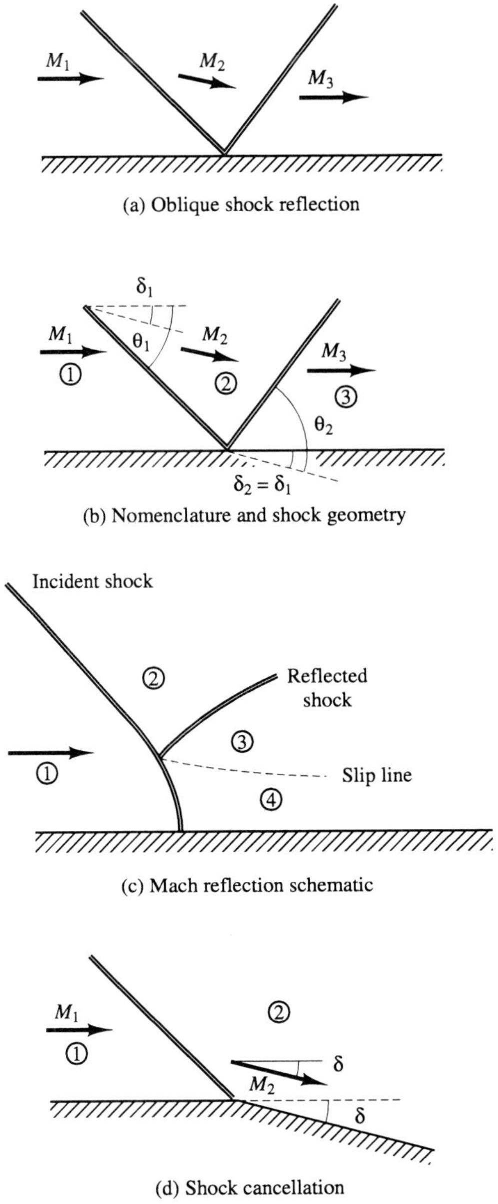

Figure 3-10 Oblique Shock Wave Reflections

The concept of oblique shock wave cancellation is also tied to flow deflection. In the regular shock reflection just described, the purpose of the reflected shock is to establish parallel flow adjacent to the plane surface. If cancellation of the shock is desired, then the need for a reflected shock wave must be removed. This can be done by altering the surface to eliminate the need for the flow to be deflected. Hence, as illustrated in Figure 3-10d, the incident shock need not lead to a reflection if, at the point of intersection of the incident shock and the wall, the wall is altered to be parallel to the flow in region 2. The altered wall removes the need for a reflection; region 3 is no longer present, and the flow is uniform downstream of the incident oblique shock wave. These concepts are used next, in Example 3-3.

Example 3-3

An oblique shock wave makes an angle of 30° with a uniform flow of air at $M_1 = 3$. If the shock is incident on a plane wall, compute the Mach number and the change in static and stagnation pressures across the system. What wall configuration should be used if no reflected shock is desired?

Solution. COMPQ will be used. The shock wave angle and upstream Mach number are known for the incident shock. Using case **S** of option **OS** produces the following results:

$$M_2 = 2.3673$$

$$\delta = 12.7735^\circ$$

$$\frac{P_2}{P_1} = 2.4583$$

$$\frac{P_{o2}}{P_{o1}} = 0.9298.$$

The reflected shock must turn the flow parallel to the plane surface. The reflected shock solution is computed using case **D** of option **OS** with the following output:

$$M_3 = 1.8586$$

$$\frac{P_3}{P_2} = 2.1178$$

$$\frac{P_{o3}}{P_{o2}} = 0.9584,$$

from which we derive the following results:

$$\frac{P_3}{P_1} = \frac{P_3}{P_2}\frac{P_2}{P_1} = (2.1178)(2.4583) = 5.2062$$

$$\frac{P_{o3}}{P_{o1}} = \frac{P_{o3}}{P_{o2}}\frac{P_{o2}}{P_{o1}} = (0.9584)(0.9298) = 0.8911.$$

Figure 3-11a contains a representation of the reflected shock wave solution. The reflected shock angle is measured from 12.77° below the plane surface.

If no shock reflection is desired, then the surface must be altered in such a fashion that it is parallel to the flow direction downstream of the incident shock wave. That can be accomplished as shown in Figure 3-11b.

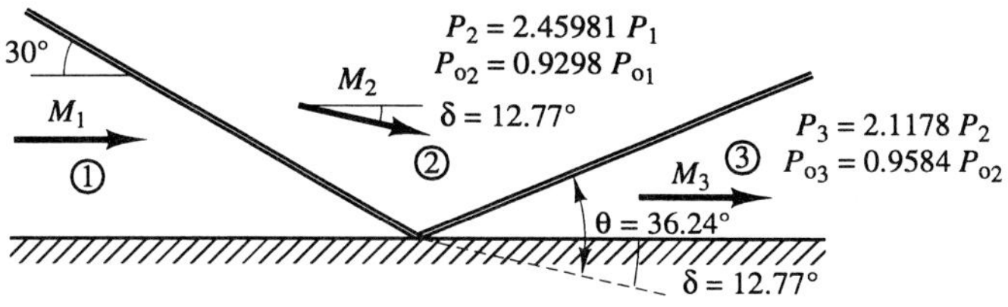

(a) Shock reflection solution

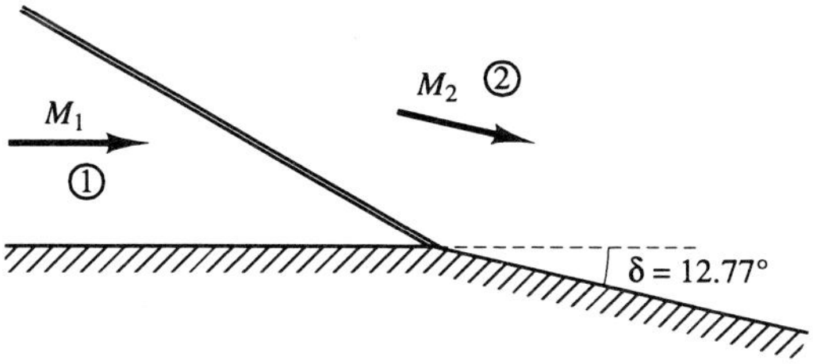

(b) Reflected shock wave cancellation

Figure 3-11 Example Problem 3-3

The preceding discussion on oblique shock wave reflections neglected viscous effects. In reality, the presence of a boundary layer on the surface complicates the interaction process. The shock waves in the reflection system can penetrate no deeper in the boundary layer than the sonic line. Below the sonic line, the flow is subsonic and information can propagate upstream, causing the boundary layer, especially if laminar, to thicken rapidly. A schematic of the interaction of an oblique shock wave with a boundary layer (whose thickness is greatly exaggerated) is illustrated in Figure 3-12. The thickening boundary layer acts as a gradual compression from which weak oblique shock waves emanate to coalesce and intersect with the incident oblique wave. Flow separation can occur if the pressure change, which results in an adverse pressure gradient, across the reflection system is large enough. Reflections with laminar boundary layers can result in large regions of separated flow. Reattachment of the separated boundary layer and thinning of the boundary layer after the interaction impose additional weak expansion and compression waves downstream of the reflected shock. These weak wave systems interact with the reflected shock. However, even if the details of the viscous interaction are neglected, the pressure recovery downstream of the reflection region is usually adequately predicted by consideration of inviscid flow.

The regular interaction of two oblique shock waves is illustrated in Figure 3-13a. Several different regions, each with uniform flow, are identified in the figure. Region 2 is downstream of the oblique shock generated by the upper wedge, and region 3 is downstream of the oblique shock generated by the lower wedge. The flow conditions, including the directions, in regions 2 and 3 are different. The flow directions in these regions are parallel to the upper and lower plane surfaces, respectively. The "refraction" of the oblique shock system results because the flow directions in regions 2 and 3 are not parallel and something must happen to achieve parallel flow between regions 4 and 5. If the flow directions in regions 4 and 5 are not parallel, then either a void is created by the diverging flow directions or the flow directions intersect. Neither situation is possible. Moreover, since regions 4 and 5 are adjacent and are separated by no bounding surface, the pressures in regions 4 and 5 must be equal.

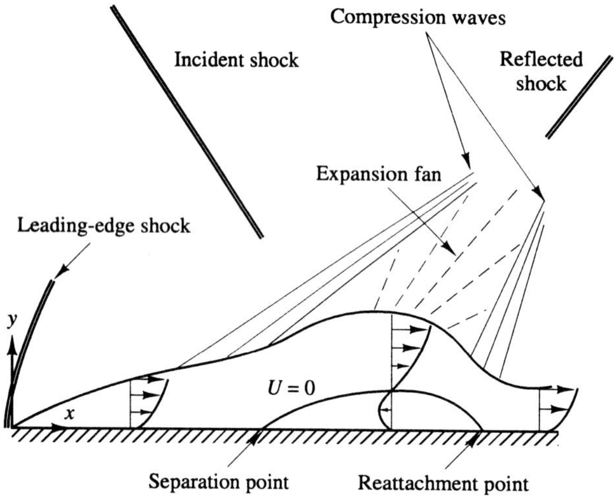

Figure 3-12 Schematic of Boundary-Layer/Oblique Shock Wave Interaction

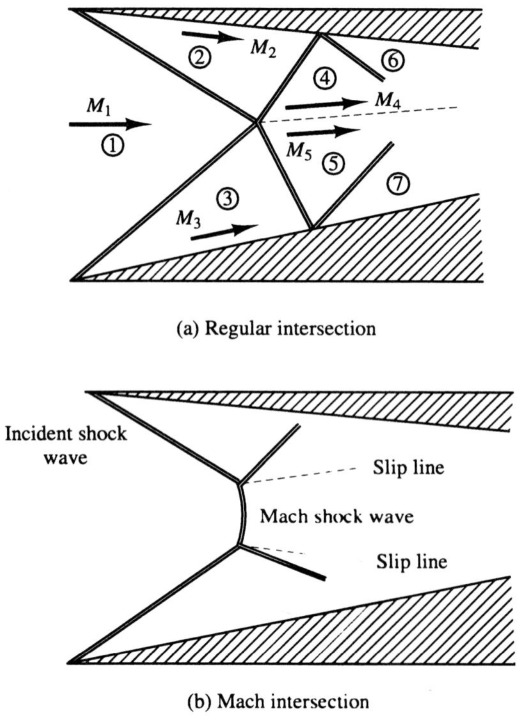

Figure 3-13 Oblique Shock Wave Intersection

The line separating regions 4 and 5 is also a slip line. Even though the flow directions and the pressures in regions 4 and 5 are equal, other physical properties, notably the entropy and temperature, may not be equal. The thermodynamic state in region 4 is determined by passage through the shock system defining region 2, whereas the state in region 5 is determined by passage through the shock system defining region 3. If the upper and lower deflections or wedge angles are equal, then the shock interaction system is symmetrical and the same states exist in regions 4 and 5. However, if the initial deflection angles are not equal, then a symmetry line is not present in the shock interaction system, and physical properties, other than the pressure, across the slip line are discontinuous. Hence, for nonsymmetrical oblique shock wave interactions, the entropy, density, temperature, Mach number, velocity, and stagnation pressure differ in regions 4 and 5. In inviscid terms we can think of the slip line as remaining distinct downstream; in reality, especially if the velocities in regions 4 and 5 differ significantly, viscous effects tend to diffuse the slip line in the downstream direction.

Figure 3-13a illustrates a regular intersection. As with the oblique shock wave reflection, the incident shocks must be weak shock solutions if a regular intersection is to take place. The deflections required in regions 2 and 3 must be less than the maximum deflection angles for attached oblique shock waves if regular intersections are to provide parallel flow in regions 4 and 5. If a regular intersection is precluded, then a Mach intersection takes place. Figure 3-13b presents a Mach intersection. The nomenclature and components of the Mach intersection are similar to the components of the Mach reflection. The Mach intersection is more complex than the regular intersection, and the curved shock waves in the system result in nonuniform flow fields.

An example is useful at this point.

Example 3-4

Solve the oblique shock wave interaction problem given in Figure 3-14a. A uniform flow of air at Mach 3.0 enters a channel defined with an upper wedge of 6° and a lower wedge of 12°.

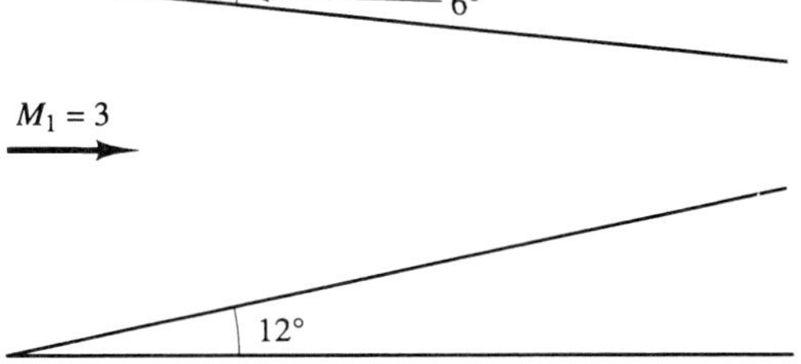

(a) Problem schematic

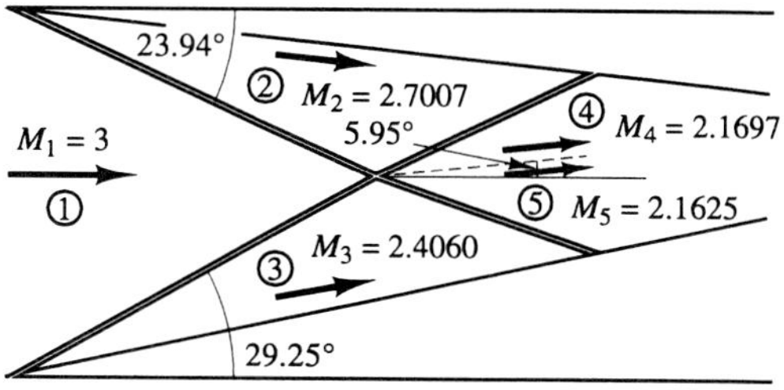

(b) Problem solution

Figure 3-14 Example Problem 3-4

Solution. COMPQ will be used. The oblique shock interaction system is sketched in Figure 3-14b. Conditions in regions 2 and 3 are found by using case **D** of option **OS**.

These conditions are

$$M_2 = 2.7007 \qquad P_2/P_1 = 1.5618$$
$$M_3 = 2.4060 \qquad P_3/P_1 = 2.3404.$$

The refracted oblique shock waves defining conditions in regions 4 and 5 must be such that both the pressures and flow directions in the regions are the same. The procedure is iterative. The iterative sequence, given in Table 3-2, is to assume a flow direction in the regions and then compute the pressures using the assumed flow direction. When the assumed flow direction results in equal pressures in regions 4 and 5, convergence is attained. The pressure ratios P_4/P_1 and P_5/P_1 are computed from the COMPQ output by

$$\frac{P_4}{P_1} = \frac{P_4}{P_2}\frac{P_2}{P_1}$$
$$\frac{P_5}{P_1} = \frac{P_5}{P_3}\frac{P_3}{P_1}.$$

TABLE 3-2 ITERATION SEQUENCE FOR OBLIQUE SHOCK WAVE INTERACTION

Flow Angle (deg)	Flow Deflection 2–4	3–5	P_4/P_2	P_4/P_1	P_5/P_3	P_5/P_1
0	6.0	12.0	1.5044	2.3496	2.0481	4.7933
+2.0	8.0	10.0	1.7106	2.6716	1.8313	4.2860
+6.0	12.0	6.0	2.1857	3.4136	1.4517	3.3976
+5.8	11.8	6.2	2.1605	3.3743	1.4689	3.4378
+5.95	11.95	6.05	2.1794	3.4038	1.4556	3.4067

The converged solution Mach numbers, the shock wave angles, and the flow direction in regions 4 and 5 are indicated on Figure 3-14b. The flow in regions 4 and 5 is inclined 5.95° toward the upper wedge. While little difference is shown in the regions' Mach numbers, variations in other properties across the slip line are more significant.

Two other interactions are of interest:

1. the confluence of two supersonic streams
2. the coalescence of two oblique shock waves to form a single oblique shock wave.

The schematic representation of the confluence of two supersonic streams is presented in Figure 3-15a. Generally, both streams must deflect such that compression processes occur unless P_1 and P_2 are quite different. The mechanism for the compression processes is the formation of two oblique shock waves. Downstream of the shock waves the pressures and flow directions in regions 3 and 4 must be the same. Thus, the shock wave angle of each of the oblique shock waves must be such that in regions 3 and 4 the flow directions are parallel and the pressures are equal. Because no curved shock waves are involved, conditions are uniform in each region. Even though the flow directions and pressures in regions 3 and 4 are the same, other properties may vary across the slip line formed at the confluence location. Given the Mach numbers, flow directions, and pressures of each stream, an iterative process similar to that used in Example 3-4 is necessary to define the deflection angles (and hence the shock wave angles) required.

The coalescence of two oblique shock waves into a single oblique shock wave is illustrated in Figure 3-15b. A slip line, across which the flow direction and pressure are equal, is formed where the two incident oblique shock waves intersect. In order for both the pressure and flow direction in regions 4 and 5 to be equal, a weak "reflected" wave dividing regions 3 and 4 is usually necessary. Depending on the nature of the upstream conditions, the weak reflected wave can be either an expansion process or a compression shock wave.

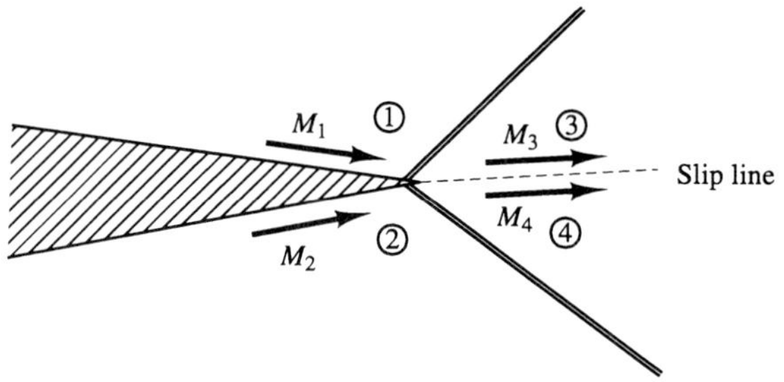

(a) Confluence of two supersonic streams

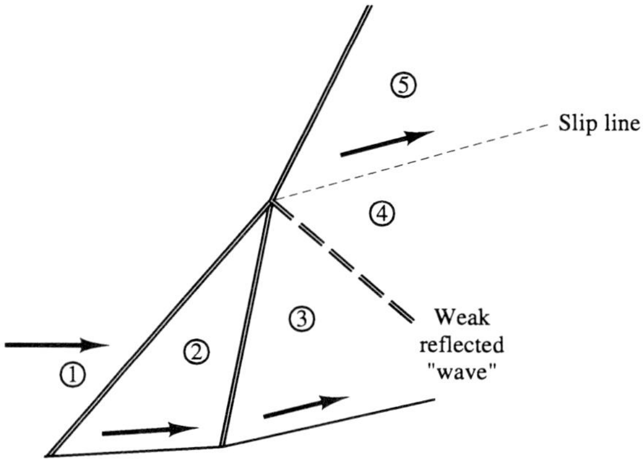

(b) Coalescence of two oblique shock waves

Figure 3-15 Other Oblique Shock Wave Interactions

3.4 SUPERSONIC FLOW OVER A CONE

The previous section examined computational techniques and aids for a variety of problems involving oblique shock waves. That development is valid only for oblique shock waves in planar, two-dimensional configurations. One frequently occurring situation in compressible flow is supersonic flow over a sharp cone—an *axisymmetric configuration*. Axisymmetric, supersonic conical flows are of great practical importance and possess significant enough differences from two-dimensional wedge flows to warrant a separate presentation.

Consider a sharp cone of half angle, δ_c, as sketched in Figure 3-16a, in a supersonic flow. An oblique shock wave, conical in shape, is attached to the vertex of the cone. The conical shock wave angle, however, is different from the oblique shock wave angle corresponding to a deflection of δ_c. The flow field between the conical shock wave and the surface of the cone is not uniform because of the three-dimensional relieving effect. As Section A-A in Figure 3-16b illustrates, the axisymmetric nature of the conical flow field results in an increased flow cross-sectional area as the distance from the axis is increased. This increased area provides additional expansion in the radial direction and is responsible for the relieving effect in axisymmetric flows when compared to planar flows.

A result of the conical flow field is that streamlines are *curved behind the conical shock wave*, whereas streamlines are *straight and parallel behind oblique shock waves*. The change of flow direction for the conical flow and the uniformity of the flow direction for the wedge flow are illustrated by the streamlines in Figure 3-16c. Hence, in conical flows an additional compression takes place between the conical shock wave and the surface of the cone. The oblique shock relations do hold locally across a conical shock wave; thus, given a conical shock wave angle, the components and properties immediately downstream of the shock can be computed. The direction of flow and properties are different on the surface of the cone than immediately downstream of the conical shock wave. The flow field between the conical shock wave and the cone surface is then a two-dimensional flow field and is more difficult to treat than the corresponding uniform flow field behind a plane, oblique shock wave.

Experimental data confirm the validity of the following assumptions made in the analysis of conical supersonic flow:

1. The shock wave is attached at the vertex of the cone.
2. The shock wave is conical in shape.
3. All flow properties are constant along radial lines, that is, along lines of constant θ. See Figure 3-17.
4. The flow field between the shock wave and the cone surface is isentropic.

Taylor and Maccoll (1933), Chapman and Walker (1971), Shapiro (1953), and Anderson (1990) all have detailed developments of the equations for conical flow. For our purposes a brief overview will be sufficient.

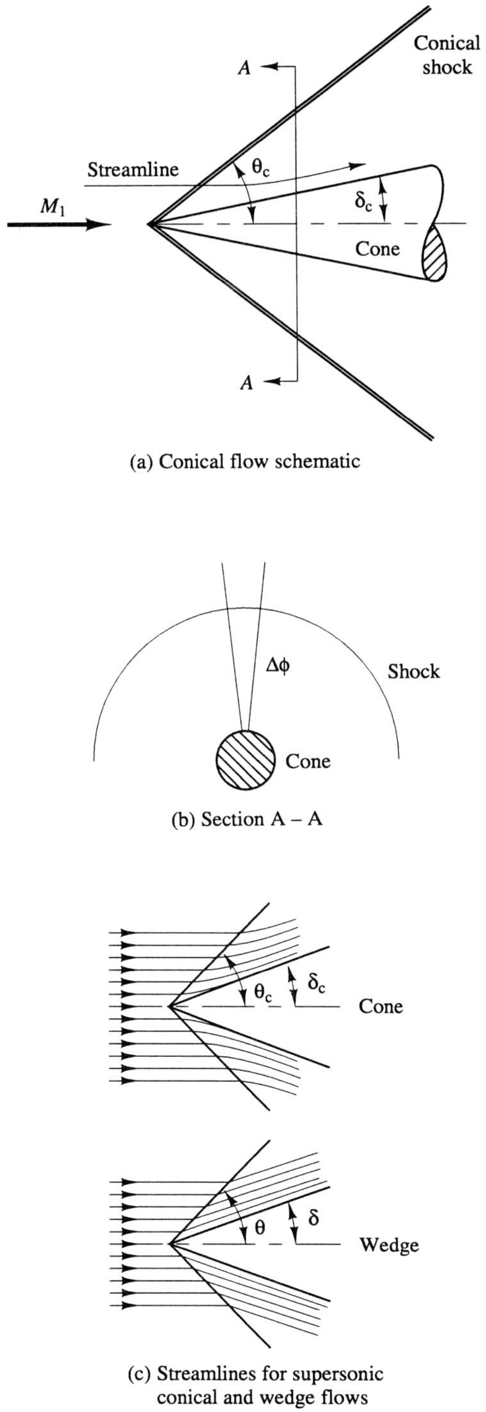

Figure 3-16 Conical Flow Nomenclature and Schematic. *Adapted from Zucrow and Hoffman.*

As we shall see in Chapter 9, one of the ramifications of a flow with constant entropy and constant stagnation enthalpy is that

$$\nabla \times \overline{V} = 0, \tag{3-31}$$

where ∇ is the del operator from vector analysis. Flows with $\nabla \times \overline{V} = 0$ are termed *irrotational*. For the case of conical flow (no circumferential or radial dependence) and spherical coordinates, the condition of irrotationality requires that

$$V_\theta = \frac{dV_r}{d\theta}, \tag{3-32}$$

where V_r is the radial component of velocity and V_θ is the tangential component of velocity. See Figure 3-17 for the velocity components and coordinate system definitions.

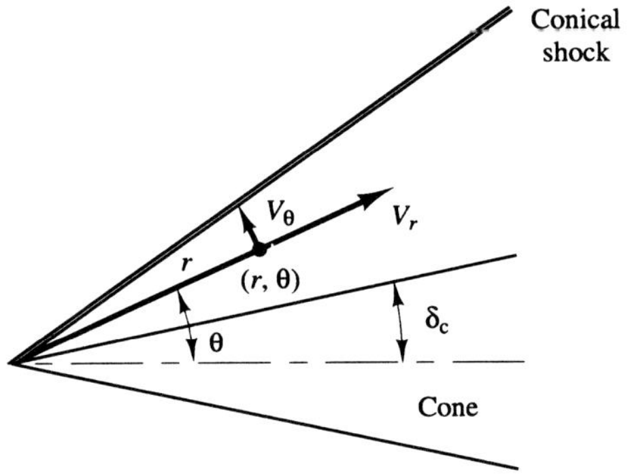

Figure 3-17 Velocity Components and Coordinate System

The continuity equation (see Chapter 9 for details) for steady compressible flow is

$$\nabla \cdot (\rho \overline{V}) = 0, \tag{3-33}$$

which for the special case of conical flow and spherical coordinates can be expressed as

$$2\rho V_r + \rho V_\theta \cot\theta + \rho\frac{dV_\theta}{d\theta} + V_\theta\frac{d\rho}{d\theta} = 0. \tag{3-34}$$

With no radial dependence, the only independent variable is θ: hence the use of the total derivatives.

The momentum equation for an inviscid flow has the form

$$dP = -\rho V dV. \tag{3-35}$$

This form of the momentum equation is called *Euler's equation* and is developed and discussed in Chapter 9. For the isentropic flow between the conical shock wave and the surface,

$$\left(\frac{\partial P}{\partial \rho}\right)_s = \frac{dP}{d\rho} = a^2. \tag{3-36}$$

The energy equation simply states that the stagnation enthalpy is constant; hence

$$h_o = h + \frac{V^2}{2} = \frac{V_{\max}^2}{2}. \tag{3-37}$$

The total velocity is

$$V = \sqrt{V_\theta^2 + V_r^2}, \tag{3-38}$$

which when combined with Eqs. 3-37 and 3-38, yields

$$\frac{V}{V_{\max}} = \left[\frac{2}{(\gamma - 1)M^2} + 1\right]^{-1/2}. \tag{3-39}$$

When Eqs. 3-32, 3-34, 3-35, and 3-36 are combined, the differential equation expressed in terms of V_r becomes

$$\begin{aligned} &\frac{\gamma - 1}{2}\left[V_{\max}^2 - V_r^2 - \left(\frac{dV_r}{d\theta}\right)^2\right]\left[2V_r + \frac{dV_r}{d\theta}\cot\theta + \frac{d^2 V_r}{d\theta^2}\right] \\ &\qquad - \frac{dV_r}{d\theta}\left[V_r\frac{dV_r}{d\theta} + \frac{dV_r}{d\theta}\frac{d^2 V_r}{d\theta^2}\right] = 0. \end{aligned} \tag{3-40}$$

Let

$$V' \equiv \frac{V}{V_{\max}} \tag{3-41a}$$

$$V_r' \equiv \frac{V_r}{V_{\max}} \tag{3-41b}$$

$$V_\theta' \equiv \frac{V_\theta}{V_{\max}}. \tag{3-41c}$$

The most convenient form of Eq. 3-40 then results:

$$\begin{aligned} &\frac{\gamma - 1}{2}\left[1 - V_r'^2 - \left(\frac{dV_r'}{d\theta}\right)^2\right]\left[2V_r' + \frac{dV_r'}{d\theta}\cot\theta + \frac{d^2 V_r'}{d\theta^2}\right] \\ &\qquad - \frac{dV_r'}{d\theta}\left[V_r'\frac{dV_r'}{d\theta} + \frac{dV_r'}{d\theta}\frac{d^2 V_r'}{d\theta^2}\right] = 0. \end{aligned} \tag{3-42}$$

This equation is frequently referred to as the *Taylor-Maccoll equation.*

Equation 3-42 is valid between the conical shock wave and the surface of the cone. It is an ordinary, nonlinear, second-order differential equation. Appropriate boundary conditions are from the physical situation as illustrated in Figure 3-16a. On the surface of the cone the tangential component of the velocity must vanish; that is,

$$V_\theta(\delta_c) = 0. \tag{3-43}$$

The Mach number and property relations, Eqs. 3-10, across an oblique shock wave also hold across a conical shock wave of the same angle; so given the upstream Mach number and the conical shock wave angle, conditions are known immediately behind the shock wave.

The complexity of the differential equation precludes a closed-form solution. A number of different numerical schemes have been used to solve the conical flow field. Program CONEQ uses a fourth-order Runge-Kutta method to solve the differential equation. Any solution technique must start either with a given cone angle and surface Mach number or with a given shock wave angle and upstream Mach number, and integrate to locate, respectively, the shock wave angle and freestream Mach number or the cone angle and the surface Mach number. CONEQ solves directly for the cone angle δ_c and surface Mach number given the freestream Mach number and conical shock angle. If the cone angle and freestream Mach number are specified—a useful pair of input quantities—the program iterates to find the shock wave angle and the surface conditions. CONEQ is more complex than any element of COMPQ and generally requires more execution time. For our purposes we will use CONEQ in much the same manner as the oblique shock option, **OS**, of COMPQ.

The greater complexity of CONEQ requires some examination of its capabilities and input and output structures. The cases available from CONEQ are listed in Figure 3-18. Basically, two inputs, either two angles or an angle and a Mach number, are required for all cases except **MC**. The maximum cone angle for an attached shock wave at a specified Mach number is computed for case **MC**. The output from a typical execution of CONEQ, case **S**, is presented in Figure 3-19. For cases **S**, **C**, or **B**, the output consists of the information on states immediately behind

```
TAYLOR-MACCOLL CONICAL FLOW CALLED FOR

INPUT PARAMETERS ARE:
      M1 AND SHOCK ANGLE          S
      M1 AND CONE ANGLE           C
      SHOCK AND CONE ANGLES       B
      MAX CONE ANGLE FOR M1       MC
      FINISHED                    FI
THE SPECIFIC HEAT RATIO IS        1.4
```

Figure 3-18 CONEQ Menu

```
ENTER THE DESIRED LETTER? s
ENTER THE UPSTREAM MACH NUMBER, M1? 3.0
ENTER THE SHOCK WAVE ANGLE IN DEGREES? 40.0

CONICAL SUPERSONIC FLOW USING THE TAYLOR-MACCOLL EQUATION

MACH NUMBER =   3.0000       SHOCK ANGLE =  40.0000       CONE ANGLE =  30.1959

BEHIND CONICAL SHOCK: M2           T2/T1        P2/P1        DELS/CP      DELTA
                      1.8942       1.6302       4.1717       0.0806      21.8461

ON CONE SURFACE:      MS           TS/T1        PS/P1        RHOS/RHO1    P. COEF.
                      1.8218       1.6829       4.6637       2.7712       0.5815

STAGNATION TEMPERATURE IS CONSTANT.  P02/P01 = P0S/P01 =   0.7542

ARE ADDITIONAL SUPERSONIC CONE FLOW SOLUTIONS NEEDED?
```

Figure 3-19 Typical CONEQ Output

the conical shock wave and on the surface of the cone. The stagnation pressure ratio across the conical shock wave is also indicated on the output. Since the flow field is isentropic between the conical shock wave and the cone surface, the stagnation pressure is constant in this region. On the surface, both the pressure ratio and the pressure coefficient are given.

CONEQ can be used much like the **OS** option of COMPQ to investigate characteristics of conical shock wave solutions. Like the wedge flow, conical flow admits two solutions: a strong and a weak. The weak conical shock solution is the one always observed in the absence of any intervention; the strong conical shock solution has only been observed on truncated cones by independently increasing the pressure near the base of the cone. CONEQ computes the weak solution only as the likelihood of the strong shock appearing naturally is quite small.

Angles larger than the maximum cone semivertex angle for an attached conical shock wave are associated with shock wave structures and flow fields that are not conical. The maximum cone semivertex angle for an attached conical shock wave is often called the *conical shock detachment angle*. For cone angles greater than the maximum for a given freestream Mach number, a detached, curved shock wave system, similar to that for a wedge, is formed. The maximum wedge angle for an attached, planar oblique shock is called the *wedge shock detachment angle*. Because

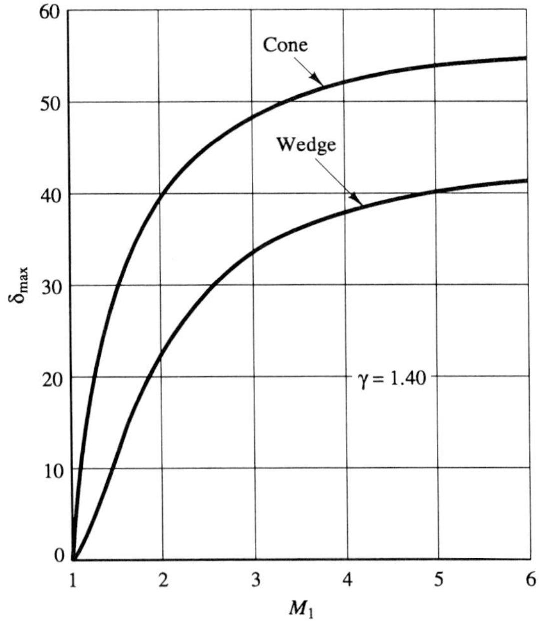

Figure 3-20 Comparison of Maximum Wedge and Cone Angles (γ = 1.4). *Adapted from Zucrow and Hoffman (1976).*

of the three-dimensional relieving effect associated with an axisymmetric flow versus a planar flow, the conical detachment angle is larger than the wedge detachment angle at a given upstream Mach number. Case **MC** of CONEQ computes the maximum cone angle for an attached conical shock wave with the freestream Mach number and the ratio of specific heats as the inputs. Figure 3-20, taken from Zucrow and Hoffman (1976), contains graphical representations of the maximum cone semivertex angle and the maximum wedge angle for attached shock waves for $\gamma = 1.4$ as a function of upstream Mach number.

Computational aids are available for supersonic conical flows. In addition to CONEQ, Figures C-4 through C-6 of Appendix C contain graphs that present conical shock wave angle, surface Mach number, and surface pressure coefficient as a function of cone angle. The maximum cone angle for an attached conical shock wave is also indicated. Consider the following examples.

Example 3-5

A 20° half-angle cone is placed in an airstream flowing at $M_1 = 3$. If the freestream static properties are 10 kPa and 200 K, find the pressure and Mach number on the surface of the cone.

Solution. CONEQ will be used. The upstream (freestream) Mach number and the cone angle are known. Case **C** of CONEQ is then appropriate. The output is shown in Figure 3-21a, from which the desired cone surface Mach number is 2.290. The solution is illustrated in Figure 3-21b. The initial deflection as the flow crosses the conical shock wave is 12.378°. During the compression process from the shock wave to the cone surface, the flow is turned away from the surface an additional 7.622°. The pressure ratio (P_s/P_1) on the surface is 2.791, from which the surface pressure is

$$P_s = 27.91 \text{ kPa}.$$

The state immediately behind ($\theta = 29.615°$) the conical shock wave corresponds to the state downstream of a plane, oblique shock wave with a shock wave angle of 29.615°. The state on the surface ($\theta_c = 20°$) corresponds to an isentropic compression from immediately behind the shock wave to the cone surface. As Figure 3-21a indicates, during that process the Mach number decreases, the pressure and temperature increase, and the entropy remains constant.

Example 3-6

At $M_1 = 2.5$, what is the difference in the maximum cone angle for an attached shock wave for a diatomic gas ($\gamma = 1.4$) and a monatomic gas ($\gamma = 1.667$)?

Solution. CONEQ will be used. Option **MC** was used to construct Table 3-3. The table shows that the smaller value of γ possesses the larger maximum cone half angle for an attached conical shock wave. This example provides some insight into the effect of the ratio of specific heats on supersonic conical flows.

```
ENTER THE DESIRED LETTER? c
ENTER THE UPSTREAM MACH NUMBER, M1? 3.0
ENTER THE CONE ANGLE IN DEGREES? 20.0


CONICAL SUPERSONIC FLOW USING THE TAYLOR-MACCOLL EQUATION

MACH NUMBER =   3.0000        SHOCK ANGLE =  29.6154        CONE ANGLE =  20.0000

BEHIND CONICAL SHOCK: M2           T2/T1        P2/P1        DELS/CP      DELTA
                      2.3871       1.3086       2.3975       0.0191      12.3783

ON CONE SURFACE:      MS           TS/T1        PS/P1        RHOS/RHO1    P. COEF.
                      2.2899       1.3667       2.7910       2.0422       0.2843

STAGNATION TEMPERATURE IS CONSTANT.  P02/P01 = P0S/P01 =  0.9352

ARE ADDITIONAL SUPERSONIC CONE FLOW SOLUTIONS NEEDED?
```

(a) Compcone output

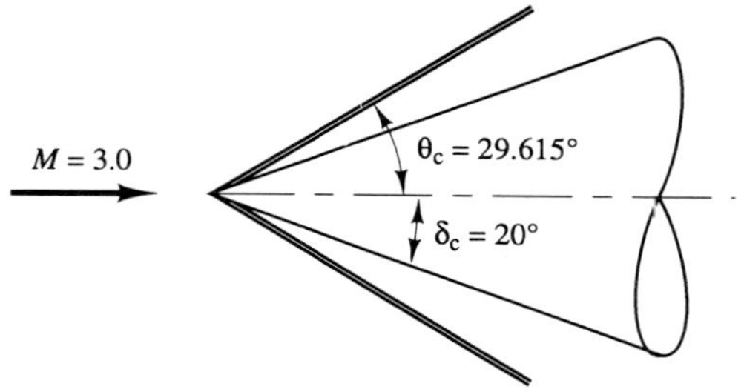

(b) Solution illustration

Figure 3-21 CONEQ Output for Example 3-5

TABLE 3-3 MAXIMUM CONE ANGLE FOR ATTACHED SHOCK WAVES FOR M = 2.5 FOR DIFFERENT RATIOS OF SPECIFIC HEATS

γ	δ_c (deg)$_{max}$	θ (deg)
1.4	46.117	70.443
1.667	41.650	68.562

As Example 3-6 illustrates, the ratio of specific heats can have a noticeable effect on conical supersonic flows. Patel et al. (1993) used a version of the software element CONEQ to study the effects of specific heats ratio on the characteristics of conical supersonic flows. They also assembled a complete set of conical supersonic flow charts, in the format of Charts C-4, C-5, and C-6 of Appendix C, for the following ratios of specific heats: 1.000001, 1.1, 1.2, 1.3, 1.4, 1.5, 1.6, and 1.67. Conical flows with different values of the ratio of specific heats are of interest since different gases have different specific heats ratios. Recall from Chapter 1 that monatomic gases, such as helium, possess the largest value of γ, 1.67; diatomic gases, such as nitrogen and oxygen, and mixtures of diatomic gases, such as air, possess a ratio of specific heats of 1.4. The smallest value of γ is near unity for gases composed of very complex molecules. The value of $\gamma = 1.000001$ (hereinafter written as 1.0^+) represents the lower limiting case considered by Patel et al.

The presentation, analogous to Figure 3-20, for ratios of specific heats of 1.0^+, 1.2, and 1.67 are given in Figure 3-22. While the same trends are evident with upstream Mach number, the smaller the value of the ratio of specific heats, the larger the shock detachment angle for a given upstream Mach number. Moreover, the smaller the value of the specific heats ratio, the smaller the difference between the cone and wedge shock detachment angles for a given upstream Mach number. The asymptotic behavior of shock detachment angles for the cone and wedge also depends on the ratio of specific heats. For example, the asymptotic nature of the $\gamma = 1.67$ behavior is evident at a much smaller upstream Mach number than is the asymptotic behavior for $\gamma = 1.0^+$.

If the conical shock detachment angle is plotted as a function of upstream Mach number for each of the ratios of specific heats, then Figure 3-23 results. As the ratio of specific heats is increased from 1.0^+ to the limiting value of 1.67, the distinctions between the behavior in increments of ratio of specific heats of 0.1 diminish.

The conical shock wave angles that correspond to the conical shock detachment angles of Figure 3-23 are also of interest. These shock wave angles are identified as θ_{max} and are plotted in Figure 3-24 as a function of upstream Mach number with the ratio of specific heats as a parameter. For upstream Mach numbers close to 1.0, the conical shock wave is nearly a normal shock wave regardless of the ratio of specific heats. As the upstream Mach number is increased from near unity, the conical shock

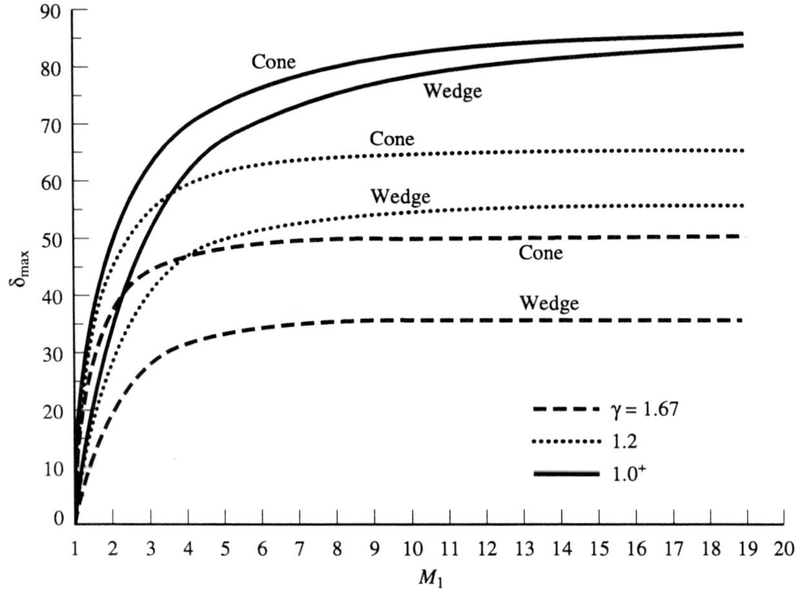

Figure 3-22 Comparison of Conical and Wedge Shock Detachment Angles for $\gamma = 1.0^+$, 1.2, and 1.67

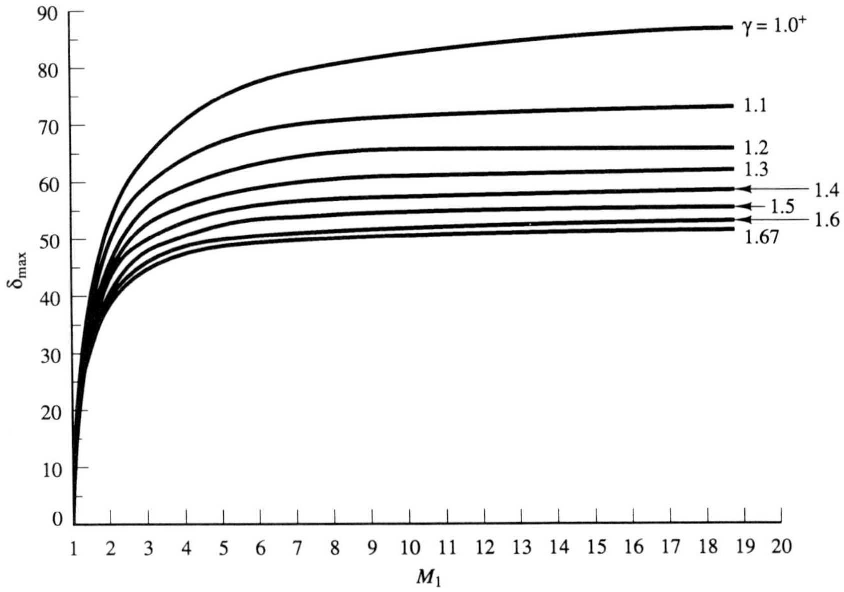

Figure 3-23 Conical Detachment Angles for Different Ratios of Specific Heats

wave angle associated with the conical shock detachment angle reaches a minimum and then increases monotonically as the Mach number increases. The ratio of specific heats is observed to have an important effect on the conical shock detachment angle and is especially significant for the lower values of the ratio.

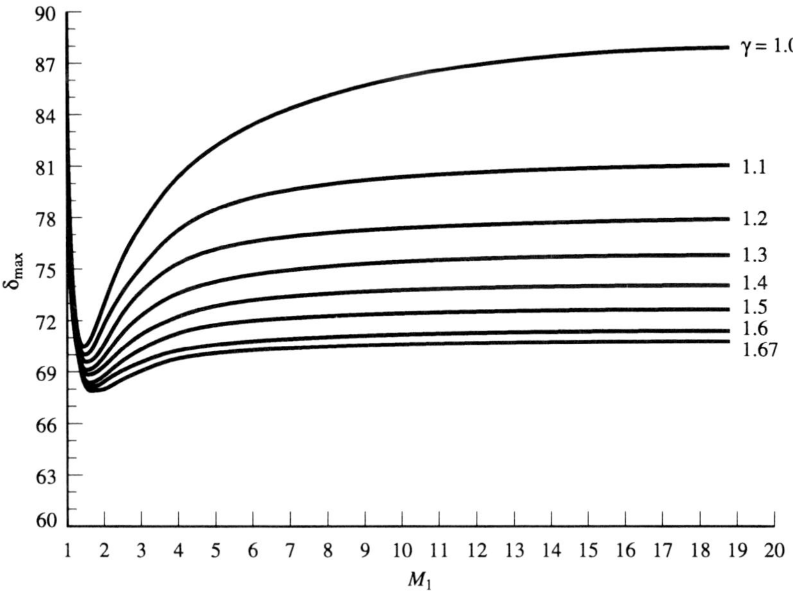

Figure 3-24 Shock Wave Angles for the Conical Detachment Angles of Figure 3-23

As mentioned earlier, the isentropic compression process that occurs from the conical shock wave to the surface of the cone results in an increase in pressure and a decrease in Mach number. For a relatively small region of shock wave angle and incident Mach numbers, the compression process starts with a supersonic Mach number and produces a subsonic Mach number on the surface, thus going from supersonic flow to subsonic flow without an additional shock wave. As an example, consider a conical flow with an incident Mach number of 1.5 and a shock angle of 58°. The output of CONEQ for this situation ($\gamma = 1.4$) is given in Figure 3-25a, and the resulting flow field is sketched in Figure 3-25b. Across the conical shock wave, the Mach number is decreased from 1.5 to 1.086. The compression process from the shock wave to the surface of the cone results in a reduction of the Mach number to 0.938. During this compression, the pressure is increased from 1.721 P_1 to 2.046 P_1, an increase of nearly 19 percent. The flow direction changes by 16.6° in the course of the compression.

```
ENTER THE DESIRED LETTER? S
ENTER THE UPSTREAM MACH NUMBER, M1? 1.5
ENTER THE SHOCK WAVE ANGLE IN DEGREES? 58.0


CONICAL SUPERSONIC FLOW USING THE TAYLOR-MACCOLL EQUATION

MACH NUMBER =    1.5000       SHOCK ANGLE =  58.0000       CONE ANGLE =  27.1203

BEHIND CONICAL SHOCK: M2          T2/T1       P2/P1       DELS/CP     DELTA
                      1.0860      1.1733      1.7212      0.0046     10.5114

ON CONE SURFACE:      MS          TS/T1       PS/P1       RHOS/RHO1   P. COEF.
                      0.9389      1.2327      2.0460      1.6598      0.6642

STAGNATION TEMPERATURE IS CONSTANT.  P02/P01 = P0S/P01 =   0.9839

ARE ADDITIONAL SUPERSONIC CONE FLOW SOLUTIONS NEEDED?
```

Figure 3-25 Conical Supersonic-to-Subsonic Flow Compression

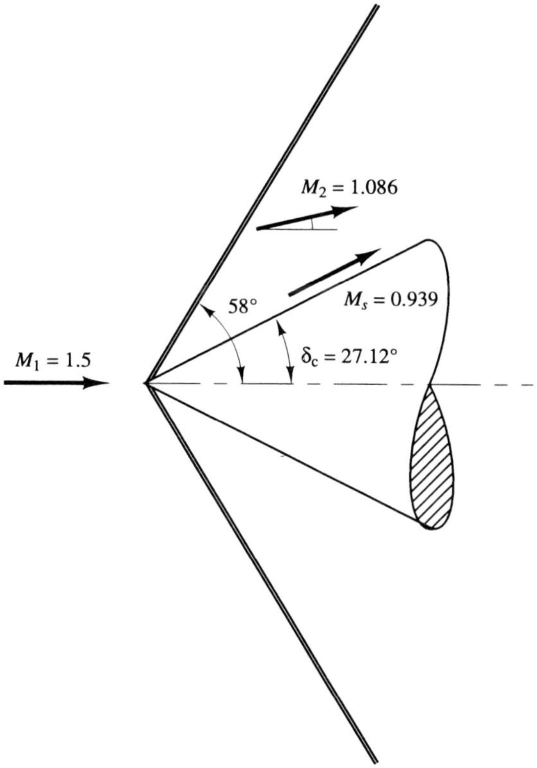

(b) Flow field illustration

Figure 3-25 (continued)

CHAPTER 3 REVIEW QUESTIONS

1. Can an oblique shock wave exist if $M_{1n} = 0.93$? Explain.
2. Sketch the flow field over a 40° wedge placed in an $M = 1.25$ flow.
3. Why are the shock wave angles for a cone and a wedge different for equal wedge and cone half-angles?
4. Why are the angle of incidence and the angle of reflection of a reflected oblique shock not equal? Is there a physical explanation?
5. How are the tangential velocities across an oblique shock wave related?
6. What are some characteristics of weak and strong oblique shock waves?
7. Contrast the flowfields behind a planar oblique shock wave and a conical shock wave.
8. For a given Mach number, why is the Mach angle the lower limiting value for the shock wave angle?
9. For a wedge of vanishing semivertex angle, what are the weak and strong shock wave angles?
10. What is a slip line?

11. Why do strong conical shock waves normally not occur in nature?
12. How are normal shock waves and oblique shock waves related?

CHAPTER PROBLEMS

1. As shown in Figure 3-26, a 15° wedge is placed in a Mach 3 stream. The static pressure and static temperature are 50 kPa and 300 K, respectively. Determine the tangential and normal components of velocity and Mach number upstream and downstream of the oblique shock wave, and compute the static and stagnation pressure ratios across the oblique shock wave. What would the wedge angle be for shock detachment?

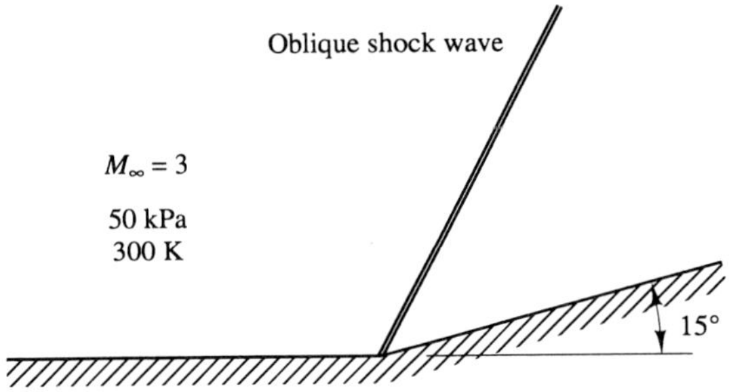

Figure 3-26

2. For a gas flow with the ratio of specific heats = 1.3, what is the turning angle above which an oblique shock wave will be detached at a Mach number of 2?
3. **(a)** For air at a Mach number of 4, the maximum wedge angle for an attached shock wave is ______________.
   (b) If the wedge angle is greater than that for part (a), sketch the resulting flow field.
4. A Mach 4 wind tunnel uses helium as the working fluid. An oblique shock wave from a wedge placed in the test section is observed to make an angle of 30° with respect to the flow direction. Find the half-angle of the wedge for this condition. What are the stagnation pressure and temperature ratios across the oblique shock wave?
5. An oblique shock wave intersects a solid boundary, as illustrated in Figure 3-27. If the solid boundary is to be turned such that no reflected shock wave will be needed, estimate the angle θ through which the boundary must be turned.

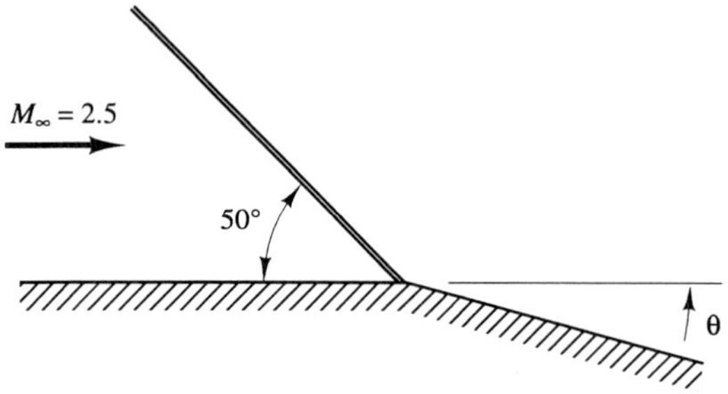

Figure 3-27

6. By observing the shock wave angle, a wedge of total included angle 30° is to be used as a "Mach meter." If the wedge axis is aligned parallel to the supersonic stream, specify the Mach number range over which such an instrument would be effective. Both upper and lower bounds of the effective range should be specified. Would a change in wedge angle help to extend the upper or lower bound of effectiveness of the meter? Hint: Plot θ vs. M.

7. A wedge of included angle 12° is placed in the center of the test section of a supersonic wind tunnel. If the test section Mach number, M_1, is equal to 2 and if the axis of the wedge is inclined 3° to the direction of the air flow, find:
   (a) the incidence and reflected angles of the oblique shock impingements on the walls (upper and lower) of the wind tunnel;
   (b) the Mach numbers and pressure ratios in the six regions delimited by the incident/reflected shock wave system.

8. Air at a Mach number of $M_1 = 2.5$ and with static conditions of $P_1 = 85$ kPa and $T_1 = 250$ K is deflected as shown in Figure 3-28. For the arrangement illustrated, determine M_2, M_3, P_2, P_3, P_{02}, and P_{03}. Sketch the oblique shock waves showing the correct angular orientation. What is the maximum value that the second deflection can have for the second oblique shock wave to be attached?

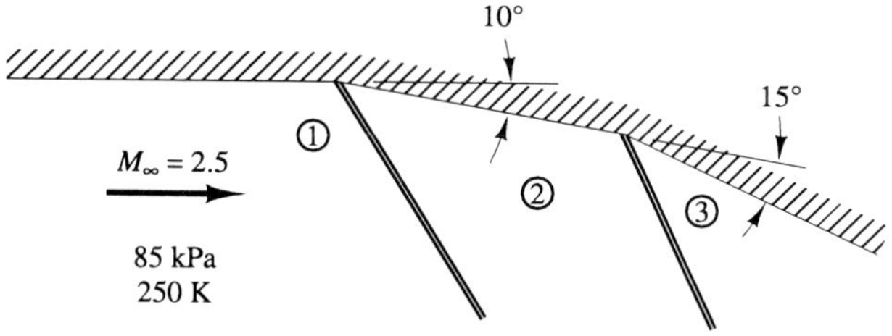

Figure 3-28

9. The reflection of an oblique shock wave from a wall is shown in Figure 3-29. If the upstream Mach number is 2.25 and the angle of the shock wave is 38°, find the Mach numbers in regions 2 and 3 and the angle of the reflected wave. Trace a streamline through the intersection.

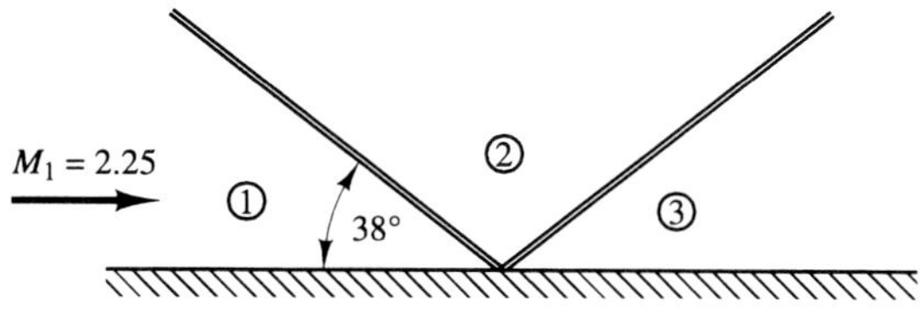

Figure 3-29

10. For the two-dimensional shock wave reflection shown in Figure 3-30, determine the pressure and Mach number in region 3 if the fluid is air. Repeat for a fluid with $\gamma = 1.67$. Plot the trace of a streamline through the shock wave system.

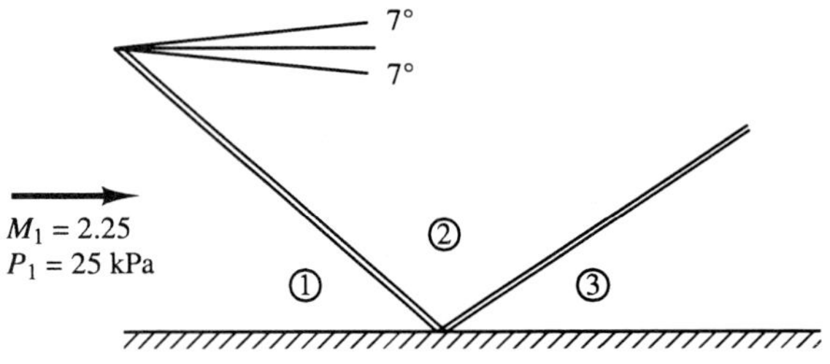

Figure 3-30

11. A cone with a 10° included angle is aligned parallel to a Mach 2.5 flow. Determine the Mach number and pressure on the surface of the cone (a) if the altitude is sea level and (b) if the altitude is 8 km.
12. The supersonic inlet sketched in Figure 3-31 contains an oblique shock wave followed by a normal shock wave. For the conditions shown in the sketch, find the stagnation temperature, the stagnation pressure, and the Mach number behind the normal shock wave (region 3 in the sketch).

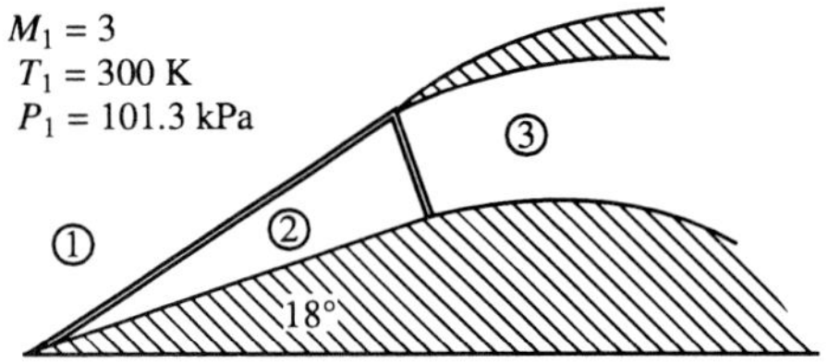

Figure 3-31

13. A two-dimensional supersonic channel in a wind tunnel has an initial Mach number of 2.5 and static conditions of 100 kPa and 300 K. If, as in Figure 3-32, a 10° wedge is placed on the one surface of the channel, calculate the static pressure, static temperature, and Mach number for the reflection process. What is the maximum wedge angle for an attached shock at the wedge?

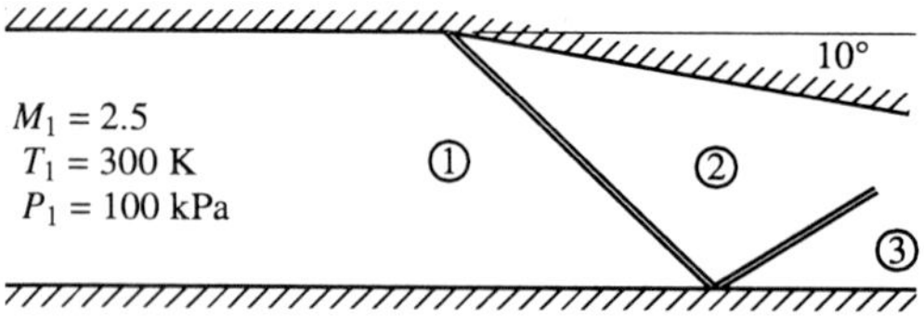

Figure 3-32

14. In a two-dimensional supersonic channel (used as a diffuser), air at an initial Mach number of 3.0 is decelerated by three oblique shock waves followed by a strong oblique shock wave; see Figure 3-33. If the flow in region 5 is in the same direction as in region 1, calculate:
    (a) the shock wave angle of each wave
    (b) the Mach number in region 5
    (c) the static pressure in region 5
    (d) the loss in stagnation pressure from region 1 to region 5.

    If the initial height of the channel is 10 cm, develop a scale drawing showing the shock waves and the flow directions.

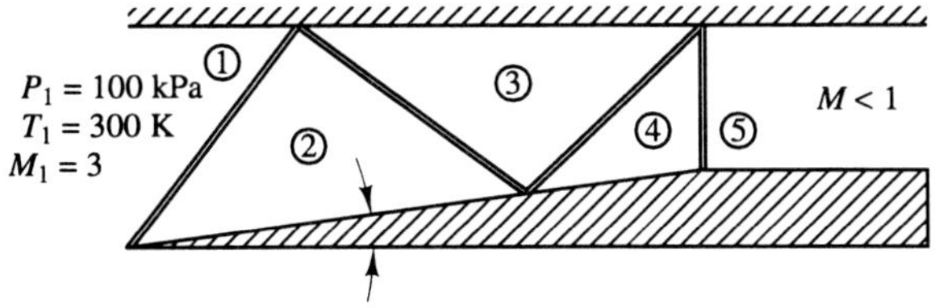

Figure 3-33

15. For the oblique shock wave intersection shown in Figure 3-34, determine the flow conditions (region 3) after the intersection with $\gamma = 1.4$.

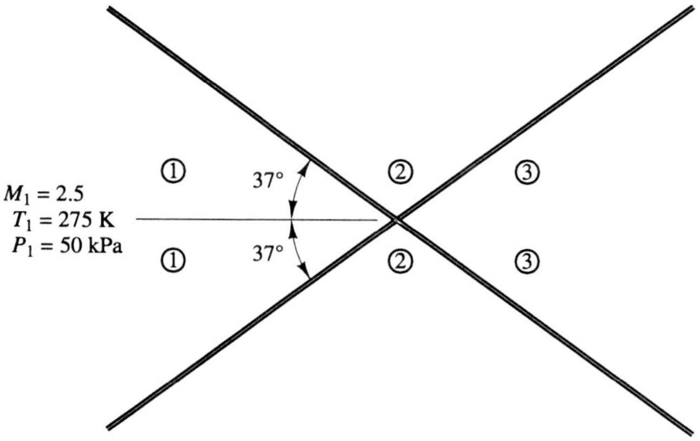

Figure 3-34

16. The flow in the exit plane of a supersonic two-dimensional nozzle is parallel and horizontal, as shown in Figure 3-35, with a Mach number of 2.5 and an exit static pressure of 50 kPa (in region 1). If the pressure of the atmosphere (the back pressure) into which the jet discharges is 101 kPa, find the static pressures in regions 2 and 3 of the figure.

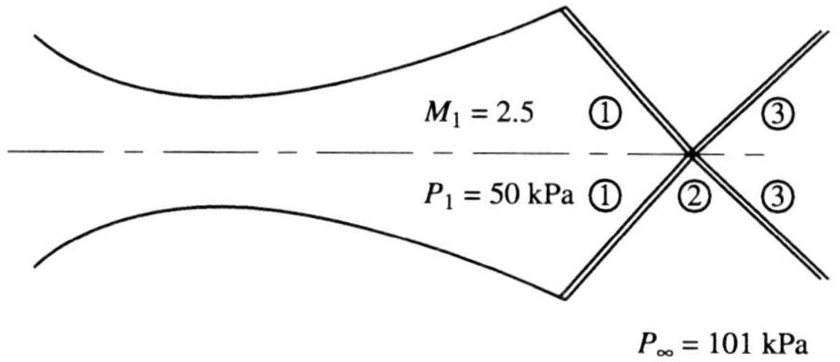

Figure 3-35

17. Alternate configurations employing oblique and normal shocks are to be evaluated. The first configuration sketched in Figure 3-36 is for a two-dimensional diffuser employing one oblique shock wave followed by a normal shock wave. The second proposed configuration in Figure 3-36 uses two oblique shock waves but still contains the normal shock wave. The total turning angle is 16 degrees for both configurations. Determine the pressures, temperatures, Mach numbers, and stagnation pressures for the subsonic state for the two configurations. Which configuration would make the better diffuser? (The purpose of a diffuser is to reduce the speed of the incoming fluid with minimum loss.)

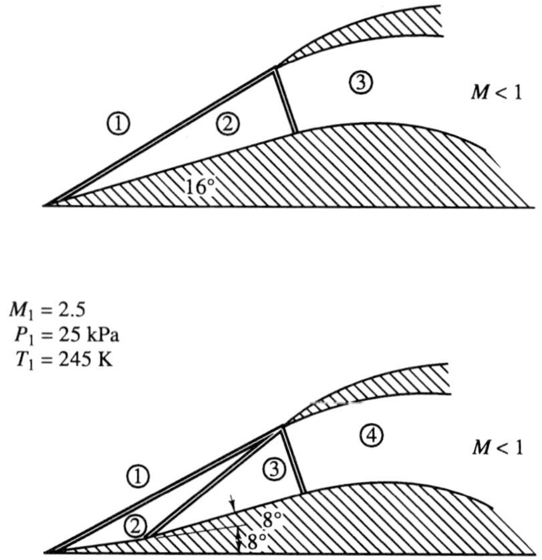

Figure 3-36

18. Proposals are made to use either a cone or a two-dimensional wedge in a Mach 3 flow. Investigate the maximum cone half-angle (semivertex angle) and the maximum wedge turning angle for attached conical and oblique shock waves. The range of ratios of specific heats is from 1.01 to 1.67.

19. Consider problem 3-12 if the inlet is asymmetrical and the inlet ramp (center section) is a cone of 18° half-angle (semivertex angle). Are the conditions (Mach number, pressure, and temperature) uniform at the upstream side of the normal shock wave? How about the conditions downstream of the normal shock wave?

20. Compare the shock detachment angles for wedges and cones over a wide range of Mach numbers if the ratio of specific heats is 1.25.

21. Rework problem 3-6 if the Mach meter is to be a cone of half-angle 15°.

Prandtl-Meyer Flows and the Shock-Expansion Procedure

4

4.1 INTRODUCTION

This chapter deals with two topics: the last of the wave processes and a supersonic application that uses our understanding of the wave processes. Prandtl-Meyer flow, an isentropic process, is the last of the wave processes we will develop. With our study of individual wave processes (normal shock, oblique shock, and Prandtl-Meyer) complete, we can use our knowledge of waves to develop especially useful, but simple, techniques to treat wing cross sections in supersonic flow.

Earlier we established that an expansion shock wave is prohibited by the second law of thermodynamics. But obviously expansions do occur, so some mechanism must be available that results in an expansion process. Because the Mach angle corresponds to the shock wave angle for an oblique shock wave of vanishing strength and no finite flow deflection, a Mach line is often called a *Mach wave*. Across the Mach wave, regions of flow are delineated, but changes in properties and directions are infinitesimal.

Consider what is required if a supersonic flow is to expand some finite angle around a curved surface. Such a situation is illustrated in Figure 4-1a. The flow is uniform upstream and downstream of the region of curvature. The figure shows the leading Mach wave associated with the incident Mach number and the trailing Mach wave associated with the downstream Mach number. Ahead of the leading Mach wave, the zone of silence precludes any effect, and aft of the trailing Mach wave, the surface curvature is constant and all Mach angles are the same. Hence, all changes must take place in the region between the leading and trailing Mach waves.

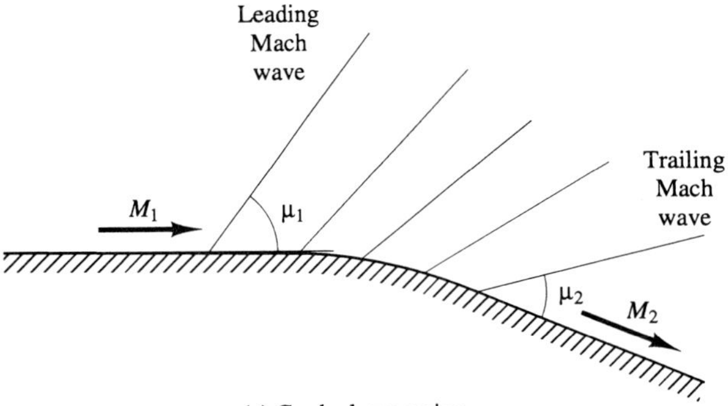

(a) Gradual expansion

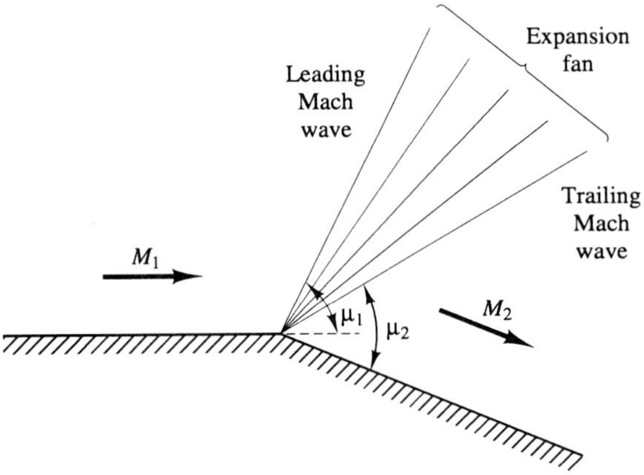

(b) Expansion corner

Figure 4-1 Expansion Process Mach Waves

Between these waves, a few of the unbounded number of Mach waves are shown. Several important concepts are illustrated here:

1. The Mach waves do not coalesce to form a finite strength wave.
2. All changes take place between the incident and the downstream Mach waves.
3. Since no dissipative mechanism is present, the process between the two bounding Mach waves is isentropic.
4. Each Mach wave sees a uniform flow.

The finite change across the expansion process is the result of the unbounded number of Mach waves contained within the incident and downstream Mach waves. This expansion process is called a *Prandtl-Meyer flow*. Prandtl-Meyer flow is valid for a gradual expansion of a supersonic flow along a curved surface as shown in Figure 4-1a or for the expansion of a supersonic flow around a corner as shown in Figure 4-1b. The region across which the expansion takes place is bounded by the leading and trailing Mach waves as illustrated in the figure. The term *Prandtl-Meyer expansion fan* is often used to describe the fan-like region defined by the decreasing Mach angles of the Mach waves associated with the expansion process. Prandtl-Meyer flow is restricted to supersonic flow since the Mach angle is defined only for supersonic flow.

4.2 EXPANSION WAVE ANALYSIS

Across a single Mach wave (see Figure 4-2) the flow is deflected by an amount $d\upsilon$ and other properties change by infinitesimal amounts. Just as for any other plane wave, the tangential components of velocity across the wave must be equal; that is

$$V_{1_t} = V_{2_t}, \tag{4-1}$$

from which

$$V \cos\mu = (V + dV)\cos(\mu + d\upsilon). \tag{4-2}$$

Using the identity for the cosine of the sum of two angles, this becomes

$$V \cos\mu = (V + dV)(\cos\mu \cos d\upsilon - \sin\mu \sin d\upsilon). \tag{4-3}$$

But since $d\upsilon$ is an infinitesimal quantity,

$$\cos d\upsilon \sim 1 \tag{4-4}$$

and

$$\sin d\upsilon \sim d\upsilon,$$

so that

$$V \cos\mu = (V + dV)(\cos\mu - d\upsilon \sin\mu). \tag{4-5}$$

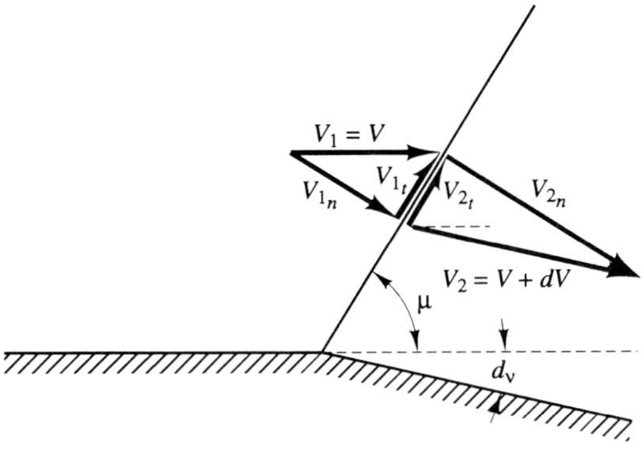

Figure 4-2 Flow Deflection across a Mach Wave (Greatly Exaggerated)

If we neglect the term containing $d\nu dV$ in comparison to terms containing only $d\nu$ or dV, Eq. 4-5 reduces to

$$\frac{dV}{V} = \tan\mu\, d\nu. \tag{4-6}$$

The Mach angle is defined as

$$\sin\mu = \frac{1}{M}. \tag{1-56}$$

In terms of a right triangle, this relationship can be construed as shown in Figure 4-3. Then

$$\tan\mu = \frac{1}{\sqrt{M^2 - 1}}. \tag{4-7}$$

Finally,

$$\frac{dV}{V} = \frac{1}{\sqrt{M^2 - 1}} d\nu. \tag{4-8}$$

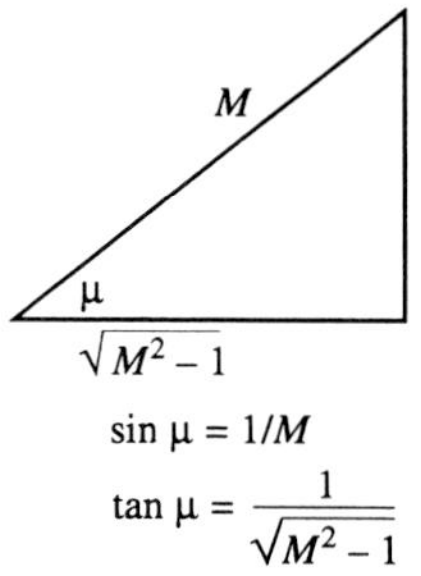

Figure 4-3 Mach Angle Relationships

Equation 4-8 represents an ordinary differential equation, but in order to be integrated the velocity terms must be expressed as functions of the Mach number. The velocity cast as the Mach number and speed of sound becomes

$$V = M\sqrt{\gamma RT}. \tag{4-9}$$

Taking the natural log yields

$$\ln V = \ln M + \tfrac{1}{2}\ln(\gamma RT). \tag{4-10}$$

After we differentiate, the following expression for dV/V results

$$\frac{dV}{V} = \frac{dM}{M} + \frac{dT}{2T}. \tag{4-11}$$

The relationship between stagnation and static temperature is

$$T_o = T\left(1 + \frac{\gamma - 1}{2} M^2\right). \tag{4-12}$$

Taking the natural log of Eq. 4-12 and differentiating gives

$$\frac{dT_o}{T_o} = \frac{dT}{T} + \frac{(\gamma - 1)MdM}{1 + \frac{\gamma - 1}{2}M^2}. \tag{4-13}$$

In isentropic flow, however, T_o is a constant, so that

$$\frac{dT}{T} = -\frac{(\gamma - 1)MdM}{1 + \frac{\gamma - 1}{2}M^2}. \tag{4-14}$$

Substituting Eq. 4-14 into Eq. 4-11 produces

$$\frac{dV}{V} = \frac{1}{1 + \frac{\gamma - 1}{2}M^2} \cdot \frac{dM}{M}. \tag{4-15}$$

Lastly, combining Eqs. 4-15 and 4-8 and simplifying produces

$$d\upsilon = \frac{\sqrt{M^2 - 1}}{1 + \frac{\gamma - 1}{2}M^2} \cdot \frac{dM}{M}, \tag{4-16}$$

which is an ordinary differential equation relating the deflection angle and the Mach number.

Integrating Eq. 4-16 (see Peirce (1929)), or the symbolic manipulation program DERIVE (1990), with the limits of υ_2 and υ_1 for M_2 and M_1 yields

$$\upsilon_2 - \upsilon_1 = \sqrt{\frac{\gamma + 1}{\gamma - 1}} \tan^{-1}\sqrt{\frac{\gamma - 1}{\gamma + 1}(M^2 - 1)} - \tan^{-1}\sqrt{M^2 - 1}\Bigg|_{M_1}^{M_2}. \tag{4-17}$$

If we take $\upsilon_1 = 0$ at $M_1 = 1$, and let $\upsilon_2 = \upsilon$ at $M_2 = M$, then

$$\upsilon = \sqrt{\frac{\gamma + 1}{\gamma - 1}} \tan^{-1}\sqrt{\frac{\gamma - 1}{\gamma + 1}(M^2 - 1)} - \tan^{-1}\sqrt{M^2 - 1}. \tag{4-18}$$

Because Eq. 4-18 is derived for Prandtl-Meyer flow, υ is usually called the *Prandtl-Meyer angle*. The Prandtl-Meyer angle, υ, represents the physical angle through which a stream initially at $M = 1$ must be expanded to reach a supersonic Mach number M. This is the situation shown in Figure 4-4a. For the sonic upstream flow, the Mach angle is 90° and the Mach wave is perpendicular to the flow. After the flow is deflected through an angle υ, the Mach angle μ is measured from the new flow direction associated with the downstream Mach number.

The real importance of Eq. 4-18 is illustrated in Figure 4-4b, which shows that the deflection required to expand a Mach number M_1 to M_2 is the difference $\upsilon_2 - \upsilon_1$. Figure 4-4b demonstrates that any supersonic Mach number can be envisioned as an expansion from Mach 1. A Mach number M_1, which can be conceived as having been expanded from a sonic flow, is deflected through an angle υ_1. The leading and trailing Mach waves are shown in Figure 4-4b. The Mach number M_1 can be imagined as an expansion of υ_1 from a sonic flow, and the Mach number M_2 as an expansion of υ_2 from the same sonic flow. The difference in the Prandtl-Meyer angles for M_1 and M_2 then corresponds to the actual deflection required from M_1 to M_2. Since the expansion process is isentropic, the isentropic relations can be used to relate the physical properties of the upstream and downstream states.

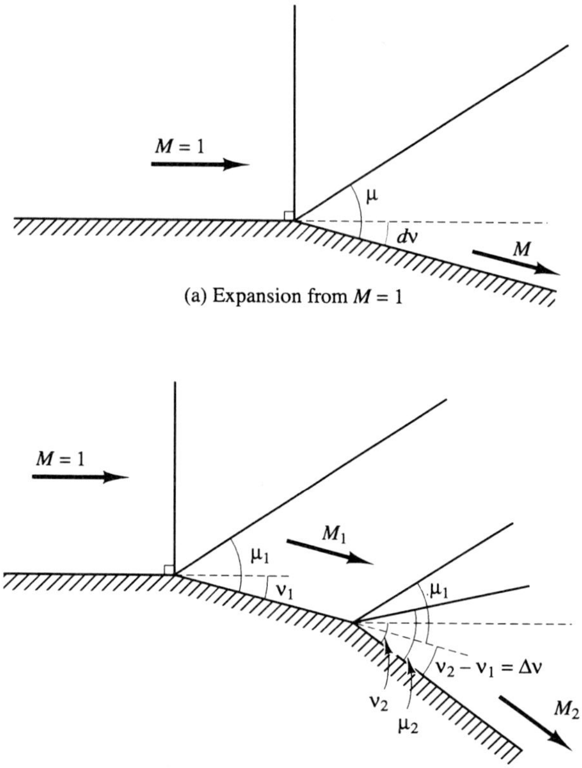

(a) Expansion from $M = 1$

(b) Expansion from M_1 to M_2

Figure 4-4 Expansion Geometry and Nomenclature

4.3 PRANDTL-MEYER APPLICATIONS

Computational aids are available for Prandtl-Meyer flow. The Prandtl-Meyer function, Eq. 4-18, and the Mach angle are tabulated in Table D-1 for $\gamma = 1.4$. The Prandtl-Meyer option of COMPQ is **PM**. Cases available in the **PM** option will be discussed after the next example.

Example 4-1

A supersonic flow initially at $M = 2$ and with static properties of 400 R and 10 psia is expanded around a 10°-sharp corner. Find the downstream Mach number and properties and construct a schematic of the expansion fan.

Solution. The situation is illustrated in Figure 4-5. From Table D-1 and $M_1 = 2.0$,

$$\upsilon_1 = 26.38°.$$

Since

$$\Delta\upsilon = \upsilon_2 - \upsilon_1$$

$$\upsilon_2 = \Delta\upsilon + \upsilon_1,$$

then

$$\upsilon_2 = 10° + 26.38° = 36.38°.$$

Again using Table D-1 and $\upsilon_2 = 36.38$, we find

$$M_2 = 2.385.$$

Since the expansion is isentropic,

$$P_2 = \frac{P_2}{P_o}\frac{P_o}{P_1}P_1 = \frac{0.070}{0.127}P_1 = 5.477 \text{ psia},$$

and

$$T_2 = \frac{T_2}{T_o}\frac{T_o}{T_1}T_1 = \frac{0.4678}{0.5556}T_1 = 336.8° R.$$

The Mach angles are

$$\mu_1 = 30°$$

$$\mu_2 = 24.8°.$$

The expansion fan then covers an angle of

$$30° - (24.8° - 10°) = 15.2°.$$

The various Mach numbers and Mach angles and the fan angle are also illustrated in Figure 4-5.

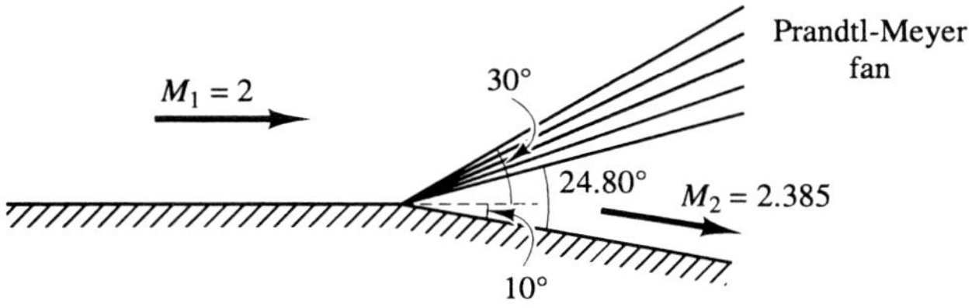

Figure 4-5 Example 4-1

The solution to this example demonstrates the typical steps required in solving a Prandtl-Meyer flow problem. The upstream and downstream Mach numbers, the Mach angles, and the deflection angle are related via the Prandtl-Meyer angles as tabulated in Table D-1 of Appendix D. The physical properties are evaluated by using the isentropic relations.

The **PM** option of COMPQ offers additional computational aids for Prandtl-Meyer flow. The menu showing the various cases for the **PM** option is given in Figure 4-6.

```
PRANDTL-MEYER PROCESS CALLED FOR
INPUT PARAMETERS ARE:
      MACH NUMBER              M
      PRANDTL-MEYER ANGLE      P
      MACH ANGLE               A
      TURNING SOLUTION         TS
      PRESSURE RATIO           PR
```

Figure 4-6 COMPQ Option **PM** Menu

The first three cases, **M**, **P**, and **A**, simply reproduce Table D-1. Given either the Mach number, the Prandtl-Meyer angle, or the Mach angle, these cases compute the remaining two parameters. The last two cases, **TS** and **PR**, are actually processes that combine a number of arithmetical operations. Case **TS** accomplishes exactly what Example 4-1 demonstrates. Execution of case **TS** for the conditions of Example 4-1 yields the output shown in Figure 4–7. The output also reminds us that the process is adiabatic—hence no change in stagnation temperature—and reversible—hence no change in stagnation pressure. Case **PR** solves the details of the process given the upstream Mach number and the static pressure ratio across the Prandtl-Meyer flow. The **TS** and **PR** cases of option **PM** of COMPQ, in conjunction with the means to specify the specific heats ratio, make this option quite useful.

```
ENTER THE DESIRED LETTER? TS

ENTER THE UPSTREAM MACH NUMBER AND THE TURNING ANGLE? 2.0,10.0

PRANDTL-MEYER TURNING SOLUTION
      M1            M2          ANGLE(DEG)         T2/T1          P2/P1
    2.0000        2.3849         10.0000           0.8421         0.5480

THE PROCESS IS ISENTROPIC; HENCE NO CHANGE IN STAGNATION PROPERTIES.
```

Figure 4-7 COMPQ Solution of Example 4-1

Additional information can be gleaned from Prandtl-Meyer flow and Eq. 4-18. Examine the limiting value of υ as the Mach number increases without bound; that is,

$$\begin{aligned}\upsilon_{max} &= \lim_{M\to\infty} \upsilon \\ &= \sqrt{\frac{\gamma+1}{\gamma-1}}\frac{\pi}{2} - \frac{\pi}{2},\end{aligned} \tag{4-19}$$

which becomes for $\gamma = 1.4$

$$\upsilon_{max} = 130.45°. \tag{4-20}$$

A literal interpretation suggests that from a sonic speed ($M = 1.0$), a flow deflection angle of 130.45° will result in an unbounded Mach number and vanishing values of the static temperature and pressure. Schematically, the situation would appear as in Figure 4-8. The limit predicted by Eq. 4-19 is of more philosophical than practical interest since long before the pressure and temperature are reduced to zero, the assumption of a calorically perfect gas would have broken down, but turning angles greater than 90° are possible. Nonetheless it does indicate some limiting value of the flow deflection into a low-pressure region. We shall pursue more about this type of expansion process in Chapter 6.

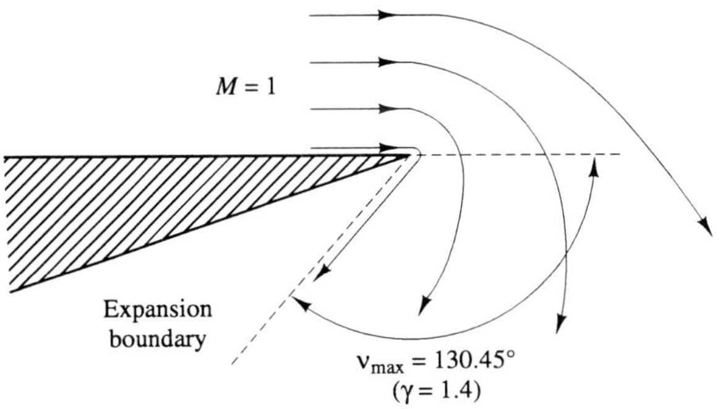

Figure 4-8 Prandtl-Meyer Expansion with Maximum Deflection

Another interesting Prandtl-Meyer flow is the gradual compression as illustrated in Figure 4-9. If in a gradual expansion the Mach waves diverge, then in a gradual compression the Mach waves converge and then coalesce to form an oblique shock wave away from the surface. (This process is schematically illustrated in the figure.) In the neighborhood of the curved surface, the compression is isentropic since the flow is undergoing a Prandtl-Meyer flow process. This is another of the few examples of an isentropic compression.

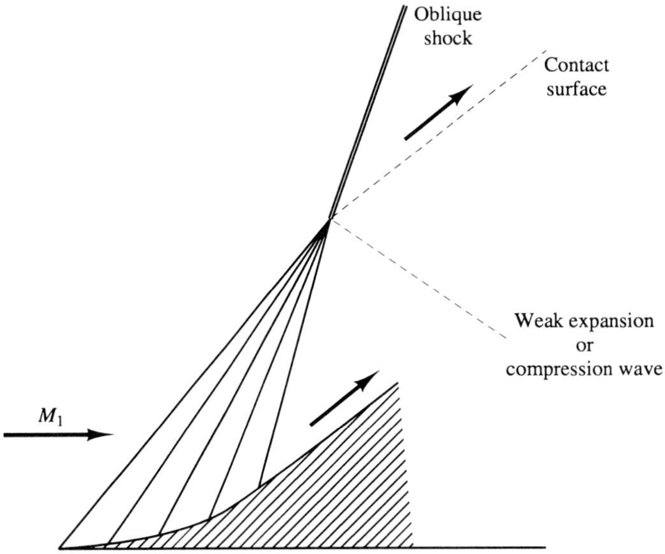

Figure 4-9 Isentropic Compression

The region of intersection and coalescence of the Mach waves depends on the specific configuration (geometrical arrangement) of the curved surface. Generally, the Mach waves coalesce in a relatively narrow region so that a relatively distinct contact surface is formed. The contact surface separates the downstream flow into regions associated either with passage through the oblique shock or with passage through the compression fan. As with the intersection of two oblique shock waves to form a single oblique shock, a weak expansion region or compression wave is formed where the Mach waves coalesce. The oblique shock wave must deflect the flow through the same angle as the gradual compression.

During our discussion of oblique shock waves we examined reflections and intersections involving them. Unless we wish to approximate expansion processes by finite amplitude discrete waves (see Zucrow and Hoffman (1976) for such a presentation), reflections and intersections of expansion waves are true two-dimensional phenomena whose proper treatment must be deferred until a later chapter. For example, the reflection of an expansion fan from a wall, as shown schematically in Figure 4-10, involves interactions between incident and reflected Mach waves.

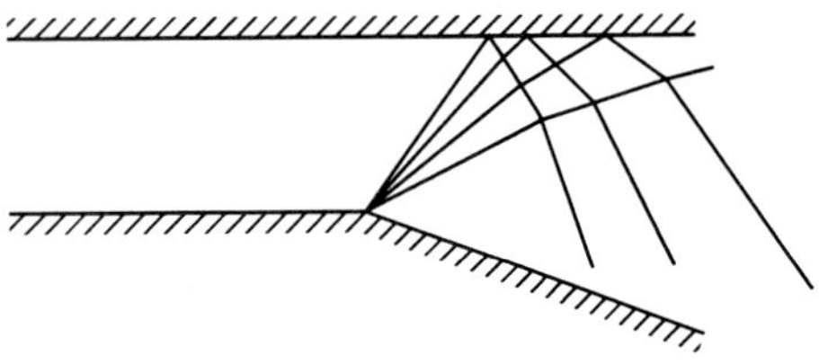

Figure 4-10 Expansion Fan Reflection Schematic

4.4 THE SHOCK-EXPANSION PROCEDURE

Having completed our initial examination of the various wave processes, we can use our understanding of them to develop some relatively simple, but quite useful, procedures for analyzing wing sections in supersonic flow. An aircraft in flight is usually held aloft by a lifting force that acts on the wing. A wing cross section (an example of which is shown in Figure 4-11), is an *airfoil* and is the fundamental aerodynamic component of an aircraft.

Airfoil nomenclature is also illustrated in Figure 4-11. The *leading* and *trailing edges* are the front and back points on the airfoil, respectively. The straight line that connects the leading and trailing edges is known as the *chord line*, while the *chord*, c, is the straight-line distance between the leading and trailing edges. A curve that is equidistant everywhere between the lower and upper surfaces can be drawn from the leading to the trailing edge. This curve is the *camber line* and is described by its distance, C, from the chord line. When the camber line is straight, the airfoil is symmetrical, or *uncambered*; otherwise, it is asymmetrical, or *cambered.* Camber is positive when the camber line is above the chord line. The distance between the lower and upper surfaces, perpendicular to the camber line, is the *thickness*, t. Camber and thickness are functions of chordwise position along the airfoil.

The angle between the freestream direction and the chord line is the angle of attack, α. Cambered airfoils generate lift at $\alpha = 0$. There is, however, an angle α_{l_0},

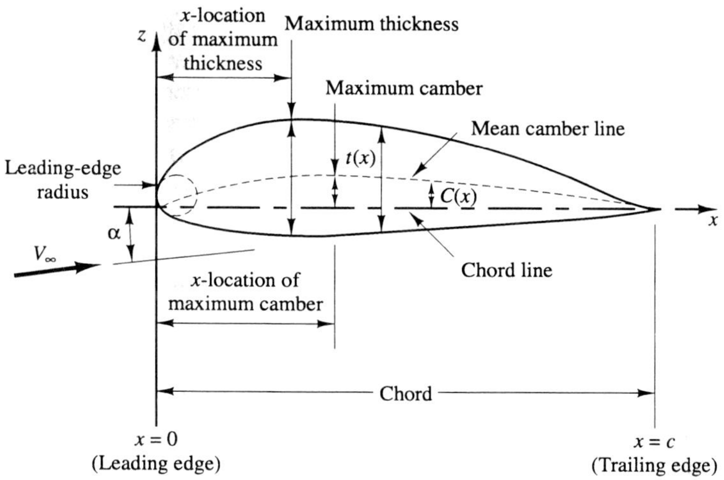

Figure 4-11 Airfoil Nomenclature

the zero lift angle, at which a cambered airfoil has no lift. Positive camber produces a negative value of α_{l_0}; increasing the amount of camber decreases α_{l_0}.

The airfoils used in most aircraft wing applications are relatively thin, have small camber, and usually operate at small angles of attack. The maximum thickness-to-chord ratio, t_{max}/c, is usually less than 0.2 and is often less than 0.1. The maximum camber, C_{max}/c, is typically less than 0.05. At cruise, α is generally only a few degrees and it rarely exceeds 20°, even in rather severe maneuvers. There are important exceptions, however. In this chapter we will restrict our discussion to airfoils in supersonic flow.

Because an airfoil is strictly a two-dimensional object, we refer to the forces and moments acting on it as *per-unit span*, where span is the straight-line distance between the tips of a wing. A prime superscript is used to designate "per-unit span" so that we write L', D', and M' for lift, drag, and moment per-unit span, respectively. A nose-up moment is considered positive; positively cambered lifting airfoils usually have negative (nose-down) moments. The magnitude of the pitching moment depends on the location on the airfoil to which it is referenced. Figure 4-12 defines the positive conventions for these forces and moment.

It is useful to present airfoil forces and moments in nondimensional forms. We define the *sectional force* and moment coefficients as

$$C_l = \frac{L'}{\frac{1}{2}\rho_\infty V_\infty^2 c} \tag{4-21a}$$

$$C_d = \frac{D'}{\frac{1}{2}\rho_\infty V_\infty^2 c} \tag{4-21b}$$

$$C_m = \frac{M'}{\frac{1}{2}\rho_\infty V_\infty^2 c^2}. \tag{4-21c}$$

The lowercase subscript denotes that these coefficients are per-unit span.

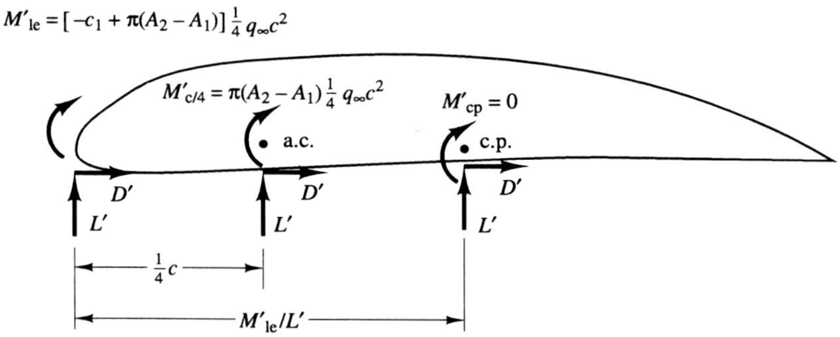

Figure 4-12 Airfoil Force and Moment Systems

The force and moment system acting on an airfoil is commonly represented in one of the three ways shown in Figure 4-12. Each system has a reference point at which the lift, drag, and moment are concentrated. In each system the magnitude of the lift and drag forces are the same; the moment varies, however, as the reference location changes. Three commonly used reference locations are the *leading edge* ($x = 0$), the *aerodynamic center* ($x = x_{ac}$), and the *center of pressure* ($x = x_{cp}$). The aerodynamic center is the location along the chord about which the pitching moment is independent of angle of attack. The center of pressure is the location along the chord line about which the pitching moment is zero. A force and moment balance on the airfoil then gives

$$x_{cp} = -\frac{M'_{le}}{L'}. \tag{4-21d}$$

Airfoils designed to operate under conditions where the freestream is supersonic are generally quite thin with relatively sharp leading edges. Bryson (1951) used a flat plate with a 10° half-angle wedge as the leading edge to study details of transonic and supersonic flow fields in the vicinity of the wedge/plate. Optical measurement techniques were used to infer the density of the gas in the region close to the plate, and lines of constant density ratios were generated. Figure 4-13 presents the density contours about the wedge/plate for two freestream Mach numbers, M_∞, 1.207 and 1.465. For the contours of constant density ratio, ρ/ρ_o, shown, $\rho/\rho_o = 0.634$ corresponds to the ratio for sonic flow ($M = 1$).

At $M_\infty = 1.207$ (Figure 4-13a) the turning angle presented to the freestream by the wedge is greater than the maximum turning angle for an attached oblique shock wave. Consequently the shock detaches and forms a nearly normal bow shock ahead of the wedge. As indicated by the density ratio values, the flow downstream of the shock is subsonic. The flow stagnates as it approaches the apex of the wedge and then accelerates along the wedge surface, remaining subsonic, however, until the corner. At the corner the flow accelerates to sonic speed and then experiences a Prandtl-Meyer expansion to a Mach number of 1.36 (based on the 0.459 density ratio). Somewhere downstream, beyond the end of this figure, the flow must slightly decelerate essentially back to the freestream conditions.

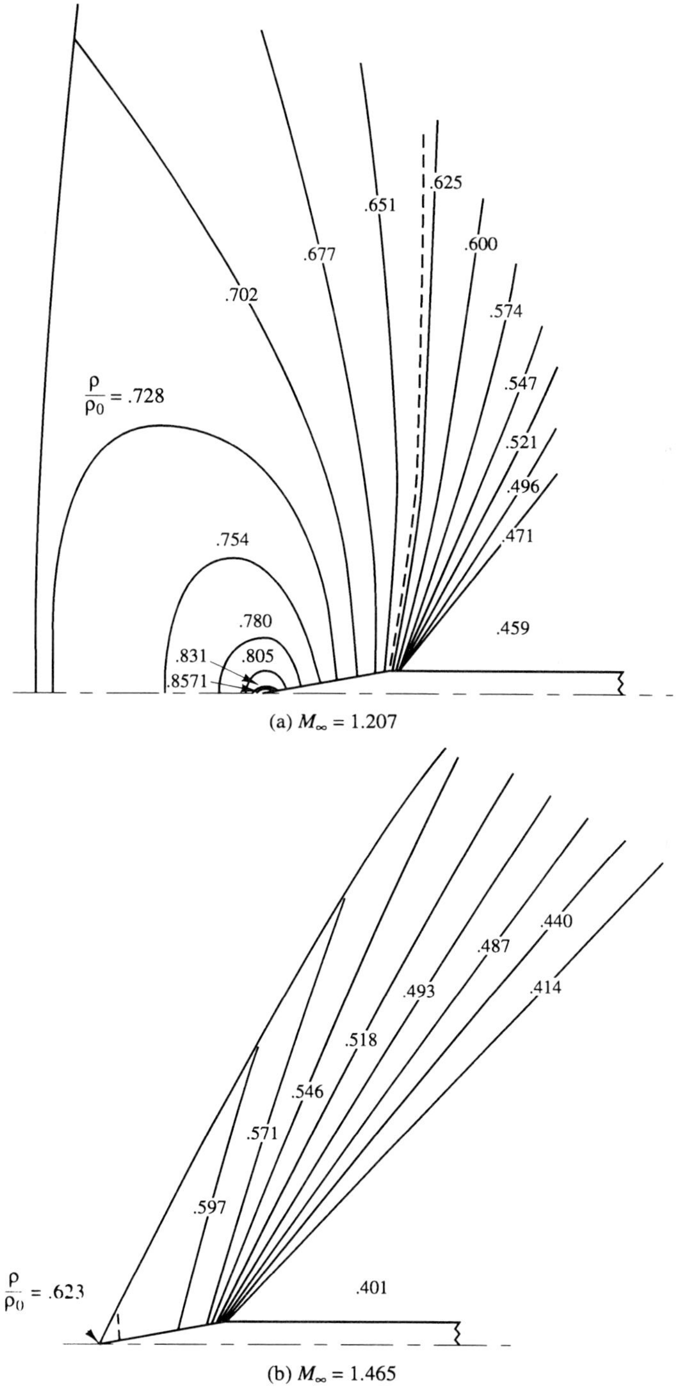

Figure 4-13 Supersonic Wedge/Plate Flowfield (Bryson, 1951)

In Figure 4-13(b), the sonic line is nearly at the apex and the shock is nearly attached. From the apex to the corner there is little change in density. Remember that for attached oblique shock waves on a wedge, the flow is uniform downstream of the shock. That is approximately the situation here. Using the oblique shock option (**OS**) of COMPQ, or referring to Appendix C, we find that the minimum freestream Mach number for an attached shock on a 10° semi-angle wedge is 1.42. Theoretically, therefore, the shock in Figure 4-13b should be fully attached since M_∞ is greater than this minimum. The differences between this theoretical solution and the actual flow are primarily a result of viscous effects on the wedge and plate surfaces. These differences are small, however, and the overall flow closely resembles the theoretical prediction.

Methods for analyzing supersonic flow past airfoils with sharp leading edges and attached shock waves are relatively simple. The *shock-expansion method* is a particularly accurate technique that is also quite easy to use. It is restricted, however, to supersonic regions of a flow.

As supersonic flow progresses over the surface of an airfoil and experiences changes in direction, compression and expansion waves will be produced. Changes in pressure and Mach number across these waves are easily determined using oblique shock and Prandtl-Meyer process calculations. This is the basis of the shock-expansion method. There is one difficulty, however, because some of the waves created at the surface will eventually intersect somewhere above or below the airfoil. Reflected waves will be produced at these intersections as indicated in Figure 4-14 and will be directed back toward the surface where they will affect the flow conditions. The shock-expansion method neglects the influence of these reflected waves on the flow and surface pressures. The pressure at each point along the airfoil is found by direct calculation using the waves produced at the surface as the direction changes. The resulting pressure distribution is then integrated to give the lift and wave drag of the airfoil.

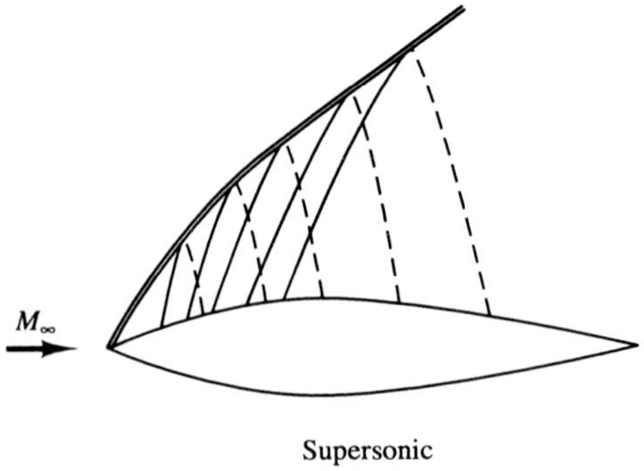

Figure 4-14 Reflection of Expansion Waves from the Leading-Edge Shock of a Supersonic Airfoil

The lift force is perpendicular to the freestream velocity (or the freestream Mach number), and the drag force is parallel to the freestream. These forces are simply the freestream perpendicular and the freestream parallel components, respectively, of the net pressure force acting on the airfoil. The surface pressures depend on the angle through which the flow turns, while the force components depend on the inclination of each segment with respect to horizontal. The following example illustrates the application of the shock-expansion method.

Example 4-2

Determine the lift and wave drag coefficients of the diamond (double-wedge) airfoil shown in Figure 4-15 for $M_\infty = 3$.

Solution. The airfoil of interest, illustrated in Figure 4-15, is a symmetrical, double-wedge configuration composed of 4° wedges. At the specified angle of attack of 6°, a 2° expansion must occur over the top surface of the leading edge and a 10° compression over the bottom surface at the leading edge. The half-chord position requires 8° expansions. Because the flow is supersonic with no upstream influences, the solutions of the oblique shock wave structures at the trailing edge are not needed for the computations associated with the airfoil lift, drag, and moment. A table will organize the solution. The oblique shock wave option of COMPQ and the Prandtl-Meyer turning angle process of COMPQ were used to obtain salient features of the states over the airfoil. Table 4-1 provides a summary of the computed results. The angles were obtained by inspection of the problem geometry. The values of M and P/P_1 were determined using COMPQ.

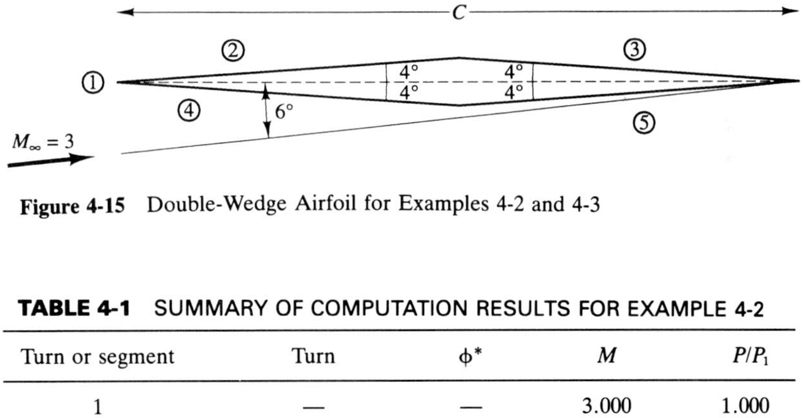

Figure 4-15 Double-Wedge Airfoil for Examples 4-2 and 4-3

TABLE 4-1 SUMMARY OF COMPUTATION RESULTS FOR EXAMPLE 4-2

Turn or segment	Turn	ϕ*	M	P/P_1
1	—	—	3.000	1.000
1 to 2	2° exp	—	—	—
2	—	88°	3.106	0.854
2 to 3	8° exp	—	—	—
3	—	80°	3.579	0.4311
1 to 4	10° shock	—	—	—
4	—	260°	2.505	2.055
4 to 5	8° exp	—	—	—
5	—	268°	2.872	1.167

* ϕ is measured counterclockwise from the $+x$-axis.

The lift and drag forces are obtained from Eq. 1-30b, which here reduces to a simple summation:

$$\vec{F}' = -\int_A P\vec{n}dA = -1\int_s P\vec{n}ds = -\sum_{i=2}^{5}(P\vec{n}s)_i = -P_1\sum_{i=2}^{5}\left(\frac{P}{P_1}\vec{n}s\right)_i.$$

The surface length, which is the same for each segment, is

$$s = \frac{\frac{c}{2}}{\cos 4^\circ}.$$

With this the force vector is

$$\vec{F}' = -P_1\frac{\frac{c}{2}}{\cos 4^\circ}\left(\frac{P_2}{P_1}\vec{n}_2 + \frac{P_3}{P_1}\vec{n}_3 + \frac{P_4}{P_1}\vec{n}_4 + \frac{P_5}{P_1}\vec{n}_5\right),$$

and the lift and drag components are

$$L' = -\frac{P_1 c}{2\cos 4^\circ}\left(\frac{P_2}{P_1}\sin 88^\circ + \frac{P_3}{P_1}\sin 80^\circ + \frac{P_4}{P_1}\sin 260^\circ + \frac{P_5}{P_1}\sin 268^\circ\right)$$

and

$$D' = -\frac{P_1 c}{2\cos 4^\circ}\left(\frac{P_2}{P_1}\cos 88^\circ + \frac{P_3}{P_1}\cos 80^\circ + \frac{P_4}{P_1}\cos 260^\circ + \frac{P_5}{P_1}\cos 268^\circ\right).$$

Evaluating these components gives

$$L' = 0.958 P_1 c$$

and

$$D' = 0.147 P_1 c.$$

The lift and drag coefficients per unit span are

$$C_l = \frac{L'}{q_\infty c}$$

and

$$C_d = \frac{D'}{q_\infty c},$$

and the dynamic pressure is

$$q_\infty = \tfrac{1}{2}\rho_\infty V_\infty^2 = \tfrac{1}{2}\gamma P_\infty M_\infty^2 = \tfrac{1}{2}\gamma P_1 M_1^2$$

for a perfect gas. So, the final results are

$$C_l = 0.152$$

and

$$C_d = 0.0233.$$

The lift-to-drag ratio, $L'/D' = C_l/C_d$ is 6.5.

Lift-to-drag ratios for well-designed subsonic airfoils at low, subsonic Mach numbers can be as high as 100 or more. The diamond airfoil in Example 10-4 is actually a reasonably efficient supersonic profile. However, its lift-to-drag ratio is quite small compared to what might be achieved at low speeds. Also, note that the calculation above neglects friction; the drag will be higher and L'/D' ratio will be even smaller when friction is included. Even though the flow is inviscid, a drag term results since the creation of waves is a source of supersonic, inviscid drag on an airfoil. This process imposes a severe penalty as shown by Example 10-4.

As Example 4-2 illustrates, the inviscid flowfield solution for a double-wedge airfoil in supersonic flow uses just the oblique shock wave and Prandtl-Meyer flow processes with which we are familiar. COMPQ is completely adequate to solve such problems providing the user is willing to assemble the information. Such an assembly is, however, somewhat tedious and can be arithmetically intensive if a number of cases are of interest. Using the same features as coded into COMPQ, the software element SHEX (for "Shock-Expansion") on the accompanying disk analyzes supersonic flow past an airfoil whose surface is constructed from straight-line segments. The freestream Mach number must be high enough for all shocks to be attached. The following example demonstrates the use of SHEX, which is best utilized by examples.

Example 4-3

Use the SHEX program to determine the lift and wave drag coefficients of the diamond (double-wedge) airfoil shown in Figure 4-15 for $M_\infty = 3$.

Solution. Figure 4-16 is a copy of the SHEX input and output for the configuration of Figure 4-15. Notice that the surface angles are measured with respect to the chord line. Positive angles correspond to compression turns and negative angles correspond to expansions as viewed with respect to the chord line. The program prompts the user for the input information it requires. This solution agrees exactly with the results in Example 4-2.

The shock-expansion procedure can be used on airfoils with configurations that are more complex than wedges and planes. Supersonic airfoils consisting of circular arcs or circular-arc segments are frequently used and are often called *biconvex* or *circular-arc airfoils*. A typical circular-arc airfoil configuration is shown in Figure 4-17. The shock-expansion procedure can be used to analyze the flow over the complete circular-arc airfoil if the flow is supersonic over the entire surface of the airfoil. This condition requires that the shock structure associated with the leading edges of the upper and lower surfaces be attached oblique shock waves. In fact, the procedure can be used over portions of airfoil surfaces in supersonic flow provided that the location of the sonic points are known. Analysis of an airfoil with a portion of the surface in subsonic flow is beyond the scope of this textbook and will not be pursued further.

```
ENTER THE RATIO OF SPECIFIC HEATS? 1.4

ENTER THE NUMBER OF SEGMENTS ON THE TOP OF THE AIRFOIL? 2
FOR EACH SEGMENT ENTER THE SURFACE ANGLE AND X/C STARTING LOCATION
? 4,0
? -4,.5

  SEGMENT         ANGLE    X/C LOCATION
     1            4.000       0.0000
     2           -4.000       0.5000
ARE THESE CORRECT (Y OR N)? Y

ENTER THE NUMBER OF SEGMENTS ON THE BOTTOM OF THE AIRFOIL? 2
FOR EACH SEGMENT ENTER THE SURFACE ANGLE AND X/C STARTING LOCATION
? 4,0
? -4,.5

  SEGMENT         ANGLE    X/C LOCATION
     1            4.000       0.0000
     2           -4.000       0.5000
ARE THESE CORRECT (Y OR N)? Y

ENTER THE FREESTREAM MACH NUMBER? 3

ENTER THE ANGLE OF ATTACK? 6



SHOCK EXPANSION SUPERSONIC AIRFOIL PROGRAM

MINF             ALPHA             GAMMA
 3                 6                1.4

TOP SURFACE OF THE AIRFOIL
 SEGMENT       ANGLE     X/C START     M         T/TINF     P/PINF
      1        4.000      0.0000     3.106       0.9559     0.8539
      2       -4.000      0.5000     3.578       0.7863     0.4312

BOTTOM SURFACE OF THE AIRFOIL
 SEGMENT       ANGLE     X/C START     M         T/TINF     P/PINF
      1        4.000      0.0000     2.505       1.2418     2.0553
      2       -4.000      0.5000     2.872       1.0566     1.1675

     CL          CD           L/D         XCP       YCP
   0.1522    0.0233         6.53        0.440     -.006


ANOTHER ANGLE OF ATTACK AT THE SAME MACH NUMBER?
```

Figure 4-16 SHEX Solution for Example 4-3

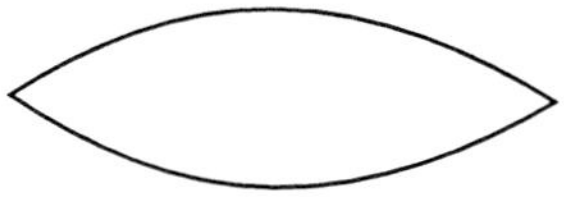

Figure 4-17 Circular-Arc Airfoil

Downstream of the oblique shock wave attached to the leading edge of a biconvex airfoil, the surface orientation is such that expansion occurs over the entire surface of the airfoil. An expansion occurs since the maximum slope of the airfoil occurs at the leading edge of such configurations. An examination of Figure 4-17 will verify that in supersonic flow, expansions will occur over both surfaces.

A schematic of the upper surface of a biconvex configuration is shown in Figure 4-18. The thickness, t, the chord, c, and the radius, R, are indicated on the figure. The x-axis is horizontal, and the z-axis is vertical. Some analytic geometry leads to the following dimensional and dimensionless expressions describing the upper surface:

$$z = -(R - \tfrac{1}{2}t) + \sqrt{R^2 - (x - \tfrac{1}{2}c)^2};$$
$$\frac{z}{c} = -\left(\frac{R}{c} - \frac{1}{2}\frac{t}{c}\right) + \sqrt{\left(\frac{R}{c}\right)^2 - \left(\frac{x}{c} - \frac{1}{2}\frac{c}{t}\right)^2}. \tag{4-22}$$

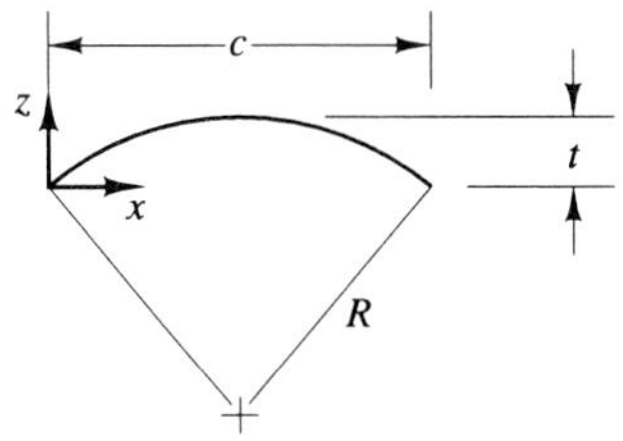

Figure 4-18 Circular-Arc Airfoil Nomenclature

Here t is the maximum thickness and R is the radius of the arc:

$$R = \frac{c^2 + t^2}{4t};$$
$$\frac{R}{c} = \frac{1 + \left(\frac{t}{c}\right)^2}{4\frac{t}{c}}. \tag{4-23}$$

The slope of the upper surface is

$$\frac{dz}{dx} = -\frac{x - \frac{1}{2}c}{\sqrt{R^2 - (x - \frac{1}{2}c)^2}} = \frac{d\frac{z}{c}}{d\frac{x}{c}} = -\frac{\frac{x}{c} - \frac{1}{2}}{\sqrt{\left(\frac{R}{c}\right)^2 - \left(\frac{x}{c} - \frac{1}{2}\right)^2}}, \tag{4-24}$$

and the angle of the surface with respect to the horizontal axis is

$$\theta = \tan^{-1}\left(\frac{dz}{dx}\right) = \tan^{-1}\left(\frac{d\frac{z}{c}}{d\frac{x}{c}}\right). \tag{4-25}$$

Equation 4-25 can be used to compute $\theta(x/c)$ for the circular-arc airfoil. Once the surface angle as a function of the dimensionless chord distance, x/c, is known, the input for SHEX is the same as that for the earlier wedge airfoil configurations. Consider the following example.

Example 4-4

Use SHEX to determine features of the flow about a circular-arc airfoil of maximum thickness $0.1c$ at angles of attack of 0° and 2° in an $M_\infty = 1.5$ flow of air.

Solution. For the given airfoil, the maximum thickness is $0.1c$ and so

$$\frac{z}{c} = -2.475 + \sqrt{6.126 - \left(\frac{x}{c}\right)^2 + \frac{x}{c}}.$$

The surface angle as a function of the chord position is determined using the above equation and appears in Table 4-2.

TABLE 4-2 SURFACE ANGLE VERSUS CHORD LOCATION FOR EXAMPLE 4-4

x/c	θ (deg.)
0.00	6.87
0.10	5.49
0.20	4.11
0.30	2.74
0.40	1.37
0.50	0.00
0.60	−1.37
0.70	−2.74
0.80	−4.11
0.90	−5.49
1.00	−6.87

The SHEX output for the 0° angle of attack is given in Figure 4-19. The Mach number immediately behind the leading-edge oblique shock is 1.254, and the flow accelerates to a Mach number of 1.683, as it traverses the airfoil surfaces. The airfoil is symmetrical and at 0° angle of attack the lift is zero but drag is present. The 2° angle of attack results are reported in Figure 4-20. The turning angle associated with the lower surface produces a Mach number behind the leading-edge oblique shock of 1.17 as compared to 1.33 for the upper surface. The flow accelerates to a Mach number of 1.753 at the trailing edge of the upper surface and to a Mach number of 1.611 at the trailing edge of the lower surface. Lift is produced in this case, and the *L*/*D* ratio is 3.42—not a very large value.

MINF	ALPHA	GAMMA
1.5	0	1.4

TOP SURFACE OF THE AIRFOIL

SEGMENT	ANGLE	X/C START	M	T/TINF	P/PINF
1	6.870	0.0000	1.254	1.1030	1.4039
2	5.490	0.1000	1.304	1.0814	1.3101
3	4.110	0.2000	1.355	1.0606	1.2239
4	2.740	0.3000	1.403	1.0405	1.1445
5	1.370	0.4000	1.450	1.0208	1.0703
6	0.000	0.5000	1.497	1.0014	1.0008
7	-1.370	0.6000	1.543	0.9822	0.9354
8	-2.740	0.7000	1.589	0.9633	0.8738
9	-4.110	0.8000	1.636	0.9445	0.8517
10	-5.490	0.9000	1.683	0.9258	0.7605

BOTTOM SURFACE OF THE AIRFOIL

SEGMENT	ANGLE	X/C START	M	T/TNIF	P/PNIF
1	6.879	0.0000	1.254	1.1030	1.4039
2	5.490	0.1000	1.304	1.0814	1.3101
3	4.110	0.2000	1.355	1.0606	1.2239
4	2.740	0.3000	1.403	1.0405	1.1445
5	1.370	0.4000	1.450	1.0208	1.0703
6	0.000	0.5000	1.497	1.0014	1.0008
7	-1.370	0.6000	1.543	0.9822	0.9354
8	-2.740	0.7000	1.589	0.9633	0.8738
9	-4.110	0.8000	1.636	0.9445	0.8157
10	-5.490	0.9000	1.683	0.9258	0.7605

CL	CD	L/D	XCP	YCP
0.0000	0.0340	0.00	0.000	0.000

Figure 4-19 SHEX Output for Example 4-4, $\alpha = 0$

MINF	ALPHA	GAMMA
1.5	2	1.4

TOP SURFACE OF THE AIRFOIL

SEGMENT	ANGLE	X/C START	M	T/TINF	P/PINF
1	6.870	0.0000	1.330	1.0711	1.2698
2	5.490	0.1000	1.379	1.0505	1.1866
3	4.110	0.2000	1.427	1.0305	1.1092
4	2.740	0.3000	1.474	1.0109	1.0373
5	1.370	0.4000	1.520	0.9916	0.9697
6	0.000	0.5000	1.567	0.9726	0.9061
7	-1.370	0.6000	1.613	0.9538	0.8462
8	-2.740	0.7000	1.659	0.9351	0.7896
9	-4.110	0.8000	1.706	0.9166	0.7362
10	-5.490	0.9000	1.753	0.8980	0.6854

BOTTOM SURFACE OF THE AIRFOIL

SEGMENT	ANGLE	X/C START	M	T/TNIF	P/PNIF
1	6.879	0.0000	1.170	1.1385	1.5604
2	5.490	0.1000	1.226	1.1148	1.4499
3	4.110	0.2000	1.279	1.0927	1.3516
4	2.740	0.3000	1.329	1.0717	1.2627
5	1.370	0.4000	1.377	1.0512	1.1805
6	0.000	0.5000	1.425	1.0313	1.1041
7	-1.370	0.6000	1.472	1.0118	1.0325
8	-2.740	0.7000	1.518	0.9925	0.9653
9	-4.110	0.8000	1.565	0.9734	0.9020
10	-5.490	0.9000	1.611	0.9545	0.8419

CL	CD	L/D	XCP	YCP
0.1330	0.0389	3.42	0.444	-.001

Figure 4-20 SHEX Output for Example 4-4, $\alpha = 2°$

4.5 CONCLUSION

In this chapter we developed and explored Prandtl-Meyer flow and examined in some detail the shock-expansion method for treating airfoils in supersonic flow. Prandtl-Meyer flow was the last topic treated in our development of waves in compressible flow. The shock-expansion section in this chapter illustrated how understanding of various flows can be utilized to provide a procedure to compute lift and drag on airfoils in supersonic flow. Additionally, some of the experimental results presented and examined in Section 4.4 demonstrated the validity of oblique shock and Prandtl-Meyer concepts. In subsequent chapters we shall find much of the material presented in the first four chapters to be useful in internal flows, such as flow of a compressible fluid in converging–diverging nozzles. Part III of this textbook develops and explores in some detail the one-dimensional, compressible internal flows.

CHAPTER 4 REVIEW QUESTIONS

1. Does stagnation pressure increase, decrease, or remain the same across a Prandtl-Meyer expansion process? Why? Answer this same question for a Prandtl-Meyer compression process.
2. As $M \to \infty$ (and $T \to 0$), what happens to the Prandtl-Meyer angle? Discuss some limitations of the answer.
3. Is a Prandtl-Meyer compression possible? Justify your answer.
4. Sketch a surface that results in an isentropic compression.
5. Sketch the shock and expansion wave patterns expected for supersonic flow over the shape shown in Figure 4-21.

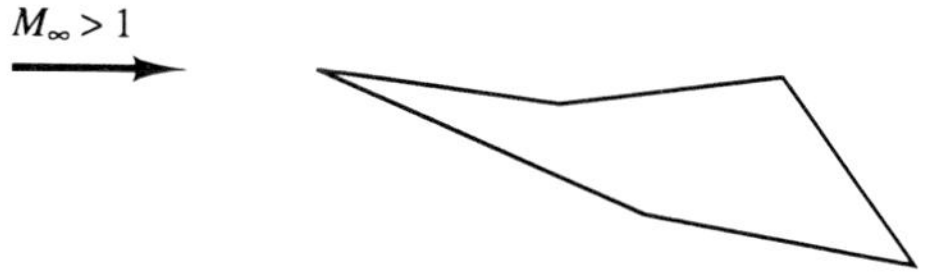

Figure 4-21

6. How does the stagnation temperature vary across a Prandtl-Meyer process?
7. Sketch the wave forms (oblique shock or Prandtl-Meyer) that would occur on the surface of the profile in a supersonic flow shown in Figure 4-22.

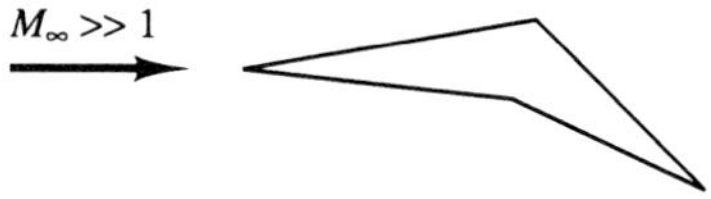

Figure 4-22

8. Why do the trailing-edge oblique shock waves not influence the lift and drag of a double-wedge airfoil in a supersonic flow?
9. Describe the Mach number and pressure distributions over the upper and lower surfaces of the biconvex airfoil in a supersonic flow shown in Figure 4-23.

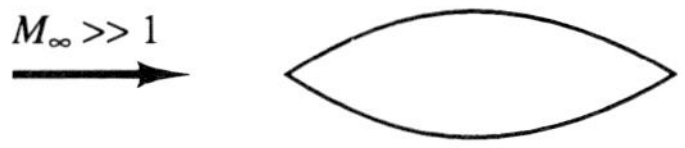

Figure 4-23

10. Sketch and label the important nomenclature of an airfoil.
11. Describe the ways in which the net forces on an airfoil section can be described in terms of lift, drag, and moment.

12. Why would the shock-expansion procedure not work near the leading edge for the circumstances of Figure 4-13a?
13. Sketch the oblique shock wave structures and the Mach lines over the top and bottom surfaces of a biconvex airfoil in supersonic flow.

CHAPTER PROBLEMS

1. Air at Mach 2.5, with stagnation conditions of 5 MPa and 1000 K, expands around a 15° convex corner. Determine the Mach number, the stagnation conditions, and the static conditions downstream of the expansion.
2. Devise a computer program subroutine that will yield values of υ versus M for Prandtl-Meyer flow with the ratio of specific heats, γ, as an input parameter and either the Prandtl-Meyer angle or the Mach number as the input variable. Repeat for μ versus M with γ as the parameter. Verify the results using COMPQ. How are these software elements related to option **PM** of COMPQ?
3. Air initially at a Mach number of 3.5 expands around a 12° compression corner and then is compressed about a 12° turning angle. The arrangement is sketched in Figure 4-24. Determine the Mach number and the static and stagnation pressures for states 2 and 3 if the static pressure at state 1 is 3 psia. Why are conditions at state 3 not equal to conditions at state 1?

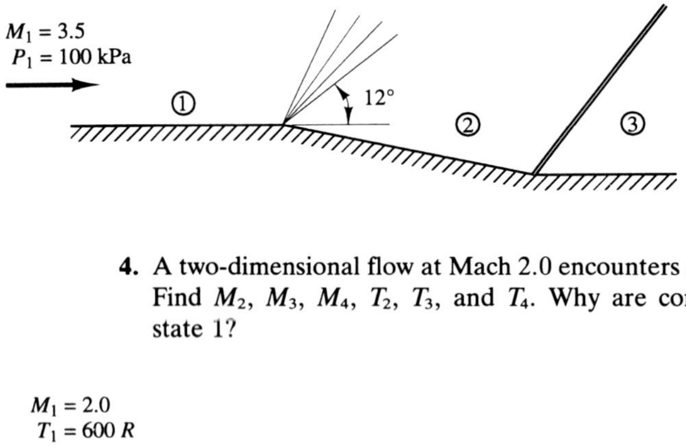

Figure 4-24

4. A two-dimensional flow at Mach 2.0 encounters the configuration shown in Figure 4-25. Find M_2, M_3, M_4, T_2, T_3, and T_4. Why are conditions at state 4 not the same as at state 1?

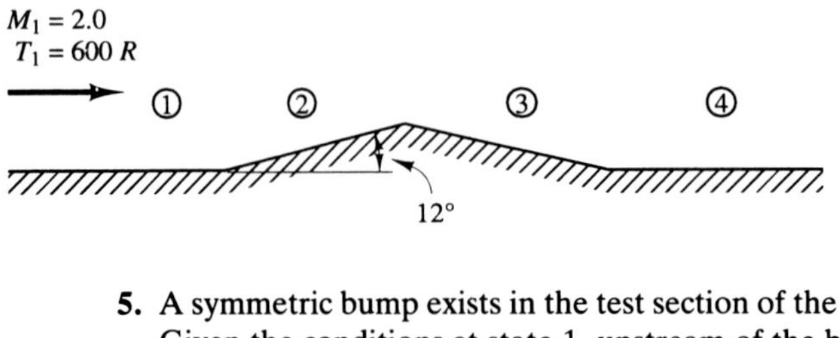

Figure 4-25

5. A symmetric bump exists in the test section of the supersonic wind tunnel in Figure 4-26. Given the conditions at state 1, upstream of the bump, find the conditions (P, P_o, T, T_o, and M) for states 3 and 4. Why are conditions at state 4 not equal to conditions at state 1?

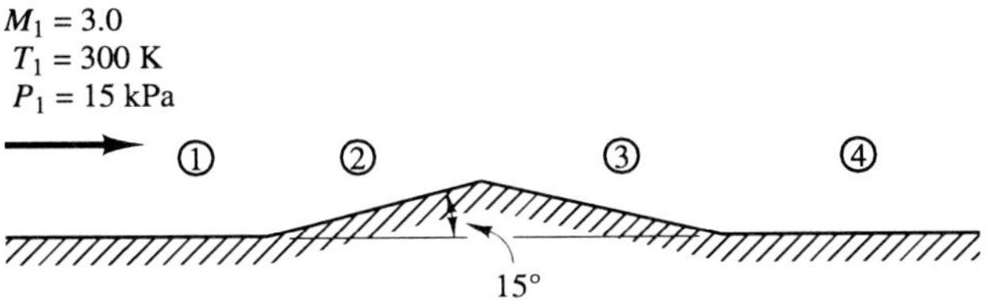

Figure 4-26

6. An air flow initially at $M = 4$, 100 F, and 2 psia passes over the configuration shown in Figure 4-27. Find the conditions (Mach number, static and stagnation temperature and pressure) in region c. How do states a and c compare?

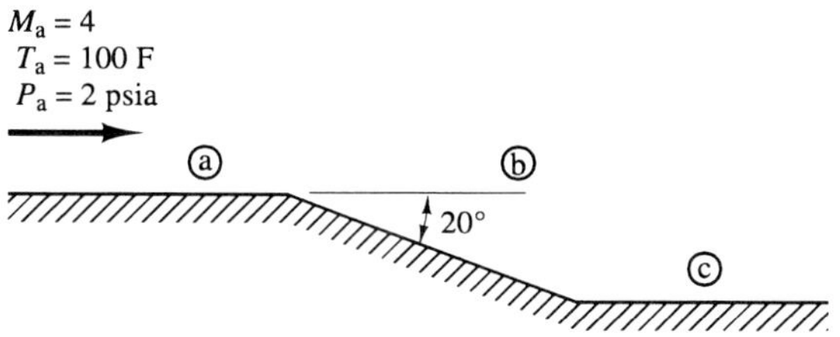

Figure 4-27

7. Sketch an arrangement to reduce the Mach number of a stream of air from 2.5 to 1.5 using a Prandtl-Meyer flow. How would the stagnation pressure losses differ for the Prandtl-Meyer arrangement if an oblique shock wave were used? How would the angles through which the flow must be turned differ for the two arrangements? What happens when the Mach lines (waves) merge?

8. A thin two-dimensional flat plate is placed in an air flow at a Mach number of 2.25. The plate orientation is such that an oblique shock, as illustrated in Figure 4-28, makes an angle of 40° with respect to the horizontal axis. On the upper side an expansion occurs at the leading edge.

(a) Find the angle the plate makes with the freestream.

(b) What is the pressure on the lower surface of the plate?

(c) What is the pressure on the upper surface of the plate?

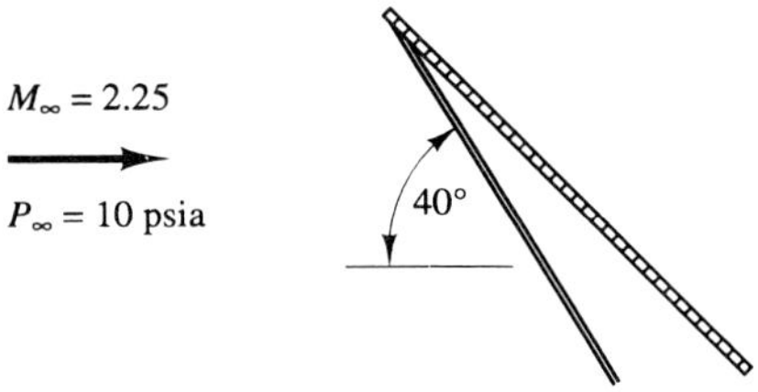

Figure 4-28

9. A nozzle attached to a reservoir containing air at 20 atm produces a uniform Mach 2.3 flow at the exit. The expansion from the exit can be considered isentropic. The flow conditions illustrated in Figure 4-29 require a Prandtl-Meyer expansion fan at the nozzle exit. Find the direction of the flow after it passes through the Prandtl-Meyer fan. What effect does this turning angle have on the net axial thrust force exerted by the fluid on the nozzle?

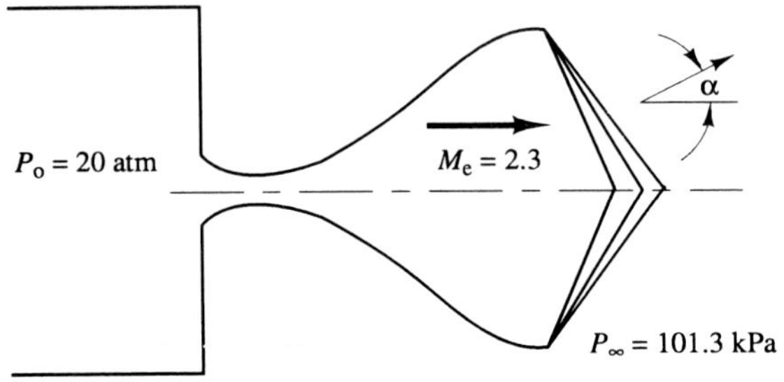

Figure 4-29

10. For the configuration shown in Figure 4-30, a compression occurs at the trailing edge, with the resultant flows in regions a and b parallel. Describe the relationship between the pressure, velocity, and entropy in regions a and b.

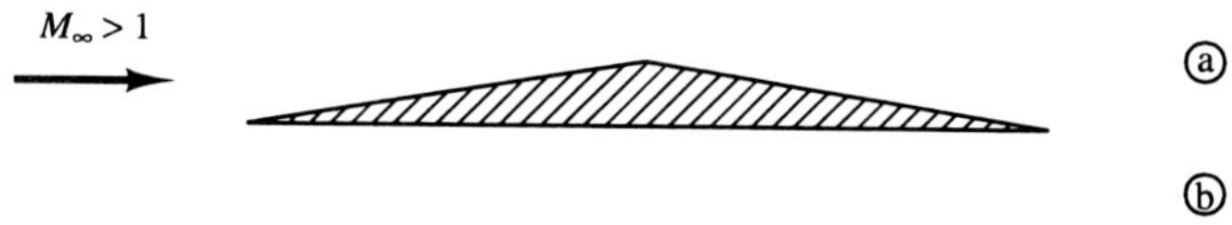

Figure 4-30

11. Generate plots of C_l, C_d, and L/D ratio as the angle of attack is varied from 0 to 15 degrees for the airfoil of Example 4-2. Indicate conditions where a detached oblique shock wave would exist.

12. Plot C_l and C_d ($0° < \alpha < 15°$) for the airfoil of Example 4-2 as the Mach number varies from $M_\infty = 2$ to $M_\infty = 5$. Indicate conditions where a detached oblique shock wave would exist.

13. Air at a Mach number of 3 flows over a flat plate at an angle of attack of 7°. Compute the Mach numbers and pressure ratios on the top and bottom surfaces. Calculate the lift and drag coefficients and the L/D ratio for the flat plate. Compute using COMPQ and verify using SHEX.

14. What happens to the lift and drag coefficients and the L/D ratio if the last $0.25c$ of the plate described in Problem 4-13 is deflected downward by 6°?

15. If $t = 0.075c$, compute C_l, C_d, and L/D for Figure 4-31's profile shapes at $M_\infty = 2$ and 0° angle of attack.

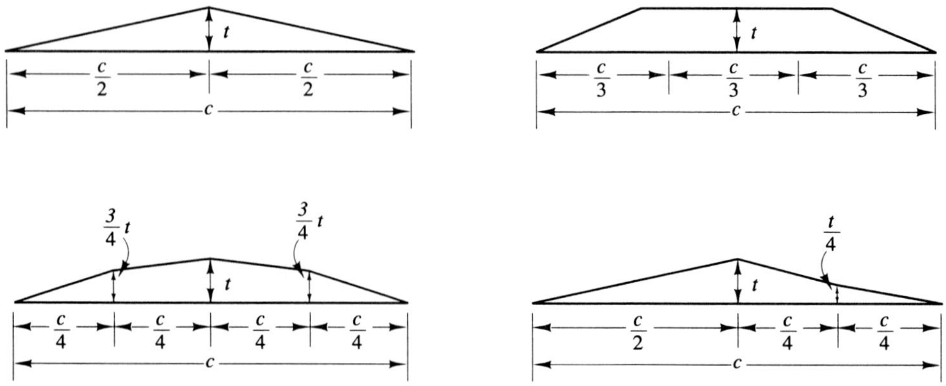

Figure 4-31 *Adapted from Saad.*

16. Repeat the computations of Problem 4-15 for an angle of attack, α, of 4°.
17. At $M_\infty = 2.5$, find C_l, C_d, and L/D for the profile shape in Figure 4-32.

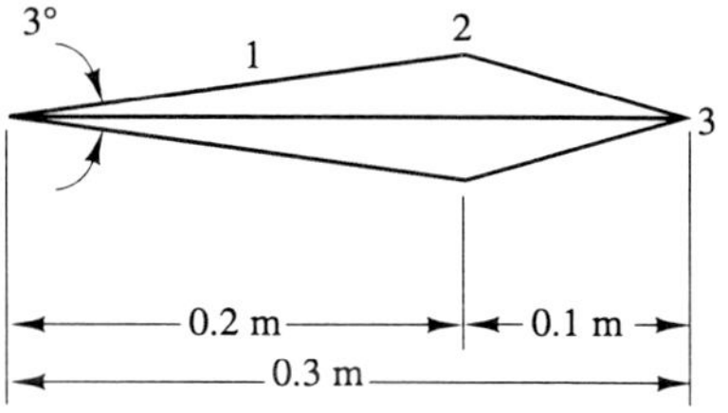

Figure 4-32 *Adapted from Saad.*

18. Generates plots of C_l, C_d, and L/D as a function of angle of attack for a $0.1c$ thickness circular-arc airfoil in a Mach 2 flow.
19. Generates plots of C_l, C_d, and L/D as a function of angle of attack for a $0.05c$ thickness circular-arc airfoil in a Mach 1.75 flow.
20. At $M_\infty = 2.5$, compare C_l, C_d, and L/D at $\alpha = 5°$ for biconvex airfoils of various thicknesses ($0.01 < t/c < 0.10$).
21. At $\alpha = 5°$, compare C_l, C_d, and L/D for $t/c = 0.05$ biconvex airfoils at various supersonic Mach numbers ($M_\infty < 5.0$).

Part III

ONE-DIMENSIONAL SIMPLE AND GENERALIZED COMPRESSIBLE FLOWS

Simple One-Dimensional Flow Formulations

5

5.1 INTRODUCTION

Compressible flow consists of more than just compression and expansion waves or processes. In Part II we focused on waves and processes associated with waves. Now we turn our attention to simple one-dimensional flows. "Simple" requires a little explanation. In an internal compressible flow we could envision area change, friction, heat transfer, and mass flow rate change as affecting the state of the flow at a location. If only one of the aforementioned is occurring (area change in the absence of friction, heat transfer, and mass flow rate change, for example), then the flow is termed *simple*. Simple flows are important for a number of reasons:

1. Many real flows can be adequately modeled using simple flow concepts.
2. Property and Mach number relations for simple flows of perfect gases can usually be expressed in closed form.
3. Each effect, such as area change, can be analyzed and understood in the absence of any other effects.
4. Quite simply, tradition.

The first reason is by far the most important; we shall see that simple flows can provide estimates of engineering accuracy for a number of important situations.

The philosophy of the mode of presentation of simple flows in this book differs from that in more traditional textbooks. The same material will be covered, but the manner of the coverage will be different. We shall first develop a well-posed system of equations that are applicable for generalized one-dimensional compressible flows that include area change, friction, heat transfer, and mass flow rate change. Each simple flow will then be viewed and analyzed as a special case of the generalized flow. The primary reasons for this sequence of presentation are that the relationships of the simple flows to the generalized flow are more readily apparent and also that the derivations are simpler and more compact overall. That the simple flows are special cases of the generalized flow is not a new notion; what is new here is the presentation sequence of generalized to simple as the theme.

5.2 DEVELOPMENT OF THE EQUATIONS FOR GENERALIZED ONE-DIMENSIONAL COMPRESSIBLE FLOW

A differential control volume for a generalized quasi–one-dimensional flow is presented in Figure 5-1. The flow is called *quasi–one-dimensional* because variations in the y-direction at a given x-location are presumed to be negligible. An implication of the quasi–one-dimensional assumption is that the area variation is such that the y-velocity component is also negligibly small. From this point on, the prefix "quasi" will no longer be used but will be understood within the framework of the assumptions. Schematic indications of shaft work, heat transfer, mass addition, area change, wall friction, and drag are given. Since this is a differential control volume, the results of the analysis will be a system of differential equations. The general plan is to invoke an equation of state (for a perfect gas), apply the conservation equations (mass, momentum, and energy), and utilize definitions to form a system of differential equations applicable for generalized, steady, one-dimensional compressible flow. The system of equations will form the foundation of the treatment of simple flows in the next two chapters and that of the more complex generalized flows in Chapter 8.

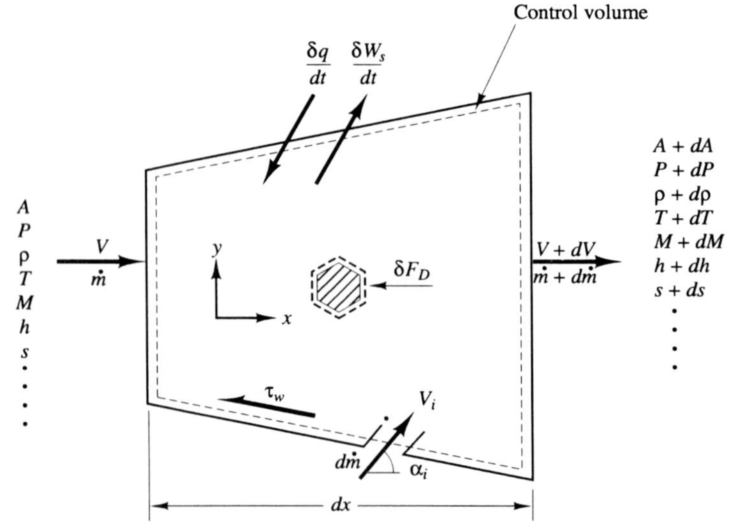

Figure 5-1 Generalized Flow Schematic and Control Volume

The system of equations is derived and assembled in the following development:

1. Equation of state. The perfect gas law is

$$P = \rho RT. \tag{1-8}$$

Taking the natural log of the above expression results in

$$\ln P = \ln \rho + \ln R + \ln T, \tag{5-1}$$

and differentiating yields

$$\frac{dP}{P} = \frac{d\rho}{\rho} + \frac{dT}{T}. \tag{5-2}$$

2. Mach number definition. The Mach number is defined as

$$M = \frac{V}{a} = \frac{V}{\sqrt{\gamma RT}}. \tag{1-40}$$

Squaring the preceding equation produces

$$M^2 = \frac{V^2}{\gamma RT}. \tag{5-3}$$

The same sequence is followed; the equation is expressed in natural log form and differentiated to give

$$\frac{dM^2}{M^2} = 2\frac{dV}{V} - \frac{dT}{T}. \tag{5-4}$$

Equation 5-4 could have been expressed in M instead of M^2; however, the analysis is easier if M^2 instead of M is consistently used throughout.

3. Conservation of mass. Because the mass flow rate is a variable of interest, the expression

$$\dot{m} = \rho AV \tag{5-5}$$

will be used instead of a formal application of the integral form of conservation of mass. Taking the natural log of Eq. 5-5 and differentiating results in

$$\frac{d\dot{m}}{\dot{m}} = \frac{d\rho}{\rho} + \frac{dA}{A} + \frac{dV}{V}. \tag{5-6}$$

4. Conservation of linear momentum. The x-component of the integral form of conservation of linear momentum for a steady-state situation is

$$\sum F_x = \int_{cs} \rho V_x \overline{V} \cdot \overline{dA}. \tag{5-7}$$

In addition to accounting for pressure acting on the boundaries of the control volume, allowance is also made for shear (friction), τ_w, acting over the surface area

A_w and for other drag forces, dF_D, acting in the x-direction. Figure 5-1 illustrates the injected mass, $d\dot{m}$, as entering the control volume with a velocity V_i at an angle α_i with respect to x-coordinate direction. As later developments will show, the injected fluid can possess a stagnation enthalpy h_{o_i} different from the main flow. The injected fluid is also assumed to have the same γ and R values as the main flow. The x-direction momentum flux carried by the injected mass, $d\dot{m}$, is $V_i \cos\alpha_i \, d\dot{m}$, where α_i is defined as shown in Figure 5-1. Application of Eq. 5-7 to the control volume becomes

$$PA - (P + dP)(A + dA) + (P + dP/2)dA - dF_D - \tau_w A_w = (\dot{m} + d\dot{m})(V + dV) - \dot{m}V - d\dot{m}V_i \cos\alpha_i. \tag{5-8}$$

Expanding, dividing by PA, dropping terms involving the quadratic or higher products of differential quantities, and neglecting terms like $d\dot{m}dV$ in comparison with terms like $\dot{m}dV$ yields

$$\frac{dP}{P} + \frac{dF_D}{PA} + \frac{\tau_w A_w}{PA} + \frac{\dot{m}}{PA}dV + \frac{V}{PA}d\dot{m} - \frac{d\dot{m}V_i \cos\alpha_i}{PA} = 0. \tag{5-9}$$

The wall shear stress due to friction is expressed in terms of the *Fanning friction factor*:

$$f = \frac{\tau_w}{\frac{1}{2}\rho V^2}. \tag{5-10}$$

Discussion of and expressions for the Fanning friction factor, which is equal to the skin friction coefficient, and the Darcy-Weisbach friction factor are presented in Appendix E. The term involving τ_w then becomes

$$\frac{\tau_w A_w}{PA} = \frac{\gamma M^2}{2}\frac{4f}{D_h}dx, \tag{5-11}$$

where

$$\frac{1}{2}\rho V^2 = \frac{P}{2RT}M^2\gamma RT = \frac{\gamma P M^2}{2}. \tag{5-12}$$

The hydraulic diameter D_h is defined in terms of the cross-sectional area A and the wetted perimeter P_w as

$$D_h = \frac{4A}{P_w}, \tag{5-13}$$

and the surface area A_w is $P_w\,dx$. In similar fashions,

$$\frac{\dot{m}}{PA}dV = \gamma M^2 \frac{dV}{V}; \tag{5-14a}$$

$$\frac{V}{PA}d\dot{m} = \gamma M^2 \frac{d\dot{m}}{\dot{m}}; \tag{5-14b}$$

$$\frac{V_i \cos\alpha_i}{PA}d\dot{m} = \gamma M^2 \frac{V_i \cos\alpha_i}{V}\frac{d\dot{m}}{\dot{m}}. \tag{5-14c}$$

Usually $(V_i \cos \alpha_i)/V$ is identified as r_i. Inserting Eqs. 5-14 in Eq. 5-9 and rearranging produces

$$\frac{dP}{P} + \frac{\gamma M^2}{2}\left(4f\frac{dx}{D_h} + \frac{2dF_D}{\gamma M^2 PA} - 2r_i\frac{d\dot{m}}{\dot{m}}\right) + \gamma M^2\frac{dV}{V} + \gamma M^2\frac{d\dot{m}}{\dot{m}} = 0. \tag{5-15}$$

5. Conservation of energy. The integral form of conservation of energy in the steady state is

$$\frac{\delta q}{dt} - \frac{\delta w_s}{dt} = \int_{cs} \rho\left(h + \frac{V^2}{2}\right)\bar{V} \cdot \overline{dA}. \tag{1-40c}$$

Applied to the control volume in Figure 5-1, the equation becomes

$$\frac{\delta q}{dt} - \frac{\delta w_s}{dt} = -\dot{m}\left(h + \frac{V^2}{2}\right) + (\dot{m} + d\dot{m})\left[h + dh + \frac{(V + dV)^2}{2} - d\dot{m}\left(h_i + \frac{V_i^2}{2}\right)\right]. \tag{5-16}$$

Expanding, dividing by $\dot{m}$, and neglecting terms involving quadratic or higher products of differentials, Eq. 5-16 reduces to

$$\delta Q - \delta W_s = dh + d(V^2/2) + \frac{d\dot{m}}{\dot{m}}\left(h - h_i + \frac{V^2 - V_i^2}{2}\right). \tag{5-17}$$

Further rearrangement of Eq. 5-17 produces

$$\delta Q - \delta W_s - \frac{d\dot{m}}{\dot{m}}\left(h - h_i + \frac{V^2 - V_i^2}{2}\right) = d\left(h + \frac{V^2}{2}\right) = dh_o. \tag{5-18}$$

For a calorically perfect gas, $dh_o = c_p\, dT_o$, so that

$$\frac{\delta Q - \delta W_s - \dfrac{d\dot{m}}{\dot{m}}\left(h - h_i + \dfrac{V^2 - V_i^2}{2}\right)}{c_p T_o} = \frac{dT_o}{T_o}. \tag{5-19}$$

The form of Eq. 5-19 demonstrates that the combined effects of heat transfer, shaft work, and the enthalpy change associated with mass addition are equivalent to change in the stagnation temperature. The stagnation temperature can then be used in place of the individual component contributions so that via the relationship between static and stagnation temperature (Eq. 1-66):

$$\frac{dT_o}{T_o} = \frac{dT}{T} + \frac{\dfrac{\gamma - 1}{2}M^2}{1 + \dfrac{\gamma - 1}{2}M^2}\,\frac{dM^2}{M^2}. \tag{5-20}$$

6. Impulse function. The *impulse function* is defined as

$$I = PA + \dot{m}V = PA(1 + \gamma M^2). \tag{5-21}$$

Differentiating the natural log of Eq. 5-21 yields

$$\frac{dI}{I} = \frac{dP}{P} + \frac{dA}{A} + \frac{\gamma M^2}{1 + \gamma M^2}\frac{dM^2}{M^2}. \tag{5-22}$$

7. Stagnation pressure. The relationship between static and stagnation pressure is

$$P_o = P\left(1 + \frac{\gamma - 1}{2}M^2\right)^{\gamma/(\gamma-1)}, \tag{1-70}$$

which can be expressed as

$$\frac{dP_o}{P_o} = \frac{dP}{P} + \frac{\frac{1}{2}\gamma M^2}{1 + \frac{\gamma - 1}{2}M^2}\frac{dM^2}{M^2}. \tag{5-23}$$

8. Entropy (second law of thermodynamics). In Chapter 1 we developed the following expression for the change in entropy for a process:

$$\frac{ds}{c_p} = \frac{dT}{T} - \frac{\gamma - 1}{\gamma}\frac{dP}{P}. \tag{1-23b}$$

The eight equations developed in the previous paragraphs are written as a system and repeated for clarity here.

$$\frac{dP}{P} = \frac{d\rho}{\rho} + \frac{dT}{T} \tag{5-24a}$$

$$\frac{dM^2}{M^2} = 2\frac{dV}{V} - \frac{dT}{T} \tag{5-24b}$$

$$\frac{d\dot{m}}{\dot{m}} = \frac{d\rho}{\rho} + \frac{dA}{A} + \frac{dV}{V} \tag{5-24c}$$

$$\frac{dP}{P} + \frac{\gamma M^2}{2}\left(4f\frac{dx}{D_h} + \frac{2dF_D}{\gamma M^2 PA} - 2\frac{d\dot{m}}{\dot{m}}r_i\right) + \gamma M^2\frac{dV}{V} + \gamma M^2\frac{d\dot{m}}{\dot{m}} = 0 \tag{5-24d}$$

$$\frac{dT_o}{T_o} = \frac{dT}{T} + \frac{\frac{\gamma - 1}{2}M^2}{1 + \frac{\gamma - 1}{2}M^2}\frac{dM^2}{M^2} \tag{5-24e}$$

$$\frac{dI}{I} = \frac{dP}{P} + \frac{dA}{A} + \frac{\gamma M^2}{1 + \gamma M^2}\frac{dM^2}{M^2} \tag{5-24f}$$

$$\frac{dP_o}{P_o} = \frac{dP}{P} + \frac{\frac{1}{2}\gamma M^2}{1 + \frac{\gamma - 1}{2}M^2}\frac{dM^2}{M^2} \tag{5-24g}$$

$$\frac{ds}{c_p} = \frac{dT}{T} - \frac{\gamma - 1}{\gamma}\frac{dP}{P} \tag{5-24h}$$

The variables from Eqs. 5-24 are $d\dot{m}/\dot{m}$, $d\rho/\rho$, dA/A, dV/V, dM^2/M^2, dP/P, dT/T, dT_o/T_o, dI/I, dP_o/P_o, ds/s, and $(4f\,dx/D_h + 2dF_D/\gamma M^2 PA - 2r_i\,d\dot{m}/\dot{m})$. Thus, the system of eight equations contains twelve variables. In order for the system to be solvable, the number of unknowns must be equal to the number of equations. We have used an equation of state and the conservation principles as well as a number of definitions in formulating Eqs. 5-24. No additional physical principles are available, and any additional definition simply adds one new variable and one new equation.

The system is solvable if four of the variables are taken as known or prescribed quantities with the remaining eight being the unknowns. We shall select the following as known or prescribed variables:

1. dA/A
2. dT_o/T_o
3. $d\dot{m}/\dot{m}$
4. $4f\,dx/D_h + 2dF_D/\gamma M^2 PA$.

Hereinafter these will be called *driving potentials* since they are assumed known and since they define the characteristics of the remaining eight variables. The unknown variables are fittingly called *flow properties* since they all involve some property of the one-dimensional flow.

In order for the technique to be tractable, the variations of the driving potentials must be known as functions of the independent variable x. Thus, it is appropriate to examine the driving potentials to ascertain if they are indeed quantities that are reasonable to be known in a problem statement. The variable A, or $A(x)$, defines the cross-sectional area distribution and is known if the shape of the duct is defined. The variable $\dot{m}$, or $\dot{m}(x)$, accounts for any mass addition and is known if the mass flow rate as a function of x is specified. The variable T_o, $T_o(x)$, addresses changes in the stagnation temperature, which according to Eq. 5-19 can result from work, heat interaction, and/or mass injection. The last expression, $4f\,dx/D_h + 2dF_D/\gamma M^2PA$, characterizes friction $(4f\,dx/D_h)$ and/or other drag-producing $(2dF_D/\gamma M^2PA)$ processes. Wall friction is accounted for by the friction factor, f, which can be specified as $f(x)$. Objects placed in a flowing stream can result in additional drag that can be accounted for by the F_D term. For the situations of interest to us, wall friction will be the only drag-producing term.

In some instances, particularly with mass addition, a driving potential can be computed with auxiliary expressions simultaneously with the remainder of the solution. Other choices of driving potentials could be made, but for our purposes these conform to the simple flows that are traditionally discussed in compressible flow.

Cast in matrix form, Eqs. 5-24 become matrix Eq. 5-25,

$$\begin{bmatrix} 1 & -1 & -1 & 0 & 0 & 0 & 0 & 0 \\ 0 & 0 & 1 & -2 & 1 & 0 & 0 & 0 \\ 0 & 1 & 0 & 1 & 0 & 0 & 0 & 0 \\ 1 & 0 & 0 & \gamma M^2 & 0 & 0 & 0 & 0 \\ 0 & 0 & 1 & 0 & \dfrac{\frac{\gamma-1}{2}M^2}{\psi} & 0 & 0 & 0 \\ -1 & 0 & 0 & 0 & -\dfrac{\gamma M^2}{1+\gamma M^2} & 0 & 1 & 0 \\ -1 & 0 & 0 & 0 & -\dfrac{\gamma M^2/2}{\psi} & 1 & 0 & 0 \\ \dfrac{\gamma-1}{\gamma} & 0 & -1 & 0 & 0 & 0 & 0 & 1 \end{bmatrix} \begin{bmatrix} \dfrac{dP}{P} \\ \dfrac{d\rho}{\rho} \\ \dfrac{dT}{T} \\ \dfrac{dV}{V} \\ \dfrac{dM^2}{M^2} \\ \dfrac{dP_0}{P_0} \\ \dfrac{dI}{I} \\ \dfrac{ds}{c_p} \end{bmatrix} = \begin{bmatrix} 0 \\ 0 \\ \dfrac{d\dot{m}}{\dot{m}} - \dfrac{dA}{A} \\ E + F \\ \dfrac{dT_0}{T_0} \\ \dfrac{dA}{A} \\ 0 \\ 0 \end{bmatrix} \tag{5-25}$$

where

$$E = -\frac{\gamma M^2}{2}\left(4f\frac{dx}{D_h} + \frac{2}{\gamma M^2}\frac{dF_D}{PA}\right), \tag{5-26}$$

$$F = -\gamma M^2(1 - r_i)\frac{d\dot{m}}{\dot{m}}, \tag{5-27}$$

$$\psi = 1 + \frac{\gamma - 1}{2}M^2. \tag{5-28}$$

Equation 5-25 is in the classic linear system form with the 8 × 8 matrix representing the coefficient matrix, the column matrix of differential properties representing the unknowns, and the column matrix of the differential driving potentials representing the knowns. With the symbolic inverse of the coefficient matrix available, the solution of Eq. 5-25 could be expressed in symbolic form for each dependent differential property in terms of the differential driving potentials and the inverse matrix elements. The symbolic inverse, generated by the symbolic manipulation language MACSYMA, was presented by Hodge (1990). The symbolic solution is given in matrix form as Eq. 5-29.

$$\begin{bmatrix} \dfrac{dP}{P} \\[2mm] \dfrac{d\rho}{\rho} \\[2mm] \dfrac{dT}{T} \\[2mm] \dfrac{dV}{V} \\[2mm] \dfrac{dM^2}{M^2} \\[2mm] \dfrac{dP_0}{P_0} \\[2mm] \dfrac{dI}{I} \\[2mm] \dfrac{ds}{c_p} \end{bmatrix} = \begin{bmatrix} \dfrac{\gamma M^2}{M^2 - 1} & -\dfrac{\gamma(\gamma - 1)M^4}{2(M^2 - 1)} & \dfrac{\gamma M^2}{M^2 - 1} & -\dfrac{1 + M^2(\gamma - 1)}{M^2 - 1} & \dfrac{\gamma M^2 \psi}{M^2 - 1} & 0 & 0 & 0 \\[2mm] \dfrac{1}{M^2 - 1} & -\dfrac{(\gamma - 1)M^2}{2(M^2 - 1)} & \dfrac{M^2}{M^2 - 1} & -\dfrac{1}{M^2 - 1} & \dfrac{\psi}{M^2 - 1} & 0 & 0 & 0 \\[2mm] \dfrac{(\gamma - 1)M^2}{M^2 - 1} & -\dfrac{(\gamma - 1)M^2(\gamma M^2 - 1)}{2(M^2 - 1)} & \dfrac{(\gamma - 1)M^2}{M^2 - 1} & -\dfrac{(\gamma - 1)M^2}{M^2 - 1} & \dfrac{(\gamma M^2 - 1)\psi}{M^2 - 1} & 0 & 0 & 0 \\[2mm] -\dfrac{1}{M^2 - 1} & \dfrac{(\gamma - 1)M^2}{2(M^2 - 1)} & -\dfrac{1}{M^2 - 1} & \dfrac{1}{M^2 - 1} & -\dfrac{\psi}{M^2 - 1} & 0 & 0 & 0 \\[2mm] -\dfrac{2\psi}{M^2 - 1} & \dfrac{(\gamma M^2 - 1)\psi}{M^2 - 1} & -\dfrac{2\psi}{M^2 - 1} & \dfrac{2\psi}{M^2 - 1} & -\dfrac{(\gamma M^2 + 1)\psi}{M^2 - 1} & 0 & 0 & 0 \\[2mm] 0 & \dfrac{\gamma M^2}{2} & 0 & 1 & -\dfrac{\gamma M^2}{2} & 0 & 1 & 0 \\[2mm] \dfrac{\gamma M^2}{\gamma M^2 + 1} & \dfrac{\gamma M^2}{\gamma M^2 + 1} & \dfrac{\gamma M^2}{\gamma M^2 + 1} & \dfrac{1}{\gamma M^2 + 1} & 0 & 1 & 0 & 0 \\[2mm] 0 & -\dfrac{(\gamma - 1)M^2}{2} & 0 & -\dfrac{\gamma - 1}{\gamma} & \psi & 0 & 0 & 1 \end{bmatrix} \begin{bmatrix} 0 \\[2mm] 0 \\[2mm] \dfrac{d\dot{m}}{\dot{m}} - \dfrac{dA}{A} \\[2mm] E + F \\[2mm] \dfrac{dT_0}{T_0} \\[2mm] \dfrac{dA}{A} \\[2mm] 0 \\[2mm] 0 \end{bmatrix} \tag{5-29}$$

Equation 5-29 represents eight differential equations, one for each of the unknowns or flow properties, expressed in terms of the driving potentials and the Mach number. Because the only independent variable is x, all differentials can be expressed in x-derivative form, $dM^2/dx, dM(x)^2/dx$, for example. The driving potentials are likewise expressible as functions of x. One of the most useful of the eight differential equations is for the Mach number. The differential equation representing the Mach number variation in generalized one-dimensional compressible flow is the solution of the matrix in Eq. 5-29 for dM^2/M^2 and is expressed in derivative form as

$$\frac{1}{M^2}\frac{dM^2}{dx} = -\frac{2\psi}{1-M^2}\frac{1}{A}\frac{dA}{dx} + \frac{(1+\gamma M^2)\psi}{1-M^2}\frac{1}{T_o}\frac{dT_o}{dx} + \frac{\gamma M^2\psi}{1-M^2}\left(4\frac{f}{D_h} + \frac{1}{\gamma PAM^2}\frac{dF_D}{dx} - 2r_i\frac{1}{\dot{m}}\frac{d\dot{m}}{dx}\right) + \frac{2(1+\gamma M^2)\psi}{1-M^2}\frac{1}{\dot{m}}\frac{d\dot{m}}{dx}. \tag{5-30}$$

Other differential equations can be obtained for the remaining seven flow properties. All the differential equations obtained from this procedure will be similar in format in that the left-hand side will contain the first derivative of the flow property, and the right-hand side will contain Mach number expressions multiplied by the derivative form of each of the driving potentials. These Mach number expressions which appear as coefficients of the driving potentials are called *influence coefficients* since they determine the influence of each driving potential on the change in each flow property.

Rather than presenting the solution of Eq. 5-25 in terms of differential equations, the usual procedure is to present the results in a table of influence coefficients. Table 5-1 represents the influence coefficients for the generalized one-dimensional flow examined in this chapter. Each of the differential equations of the flow properties (the unknowns in Eq. 5-29) are readily obtained from the table; for example, Eq. 5-30 is just the first-row influence coefficients multiplied by the indicated driving potentials. We shall extensively pursue generalized one-dimensional compressible flow using the influence coefficients in Chapter 8, but the simple flows will be developed in detail in the next two chapters.

5.3 SIMPLE FLOWS

A simple compressible flow is a flow in which only a single driving potential is involved. Based on the generalized one-dimensional analysis, which named four driving potentials, four simple flows can be defined:

1. simple area change, usually called isentropic flow
2. simple friction, usually called Fanno flow
3. simple heat addition or rejection (diabatic flow), usually called Rayleigh flow
4. simple mass addition.

We shall examine each of these in detail in the next two chapters of this book.

TABLE 5-1 INFLUENCE COEFFICIENTS FOR THE GENERALIZED STEADY ONE-DIMENSIONAL FLOW OF A PERFECT GAS

Change in flow property	Driving Potentials			
	$\frac{dA}{A}$	$\left[\left(\frac{4f\,dx}{D_h}\right) + \frac{2dF_D}{\gamma M^2 pA}\right]$	$\frac{dT_o}{T_o}$	$\frac{d\dot{m}}{\dot{m}}$
$\frac{dM^2}{M^2}$	$-\frac{2\psi}{1-M^2}$	$\frac{\gamma M^2\psi}{1-M^2}$	$\frac{(1+\gamma M^2)\psi}{1-M^2}$	$\frac{2\psi[(1+\gamma M^2) - r_i\gamma M^2]}{1-M^2}$
$\frac{dP}{P}$	$\frac{\gamma M^2}{1-M^2}$	$-\frac{\gamma M^2[1+(\gamma-1)M^2]}{2(1-M^2)}$	$-\frac{\gamma M^2\psi}{1-M^2}$	$-\frac{\gamma M^2[2\psi(1-r_i)+r_i]}{1-M^2}$
$\frac{d\rho}{\rho}$	$\frac{M^2}{1-M^2}$	$-\frac{\gamma M^2}{2(1-M^2)}$	$-\frac{\psi}{1-M^2}$	$-\frac{[(\gamma+1)M^2 - r_i\gamma M^2]}{1-M^2}$
$\frac{dT}{T}$	$\frac{(\gamma-1)M^2}{1-M^2}$	$-\frac{\gamma(\gamma-1)M^4}{2(1-M^2)}$	$\frac{(1-\gamma M^2)\psi}{1-M^2}$	$-\frac{(\gamma-1)M^2[(1+\gamma M^2) - r_i\gamma M^2]}{1-M^2}$
$\frac{dV}{V}$	$-\frac{1}{1-M^2}$	$\frac{\gamma M^2}{2(1-M^2)}$	$\frac{\psi}{1-M^2}$	$\frac{[(1+\gamma M^2) - r_i\gamma M^2]}{1-M^2}$
$\frac{dP_o}{P_o}$	0	$-\frac{\gamma M^2}{2}$	$-\frac{\gamma M^2}{2}$	$-\gamma M^2(1-r_i)$
$\frac{ds}{C_p}$	0	$\frac{(\gamma-1)M^2}{2}$	ψ	$(\gamma-1)M^2(1-r_i)$

One of the stated reasons for examining simple flows was that closed-form expressions could be found for most situations. Consider, for example, the relation between the static pressure and the Mach number for simple flow heat addition (or rejection). From Table 5-1, the influence coefficient expression for the just-mentioned case is

$$\frac{dP}{P} = -\frac{\gamma M^2 \psi}{1 - M^2}\frac{dT_o}{T_o}, \tag{5-31}$$

and for the Mach number,

$$\frac{dM^2}{M^2} = \frac{(1 + \gamma M^2)\psi}{1 - M^2}\frac{dT_o}{T_o}, \tag{5-32}$$

from which

$$\frac{dT_o}{T_o} = \frac{1 - M^2}{(1 + \gamma M^2)\psi}\frac{dM^2}{M^2}. \tag{5-33}$$

Eliminating dT_o/T_o from Eq. 5-31 yields

$$\frac{dP}{P} = \frac{-\gamma}{1 + \gamma M^2} dM^2. \tag{5-34}$$

Equation 5-34 is a variable separable differential equation involving only the pressure and the Mach number. From previous chapters we know that sonic flow, $M = 1.0$, represents an important condition in compressible flow. If we adopt the convention that the superscript "*" represents a property evaluated at $M = 1$, Eq. 5-34 can then be integrated as

$$\int_{P^*}^{P} \frac{dP}{P} = \int_{1}^{M^2} \frac{-\gamma}{1 + \gamma M^2} dM^2, \tag{5-35}$$

where P^* represents the static pressure at a point in the flow where $M = 1$. The evaluation of Eq. 5-35 yields

$$\frac{P}{P^*} = \frac{\gamma + 1}{1 + \gamma M^2}, \tag{5-36}$$

which is the desired relationship between Mach number and pressure in a one-dimensional generalized flow with simple heat addition or rejection.

This same procedure can be applied for every property ratio for each of the simple flows to find closed-form expressions. Closed-form expressions for each of the property ratios for each of the simple flows are presented in Table 5-2. In general the Mach number-driving potential relationships for each of the simple flows can be obtained by integration of the first line of the influence coefficients in Table 5-1. The other property ratios that are expressed in terms of Mach number are obtained by following the substitution procedure as in Eqs. 5-31 to 5-36. In Table 5-2 the quantities with the superscript "*" are evaluated at the point where $M = 1$. Generally, conditions for various properties at the point where $M = 1$ will differ for different processes. For example, P^* is different for heat addition and friction processes.

TABLE 5-2 EQUATIONS FOR SIMPLE ONE-DIMENSIONAL FLOW OF A PERFECT GAS

	Simple area change	Simple frictional flow	Simple diabatic flow	Simple mass addition, $r_i = 0$
$\frac{A}{A^*}$	$\frac{1}{M}\left[\left(\frac{2}{\gamma+1}\right)\psi\right]^{(\gamma+1)/2(\gamma-1)}$	1	1	1
$\frac{T_o}{T_o^*}$	1	1	$\frac{2(\gamma+1)M^2\psi}{(1+\gamma M^2)^2}$	1
$\frac{4fL^*}{D_h}$	0	$\frac{1-M^2}{\gamma M^2}+\left(\frac{\gamma+1}{2\gamma}\right)\ln\left\{M^2\left[\left(\frac{2}{\gamma+1}\right)\psi\right]^{-1}\right\}$	0	0
$\frac{m}{\dot{m}^*}$	1	1	1	$\frac{M\sqrt{2(\gamma+1)\psi}}{1+\gamma M^2}$
$\frac{P}{P^*}$	$\left[\left(\frac{2}{\gamma+1}\right)\psi\right]^{-\gamma/(\gamma-1)}$	$\frac{1}{M}\left[\left(\frac{2}{\gamma+1}\right)\psi\right]^{-1/2}$	$\frac{\gamma+1}{1+\gamma M^2}$	$\frac{\gamma+1}{1+\gamma M^2}$
$\frac{\rho}{\rho^*}$	$\left[\left(\frac{2}{\gamma+1}\right)\psi\right]^{-1/(\gamma-1)}$	$\frac{1}{M}\left[\left(\frac{2}{\gamma+1}\right)\psi\right]^{1/2}$	$\frac{1+\gamma M^2}{(\gamma+1)M^2}$	$\frac{2\psi}{1+\gamma M^2}$
$\frac{T}{T^*}$	$\left[\left(\frac{2}{\gamma+1}\right)\psi\right]^{-1}$	$\left[\left(\frac{2}{\gamma+1}\right)\psi\right]^{-1}$	$\frac{(\gamma+1)^2M^2}{(1+\gamma M^2)^2}$	$\left[\left(\frac{2}{\gamma+1}\right)\psi\right]^{-1}$
$\frac{V}{V^*}$	$M\left[\left(\frac{2}{\gamma+1}\right)\psi\right]^{-1/2}$	$M\left[\left(\frac{2}{\gamma+1}\right)\psi\right]^{-1/2}$	$\frac{(\gamma+1)M^2}{1+\gamma M^2}$	$M\left[\left(\frac{2}{\gamma+1}\right)\psi\right]^{-1/2}$
$\frac{P_o}{P_o^*}$	1	$\frac{1}{M}\left[\left(\frac{2}{\gamma+1}\right)\psi\right]^{(\gamma+1)/2(\gamma-1)}$	$\frac{\gamma+1}{1+\gamma M^2}\left[\left(\frac{2}{\gamma+1}\right)\psi\right]^{\gamma/(\gamma-1)}$	$\frac{\gamma+1}{1+\gamma M^2}\left[\left(\frac{2}{\gamma+1}\right)\psi\right]^{\gamma/(\gamma-1)}$
$\frac{I}{I^*}$	$\frac{1+\gamma M^2}{M\sqrt{2(\gamma+1)\psi}}$	$\frac{1+\gamma M^2}{M\sqrt{2(\gamma+1)\psi}}$	1	1
$\frac{s-s^*}{c_p}$	0	$\ln\left\{M^{(\gamma-1)/\gamma}\left[\left(\frac{2}{\gamma+1}\right)\psi\right]^{-(\gamma+1)/2\gamma}\right\}$	$\ln\left[M^2\left(\frac{\gamma+1}{1+\gamma M^2}\right)^{(\gamma+1)/\gamma}\right]$	$\ln\left\{\left[\left(\frac{2}{\gamma+1}\right)\psi\right]^{-1}\left(\frac{1+\gamma M^2}{\gamma+1}\right)^{(\gamma-1)/\gamma}\right\}$

Adapted from Zucrow and Hoffman.

As will be demonstrated in the next two chapters, the integration of each of the property relations for each of the simple flows is not just an exercise in mathematical manipulation; each integrated property relation in Table 5-1 expresses the ratio of the property at a Mach number M to the property at the point where the Mach number is unity. Equation 5-36, for example, will be used to relate the effect of simple heat addition or rejection on the static pressure for that simple flow. Hence, by reference to Table 5-2 we have the solutions to the four simple compressible flows obtained from the generalized one-dimensional analysis. The discussions and developments of concepts related to the four simple flows can then start with the expressions from Table 5-2.

One of the reasons for approaching the simple flows in this fashion was to demonstrate the relationships between the simple flows: the relationship of the simple flows to the generalized is evident. Another reason for approaching simple flows in this manner was the rapidity with which expressions for various simple flow relations could be obtained. Most compressible flow textbooks obtain the expressions shown in Table 5-2 by considering each simple flow as a separate entity, drawing a control volume valid for only that flow, applying the conservation statements and various definitions, reducing the system to an integrable form, and finally presenting the property relations. Considering each simple flow as a separate entity in deriving the governing equations is thus a more laborious effort. The behavior and characteristics of each simple flow must, however, be considered separately.

5.4 OTHER GENERALIZED FORMULATIONS

The distinction between a calorically perfect gas (constant specific heats and $P = \rho RT$) and a thermally perfect gas ($P = \rho RT$) was pointed out in Chapter 1. Calorically perfect gases form the basis of treatment of compressible flows in Chapters 1 through 10; Chapter 11 examines in some detail compressible flow processes of thermally perfect gases and presents some introductory materials on chemically reacting flows. The developments so far in this chapter have been restricted to calorically perfect gas processes.

As is evident from Table 5-2, the property ratios for simple processes of a calorically perfect gas can be expressed in closed form. The generalized one-dimensional compressible flow analysis can be extended to compressible flow processes with thermally perfect gases. Shapiro (1953) presents the resulting influence coefficient formulation for both variable specific heats and variable molecular weight. Such a formulation adds two driving-potential–like variables: the variations with x of the molecular weight and the variation with x of the ratio of specific heats. The influence coefficient entries for the four traditional driving potentials remain unchanged for this more involved formulation. Table 5-3 presents the additions needed to consider variable ratios of specific heats and variable molecular weight.

Although the general form of the influence coefficients for variable specific heats and variable molecular weight is similar, an important difference is present. That difference precludes the formulation of "simple" flows for thermally perfect gases in the same sense as simple flows for calorically perfect gases. As we shall see

TABLE 5-3 INFLUENCE COEFFICIENT ADDITIONS FOR VARIABLE MOLECULAR WEIGHT AND RATIO OF SPECIFIC HEATS

	Driving potentials	
Change in flow property	$\frac{dMW}{MW}$	$\frac{d\gamma}{\gamma}$
$\frac{dM^2}{M^2}$	$-\frac{1+\gamma M^2}{1-M^2}$	-1
$\frac{dP}{P}$	$\frac{\gamma M^2}{1-M^2}$	0
$\frac{d\rho}{\rho}$	$\frac{1}{1-M^2}$	0
$\frac{dT}{T}$	$\frac{(\gamma-1)M^2}{1-M^2}$	0
$\frac{dV}{V}$	$-\frac{1}{1-M^2}$	0
$\frac{ds}{c_p}$	0	0

Adapted from Shapiro (1953).

in Chapter 11, the molecular weight of a mixture of gases is determined by its chemical composition, which in the equilibrium chemical sense is a function of the local temperature and pressure. The specific heats are generally functions of the local temperature. But the local temperature and pressure, and hence the local chemical composition, are not known (if they were, we would not need to use influence coefficients), and so the local variations of the ratio of specific heats and molecular weights are not known. Generally, for thermally perfect compressible flow processes, numerical solutions of the differential equations formed from the influence coefficients are the only tractable approach. Hence, for the more complex thermally perfect formulation, the closed-form format of the simple flow solutions that we saw for calorically perfect gases is not possible. We will not make much use of the influence coefficient formulation for variable ratio of specific heats and variable molecular weight, but it should serve to remind us that the assumption of a calorically perfect gas is fundamental to the first ten chapters of this textbook.

5.5 CONCLUSION

This chapter provides the basic developments needed to solve and understand one-dimensional flows of calorically perfect compressible gases. The next two chapters are devoted to examining and applying the results of the simple flows as presented in Table 5.2. Chapter 8 then revisits generalized one-dimensional compressible flows and develops the necessary numerical techniques to solve such generalized problems.

CHAPTER 5 REVIEW QUESTIONS

1. What is a simple flow?
2. What is a driving potential?
3. Name the four driving potentials specified in this chapter.
4. What is simple area change flow usually called?
5. What is simple frictional flow usually called?
6. What is simple heat addition or rejection usually called?

CHAPTER PROBLEMS

1. Verify Eq. 5-2.
2. Verify Eq. 5-4.
3. Verify Eq. 5-6.
4. Verify Eq. 5-15.
5. Verify Eq. 5-18.
6. Explain what Eq. 5-19 means.
7. Verify Eq. 5-20.
8. Verify Eq. 5-22.
9. Verify Eq. 5-23.
10. List the independent variables and the dependent variables in Eqs. 5-24.
11. Using Eq. 5-25, specify differential expressions for all the dependent variables.
12. Verify that the results of Problem 5-11 are consistent with the entries of Table 5-1, the influence coefficient table.
13. For simple area change obtain the expression for A/A^* using Table 5-1, and verify the results with Table 5-2.
14. For simple area change obtain any of the property relations using Table 5-1, and verify the results with Table 5-2.
15. For simple frictional flow obtain the expression for $4fL^*/D_h$ using Table 5-1, and verify the results with Table 5-2.
16. For simple frictional flow obtain any of the property relations using Table 5-1, and verify the results with Table 5-2.
17. For simple diabatic flow obtain the expression for T_o/T_o^* using Table 5-1, and verify the results with Table 5-2.
18. For simple diabatic flow obtain any of the property relations using Table 5-1, and verify the results with Table 5-2.
19. For simple mass addition obtain the expression for $\dot{m}/\dot{m}^*$ using Table 5-1, and verify the results with Table 5-2.
20. For simple mass addition obtain any of the property relations using Table 5-1, and verify the results with Table 5-2.

Simple Area Change (Isentropic)

6

6.1 INTRODUCTION

In this and the next chapter the simple flows will be investigated. This chapter is devoted to the first of the simple flows that we shall study: simple area change, which is also called *isentropic flow*. Chapter 5 contained the development of both differential and closed-form expressions for the various simple flows. The chapters (6 and 7) devoted to the simple flows contain few derivations but instead utilize the results of Tables 5-1 and 5-2 to develop applications of the simple process concepts. Chapters 6 and 7 follow the same format:

1. The appropriate differential and algebraic equations for each simple flow are listed.
2. The ramifications of each particular driving potential are examined.
3. Inferences from the equations for each simple flow are made.
4. Applications are presented and discussed.

6.2 THE BEHAVIOR OF SIMPLE AREA CHANGE FLOW

Table 6-1(a) reproduces for convenience the influence coefficient entries from Table 5-1 for simple area change. Since the influence coefficients for both entropy and stagnation pressure are zero, the flow must be isentropic and must possess constant stagnation pressure. Table 6-1(b) reproduces the equations for simple area change from Table 5-2 and presents the requirements for the remaining driving potentials to vanish (T_o = constant; $\dot{m}$ = constant; and $f = 0$). Thus, simple area change possesses both constant stagnation pressure and temperature. These same results could have been obtained by applying the conservation equations and the various definitions to the control volume of Figure 6-1. The control volume shows only area change since heat transfer, mass addition, and friction are not present.

TABLE 6-1 SUMMARY OF RELATIONS FOR SIMPLE AREA CHANGE

(a) Differential form	(b) Property ratios	
$\frac{dM^2}{M^2} = -\frac{2\psi}{1-M^2}\frac{dA}{A}$	$\frac{A}{A^*} = \frac{1}{M}\left[\left(\frac{2}{\gamma+1}\right)\psi\right]^{(\gamma+1)/[2(\gamma-1)]}$	
$\frac{dV}{V} = -\frac{1}{1-M^2}\frac{dA}{A}$	$\frac{P}{P^*} = \left[\left(\frac{2}{\gamma+1}\right)\psi\right]^{-\gamma/(\gamma-1)}$	or $\frac{P}{P_o} = (\psi)^{-\gamma/(\gamma-1)}$
$\frac{dT}{T} = \frac{(\gamma-1)M^2}{1-M^2}\frac{dA}{A}$	$\frac{\rho}{\rho^*} = \left[\left(\frac{2}{\gamma+1}\right)\psi\right]^{-1/(\gamma-1)}$	or $\frac{\rho}{\rho_o} = (\psi)^{-1/(\gamma-1)}$
$\frac{d\rho}{\rho} = \frac{M^2}{1-M^2}\frac{dA}{A}$	$\frac{T}{T^*} = \left[\left(\frac{2}{\gamma+1}\right)\psi\right]^{-1}$	or $\frac{T}{T_o} = \psi^{-1}$
$\frac{dP}{P} = \frac{\gamma M^2}{1-M^2}\frac{dA}{A}$	$\frac{V}{V^*} = M\left[\left(\frac{2}{\gamma+1}\right)\psi\right]^{1/2}$	
$\frac{dP_o}{P_o} = 0 \rightarrow P_o$ = constant	$\frac{I}{I^*} = \frac{1+\gamma M^2}{M\sqrt{2(\gamma+1)\psi}}$	
$\frac{dI}{I} = \frac{1}{1+\gamma M^2}\frac{dA}{A}$	T_o = constant	
$\frac{ds}{c_p} = 0 \rightarrow s$ = constant	P_o = constant	
	$\dot{m}$ = constant	
	$f = 0$	
	s = constant	
	$\psi = 1 + \frac{\gamma-1}{2}M^2$	

A more useful form for the property ratios P/P_o, ρ/ρ_o, and T/T_o can be obtained from the property ratio expressions in Table 6-1(b) by evaluating the table entries at stagnation; that is,

$$\frac{P_o}{P^*} = \left(\frac{2}{\gamma+1}\right)^{-\gamma/(\gamma-1)}, \tag{6-1}$$

and forming the ratio P/P_o and using ψ from Eq. 5-28,

$$\frac{P}{P_o} = \frac{P}{P^*}\frac{P^*}{P_o} = \left(\frac{2}{\gamma+1}\psi\right)^{-\gamma/(\gamma-1)}\left(\frac{\gamma+1}{2}\right)^{-\gamma/(\gamma-1)}, \tag{6-2}$$

from which

$$\frac{P}{P_o} = (\psi)^{-\gamma/(\gamma-1)} = \left(1 + \frac{\gamma-1}{2}M^2\right)^{-\gamma/(\gamma-1)}. \tag{6-3}$$

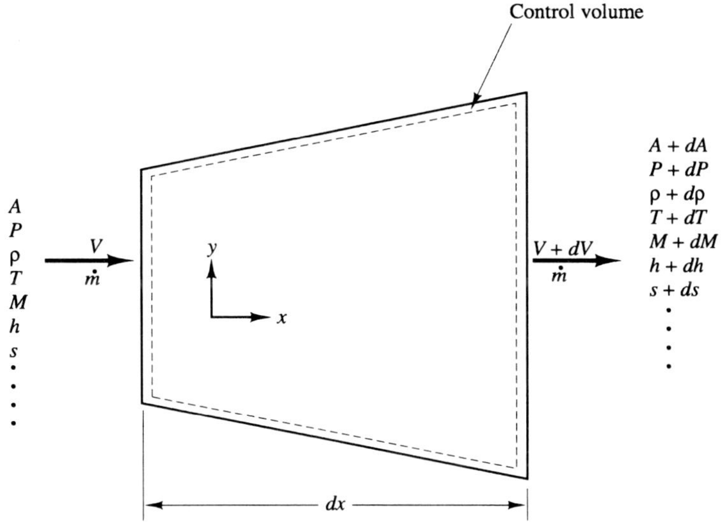

Figure 6-1 Control Volume for Simple Area Change

In similar fashions,

$$\frac{\rho}{\rho_o} = (\psi)^{-1/(\gamma-1)} = \left(1 + \frac{\gamma - 1}{2}M^2\right)^{-1/(\gamma-1)}; \tag{6-4a}$$

$$\frac{T}{T_o} = \psi^{-1} = \left(1 + \frac{\gamma - 1}{2}M^2\right)^{-1}. \tag{6-4b}$$

But Eqs. 6-3 and 6-4a are merely the isentropic relations discussed in Chapter 1. However, whereas the isentropic relations were discussed in the context of relating stagnation and static properties at a point in Chapter 1, in this development the isentropic relations apply throughout the entire flow field. This is a very significant distinction as Eqs. 6-3 and 6-4 can be applied between different points of the flow field with simple area change.

We start our development of isentropic flow by considering the differential form of the Mach number area relation as given in Table 6-1(a):

$$\frac{dM^2}{M^2} = -\frac{2\psi}{1 - M^2}\frac{dA}{A}. \tag{6-5}$$

Combining this with the differential form of the static pressure relation results in

$$\frac{dP}{P} = \frac{\gamma M^2}{(1 - M^2)}\frac{dA}{A} = \frac{-\gamma M^2}{1 - M^2}\frac{(1 - M^2)}{2\psi}\frac{dM^2}{M^2} = -\frac{\gamma M^2}{2\psi}\frac{dM^2}{M^2}. \tag{6-6}$$

Equation 6-6 shows that a decrease in Mach number leads to an increase in static pressure and conversely that an increase in Mach number leads to a decrease in static pressure. Since Eq. 6-5 contains the term $1 - M^2$, which changes sign at $M = 1$, the effect of area change on Mach number is a function of Mach number. If the flow is subsonic, then $dM^2 \propto -dA$; and if the flow is supersonic, then $dM^2 \propto +dA$. This behavior leads to very important differences between subsonic and supersonic simple area change flow. For a subsonic Mach number, a decrease in area results in an increase in Mach number, while an increase in area results in a decrease in Mach number. Supersonic flow has the opposite effect; a decrease in area results in a decrease in Mach number, and an increase in area results in an increase in Mach number. If $M = 1$, then the denominator of Eq. 6-5 vanishes, and the rate of change of Mach number is unbounded if the value of the numerator is finite. This is not possible because we accept that unbounded rates of change cannot physically occur. Thus, at the sonic point, the location where $M = 1$, dA must also vanish if an unbounded change in Mach number is to be prevented at that location. Hence, for a given isentropic flow, the $M = 1$ location will be either at the area minimum or the area maximum. Since the area can be unbounded with no maximum generally existing, if $M = 1$ occurs, it must be at the minimum area. The minimum area in a passage is called the *throat*. This does not say that the Mach number at the throat must be 1, only that a Mach number of 1, if it occurs in an isentropic flow, must occur at the throat. The effects of decreasing and increasing area on Mach number or pressure are summarized in Figure 6-2.

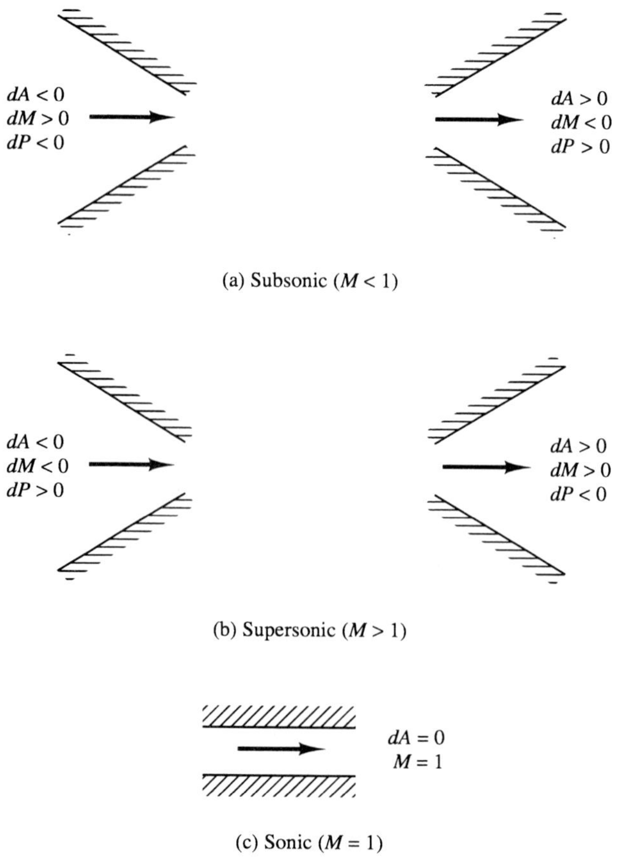

Figure 6-2 Mach Number and Pressure Behavior for Simple Area Change

The mass flow rate, $\dot{m}$, is usually written as ρAV; hence, it follows that for a given mass flow rate, the product ρV is a maximum at the minimum area. Moreover, for an isentropic flow the largest Mach number that can occur at the throat is 1. A Mach number larger than 1 can occur only in or after a passage with a cross-sectional area larger than the throat. Thus, we conclude that when the Mach number is equal to 1 in the minimum area, or throat, the mass flow rate is a maximum. This is a very important conclusion and will be extensively used. The flow is said to be *choked* if the $M = 1$ at the minimum area. Choked flow represents the maximum mass flow rate, for a given set of stagnation conditions, that can pass through a given throat.

The idea that $M = 1$, if it occurs, must occur at a minimum area raises an interesting question. Suppose that a flow is established such that $M = 1$ at the throat, and further suppose that the passage geometry is altered to result in a smaller area than the original throat. What happens? Since the minimum area is now smaller than the original throat, M cannot equal 1 at the original throat area; $M = 1$ must occur at the new minimum area and the passage flowfield would be altered such that $M = 1$ at the new minimum area. Because the new minimum area is less than the original, the mass flow rate through the new minimum area would be less than through the original minimum area, and the local Mach number distribution within the passage would also be altered from the original. Problem 6-1 explores this question quantitatively and should be understood.

When evaluated at the sonic point, Eq. 6-5 produces the indeterminate form 0/0, and l'Hospital's rule can be applied with the following results:

$$\left(\frac{dM}{dx}\right)_{M=1} = \pm\frac{\gamma + 1}{4A}\frac{d^2A}{dx^2}. \tag{6-5a}$$

Any passage downstream of the throat must diverge since the throat represents the minimum area passage. Equation 6-5b shows two possibilities downstream of a choked throat: $dM/dx > 0$ or $dM/dx < 0$. If $dM/dx > 0$, then the flow is accelerated to supersonic velocities; if $dM/dx < 0$, then the flow is decelerated to subsonic velocities.

Piecing together the information from the previous paragraphs, we obtain the following scenario for achieving supersonic flow. A converging–diverging nozzle is connected to a reservoir with stagnation conditions P_o and T_o. The situation is schematically illustrated in Figure 6-3. The fluid is essentially at rest in the reservoir so that it enters the converging portion of the nozzle subsonically and is accelerated in that portion of the nozzle. The sonic point must be at the minimum area, the throat. The flow enters the diverging portion of the nozzle at $M = 1$ and is accelerated supersonically in the diverging portion of the nozzle. Only a converging–diverging nozzle can isentropically accelerate the flow from rest to supersonic Mach numbers. The static pressure decreases monotonically from the reservoir to the exit of the nozzle. The exit exhausts into a back pressure, P_b. The back pressure is the ambient pressure into which the nozzle exhausts. Later in this chapter the effects of back pressure on converging–diverging nozzle flows will be examined. In much of the compressible flow literature, a converging–diverging nozzle is called a *de Laval nozzle*, after the Swedish engineer, Carl de Laval.

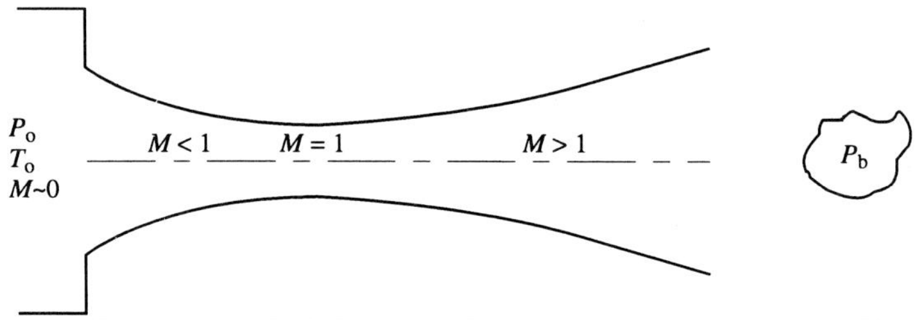

Figure 6-3 Converging–Diverging Nozzle

The preceding discussion indicates the importance in simple area change of the location of $M = 1$. In a de Laval nozzle the $M = 1$ location must be at the minimum area. Hence, for a given isentropic flow the area at the sonic location is unique throughout the flow field and can be regarded as a property of that isentropic flow and referenced throughout that isentropic flow. Evaluation of the property ratios, given in Table 6-1(b), at the $M = 1$ location yields, with the values for $\gamma = 1.4$ indicated in parentheses,

$$\frac{P^*}{P_o} = \left(\frac{2}{\gamma + 1}\right)^{\gamma/(\gamma-1)} (0.5283 \text{ for } \gamma = 1.4); \tag{6-7}$$

$$\frac{\rho^*}{\rho_o} = \left(\frac{2}{\gamma + 1}\right)^{1/(\gamma-1)} (0.6339 \text{ for } \gamma = 1.4); \tag{6-8}$$

$$\frac{T^*}{T_o} = \frac{2}{\gamma + 1} (0.8333 \text{ for } \gamma = 1.4). \tag{6-9}$$

The behavior of the simple area change property ratios and the area ratio A/A^* are plotted in Figure 6-4. While this figure is not very useful for numerical purposes, it does serve as a convenient illustration for the behavior of the various property and area ratio expressions for isentropic flow. As the Mach number increases, the pressure, density, and temperature ratios—expressed as the quotient of the stagnation divided by static property—all increase monotonically and become unbounded as the Mach number increases without bound. The impulse function also increases monotonically for supersonic flow but approaches an asymptote at elevated Mach numbers. The area ratio, A/A^*, however, is double-valued, exhibiting a subsonic and a supersonic root. This should not be a surprise, since in the converging–diverging nozzle discussion we saw that subsonic flow and supersonic flow occurred at the same area ratio so that the double-valuedness was apparent.

Computational aids similar to those already examined are available for isentropic flow. Appendix A and option **IS** of program COMPQ have already been examined in Chapter 1 in conjunction with stagnation and static property relations. Based on previous discussions in this section, these computational aids can also be used for isentropic flow field calculations. In the earlier examination of isentropic computational aids, the area ratio A/A^* was not used. We now know that it is the ratio of the area at a specified Mach number to the area where $M = 1$.

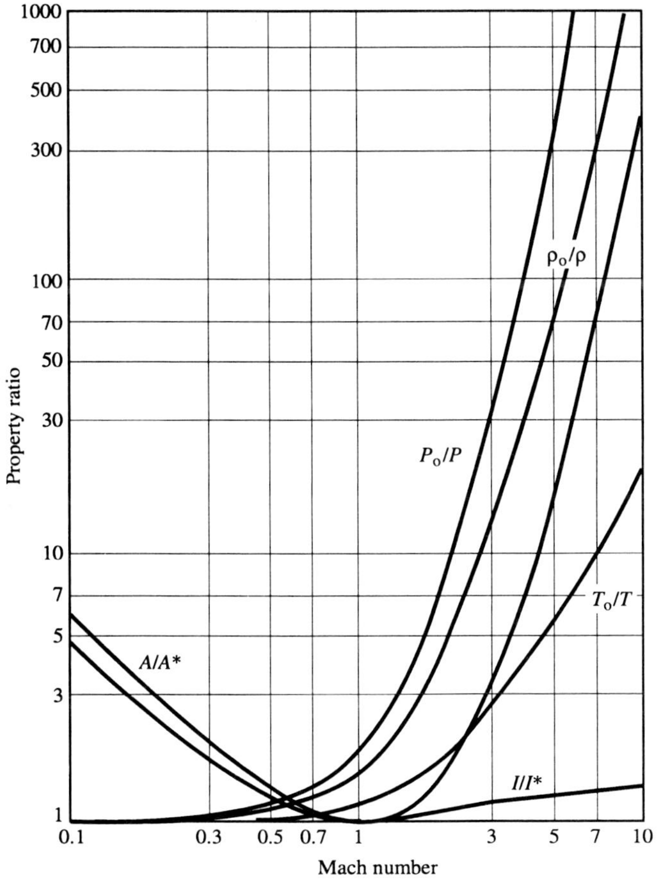

Figure 6-4 Isentropic Property Relations for $\gamma = 1.4$

The COMPQ menu for option **IS** is presented in Figure 6-5. Cases **P**, **M**, **A**, and **T** have as input the pressure ratio, the Mach number, the area ratio, or the temperature ratio, respectively, and have as output the remaining isentropic quantities. These cases basically replicate and are used in the same fashion as the entries in Appendix A. Since case **A** is double-valued, both the subsonic and supersonic Mach numbers and the corresponding property ratios are computed and displayed.

```
ISENTROPIC PROCESS CALLED FOR
INPUT PARAMETERS ARE:
      PRESSURE RATIO          P
      MACH NUMBER             M
      AREA RATIO              A
      TEMPERATURE RATIO       T
      AREA CHANGE PROCESS    AC
      M1-M2 AREA REQUIRED    AR

ENTER THE DESIRED LETTER?
```

Figure 6-5 Menu for Option **IS**

```
ISENTROPIC PROCESS CALLED FOR
INPUT PARAMETERS ARE:
      PRESSURE RATIO            P
      MACH NUMBER               M
      AREA RATIO                A
      TEMPERATURE RATIO         T
      AREA CHANGE PROCESS      AC
      M1-M2 AREA REQUIRED      AR

ENTER THE DESIRED LETTER? A

ENTER THE AREA RATIO? 2.5

    MACH       T RATIO      P RATIO      A RATIO    IMPULSE RATIO
   0.2395       0.9887       0.9608       2.5000        2.0468


    MACH       T RATIO      P RATIO      A RATIO    IMPULSE RATIO
   2.4427     4.559E-01    6.399E-02    2.500E+00       1.1801

STAGNATION PRESSURE AND TEMPERATURE ARE CONSTANT FOR ISENTROPIC FLOW.

ARE ADDITIONAL ISENTROPIC RELATIONS DESIRED?
```

Figure 6-6 Input/Output for Case **A** of Option **IS**

Case **A** for an area ratio of 2.5 produces the output given in Figure 6-6, where the subsonic and supersonic Mach numbers and property ratios are shown. A reminder that stagnation properties are constant for an isentropic flow is also given as part of the output.

Cases **AC** and **AR** represent processes in which several arithmetical operations are combined. Case **AC** calculates the Mach number M_2 and property ratios given an initial Mach number M_1 and area change ratio A_2/A_1. Case **AR** computes the area change ratio A_2/A_1 required in a process with initial Mach number M_1 and final Mach number M_2. Example 6-1 illustrates the use of many of the computational aids.

Example 6-1

A supersonic flow is expanded through an area change ratio $A_2/A_1 = 2$ from an initial Mach number of 1.5. Find the final Mach number and the pressure and temperature ratios for the process.

Solution. This example will be worked using the individual cases **P**, **M**, **A**, and **T** of option **IS** and then reworked using the case **AC** for the process. Figure 6.7a presents a schematic representation of the problem. The throat, although not present in the problem description, can be imagined at either or both ends of the diverging passage. Remember that for an isentropic flow, the throat area is an invariant property of the flow field and can be referenced at any point in the flow field. The solution using the individual cases of COMPQ then starts with the observation

$$\frac{A_2}{A^*} = \frac{A_2}{A_1}\frac{A_1}{A^*}.$$

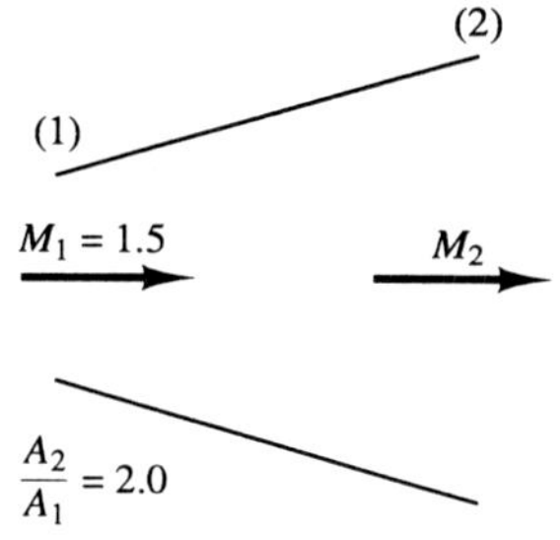

(a) Problem statement

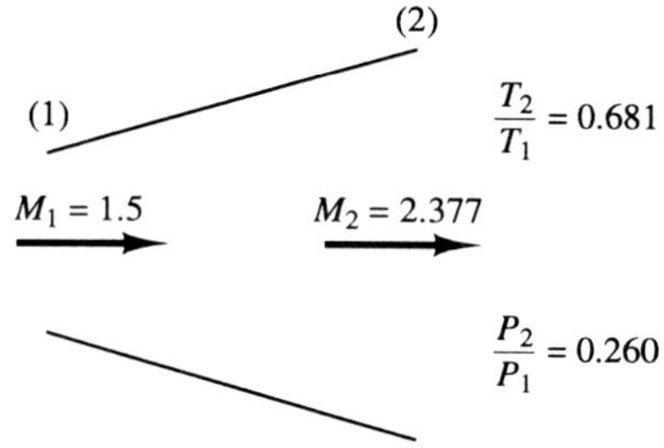

(b) Problem solution

Figure 6-7 Example Problem 6-1

For a Mach number of 1.5, the isentropic option of COMPQ yields

$$A_1/A^* = 1.1762$$
$$T_1/T_o = 0.6897$$
$$P_1/P_o = 0.2724$$

so that

$$\frac{A_2}{A^*} = 2(1.1762) = 2.3524,$$

from which

$$M_2 = 0.2558,\ 2.3768.$$

Since the flow is initially supersonic, the supersonic root will be chosen. For $M_2 = 2.3768$,

$$T_2/T_o = 0.4695;$$
$$P_2/P_o = 0.0709.$$

Finally, since the flow field is isentropic,

$$\frac{T_2}{T_1} = \frac{T_2}{T_o}\frac{T_o}{T_1} = \frac{T_2/T_o}{T_1/T_o} = \frac{0.4695}{0.6897} = 0.6807;$$

$$\frac{P_2}{P_1} = \frac{P_2}{P_o}\frac{P_o}{P_1} = \frac{P_2/P_o}{P_1/P_o} = \frac{0.0709}{0.2724} = 0.2603.$$

The completed solution is shown in Figure 6-7b.

An alternate, and preferred, approach is to use case **AC** of the **IS** option because it treats the process of an area change. The input and output for the solution to this problem using case **AC** are reproduced in Figure 6-8. The results of the two methods agree to three decimal places. Case **AC** uses the methodology and steps explained in the previous paragraph for the process computations.

```
ENTER THE DESIRED LETTER? AC

ENTER M1 AND A2/A1 ? 1.5,2.0

AREA CHANGE PROCESS IN ISENTROPIC FLOW
   M1            M2           A2/A1          T2/T1          P2/P1
 1.5000        2.3768        2.0000         0.6808         0.2604

THE PROCESS IS ISENTROPIC; HENCE NO CHANGE IN STAGNATION PROPERTIES.
```

Figure 6-8 COMPQ Solution for Example 6-1

The idea of choked flow—$M = 1$ at the minimum area of an isentropic flow—has already been discussed. Consider evaluating the mass flow rate at a choked throat; then

$$\dot{m}^* = \rho^* A^* V^* = \frac{P^*}{RT^*} A^*(1)\sqrt{\gamma R T^*} = P^* A^* \sqrt{\frac{\gamma}{RT^*}}, \tag{6-10}$$

since the $M = 1$. From Eqs. 6-7 and 6-9, the mass flow can be cast in terms of the stagnation properties as

$$\dot{m}^* = \left(\frac{2}{\gamma + 1}\right)^{\gamma/(\gamma-1)} P_o A^* \sqrt{\gamma/R}\left(\frac{\gamma + 1}{2}\right)^{1/2} \frac{1}{\sqrt{T_o}}, \tag{6-11}$$

which reduces to

$$\dot{m}^* = \left(\frac{2}{\gamma + 1}\right)^{(\gamma+1)/[2(\gamma-1)]} \sqrt{\gamma/R}\,\frac{A^* P_o}{\sqrt{T_o}}. \tag{6-12}$$

This very informative equation explicitly shows that for a choked flow the mass flow rate is directly proportional to the throat area and the stagnation pressure and is inversely proportional to the square root of the stagnation temperature. The remaining terms are constants for a given gas. Thus, for a given minimum area and stagnation conditions, the maximum mass flow rate can be determined from Eq. 6-12.

A normal shock wave standing in a diverging passage is a common occurrence in compressible flow. Such a situation is schematically illustrated in Figure 6-9a. Upstream of the normal shock wave, the flow field is isentropic with an entropy of s_1. Downstream of the shock wave, the flow field is isentropic with an entropy of s_2. Only across the shock—a very thin region—does the entropy change. The stagnation pressure downstream of the shock wave is lower than that upstream of the shock

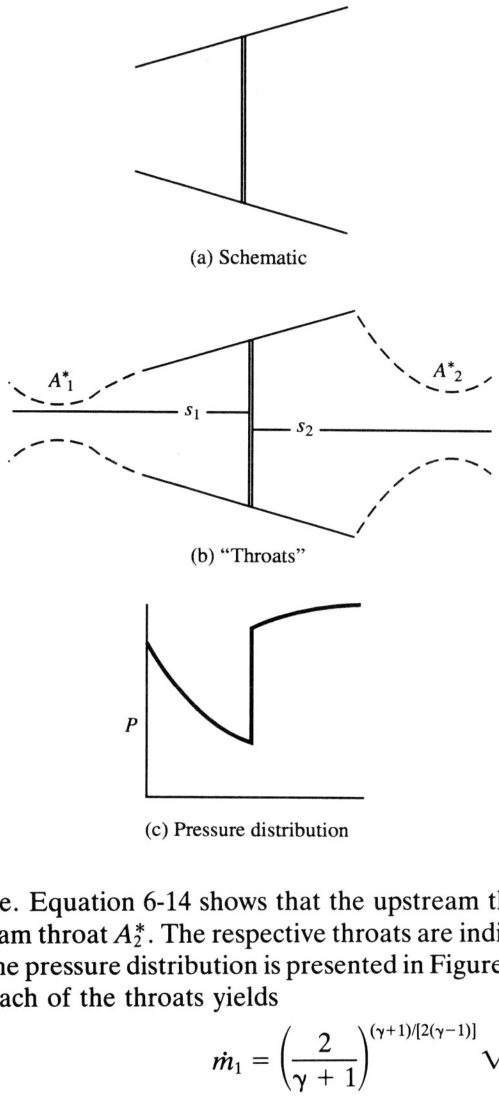

Figure 6-9 Normal Shock in Diverging Passage

wave. Equation 6-14 shows that the upstream throat A_1^* is smaller than the downstream throat A_2^*. The respective throats are indicated in Figure 6-9b, while a sketch of the pressure distribution is presented in Figure 6-9c. Evaluating the mass flow rate at each of the throats yields

$$\dot{m}_1 = \left(\frac{2}{\gamma + 1}\right)^{(\gamma+1)/[2(\gamma-1)]} \sqrt{\gamma/R}\,\frac{A_1^* P_{o_1}}{\sqrt{T_o}},$$
$$\dot{m}_2 = \left(\frac{2}{\gamma + 1}\right)^{(\gamma+1)/[2(\gamma-1)]} \sqrt{\gamma/R}\,\frac{A_2^* P_{o_2}}{\sqrt{T_o}}, \tag{6-13}$$

or

$$\frac{A_1^*}{A_2^*} = \frac{P_{o_2}}{P_{o_1}}. \tag{6-14}$$

The throat area ratio required for the shock in a diverging channel situation is directly dependent on the stagnation pressure ratio across the normal shock system.

An ancillary conclusion is that downstream of the normal shock wave any subsequent area reduction must not be less than A_2^*. If a passage with an area less than A_2^* downstream of the normal shock wave were considered, then the mass flow rate of the entire system would have to be reduced, thereby leading to an inconsistency in the original problem statement. We shall use this concept later when considering nozzle flow applications.

6.3 SIMPLE AREA CHANGE (ISENTROPIC) BASIC APPLICATIONS

The previous section developed some useful concepts and computational aids for isentropic flow. This section applies those to problems of interest in one-dimensional compressible flow.

Consider a converging nozzle attached to a reservoir with stagnation conditions P_o and T_o and exhausting into a variable back pressure P_b. The arrangement is illustrated in Figure 6-10a. If the back pressure is equal to the stagnation pressure, then no flow occurs and as indicated in Figure 6-10b, the pressure is constant throughout the nozzle. A slight decrease in back pressure induces a flow through the nozzle. The flow rate must be such that the pressure in the nozzle exit plane is equal to the back pressure. This is the same exit plane condition that exists in an incompressible flow.

As the back pressure is decreased further (see Figure 6-10b for a schematic illustration), the mass flow rate increases and the pressure continues to decrease to keep the exit plane pressure equal to the back pressure. The increasing mass flow rates corresponding to the decreasing back pressures are shown in Figure 6-10c. Finally, the back pressure is reduced low enough so that the exit plane Mach number is equal to 1. When this occurs, Mach one at the minimum area, the nozzle flow is choked. Since the exit plane pressure equals the back pressure just as choked flow is reached, the maximum back pressure that produces choked flow then becomes

$$P_b^* = \left(\frac{2}{\gamma + 1}\right)^{\gamma/(\gamma-1)} P_o \ (0.5283\ P_o \text{ for } \gamma = 1.4). \tag{6-15}$$

What happens if the back pressure is reduced below P_b^*? When the back pressure is P_b^*, the fluid has reached sonic velocity in the nozzle exit plane. Any further reduction of back pressure will have no effect on either the mass flow rate or the pressure distribution in the converging nozzle. This is a startling statement, but consider what we have already established about converging nozzle flows. The Mach number can never exceed 1 at the exit plane of a converging nozzle since the exit plane represents the minimum area. Thus any further reductions in back pressure cannot have any effect on the exit plane Mach number because a sound wave cannot propagate upstream against the sonic velocity at the minimum area. With the pressure distribution within the nozzle set by the reservoir stagnation conditions, the nozzle profile (area distribution), and the exit plane Mach number,

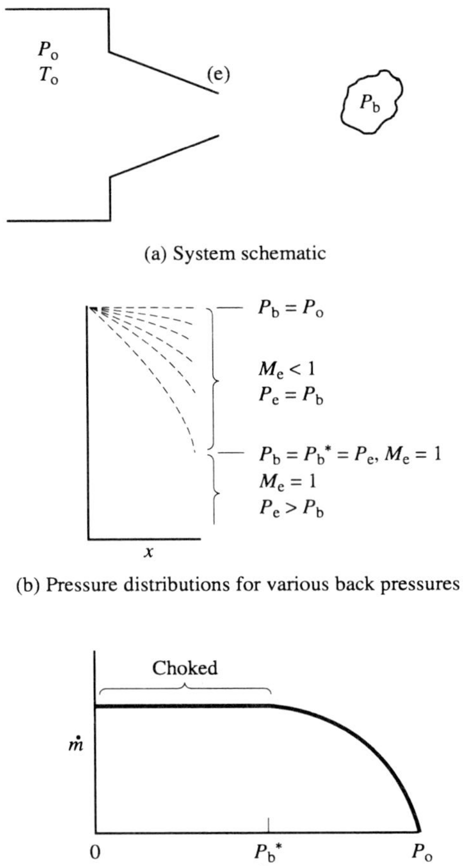

(a) System schematic

(b) Pressure distributions for various back pressures

(c) Mass flow for various back pressures

Figure 6-10 Converging Nozzle Behavior

no changes occur in the mass flow rate or the pressure distribution if the nozzle exit plane Mach number remains 1. The behavior of mass flow rate for back pressures less than P_b^* is illustrated in Figure 6-10c. For the converging nozzle in a compressible flow, the exit plane conditions then become

$$P_e = P_b \text{ for } M_e < 1 \tag{6-16}$$

for subsonic exit Mach numbers, and

$$P_e \geq P_b \text{ for } M_e = 1 \tag{6-17}$$

for a sonic exit Mach number. This behavior is dramatically different from that expected for an incompressible flow.

As long as the converging nozzle is not choked, the exit plane and back pressures are equal and the exit flow leaves the nozzle as a subsonic jet. A back pressure less than P_b^* results in an exit plane pressure greater than the back pressure and sonic conditions in the exit. Outside the nozzle, the flow then expands and the pressure is reduced to the back pressure value. This expansion is accomplished by a Prandtl-Meyer process in the vicinity of the nozzle exit plane. The exit plane flow is initially turned outward, as illustrated in Figure 6-11. The reflection of the expansion fan from the center line eventually limits the extent of the expansion. Other aspects of the structure of exhaust plumes will be discussed later in this chapter.

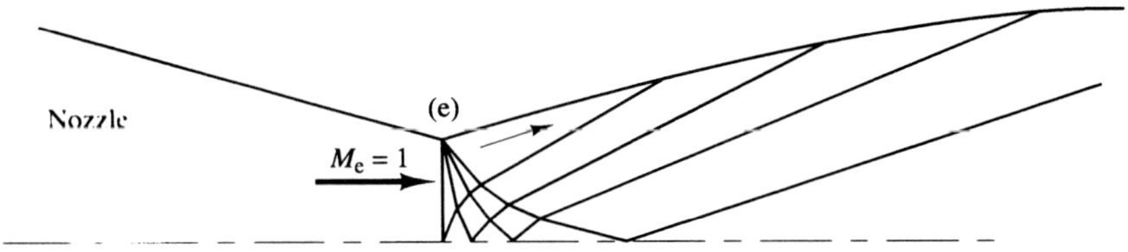

Figure 6-11 Illustration of Expansion from a Choked Converging Nozzle ($P_b < P_b^*$)

Example 6-2

A converging nozzle with an exit area of 0.05 m^2 is connected to a reservoir containing air with a stagnation pressure of 600 kPa and a stagnation temperature of 27° C. Find the exit plane pressure, the exit plane Mach number, and the mass flow rate for back pressures of 400 kPa, 200 kPa, and 100 kPa.

Solution. The problem is sketched in Figure 6-12. The first step is to find the maximum back pressure, P_b^*, that will allow choked flow. This occurs when the exit Mach number is 1 and the back pressure is equal to the exit plane pressure; hence

$$P_b^* = 600.0(0.5283) = 316.0 \text{ kPa}.$$

For back pressure greater than 316.0 kPa, the converging nozzle will be unchoked, will possess an exit plane Mach number less than 1, and will have an exit plane pressure equal to the back pressure. If the back pressure is less than 316.0 kPa, the nozzle will be choked; the exit Mach number will be 1; and the exit plane pressure will be 316.0 kPa. When the nozzle is choked, further reductions in back pressure do not affect the flow field in the nozzle. We will now examine each back pressure separately.

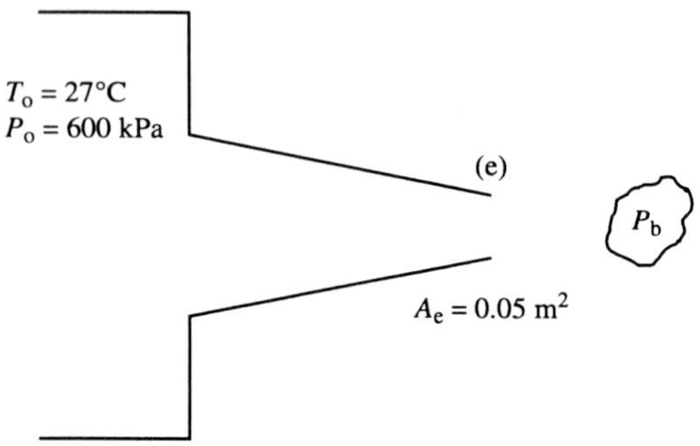

Figure 6-12 Example Problem 6-2

(a) $P_b = 400$ kPa. The nozzle will not be choked because the back pressure is greater than the maximum permitted for choked flow in the nozzle. Since the exit Mach number is less than 1, the exit pressure is equal to the back pressure. Also, since the flow is isentropic,

$$\frac{P_e}{P_o} = \frac{P_b}{P_o} = \frac{400.0}{600.0} = 0.6667,$$

from which case **P** of option **IS** of COMPQ yields

$$M_e = 0.7836$$

$$\frac{T_e}{T_o} = 0.8906.$$

The mass flow rate is

$$\dot{m} = \rho_e A_e V_e,$$

with

$$\rho_e = \frac{P_e}{RT_e} = \frac{400{,}000}{287.0(0.8906)(300.0)} = 5.2164 \text{ kg/m}^3,$$

and

$$V_e = M_e\sqrt{\gamma R T_e} = 0.7836[(1.4)(287.0)(0.8906)(300.0)]^{1/2}$$
$$= 256.7 \text{ m/s}.$$

Thus,

$$\dot{m} = (5.2164)(0.05)(256.7) = 66.95 \text{ kg/s}.$$

(b) $P_b = 200$ kPa. The flow is choked since the back pressure is less than P_b^*. Hence, conditions in the nozzle exit plane are sonic and

$$\dot{m} = \rho_e A_e V_e = \rho^* A^* V^*,$$

from which

$$\rho^* = \frac{(0.5283)(600{,}000)}{(287.0)(0.8333)(300)} = 4.4181 \text{ kg/m}^3$$

and

$$V^* = M_e\sqrt{\gamma R T_e} = 1[(1.4)(287.0)(0.8333)(300)]^{1/2} = 316.9 \text{ m/s}.$$

Finally,

$$\dot{m} = (4.4181)(0.05)(316.9) = 70.0 \text{ kg/s}.$$

(c) $P_b = 100$ kPa. The flow is choked. Since the nozzle exit conditions are unchanged from part (b), the exit pressure, the exit Mach number, and the mass flow rate for this back pressure are the same as for part (b).

This is an important example. It illustrates how to determine if a converging nozzle is choked, and it reinforces the concept that once the nozzle is choked, further reductions in back pressure have no effect on the flow field within the nozzle.

Example 6-2 treated the concept of a spatially isentropic flow to arrive at the various answers. Flows that are isentropic temporally as well as spatially can also be treated. An important example of such a flow is the unsteady blowdown of a pressurized tank. Figure 6-13 illustrates a pressurized tank of volume *Vol*, initially containing a gas at P_{o_i} and T_{o_i} venting through a converging nozzle of area A^* into a back pressure P_b. The process is unsteady since as the blowdown proceeds, the pressure in the tank decreases. Since the pressure in the tank is a function of time, the mass flow rate of the converging nozzle is also a function of time. By assuming the process to be quasi-steady, we can develop expressions relating the tank pressure and time. The *quasi-steady assumption* presumes that no serious error will be introduced if at a given instant of time the nozzle flow field is taken to be that corresponding to the instantaneous pressure in the tank. This assumption is valid if the pressure in the tank is not changing rapidly with respect to time. The quasi-steady assumption will thus be acceptable if the mass flow rate out of the converging nozzle is small compared to the mass of the gas within the tank.

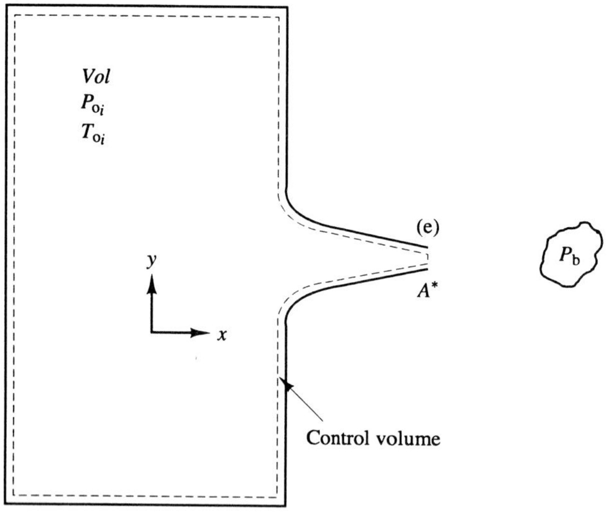

Figure 6-13 Tank Blowdown Schematic

The control volume, shown as the dotted line in Figure 6-13, is taken as the mass within the tank. The control surface then becomes the inside of the tank and the nozzle as well as the nozzle exit plane. The only mass crossing the control surface crosses at the exit plane of the nozzle. Again, conservation of mass is

$$0 = \frac{\partial}{\partial t}\int_{cv} \rho d Vol + \int_{cs} \rho \overline{V} \cdot \overline{dA}. \tag{1-33}$$

Applied to the control volume, this equation becomes

$$0 = \frac{d}{dt}\left(\frac{P_o}{RT_o} Vol\right) + \rho_e A_e V_e, \tag{6-18}$$

where uniform conditions in the tank, isentropic flow within the nozzle, and one-dimensional flow across the exit plane are assumed. If the analysis is limited to choked flow, then

$$\rho_e A_e V_e = \rho^* A^* V^* = \dot{m}, \tag{6-19}$$

and the mass flow rate can be expressed as

$$\dot{m} = \rho^* A^* V^* = \frac{P^*}{RT^*} A^* 1\sqrt{\gamma RT^*} = \sqrt{\gamma/(RT^*)} A^* P^*. \tag{6-20}$$

Let's recall conditions for choked flow:

$$P^* = \left(\frac{2}{\gamma + 1}\right)^{\gamma/(\gamma-1)} P_o \tag{6-21}$$

and

$$T^* = \frac{2}{\gamma + 1} T_o. \tag{6-22}$$

The mass flow rate then becomes

$$\dot{m} = \sqrt{\gamma/R} A^* \sqrt{\frac{\gamma + 1}{2} \frac{1}{T_o}} \left(\frac{2}{\gamma + 1}\right)^{\gamma/(\gamma-1)} P_o, \tag{6-23a}$$

which reduces to

$$\dot{m} = \sqrt{\gamma/R} \left(\frac{2}{\gamma + 1}\right)^{(\gamma+1)/2(\gamma-1)} \frac{A^* P_o}{\sqrt{T_o}}. \tag{6-23b}$$

Since the tank volume and the gas constant R are not functions of time, Eq. 6-18 reduces to the first-order differential equation

$$\frac{Vol}{R} \frac{d}{dt}\left(\frac{P_o}{T_o}\right) = -\sqrt{\gamma/R} \left(\frac{2}{\gamma + 1}\right)^{(\gamma+1)/2(\gamma-1)} \frac{A^* P_o}{\sqrt{T_o}}. \tag{6-24}$$

In order to integrate this equation, a relationship between P_o and T_o must be known. As will be verified later, the isothermal and the isentropic are two meaningful processes relating the pressure and temperature of the gas within the tank. We shall consider each in turn.

The isothermal process will be examined first. If the flow is isothermal, then $T_o = T_{o_i}$. This process results in a relatively simple analysis since Eq. 6-24 becomes

$$\frac{Vol}{R} \frac{1}{T_o} \frac{dP_o}{P_o} = -\sqrt{\gamma/R} \left(\frac{2}{\gamma + 1}\right)^{(\gamma+1)/2(\gamma-1)} \frac{A^*}{\sqrt{T_o}} dt. \tag{6-25}$$

If we take the initial conditions to be P_{o_i} when $t = 0$ and specify the tank pressure to be P_o at time t, Eq. 6-25 can be integrated to yield

$$t = -\frac{Vol}{A^*} \left[\frac{1}{\gamma R T_o} \left(\frac{\gamma + 1}{2}\right)^{(\gamma+1)/(\gamma-1)}\right]^{1/2} \ln\left(\frac{P_o}{P_{o_i}}\right). \tag{6-26}$$

This expression can be used to relate time to the instantaneous pressure in the tank. An example will be presented later.

The isentropic assumption will be considered next. In order to arrive at Eq. 6-26, a spatially isentropic flow was assumed. A temporal, or time-based, isentropic process will be assumed to occur in the tank during the tank blowdown. Thus, in the tank the initial stagnation conditions are related to the time-dependent stagnation conditions by

$$\frac{P_o}{P_{o_i}} = \left(\frac{T_o}{T_{o_i}}\right)^{\gamma/(\gamma-1)}, \tag{6-27}$$

from which

$$T_o = T_{o_i}\left(\frac{P_o}{P_{o_i}}\right)^{(\gamma-1)/\gamma}. \tag{6-28}$$

Within the time derivative, P_o and T_o appear. Working on just that term,

$$\frac{d}{dt}\left(\frac{P_o}{T_o}\right) = \frac{d}{dt}\left[P_o \frac{1}{T_{o_i}}\left(\frac{P_{o_i}}{P_o}\right)^{(\gamma-1)/\gamma}\right], \tag{6-29}$$

which can finally be reduced to

$$\frac{d}{dt}\left(\frac{P_o}{T_o}\right) = \frac{1}{\gamma}\frac{P_{o_i}^{(\gamma-1)/\gamma}}{T_{o_i}} P_o^{(1-\gamma)/\gamma}\frac{dP_o}{dt}. \tag{6-30}$$

In the same fashion, $P_o/\sqrt{T_o}$ becomes

$$\frac{P_o}{\sqrt{T_o}} = \frac{P_{o_i}^{(\gamma-1)/2\gamma}}{\sqrt{T_{o_i}}} P_o^{(\gamma+1)/2\gamma}. \tag{6-31}$$

Using Eqs. 6-30 and 6-31, the differential equation appears as

$$\frac{Vol}{\sqrt{RT_{o_i}}\,\gamma} P_{o_i}^{(\gamma-1)/2\gamma} P_o^{(1-3\gamma)/2\gamma}\, dP_o = -\gamma\left(\frac{2}{\gamma+1}\right)^{(\gamma+1)/2(\gamma-1)} A^* dt. \tag{6-32}$$

As before, integrating with the initial condition of P_{o_i} at $t = 0$ and the boundary condition of P_o at t, Eq. 6-32 becomes

$$t = \frac{2Vol[(P_o/P_{o_i})^{(1-\gamma)/2\gamma} - 1]}{(\gamma - 1)A^*\sqrt{\gamma RT_{o_i}\left(\frac{2}{\gamma+1}\right)^{(\gamma+1)/(\gamma-1)}}}. \tag{6-33}$$

Equation 6-33 was derived from consideration of two isentropic processes: (1) spatially in the nozzle at any instant of time and (2) temporally in the tank as a function of time.

The choked phase of the blowdown with either an isothermal or isentropic process has been analyzed, and the results are presented in Eqs. 6-26 and 6-33. Obviously these expressions yield different results. Which is better? To answer that question consider Figure 6-14. This figure, taken from Saad (1992), shows the time–pressure ratio relationship for both the isothermal and isentropic processes and some experimental results from the blowdown of a pressurized tank through an orifice. One reason for choosing the isothermal and the isentropic processes for the blowdown analyses is that they bound the actual process.

The isentropic process can occur only if the flow is adiabatic. Initially this is a good assumption because little time has passed and the heat transfer coefficient is low in the free convective regime, so that little energy transfer to the gas takes place. As the blowdown proceeds, the convective heat transfer coefficient is increased since the difference in temperature between the tank wall and the gas becomes larger. The combination of more time for heat transfer to occur, larger convective heat transfer coefficients, and larger differences between the gas and tank wall temperatures moves the process away from the isentropic and toward the isothermal. The experimental data confirm this as the initial pressure–time history follows the isentropic but then tends to deviate and move toward the isothermal as time passes.

Figure 6-14 provides another important point. As the pressure decreases during the blowdown process, so must the temperature; see Eq. 6-28. Thus an isentropic or nearly isentropic tank blowdown results in a decrease in the stagnation temperature of the gas inside the tank. If large pressure changes occur during the blowdown, then large temperature changes will also occur. A tank filled with moist, compressed air, for example, can discharge snow from the condensation of the water vapor within the air if discharged over a large pressure, and hence temperature, range. For this reason many blowdown applications specify that the air (or any other gas) be dry in the sense of containing little water vapor.

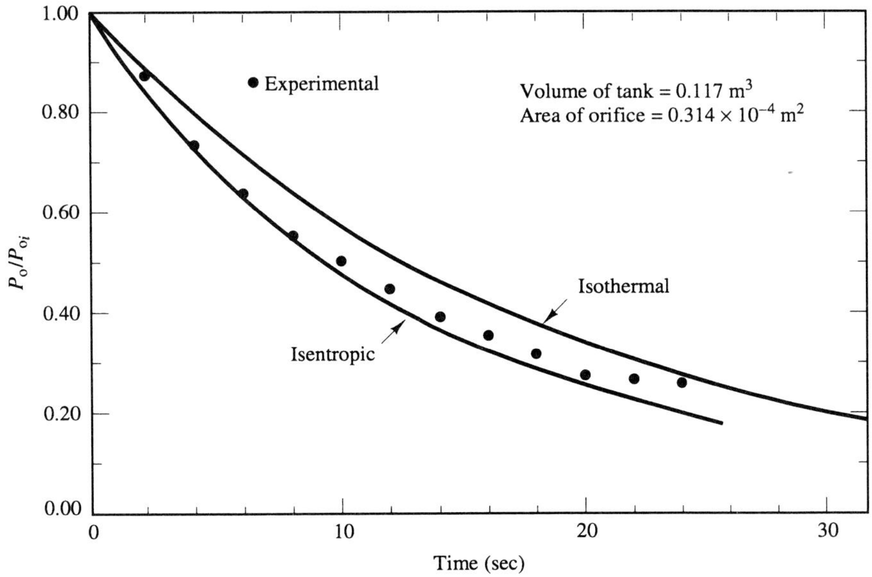

Figure 6-14 Pressure–time History during the Blowdown of a Pressurized Tank through an Orifice. *Adapted from Saad (1985).*

Compared with converging nozzles, converging–diverging nozzles present a more complex array of possible states. Converging–diverging nozzles, for example, have already been mentioned in the context of accelerating flows to supersonic speeds, something a converging nozzle alone cannot accomplish. As with converging nozzles, the starting point for understanding converging–diverging nozzles is the nozzle flow field behavior with respect to back pressure.

Consider a converging–diverging nozzle connected to a reservoir with fixed stagnation conditions P_o and T_o and exhausting into a variable back pressure P_b. The nozzle exit area is A_e, and the throat area is A^*. The situation is illustrated in Figure 6-15a. The pressure distributions down the nozzle for various back pressures are sketched in Figure 6-15b. Notice the similarity between Figures 6-10 and 6-15.

When the back pressure, P_b, is equal to the reservoir pressure, no flow occurs. If the back pressure is reduced slightly, say to P_{b_a}, flow is induced in the nozzle. The pressure distribution associated with a back pressure P_{b_a} is sketched in Figure 6-15b.

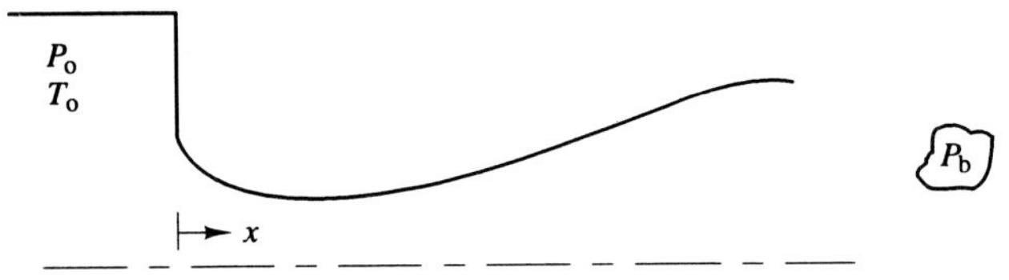

(a) Schematic

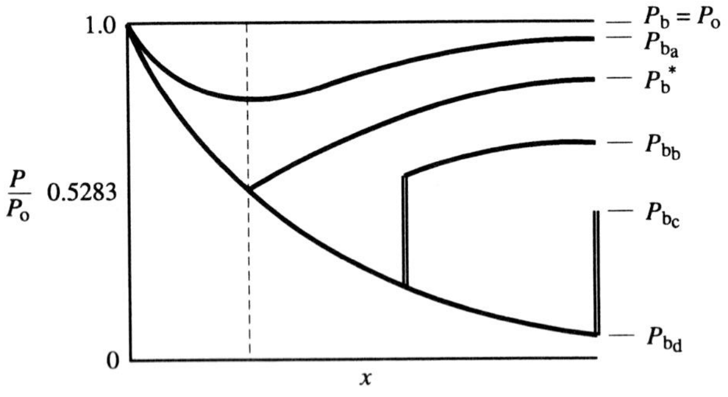

(b) Pressure distributions for various back pressures

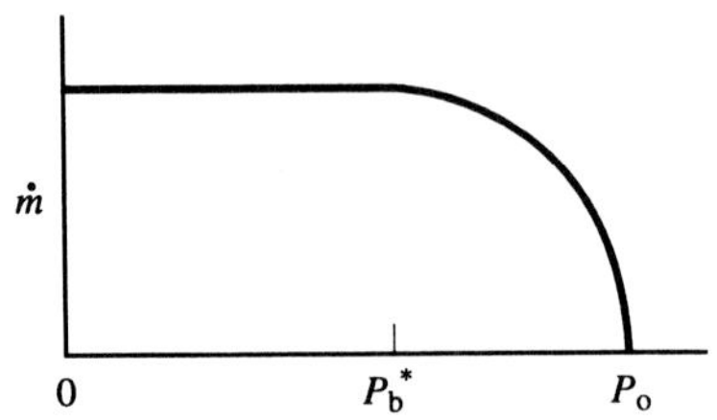

(c) Mass flow rate as a function of back pressure

Figure 6-15 Converging–Diverging Nozzle Behavior

The pressure in the converging portion of the nozzle decreases as the area decreases since the Mach number is increasing. In the diverging portion, the Mach number decreases and the pressure increases as the area increases. The highest Mach number occurs at the minimum area, which is also the point of minimum pressure. The back pressure P_{b_a} is high enough that subsonic flow exists throughout the nozzle. The exhaust plume shape is illustrated in Figure 6-16a.

As with the converging nozzle, there exists a maximum back pressure, P_b^*, for which the throat is choked. This is indicated as P_b^* in Figure 6-15b. This back pressure

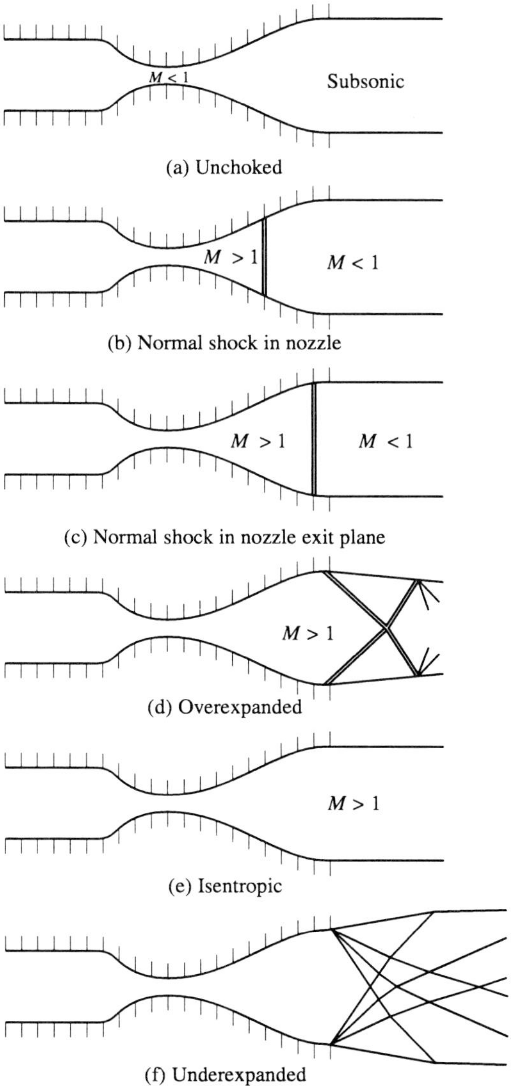

Figure 6-16 Exhaust Plume Shapes for Various Conditions

results in sonic flow in the throat and subsonic isentropic flow elsewhere. Any further reduction in back pressure will have no effect on the mass flow rate or the pressure distribution in the converging portion of the nozzle. The effect of back pressure on mass flow rate is sketched in Figure 6-15c. A converging–diverging nozzle chokes at a higher back pressure than a converging nozzle because the pressure increases for subsonic flow in the diverging segment of the nozzle.

The back pressure P_{b_d} results in a flow that is isentropic throughout the nozzle but supersonic within the diverging segment. Thus, back pressures P_b^* and P_{b_d} are both associated with choked isentropic flow in the nozzle, but the lower back pressure produces a supersonic expansion rather than a subsonic compression in the diverging portion. The exhaust plume shape for this condition is sketched in Figure 6-16e. This is sometimes called the *design condition* for a nozzle since the flow is fully expanded to the back pressure value in the absence of any interior or exterior wave structure. In propulsion the condition is called the *optimum expansion* for the same reasons.

Reduction in back pressure below P_{b_d} has no effect on the Mach number or pressure distribution within the nozzle since the flow is supersonic in the diverging portion. The pressure and Mach number in the exit plane are not functions of back pressure for $P_b < P_{b_d}$. Thus, in converging–diverging nozzle flow, P_e can be greater than P_b if the nozzle exit plane Mach number is supersonic. The flow adjusts from P_e to P_b outside the nozzle. Since the nozzle exit plane pressure is greater than the back pressure, this flow condition is often called *underexpanded*. Expansion waves occurring outside the nozzle expand the flow to the back pressure. This behavior is similar to that of a choked converging nozzle with a back pressure lower than P_b^* for the converging nozzle. Figure 6-16f illustrates the expansion process.

Perhaps the most interesting regime of nozzle flow occurs when $P_b^* < P_b < P_{b_d}$. As the back pressure is reduced from P_b^* (see Figure 6-15b), normal shock waves occur within or at the exit of the nozzle. The back pressure P_{b_b}, for example, produces a flow that is supersonic from the throat to the normal shock and subsonic from the normal shock to the nozzle exit plane. The exit Mach number is subsonic and the shock location is such that $P_e = P_b$. Indeed, for a subsonic exit Mach number, the shock wave must be located so that P_e is equal to P_b. The exhaust plume for this shape is schematically illustrated in Figure 6-16b. As the back pressure is decreased from P_b^*, the normal shock is pulled farther and farther into the diverging segment of the nozzle. Finally, at a back pressure indicated as P_{b_c} in Figure 6-15b, a normal shock stands in the nozzle exit plane. Upstream of the shock wave in the exit plane, the Mach number is supersonic; downstream, the Mach number is subsonic. The pressure increase across the exit plane shock wave raises the pressure to the back pressure. The plume structure for this case is presented in Figure 6-16c.

Back pressures in the range $P_{b_c} < P_b < P_{b_d}$ result in oblique shock waves attached to the lip of the nozzle exit. For back pressures in this range, the Mach number and pressure distributions within the nozzle are not affected by the back pressure, although the oblique shock structure attached to the nozzle lip is affected. The pressure in the nozzle exit plane is less than the back pressure, so this condition is often called an *overexpanded nozzle flow*. One effect of the oblique shock wave is to compress the exit flow and raise its pressure to the back pressure. An oblique shock structure typical of the underexpanded condition is illustrated in Figure 6-16d.

The representations of the plume behavior as depicted in Figure 6-16 are idealized to the extent that they ignore viscous effects. In actuality the viscous effects would result in the nozzle exhaust entraining ambient air, jet entrainment, and the demarcation between the plume and the surroundings would not be maintained as the distinct lines indicated in the figure. Nonetheless, the features are qualitatively correct.

The preceding converging–diverging nozzle concepts can be easily codified, as in Figure 6-17, by realizing that three key back pressures separate the nozzle exit conditions into four regions. In order of decreasing values, the three key back pressures are

1. the maximum for choked flow
2. the pressure required for a normal shock in the nozzle exit plane
3. the pressure required for supersonic nozzle flow with no wave structure exterior to the exit plane.

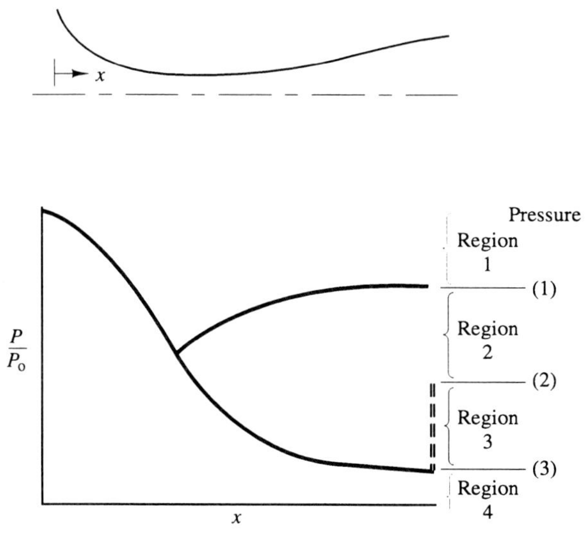

Figure 6-17 Key Pressures and Regions for Converging–Diverging Nozzle Flows

These are indicated and labeled as (1), (2), and (3) in Figure 6-17. The four regions are also shown in the figure. Region 1 is associated with subsonic flow throughout the nozzle, region 2 with a normal shock wave within the nozzle, region 3 with an oblique shock wave at the lip of the nozzle, and region 4 with an expansion wave at the lip of the nozzle. Table 6-2 summarizes attributes of the various key pressures and regions. Some examples are in order.

TABLE 6-2 CONVERGING–DIVERGING NOZZLE ATTRIBUTES

Key pressure	Region	Throat	Exit Mach no.	Exit pressure	Other
	1	$M < 1$	$M_e < 1$	$P_e = P_o$	Isentropic
1		$M = 1$	$M_e < 1$	$P_e = P_b$	min P_o or max P_b^* for choked flow
	2	$M = 1$	$M_e < 1$	$P_e = P_b$	Shock in nozzle
2		$M = 1$	$M_{e_1} > 1$	$P_{e_1} < P_b$	Shock in exit (1) upstream of shock
			$M_{e_2} < 1$	$P_{e_2} = P_b$	(2) downstream of shock
	3	$M = 1$ (overexpanded nozzle)	$M_e > 1$	$P_e < P_b$	Oblique shock at exit plane
3		$M = 1$ (optimum expansion)	$M_e > 1$	$P_e = P_b$	Isentropic everywhere
	4	$M = 1$ (underexpanded nozzle)	$M_e > 1$	$P_e > P_b$	Expansion wave at exit plane

Example 6-3

A converging–diverging nozzle with an area ratio of 3.0 is attached to a reservoir with stagnation conditions of 500 kPa and 300 K. Identify the key pressures and back pressure ranges for each region. Air is the fluid.

Solution. The solution was obtained using COMPQ. A sketch of the problem is provided in Figure 6-18a. Key pressures (1) and (3) correspond to the two isentropic conditions. Using A/A^* as input for case **A** of option **IS** yields

$$A_e/A^* = 3.0 \qquad M_e = 0.1974 \qquad M_e = 2.6374$$

$$T_e/T_o = 0.9923 \qquad T_e/T_o = 0.4182$$

$$P_e/P_o = 0.9732 \qquad P_e/P_o = 0.0473$$

from which the key pressures are

$$P_1 = 486.6 \text{ kPa}$$

$$P_3 = 23.65 \text{ kPa}.$$

The situation for key pressure (2) is illustrated in Figure 6-18b. The normal shock occurs at a Mach number of 2.6374. Case **U** of option **NS** of COMPQ produces

$$M_1 = 2.6374 \qquad M_2 = 0.5007$$

$$P_2/P_1 = 7.9485.$$

The pressure downstream of the normal shock wave becomes

$$P_{2_e} = P_b = \frac{P_{2_e}}{P_{1_e}}\frac{P_1}{P_o}P_o = (7.9485)(0.0473)(500) = 188.0 \text{ kPa}.$$

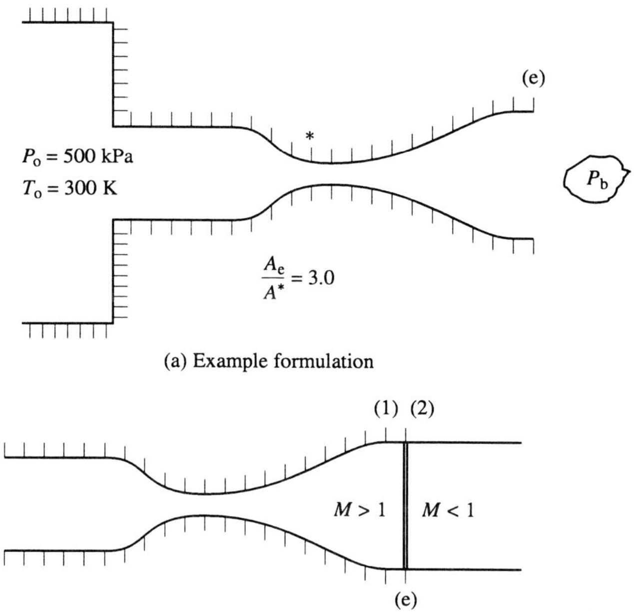

(a) Example formulation

(b) Shock in nozzle exit plane

Figure 6-18 Example Problem 6-3

This is also the back pressure since the Mach number downstream of the shock wave is subsonic. Table 6-3 presents the back pressure ranges associated with the various regions. Given any back pressure for the conditions of this problem, the nozzle flow region can be determined by consulting Table 6-3.

TABLE 6-3 BACK PRESSURE RANGES FOR REGIONS

Region	Back pressure range (kPa)
1	$P_b > 486.6$
2	$188.0 < P_b < 486.6$
3	$23.65 < P_b < 188.0$
4	$P_b < 23.65$

Example 6-4

For the nozzle and conditions of Example 6-3, find the exit Mach numbers and sketch the salient characteristics of the flow field for back pressures of 100 kPa and 10 kPa.

Solution.

(a) Back pressure of 100 kPa

From Table 6-3, a back pressure of 100 kPa lies in region 3. Thus the nozzle is choked, the flow is supersonic in the diverging portion, and an oblique shock is attached to the lip of the nozzle. The flow field is sketched in Figure 6-19a. From Example 6-3, the pressure in the nozzle exit plane is 23.65 kPa with an exit Mach number of 2.6374. Across the oblique shock the pressure must increase from 23.65 kPa to 100 kPa, a pressure ratio of 4.2283. Using case **PR** of option **OS** of COMPQ, the oblique shock angle and the deflection are

$$M_c = 2.6374 \qquad \theta = 47.38° \qquad \delta = 24.52°.$$

Figure 6-19b illustrates the shock angle and the deflection as well as the initial plume angle of the flow as it leaves the nozzle.

(b) Back pressure of 10 kPa

Table 6-3 shows that this back pressure is in region 4. Hence the nozzle is choked, the flow is supersonic in the diverging portion, and an expansion fan is attached to the lip of the nozzle. The flow field is indicated in Figure 6-19c. The exit plane Mach number and pressure are 2.6374 and 23.65 kPa, respectively. The expansion fan must expand the flow from 23.65 kPa to 10.0 kPa. The pressure ratio for the expansion is 0.4228, from which case **PR** of option **PM** of COMPQ yields

$$\Delta\nu = 11.359°$$

$$M_b = 3.208.$$

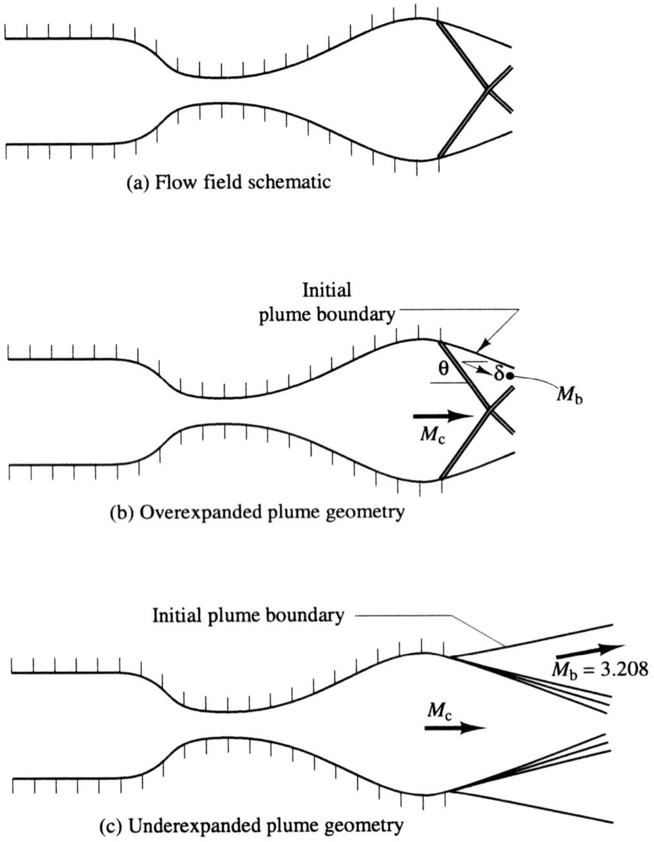

Figure 6-19 Example Problem 6-4

These are shown in Figure 6-19c. This example demonstrates how to use an understanding of the key pressures to ascertain the behavior of a nozzle with a specified back pressure. Conditions in regions 3 and 4 are simple to treat since the exit plane conditions are given by key pressure 3. Region 2 problems involving shock locations for given exit conditions are more involved and additional considerations are needed.

Problems associated with normal shocks occurring in diverging portions of nozzles will be examined next. A general schematic of the arrangement is shown in Figure 6-20. In inviscid flows with normal shock waves in diverging passages, the pressure increases discontinuously across the shock but the flow continues to fill the passage. State i can be any supersonic state of the flow; in many instances it will correspond to the actual throat of a nozzle. With conditions (Mach number and properties) at state i known, three different problem specifications are likely:

1. Given the shock location, find the Mach number and properties at the nozzle exit.
2. Given the exit Mach number, M_e, find the shock wave location and the exit properties.
3. Given the exit pressure, find the exit Mach number and properties.

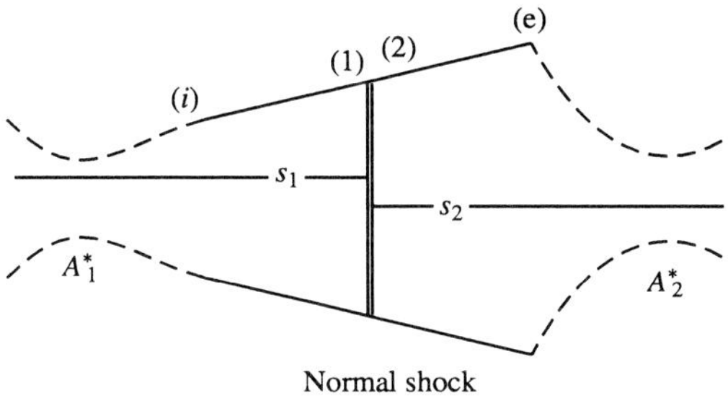

Figure 6-20 Normal Shock in Nozzle Nomenclature

Problems of the first type are readily solved since the shock location determines the upstream Mach number. Conditions at state 1 are found, the shock wave is crossed, and then conditions in the exit plane are found.

The second type of problems at first appear somewhat intractable, but consider the sonic locations that are associated with the upstream and downstream flows. These are the circumstances of Figure 6-9a. Recall Eq. 6-14:

$$\frac{A_1^*}{A_2^*} = \frac{P_{o_2}}{P_{o_1}}, \tag{6-14}$$

from which we derive

$$\frac{P_{o_2}}{P_{o_1}} = \frac{A_1^*}{A_i}\frac{A_i}{A_e}\frac{A_e}{A_2^*} = \frac{A_e/A_2^*}{A_i/A_1^*}\frac{A_i}{A_e}. \tag{6-34}$$

The inlet and exit Mach numbers as well as the area change ratio are known, and so the stagnation pressure ratio can be computed directly. From the stagnation pressure ratio, the incident Mach number M_1 can be obtained. Once M_1 is known, all of the remaining quantities of interest can be found.

The third type of problem specifications, in which the back pressure is known, are a little more involved than the second type of problems. This final type of problems can be solved either iteratively or by evaluating a closed-form expression. The iterative procedure will be demonstrated in Example 6-5. For steady-state flow,

$$\rho_i V_i A_i = \rho_e V_e A_e, \tag{6-35}$$

which can be written as

$$\frac{P_i}{RT_i} M_i \sqrt{\gamma R T_i} A_i = \frac{P_e}{RT_e} M_e \sqrt{\gamma R T_e} A_e. \tag{6-36}$$

With conditions at the inlet and pressure at the exit known, Eq. 6-36 reduces to

$$\frac{P_i}{P_e} \frac{A_i}{A_e} \frac{M_i}{\sqrt{T_i/T_o}} = \frac{M_e}{\sqrt{T_e/T_o}}. \tag{6-37}$$

Naming the constants on the left-hand side of Eq. 6-37 as C, squaring, and solving the resulting quadratic equation for M_e^2 yields

$$M_e^2 = \frac{-1 \pm [1 + 2(\gamma - 1)C^2]^{1/2}}{\gamma - 1}. \tag{6-38}$$

Example 6-5 will use Eq. 6-38.

Example 6-5

Given the conditions of Example 6-3, find the exit Mach number and pressure and sketch the flow field for a back pressure of 250 kPa.

Solution. Table 6-3, constructed in Example 6-3, will be used here to determine the region associated with a back pressure of 250 kPa. A back pressure of 250 kPa places the nozzle flow in region 2, a normal shock standing in the diverging section.

This example will be solved first by using Eq. 6-38 and then by using an iterative process. The "inlet" for the diverging nozzle will be taken as the actual throat of the nozzle. The term C then becomes

$$C = \frac{P^*}{P_e} \frac{A^*}{A_e} \frac{1}{\sqrt{T^*/T_o}} = \frac{(0.5283)(500)}{250.0} \frac{1}{3.0} \frac{1}{\sqrt{.8333}} = 0.38582.$$

M_e^2 is obtained from Eq. 6-36:

$$M_e^2 = \frac{-1 \pm [1 + 2.0(1.4 - 1)0.38582^2]^{1/2}}{1.4 - 1},$$

so that

$$M_e^2 = 0.1447, -5.1447.$$

With the exit Mach number and static pressure known,

$$P_{o_2} = \frac{P_{o_2}}{P_2} P_2 = \frac{P_2}{P_2/P_{o_2}} = \frac{250}{0.9050} = 276.24 \text{ kPa},$$

so that

$$\frac{P_{o2}}{P_{o_1}} = \frac{276.24}{500} = 0.55249.$$

Option **NS** of COMPQ yields

$$M_2 = 0.5263$$
$$M_1 = 2.3709$$
$$P_2/P_1 = 6.3914.$$

The shock is located at

$$\frac{A_1}{A_1^*} = \frac{A_1}{A_i} = 2.3396.$$

The iterative process requires a series of region 2 computations. The usual procedure is to assume an upstream Mach number M_1 (or an area ratio A_1/A_1^*) from which P_{o_2}/P_{o_1} can be found. Then

$$\frac{A_e}{A_2^*} = \frac{A_e}{A_1^*}\frac{A_1^*}{A_2^*} = \frac{A_e}{A_i}\frac{P_{o2}}{P_{o_1}},$$

which defines the exit Mach number M_e and P_e/P_{o2}. Finally, P_e/P_{o_1} can be expressed as

$$\frac{P_e}{P_{o_1}} = \frac{P_e}{P_{o_2}}\frac{P_{o2}}{P_{o_1}}.$$

When the assumed value of M_1 yields the correct value of P_e/P_{o_1}, which is known, the iteration has converged. For this problem, the iterative sequence is given in Table 6-4. The iteration yields the same results as evaluation of Eq. 6-38. Once the Mach number is found by the iterative process, the remaining parameters are computed as in the first solution.

TABLE 6-4 ITERATIVE SEQUENCE FOR EXAMPLE 6-5

M_1	P_{o_2}/P_{o_1}	A_e/A_2^*	M_e	P_2/P_{o_2}	P_e/P_{o_1}
1.30	0.9794	2.9382	0.2018	0.9720	0.9520
2.00	0.7209	2.1617	0.2804	0.9469	0.6830
2.32	0.5745	1.7235	0.3630	0.9130	0.5295
2.48	0.5071	1.5213	0.4227	0.8844	0.4485
2.40	0.5401	1.6203	0.3909	0.9000	0.4860
2.36	0.5572	1.6716	0.3765	0.9068	0.5053
2.38	0.5486	1.6458	0.3836	0.9035	0.4957
2.37	0.5529	1.6587	0.3800	0.9052	0.5005

The preceding discussion involving converging–diverging nozzle flows in regions 2 and 3 has been based on the assumption of inviscid flow and has not addressed what sequence of events are needed in order to obtain some of the postulated nozzle flow patterns. For example, in region 2 a normal shock wave has been shown in the nozzle divergence in the inviscid one-dimensional sense; that is, as providing a step increase in pressure at some point along the nozzle but not inducing flow separation from the nozzle. In actuality, operation in region 2 almost always leads to flow separation so that the flow no longer continues to fill the nozzle channel. A typical region 2, symmetrically separated flow field is illustrated in Figure 6-21a. As shown in the figure, the separated region does not reattach to the nozzle wall. The resulting deflection of the stream away from the wall results in the oblique shock wave intersection pattern indicated in the figure. The actual flow field, qualitatively sketched in Figure 6-21, resulting from operation in region 2, differs significantly from that postulated by inviscid considerations, qualitatively sketched in Figure 6-15. The essential differences between the actual and the postulated flow fields are caused by viscous effects. In regions of favorable pressure gradients ($dP/dx < 0$), the boundary layer is well behaved and, for the usual Reynolds number range, is confined to a small region near the surface. Viscosity has little effect on the postulated flow field for such accelerating flows, and the inviscid results from the one-dimensional analysis are quite adequate for many engineering purposes. In regions of adverse pressure gradients ($dP/dx > 0$), the boundary layer thickens rapidly and is susceptible to separation. Were the flow field to be as postulated in the inviscid analysis, the pressure gradient across the normal shock wave intersecting the nozzle walls would be quite large in the neighborhood of the intersection location, and separation—a viscous-induced phenomenon—would occur. The actual flow field would be qualitatively as sketched in Figures 6-21a or c.

Experimental data demonstrate that the nozzle wall pressures follow the isentropic values for the local area ratios until the separation point is reached. The nozzle wall pressures downstream of the separation point tend to be constant at values close to the back pressure. Figure 6-21b illustrates the typical nozzle wall pressure distribution occurring in a symmetrically separated flow. The constant pressure felt by the nozzle wall downstream of the separation point is evident. Thompson (1972) suggests to a reasonable estimate that the static pressure at the separation location is about 0.4 of the back pressure.

The separation pattern may not be symmetrical with respect to the nozzle wall, and thus the flow can separate from one side and remain attached to the other. A typical nonsymmetrical separated flow pattern is presented in Figure 6-21c. Obviously, proper treatment of separated flows in nozzles requires consideration of two- or three-dimensional effects as well as viscous effects. Such treatment is beyond the scope of this text.

The flow pattern postulated for region 3 conditions warrants some additional examination. If a converging–diverging nozzle test run were initiated with the reservoir and back pressures set for a flow pattern in region 3, which should result in an oblique shock standing in the nozzle exit plane, a separated flow pattern would be observed. The only way to obtain the flow pattern identified with region 3 is to start the nozzle with a sufficiently large reservoir pressure to place the flow in region 4 and then to decrease the reservoir pressure, or increase the back pressure. As the pressure required for region 4 flow passes through the conditions for key pressure

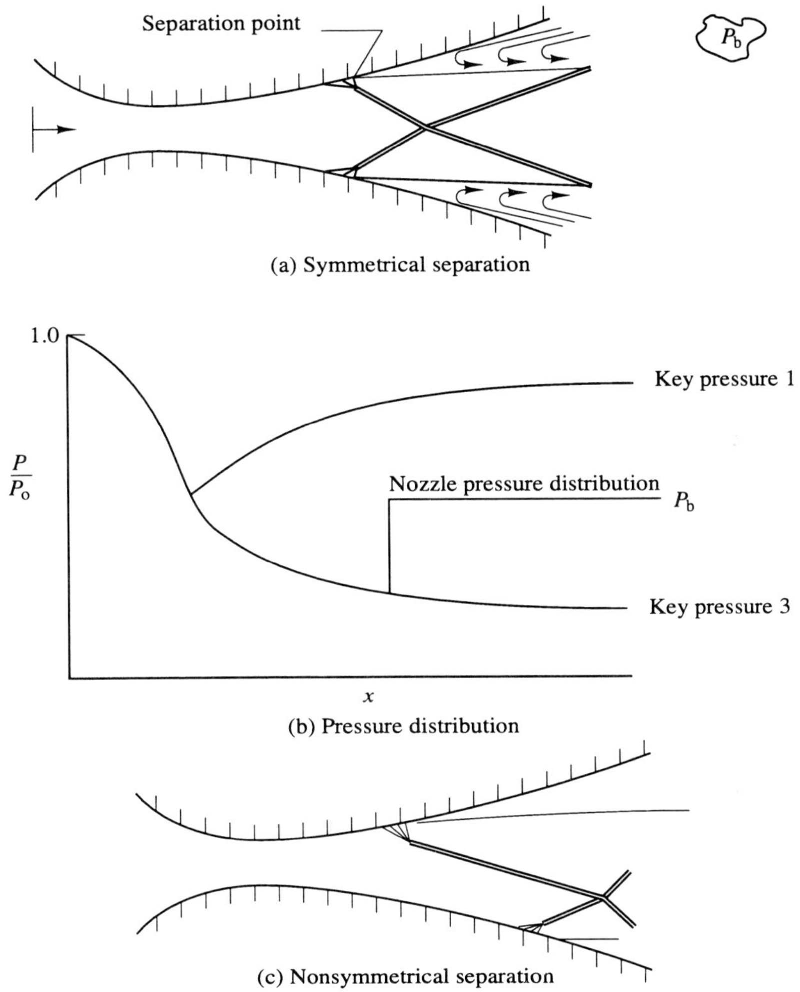

Figure 6-21 Separated Nozzle Flows

3 and into region 3 flow, the condition of an oblique shock standing in the nozzle exit plane would be obtained. We should pay particular attention to steady-state isentropic flows, since some steady-state flow conditions can only be obtained via transient approaches.

6.4 SUPERSONIC WIND TUNNELS

A supersonic wind tunnel is a facility for producing, in a laboratory, conditions that can be used to simulate flight at supersonic Mach numbers. The use of converging–diverging nozzles to produce supersonic flow has been discussed. Zucrow and

Hoffman (1976) identify the following four general arrangements for supersonic wind tunnels:

1. high-pressure gas storage tanks for blowdown wind tunnels
2. a compressor and diffuser for continuous-duty wind tunnels
3. a vacuum tank for atmospheric-entry wind tunnels
4. shock tubes for high-enthalpy wind tunnels.

In this section we shall be concerned only with types 1 and 2 since the majority of wind tunnels fall into these categories.

The blowdown process was discussed in terms of choked flow in converging nozzles. Much of that discussion and development is useful for blowdown wind tunnel operation. The typical blowdown wind tunnel consists of a pressurized tank feeding a converging–diverging nozzle. The typical blowdown wind tunnel is schematically illustrated in Figure 6-22a. The purpose of the wind tunnel is to develop a specified Mach number, determined by the test section area ratio, in the test section. So long as the pressure in the tank is sufficient to provide region 3 or region 4 flow, the test section operates at the specified or design Mach number. When the

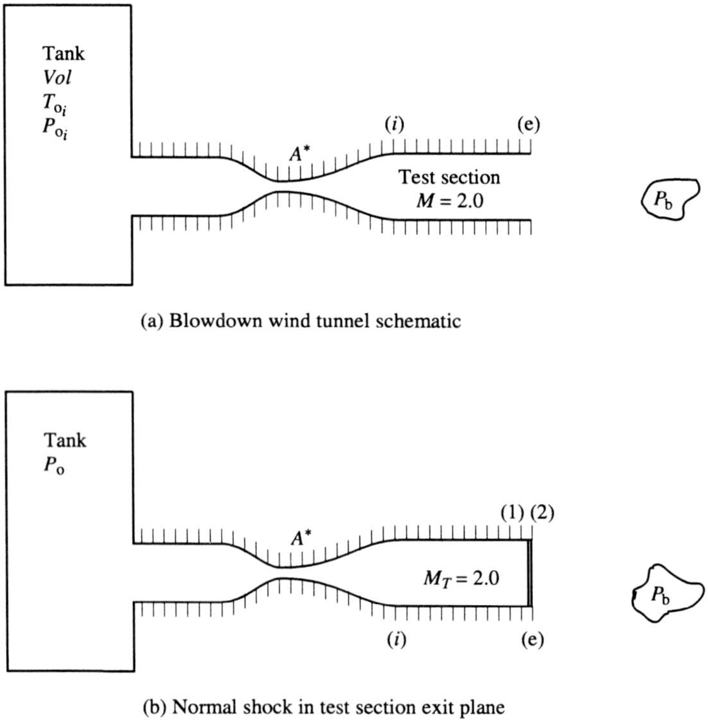

Figure 6-22 Blowdown Wind Tunnel and Normal Shock Wave

reservoir pressure becomes too low with respect to the back pressure, region 2 flow is initiated and the test section no longer operates supersonically. Two of the simpler options for prolonging acceptable test section conditions are to discharge the test section into an initially evacuated tank and/or to provide a diffuser between the test section and the back pressure. Both of these options result in lower reservoir pressure for region 3 or 4 flow. The vacuum tank essentially provides a lower back pressure for the system. This lower back pressure translates into longer testing times for a given pressure at the start of the blowdown process. The next examples explore the effectiveness of diffusers.

Example 6-6

A blowdown wind tunnel exhausting to atmospheric pressure (14.7 psia) is to be designed. The test section cross-sectional area is specified to be 1 ft^2, and the desired test section Mach number is 2.0. The supply tank can be pressurized to 150 psia and heated to 150° F. Determine the throat area and supply tank volume required for a testing time of 30.0 s.

Solution. The blowdown wind tunnel will appear as in Figure 6-22a. The test section extends from station i to station e. For simple area change, the Mach number is constant through the test section since the area is constant. Useful operation occurs when the flow is isentropic from the tank through the test section and supersonic within the test section. Using case **M** of option **IS** of COMPQ yields the test section area ratio

$$\frac{A_T}{A^*} = 1.6875$$

and the test section pressure ratio

$$\frac{P_T}{P_o} = 0.12780.$$

The required supersonic Mach number will be maintained within the test section until the tank pressure drops so low that a normal shock wave sits at the test section exit, station e. Any further reduction in the tank pressure will result in the normal shock moving into the test section or the diverging portion of the nozzle. Testing must be terminated at this point since the flow is no longer supersonic within the test section. The tank pressure corresponding to the condition of the normal shock wave in the exit plane can be calculated by reference to Figure 6-22b. For a test section Mach number of 2.0, the Mach number downstream of the normal shock wave is

$$M_2 = 0.5774,$$

with the pressure ratio across the shock wave of

$$\frac{P_2}{P_1} = 4.500.$$

We assume the process to be quasi-steady, and so the tank pressure for this condition becomes

$$P_o = \frac{P_o}{P_1}\frac{P_1}{P_2}\frac{P_2}{P_b}P_b = \frac{1}{0.1278}\frac{1}{4.5}14.7 = 25.561 \text{ psia}.$$

The blowdown of a pressurized tank through a choked throat has already been examined and analyzed for a converging nozzle. That analysis is valid for a converging–diverging nozzle as long as the throat of the nozzle remains choked. In the earlier development of the tank blowdown, experimental data suggested that the actual process in the tank is bounded by the isentropic and the isothermal processes. The data indicate that the isentropic process results in reaching a given pressure ratio, P_o/P_{o_i}, faster than the isothermal process. Thus the volume required for a specified testing time will be larger for the isentropic than for the isothermal; see Figure 6-14. The required volume computed using the isentropic tank process will be conservative. The time required to reach a given pressure ratio for the isentropic tank process is given by Eq. 6-33, which is repeated here:

$$t = \frac{2Vol[(P_{o_f}/P_{o_i})^{(1-\gamma)/2\gamma} - 1]}{(\gamma - 1)A^*\sqrt{\gamma R T_{o_i}}\sqrt{\left(\frac{2}{\gamma + 1}\right)^{(\gamma+1)/(\gamma-1)}}}. \tag{6-33}$$

The required volume becomes

$$Vol = \frac{t(\gamma - 1)A^*\sqrt{\gamma R T_{o_i}}\sqrt{\left(\frac{2}{\gamma + 1}\right)^{(\gamma+1)/(\gamma-1)}}}{2[(P_{o_f}/P_{o_i})^{(1-\gamma)/2\gamma} - 1]}.$$

For a final tank pressure of 25.561 psia, the volume is

$$Vol = \frac{(30)(1.4 - 1)(0.5926)[(1.4)(1716)(610)]^{1/2}\sqrt{\left(\frac{2}{1.4 + 1}\right)^{(1.4+1)/(1.4-1)}}}{2[(25.56/150)^{(1 - 1.4)/2(1.4)} - 1]}$$

$$= 8660 \text{ ft}^3.$$

Had the computations been made using the assumption of an isothermal process, Eq. 6-26 would have been used with the resulting tank volume requirement of 7038 ft^3. A tank volume of 8660 ft^3 and a throat area of 0.5926 ft^2 will meet the requirements.

Example 6-7

If the system specified in Example 6-6 has a diffuser of area ratio $3.375(A_d/A^*)$ attached to the end of the test section, what is the testing time?

Solution. The results of Example 6-6 will be used. The system with the diffuser is shown schematically in Figure 6-23a. As in the previous example, testing can continue until a normal shock wave is located at the end of the test section. The previous example indicated these values:

$$Vol = 8660 \text{ ft}^3$$

$$A^* = 0.5926 \text{ ft}^2, \qquad A_T = 1.0 \text{ ft}^2$$

$$P_T/P_o = 0.12780$$

$$M_2 = 0.5774$$

$$P_2/P_1 = 4.5.$$

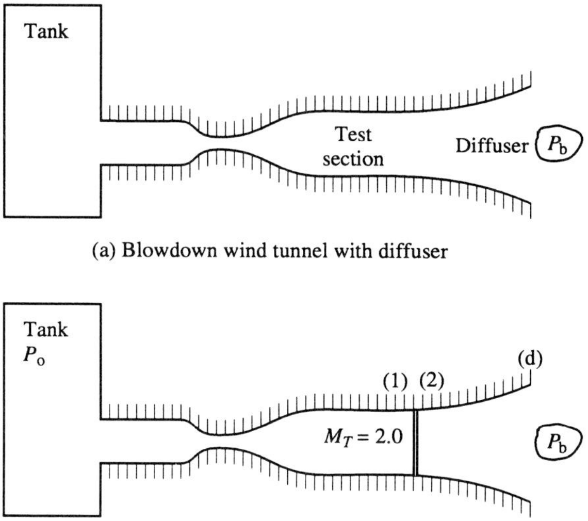

(a) Blowdown wind tunnel with diffuser

(b) Normal shock in test section exit plane

Figure 6-23 Example 6.7

The diffuser-to-test-section exit ratio is $A_d/A_T = 3.375/1.6875 = 2.0$. Using case **AC** of option **IS** of COMPQ with an initial Mach number of 0.5774 and an area ratio of 2.0 yields

$$M_d = 0.2470$$

$$P_d/P_2 = 1.202.$$

Then the reservoir pressure corresponding to the condition shown in Figure 6-23b becomes

$$P_o = \frac{P_o}{P_1}\frac{P_1}{P_2}\frac{P_2}{P_d}\frac{P_d}{P_b}P_b = \frac{1}{1.278}\frac{1}{4.5}\frac{1}{1.202}(1)(14.7) = 21.27 \text{ psia.}$$

Again, the isentropic tank process will be followed so that

$$t = \frac{(2)(8660)[(21.27/150)^{(1-1.4)/2(1.4)} - 1]}{(1.4-1)(0.5926)\sqrt{(1.4)(1716)(610)}\sqrt{\left(\frac{2}{1.4+1}\right)^{(1.4+1)/(1.4-1)}}}$$

$$= 33.55 \text{ s.}$$

The addition of the diffuser to the test section results in an increase of 3.55 s in testing time. This is slightly more than a 10 percent increase. The diffuser is helpful because it has a lower exit Mach number than the test section. The higher pressure ratio for the lower Mach number permits a lower reservoir (tank) pressure for the normal shock wave to stand in the test section exit plane.

Subsonic flow in the diffuser results in an adverse (increasing) pressure gradient down the diffuser. In an inviscid analysis, such as simple area change, viscous effects are neglected and the flow possesses no sensitivity to adverse pressure gradients. In reality, if the rate of increase of area in the diffuser is too large, flow separation will occur because of the viscous effects. Because of the flow separation possibilities in subsonic diffusers, they are usually constructed with relatively small area change rates; circular diffusers may have only a 7 or 8 degree divergence (slope) to inhibit separation.

Continuous-flow supersonic wind tunnels comprise the other class of wind tunnels that we shall examine. The continuous-flow wind tunnels are used where long testing times are required. These wind tunnels are basically closed loop devices that contain compressors and heat exchangers. A continuous-flow wind tunnel is schematically illustrated in Figure 6-24a. In order for the tunnel to have continuous flow, the compressors must be large enough to overcome all steady-state losses in the system and to provide for start up. Heat exchangers must be provided to cool the gas since the gas (usually air) is heated by the compressors.

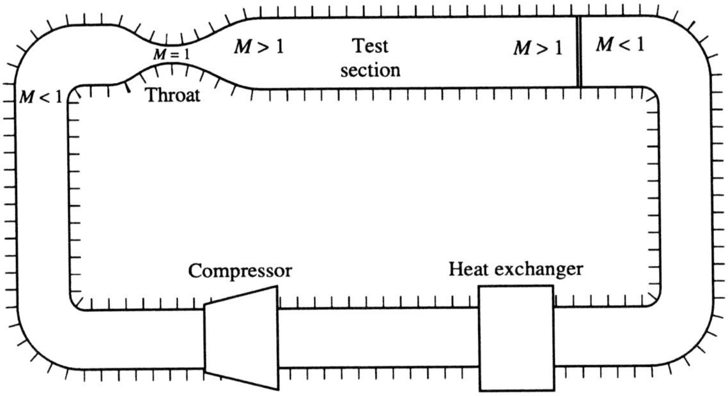

(a) Continuous wind tunnel (no diffuser)

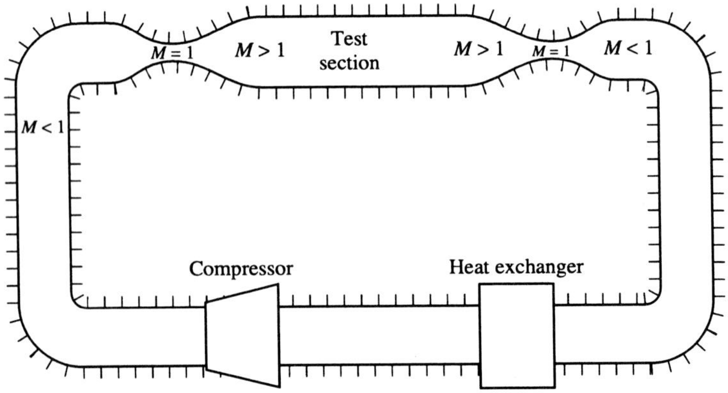

(b) Continuous wind tunnel with diffuser

Figure 6-24 Continuous Wind Tunnel Schematics

The steady-state power required by the compressors can be minimized by minimizing the losses that the compressors must make up. A significant component of the losses can be the loss in stagnation pressure as the flow is decelerated, via a shock wave, from supersonic to subsonic speeds, as it must be prior to entering the compressors. This stagnation pressure loss can be reduced by decreasing the Mach number of the shock wave that terminates the supersonic flow. One conceptual scheme is represented by Figure 6-24b, which shows a supersonic diffuser attached to the test section. The diffuser is basically an additional converging–diverging section that is used to reduce the incoming supersonic flow to subsonic flow without passage through a shock wave.

How is such a flow condition established? Consider the startup sequence shown in Figure 6-25. As power is initially applied to the compressor during the startup phase, the flow is subsonic through the entire loop as indicated by Figure 6-25a. Further increases in power to the compressor lead to the flow's choking at the minimum area, with the flow subsonic elsewhere. The wind tunnel Mach number distribution at initial choking is schematically illustrated in Figure 6-25b. Any further increase in power to the compressor will result in establishing supersonic flow in a region downstream of the first throat. Additional increases in power to the compressor cause the normal shock to move farther into the diverging section, as in Figure 6-25c. Eventually the normal shock will be positioned at the exit of the diverging section, which is the test section inlet. This condition is illustrated in Figure 6-25d. The supersonic region will be terminated by a normal shock wave across which stagnation pressure is lost. The diffuser throat (the second throat) must be larger than the first throat. The largest stagnation pressure loss associated with the startup procedure will occur when the normal shock wave is in the test section. The diffuser throat must be large enough to pass the required mass flow rate even with the reduced stagnation pressure downstream of the shock wave. The first throat and the diffuser throat are then related as in Eq. 6-14:

$$\frac{A_2^*}{A_1^*} = \frac{P_{o_1}}{P_{o_2}}. \tag{6-14}$$

For simple area change flows, any further increase in power to the compressor will result in the shock wave moving into the converging section of the diffuser. Because the shock wave cannot remain in a converging section, it moves through the diffuser throat, the second throat, and is said to be *swallowed*. The swallowed normal shock condition is indicated in Figure 6-25e. Since the normal shock wave has been swallowed, both throats now possess the same stagnation pressure. If the diffuser throat area is variable, it can now be reduced to the same area as the first throat. That condition is illustrated in Figure 6-25f. Increasing the back pressure pushes the shock wave into the diffuser throat where the $M = 1$ and the shock-free deceleration conditions of Figure 6-24b are obtained. If the diffuser throat is not variable, then the back pressure can be increased (by reducing the power to the compressor) to move the shock in the diverging section of the diffuser close to the throat. The limit on this shock movement is to place the normal shock in the diffuser throat, as that

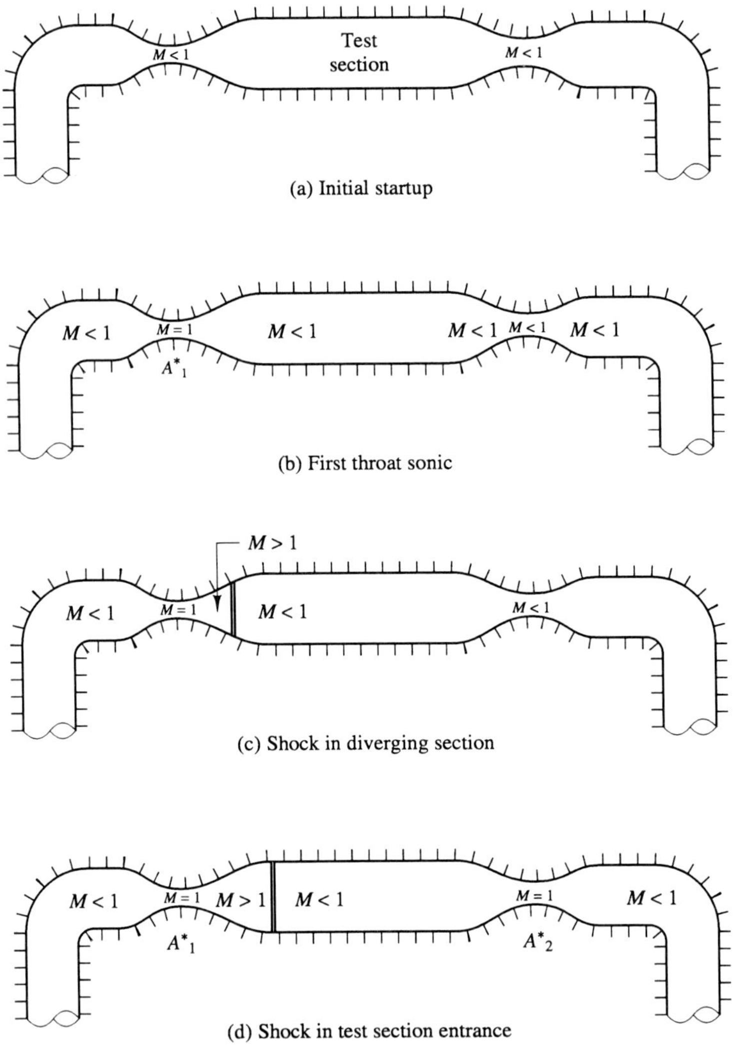

Figure 6-25 Wind Tunnel Startup Sequence

position represents the lowest Mach number possible into the shock wave for a fixed throat diffuser. Figure 6-25g schematically illustrates wind tunnel operation for a fixed-area diffuser with the minimum normal shock wave Mach number.

Generally, because some fluctuations are always present in the flow, prudence suggests that the shock be placed slightly downstream of the diffuser throat for the fixed-area diffuser. For the same reason, for the variable-area diffuser, rather than having subsonic flow downstream of the diffuser throat, a weak shock wave in the diverging section close to the diffuser throat is usually preferred. These arrange-

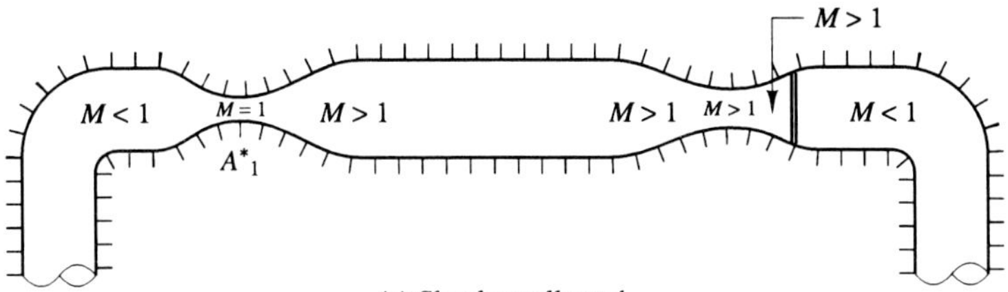

(e) Shock swallowed

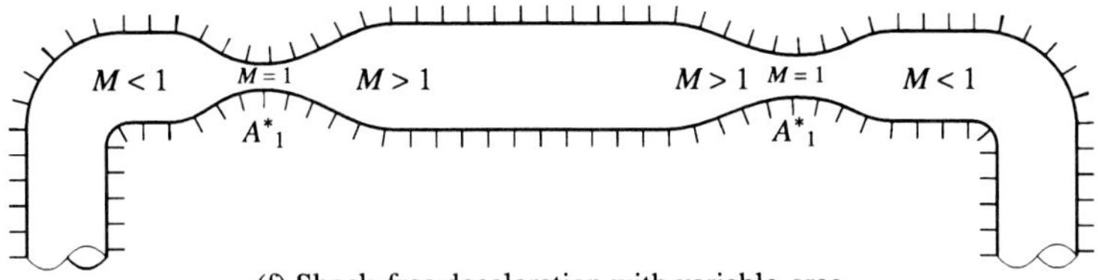

(f) Shock-free deceleration with variable-area diffuser throat

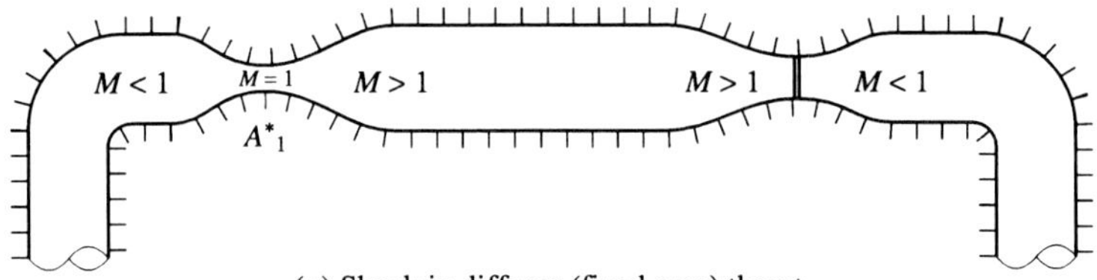

(g) Shock in diffuser (fixed-area) throat

Figure 6-25 (continued)

ments permit some flow fluctuations to occur and the swallowed shock wave to remain in the diffuser divergent section. Otherwise instabilities or fluctuations in the flow could cause the diffuser to become unchoked, and the entire starting process would need to be repeated.

One of the key features of the shock swallowing process is the movement of the shock wave through the converging section of the diffuser. The statement was made that a normal shock wave could not reside in a converging segment of a nozzle. Consider, as illustrated in Figure 6-26, a normal shock wave in a converging passage. If a fluctuation in the flow moves the shock wave farther into the passage, the upstream Mach number of the shock is decreased. The decrease in the upstream Mach number causes an increase in stagnation pressure over the original, and the increase in stagnation pressure allows more mass to pass through a given area. Since mass is being depleted at a faster rate between the shock and the diffuser throat, the shock wave is pulled farther into the converging passage. If, on the other hand, the flow fluctuation causes the normal shock wave to move out (to a larger area) of the passage, the upstream Mach number increases. The increased Mach number results in a stronger shock wave with a larger decrease in stagnation pressure. The decrease in stagnation pressure reduces the mass flow rate that can pass through a given area. The decrease in mass flow rate between the shock wave and the diffuser throat pushes the shock wave farther out in the divergence since mass tends to accumulate between the shock wave and the diffuser throat. Thus a shock wave in a converging passage exhibits the attribute of instability since movement either into or out of the passage tends to produce movement away from the original position. The normal shock wave will thus be either swallowed or disgorged, depending on the back pressure into which the diffuser exhausts.

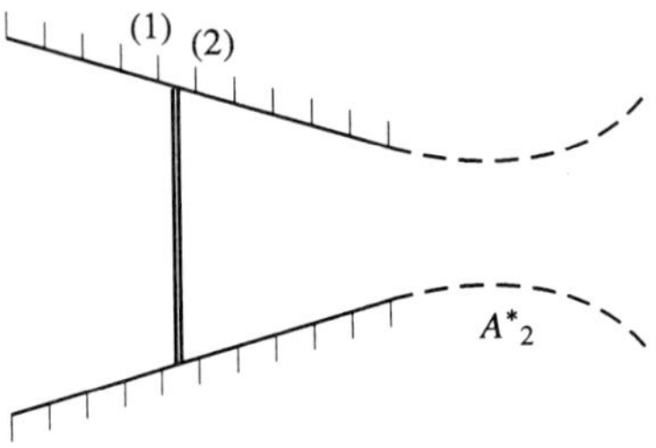

Figure 6-26 Shock in Converging Passage

Example 6-8

A continuous-duty supersonic wind tunnel is to be designed. The test section specifications are a Mach number of 2, a static pressure of 40 kPa, a static temperature of 250 K, and an area of 0.5 m^2. Determine the stagnation conditions required upstream of the test section, the mass flow rate required, and the throat area required. During the startup process, what is the maximum stagnation pressure loss across the shock system? If a constant-area diffuser is used, what is the minimum diffuser throat area? Sketch the steady-state operating conditions for the constant-area diffuser. If a variable-area diffuser is used, explain the startup sequence required to achieve shock-free operation.

Solution. COMPQ will be used. A sketch of the wind tunnel with the constant-area diffuser is given in Figure 6-27a. For a test section Mach number of 2.0, the following values are required:

$$A_T/A_1^* = 1.6875$$

$$P_T/P_{o_1} = 0.1278$$

$$T_T/T_{o_1} = 0.5556.$$

The throat area and stagnation conditions required are then

$$A_1^* = A_T/1.6875 = 0.2963 \text{ m}^2$$

$$P_{o_1} = P_T/0.1278 = 313.0 \text{ kPa}$$

$$T_{o_1} = T_T/0.5556 = 450 \text{ K}$$

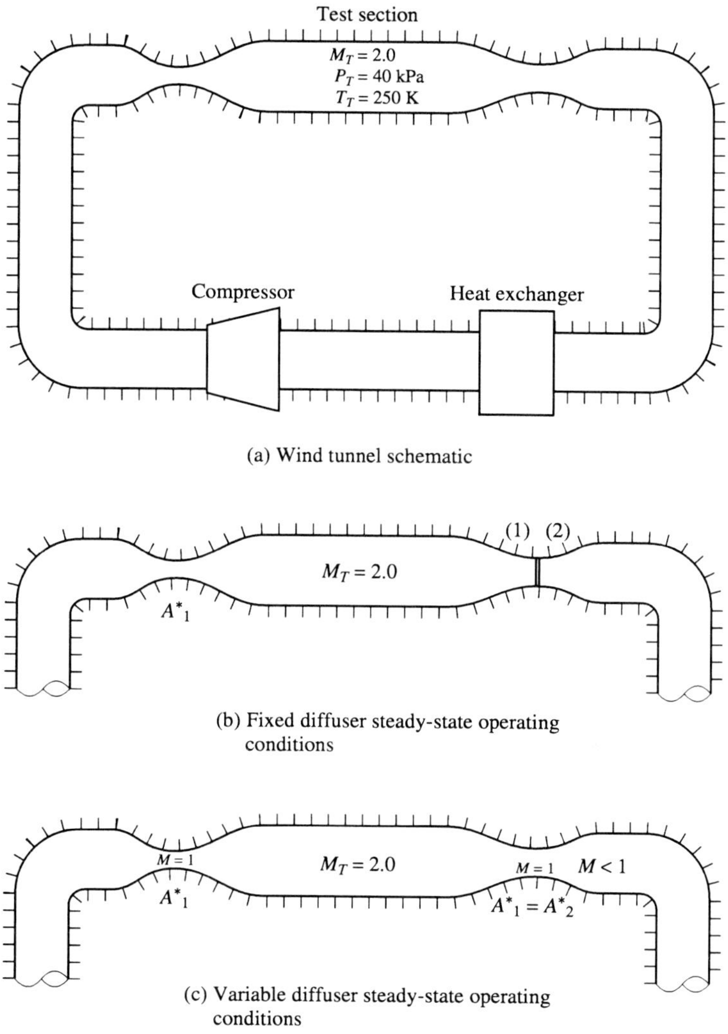

Figure 6-27 Example Problem 6-8

The mass flow rate for a choked throat, using Eq. 6-12, is

$$\begin{aligned} \dot{m} &= \left(\frac{2}{\gamma + 1}\right)^{(\gamma+1)/2(\gamma-1)} \sqrt{\gamma/R}\, \frac{A_1^* P_{o_1}}{\sqrt{T_{o_1}}} \\ &= \left(\frac{2}{1.4 + 1}\right)^{(1.4+1)/2(1.4-1)} \sqrt{1.4/287}\, \frac{0.2963(313.0)}{\sqrt{450}} \\ &= 0.1767 \text{ kg/s.} \end{aligned}$$

The maximum stagnation pressure loss during startup will occur when the normal shock wave is positioned in the test section and will occur at the test section design Mach number of 2.0. For an upstream Mach number of 2.0, option **NS** of COMPQ yields

$$\frac{P_{o2}}{P_{o_1}} = 0.7209.$$

The maximum loss in stagnation pressure is

$$\begin{aligned}\Delta P_o &= P_{o_2} - P_{o_1} = P_{o_1}(P_{o_2}/P_{o_1} - 1)\\ &= P_{o_1}(0.7209 - 1) = -0.2791 P_{o_1}\\ &= -86.36 \text{ kPa}.\end{aligned}$$

The constant-area diffuser requires a throat large enough to pass the mass flow rate with the maximum stagnation pressure loss; recall Eq. 6-14:

$$\frac{A_2^*}{A_1^*} = \frac{P_{o1}}{P_{o_2}}.$$

The minimum diffuser throat area then becomes

$$A_2^* = \frac{A_1^*}{P_{o_2}/P_{o_1}} = \frac{0.2963}{0.7209} = 0.4110 \text{ m}^2.$$

For a constant-area diffuser, steady-state operating conditions are illustrated in Figure 6-27b. The shock appearing in the diffuser throat occurs at an area ratio of

$$\frac{A}{A_1^*} = \frac{0.4110}{0.2963} = 1.3874,$$

so that the upstream Mach number is

$$M_1 = 1.7509,$$

and the stagnation pressure loss becomes

$$\frac{P_{o2}}{P_{o_1}} = 0.8342.$$

If a variable throat diffuser is used, the throat must initially possess at least an area of 0.4110 m^2 in order to swallow the normal shock wave. The power to the compressor would be increased until the shock wave was swallowed. Once the shock wave is swallowed, the variable-area throat would be reduced to the area of the first throat, 0.2963 m^2. Simultaneously, the back pressure would be increased to place the shock at the diffuser throat. Ideally the limit for a variable throat diffuser would be shock-free flow as illustrated in Figure 6-27c.

The solution to this example problem considered inviscid flow. In reality, viscous effects would add to the output requirements of the compressor, but the basic mode of operation would be as described.

An example of an older wind tunnel with several additional features is the NASA/Langley Research Center Continuous-Flow Hypersonic Tunnel. A NASA illustration (Baals and Corliss, 1981) representing the dominant tunnel features in schematic form is reproduced in Figure 6-28. Planning for this facility was initiated in 1958, and since the early 1960s the wind tunnel has been heavily used for heat transfer and aerodynamic studies. The Continuous-Flow Hypersonic Tunnel is connected to a pressurized-air storage system, a vacuum tank, a heater, a bank of six compressors, and a heat exchanger. Normal operating procedure is to start the tunnel with the pressurized air flowing into the vacuum sphere, and then to divert the flow through the compressors for continuous operation. The heater raises the stagnation enthalpy to the desired level for the test conditions. The diffuser is used to reduce the loss in shock wave stagnation pressure by placing the shock wave terminating the supersonic flow at a low Mach number. Depending on the mass flow rate and the hypersonic Mach number desired in the test section, all six of the compressors may not be required, and so some can be bypassed. In spite of the added complexity of this wind tunnel, the principles discussed earlier are sufficient to understand the operating modes of the wind tunnel.

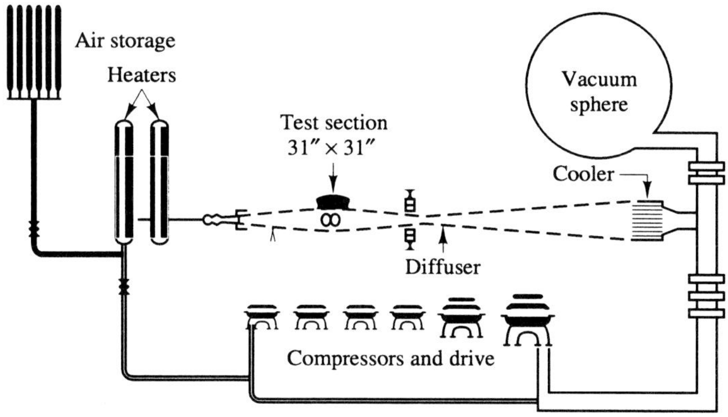

Figure 6-28 NASA/Langley Continuous-Flow Hypersonic Tunnel

The use of heaters in hypersonic wind tunnels warrants additional consideration. The test section temperature for a constant stagnation temperature wind tunnel can be computed by using the ratio of the static-to-stagnation temperature at the test section Mach number. For example, for a Mach six test section with air as the working fluid, $T/T_0 = 0.12195$. This corresponds to a test section temperature of 37 K for air initially at 300 K. But the liquefaction temperature of air starts at about 90 K for oxygen and at a slightly lower temperature for nitrogen. Thus, to avoid problems associated with liquefaction, hypersonic wind tunnels are normally heated.

Thus far in our discussion of supersonic wind tunnels we have concentrated on steady-state and startup features of the flow conditions. An equally important consideration for continuous-flow supersonic wind tunnels is the power required for both startup and continuous operations. Steady-state operation of a typical supersonic wind tunnel, as sketched in Figure 6-29, requires that the compressor make up all the pressure losses from the compressor exit through the tunnel circuit to the compressor inlet. The heat exchanger is necessary because compression of a gas

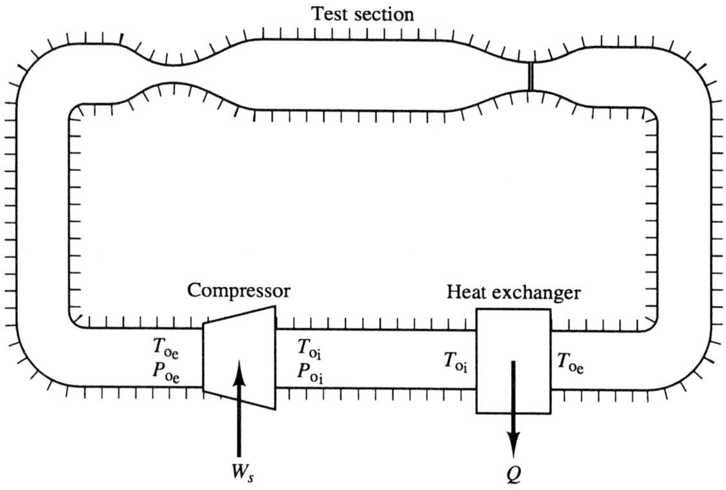

Figure 6-29 Nomenclature for Wind Tunnel Power Required Calculation

always results in an increase in temperature of the gas, and steady-state operation of the systems requires that the power input by the compressor be equal to the heat rejection rate of the heat exchanger. The pressure losses result primarily from skin friction due to the wind tunnel wall surfaces, stagnation pressure losses across shock waves, and pressure drop across the heat exchanger. If the compressor were isentropic, the limiting case of the best possible process, then the discharge temperature T_{o_e}, would be given by the isentropic relation

$$\frac{T_{o_e}}{T_{o_i}} = \left(\frac{P_{o_e}}{P_{o_i}}\right)^{(\gamma-1)/\gamma}, \tag{6-39a}$$

with the power input to the air

$$W_s = \dot{m}(h_{o_e} - h_{o_i}) = \dot{m}\, c_p(T_{o_e} - T_{o_i}). \tag{6-39b}$$

Equations 6-39 can be manipulated to yield

$$\begin{aligned} W_s &= \dot{m}\, c_p\, T_{o_i}(T_{o_e}/T_{o_i} - 1) \\ &= \dot{m}\, c_p\, T_{o_i}\left[\left(\frac{P_{o_e}}{P_{o_i}}\right)^{(\gamma-1)/\gamma} - 1\right]. \end{aligned} \tag{6-40}$$

The rate of heat rejection by the heat exchanger is

$$Q = \dot{m}(h_{o_e} - h_{o_i}) = \dot{m}\, c_p(T_{o_e} - T_{o_i}). \tag{6-41}$$

Equation 6-40 addresses the power delivered by an isentropic compressor in addition to establishing that the heat exchanger should be placed before the compressor since the power required from a compressor is directly proportional to the inlet temperature of the fluid. Thus, cooling the air prior to compressing it requires less power than compressing the air prior to cooling it. Moreover, since less power is delivered to the fluid, less heat rejection is required from the heat exchanger.

The discussion in the preceding paragraphs was centered about the assumption of an isentropic compressor. Compressors that are not isentropic can be easily incorporated into the analysis by using an efficiency to relate the behavior of the actual compressor when compared with that of a reference. The most commonly used reference condition for compressor efficiency is the isentropic, and the efficiency that results is called the *isentropic compression efficiency*. Consider, as in Figure 6-30, the *h–s* diagram for the development of the isentropic efficiency concept. The initial stagnation state of the gas entering the compressor is labeled as "i," and the actual discharge state of the compressor is labeled as "e." The reference process is the isentropic process required to increase the pressure to the actual exit pressure. On the *h–s* diagram, this reference process is represented as a vertical line with the exit state indicated as "s." Because of the reversibilities of the compression process, the actual discharge stagnation enthalpy is higher than that of the isentropic process at the same discharge pressure. Remember, the actual discharge state is at "e"; state "s" is just the isentropic reference. The isentropic compression efficiency is defined as the ratio of the isentropic change in stagnation enthalpy to the actual change in stagnation enthalpy to affect the increase in pressure or

$$\eta_c = \frac{h_{o_s} - h_{o_i}}{h_{o_e} - h_{o_i}} = \frac{T_{o_s} - T_{o_i}}{T_{o_e} - T_{o_i}} = \frac{T_{o_i}(T_{o_s}/T_{o_i} - 1)}{T_{o_e} - T_{o_i}}. \tag{6-42}$$

Since the process "i" in "s" is isentropic,

$$\frac{T_{o_s}}{T_{o_i}} = \left(\frac{P_{o_e}}{P_{o_i}}\right)^{(\gamma-1)/\gamma}, \tag{6-43}$$

and following the development sequence of Eqs. 6-39 and 6-40, the isentropic efficiency can be expressed as

$$\eta_c = \frac{T_{o_i}[(P_{o_e}/P_{o_i})^{(\gamma-1)/\gamma} - 1]}{T_{o_e} - T_{o_i}}. \tag{6-44}$$

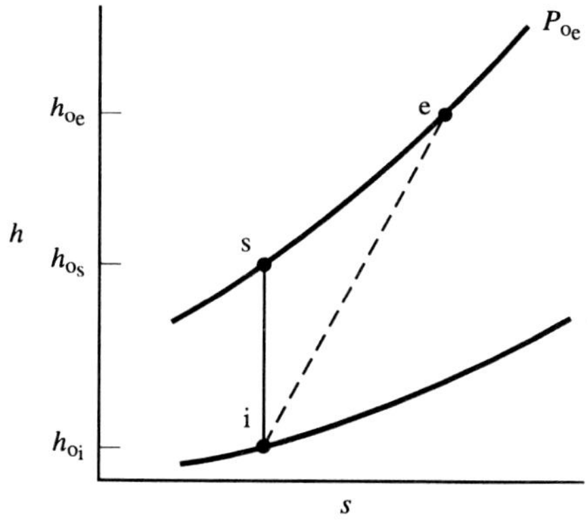

Figure 6-30 *h–s* Diagram for Isentropic Compression Efficiency

The power delivered to the fluid then takes the form

$$W_s = \dot{m}\, c_p(T_{o_e} - T_{o_i}) = \frac{\dot{m}\, c_p\, T_{o_i}}{\eta_c}\left[\left(\frac{P_{o_e}}{P_{o_i}}\right)^{(\gamma-1)/\gamma} - 1\right]. \tag{6-45}$$

When cast as in Eq. 6-45, the reversibilities of the compressor result in an increase in the power delivered to the fluid over that required by the isentropic process. Because more power is delivered to the fluid, more heat must be rejected by the heat exchanger; then

$$Q = \dot{m}\, c_p(T_{o_e} - T_{o_i}). \tag{6-46}$$

This appears identical to Eq. 6-41, but in Eq. 6-41 the exit temperature was based on an isentropic compression process, so $T_{o_e} = T_{o_s}$ for that expression. With the flow adiabatic except for the heat exchanger, the stagnation temperature is T_{o_e} everywhere except for the duct connecting the heat exchanger and the compressor. The heat exchanger heat rejection rate is taken as the power delivered to the fluid by the compressor; thus

$$\dot{m}\, c_p(T_{o_e} - T_{o_i}) = \frac{\dot{m}\, c_p\, T_{o_i}}{\eta_c}\left[\left(\frac{P_{o_e}}{P_{o_i}}\right)^{(\gamma-1)/\gamma} - 1\right], \tag{6-47}$$

which can be solved for the compressor inlet temperature (which is also the heat exchanger outlet temperature), T_{o_i}:

$$T_{o_i} = \frac{T_{o_e}}{1 + [(P_{o_e}/P_{o_i})^{(\gamma-1)/\gamma} - 1]/\eta_c}. \tag{6-48}$$

The last few equations relate to the power delivered to the fluid, not to the power requirements of the compressor. The overall efficiency η is defined as

$$\eta = \frac{W_s}{\text{Power}} \tag{6-49}$$

and is used to relate the compressor power input to the compressor power output (power delivered to the fluid). Consider the following example.

Example 6-9

Estimate the power required to start and run the wind tunnel of Example 6-8 if the constant-area diffuser is used. What would be the power if no diffuser were present? The isentropic compression efficiency is 83 percent.

Solution. The results of Example 6-8 provide most of the information needed to compute the power requirements. For the purposes of this problem we shall ignore all losses except those associated with shock waves. The maximum power delivery occurs during the startup process when the shock wave is in the test section. For the wind tunnel under consideration in this problem, the upstream Mach number is 2 and the stagnation pressure loss is 86.36 kPa. The compressor discharge pressure is 313 kPa, and the compressor inlet, or suction, pressure is 225.64 kPa (313 kPa − 86.36 kPa). If the heat exchanger is placed upstream of the compressor, then the compressor inlet temperature can be computed from Eq. 6-48 as

$$T_{o_i} = \frac{450}{1 + \dfrac{(313/225.69)^{.2857} - 1}{.83}} = 402.5 \text{ K}.$$

The power delivered to the fluid using Eq. 6-45 then has the value

$$W_s = 0.1767 \text{ kg/s } 1003.5 \frac{\text{J}}{\text{kg K}} 402.5 \text{ K} \frac{1}{.83}\left[\left(\frac{313}{225.64}\right)^{.2857} - 1\right]$$

$$= 8429 \text{ W} = 8.429 \text{ kW}.$$

Steady-state operation with a fixed diffuser is achieved by placing the shock wave near the diffuser throat. For the wind tunnel of Example 6-8 the upstream Mach number for this condition is 1.7509 with the corresponding loss in stagnation pressure of 49.08 kPa. The compressor inlet temperature is then

$$T_{o_i} = \frac{450}{1 + [(313/263.92)^{.2857} - 1]/.83} = 424.5 \text{ K},$$

and the power delivered to the fluid is calculated by

$$W_s = (0.1767)(1003.5)\frac{424.5}{.83}\left[\left(\frac{313}{263.92}\right)^{.2857} - 1\right] = 4.525 \text{ kW}.$$

For steady-state operations, the diffuser permits the power delivered to the fluid to be reduced by 46 percent. Had the heat exchanger been placed downstream of the compressor, the startup power delivered would be 9.423 kW and the steady-state power delivered would be 4.8 kW.

6.5 SUBSONIC AND SUPERSONIC INLETS

Two of the prime requirements for proper, efficient operation of an air-breathing engine are that the ambient air delivered to it be at a subsonic Mach number, usually not exceeding 0.4, and that the flow pattern at the entrance to the engine be fairly uniform. Since most gas turbine (turbojet, turbofan, or turboprop) or ramjet-powered aircraft fly at Mach numbers considerably greater than 0.4, the flow must usually be decelerated or diffused. Diffusion results in an adverse pressure gradient, is generally difficult to accomplish without some loss in stagnation pressure, and can lead to separated flow. The inlet is the device that must meet the Mach number and uniform flow requirements with the minimum loss of stagnation pressure. For a supersonic vehicle, shock waves are almost invariably encountered in the diffusion process. The design of inlets and their integration with engines and airframes are of vital importance if high-performance, efficient flight vehicles are to result. Inlet design and inlet/engine/airframe integration are well-developed subjects with extensive literatures. In this section we shall investigate, in the one-dimensional sense, fundamental concepts and ideas concerning subsonic and supersonic inlets.

Inlets are broadly divided into subsonic and supersonic, depending on the flight Mach number of the vehicle. For either subsonic or supersonic inlets, the freestream air entering the inlet is called the *captured air*. Far upstream of an inlet the area defined by the streamlines enclosing the captured air is called the *capture area*, A_o. The capture area definition is illustrated in Figure 6-31. If the inlet entrance area, A_i, is greater than the capture area, then $\rho_\infty V_\infty(A_o - A_i)$ represents flow that is spilled over the inlet nacelle. This spilled-over flow is decelerated upstream of the entrance to the inlet and then accelerated as it passes over the leading edge of the inlet cowl. Following this region of acceleration, the spilled-over flow is then decelerated as it traverses the nacelle body. Excessive deceleration upstream of the inlet is often associated with significant acceleration over the inlet leading edge followed by sufficient deceleration to induce flow separation on the nacelle cowl. This separation can result in excessive nacelle drag. Spillover can also lead to problems in inlets operating at high subsonic Mach numbers since cowl leading-edge acceleration can result in locally supersonic flow. Deceleration of such a local supersonic flow is almost always associated with shock waves that result in flow separation on the nacelle body and excessive nacelle drag. Inlets for supersonic flight Mach numbers are invariably involved with shock waves and the attendant loss of stagnation pressure. Spillover in supersonic inlets can also be a problem that results in increased inlet nacelle drag.

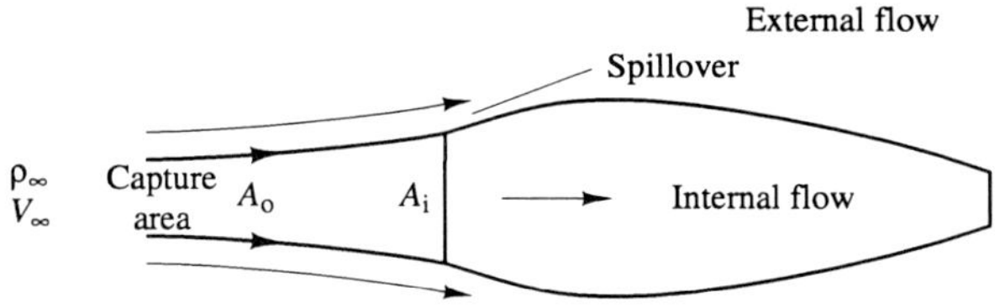

Figure 6-31 Inlet Nomenclature

In a subsonic inlet the velocity of the air entering the inlet—the captured air—is subsonic throughout its flow path. Subsonic inlets can be subdivided into external compression and internal compression inlets. Schematic illustrations of the two basic types of subsonic inlets and their capture areas are presented in Figure 6-32. The subsonic external compression inlet, Figure 6-32a, is composed of a constant-area duct enclosed by a contoured cowl. Once the captured air enters the constant-area duct, its velocity is constant. Any diffusion that takes place must occur before the air enters the duct; hence, the name external compression.

In an external compression inlet, the diffusion process is isentropic, but the air deflected by the cowl—the spillover—is at a higher Mach number than the freestream, so an external drag is produced. Local Mach numbers on the cowl of an external compression inlet should be less than Mach 1 to prevent the formation of shock waves on the cowl. Generally, external compression subsonic inlets are not attractive for high subsonic Mach number applications because of the possibility of local Mach numbers greater than 1.

Subsonic internal compression inlets, such as illustrated in Figure 6-32b, accomplish their diffusion internally in a diverging area duct. As no diffusion occurs

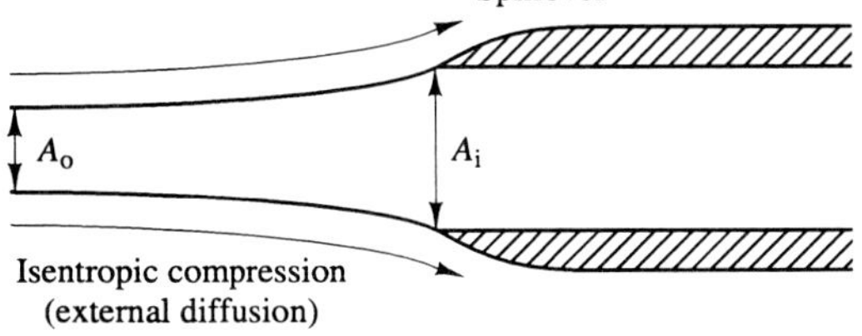

(a) External compression

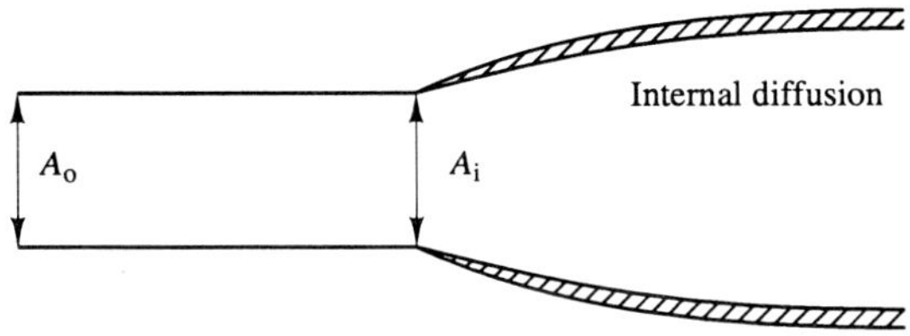

(b) Internal compression

Figure 6-32 Subsonic Inlets

outside the inlet, the capture area is equal to the inlet entrance area. Zucrow and Hoffman (1976) suggests that the divergence angle for cylindrical internal compression inlets be no more than 5 to 7 degrees if separation is to be avoided. Such low-divergence angles yield inlets that tend to be too heavy and too long for many applications. Design of internal compression inlets thus becomes a compromise between length, weight, and stagnation pressure loss. Additional details on subsonic inlet design are provided by Seddon and Goldsmith (1985) and by Hill and Peterson (1992).

Supersonic inlets, because they invariably involve shock waves, are generally more difficult to design or operate than subsonic inlets. Supersonic inlets can be broadly divided into normal shock, oblique shock, and converging–diverging inlets. Normal shock inlets are similar in appearance to internal compression subsonic inlets. Inlets involving oblique shock waves are classified as external compression or mixed compression, depending on where the diffusion of supersonic flow is accomplished. Converging–diverging configurations can also be used as supersonic inlets.

The normal shock inlet is sometimes called a *pitot inlet* because of its resemblance to a pitot tube. The normal shock inlet is designed to operate with a normal shock attached to the lip of the inlet entrance. This operating condition is illustrated in Figure 6-33a. If the back pressure is increased, the normal shock moves out in front of the inlet, and the conditions of Figure 6-33b result. A decrease in back pressure causes the shock to be pulled into the inlet, where a weak expansion or compression wave system is generated ahead of the normal shock. The supersonic region is eventually terminated by a relatively strong normal shock wave. For the sake of clarity, only the normal shock is shown in Figure 6-33c. Because of excessive stagnation pressure losses with increasing flight Mach numbers, the normal shock inlet is not attractive for use above Mach numbers greater than 1.8.

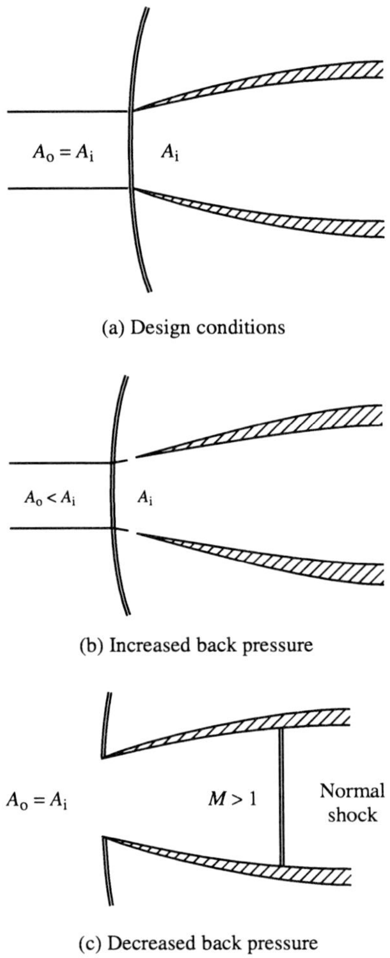

Figure 6-33 Normal Shock or Pitot Inlet Flow Configuration

Oblique shock inlet nomenclature is presented in Figure 6-34, which schematically illustrates a mixed-compression inlet. The inlet is composed of a cowl and a center body. The configuration is called mixed because part of the compression takes place external to the inlet and part takes place within the inlet. The flow field consists of several oblique shock waves and a normal shock wave that finally terminates the supersonic region. Minimum stagnation pressure loss through the normal shock will occur if the shock is positioned at the throat of the inlet. The desired operating condition is to place the first oblique shock wave so that it impinges on the lip of the cowl. With the inlet operating as illustrated in Figure 6-34, the capture area is maximized and the spillover is eliminated. More details on supersonic inlets will be considered in the following.

External compression supersonic inlets will be examined first. Such an inlet is represented schematically in Figure 6-35a. In this arrangement, the throat is located at the cowl lip. The flow field shown in the figure, oblique shock impinging on the

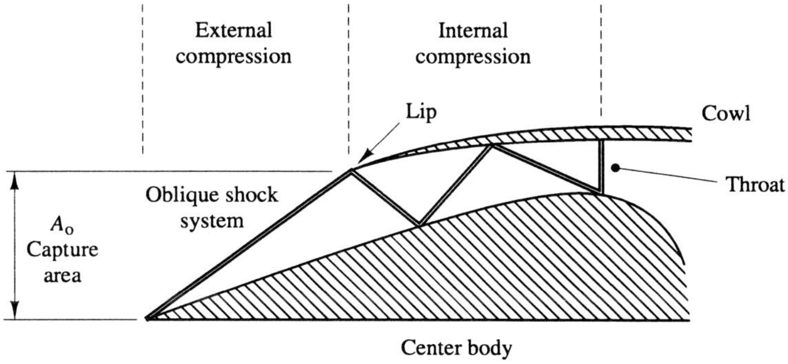

Figure 6-34 Supersonic Inlet Nomenclature

lip and the normal shock at the throat, represents the desired conditions at the design Mach number. Diffusion within the inlet then takes place subsonically. This flow field arrangement, with the normal shock at the inlet throat, is called *critical inlet operation*.

The wedge (or cone) angle of the center body and the Mach number determine the oblique shock wave angle. The location of the normal shock is fixed by the oblique shock wave angle and the back pressure the engine imposes on the inlet exit. If the back pressure is increased from the critical operating condition, the normal

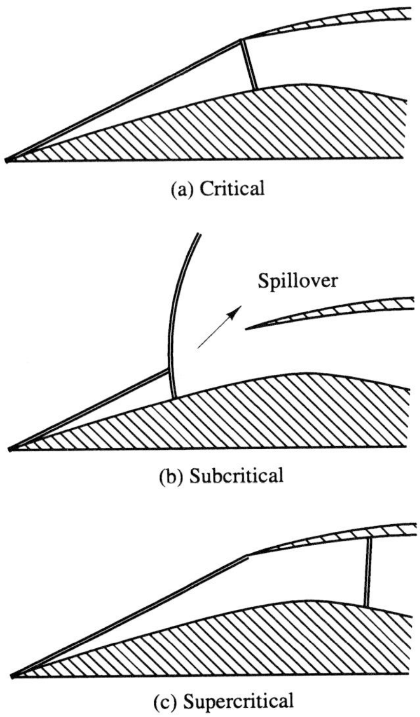

Figure 6-35 External Compression Inlet Operating Conditions

shock wave moves off the lip of the cowl and onto the exterior portion of the center body. This flow condition, which is illustrated in Figure 6-35b, is called *subcritical*. Spillover generally occurs, and the drag for subcritical operation can be large. Especially at low flow rates, the shock wave system on the external portion of the center body can exhibit marked instabilities which lead to oscillations in the shock locations. This is called *inlet buzz* and is a problem that often occurs in subcritical inlet operation. Subcritical operation is generally avoided.

If the back pressure is decreased, then the normal shock is pulled into the diverging passage of the inlet. Inlet operation is termed supercritical when the normal shock is positioned downstream of the throat in the diverging passage. Supercritical operation is illustrated in Figure 6-35c. Most external compression inlets are operated in a slightly supercritical mode since this prevents freestream fluctuations from causing subcritical flow.

If the external compression inlet geometry is fixed, then operation at off-design flight Mach numbers must be considered. If the Mach number is less than the design Mach number, then the oblique wave shock angle is increased, and the oblique shock no longer impinges on the cowl lip. A normal shock usually appears upstream of the cowl lip, and the flow field becomes similar to that in Figure 6-35b. The oblique shock wave does not intersect the inlet cowl; the capture area is reduced, and spillage between the oblique/normal shock system and the inlet cowl occurs. If the Mach number is larger than the design Mach number, the shock wave angle is reduced and the shock wave impinges on the inside of the inlet. No spillage occurs in this flow configuration, but an additional shock structure, which may induce flow separation, appears inside the inlet.

One alternative to off-design operation of a fixed external compression inlet is to provide for translation of the center body. If the center body is moved forward or aft as the flight Mach number is increased or decreased, the oblique shock wave can be positioned to impinge on the lip of the inlet. Figure 6-36 illustrates how movement of the center body can be used to maintain the desired inlet flow conditions for off-design Mach numbers. Although portrayed as located near the lip of the inlet, the location of the normal shock waves would be determined in part by the back pressure imposed on the inlet.

Some other ideas that are useful for external compression inlets were introduced in Chapter 3, such as the use of multiple oblique shock waves to reduce the stagnation pressure loss for a given deflection angle. An external compression inlet that utilizes multiple oblique shock waves followed by a normal shock wave is indicated in Figure 6-37. A single oblique shock inlet is adequate for flight Mach numbers less than about 2.5, but for Mach numbers greater than that, multiple external oblique shock waves are attractive for preventing excessive inlet stagnation pressure loss. For a given center body deflection angle, the multiple oblique shock configuration has a lower stagnation pressure loss than the single oblique shock design.

An isentropic supersonic compression can be obtained by providing a center body with a contour that gradually turns the flow. The isentropic spike inlet is schematically illustrated in Figure 6-38. Isentropic compression using a Prandtl-

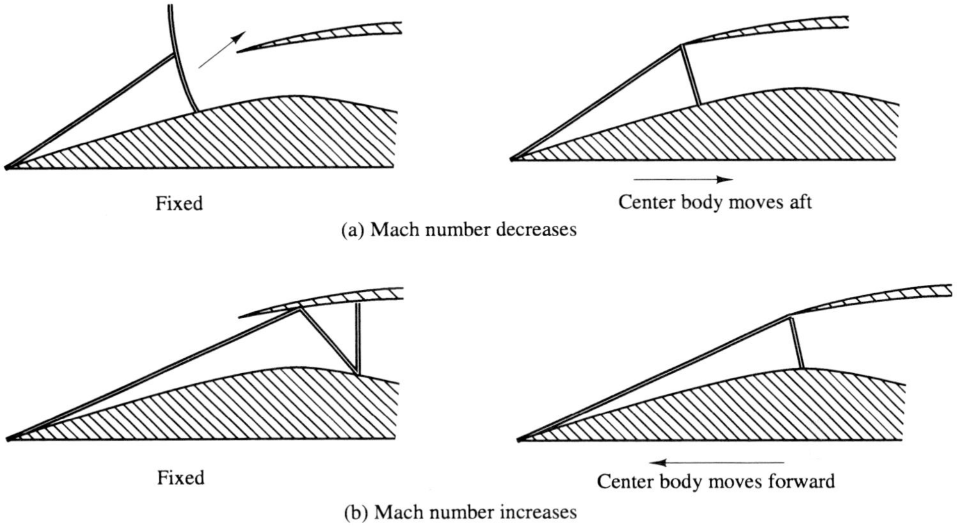

Figure 6-36 Center Body Translation for Off-Design Mach Numbers

Meyer compression fan was discussed in Chapter 3. Selected Mach waves from the Prandtl-Meyer compression process are shown in the figure. Depending on the local radii of curvature distribution of the center body, the Mach waves may converge at the lip of the inlet, as in Figure 6-38, or may coalesce and form an oblique shock wave near the lip of the cowl. The situation at the design Mach number of the inlet of Figure 6-38 permits an isentropic supersonic compression to occur as the flow traverses the Prandtl-Meyer compression region. Note, however, that regardless of the inlet compression process, a shock wave system is likely to be present on the external surface of the inlet cowl. Although the isentropic inlet is attractive at the design Mach number, off-design performance is poor and the concept has had little use.

In spite of the favorable performances multiple external oblique shock inlets (above Mach 2.5) and isentropic compression inlets offer, few existing aircraft utilize these configurations. A variety of reasons exist for their nonuse. These inlet configurations possess poor off-design performances; at flight Mach numbers greater than

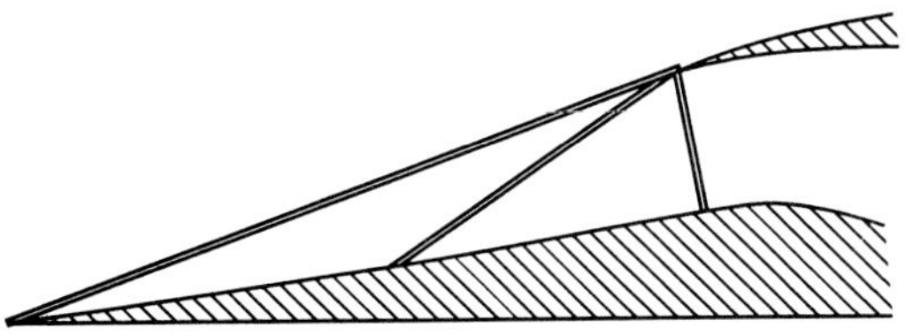

Figure 6-37 Multiple Shock External Compression Inlet

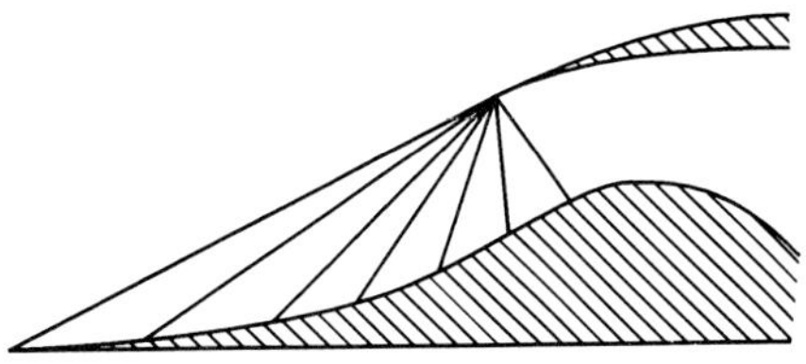

Figure 6-38 Isentropic External Compression Inlet

or less than the design Mach number, the shock/Mach waves patterns degrade performance. Variable geometry inlets can be used to alleviate the performance degradation these configurations suffer, but the increase in complexity is significant.

Inlets designed for efficient operation at high supersonic Mach numbers often diffuse the supersonic flow inside the inlet with a series of oblique shock waves. Such a mixed compression device was shown in Figure 6-34. Much of the discussion about external supersonic inlets is appropriate for the external compression portions of mixed compression supersonic inlets. High-performance aircraft can use mixed compression inlet configurations that contain multiple external oblique shock waves followed by multiple internal oblique shock waves with a final normal shock terminating the supersonic region.

A good example of a high supersonic inlet is that of the XB-70 research aircraft. Originally designed in the mid-1960s as a high-altitude Mach 3 strategic bomber, only two XB-70 aircraft were completed, and those were used as research vehicles. A three-view drawing (Gallagher, 1970) of the aircraft is presented in Figure 6-39a, where the dimensions are indicated in feet with the metric equivalents in parentheses. Considering the technology of its day, a design cruise speed of Mach 3 at 70,000 to 80,000 ft represented an awesome task. A variable geometry inlet system was necessary since air had to be delivered to the XB-70's six engines at altitudes from sea level to 80,000 ft and at speeds ranging from takeoff to cruise at Mach 3. The overall inlet arrangement, with salient features indicated, is shown schematically in Figure 6-39b as abstracted from Martin (1970). The ramp positions are illustrated for takeoff and cruise, and the region of boundary-layer bleed is indicated by the cross-hatch. Although we have examined inlets from an inviscid point of view, viscous effects can be dramatic and can result in significant performance degradation. Such effects can be alleviated and/or partially controlled by boundary-layer bleed, which removes and dumps overboard air contained within the boundary layer. Bypass doors are also indicated in the figure. Inlet systems of virtually all supersonic aircraft contain bypass doors that can be used to dump overboard air in the inlet not needed by the engine. The bypass feature is needed if the engine mass flow rate is to be matched to the inlet size, aircraft speed, fuel flow rate, and freestream density.

The variable geometry feature permits efficient subsonic, low supersonic, and high supersonic inlet operation. Figure 6-39c illustrates the ramp positions for subsonic flight. The flow is completely subsonic in this mode. At low supersonic speeds, the ramp positions, Figure 6-39d, are such that deceleration to subsonic speed is accomplished by a series of oblique shock waves followed by a normal shock, all external to the inlet. Hence, in the low supersonic regime, the inlet operates as an external compression inlet. As Figure 6-39e shows, at high supersonic Mach numbers the inlet operates in the mixed compression mode. Deceleration to subsonic flow is accomplished by a system of oblique shock waves, both outside and inside the inlet, followed by a normal shock wave standing downstream of the inlet throat. Lower-than-normal performance is schematically illustrated by the normal shock wave occurring at a larger area ratio location. The decreased performance of the inlet results from the greater loss of stagnation pressure because of the higher-than-normal upstream Mach number of the shock wave. Various pressure ratios,

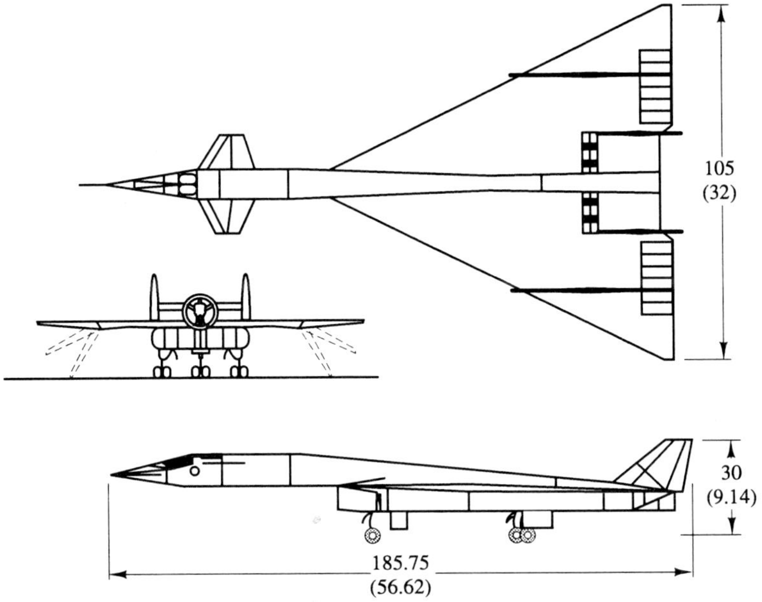

(a) XB-70 Aircraft three-view *Gallagher* (*1970*)

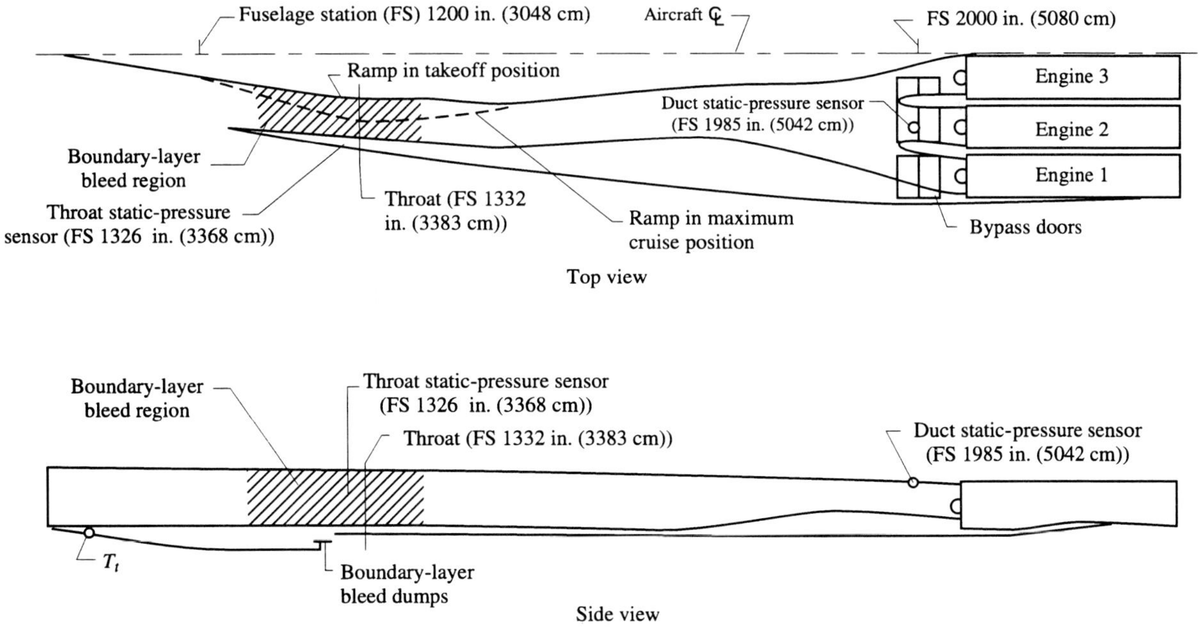

(b) Inlet arrangement *Martin* (*1970*)

Figure 6-39 XB-70 Inlet Geometry and Operation

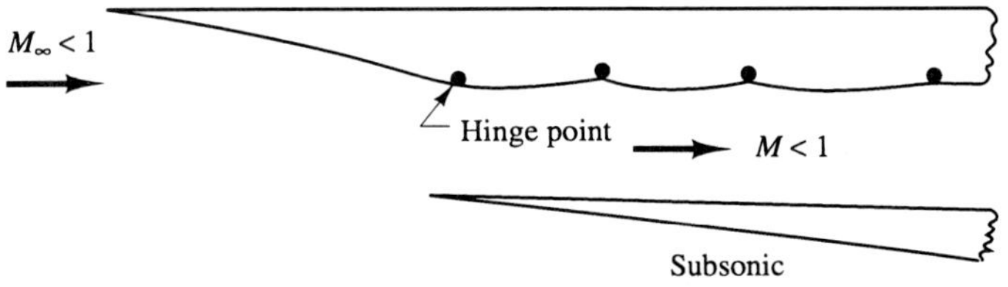

(c) Subsonic operation *Martin* (*1970*)

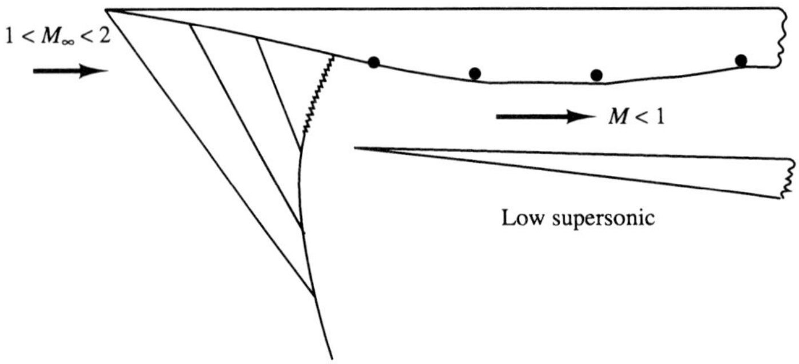

(d) Low supersonic operation *Martin* (*1970*)

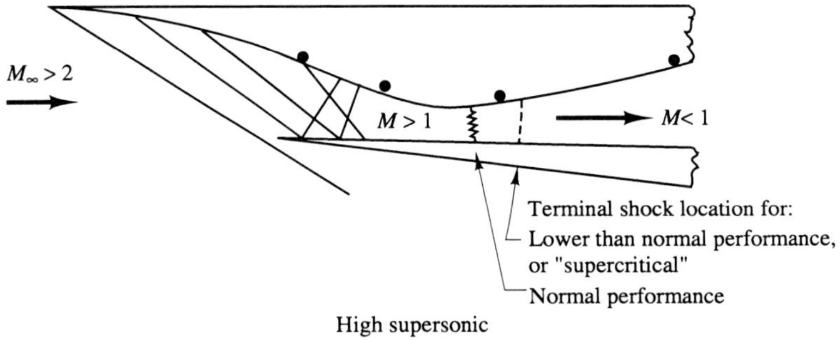

(e) High supersonic operation *Martin* (*1970*)

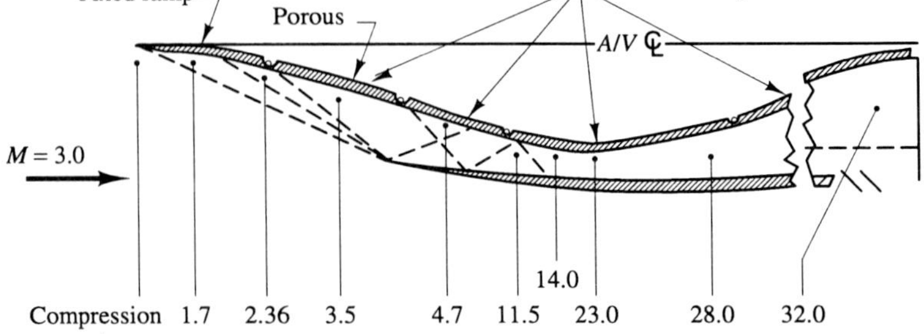

(f) Compression ratio distribution

Figure 6-39 (continued)

expressed as the compression ratio, P_{local}/P_∞, for design operation at Mach 3 are given in Figure 6-39f. After deceleration is complete, the compression ratio immediately upstream of the engine is 32. This value is typical for high supersonic aircraft; another example is the MiG 25, a Mach 2.8 fighter aircraft of the former USSR, which possesses an inlet system compression ratio of near 30. Thus, at high supersonic speeds, the inlet provides significant compression for the engine system.

High compression ratios are always associated with high supersonic inlets since low compression ratios at high supersonic speeds can occur only if the inlet is not efficient. The range of feasible compression ratios for supersonic inlets can be established by considering an isentropic compression inlet as the most efficient and a pitot inlet as the least efficient. The compression ratio across a pitot inlet can be estimated by considering passage through a normal shock wave at the freestream Mach number followed by an isentropic compression to a Mach number of 0.4, a reasonable goal for a typical inlet. The compression ratio across an isentropic inlet assumes an isentropic compression from the freestream Mach number to 0.4. The range of feasible inlet compression ratios for a Mach 3 aircraft is from near 10 for the pitot inlet to near 33 for the isentropic inlet. The XB-70's value of 32 (see Figure 6-39f) compares quite favorably and is indicative of the overall efficiency of the inlet.

Our study of inlets has been based on one-dimensional considerations and has neglected viscous effects. The primary purpose of an inlet is to decelerate the air coming to the compressor section of a jet engine. Deceleration always results in an adverse pressure gradient since the static pressure increases as the velocity decreases. Flows with adverse pressure gradients, where $dP/dx > 0$, are more difficult to handle than are accelerating flows since adverse pressure gradients greatly exacerbate viscous effects and generally result in increased boundary-layer thickness and susceptibility to separation. Boundary-layer bleed, as mentioned earlier, is one way of alleviating viscous effects. Nonetheless, viscous effects and two- and three-dimensional effects are important in inlet operation. This short presentation on inlets has been merely an introduction to the subject, more accurate in the qualitative sense than in the quantitative. Seddon and Goldsmith (1985) presents a more comprehensive and quantitative treatment.

Example 6-10

Compare the stagnation pressure loss for a flight Mach number of 3.5 for a pitot (normal shock) inlet and for an external compression inlet with a total turning angle of 20° using one, two, or four oblique shock waves followed by a normal shock.

Solution. The inlet configurations are shown in Figure 6-40. Options **NS**, **OS**, and **PM** of COMPQ were used to construct Tables 6-5a and b, which summarize solution parameters. Table 6-5a presents the stagnation pressure ratios after each oblique and each normal shock and the overall stagnation pressure ratio for oblique shock(s)/normal shock system. The pitot inlet stagnation pressure ratio of 0.2129 is unacceptable and points out the inefficiency of such an inlet at high supersonic Mach numbers. The overall stagnation pressure ratio increases as the number of oblique shock waves increases. For comparison, the case of an isentropic compression through 20° followed by a normal shock was also computed. This is the limiting value expected as the number of oblique

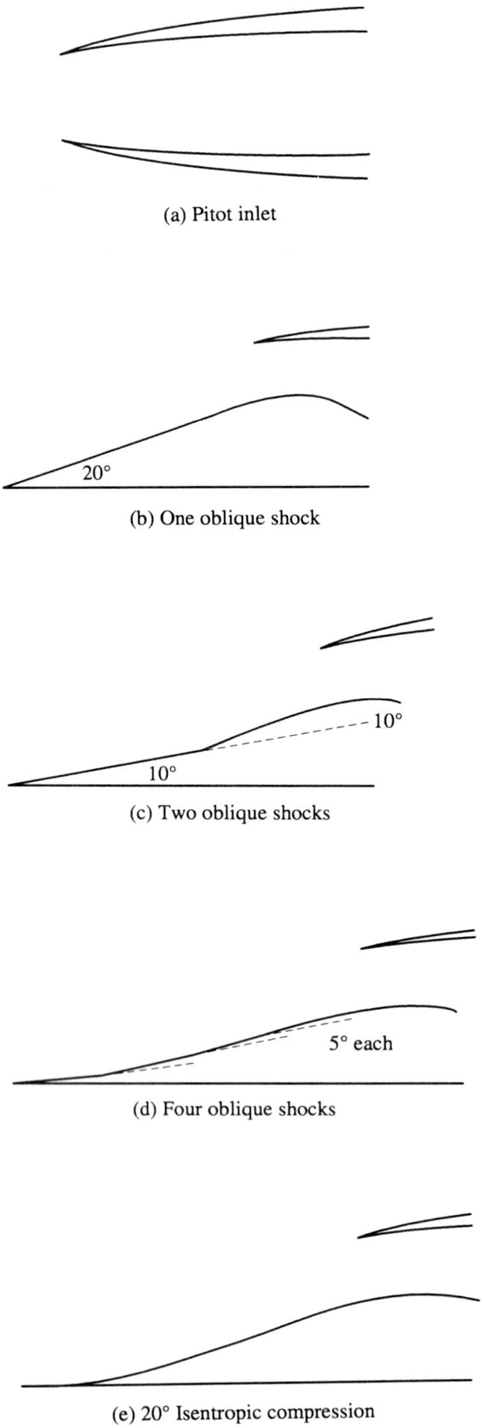

Figure 6-40 Inlet Configurations for Example Problem 6-10

TABLE 6-5a EXAMPLE 6-10 STAGNATION PRESSURE RATIOS

Deflection angle	Oblique shock 1	2	3	4	Normal shock	Stagnation pressure ratio
					0.2129	0.2129
20°	0.7266				0.5840	0.3982
10°	0.9463	0.9658			0.5290	0.4835
5°	0.9921	0.9937	0.9950	0.9960	0.5142	0.5024
20°	Isentropic compression				0.5091	0.5091

TABLE 6-5b EXAMPLE 6-10 MACH NUMBER VALUES

Initial Mach	Deflection angle	After oblique shock 1	2	3	4	Normal shock
3.5						0.4512
3.5	20°	2.2985				0.5346
3.5	10°	2.9045	2.4266			0.5203
3.5	5°	3.1984	2.9288	2.6852	2.4625	0.5167
3.5	20°	Isentropic compression			2.4750	0.5154

shock waves increases without bound. The four oblique shock waves/normal shock wave case yields a stagnation pressure ratio of 0.5024, while the isentropic compression followed by a normal shock wave gives 0.5091. Hence, for this condition four oblique shock waves is a practical limit beyond which little return can be expected. The Mach numbers behind each shock wave are presented in Table 6-5b. The Mach number after the isentropic compression is 2.475.

The stagnation pressure ratio can be increased by specifying a larger total deflection angle (more than 20°) and providing more oblique shocks, but if the flow is turned too much the inlet cross-sectional area and drag increase excessively.

Proper inlet design and operation are necessary for flight vehicles operating supersonically, and the higher the Mach number the more important inlet considerations become. The Lockheed SR-71 Blackbird, with sustained cruise Mach numbers in excess of 3, is probably the best-known example of such a vehicle. The SR-71 is equipped with a mixed compression inlet that includes a "spike" that can be translated longitudinally for more efficient inlet operation. Figure 6-41, from Drendel (1982), shows a three-view sketch of the SR-71. The two spike inlets are dominant visual features of the aircraft. Bleed doors located in the engine nacelle can be opened to increase the air mass flow rate through the inlet. At the time of the design and flight tests of the SR-71, during the early 1960s, analysis and design methods were not available that could reliably predict the needed spike location and bleed air door positions for the starting and efficient operation of the inlet system for various supersonic Mach numbers. The spike location and bleed air door positions were determined as part of the flight test program. Indeed, inlet operation was considered one of the biggest problems encountered in that program.

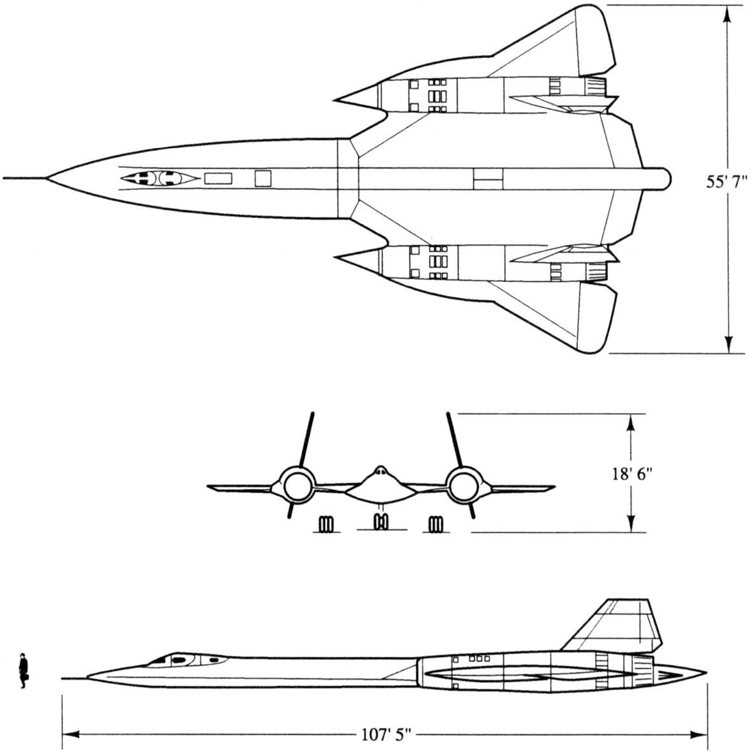

Figure 6-41 Three-View Sketch of the SR-71. *Adapted from Lockheed (1981).*

Crickmore (1986) provides a summary of the inlet operation problems encountered in the early phases of flight test for this aircraft. Flight at high Mach numbers required that the inlet spike position schedule—the spike location versus Mach number—be programmed accurately if excess fuel consumption and inlet unstarts were to be avoided. Nearly 70 flights were required to establish acceptable spike position/door settings for the SR-71 inlet. The sequence started at Mach 2 and progressed in Mach number increments of 0.1 until reaching Mach 3.2. During these flights the usual procedure was to increase the Mach number by 0.1 and manually select the spike position. If that spike location appeared to give efficient operation, as confirmed by flight test data analysis, that spike position location was incorporated into the schedule and the next Mach number increment was added. In many instances, however, the attempt to increase the Mach number would result in inlet unstart, often accompanied by harsh yawing in the direction of the unstarted inlet. The inlet unstart resulted in reduced thrust and lowered the flight Mach number. Inlet restart was accomplished by opening the bleed air doors to increase the air flow rate and swallow the shock. After inlet restart, the original Mach number was again attained and a new spike position tried. Eventually, the most efficient spike locations were determined for the Mach number range and incorporated into a control system that automatically positioned the spike as a function of Mach number. The SR-71 enjoyed a long (more than 25 years) and successful operational career and has probably accumulated more high supersonic Mach number (3 and above) flight time than any other flight vehicle.

Converging–diverging configurations can also be used as supersonic inlets. When used as an inlet, the converging–diverging passage operates in the same fashion as a diffuser on a supersonic wind tunnel. In steady state, the converging–diverging inlet would operate shock-free and reduce the incoming supersonic Mach number to a subsonic value. This mode of operation is illustrated in Figure 6-42a. As with supersonic wind tunnel diffusers, the problem is how to achieve the flow conditions indicated in the figure.

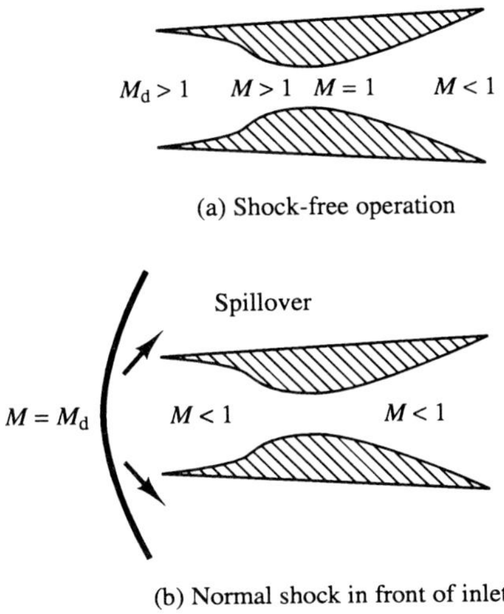

Figure 6-42 Converging–Diverging Inlet Operation

A fixed area converging–diverging supersonic inlet if accelerated to the design Mach number from rest would possess the flow conditions given in Figure 6-42b. A normal shock wave would stand in front of the inlet. Since the stagnation pressure decreases across the normal shock, the inlet throat is no longer able to pass the design mass flow rate. As a result, some spillover would occur. The procedure for attaining the shock-free flow conditions shown in Figure 6-42a is called *inlet startup*. The two methods available for starting such an inlet are: (1) overspeeding and (2) area change. The following example problem illustrates both starting procedures.

Example 6-11

A converging–diverging inlet operating in a shock-free mode is to be used at a Mach number of 1.75. Find the throat-to-entrance area ratio required at the design Mach number, the overspeed Mach number at which the inlet can be started, and the throat-to-entrance area ratio necessary for inlet startup by area change.

Solution. The desired inlet conditions at the design Mach number are depicted in Figure 6-43a. The area ratio A_i/A^* required for the freestream Mach number of 1.75 is from the **IS** option of COMPQ:

$$\frac{A_i}{A^*} = 1.3865.$$

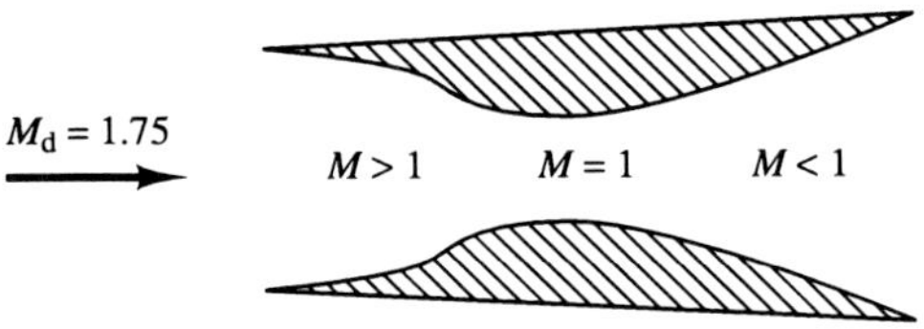

(a) Desired operating conditions

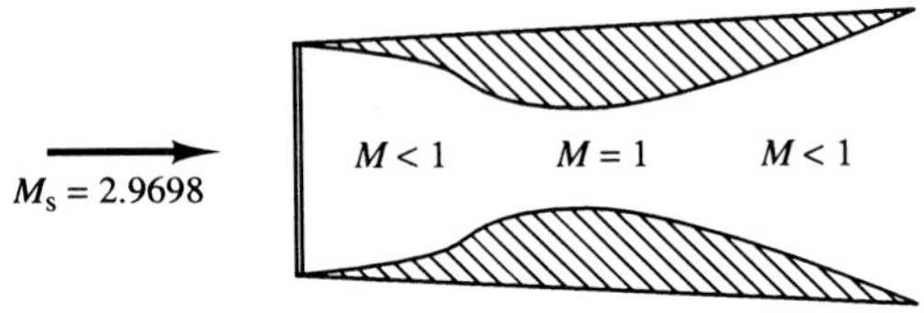

(b) Conditions at M_s

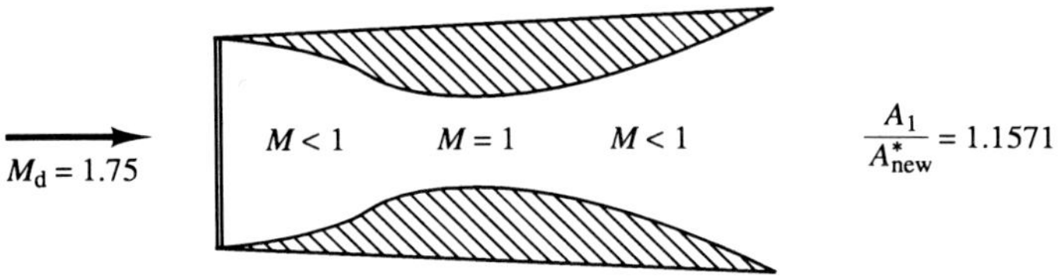

(c) Throat area increased

Figure 6-43 Example 6-11. *Adapted from Zucrow and Hoffman.*

If the inlet were accelerated from rest to the design Mach number, a shock wave would stand in front of the inlet, as in Figure 6-42b. By overspeeding the inlet, the shock can be moved to the lip of the inlet, and the situation becomes that shown in Figure 6-43b. For the area ratio of 1.3685, the Mach number downstream of the normal shock wave becomes

$$M_2 = 0.4770,$$

for which (using option **NS** of COMPQ)

$$M_1 = M_s = 2.9698.$$

Any further increase in the Mach number will lead to the shock being swallowed. Once the shock is swallowed, the vehicle can be slowed to the design Mach number of 1.75 and the desired shock-free inlet operation, Figure 6-42a, will be obtained. Overspeeding is not a practical technique for starting the inlet because of the extra thrust required to accelerate the vehicle to the Mach number required to start the inlet. Additionally, overspeeding to start a converging–diverging inlet is limited to design Mach numbers less than approximately 2.

Converging–diverging inlet startup by area change is a viable procedure that suffers only from mechanical complexity. If the throat area of the converging–diverging inlet is increased, A_i/A^* will be decreased, more mass flow can pass through the inlet, and the shock wave will be drawn closer to the lip of the inlet. The throat area can be increased enough to move the shock to the lip of the inlet. Such a condition is illustrated in Figure 6-43c. If the throat area is made incrementally larger than is required to place the shock wave at the lip of the inlet, then the shock will be swallowed. Once the shock is swallowed, the area ratio of the inlet can be reduced to the design Mach number value and the back pressure increased to place a normal shock wave of vanishing strength ($M_1 = 1.0$) at the throat.

For the condition illustrated in Figure 6-43c, the inlet throat is sonic. With a design Mach number of 1.75, the downstream Mach number is

$$M_2 = 0.6281,$$

and the area ratio required for sonic flow at the throat becomes

$$\frac{A_i}{A^*} = 1.1571.$$

Any further increase in throat area will result in the shock being swallowed. Once the shock is swallowed, the inlet throat area can be returned to the design value. If the back pressure is increased, the swallowed shock can be placed at the throat, giving the desired shock-free inlet conditions of Figure 6-43a. Because freestream fluctuations can easily result in disgorging of the shock at the throat, unstarting of the inlet, prudence suggests that the converging–diverging inlet be operated with the normal shock slightly downstream of the throat.

A convenient summary of both overspeeding and area change procedures for converging–diverging inlet startup is provided in Figure 6-44. The upper curve, labeled "shock wave attached," represents the reciprocal area ratio required, A^*/A_i, for a normal shock to be attached to the lip of the inlet. The lower curve, labeled "flow is supersonic throughout," represents the reciprocal area ratio required for isentropic flow through a converging–diverging inlet.

As an example, consider the requirements for an inlet at a design Mach number of 1.75. Point "a" on the figure represents the required reciprocal area ratio for shock-free flow. The ratio required for $M = 1.75$ is 0.72, which yields an area ratio of 1.39. If the inlet is to be started by area change, then the Mach number remains constant and the area ratio must be changed. Traversing from "a" to "o" at constant Mach number in Figure 6-44 yields the reciprocal area ratio required to position a normal shock at the inlet. With A^*/A_i read as 0.87, the required area ratio is 1.149. Within the accuracy of reading the figure, these numbers are in agreement with Example 6-11.

If the inlet is to be started by overspeeding, then the Mach number is increased at constant-area ratio. The overspeed Mach number is indicated as "c" on Figure 6-44. Traversing from "a" to "c" yields the required overspeed Mach number of 3.0, which agrees within reading accuracy with the results of Example 6-11. The asymptotic value of the upper curve is shown in Figure 6-44. For a Mach number greater

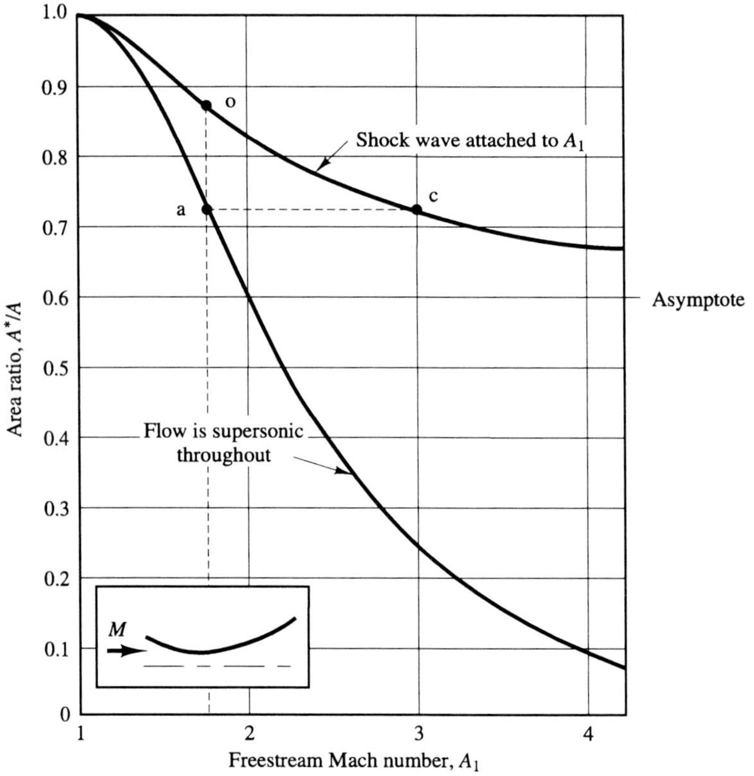

Figure 6-44 Reciprocal Area Ratio Relationships for Attached and Swallowed Normal Shock Waves for Supersonic Converging–Diverging Inlets

than about 2, starting the inlet by overspeeding is not possible (intersection with the upper curve will not occur).

In reality, the converging–diverging inlet is not attractive since it possesses very poor off-design operating characteristics. Even if inlet starting is achieved by area change, the mechanical requirements for varying the internal geometry of the inlet are considerable.

A single parameter is often used to characterize the overall performance or efficiency of an inlet. The two most common such parameters are the *isentropic diffuser efficiency* and the *total pressure ratio*. The total pressure ratio is the ratio of the stagnation pressure of the air delivered by the inlet to the stagnation pressure of the freestream. In much of the literature the total pressure ratio is represented by r_o and is defined as

$$r_o = \frac{P_{o_2}}{P_{o_\infty}}. \tag{6-50}$$

The total pressure ratio thus simply measures the effectiveness of the inlet in preserving the stagnation pressure of the entering air. Much of our discussion of inlets has used the total pressure ratio as a quantitative indicator of the effectiveness of the inlet diffusion process. The isentropic diffuser efficiency, η_d, is frequently used as the parameter of merit for supersonic inlets. Since both the total pressure ratio and the isentropic diffuser efficiency appear in the literature, an examination of the relationship between the two is appropriate. The isentropic diffuser efficiency is based on an isentropic process as the reference and is similar in concept to the isentropic compression efficiency discussed earlier in this chapter. Figure 6-45, an h–s diagram for an inlet diffusion process, is helpful in developing the isentropic inlet efficiency concept. Air at freestream conditions, in particular with freestream stagnation conditions of h_{o_∞} and P_{o_∞}, enters the inlet and is diffused to state 2 with stagnation conditions of h_{o_2} and P_{o_2}. Since the diffusion process is adiabatic, the stagnation enthalpy is unchanged throughout the process; hence, $h_{o_\infty} = h_{o_2}$. However, if the inlet process were isentropic, then compression to P_{o_2} would be to state 2′ in Figure 6-45 and would require only a stagnation enthalpy of h_{o_2}. The isentropic diffuser efficiency is defined as the ratio of the isentropic enthalpy change, $h_{o_{2'}} - h_\infty$, to the actual change, $h_{o_2} - h_\infty$, required to compress the flow to P_{o_2} or

$$\eta_d = \frac{h_{o_{2'}} - h_\infty}{h_{o_2} - h_\infty} = \frac{T_{o_{2'}} - T_\infty}{T_{o_2} - T_\infty} = \frac{T_{o_{2'}}/T_\infty - 1}{T_{o_2}/T_\infty - 1}. \tag{6-51}$$

But the process is isentropic from the freestream to the state designated as 2′; thus

$$\frac{T_{o_{2'}}}{T_\infty} = \left(\frac{P_{o_2}}{P_\infty}\right)^{(\gamma-1)/\gamma} \qquad T_{o_2} = T_\infty\left(1 + \frac{\gamma - 1}{2} M_\infty^2\right). \tag{6-52}$$

Hence, Eq. 6-51 can be expressed as

$$\eta_d = \frac{(P_{o_2}/P_\infty)^{(\gamma-1)/\gamma} - 1}{[(\gamma - 1)/2]M_\infty^2}. \tag{6-53}$$

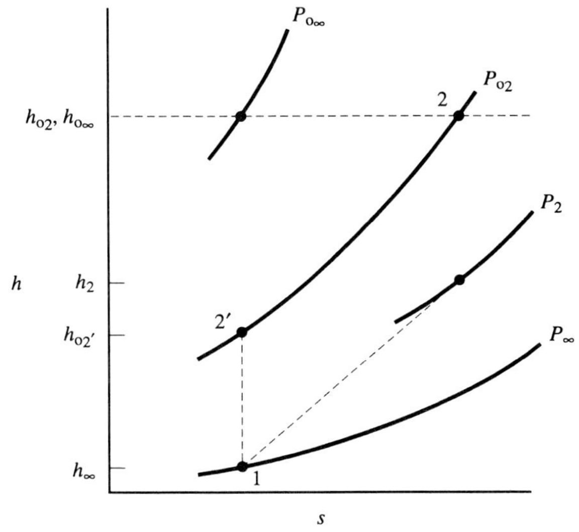

Figure 6-45 h–s Diagram for Isentropic Diffuser Efficiency Derivation

Solving Eq. 6-53 for P_{o_2}/P_∞ results in

$$\frac{P_{o_2}}{P_\infty} = \left(1 + \frac{\gamma - 1}{2}\eta_d M_\infty^2\right)^{\gamma/(\gamma-1)}. \tag{6-54}$$

Then

$$\frac{P_{o_2}}{P_\infty} = \frac{P_{o_2}}{P_{o_\infty}}\frac{P_{o_\infty}}{P_\infty} = r_o\left(1 + \frac{\gamma - 1}{2}M_\infty^2\right)^{\gamma/(\gamma-1)}, \tag{6-55}$$

recognizing the definition of the total pressure ratio, r_o. Using Eq. 6-54 for P_{o_2}/P_∞ in Eq. 6-55 permits r_o to be cast as

$$r_o = \left[\frac{1 + \frac{\gamma - 1}{2}\eta_d M_\infty^2}{1 + \frac{\gamma - 1}{2}M_\infty^2}\right]^{\gamma/(\gamma-1)}, \tag{6-56}$$

or solving for the isentropic diffuser efficiency:

$$\eta_d = \frac{\left[1 + \frac{\gamma - 1}{2}M_\infty^2\right]r_o^{(\gamma-1)/\gamma} - 1}{\frac{\gamma - 1}{2}M_\infty^2}. \tag{6-57}$$

This expression relates the total pressure ratio and the isentropic diffuser efficiency as a function of Mach number and ratio of specific heats in the functional form

$$\eta_d = f(\gamma, M_\infty, r_o). \tag{6-58}$$

Although the values of r_o and η_d differ, both monotonically decrease from unity as inlet performance degrades. Moreover, given the value of one, the other is readily available from either Eq. 6-56 or Eq. 6-57.

Example 6-12

Calculate the isentropic diffuser efficiency of the multiple oblique inlet described in Example 6-11.

Solution. Table 6-5 contains summary information for the inlet with the various oblique and normal shock arrangements. The solution to this example will be presented in Table 6-6. The values of r_o have already been calculated for the various options and are available in Table 6-5. For a freestream Mach number of 3.5 and a ratio of specific heats of 1.4, Eq. 6-57 becomes

$$\eta_d = \frac{\left(1 + \frac{\gamma - 1}{2}M_\infty^2\right)r_o^{(\gamma-1)/\gamma} - 1}{\frac{\gamma - 1}{2}M_\infty^2}$$

$$= \frac{3.45r_o^{.2857} - 1}{2.45}.$$

TABLE 6-6 INLET EFFICIENCIES FOR EXAMPLE 6-11

Oblique shocks	Normal shocks	r_o	η_d
0	1	0.2129	0.4970
1	1	0.3982	0.6743
2	1	0.4835	0.7360
4	1	0.5024	0.7486

Equation 6-57, for a given Mach number and gas, thus yields a one-to-one correspondence between the total pressure ratio and the isentropic diffuser efficiency. The isentropic diffuser efficiencies can be calculated directly from the preceding equation since the values of the total pressure ratios are known from Example 6-11. The isentropic diffuser efficiencies are somewhat larger than the total pressure ratios, so to avoid confusion, we must be aware of which measure of merit is being used to quantify inlet performance.

6.6 CONCLUSION

This chapter has examined the principles of, and developed methods, computational aids, and procedures for, simple area change or simple isentropic flow. Even though one-dimensionality was assumed, the results, especially for high Reynolds number flows, are acceptable for preliminary purposes. Even for multidimensional flows, the principles of operation of nozzles and inlets developed in this chapter are phenologically correct. A good understanding of simple area change is necessary if compressible flow is to be mastered.

CHAPTER 6 REVIEW QUESTIONS

1. A supersonic flow expands indefinitely in a diverging channel. Does the flow velocity approach a finite limit, or does it continue to increase without bounds? Assume a perfect gas with constant specific heats.
2. What is choked flow?
3. For an isentropic flow, must the Mach number always be 1.0 at the minimum area? Explain.
4. A gas stored in a reservoir is discharged through a converging nozzle. For a constant back pressure, sketch the variation of mass flow rate with reservoir pressure. What if the nozzle is converging–diverging?
5. What must the relationship be between the nozzle exit plane pressure and the back pressure for the following situations:
   (a) $M_e = 1.0$;
   (b) $M_e < 1.0$;
   (c) $M_e > 1.0$.

6. If the nozzle exit Mach number exceeds 1, what is the relationship between the nozzle exit and back pressures?
7. Sketch $P(x)$ for the situations in Figure 6-46. Assume isentropic flow except for the normal shock waves.

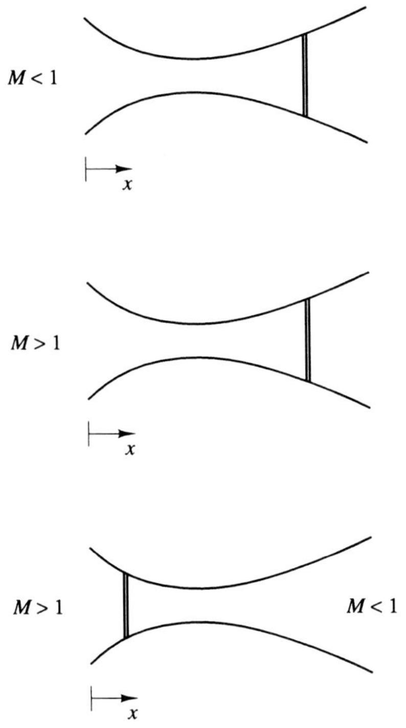

Figure 6-46

8. Sketch $P(x)$ for the condition shown in Figure 6-47.

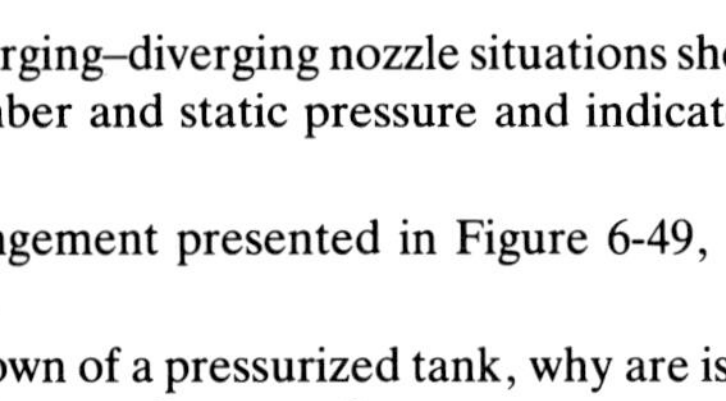

Figure 6-47

9. For the converging–diverging nozzle situations shown in Figure 6-48, sketch the variations of Mach number and static pressure and indicate whether the arrangement is stable or unstable.
10. For the arrangement presented in Figure 6-49, sketch the Mach number and pressure distributions.
11. In the blowdown of a pressurized tank, why are isothermal and isentropic processes good bounds for the actual process?

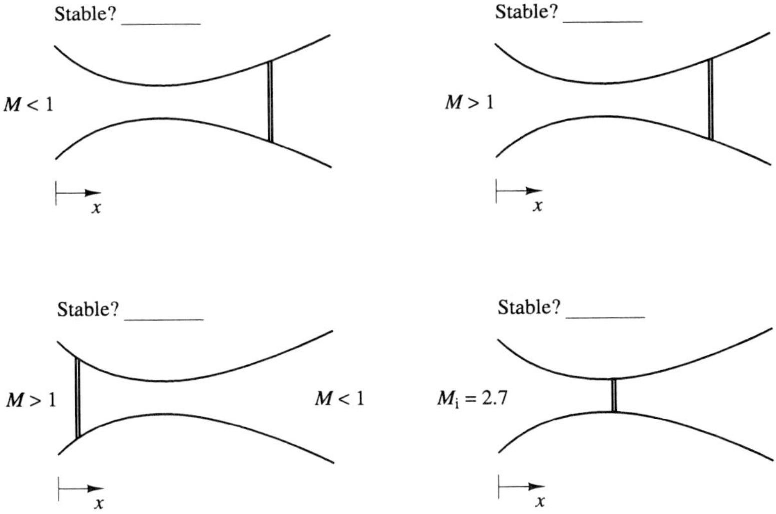

Figure 6-48

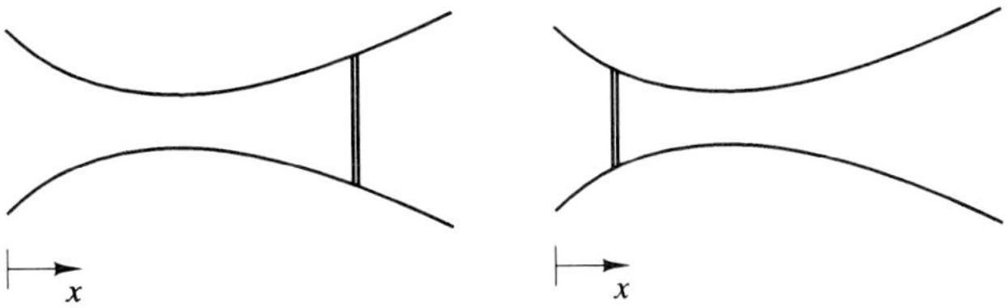

Figure 6-49

12. In the blowdown of a pressurized tank, sketch the time–temperature histories of the stagnation temperature inside the tank for an isothermal process and for an isentropic process inside the tank during the blowdown.
13. What advantage does a multiple shock inlet possess over a single shock inlet?
14. In an inlet arrangement, why are several oblique shock waves preferred over a single normal shock wave?
15. Why is overspeeding to start an inlet not a practical idea?
16. In a continuous-duty supersonic wind tunnel, why is a diffuser attractive?
17. In a blowdown supersonic wind tunnel, why is a diffuser attractive?
18. Explain how a continuous-duty supersonic wind tunnel with a variable area diffuser would be started.

CHAPTER PROBLEMS

1. Air is flowing at a Mach number of 0.5 in a two-dimensional channel at a location where the area is 1 ft^2. At this location the static pressure is 10 psia and the static temperature is 500 R.

(a) Calculate the mass flow rate through the channel.
(b) What percent reduction in area would be necessary to increase the flow Mach number to 0.75?
(c) What reduction in area would be necessary to reach a Mach number of 1.0?
(d) What would happen if the area were reduced to a value smaller than that in part (c)?

2. A converging nozzle is attached to a reservoir. If the back pressure (P_b) is 101.3 kPa, what is the minimum reservoir pressure required for the flow to be choked? The reservoir temperature is 300 K.

3. At a cross-sectional area of 1 m^2, the Mach number is 1.75. If the flow is isentropic, what is the Mach number at a flow area of 1.5 m^2? Work both with and without using case **AC** of COMPQ option **IS**.

4. As shown in Figure 6-50, a normal shock wave stands in a diverging channel. The inlet Mach number is 2.5 and the exit Mach number is 0.5. The area ratio, A_{exit}/A_{inlet}, is equal to 2.0 for the channel. Find the location of the shock wave in the channel in terms of the area ratio. What are P_2/P_1 and P_e/P_i for the situation illustrated in the figure?

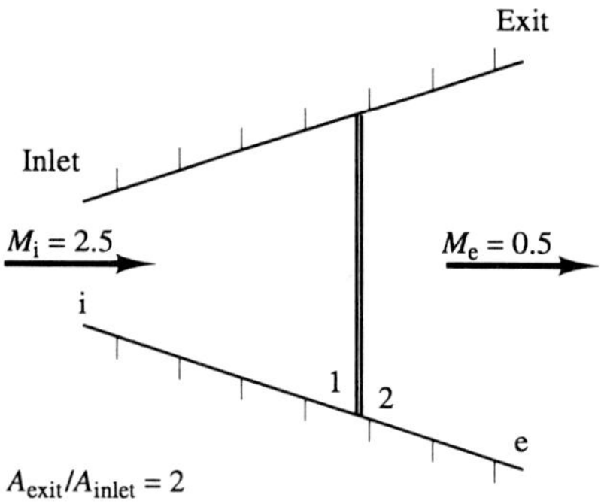

Figure 6-50

5. Compute the pressures required at station b of the supersonic diffuser for the two different shock wave locations illustrated in Figure 6-51. The area ratio of the diffuser is $2.5(A_b/A_a)$. For the two conditions shown, find the losses in stagnation pressure. Which would be the preferred arrangement?

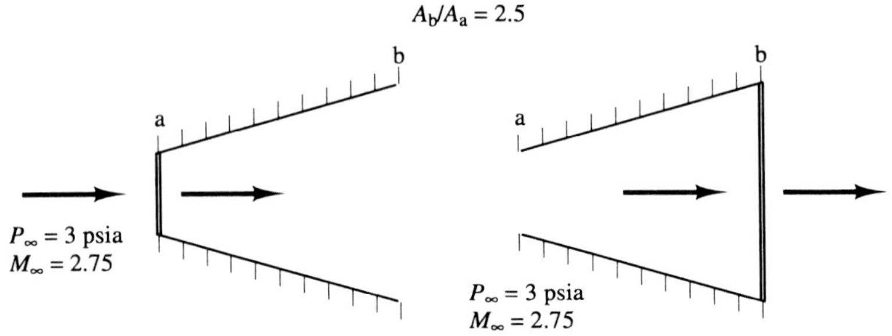

Figure 6-51

6. A converging–diverging nozzle with an A/A^* of 4.0 is supplied with air at a stagnation temperature of 650 R. If the back pressure is 14.696 psia, find

(a) the reservoir pressure for key pressure 1 (isentropic #1);
(b) the reservoir pressure for key pressure 3 (isentropic #2);
(c) the reservoir pressure for a normal shock to stand in the nozzle exit plane;
(d) the reservoir pressure for a normal shock to stand at the $A/A^* = 2.0$ location;
(e) for reservoir pressures greater than that required for key pressure 3 (isentropic #2), sketch the flow field at the nozzle exit plane.

7. A reservoir (containing air) with a stagnation pressure of 1000 kPa feeds a converging–diverging nozzle with an area ratio of 2.6367. What is the supersonic Mach number in the nozzle exit plane? Sketch the flow field for a back pressure of 250 kPa.

8. What is the exit Mach number in a converging nozzle connecting a reservoir at 1000 kPa and a back pressure of 700 kPa?

9. A converging–diverging nozzle of area ratio 2.75 is connected to a pressurized tank. The nozzle exhausts into a back pressure of 101 kPa. What tank stagnation pressure is necessary for isentropic #2 to exist? What tank stagnation pressure is necessary for a normal shock wave to stand in the nozzle exit plane?

10. A converging–diverging nozzle with an A/A^* of 4.0 is supplied with air at a stagnation temperature of 1000 K. If the back pressure is 101 kPa, find
(a) the reservoir pressure for key pressure 1 (isentropic #1);
(b) the reservoir pressure for key pressure 3 (isentropic #2);
(c) the reservoir pressure for a normal shock to stand in the nozzle exit plane;
(d) the reservoir pressure for a normal shock to stand in the nozzle at the $A/A^* = 2.0$ location;
(e) for reservoir pressures greater than that required for key pressure 3 (isentropic #2), sketch the flow field at the nozzle exit plane.

11. A converging–diverging nozzle with an area ratio of 4.25 is attached to a reservoir containing nitrogen that exhausts to atmospheric pressure. Determine
(a) the range of reservoir pressures over which a normal shock will appear in the nozzle;
(b) the value of reservoir pressure for the nozzle to be perfectly expanded with supersonic flow at the exit plane;
(c) the minimum reservoir pressure to produce sonic flow.

12. A converging nozzle of exit area of 2.5 cm^2 is used to discharge air from a reservoir to the atmosphere. If the air in the reservoir is at 350 R, determine the flow rate for reservoir pressures of 120 kPa, 140 kPa, 200 kPa, and 1 MPa. In each case, what is the increase in mass flow if the back pressure is reduced from 101 kPa to 0 kPa?

13. A converging–diverging nozzle with exit and throat areas of 10 in^2 and 5 in^2, respectively, is used to vent a large chamber. For a back pressure of 14.7 psia and a chamber temperature of 550 R, compute the following:
(a) minimum chamber pressure to choke the nozzle;
(b) mass flow rate for a chamber pressure of 60 psia;
(c) mass flow rate for a chamber pressure of 30 psia.

14. A converging nozzle of rectangular cross section is constructed with a constant width of 2.5 in. As shown in Figure 6-52, the top and bottom walls of the channel are tapered at an angle of 10°. At station 1 the nozzle possesses a square cross section. If stagnation conditions are 300 psia and 600 R and the Mach number at station 1 is 0.3, construct plots of $M(x)$, $P(x)$, and $T(x)$. What is the maximum permissible nozzle length? Is an assumption of one-dimensional flow justified?

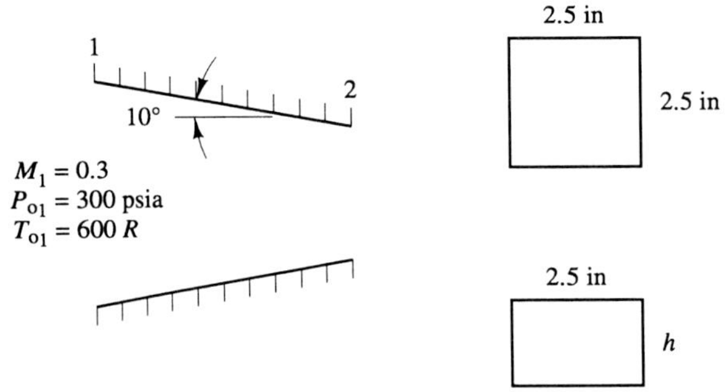

Figure 6-52

15. A gas with a molecular weight of 20 kg/kg-mol has $c_p = 2.0$ kJ/kg K. If the gas is expanded from a large reservoir (2 MPa and 700 K) through a converging–diverging nozzle with an area ratio of 3.5, find the nozzle exit velocity if the back pressure is very low.

16. A converging nozzle, supplied from an air reservoir maintained at 600 R and 250 psia, exhausts into a back pressure of 10 psia. Determine the nozzle exit plane pressure and Mach number for the above conditions. Determine the flow direction and Mach number after the air exits the nozzle. Sketch the process in correct proportions, indicating important features of the flow outside the nozzle.

17. Find the nozzle exit plane velocity and static pressure for a converging–diverging nozzle of area ratio 3.25 connected to a reservoir maintained at 300 kPa and 500 K. The back pressure is 101 kPa.

18. A converging–diverging nozzle, designed to operate isentropically with air at an exit Mach number of 1.75, is connected to a reservoir with stagnation conditions of 4 MPa and 600 K. Compute the following:
(a) the maximum back pressure to choke nozzle;
(b) the flow rate for a back pressure of 101 kPa;
(c) the flow rate for a back pressure of 1 MPa.
The nozzle exit area is 0.15 m^2.

19. A converging–diverging nozzle with an area ratio of 2.75 is attached to a reservoir containing air at 250 psia and 500 R.
(a) Find the three key back pressures.
(b) For a back pressure (P_b) of 75 psia, sketch the nozzle flow field including the region just downstream of the nozzle exit plane.
(c) Sketch the nozzle flow field including the region just downstream of the nozzle exit if the nozzle exhausts into a vacuum ($P_b = 0$).

20. Air in a reservoir maintained at 10 MPa is discharged through a converging–diverging nozzle with an area ratio of 3.0. The flow field in the nozzle exit plane contains a Prandtl-Meyer expansion fan that turns the flow through an angle of 10°. For this condition, find the back pressure into which the nozzle is exhausting.

21. An oxygen reservoir is vented by a converging nozzle. If the exit area of the converging nozzle is 1.75 cm^2, and if conditions inside the reservoir are 350 kPa and 1000 K, find the nozzle mass flow rate for back pressures of 0, 100, 200, and 300 kPa.

22. A converging–diverging nozzle of area ratio 2.5 is connected to a large reservoir and exhausts into a region of back pressure P_b. When a normal shock is located in the nozzle at an area equal to 1.75 times the throat area, find the percent decrease in back pressure that would be required to place the normal shock wave in the exit plane of the converging–diverging nozzle.

23. A normal shock is positioned in a converging–diverging nozzle at an area equal to 1.25 times the throat area. If the area ratio of the nozzle is 2.5, determine the ratio of the nozzle exit velocity for this case to that for supersonic flow in the nozzle exit plane. Plot this ratio as a function of the ratio of specific heats.

24. A converging nozzle discharges into a zero back pressure. If air is the fluid and if the reservoir pressure is 50 psia, sketch the initial flow direction for the exhaust plume.

25. A converging nozzle is fed by a tank with a stagnation pressure of 150 kPa. If the back pressure into which the tank exhausts is 101 kPa, what is the exit Mach number? Repeat for a back pressure of 10 kPa.

26. A converging–diverging nozzle with a throat area of 40 cm^2 and an exit area of 250 cm^2 is connected to a helium reservoir maintained at 300 K and 350 kPa. Find the range of back pressures for which a normal shock will appear in the nozzle. If the nozzle exhausts into a vacuum, sketch the initial plume angle at the nozzle exit plane.

27. A nitrogen reservoir is charged to a pressure of 200 psia at a temperature of 550 R. For a converging–diverging nozzle of area ratio 3.75, determine the back pressure required for a normal shock wave to be located at a nozzle area equal to twice the throat area. For this condition, compute the pressure and velocity in the nozzle exit plane. What back pressure would be required for a reduction in the nozzle mass flow rate for the condition described in the first part of this problem?

28. Determine the thrust of a rocket nozzle designed to operate isentropically (with key pressure 3) at 45,000 ft with a chamber pressure of 350 psia. The combustion products of the gas in the rocket can be assumed to be a calorically perfect gas with a stagnation (chamber) temperature of 3000 R, $\gamma = 1.27$, and $MW = 23$ lb_m/lb_m-mol). The exit area of the converging–diverging nozzle is 1.25 ft^2. Compute the thrust at sea level and at 300,000 ft altitude for this engine. Sketch the flow field at the nozzle exit for each back pressure.

29. In Figure 6-53, the exit-area-to-throat-area ratio of the nozzle is 4, and the normal shock wave is located at an area ratio of 2. The stagnation conditions in the reservoir are 500 kPa and 500 K.

(a) Find the back pressure for the situation shown.
(b) What is the loss in stagnation pressure through the nozzle?
(c) What is the maximum back pressure for which the nozzle will choke?
(d) Sketch the flow field at the nozzle exit if the back pressure is zero.

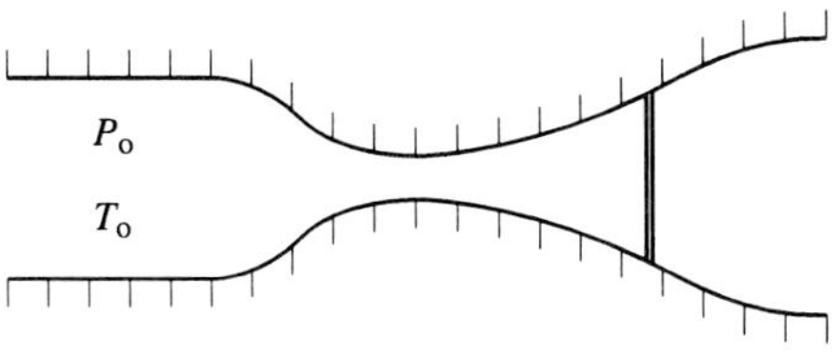

Figure 6-53

30. For the system shown in Figure 6-54, $M_a = 1.75$, $A_a = 2$ in^2, throat area $= 1.0$ in^2, area at the shock wave $= 1.75$ in^2, and exit area $= 2.5$ in^2. If the working fluid is a calorically perfect gas with the ratio of specific heats, γ, of 1.275, calculate the following:

(a) throat Mach number;
(b) exit Mach number;
(c) ratio of exit static pressure to static pressure at a;
(d) ratio of exit stagnation pressure to stagnation pressure at a.

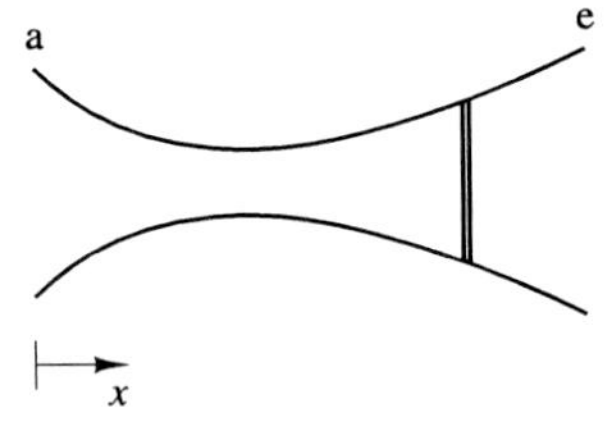

Figure 6-54

31. For the flow situation depicted in the converging–diverging diffuser of Figure 6-55, find the diffuser exit conditions (velocity, Mach number, static temperature, and static pressure) if the inlet conditions (station i) are $M = 2.8$, $P = 5$ psia, and $T = 400\ R$.

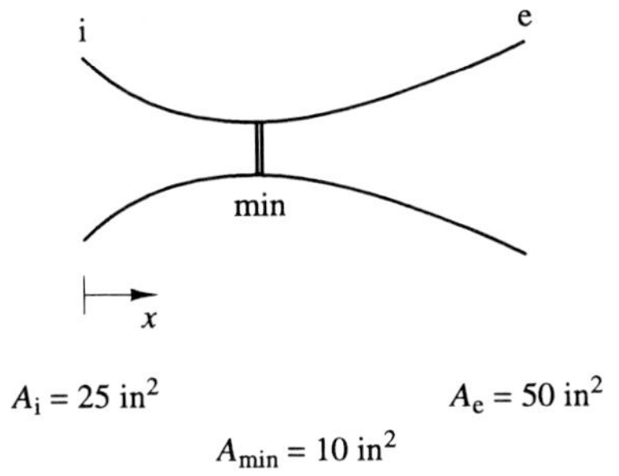

Figure 6-55

32. Because of fuel and oxidant flow rate variations in a rocket engine, the stagnation pressure at the inlet of the converging–diverging nozzle varies as $P_o = 350.0[1 + 0.05\sin(\pi/2)t]$, where t is in seconds and P_o is in psia. If the nozzle exit area is 2 ft^2 and the nozzle area ratio is 5, compute the resultant variations in nozzle mass flow rate, nozzle exit plane velocity, and nozzle exit plane static pressure. How does the thrust vary with time? The combustion products can be modeled as a calorically perfect gas with a ratio of specific heats of 1.3 and a molecular weight of 19 lb_m/lb_m-mol. The back pressure is low enough to permit supersonic flow throughout the diverging portion of the nozzle. State and defend your assumptions.

33. A 40-m^3 spherical tank contains air at 750 kPa and 300 K. A 5-cm diameter rupture diaphragm provided as a safety feature ruptures, causing the tank to vent to atmospheric pressure. Estimate the time required for the tank pressure to drop to 375 kPa if the actual process in the tank is bounded by the isothermal and isentropic.

34. Develop a computer program to predict the pressure history of the blowdown situation in Problem 33 from initial rupture of the diaphragm until the tank pressure has dropped to 1.1 atm.

35. Air is stored in a tank of volume *Vol* and initial stagnation pressure of P_{o_i} and initial stagnation temperature of T_o. The gas is discharged through a converging–diverging nozzle of exit-to-throat area ratio 2.75. The throat area is A^*.

(a) Develop your own original expression for the mass flow rate (for choked flow) as a function of tank stagnation conditions.

(b) If the discharge process is isothermal and choked, develop (not copy) an expression for the time required for the tank pressure to drop to $P_{o_{final}}$.

36. A converging nozzle is attached to a large tank containing air at 400 kPa and 300 K. If the back pressure is 101.3 kPa and the inlet and exit areas of the nozzle are 200 cm^2 and 70 cm^2, respectively, find the force of the air on the nozzle. Figure 6-56 illustrates the arrangement.

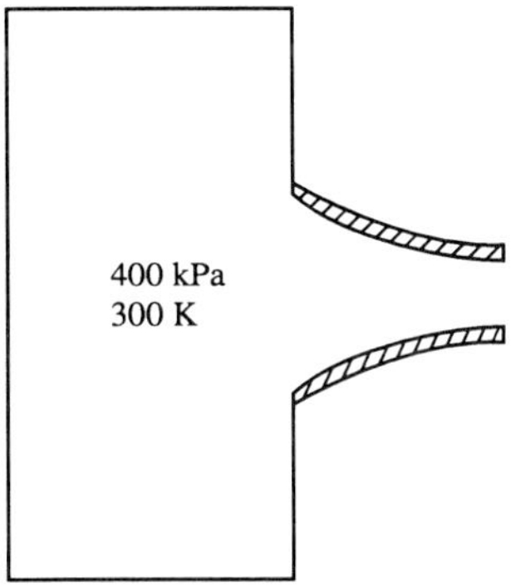

Figure 6-56

37. For the converging–diverging nozzle/tank arrangement shown in Figure 6-57, calculate the range of back pressures for which $P_e > P_b$, the range of back pressures for which $P_e < P_b$, and the range of back pressures for which the nozzle is choked.

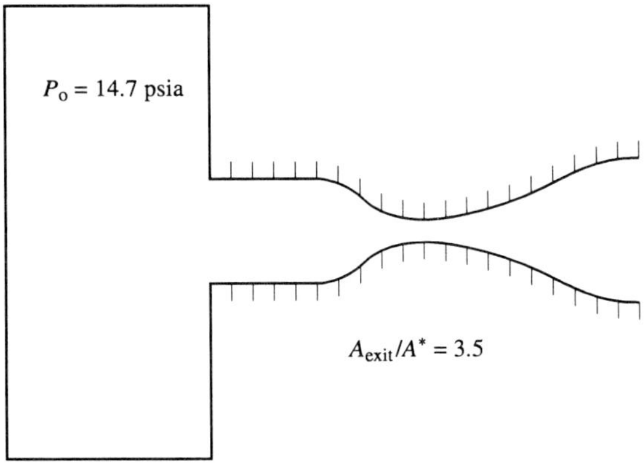

Figure 6-57

38. The cabin of a transport plane flying at an altitude of 10 km is maintained at a pressure of 85 kPa and a temperature of 22° C. The fuselage is cylindrically shaped with a diameter of 3 m and a length of 45 m. If the skin of the fuselage is pierced such that a converging nozzle-shaped hole of exit diameter 1.2 cm is formed, calculate the time for the cabin pressure to decrease to one-half of the initial value.

39. A storage tank contains helium at 70 psia and 70° F, as Figure 6-58 shows. A converging–diverging nozzle with an area ratio of 4 and a throat area of 0.25 ft^2 is attached to the tank.

Find the maximum mass flow possible through the nozzle and the range of back pressures over which the mass flow can be attained. Repeat for oxygen.

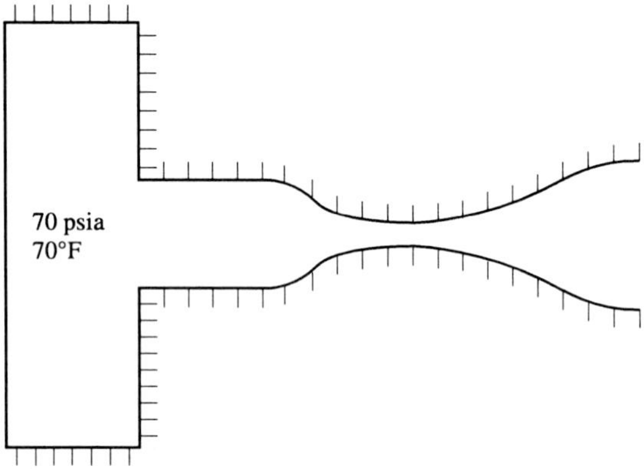

Figure 6-58

40. A pressurized tank is filled with air at 250 kPa and 300 K. A converging–diverging nozzle with a throat area of 3 m^2 and an area ratio of 3.5 is attached to the tank. As illustrated in Figure 6-59 for a certain back pressure, a normal shock wave stands in the diverging portion of the nozzle at an area of 6 m^2. For these conditions, determine the following:
(a) A^* for flow from inlet to shock;
(b) A^* for flow from shock to exit;
(c) Mach number at nozzle exit plane;
(d) stagnation pressure at nozzle exit plane;
(e) exit plane static pressure and velocity.

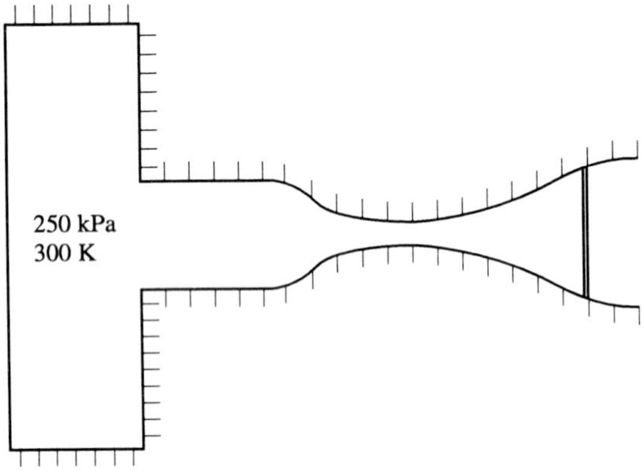

Figure 6-59

41. A jet plane flying supersonically at a Mach number of 1.75 utilizes a diverging passage as a diffuser (see Figure 6-60). If the ambient conditions are 80 kPa and 250 K, estimate the back pressure range over which a normal shock will appear in the diffuser. $A_{exit}/A_{inlet} = 2.25$ for the diffuser.

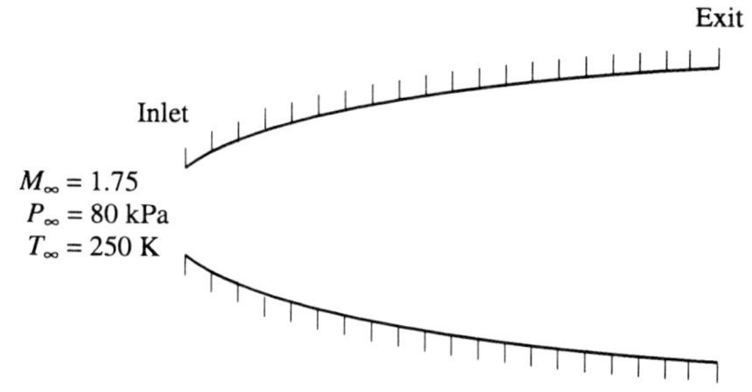

Figure 6-60

42. A converging–diverging nozzle designed to produce a Mach 2 flow at the exit plane is connected to a reservoir containing air at 400 psia. For an ambient pressure of 14.696 psia, the nozzle is underexpanded and a Prandtl-Meyer fan will be attached to the lip of the nozzle. Find the angle, indicated in Figure 6-61, through which the Prandtl-Meyer expansion will turn the flow. Sketch the flow field immediately downstream of the nozzle exit plane; include the initial and final expansion fan angles. What effect will this turning angle have on the thrust associated with the nozzle?

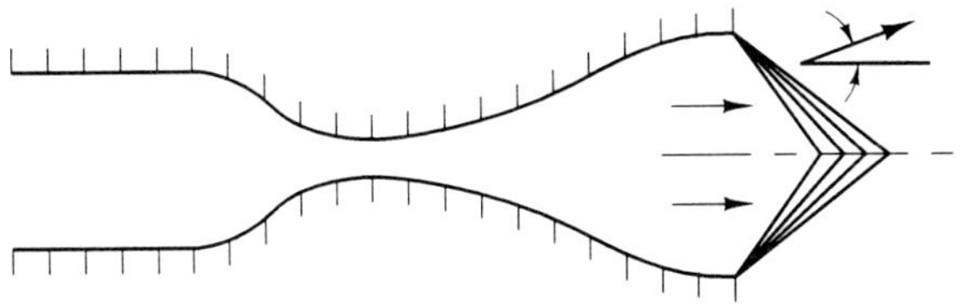

Figure 6-61

43. A converging–diverging nozzle (shown in Figure 6-62) with an area ratio of 2.75 is connected to a reservoir with stagnation conditions of 150 psia and 530 R. Determine the back pressure required for a normal shock wave to appear in the nozzle exit plane.

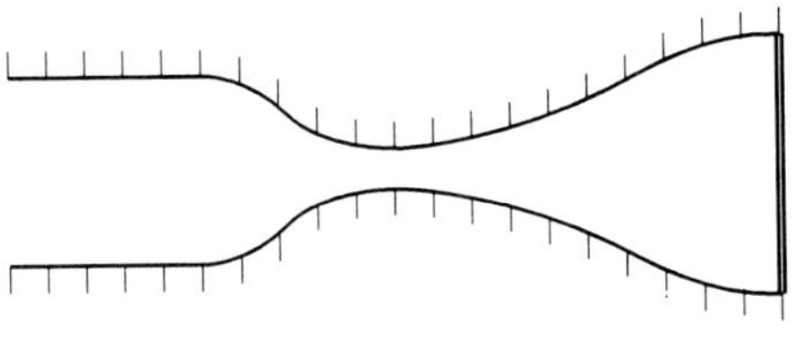

Figure 6-62

44. A converging–diverging nozzle with a design exit Mach number of 2.2 is connected to a reservoir pressurized to 550 kPa and vented to an ambient pressure of 101.3 kPa. For these circumstances the nozzle is overexpanded and results in an exhaust plume structure as sketched in Figure 6-63. Determine the flow direction, static pressure, and Mach number in regions R_1, R_2, and R_3 of the figure.

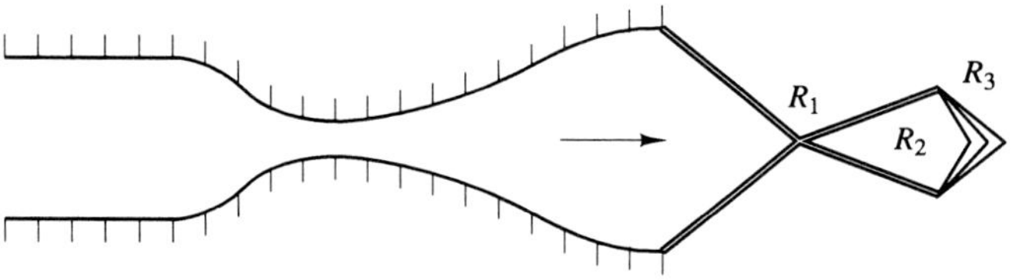

Figure 6-63

45. A supersonic wind tunnel, arranged as in Figure 6-64, consists of a converging–diverging nozzle venting into a vacuum tank. The test section cross-sectional area is 75 cm^2, and the test section design Mach number is 2.5. The vacuum tank is evacuated to near-zero pressure, the upstream valve is opened, and the run begins. If the tank volume is 5 m^3, how long can a run last before a shock wave is forced into the test section? What tank volume would be required for a running time of 45 s? Atmospheric conditions are 101 kPa and 27° C. Hint: What is the temperature of the air in the tank?

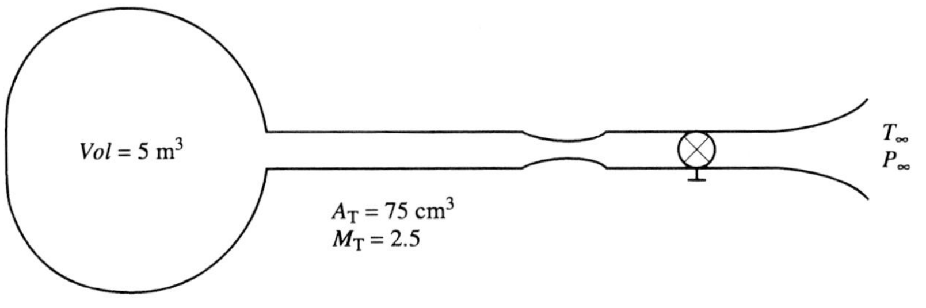

Figure 6-64

46. A wind tunnel (see Figure 6-65) with a test section Mach number of 2.75 uses a fixed-area diffuser. If the test section area is 10 ft^2, what are the areas of the first throat and the diffuser throat?

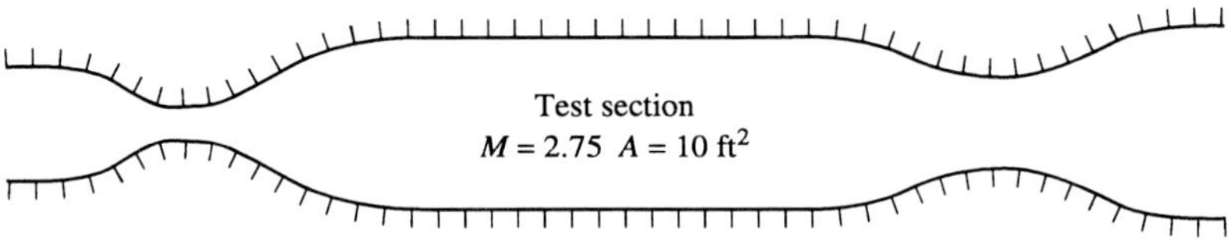

Figure 6-65

47. A supersonic wind tunnel with a test section Mach number of 2.2 is equipped with a fixed-area converging–diverging diffuser. If the test section cross-sectional area is 0.1 m^2, find the minimum area of the diffuser throat.

48. A converging–diverging inlet designed for operation at Mach = 3 is to be started by area variation. How much must the throat area be changed in order to start the inlet?

49. A jet plane flying supersonically at a Mach number of 1.75 utilizes a diverging passage as a diffuser (see Figure 6-66). If the ambient conditions are 8 psia and 430 R, estimate the back pressure range over which a normal shock will appear in the diffuser. $A_{exit}/A_{inlet} = 2.25$ for the diffuser. What is the percentage change in mass flow rate for the range of back pressures?

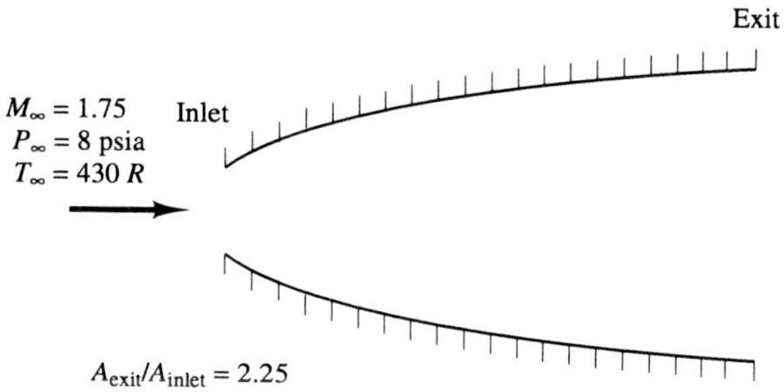

Figure 6-66

50. As shown in Figure 6-67, a supersonic wind tunnel with a test section Mach number of 2.0 using helium as the working fluid has a fixed-area diffuser. If the upstream throat area is 0.2 m^2, what is the minimum diffuser throat area required for startup?

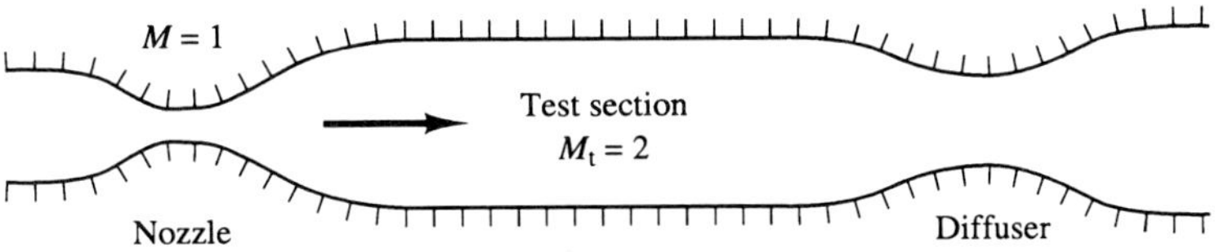

Figure 6-67

51. A converging–diverging supersonic diffuser is to be used at Mach 3.0. The diffuser utilizes a variable throat area to swallow the starting shock. What percentage increase in diffuser throat area as compared with the first throat area will be necessary if the working fluid is helium?

52. A supersonic wind tunnel with a test section Mach number of 3.0 is equipped with a converging–diverging diffuser. Find the minimum area of the diffuser (throat) required to swallow the starting shock if the nozzle throat area is 2 ft^2.

53. A supersonic wind tunnel with a test section Mach number of 2.2 is equipped with a fixed-area converging–diverging diffuser.

(a) If the test section cross-sectional area is 1.5 m^2, find the minimum area of the diffuser throat.

(b) After the shock is swallowed, what would the diffuser throat Mach number be?

54. A supersonic wind tunnel with a test section Mach number of 3.0 is driven by a compressor with an output pressure of 400 kPa and an output temperature of 400 K. If the diffuser is of the fixed-area type, find the throat areas of the first and second (diffuser) throats if the test section area is 1 m^2.

55. As illustrated in Figure 6-68, a supersonic wind tunnel is equipped with a converging–diverging diffuser. If the area ratio of the nozzle is 2.71, find the area ratio A_3/A_1 necessary to produce sonic flow in the diffuser throat, with a normal shock wave at A_2. Find the percent of loss in stagnation pressure for this flow. Determine the percent of loss in stagnation pressure when a normal shock wave appears at A_3.

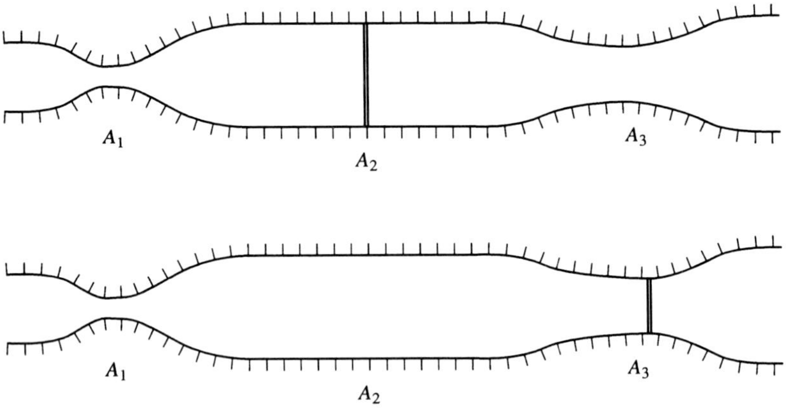

Figure 6-68

56. The supersonic inlet sketched in Figure 6-69 contains an oblique shock wave followed by a normal shock wave. For the conditions shown in the sketch, find the stagnation temperature, the stagnation pressure, and the Mach number behind the normal shock wave (region 3 in the sketch).

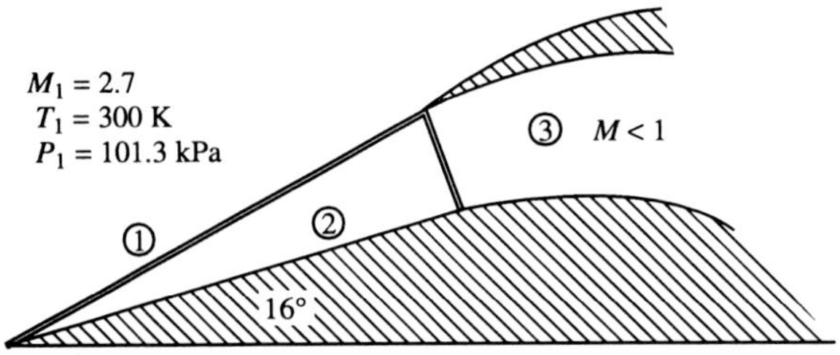

Figure 6-69

57. Two options for a supersonic inlet designed to handle 15 kg/s and operate at a Mach number of 2.0 at 8000 m altitude are to be investigated. The general configuration is sketched in Figure 6-70. Estimate the size required for such inlets if a 14° included angle wedge is to be used and if a 14° included angle cone is to be used.

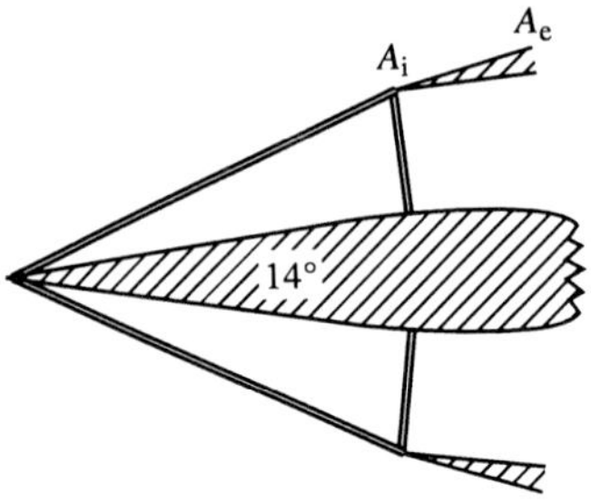

Figure 6-70

58. For the two-dimensional diffuser shown in Figure 6-71, determine the velocity, static pressure, stagnation pressure, and static temperature downstream of the normal shock wave.

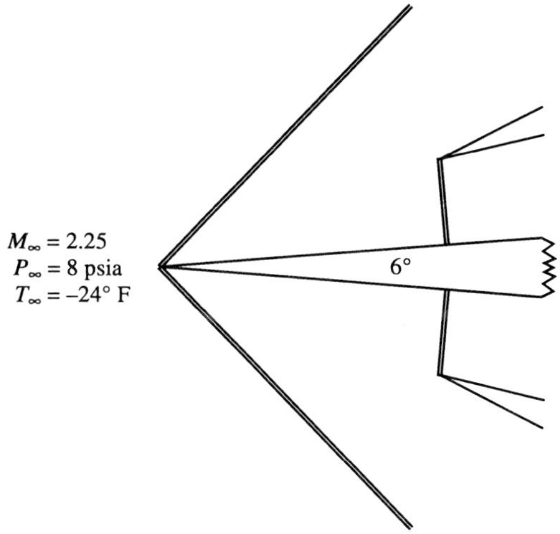

Figure 6-71

59. A preliminary concept study is to be performed for a two-dimensional supersonic inlet designed for operation at Mach 2.5. As shown in Figure 6-72, the flow is to be decelerated by a series of oblique shock waves followed by a normal shock wave. Using the loss of stagnation pressure as a figure of merit, investigate the cases of one, two, three, and four oblique shocks. The total turning angle for all the oblique shock waves must be 24° and for a given number of oblique shock waves, the turning angle associated with each shock must be the same. What would be the minimum loss for a large number of oblique shock waves for such a system? What problems might be encountered in using such a system?

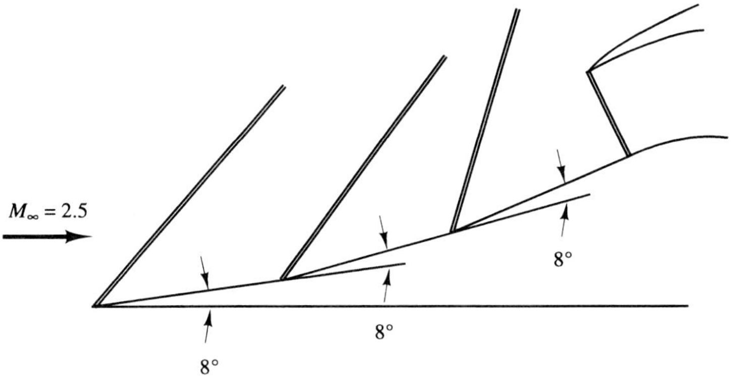

Figure 6-72

60. As shown in Figure 6-73, oblique shock waves appear at the exit of an overexpanded supersonic nozzle. Determine the stagnation pressure in the chamber when the nozzle back pressure is 14.696 psia. The nozzle exit-to-throat area is 2.75. If the chamber stagnation temperature is 700 R, calculate the nozzle exit velocity.

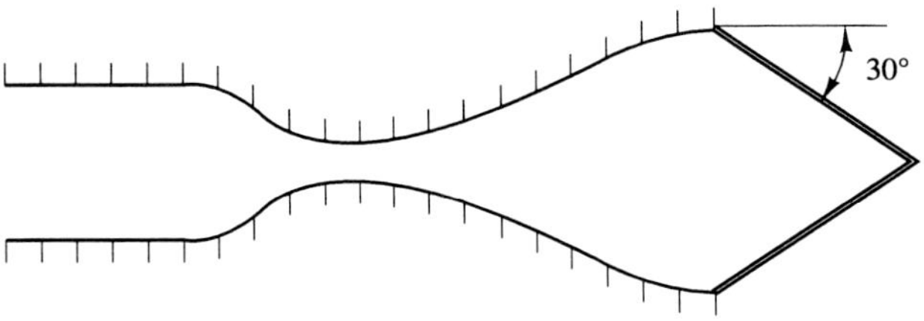

Figure 6-73

61. A converging–diverging nozzle with a design exit Mach number of 2.75 operates in an overexpanded mode between a back pressure of 80 kPa and a reservoir pressure of 280 kPa, as shown in Figure 6-74. Compute the flow direction and flow Mach number in region R if air is the working fluid.

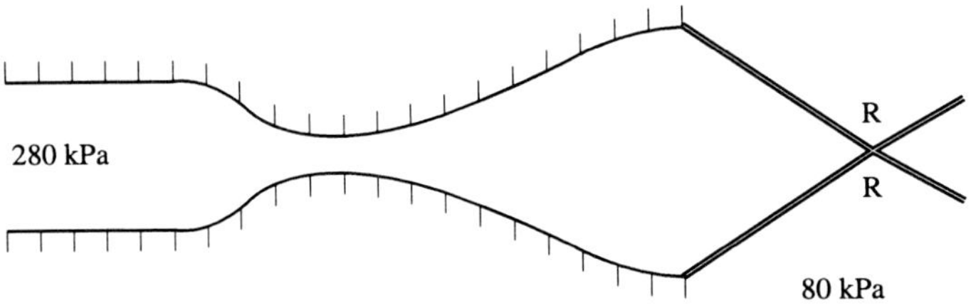

Figure 6-74

62. Two converging–diverging nozzles are in series as shown in Figure 6-75. If air is the fluid, find the area ratio A_3/A_2 necessary to produce sonic flow at the second throat in the test section. Find the loss in stagnation pressure for this situation. At another condition, a shock appears at the second throat. Find the loss in stagnation pressure for this condition.

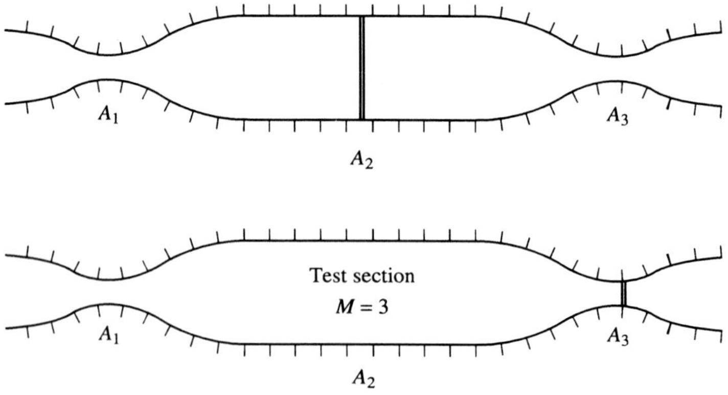

Figure 6-75

63. Write a computer program that will yield values of T/T_o, P/P_o, and A/A^* for isentropic flow of a perfect gas with constant $\gamma = 1.35$. Use Mach number increments of 0.05 over the range $M = 0$ to $M = 5.0$. Verify using COMPQ.

Other Simple Flows

7

7.1 INTRODUCTION

In Chapter 5, the generalized one-dimensional formulation was developed by identifying four driving potentials: area change, friction, heat addition or rejection, and mass addition. Flows in which only a single driving potential is present were labeled simple flows. Because Mach number and property relations for simple flows can be expressed in closed form (see Table 5-2 for a compact presentation), such flows are usually treated individually. Simple area change was explored and developed in Chapter 6. The remaining three simple flows will be examined in this chapter.

Flow with friction is called Fanno flow, and flow with heat addition or rejection is called Rayleigh flow. Simple mass addition has not been associated with any particular individual. The same basic approach will be used in the study of these three simple flows as was used in the discussion of simple area change.

7.2 SIMPLE FRICTIONAL FLOW (FANNO FLOW)

If the only driving potential is friction, simple frictional flow, or *Fanno flow*, is defined. Table 7-1a reproduces for convenience the influence coefficient entries from Table 5-1 for Fanno flow. The independent variable is given as $4\,fdx/D_h$, where f is the Fanning friction factor (defined by Eq. 5-11) and D_h is the hydraulic diameter (defined by Eq. 5-13). Since no additional drag forces are considered for the development in this section, dF_d is taken as zero. The closed-form expressions for Fanno flow from Table 5-2 are restated in Table 7-1b. The entropy change is referenced to the sonic state, so $\Delta s = s - s^*$. As with simple area change, the starred variables denote conditions at the sonic location. Fanno flow is thus flow in an adiabatic (T_o = constant), constant-area duct (A = constant) with no mass addition ($\dot{m}$ = constant), but with friction. If the conservation equations and various definitions are applied to an elemental control volume of constant area, stagnation temperature, and mass flow rate, the expressions of Table 7-1a result.

TABLE 7-1 SUMMARY OF RELATIONS FOR FANNO FLOW

(a) Differential forms	(b) Property ratios
$\frac{dM^2}{M^2} = \frac{\gamma M^2 \psi}{1 - M^2} \frac{4f}{D_h} dx \qquad \psi = 1 + \frac{\gamma - 1}{2} M^2$	$4\frac{fL^*}{D_h} = \frac{1 - M^2}{\gamma M^2} + \left(\frac{\gamma + 1}{2\gamma}\right) \ln\left\{M^2\left[\left(\frac{2}{\gamma + 1}\right)\psi\right]^{-1}\right\}$
$\frac{dV}{V} = \frac{\gamma M^2}{2(1 - M^2)} \frac{4f}{D_h} dx$	$\frac{P}{P^*} = \frac{1}{M}\left[\left(\frac{2}{\gamma + 1}\right)\psi\right]^{-1/2}$
$\frac{dT}{T} = -\frac{\gamma(\gamma - 1)M^4}{2(1 - M^2)} \frac{4f}{D_h} dx$	$\frac{\rho}{\rho^*} = \frac{1}{M}\left[\left(\frac{2}{\gamma + 1}\right)\psi\right]^{-1/2}$
$\frac{d\rho}{\rho} = -\frac{\gamma M^2}{2(1 - M^2)} \frac{4f}{D_h} dx$	$\frac{T}{T^*} = \left[\left(\frac{2}{\gamma + 1}\right)\psi\right]^{-1}$
$\frac{dP}{P} = -\frac{\gamma M^2[1 + (\gamma - 1)M^2]}{2(1 - M^2)} \frac{4f}{D_h} dx$	$\frac{V}{V^*} = M\left[\left(\frac{2}{\gamma + 1}\right)\psi\right]^{-1/2}$
$\frac{dP_o}{P_o} = -\frac{\gamma M^2}{2} \frac{4f}{D_h} dx$	$\frac{P_o}{P_o^*} = \frac{1}{M}\left[\left(\frac{2}{\gamma + 1}\right)\psi\right]^{(\gamma+1)/2(\gamma-1)}$
$\frac{dI}{I} = -\frac{\gamma M^2}{2(1 + \gamma M^2)} \frac{4f}{D_h} dx$	$\frac{I}{I^*} = \frac{1 + \gamma M^2}{M\sqrt{2(\gamma + 1)\psi}}$
$\frac{ds}{c_p} = \frac{(\gamma - 1)M^2}{2} \frac{4f}{D_h} dx$	$\frac{\Delta s}{c_p} = \ln\left\{M^{(\gamma-1)/\gamma}\left[\left(\frac{2}{\gamma + 1}\right)\psi\right]^{-(\gamma+1)/2\gamma}\right\}$
	$A = A^* = \text{constant}$
	$T_o = T_o^* = \text{constant}$
	$\dot{m} = \dot{m}^* = \text{constant}$
	$\psi = 1 + \frac{\gamma - 1}{2} M^2$

The only expression with a new functional form appearing in Table 7-1b is the first entry, the one that relates $4fL^*/D_h$ to the Mach number. The variable L refers to the duct length, and so L^* refers to the length of duct required to achieve Mach 1 from a Mach number M. This concept will be fully discussed later in this section. The distance down the duct then effectively becomes the independent variable in Fanno flow. The influence coefficients for entropy and stagnation pressure show that entropy increases and stagnation pressure decreases in the positive x-direction in a

Fanno flow. Figure 7-1 schematically represents the conditions and nomenclature for Fanno flow. Fanno flow is adiabatic but irreversible, which is not surprising when one considers the effects of friction.

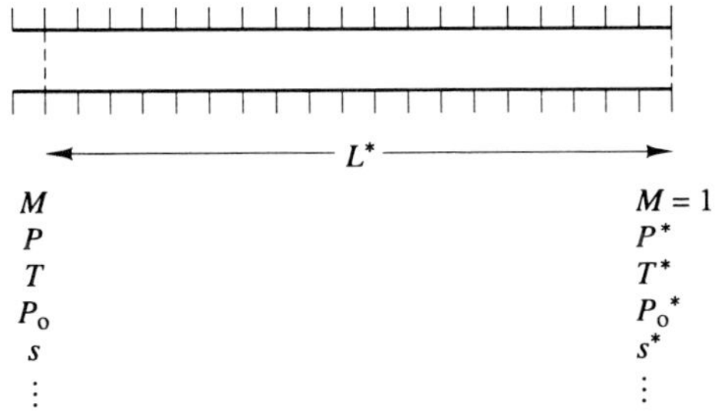

Figure 7-1 Fanno Flow Nomenclature

The friction factor was taken as constant in the simple frictional flow results derived in Chapter 5 and presented in Table 5-2. Since those equations serve as the basis for Fanno flow, the friction factor, f, over a reach of Fanno flow is likewise taken as constant. Thus, the f in the Fanno term, $4fL/D_h$, is usually viewed as the average value

$$\bar{f} = \frac{1}{L_1 - L_2}\int_{L_1}^{L_2} f\,dx \tag{7-1a}$$

acting over the line length of interest. Indeed, the friction factor over a segment of fully developed, isothermal, incompressible flow in a constant-area duct is invariant along the duct, which makes the average value just the invariant value. In a compressible flow, the situation is not nearly so clear, and the question of what is the appropriate constant value to use in Fanno flow must be addressed.

The friction factor for incompressible, isothermal flow is a function of Reynolds number and relative roughness; that is, $f = f(\mathrm{Re}_D, \epsilon/D)$. The Moody diagram (Moody, 1944) is the accepted graphical representation for $f(\mathrm{Re}_D, \epsilon/D)$ for incompressible, isothermal flow. Appendix I contains a general discussion and presents several alternative approaches for evaluating f in an incompressible flow. Benedict (1980) presents an excellent discussion of friction factors including alternative expressions.

The addition of Mach number as an important dimensionless parameter in a compressible flow results in a more complex friction factor functional relationship, $f = f(\mathrm{Re}_D, \epsilon/D, M)$, than in incompressible flow. Shapiro (1953) suggests that for subsonic flows, the Mach number dependence is not important and that the incompressible value can be used. The usual procedure is to use the Moody diagram or its functional representation, as in Appendix I, and to evaluate the friction factor using the Reynolds number definition

$$\mathrm{Re}_{D_h} = \rho V D_h/\mu \tag{7-1b}$$

and the relative roughness, ϵ/D_h. For duct flows, ρV is usually written as $\dot{m}/A$, where A is the duct cross-sectional area. Since the product ρV is constant for steady-state duct flows, it can be computed at any location along the duct. The viscosity is evaluated at the mean static temperature over the region of interest. No problems arise in Fanno situations for which the Mach number range is small, since the fluid static temperature and viscosity are nearly constant. Some experimental evidence supports this concept for subsonic flows.

Supersonic flow that can be adequately modeled as Fanno flow is relatively difficult to establish and, as we shall see, is very limited in the duct length along which it can be maintained. The limited duct length associated with supersonic Fanno flow precludes establishing fully developed conditions. In supersonic duct flows the fluid properties are changing so rapidly that one-dimensional representation of the process typically uses an apparent friction factor that includes wall shearing stress as well as momentum flux differences. Shapiro (1953), based on his observation of data taken prior to 1953, suggests that the friction factor decreases, perhaps by as much as 50 percent, for supersonic duct flows. Zucrow and Hoffman (1976), however, state that experiments with steam and air indicate that in well-developed turbulent flow the friction factor is nearly independent of the Mach number. Emanuel (1986) points out that for supersonic duct flows no equivalent chart to the Moody diagram exists. Thus, the usual procedure is to compute the friction factor for a supersonic flow as for a subsonic flow, realizing that the range of static temperature variation may be significant. We may console ourselves in this situation by remembering that although some of the results for supersonic duct flows may not possess the accuracy that we would like, the overall behavior of supersonic duct flow is captured by the Fanno analysis.

In much of our examination of Fanno flow, we shall be more interested in understanding the nuances of Fanno flow and will simply specify the friction factor. Reality, however, is that the friction factor is not usually known and must be computed using the procedures of the previous paragraph. If the Reynolds number is to be computed, then the viscosity must be known. Many textbooks and handbooks contain tables of values of viscosity for different gases. The most widely used closed-form expression for the viscosity for gases of interest to us is that due to Sutherland (White, 1991):

$$\mu = \mu_{\text{ref}}\left(\frac{T}{T_{\text{ref}}}\right)^{1.5}\frac{T_{\text{ref}} + S}{T + S}, \tag{7-2}$$

where s is the Sutherland constant and μ_{ref} is the viscosity evaluated at the reference temperature, T_{ref}. Table 7-2 presents the Sutherland constant, T_{ref}, and μ_{ref} for several gases of interest. Sutherland's expression is valid for dilute gases of a single component substance. Air is a mixture primarily of nitrogen and oxygen with small traces of other elements and components, but nitrogen and oxygen molecules are so similar to each other that the viscosity of air can be computed accurately using the Sutherland expression.

Many tabulations of thermophysical properties of gas, particularly air, are available in the literature. For example, Brower (1990) presents temperature, density, speed of sound, viscosity, and thermal conductivity as a function of pressure

TABLE 7-2 SUTHERLAND LAW CONSTANTS FOR SELECTED GASES

Gas	T_{ref} °R	$\mu_{ref} \times 10^5$ lbm/ft-s	S °R	T_{ref} °K	$\mu_{ref} \times 10^5$ N s/m²	S °K
Air	491.6	1.1530	199	273.1	1.7160	111
N_2	491.6	1.1170	192	273.1	1.6630	107
O_2	491.6	1.2900	250	273.1	1.9190	139
H_2	491.6	0.5652	174	273.1	0.8411	97

Adapted from White (1991) and Hilsenrath et al. (1955).

for several gases, including air, nitrogen, and oxygen. Sauerwein and Dalton (1985) contains a contemporary, extensive list of references for thermophysical properties.

The path of the Fanno process on an *h–s* or *T–s* diagram is a convenient mechanism for exploring this simple flow. Although the Fanno path can be represented on property diagrams involving other variables, the *h–s* is the most useful version and is the traditional one used. The Fanno path represents the locus of states for a particular Fanno flow and is usually called the *Fanno line*. From Table 7-1b the change in entropy is

$$\frac{\Delta s}{c_p} = \ln\left\{M^{(\gamma-1)/\gamma}\left[\left(\frac{2}{\gamma+1}\right)\psi\right]^{-(\gamma+1)/2\gamma}\right\}. \tag{7-3}$$

The static and stagnation temperature are related by

$$\frac{T_o}{T} = 1 + \frac{\gamma-1}{2}M^2 \equiv \psi, \tag{1-66}$$

from which

$$M = \sqrt{\left(\frac{T_o}{T} - 1\right)\frac{2}{\gamma-1}}. \tag{7-4}$$

At the sonic point,

$$\frac{T_o}{T^*} = \frac{\gamma+1}{2}. \tag{7-5}$$

Substituting Eq. 7-4 for the Mach number and using the definition of ψ, the change in entropy, expressed in terms of the temperature ratios T/T_o and T^*/T_o, becomes

$$\frac{\Delta s}{c_p} = \ln\left\{\left(\frac{T_o}{T} - 1\right)^{(\gamma-1)/2\gamma}\left(\frac{2}{\gamma-1}\right)^{(\gamma-1)/2\gamma}\left(\frac{\gamma+1}{2}\right)^{(\gamma+1)/2\gamma}\left(\frac{T}{T_o}\right)^{(\gamma+1)/2\gamma}\right\}. \tag{7-6}$$

For a calorically perfect gas the temperature ratio is equal to the enthalpy ratio since

$$\frac{h}{h_o} = \frac{c_p T}{c_p T_o} = \frac{T}{T_o}. \tag{7-7}$$

Equation 7-6 represents a Fanno line on a dimensionless *h–s* or *T–s* diagram where the abscissa is the dimensionless entropy, $\Delta s/c_p$, and the ordinate is the dimensionless temperature or the enthalpy ratio. The dimensionless Fanno plot on a *T–s* diagram for a specific heat ratio of 1.4 is given in Figure 7-2a. Mach number as a function of $\Delta s/c_p$ for $\gamma = 1.4$ is plotted as Figure 7-2b. Although Fanno lines can be drawn on property diagrams for any two properties, unless otherwise stated in this text a Fanno line will mean the enthalpy- (or temperature) versus-entropy presentation. Enthalpy–entropy representations are usually called *Mollier diagrams*.

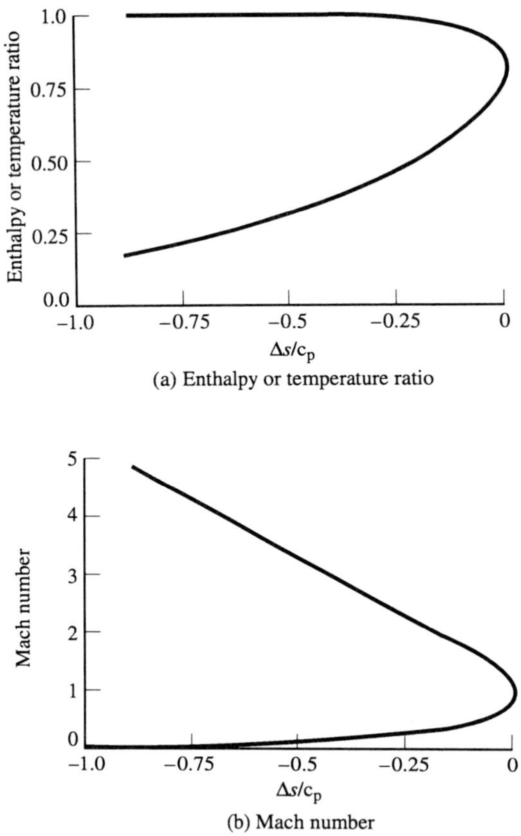

Figure 7-2 Fanno Lines for Calorically Perfect Gas ($\gamma = 1.4$)

By observation, Figure 7-2b suggests that the maximum entropy state is the sonic state. This can be verified. Cast in terms of the Mach number, the change in entropy for a Fanno flow becomes

$$\frac{\Delta s}{c_p} = \ln\left\{M^{(\gamma-1)/\gamma}\left[\left(\frac{2}{\gamma+1}\right)\left(1 + \frac{\gamma-1}{2}M^2\right)\right]^{-(\gamma+1)/2\gamma}\right\}. \tag{7-8}$$

To find the Mach number at the maximum entropy state, the derivative of $\Delta s/c_p$ or $\exp(\Delta s/c_p)$ with respect to Mach number can be taken and set equal to zero. No entropy minimum exists since both branches of the Fanno line are unbounded in the negative direction. The state at which the derivative of entropy vanishes will thus be the state of maximum entropy. The derivative becomes

$$\frac{d}{dM}(e^{\Delta s/c_p}) = \frac{\gamma-1}{\gamma}M^{-(1)/\gamma}\left[\left(\frac{2}{\gamma+1}\right)\left(1+\frac{\gamma-1}{2}M^2\right)\right]^{-(\gamma+1)/2\gamma}$$
$$+ M^{(\gamma-1)/\gamma}\left(-\frac{\gamma+1}{2\gamma}\right)\left[\left(\frac{2}{\gamma+1}\right)\left(1+\frac{\gamma-1}{2}M^2\right)\right]^{-(3\gamma+1)/2\gamma} \tag{7-7a}$$
$$\cdot\left(\frac{2}{\gamma+1}\right)\left(\frac{\gamma-1}{2}\right)2M = 0.$$

Simplifying yields

$$\frac{1}{M} = \frac{\gamma+1}{2}\frac{1}{1+\frac{\gamma-1}{2}M^2}M, \tag{7-7b}$$

which reduces to

$$M^2 = 1. \tag{7-7c}$$

For the Fanno flow of a calorically perfect gas, the state of maximum entropy is thus the sonic state, $M = 1$.

The second law of thermodynamics compels the Fanno process to move into a region of increasing entropy; hence, the process must move from left to right as indicated on the Fanno line in Figure 7-3. The sonic state occurs at the point of maximum entropy and is also shown on the figure. Subsonic Mach numbers are associated with the large temperature or enthalpy ratios, and supersonic Mach numbers with the smaller values of the ratios. Thus, the upper portion of the Fanno line corresponds to subsonic flow and the lower portion to supersonic flow. Moreover, since the sonic point cannot be traversed without moving into a region of decreasing entropy, a Fanno flow that is initially subsonic must remain subsonic, and a Fanno flow that is initially supersonic must remain supersonic. Both supersonic and subsonic Fanno flows then move toward the Mach 1 state as the duct is traversed in the direction of flow. The meaning of L^* is now clear: it is the distance required in

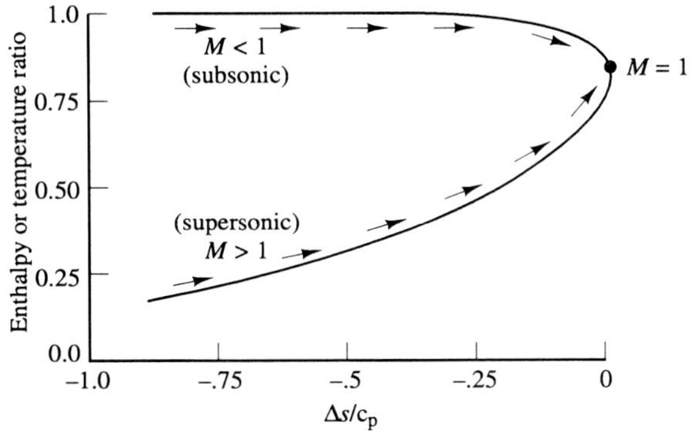

Figure 7-3 Dimensionless Fanno Line Representation ($\gamma = 1.4$)

a Fanno flow for the Mach number to become 1. As with the other simple flows, conditions at the location of Mach 1 in a Fanno flow are unique for that given flow and can be used as reference conditions to aid in calculations.

Figure 7-3 is the Fanno line representation for all Fanno flows with $\gamma = 1.4$ as expressed in the dimensionless coordinates $\Delta s/c_p$ and T/T_o (or h/h_o). If the Fanno lines for flows with different stagnation conditions and/or different mass flow rates were represented in dimensional coordinates s and T (or h), then the shapes would be similar to that of Figure 7-3, but the Fanno lines for the individual flows would be distinct. For example, in dimensional coordinates, Fanno flows with the same initial static states (T and P) but with different stagnation conditions would appear as in Figure 7-4. The initial state is the same for all three Fanno flows, but the different stagnation conditions lead to different and distinct Fanno lines. For given stagnation conditions, but with different mass flow rates, the various individual Fanno lines appear as in Figure 7-5. As the mass flow rate is decreased, the Fanno line is shifted to the right, in the direction of increasing entropy. In general, for our purposes the dimensionless Fanno line, Figure 7-3, will be used to explain various phenomena associated with Fanno flows.

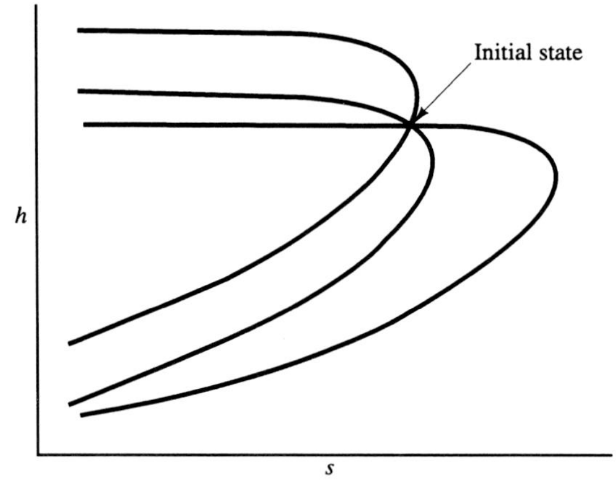

Figure 7-4 Fanno Lines with Different Stagnation Conditions but Equal Initial States

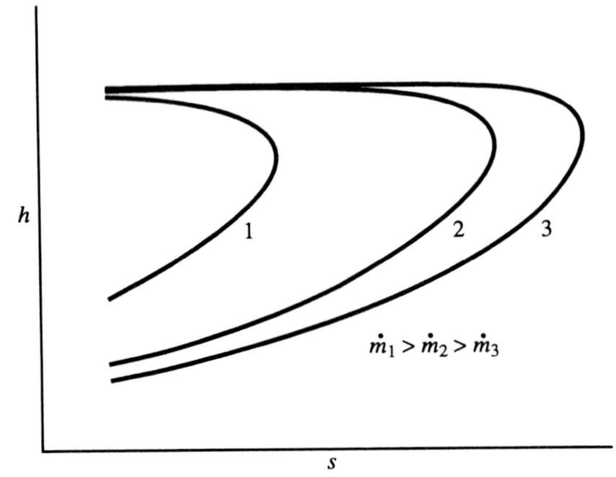

Figure 7-5 Fanno Lines with Different Mass Flow Rates but Equal Stagnation Conditions

Example 7-1

Plot the dimensional Fanno line for a flow with a stagnation temperature of 1000 R and an initial stagnation pressure of 100 psia. The initial Mach number is 0.2 for the Fanno flow.

Solution. The initial state occurs at $M = 0.2$ and is

$$M_i = 0.2$$

$$T_i = \frac{T_i}{T_o} T_o = (0.9921)(1000) = 992.1\ R$$

$$P_i = \frac{P_i}{P_{o_i}} P_{o_i} = (0.9725)(100) = 97.25 \text{ psia}.$$

The entropy of the initial state is arbitrarily taken as 0.25 Btu/lbm. Because only the change in entropy between two states is meaningful, the entropy datum could have been taken as any number; however, the choice of this number yields a Fanno plot with positive entropy at each state considered. A computer program was written to generate the enthalpy–entropy values needed for the plot. Using the Mach number as a parameter, the computation sequence was to compute the static temperature using

$$T = \frac{T_o}{1 + \dfrac{\gamma - 1}{2} M^2}$$

and the static enthalpy using

$$h = c_p T.$$

The dimensionless entropy was computed using, from Table 7-1b,

$$\frac{\Delta s}{c_p} = \ln\left\{ M^{(\gamma-1)/\gamma} \left[\left(\frac{2}{\gamma + 1} \right) \psi \right]^{-(\gamma+1)/2\gamma} \right\}. \tag{7-8}$$

The dimensional entropy was then generated from

$$s = s^* + c_p \left(\frac{\Delta s}{c_p} \right). \tag{7-9}$$

Plotting the results of the preceding calculations yields the Fanno line in Figure 7-6. The initial state is indicated on the figure.

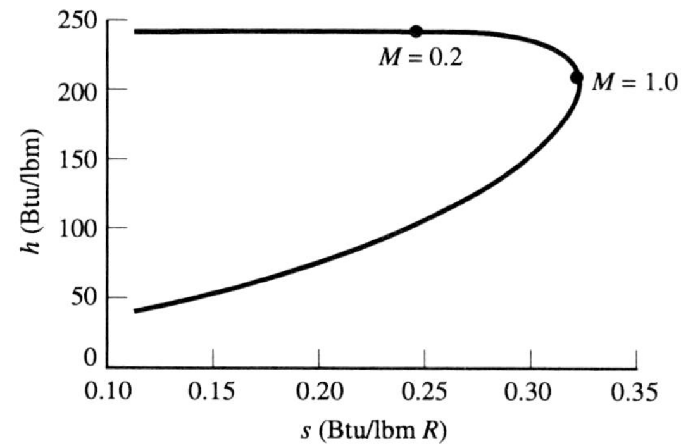

Figure 7-6 Dimensional Fanno Line for Example 7-1

As duct length is added, the process moves to the right in Figure 7-6. The flow is accelerated if M is subsonic and is decelerated if M is supersonic. Duct length can be added with no change in the mass flow rate until the sonic point, which is also indicated in the figure, is reached. If duct length beyond that associated with sonic flow of the initial Fanno line is added, the mass flow rate must be reduced, and the Fanno line corresponding to the new, smaller mass flow rate would be shifted to the right of the initial line.

Computational aids are available for Fanno flow. The expressions from Table 7-1b give the values of various property ratios and of $4fL^*/D_h$ as functions of Mach number. Tabulations of these relations for a specific heat ratio of 1.4 are presented in Appendix E. Figure 7-7 presents plots of the various property ratios and the Fanno parameter, $4fL^*/D_h$. The property ratios are all referenced to sonic conditions. Examination of the figure reveals that the Fanno parameter, which is unbounded as the Mach number approaches zero in subsonic flow, is restricted to a relatively small value for supersonic flow. This should not be surprising since

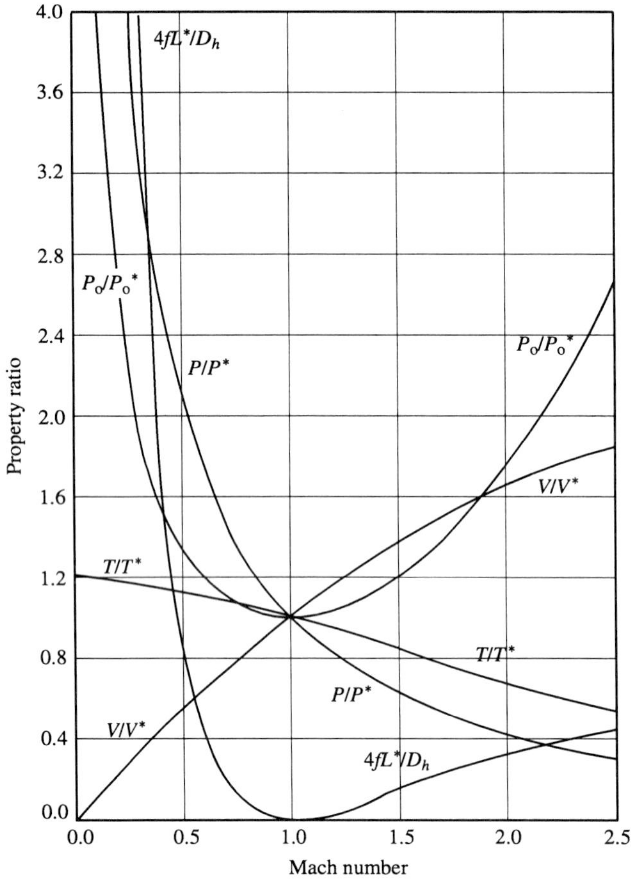

Figure 7-7 Property Ratios for Fanno Flow ($\gamma = 1.4$)

pressure drops in pipes are nearly proportional to the square of the velocity, indicating rapid changes in pressure down the duct for the high velocities encountered in supersonic flows.

Fanno flow computational aids are available under option **FA** in program COMPQ. The Fanning friction factor, defined in Eq. 5-10, is used in all COMPQ Fanno computations. The menu for option **FA** is given in Figure 7-8a. Four cases are available. Cases **M**, **F**, and **P** effectively replace the Fanno tables of Appendix E. The input for these cases is either the Mach number for case **M**, the Fanno parameter for case **F**, or the static pressure ratio for case **P**. The output for these cases are the Mach number, Fanno parameter, and various property ratios. For a specified Mach number of 0.2, the output for case **M** is given in Figure 7-8b. The output also reminds us that in Fanno flow the stagnation temperature is constant. The final case, **DA**, is used for the process involved in duct addition and will be examined in Example 7-2. COMPQ provides added versatility since the value of the specific heat ratio can be specified.

```
FANNO FLOW SELECTED
INPUT PARAMETERS ARE:
      MACH NUMBER              M
      4fL/D                    F
      P/PSTAR                  P
      DUCT ADDITION            DA
THE FANNING FRICTION FACTOR IS USED IN ALL FANNO COMPUTATIONS

ENTER THE DESIRED LETTER(S)? M
```

(a) COMPQ option **FA** menu

```
THE MACH NUMBER IS    0.2000
   4fL/D        T/T*          P/P*          PO/PO*       RHO/RHO*
  14.5333      1.1905        5.4554        2.9635        4.5826

STAGNATION TEMPERATURE IS CONSTANT FOR A FANNO FLOW.
```

(b) Case **M** output

Figure 7-8 Option **FA** Menu and Sample Output

7.3 FANNO FLOW APPLICATIONS

Duct addition is a recurring circumstance that is involved in many Fanno flow situations. In duct addition, information on the initial state, the friction factor, and the duct diameter and additional length are given; the final state after the flow traverses the duct length addition is desired. The duct addition problem is schematically illustrated in Figure 7-9. Conditions at station 1, f, and L are known. The solution technique is based on the concept that the duct location of the sonic point and the conditions at the sonic point in a given Fanno flow are unique. Thus, for state 1 the duct length required to achieve Mach 1 is L_1^*, and for state 2 the duct length required to achieve Mach 1 is L_2^*. For both states the sonic location and the

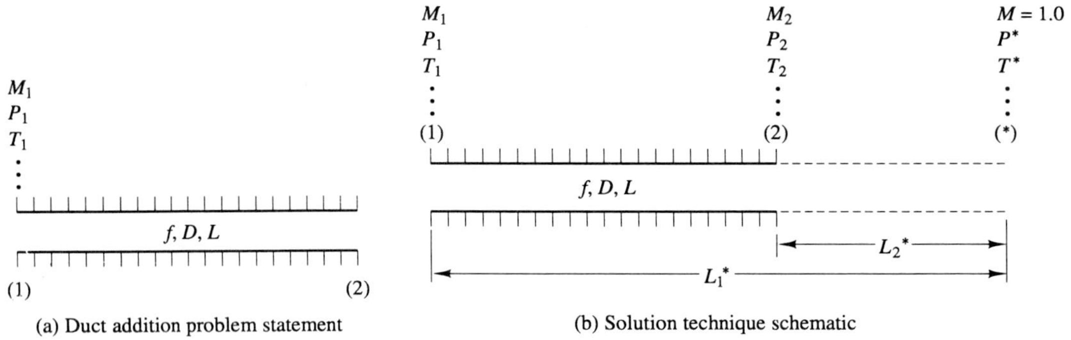

Figure 7-9 Duct Addition Problem

conditions at the sonic location must be the same. The Fanno parameters for states 1 and 2 are then related as

$$4\frac{fL_1^*}{D_h} = \left(4\frac{fL}{D_h}\right)_{\text{act}} + 4\frac{fL_2^*}{D_h}, \tag{7-10}$$

where $(4\,fL/D_h)_{\text{act}}$ is the physical contribution to the Fanno parameter made by the duct length L. Given the Fanno parameter at the inlet, we can find the Fanno parameter at the exit by using Eq. 7-10. The Mach number and property ratios at state 2 can then be found using the value of the Fanno parameter and the Fanno tables in either Appendix E or case **F** of option **FA** in COMPQ. Given the Mach number at state 2, the property ratios at state 2 can be computed by using the ratios

$$\frac{T_2}{T_1} = \frac{T_2}{T^*}\frac{T^*}{T_1} = \frac{T_2/T^*}{T_1/T^*};$$
$$\frac{P_2}{P_1} = \frac{P_2}{P^*}\frac{P^*}{P_1} = \frac{P_2/P^*}{P_1/P^*}. \tag{7-11}$$

Case **DA** of option **FA** accomplishes the duct addition process by going through the steps just outlined. Consider the following example.

Example 7-2

At the inlet to a segment of duct flow that can be modeled as a Fanno flow, the initial Mach number is 0.2 and the static pressure and temperature are 300 K and 200 kPa, respectively. If the duct is 0.1 m in diameter and the friction factor is 0.005, find the Mach number and properties after the flow has traversed 50 m of duct. Solve the duct addition problem using cases **M** and **F** as well as using the process **DA**.

Solution. This is the classical duct addition problem. The problem statement and solution schematics are presented in Figure 7-10. For circular cross sections, $D_h = D$. The path of the process is indicated on the Fanno line in Figure 7-10c. Consider the solution using cases **M** and **F**. At state 1, the stagnation conditions (case **M** of option **IS**) are

$$T_{o_1} = \frac{T_{o_1}}{T_1}T_1 = \frac{T_1}{T_1/T_{o_1}} = \frac{300}{0.9921} = 302.4 \text{ K};$$

$$P_{o_1} = \frac{P_{o_1}}{P_1}P_1 = \frac{P_1}{P_1/P_{o_1}} = \frac{200}{0.9725} = 205.7 \text{ kPa}.$$

Using case **M** of option **FA** for **M** = 0.2 yields

$$4\frac{fL_1^*}{D} = 14.533 \qquad \frac{T_1}{T^*} = 1.190$$

$$\frac{P_1}{P^*} = 5.455 \qquad \frac{P_{o_1}}{P_0^*} = 2.964.$$

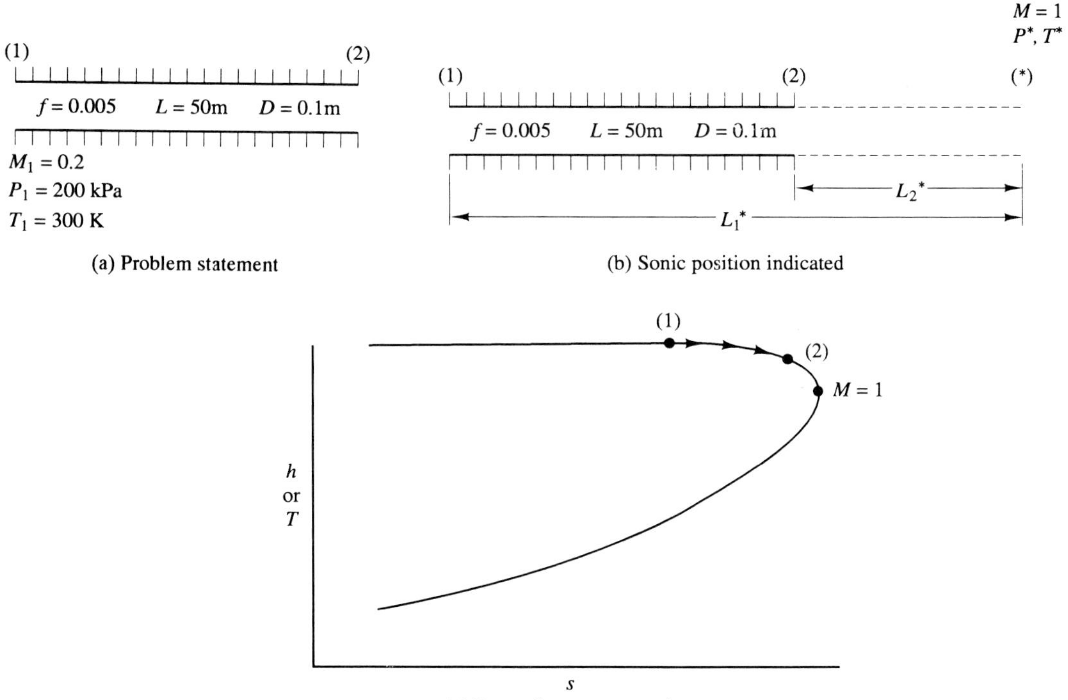

(a) Problem statement (b) Sonic position indicated

(c) Fanno line representation

```
ENTER THE DESIRED LETTER(S)? DA

ENTER THE MACH NUMBER AND THE 4fL/D INCREMENT? 0.2,10.0

     M1      (4fL/D)1   D(4fL/D)  (4fL/D)2
   0.2000    14.5333    10.0000    4.5333

AT STATE 2 THE MACH NUMBER IS   0.3178

   T2/T1      P2/P1      PO2/PO1   RHO2/RHO1
   0.9880     0.6255     0.6524     0.6331

STAGNATION TEMPERATURE IS CONSTANT FOR A FANNO FLOW.
```

Figure 7-10 Example 7-2

The actual value of $4\,fL/D$ for the duct addition is

$$\left(4\frac{fL}{D}\right)_{\text{act}} = \frac{(4)(0.005)(50)}{0.1} = 10.0,$$

so that the Fanno parameter at state 2 is

$$4\frac{fL_1^*}{D} = \left(4\frac{fL}{D}\right)_{\text{act}} + 4\frac{fL_2^*}{D};$$

$$4\frac{fL_2^*}{D} = 14.533 - 10.0 = 4.533.$$

Using case **F** of option **FA** allows us to find the Mach number and property ratios of

$$M_2 = 0.3178 \qquad \frac{T_2}{T^*} = 1.176$$

$$\frac{P_2}{P^*} = 3.413 \qquad \frac{P_{o2}}{P_o^*} = 1.934.$$

Then, the state-2-to-state-1 property ratios are

$$\frac{P_2}{P_1} = \frac{P_2}{P^*}\frac{P^*}{P_1} = \frac{P_2/P^*}{P_1/P^*} = \frac{3.413}{5.455} = 0.6257$$

$$\frac{T_2}{T_1} = \frac{T_2}{T^*}\frac{T^*}{T_1} = \frac{T_2/T^*}{T_1/T^*} = \frac{1.176}{1.190} = 0.9882$$

$$\frac{P_{o_2}}{P_{o_1}} = \frac{P_{o_2}}{P_o^*}\frac{P_o^*}{P_{o_1}} = \frac{P_{o_2}/P_o^*}{P_{o_1}/P_o^*} = \frac{1.934}{2.964} = 0.6525,$$

from which

$$P_2 = 125.1 \text{ kPa}$$

$$T_2 = 296.5 \text{ K}$$

$$P_{o_2} = 134.2 \text{ kPa}.$$

The loss in stagnation pressure is

$$\Delta P_o = P_{o_2} - P_{o_1} = -71.5 \text{ kPa}.$$

The stagnation temperature is constant at 302.4 K for the Fanno process.

The solution using case **DA**—the process solution—is presented next. Case **DA** requires as inputs the initial Mach number and the $(4\,fL/D)_{\text{act}}$ of the added duct length. Figure 7-10d contains the output for the process solution. The Mach number and the property ratios from case **DA** are

$$M_2 = 0.3178 \qquad \frac{T_2}{T_1} = 0.988$$

$$\frac{P_2}{P_1} = 0.626 \qquad \frac{P_{o2}}{P_{o_1}} = 0.652.$$

These values are the same as in the first solution technique but were obtained using only a single case of option **FA**. Case **DA** for the process of duct addition takes less effort and arithmetic than the solution using cases **M** and **F**.

Example 7-3

At the end of a segment of Fanno flow, the Mach number is 0.7, the pressure is 150 kPa, and the temperature is 300 K. If the friction factor is 0.004, and the duct diameter is 0.05, find the Mach number and the property values 25 m upstream of the duct end.

Solution. The situation is sketched in Figure 7-11a. The states on the Fanno line, as indicated in Figure 7-11b, show state 2 lying upstream of state 1. This problem is very similar to Example 7-2 except that conditions are required upstream in a duct addition formulation. The Fanno parameters are related as

$$4\frac{fL_2^*}{D} = \left(4\frac{fL}{D}\right)_{\text{act}} + 4\frac{fL_1^*}{D},$$

so that

$$4\frac{fL_1^*}{D} = 4\frac{fL_2^*}{D} - \left(4\frac{fL}{D}\right)_{\text{act}}.$$

The problem can be solved in the same fashions as Example 7-2 except that $(4fL/D_h)_{\text{act}}$ is negative, not positive. Case **DA** of option **FA** will be used. The value of $(4fL/D)$ is

$$\left(4\frac{fL}{D}\right)_{\text{act}} = \frac{(4)(0.004)(25)}{0.05} = 8.$$

Since state 2 is upstream of state 1, the value of the $(4fL/D)$ that is input for case **DA** is negative. The input and output for this "negative duct addition" problem is shown

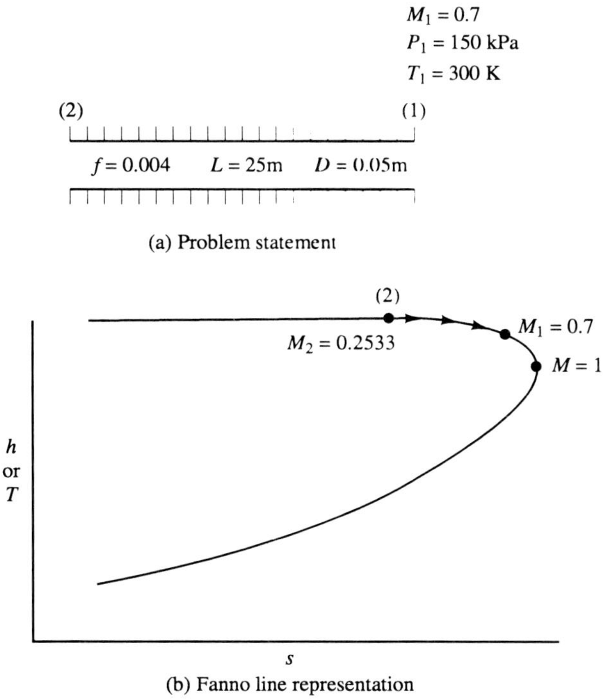

Figure 7-11 Example 7-3

```
ENTER THE DESIRED LETTER(S)? DA

ENTER THE MACH NUMBER AND THE 4fL/D INCREMENT? 0.7,-8.0
A NEGATIVE VALUE OF THE 4fL/D INCREMENT INDICATES THAT STATE 2 IS UPSTREAM OF ST
ATE 1.

     M1     (4fL/D)1   D(4fL/D)  (4fL/D)2
   0.7000     0.2081   -8.0000     8.2081

AT STATE 2 THE MACH NUMBER IS    0.2533

   T2/T1      P2/P1      PO2/PO1   RHO2/RHO1
   1.0841     2.8770     2.1688     2.6538

STAGNATION TEMPERATURE IS CONSTANT FOR A FANNO FLOW.
```

(c) Case **DA** output

Figure 7-11 (continued)

in Figure 7-11c. In using case **DA** in this sense, it is very important to note that state 2 is ahead of state 1. The conditions 25 m upstream of the end of the duct are

$$M_2 = 0.2533$$
$$P_2 = (2.877)(150) = 431.6 \text{ kPa}$$
$$T_2 = (1.084)(300) = 325.2 \text{ K}$$
$$\frac{P_{o_2}}{P_{o_1}} = 2.169.$$

The solution techniques using the individual cases **M** and **F** could have been used to solve the problem. Again, the Fanno parameter increment, $(4\,fL/D)_{\text{act}}$, would have been used in the negative sense. This example illustrates a very important aspect of Fanno flows and duct addition problem statements. The use of case **DA** going upstream in a Fanno flow is inherent in the technique and requires no additional logic in COMPQ.

7.4 FANNO FLOW APPLICATIONS INVOLVING NOZZLES

Thus far the Fanno flow discussions have involved considerations of simple frictional flows in constant-area duct segments. However, constant-area duct segments are often used in conjunction with converging or converging–diverging nozzles. Consider, as in Figure 7-12, a converging nozzle feeding a constant-area duct. The converging nozzle can be modeled as an isentropic flow, and the constant-area segment as a Fanno flow. As we shall see in Chapter 8, in nonsimple flows involving both area change and friction, area change effects usually dominate friction effects, so that the assumption of isentropic flow in the converging section is reasonable. A system involving both a converging nozzle and a constant-area duct segment must not violate any of the principles established for the behavior of either the nozzle or the duct segment.

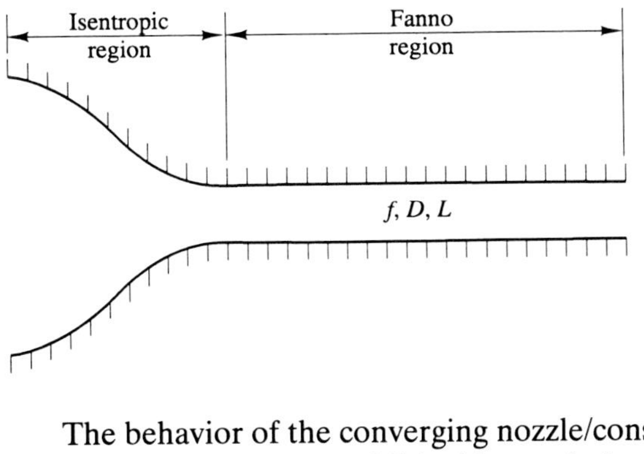

Figure 7-12 Converging Nozzle/Constant-Area Duct System

The behavior of the converging nozzle/constant-area duct system with respect to back pressure can be established in much the same fashion as for the converging nozzle. The situation is illustrated in Figure 7-13a. If the back pressure and the reservoir pressure are the same, then no flow occurs. As the back pressure is

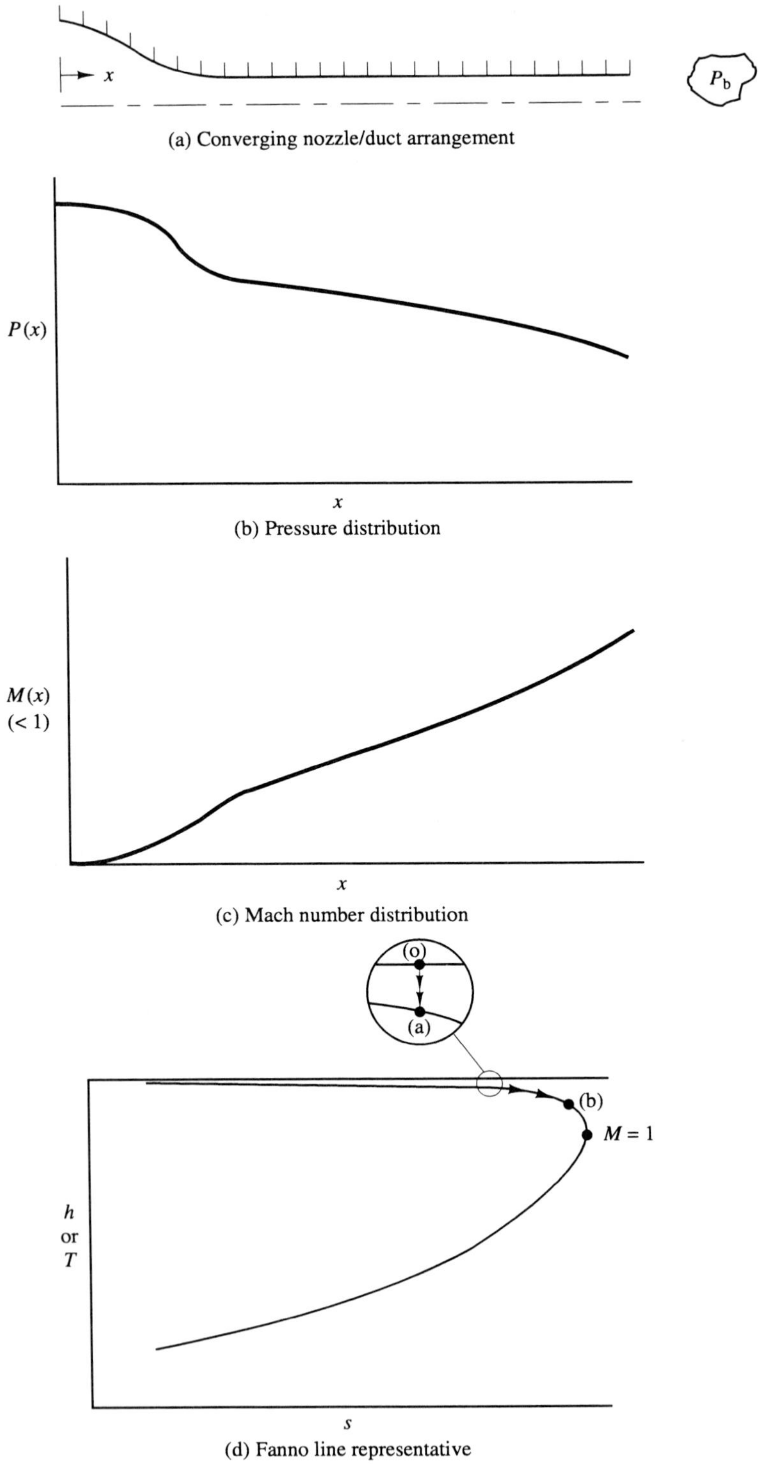

Figure 7-13 Converging Nozzle/Constant-Area Duct in Subsonic Flow

decreased, flow is induced in the system. The duct exit plane pressure must follow the same relationships as for any converging duct exit plane: if the exit Mach number is subsonic, the duct exit plane pressure must be equal to the back pressure, and if the exit Mach number is sonic (or supersonic), the duct exit plane pressure must be equal to or greater than the back pressure. A back pressure that results in a subsonic exit Mach number yields a pressure distribution as sketched in Figure 7-13b and a Mach number distribution as depicted in Figure 7-13c. The process path on the Fanno line is portrayed in Figure 7-13d. Flow in the converging nozzle is isentropic and is indicated in Figure 7-13d as the vertical line going from the stagnation state "o" to the nozzle exit state "a." The constant-area duct Fanno flow segment is indicated as from state "a" to the exit plane state "b." In both the converging nozzle segment and the constant-area duct segment, the Mach number increases and the pressure decreases. The highest Mach number in the system occurs at the duct exit.

The back pressure can be reduced low enough for the converging nozzle/constant-area duct system to choke. Since the highest Mach number in the system exists in the duct exit plane, sonic flow, if it occurs, must occur in the duct exit plane. This behavior is evident from an examination of the Fanno line. The flow cannot choke at the converging nozzle exit, because the entrance to the Fanno flow region would then occur at the point of maximum entropy, and movement along the Fanno line into regions of decreasing entropy would thereby violate the second law of thermodynamics. The only tenable path is for the flow to be subsonic in the nozzle exit plane, which is also the entrance to the Fanno flow region, and for the flow to move subsonically to the sonic point on the Fanno line.

Once the back pressure is reduced low enough for the flow to choke in the converging nozzle/constant-area duct system, further reductions in back pressure have no effect on either the mass flow rate of the system or the distributions of Mach number or properties through the system. This is exactly the behavior that the converging nozzle possessed with respect to back pressure. In Figure 7-14a the pressure distributions for a constant reservoir pressure/variable back pressure situation are shown. The pressure distributions, and hence the mass flow rates, depend on the back pressure as long as the flow is not choked. Once the flow is choked, the duct exit plane pressure can be greater than the back pressure, and the system will operate in an underexpanded mode with an expansion fan at the duct exit. The key back pressure is then the maximum back pressure, $P = P_b^*$, that results in choked flow in the system. The path of the process on the Fanno line is illustrated in Figure 7-14b. All other parameters being the same, the longer the constant-area duct section, the lower the maximum back pressure required for choked flow in the system. An example problem is in order.

Example 7-4

A converging nozzle/constant-area duct is connected to a reservoir with stagnation conditions of 100 psia and 500 R. The diameter is 0.1 ft, and the Fanning friction factor is 0.0025. For duct lengths of 0, 10, and 100 ft, find the maximum back pressure required for choked flow. What is the percentage reduction in mass flow rate caused by the two finite-length ducts?

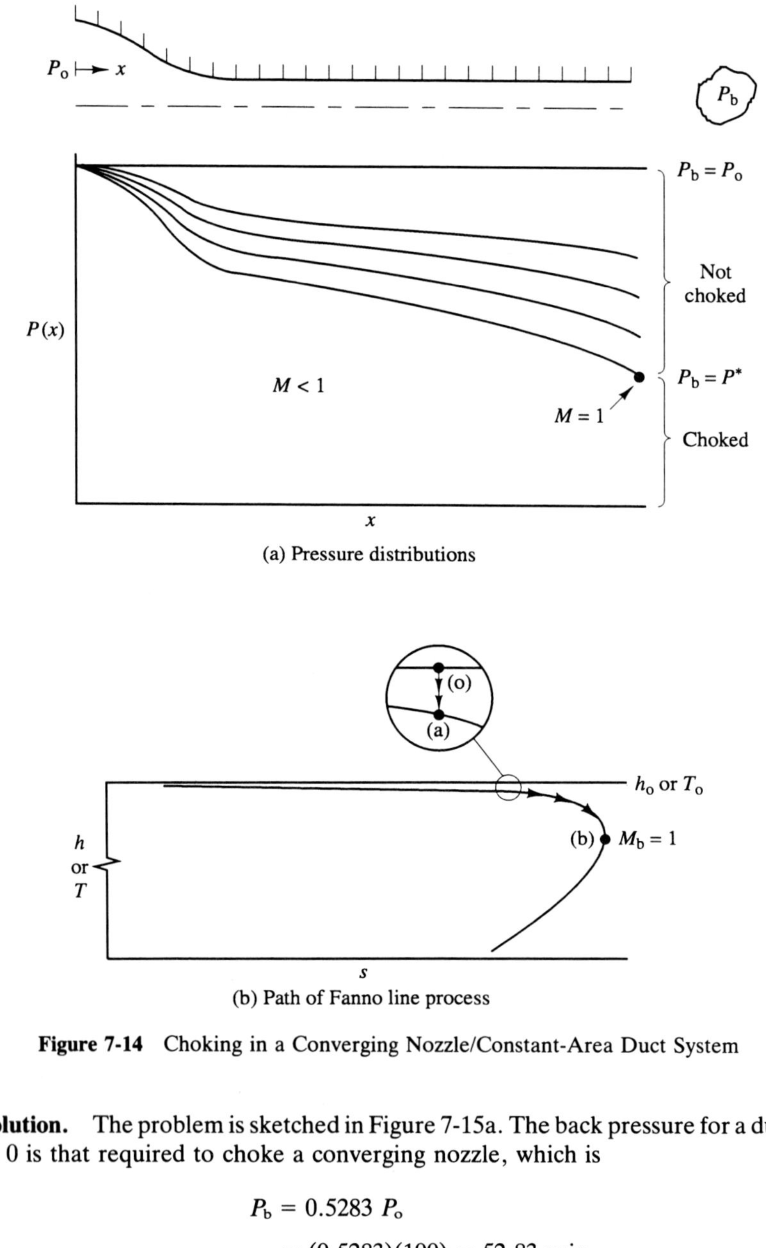

Figure 7-14 Choking in a Converging Nozzle/Constant-Area Duct System

Solution. The problem is sketched in Figure 7-15a. The back pressure for a duct length of 0 is that required to choke a converging nozzle, which is

$$P_b = 0.5283\ P_o$$
$$= (0.5283)(100) = 52.83 \text{ psia.}$$

The back pressure required to choke the system for a finite duct length can be obtained from the pressure ratio relation

$$P_b = P_1 = P^* = \frac{P_1}{P^*}\frac{P^*}{P_2}\frac{P_2}{P_{o2}}P_{o2} = \frac{P_2/P_{o2}}{P_2/P^*}P_{o2}$$

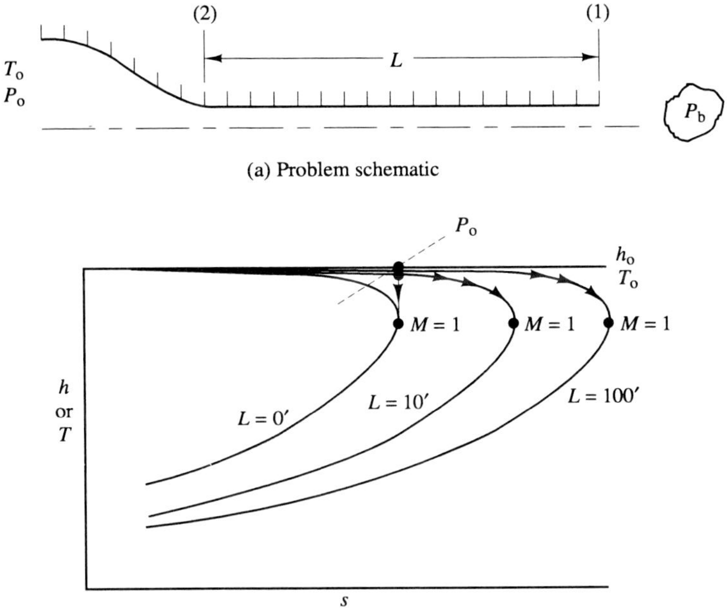

Figure 7-15 Example Problem 7-4

if cases **M** and **F** of option **FA** are used, or from

$$P_b = P_1 = P^* = \frac{P_1}{P_2}\frac{P_2}{P_{o_2}}P_{o_2} = \frac{P_2/P_{o2}}{P_2/P_1}P_{o_2}$$

if case **DA**, the duct addition option, is used. The $(4\,fL/D)_{\text{act}}$ for the two finite-length duct additions are given in Table 7-3. Case **DA** of option **FA** was used to compute the Fanno region property ratios and upstream Mach numbers for the two finite duct lengths. With the Mach number known at the duct exit and with the $(4\,fL/D)_{\text{act}}$ of the constant-area duct prescribed, the execution of case **DA** yields the condition at station 2, the converging section exit. The outputs are summarized in Table 7-3. Evaluating the mass flow rates at the exit of the converging nozzle gives

$$\dot{m} = \rho_2 A_2 V_2.$$

Using the equation of state and the definitions of Mach number and speed of sound, we can reduce this to

$$\dot{m} = \sqrt{\gamma/R}A_2\frac{P_2 M_2}{\sqrt{T_2}}.$$

TABLE 7-3 SOLUTION SUMMARY FOR EXAMPLE 7-4

L (ft)	$4\,fL/D$	M_1	M_2	P_2/P_1	T_2/T_1	P_2/P_{o_2}	T_2/T_0	P_b
0	0		1.0000			0.5283	0.8333	52.83
10	1.0	1.0	0.5087	2.100	1.141	0.8381	0.9508	39.91
100	10.0	1.0	0.2339	4.658	1.187	0.9626	0.9892	20.67

Then for the three specified duct lengths

$$\dot{m}|_{L=0'} = \sqrt{\gamma/R}A_2\frac{0.5283P_o(1)}{\sqrt{.833T_o}} = 0.5787\sqrt{\gamma/R}A_2\frac{P_o}{\sqrt{T_o}};$$

$$\dot{m}|_{L=10'} = \sqrt{\gamma/R}A_2\frac{0.8381P_o(0.5087)}{\sqrt{0.9508T_o}} = 0.4372\sqrt{\gamma/R}A_2\frac{P_o}{\sqrt{T_o}};$$

$$\dot{m}|_{L=100'} = \sqrt{\gamma/R}A_2\frac{0.9626P_o(.2339)}{\sqrt{.9892T_o}} = 0.2264\sqrt{\gamma/R}A_2\frac{P_o}{\sqrt{T_o}}.$$

The percentage changes in mass flow rate over only the converging nozzle configuration are

$$\frac{\dot{m}|_{L=10'} - \dot{m}|_{L=0}}{\dot{m}|_{L=0}} \times 100 = -24.45;$$

$$\frac{\dot{m}|_{L=100'} - \dot{m}|_{L=0}}{\dot{m}|_{L=0}} \times 100 = -60.88.$$

For a constant pressure reservoir feeding a converging nozzle, the addition of a constant-area duct to the converging nozzle requires a lower back pressure for choking to occur and reduces the mass flow rate. The former occurs because of the reduction in static pressure in the downstream direction as the constant-area duct is traversed. The longer the duct, the more the static pressure is reduced and the lower back pressure required for choking. The latter condition is caused by the friction-induced reduction in stagnation pressure at the end of the duct where choking takes place. The longer the duct, the more the loss in stagnation pressure and the lower the flow rate for given entrance stagnation conditions. In Chapter 6 we developed an expression for the mass flow rate for choked flow expressed in terms of the stagnation conditions for an isentropic flow. That expression, Eq. 6-12, is also valid for a choked Fanno flow if the stagnation conditions at the point where choking occurs are used. Hence,

$$\dot{m} = \sqrt{\gamma/R}\left(\frac{2}{\gamma + 1}\right)^{(\gamma+1)/[2(\gamma-1)]}\frac{A^*P_o^*}{\sqrt{T_o}}. \tag{7-12}$$

The above expression explicitly demonstrates that the friction-induced loss of stagnation pressure directly affects the mass flow rate for given stagnation conditions in a specified duct. The dimensional Fanno line paths for the three duct lengths are sketched in Figure 7-15b. The dimensional Fanno lines must shift to the right (in the direction of increasing entropy) as the duct length increases in order to accommodate the increase in entropy. The stagnation conditions P_o and T_o are indicated on the Fanno line sketch. Because the Fanno lines are shifted to the right for increasing duct length, the Mach number at the start of the Fanno segment decreases as the duct length is increased.

Converging nozzle/constant-area duct systems can take many different forms. For example, the converging nozzle area segment can be used as an exit nozzle feed by a constant-area duct segment. Such an arrangement is illustrated in Figure 7-16a. The constant-area duct is the Fanno flow region, and the converging nozzle is the isentropic region. The behavior of the converging nozzle exit plane pressure with respect to Mach number is unchanged from that discussed earlier. The pressure distribution and the Fanno path for a constant-area duct/converging nozzle system operating in a choked mode are given in Figures 7-16b and c. In such an arrangement, choking can occur only at the converging nozzle exit.

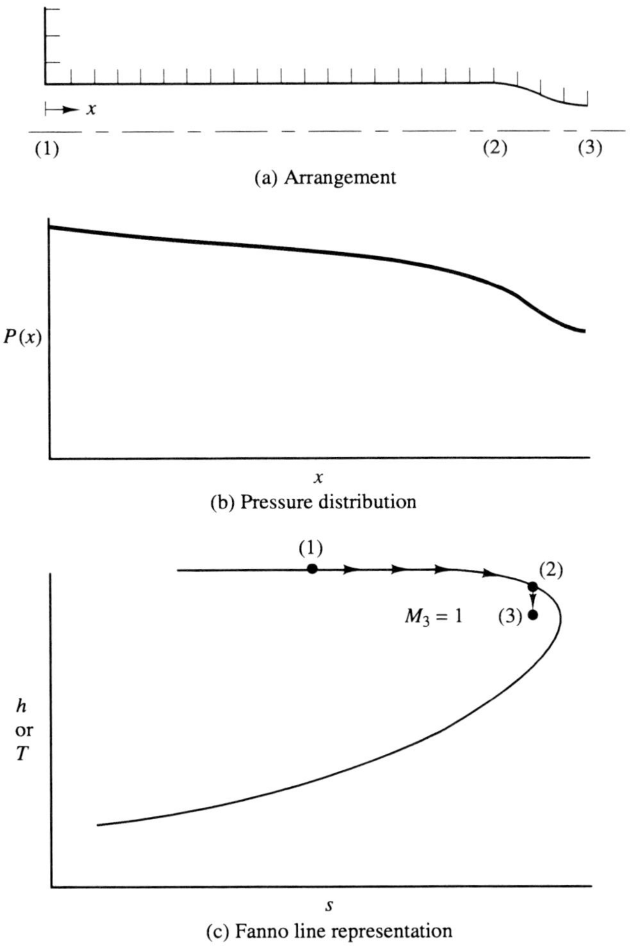

Figure 7-16 Constant-Area Duct/ Converging Nozzle System

Converging–diverging nozzles can also be used in conjunction with constant-area ducts. Usually the converging–diverging nozzle precedes the constant-area duct segment. A typical arrangement, with Fanno and isentropic regions indicated, is schematically illustrated in Figure 7-17. Since the exit Mach number for the converging–diverging nozzle can be either subsonic or supersonic, and since shock waves can occur in both the nozzle and the duct, the diversity of flow conditions possible in this arrangement is larger than that for the simpler converging nozzle/constant-area duct.

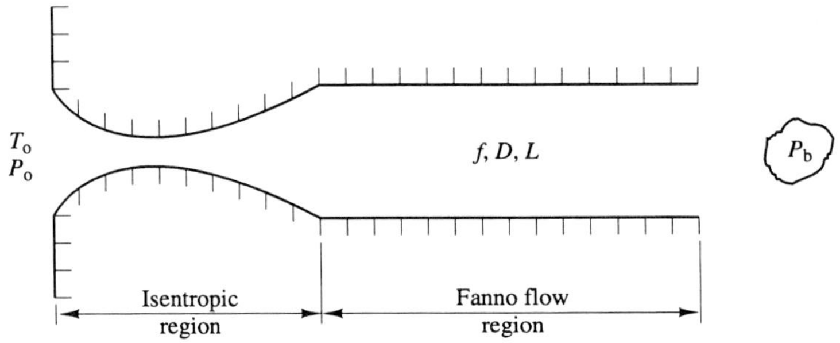

Figure 7-17 Converging–Diverging Nozzle/Constant-Area Duct Schematic

The major difference between supersonic and subsonic Fanno flow is that the permissible length of the duct segment with supersonic flow is limited while the permissible length for subsonic flow is unlimited. Consider, as in Figure 7-18a, a duct segment with a supersonic entrance Mach number. For a given entrance supersonic Mach number, the Fanno parameter, $4fL^*/D_h$, determines the maximum duct length over which supersonic flow can be maintained. If the actual duct length L for a supersonic entrance Mach number is greater than the maximum length L^* permitted for supersonic flow, then a shock wave must stand in the duct. A normal shock wave can also appear in a Fanno flow with supersonic entrance Mach number if the actual duct length is less than L^* for supersonic flow. The exact location of the shock wave will depend on the duct conditions and geometry as well as the back pressure; details will be examined later.

The situation for a normal shock appearing in the duct is illustrated in Figure 7-18a, with the pressure distribution and Fanno path presented in Figures 7-18b and c, respectively. Across the shock wave the Mach number goes from supersonic to subsonic; in addition, since the shock process is not isentropic, the entropy increases. A "jump" is made from the supersonic Fanno line region to the subsonic Fanno line region.

The overall behavior of a converging–diverging nozzle/constant-area duct system with respect to back pressure will be considered next. Envision such a system connected to a constant pressure reservoir and exhausting into a variable back pressure environment. We wish to examine the flow conditions as the back pressure is varied. When the entrance Mach number to the constant-area segment is supersonic, two possibilities exist for the flow conditions in the duct. If the actual length is less than L^* for the entrance supersonic Mach number, a normal shock wave may or may not stand in the duct. If the actual length is greater than L^* for the entrance supersonic Mach number, a normal shock wave *must* stand in the duct.

The behavior with respect to back pressure for $L < L^*$ will be examined first. The system is illustrated in Figure 7-19a. As the back pressure is decreased, flow is induced in the system. The pressure distribution for a back pressure that results in completely subsonic flow in the system is illustrated as curve "a" in Figure 7-19b, with the Fanno path sketched in Figure 7-19c. For a sufficiently low back pressure,

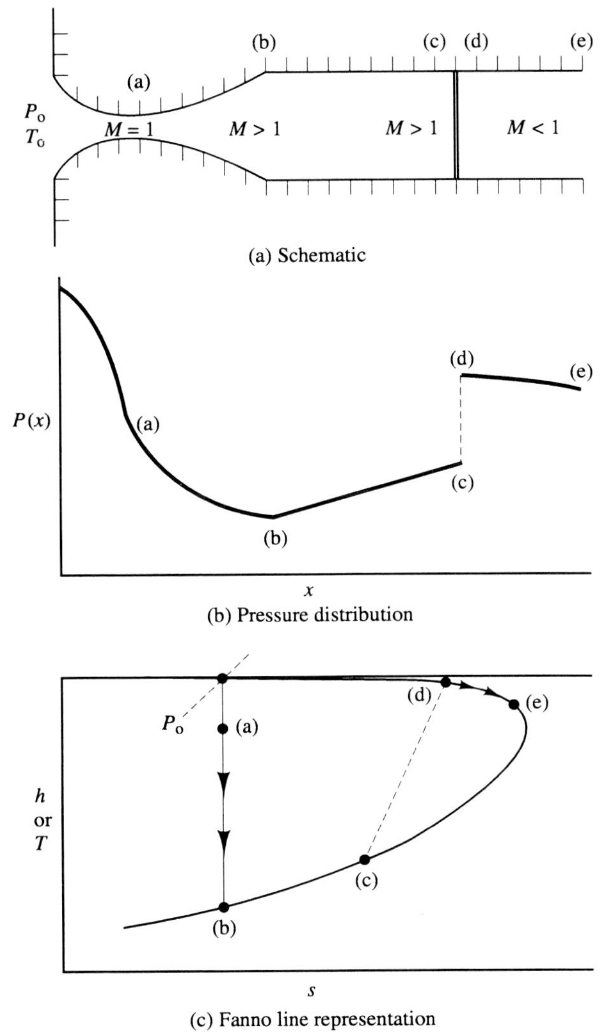

Figure 7-18 Normal Shock in Fanno Region

the flow chokes in the nozzle throat, resulting in the pressure distribution of curve "b" and the Fanno path of Figure 7-19d. Further reduction in back pressure leads to a normal shock standing in the diverging portion of the nozzle. The pressure distribution for a shock in the nozzle is indicated in curve "c" of Figure 7-19b, with the Fanno path in Figure 7-19e. A further decrease in back pressure places the shock wave at the nozzle exit plane, shown as curve "d" in Figure 7-19b and Figure 7-19(f) for its Fanno path. A further reduction in the back pressure leads to the shock standing in the duct exit plane as presented in curve "e" and Figure 7-19g.

Any further decrease in back pressure below that required to place the shock wave in the duct exit plane has no effect on the flow field in the nozzle/duct system.

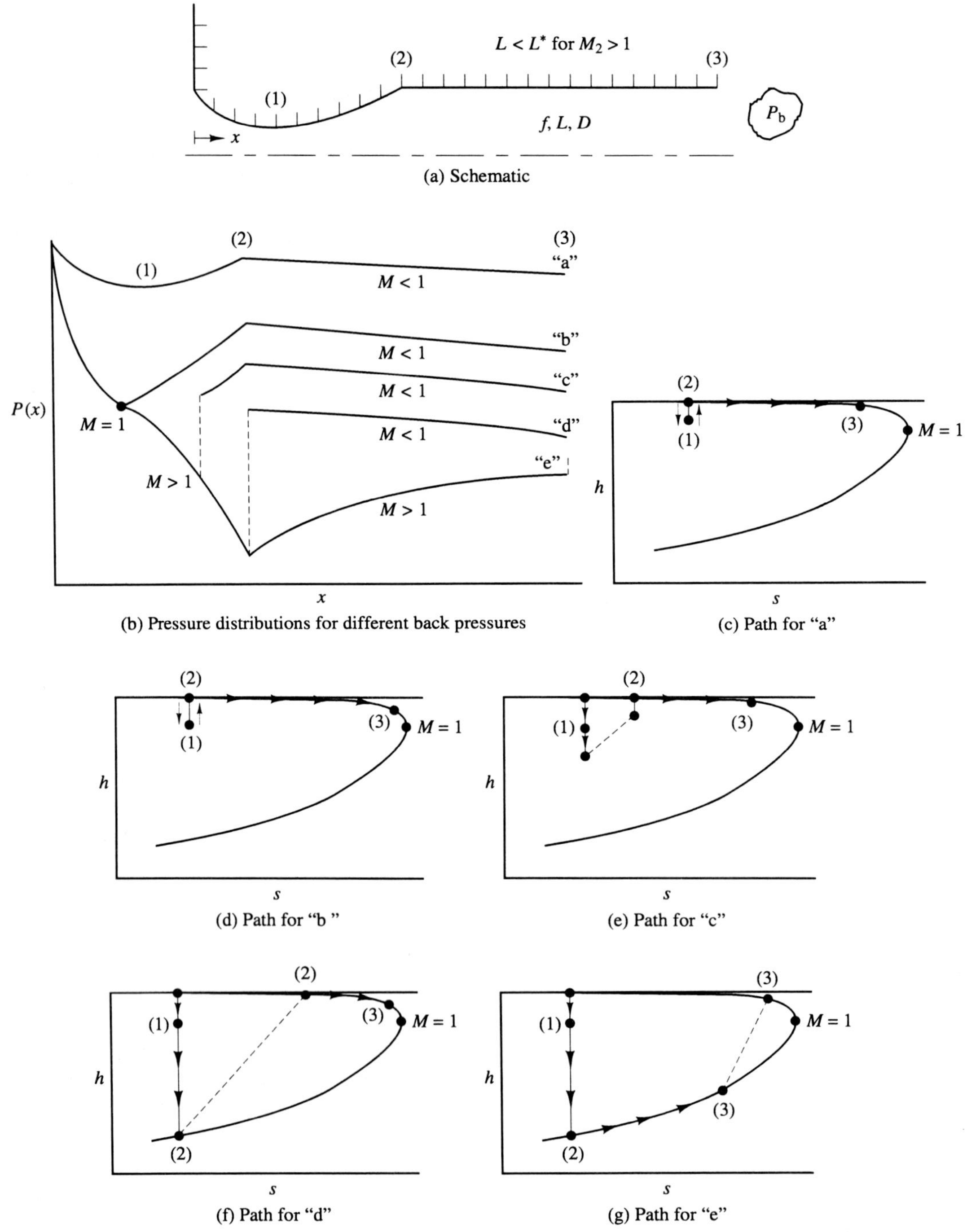

Figure 7-19 Conditions for $L < L^*$ in Supersonic FLow

Further reductions in back pressure beyond that point result in changes only in the exhaust plume flow structure outside the duct and generally follow the sequence of regions 3 and 4 for the converging–diverging nozzle behavior. A back pressure does exist for which for flow is shock-free throughout the nozzle/duct system and the exhaust plume, corresponding to the conditions of key pressure 3 of the converging–diverging nozzle sequence. Although a shock-free flow throughout the system can be generated, the flow is not isentropic because of the friction effects associated with the Fanno flow segment. The converging–diverging nozzle/constant-area duct combination obeys all the principles covered for both the simple flows involved in the series combination.

The behavior with respect to back pressure for $L > L^*$ is the second arrangement of interest with respect to back pressure effects on the flow field of the nozzle/duct system. The system is illustrated in Figure 7-20a. The behavior of the nozzle/duct system with $L > L^*$ as the back pressure is reduced follows the same basic sequence as for $L < L^*$ until the back pressure is reduced low enough to permit a normal shock wave to stand at the nozzle exit plane. These are indicated as curves "a" through "d" on Figure 7-20b with the Fanno paths shown as Figures 7-20c–f. The behavior for back pressures lower than that required to place the shock at the nozzle exit plane is different for the two duct length conditions. A further decrease in back pressure moves the shock wave into the constant-area duct segment. Because $L > L^*$, a normal shock wave must stand in the duct even if the back pressure is reduced to zero. For back pressures lower than that required to place the shock at the nozzle exit, the duct flow field will always contain a supersonic region and a subsonic region separated by a normal shock wave. For low back pressures the duct exit Mach number will also be 1 in order to permit the duct exit plane pressure to be larger than the back pressure. The exact location of the shock wave in such a system depends on the system parameters and the back pressure. Curve "e" of Figure 7-20b shows the pressure distribution for a shock wave in the duct. The corresponding Fanno path is presented in Figure 7-20g. As with the $L < L^*$ arrangement, the converging–diverging nozzle/constant-area duct combination must obey all the principles covered for both the simple flows involved in the series combination.

Example 7-5

A converging–diverging nozzle/constant-area duct system is connected to a reservoir containing air with a stagnation pressure of 100 psia. The exit-to-throat area ratio of the nozzle is 3; the constant area duct is 20 ft long and 1 ft in diameter with a friction factor of 0.005. Determine the range of back pressures for which a normal shock wave will stand in the duct.

Solution. The problem is sketched in Figure 7-21a. For an area ratio of 3, the nozzle exit supersonic Mach number is 2.6374. The Fanno parameter, $4\,fL^*/D$, for this Mach number is 0.460. The duct possesses an actual $4\,fL/D$ value of

$$\left(4\frac{fL}{D}\right)_{\text{act}} = \frac{(4)(0.005)(20)}{1} = 0.4.$$

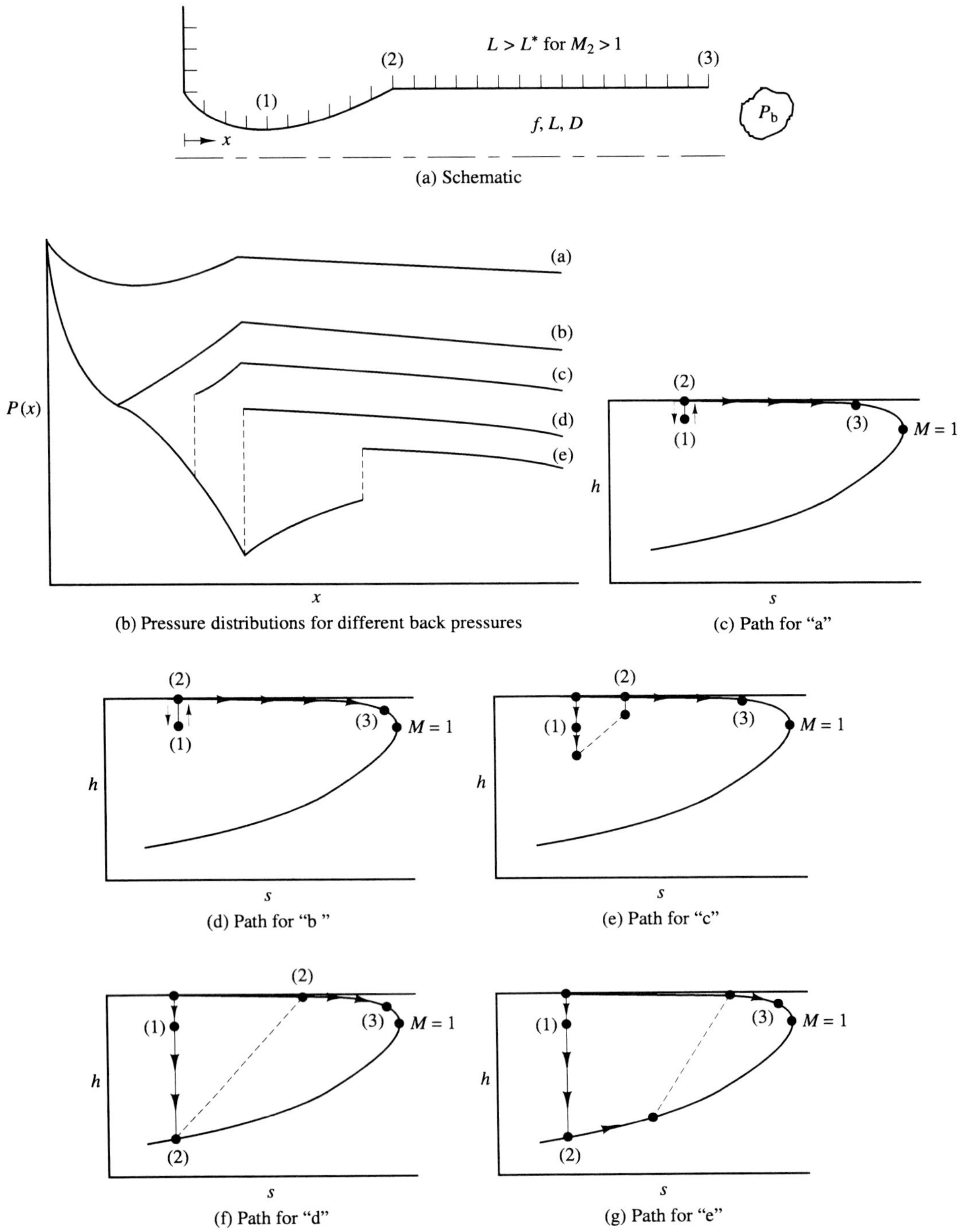

Figure 7-20 Conditions for $L > L^*$ for Supersonic Flow

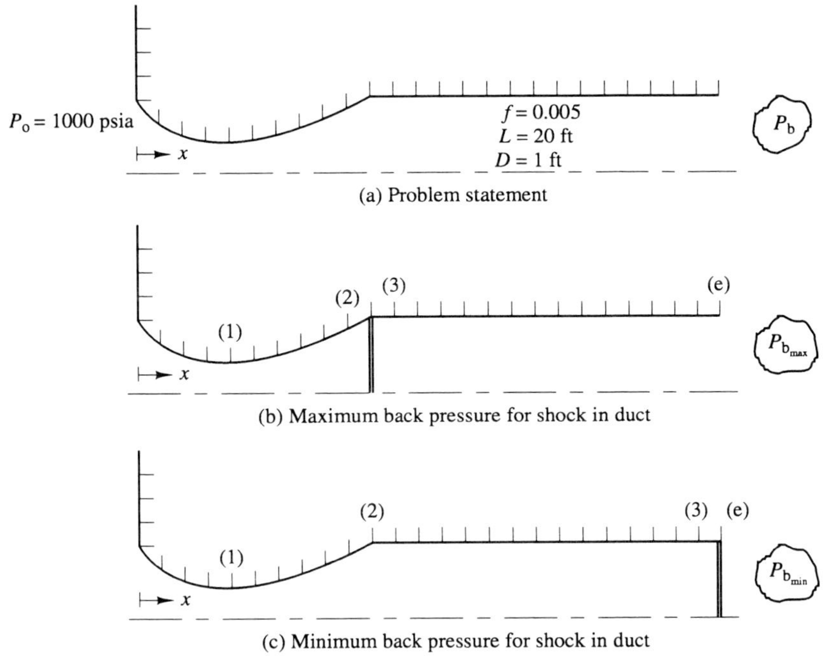

Figure 7-21 Example Problem 7-5

Thus, since $L < L^*$, the situation depicted is as in Figure 7-19. Depending on the back pressure, a normal shock wave can stand anywhere in the constant-area segment. The maximum back pressure associated with a shock in the duct will place the shock at the nozzle exit/duct entrance. This condition is illustrated in Figure 7-21b with the Fanno path shown in Figure 7-19f. The minimum back pressure will place the shock in the duct exit plane. This condition is sketched in Figure 7-21c, with the corresponding Fanno path presented in Figure 7-19g.

The maximum back pressure can be obtained from the following property ratios (see Figure 7-21b for the subscript meanings):

$$P_b = P_e = \frac{P_e}{P_3}\frac{P_3}{P_2}\frac{P_2}{P_o}P_o.$$

For the normal shock at the entrance to the Fanno region,

$$M_2 = 2.6374;$$

$$\frac{P_2}{P_o} = 0.0473.$$

Across the normal shock,

$$M_3 = 0.5007;$$

$$\frac{P_3}{P_2} = 7.9485.$$

Using case **DA** of option **FA** produces the following end-of-duct conditions:

$$M_e = 0.5619;$$

$$\frac{P_e}{P_3} = 0.8860.$$

With a subsonic Mach number, the exit plane pressure and the back pressure must be equal; thus

$$P_e = P_b = (0.886)(7.9485)(0.0473)(100) = 33.31 \text{ psia}.$$

The minimum back pressure is obtained from the following property ratios (see Figure 7-21c for the locations designated by the subscripts):

$$P_b = P_e = \frac{P_e}{P_3}\frac{P_3}{P_2}\frac{P_2}{P_o}P_o.$$

With an entrance Mach number of 2.6374 and an $(4\,fL/D)_{act}$ of 0.4, case **DA** of option **FA** yields at the duct exit plane

$$M_3 = 1.2853;$$

$$\frac{P_3}{P_2} = 2.7510.$$

Across the normal shock with an upstream Mach number of 1.2853,

$$M_e = 0.7935;$$

$$\frac{P_e}{P_3} = 1.7607,$$

from which

$$P_e = P_b = (1.7607)(2.751)(0.0473)(100) = 22.91 \text{ psia}.$$

A normal shock will appear in the constant-area duct when

$$22.91 \text{ psia} \leq P_b \leq 33.31 \text{ psia}.$$

Back pressures higher than 33.31 psia will move the shock into the diverging portion of the nozzle, while back pressures below 22.91 psia will result in either an oblique shock or an expansion fan at the duct exit. For back pressures below 22.91 psia, the flow field in the entire nozzle/duct system is not affected by the value of the back pressure.

Example 7-6

Find the range of back pressures for which a shock wave will stand in the duct for the conditions of Example 7-5 if the duct length is increased to 50 ft.

Solution. The problem is sketched in Figure 7-22a. For an area ratio of 3, the exit supersonic Mach number is 2.6374. The Fanno parameter, $4\,fL^*/D$, for this Mach number is 0.460. The duct possesses an actual $(4\,fL/D)_{act}$ value of

$$\left(4\frac{fL}{D}\right)_{act} = \frac{(4)(0.005)(50)}{1} = 1.0.$$

Thus, since $L > L^*$, the situation is as depicted in Figure 7-20. No matter how low the back pressure, a normal shock wave will always stand in the constant-area segment if the entrance duct Mach number is supersonic. The maximum back pressure associated with a shock in the duct will place the shock at the nozzle exit/duct entrance. This condition is illustrated in Figure 7-22b, with the Fanno path shown in Figure 7-20f. A normal shock wave will stand somewhere in the duct for zero back pressure. The Fanno path for zero back pressure is sketched in Figure 7-22c.

The maximum back pressure can be obtained from the following property ratios (see Figure 7-22b for the subscript meanings):

$$P_b = P_e = \frac{P_e}{P_3}\frac{P_3}{P_2}\frac{P_2}{P_o}P_o.$$

For the shock at the entrance to the Fanno region,

$$M_2 = 2.6374;$$

$$\frac{P_2}{P_o} = 0.0473.$$

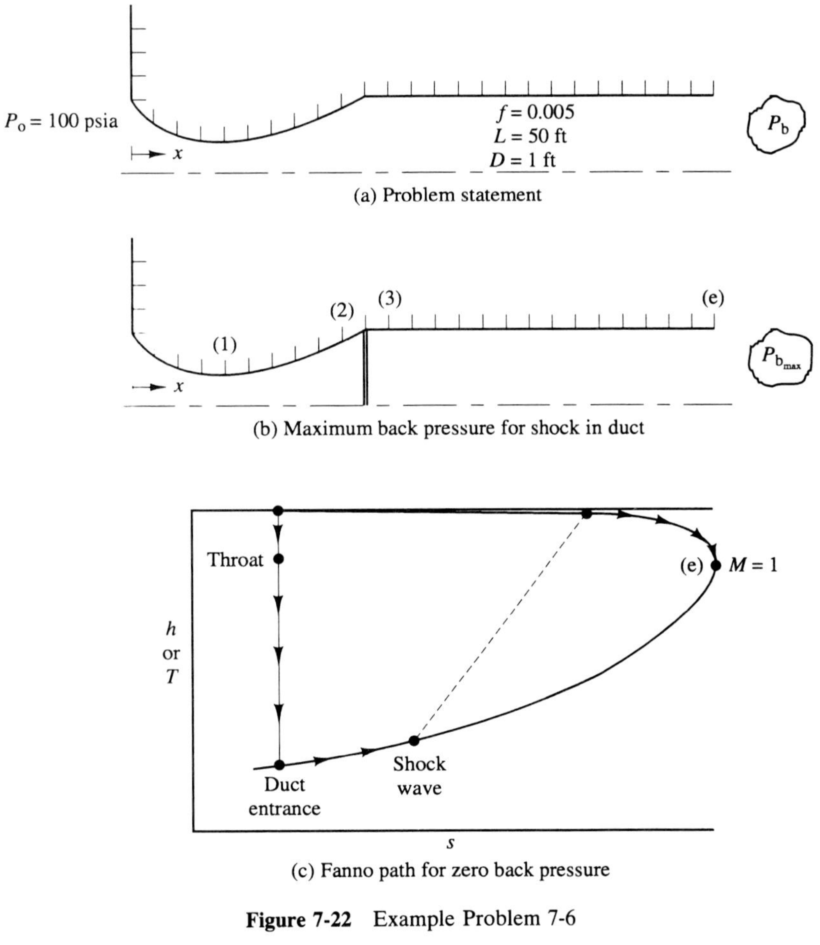

Figure 7-22 Example Problem 7-6

Across the normal shock,

$$M_3 = 0.5007;$$

$$\frac{P_3}{P_2} = 7.9485.$$

Using case **DA** of option **FA** with $M = 0.5007$ and $(4\,fL/D)_{act} = 1.0$ produces the following results:

$$M_e = 0.8105;$$

$$\frac{P_e}{P_3} = 0.5950.$$

With a subsonic Mach number, the exit plane pressure and the back pressure must be equal, and thus

$$P_b = P_e = (0.5950)(7.9485)(0.0473)(100) = 22.37 \text{ psia}.$$

A normal shock will appear in the constant-area duct when

$$P_b \leq 22.37 \text{ psia}.$$

Back pressures higher than 22.37 psia will move the shock into the diverging portion of the nozzle, while back pressures less than or equal to 22.37 psia will always result in a normal shock wave in the duct.

This and the previous example illustrate the differences in behavior for a converging–diverging nozzle/constant-area duct system according to whether the duct length is less than or greater than L^* for supersonic flow. The longer duct length associated with this example as compared with the previous example also demonstrates the effect of increasing the duct length on the back pressures for key events in the ducts' behavior.

Example 7-7

The converging–diverging nozzle/constant-area duct system shown in Figure 7-23a is exposed to a back pressure of 400 kPa. If the system is supplied from a 1000 kPa reservoir, find the location of the normal shock in the system. Sketch the path on the Fanno line.

Solution. With a back pressure of 400 kPa and a reservoir pressure of 1000 kPa, the flow will be assumed choked with a normal shock standing in the duct segment. This assumption must be verified. Important features of the flow field are labeled in Figure 7-23b. For a nozzle area ratio of 2, the supersonic exit Mach number is 2.1971. For an entrance Mach number of 2.1971,

$$4\frac{fL^*}{D} = 0.36.$$

The actual value of $4\,fL/D$ for the duct is

$$\left(4\frac{fL}{D}\right)_{act} = \frac{(4)(0.005)(20)}{1} = 0.40.$$

Since $L > L^*$, a normal shock will stand in the constant-area duct if M_2 is supersonic. The solution to the problem of finding the location of the shock is iterative. The general

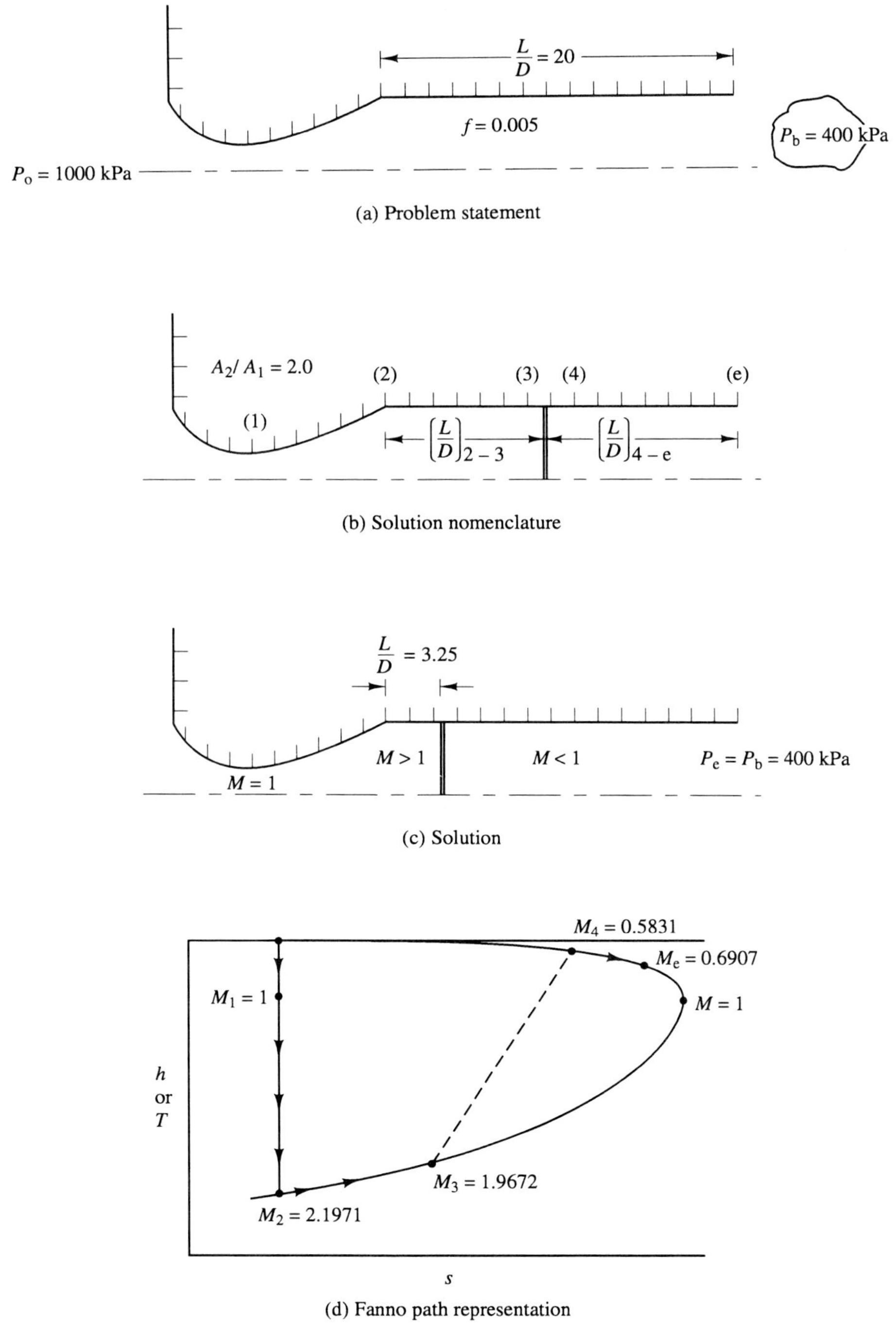

Figure 7-23 Example Problem 7-7

procedure for the iteration is to assume shock locations in terms of L/D, solve the duct addition problem for conditions at state 3, jump across the normal shock wave to state 4, and finally solve the duct addition problem for conditions at state e. The back pressure for the assumed shock wave location can be calculated using the pressure ratios

$$P_b = P_e = \frac{P_e}{P_4}\frac{P_4}{P_3}\frac{P_3}{P_2}\frac{P_2}{P_o}P_o.$$

The process has converged when the assumed shock location yields the computed back pressure to be 400 kPa. For this problem

$$(L/D)_{2-3} + (L/D)_{4-e} = 20.$$

Case **DA** for the COMPQ Fanno flow option **FA** is very useful in the iterative scheme proposed. Using the tables or the individual cases **M** and **F** more than triples the calculations involved in the process. Table 7-4 presents the iterative sequence for the solution. The duct exit Mach number is subsonic, the duct exit pressure is 400 kPa, and the back pressure is 400 kPa. Thus the original assumption of a normal shock wave standing in the duct is correct. The flow field is as sketched in Figure 7-23c with the Fanno path given in Figure 7-23d.

The preceding example problems illustrate various aspects of the behavior of converging–diverging nozzle/constant-area duct systems. In such systems the changes in pressure and the losses in stagnation pressure in the duct segment have an effect similar to back pressure changes for the nozzle segment. The next example problem explores the effect of adding duct length to a nozzle/duct arrangement exhausting into a zero back pressure.

Example 7-8

A converging–diverging nozzle/constant-area duct system is connected to a reservoir. The reservoir pressure is 100 psia, and the back pressure into which the system exhausts is zero. If the nozzle exit-to-throat area ratio is 2.005, determine the system behavior as the constant-area duct length is increased.

Solution. The situation is sketched in Figure 7-24a. Increases in duct length are reflected in increases in the value of $4fL/D$. The results of this problem will be expressed in terms of the parameter $4fL/D$. Since the back pressure is zero, no matter what the system configuration, the system must be choked. The pressure in the duct exit plane will always be finite. With a zero back pressure and a finite duct exit pressure, the duct exit Mach number must be greater than or equal to 1.

For only the nozzle (no constant-area duct attached) and a zero back pressure, the nozzle operates in an underexpanded mode: region 4 in the converging–diverging nozzle discussion of Chapter 7. For an area ratio of 2.005 and a zero back pressure, the nozzle chokes and the flow is expanded supersonically in the diverging section. The exit Mach number is

$$M_2 = 2.20,$$

and the exit pressure becomes

$$P_e = \frac{P_2}{P_o}P_o$$

$$= (0.0935)(100) = 9.35 \text{ psia}.$$

TABLE 7-4 ITERATIVE SEQUENCE FOR EXAMPLE PROBLEM 7-7

$(L/D)_{2-3}$	$(4fL/D)_{2-3}$	M_3	P_3/P_2	M_4	P_4/P_3	$(L/D)_{4-e}$	$(4fL/D)_{4-e}$	M_e	P_e/P_4	P_e (kPa)
10.00	0.200	1.5662	1.611	0.6789	2.6951	10.00	0.20	0.8251	0.804	329
5.00	0.100	1.8556	1.277	0.6045	3.8505	15.00	0.30	0.7193	0.829	383
2.50	0.050	2.0174	1.134	0.5744	4.5816	17.50	0.35	0.6856	0.836	408
3.00	0.060	1.9838	1.162	0.5802	4.4247	17.00	0.34	0.6869	0.834	403
3.50	0.070	1.9508	1.190	0.5860	4.2732	16.50	0.33	0.6945	0.833	398
3.25	0.065	1.9672	1.176	0.5831	4.3482	16.75	0.33	0.6907	0.834	400

An expansion fan is located on the lip of the nozzle exit. The flow field characteristics are sketched for this condition in Figure 7-24b.

If duct length less than L^* for a supersonic duct entrance Mach number is added, then the flow remains supersonic throughout the duct segment. The Mach number decreases as the duct is traversed. If a duct of length L^* is added, then the Mach number at the duct exit is 1, the exit pressure is greater than the back pressure, and an expansion fan is attached to the lip of the duct exit. The exit pressure for $L = L^*$ is

$$P_e = \frac{P_e}{P^*}\frac{P^*}{P_2}\frac{P_2}{P_o}P_o$$

$$= (1)\frac{1}{0.3549}(0.0935)(100) = 26.35 \text{ psia},$$

and

$$4\,fL/D = 0.3609.$$

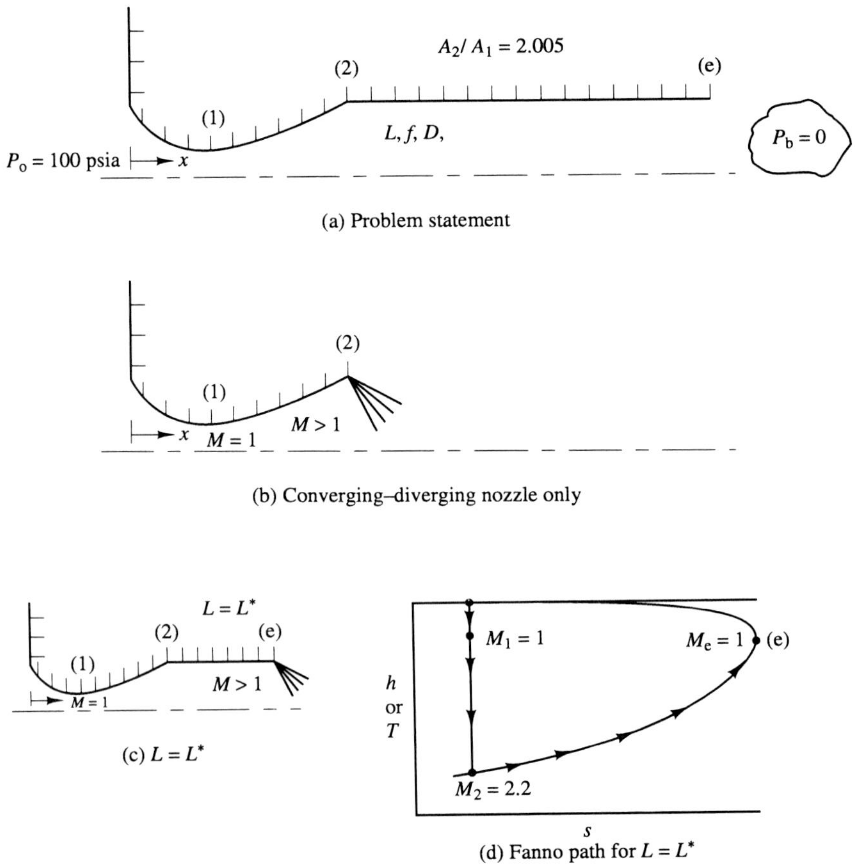

Figure 7-24 Example Problem 7-8

The salient features of the flow field for $L = L^*$ are presented in Figure 7-24c, with the Fanno path shown in Figure 7-24d. For

$$0 < 4\,fL/D < 0.3609,$$

supersonic flow exists throughout the constant-area duct.

If more duct length is added, a normal shock wave will appear in the duct section since $L > L^*$. A shock will appear in the duct until sufficient duct length is added to place the shock wave at the entrance to the constant-area duct. This condition is illustrated in Figure 7-24e, with the Fanno path represented in Figure 7-24f. For this condition,

$$M_{2_u} = 2.2,$$

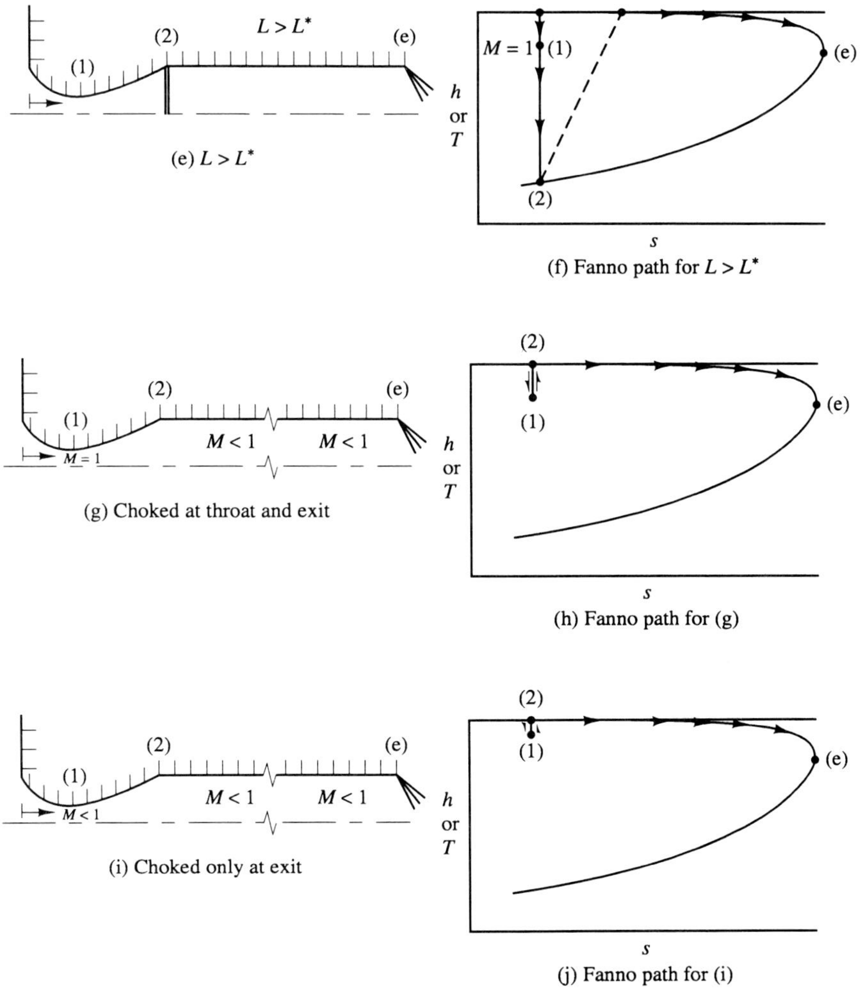

Figure 7-24 (continued)

and across the normal shock wave

$$M_{2_d} = 0.5741.$$

The Fanno parameter for $M_{2_d} = 0.5741$ is 0.603. Thus, a shock will appear in the duct for

$$0.3609 < 4\,fL/D < 0.6030.$$

Any further duct addition beyond $4fL/D = 0.6030$ will result in the shock wave moving into the diverging section of the nozzle. The longer the duct addition, the closer the shock location will be to the nozzle throat. If the duct is sufficiently long, the normal shock wave can be placed in the throat of the nozzle. At this position, the upstream Mach number is 1 and the shock wave is of vanishing strength. For the case of sonic flow at the nozzle throat and at the duct exit, the flow field features appear as in Figure 7-24g, with the Fanno path indicated in Figure 7-24h. For subsonic flow except at the nozzle throat and the duct exit, the area ratio of 2.005 yields a nozzle exit Mach number of 0.305, for which

$$4\,fL/D = 5.069.$$

Thus, a shock will stand in the diverging portion of the nozzle for

$$0.6030 < 4\,fL/D < 5.069.$$

Because area change effects are generally more significant than Fanno effects, the duct addition required to obtain this condition is much longer than the duct addition required to move the shock to the duct entrance.

If the duct length is increased so that $4fL/D > 5.069$, then the nozzle throat will unchoke, leaving only the duct exit choked. The flow field for this condition is depicted in Figure 7-24i, with the Fanno path in Figure 7-24j. For example, for a $4\,fL/D = 10$ value, the Mach number at the entrance to the duct is 0.2339. Then at the nozzle minimum area,

$$\frac{A_1}{A^*} = \frac{A_1}{A_2}\frac{A_2}{A^*}$$

$$= \frac{1}{2.005}2.5563 = 1.275,$$

so that the minimum area Mach number becomes

$$M_1 = 0.5371.$$

As long as the nozzle throat is choked, the duct length addition has no effect on the system mass flow rate. Once the nozzle throat is unchoked, the mass flow rate for the system is set by the choked flow at the duct exit. The mass flow rate for the condition of $4\,fL/D > 5.069$ is reduced from that for choked flow at the nozzle throat.

This same problem can be considered if the back pressure is not zero. The same principles hold, but the values of $4fL/D$ for the various conditions will be changed from the values for the zero back pressure. Also, if the back pressure is finite, the duct exit Mach number can be subsonic for situations in which the duct exit pressure and the back pressure are equal. Notice also that increasing the duct length has the same effect on the converging–diverging nozzle flow field as increasing the back pressure in the absence of a constant-area Fanno segment.

Example 7-9

Nitrogen is stored at 200 psia and 530 *R* in a large pressure tank. Nitrogen for a purge system is needed 200 ft from the storage tank. The required flow rate is 1.0 lbm/s at a pressure of 30 psia or more. Find the diameter of the copper tube (absolute roughness = 0.000005 ft) required to deliver the nitrogen at the desired flow rate and pressure.

Solution. This is a more realistic example problem than most of the previous Fanno examples since this problem involves finding the friction factor as part of the solution. The system is schematically illustrated in Figure 7-25. Although problems like this one require several levels of iteration, the procedure is logical. Problems of this type in incompressible fluid mechanics (Hodge, 1990) are often called *category III* problems since the diameter is not known, but a flow rate requirement is specified. The same general process will be followed as for category III problems, but the Fanno version is complicated by the possibility of choking.

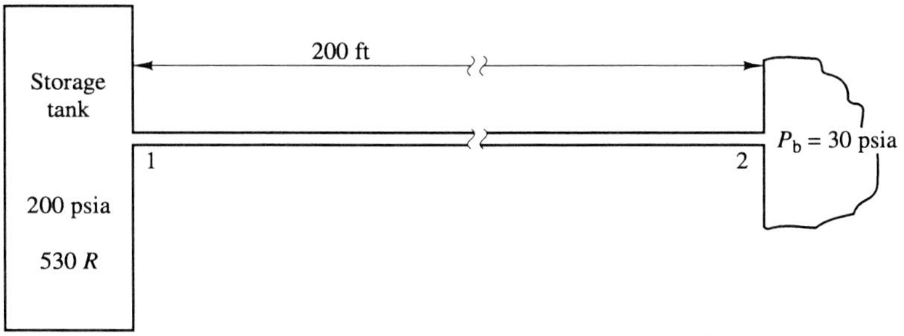

Figure 7-25 Schematic for Example 7-9

Consider the following outline of a procedure to solve this problem:

1. Assume a pipe diameter.
   (outer iteration: steps 1–7)
2. Assume a pipe exit Mach number, M_2.
   (middle iteration: steps 2–5)
3. Assume a friction factor.
   (inner iteration: steps 3–4)
4. Iterate to find the friction factor for the specified diameter and M_2.
   a. Compute inlet conditions for f and M_2.
   b. Find inlet temperature, pressure, and density.
   c. Compute Reynolds number using viscosity at mean temperature.
   d. Use Appendix I to find the f based on part 4c.
   e. Assume a new f and repeat steps (a)–(d) until f converges.
5. Compute exit pressure, P_2, from step 4 and compare with back pressure, P_b.
   a. If $P_2 > P_b$, then the flow is choked and if $M_2 = 1$, go to step 6.
   b. If $P_2 < P_b$, then the flow is not choked and M_2 must be established such that $P_b = P_2$.
   c. Assume a new M_2 and repeat steps 2–5.
   d. M_2 converges when $P_2 = P_b$; go to step 6.

6. Compute mass flow rate.
   a. If mass flow computed equals mass flow required, the problem is finished.
7. If mass flow computed does not equal mass flow required, start at step 1 with a new pipe diameter and repeat the entire process.

The procedure contains three levels of iteration: the inner is to determine the correct friction factor for the assumed M_e and pipe diameter; the middle is to determine whether or not the flow is choked; and the outer is to determine the pipe diameter.

The mean viscosity will be required in order to evaluate the Reynolds number, which is needed to obtain the friction factor. The limits of the static temperature for this problem must be between T_o (the stagnation temperature) and T^* (the static temperature at a Mach number of 1). The average of these temperatures is 477 R. Initially we shall assume that this is a reasonable mean temperature for the evaluation of the viscosity. The viscosity will be taken as 1.138×10^{-5} lbm/ft s. Either Sutherland's expression, Eq. 7-2, can be used to evaluate the viscosity or, as in this case, the viscosity can simply be read from a table (Hodge, 1990). The gas constant for nitrogen was taken as 55.15 ft-lbf/lbm R.

The solution begins by assuming a diameter: assume $ID = 1$ in. Then the relative roughness, $\epsilon/D = 0.000005/(1/12) = 0.00006$. Since this is a long pipe ($L/D = 2400$), and since the allowable pressure drop is from 200 psia to 30 psia, the flow will be assumed choked. Thus, initially $M_2 = 1$. The inner iterative sequence is started by assuming a value of the friction factor. The relative roughness is quite small, so that the friction factor is unlikely to be large; assume an initial friction factor of 0.005. This is usually a reasonable initial friction factor value. With assumed values of both the pipe inside diameter and the exit Mach number, the computational sequence to establish the correct friction factor starts with calculating $4\,fL/D$. The sequence, delineated in Table 7-5, is then to obtain the inlet Mach number, static temperature, static pressure, density, Reynolds number, and friction factor. The diameter of 1.1 in results in a flow rate of 1.122 lbm/s at a pressure of 30 psia. An additional iteration with a diameter between 1.0 and 1.1 in could be tried, but an examination of availability of copper tubes (Hodge, 1990) reveals that the nominal 1.25 BWG 16-gauge copper tube possesses an inside diameter of 1.12 in. Thus, the available copper tube would be larger than 1.1 in, and another iteration would not result in any useful information. An examination of some of the tabulated results is appropriate in order to explain in more detail the outlined iteration sequence.

The initial tube diameter chosen was 1 in with an initial guess on friction factor of 0.005 with the assumption of choked flow. The table shows that with $M_2 = 1$, the inner friction factor iteration converged with $f = 0.003125$. But for these conditions the exit pressure P_2 attained only 26.1 psia, while the back pressure was fixed at 30 psia. Thus, we conclude that the tube could not be choked and that an M_2 of less than 1 is required. This problem is somewhat simplified because of the rapid change of pressure near sonic flow. A P_2 of 30 psia is required, so that the appropriate pressure ratio, P/P^*, is $30.0/26.1 = 1.149$. The exit Mach number required is 0.8829, but the $4\,fL/D$ at this Mach number is only 0.005. The new calculation sequence would then start at $M_2 = .8829$ with the same $4\,fL/D$ change (30.00). The net result at the entrance to the pipe is that the change in the Fanno parameter of 0.005 out of a value of 30.0 has no appreciable effect on the inlet Mach number, M_1, or the flow properties. Hence, for a diameter of 1 in and an entrance Mach number of 0.1452, the mass flow rate is 0.884 lbm/s with both the exit Mach number and the friction factor converged.

TABLE 7-5 ITERATIVE SEQUENCE FOR EXAMPLE 7-9

ID (in)	M_2	f	$4fL/D$	M_1	T_1 (R)	P_1 (psi)	ρ_1 (lbm/ft^3)	a_1 (ft/s)	V_1 (ft/s)	Re	f	P_2 (psi)	P_b (psi)	$\dot{m}$ (lbm/s)	Converged on
1.0	1.0	.005	47.80	.1169	527.57	197.10	.9786	1146.0	133.95	959,858	.0032				
	1.0	.0032	30.72	.1437	527.83	197.14	.9752	1145.1	164.55	1,175,000	.003125				
	1.0	.003125	30.00	.1452	527.77	197.08	.9750	1145.0	166.25	1,186,900	.003125	26.1	30.0		f
	0.8829	.003125	30.00	.1452	527.77	197.08	.9750	1145.0	166.25	1,186,900	.003125	30.0	30.0	.884	f, M_2
1.2	1.0	.0033	26.40	.1538	527.51	196.72	.9737	1144.7	176.05	1,506,326	.00325				
	1.0	.00324	26.00	.1549	527.46	196.68	.9736	1144.7	177.31	1,516,950	.00325	27.88	30.0		f
	0.9387	.00324	26.00	.1549	527.46	196.68	.9736	1144.7	177.31	1,516,950	.00325	30.00	30.0	1.356	f, M_2
1.1	1.0	.00319	27.84	.1502	527.62	196.88	.9743	1144.9	171.96	1,349,600	.00308				
	1.0	.00308	26.88	.1526	527.56	196.78	.9739	1144.8	174.69	1,409,600	.00306	27.48	30.0		f
	0.9268	.00308	26.88	.1526	527.56	196.78	.9739	1144.8	174.69	1,409,600	.00306	30.00	30.0	1.122	f, M_2

Since the desired mass flow rate is more than that provided by the 1-in tube, the second tube diameter chosen was 1.2 in. The entries in the table associated with the 1.2-in tube follow the same sequence as for the 1-in tube. The converged results for the 1.2-in tube show a mass flow rate of 1.356 lbm/s with $M_2 = .9387$.

The required tube diameter is bracketed between 1 and 1.2 in at this point. The last iteration involves a diameter of 1.1 in. The 1.1-in diameter tube has a mass flow rate of 1.122 lbm/s with an exit Mach number of 0.9263. No further iterations are needed since the sizes of available BWG 16-gauge tubes are outside the range of 1.0 to 1.1 in.

This example problem is more involved than the previous examples since the friction factor must be determined as part of the solution. As we've seen, having to compute the friction factor as part of the solution generally requires some iteration, but the overall ideas that we have developed and examined in Fanno flow are not changed. This essentially says that if we can work in a Fanno scenario with the friction factor given, then we can work that scenario if the friction factor is to be determined as part of the solution, albeit some additional effort being required. Continued iteration to secure the converged friction factor can mask the point of many Fanno problems, so the usual procedure is to specify the friction factor. But with this problem understood, we are now in a position to work Fanno problems in which the friction factor must be determined as part of the solution.

7.5 SIMPLE HEAT ADDITION OR REJECTION (RAYLEIGH FLOW)

If the only driving potential is heat addition or rejection, simple diabatic flow—or *Rayleigh flow*—is defined. Because the only driving potential that changes in simple diabatic flow is T_o, it is sometimes called *simple T_o change*. Table 7-6a reproduces for convenience the influence coefficient entries from Table 5-1 for Rayleigh flow. The independent variable is given as dT_o/T_o. The closed-form expressions for Rayleigh flow from Table 5-2 are restated in Table 7-6b. The entropy change is referenced to the sonic state, so $\Delta s = s - s^*$. As with simple area change, the starred variables denote conditions at the sonic location. Rayleigh flow is thus flow in a frictionless ($f = dF_D = 0$), constant-area duct (A = constant) with no mass addition ($\dot{m}$ = constant), but with stagnation temperature and, hence, stagnation enthalpy change. If the conservation equations and various definitions are applied to an elemental control volume with constant area, no change in mass flow rate, and no friction, the expressions of Table 7-6a result. Integration of the entries in Table 7-6a with conditions referenced to the sonic point produce the entries of Table 7-6b.

The only expression with a new functional form appearing in Table 7.6b is the one that relates T_o/T_o^* to the Mach number. For Rayleigh flow the increase in stagnation temperature is the important parameter; no distinction is made involving the distance over which the stagnation temperature changes. Thus a step increase from T_{o_1} to T_{o_2}, or a gradual increase from T_{o_1} to T_{o_2}, yields the same results under the Rayleigh flow assumptions; the duct length over which the change in T_o is accomplished is not a consideration in Rayleigh flow. This concept will be fully discussed later in this section. The influence coefficients for entropy and stagnation pressure show that heat addition produces an increase in entropy and a decrease in

TABLE 7-6 SUMMARY OF RELATIONS FOR RAYLEIGH FLOW

(a) Differential forms	(b) Property ratios
$\dfrac{dM^2}{M^2} = \dfrac{(1+\gamma M^2)\psi}{1-M^2}\dfrac{dT_o}{T_o}$ $\quad \psi = 1 + \dfrac{\gamma-1}{2}M^2$	$\dfrac{T_o}{T_o^*} = \dfrac{2(\gamma+1)M^2\psi}{(1+\gamma M^2)^2}$
$\dfrac{dV}{V} = \dfrac{\psi}{1-M^2}\dfrac{dT_o}{T_o}$	$\dfrac{\rho}{\rho^*} = \dfrac{1+\gamma M^2}{(\gamma+1)M^2}$
$\dfrac{dT}{T} = \dfrac{(1-\gamma M^2)\psi}{1-M^2}\dfrac{dT_o}{T_o}$	$\dfrac{T}{T^*} = \dfrac{(\gamma+1)^2 M^2}{(1+\gamma M^2)^2}$
$\dfrac{d\rho}{\rho} = -\dfrac{\psi}{1-M^2}\dfrac{dT_o}{T_o}$	$\dfrac{V}{V^*} = \dfrac{(\gamma+1)M^2}{1+\gamma M^2}$
$\dfrac{dP}{P} = -\dfrac{\gamma M^2\psi}{1-M^2}\dfrac{dT_o}{T_o}$	$\dfrac{P_o}{P_o^*} = \dfrac{\gamma+1}{1+\gamma M^2}\left[\left(\dfrac{2}{\gamma+1}\right)\psi\right]^{\gamma/(\gamma-1)}$
$\dfrac{dP_o}{P_o} = -\dfrac{\gamma M^2}{2}\dfrac{dT_o}{T_o}$	$\dfrac{P}{P^*} = \dfrac{\gamma+1}{1+\gamma M^2}$
$\dfrac{dI}{I} = 0$	$\dfrac{\Delta s}{c_p} = \ln\left\{M^2\left(\dfrac{\gamma+1}{1+\gamma M^2}\right)^{(\gamma+1)/\gamma}\right\}$
$\dfrac{ds}{c_p} = \psi\dfrac{dT_o}{T_o}$	$A = A^* = \text{constant}$
	$f = 0$
	$\dot{m} = \text{constant}$
	$I = I^*$
	$\psi = 1 + \dfrac{\gamma-1}{2}M^2$

stagnation pressure while heat rejection produces a decrease in entropy and an increase in stagnation pressure. The decrease in stagnation pressure associated with heat addition is sometimes called the *Rayleigh effect* and is of considerable importance in the design of combustion systems. Figure 7-26 schematically represents the conditions and nomenclature for Rayleigh flow. This flow is sometimes called *reversible heat addition or rejection* because the only change in entropy is caused by the heat addition/rejection process. If the same amount of energy is added and then rejected (or rejected and then added), no property changes occur between the entrance and exit conditions.

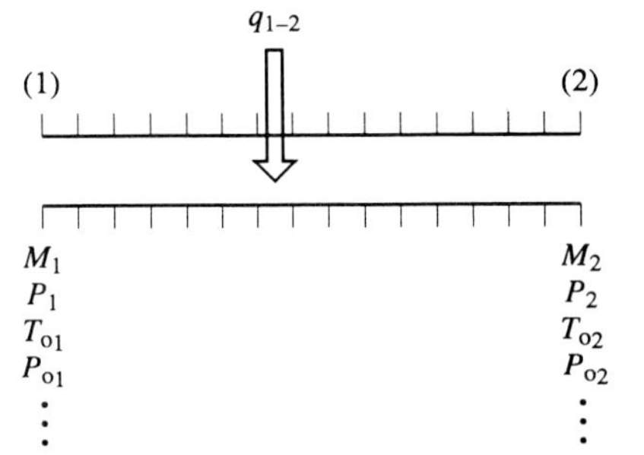

Figure 7-26 Rayleigh Flow Nomenclature

The path of the Rayleigh process on an h–s or T–s diagram is a convenient mechanism for exploring this simple flow. Although the Rayleigh path can be represented on property diagrams involving other variables, the h–s is the most useful version and is the traditional one used. The Rayleigh path represents the locus of states for a particular Rayleigh flow and is usually called the *Rayleigh line*. Table 7-6b gives the change in entropy (referenced to the sonic state) as

$$\frac{\Delta s}{c_p} = \ln\left\{M^2\left(\frac{\gamma + 1}{1 + \gamma M^2}\right)^{(\gamma+1)/\gamma}\right\}. \tag{7-13}$$

The static temperature ratio as a function of Mach number is

$$\frac{T}{T^*} = \left[\frac{(\gamma + 1)M}{1 + \gamma M^2}\right]^2. \tag{7-14}$$

At the sonic point

$$\frac{T^*}{T_o^*} = \frac{2}{\gamma + 1}. \tag{7-15}$$

The dimensionless Rayleigh line can be generated by using the Mach number as the parameter in conjunction with Eqs. 7-13 and 7-14. A single expression for entropy change as a function of the temperature ratio, T/T^*, can be derived by eliminating the Mach number, M, between Eqs. 7-13 and 7-14, but it is long and unwieldy. For a calorically perfect gas the temperature ratio is equal to the enthalpy ratio, since

$$\frac{h}{h^*} = \frac{c_p T}{c_p T^*} = \frac{T}{T^*}. \tag{7-16}$$

Equations 7-13 and 7-14 represent a Rayleigh line on a dimensionless h–s or T–s diagram where the abscissa is the dimensionless entropy, $\Delta s/c_p$, and the ordinate is the dimensionless temperature or the enthalpy ratio, T/T^* or h/h^*. The dimensionless Rayleigh plot on a T–s diagram for a specific heat ratio of 1.4 is given in Figure 7-27a. Mach number as a function of $\Delta s/c_p$ for $\gamma = 1.4$ is plotted as Figure 7-27b. Although Rayleigh lines can be drawn on property diagrams for any two properties, in this text a Rayleigh line will mean the enthalpy (or temperature) versus entropy, or Mollier diagram, presentation, unless otherwise stated.

By observation, Figure 7-27b suggests that the maximum entropy state is the sonic state. This can be verified. The change in entropy for a Rayleigh flow, cast in terms of the Mach number, is given in Eq. 7-13. To find the Mach number at the maximum entropy state, the derivative of $\Delta s/c_p$ or $\exp(\Delta s/c_p)$ with respect to Mach number can be taken and set equal to zero. No entropy minimum exists since both branches of the Rayleigh line are unbounded in the negative direction. The state at which the derivative of entropy vanishes will thus be the state of maximum entropy. The derivative becomes

$$\frac{d}{dM}e^{\Delta s/c_p} = 2M\left(\frac{\gamma + 1}{1 + \gamma M^2}\right)^{(\gamma+1)/\gamma}$$
$$+ M^2\left(\frac{\gamma + 1}{\gamma}\right)\left(\frac{\gamma + 1}{1 + \gamma M^2}\right)^{1/\gamma}\frac{(\gamma + 1)(-1)}{(1 + \gamma M^2)^2}2\gamma M = 0, \tag{7-17}$$

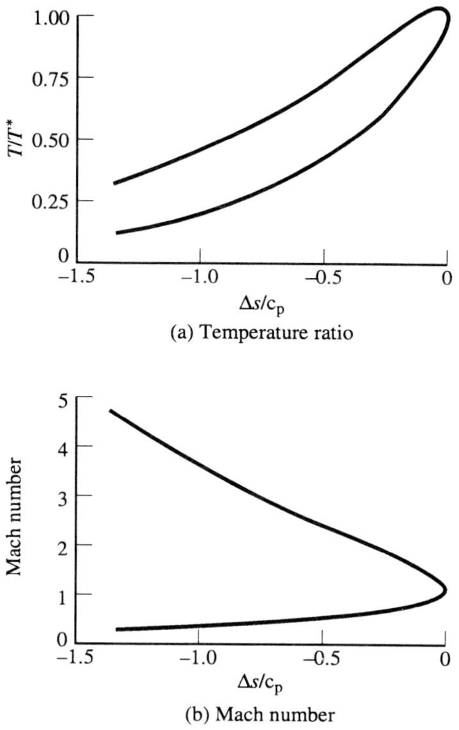

Figure 7-27 Rayleigh Lines for Calorically Perfect Gas ($\gamma = 1.4$)

which reduces to

$$M^2 = 1. \tag{7-18}$$

For the Rayleigh flow of a calorically perfect gas, the state of maximum entropy is thus the sonic state, $M = 1$. The Rayleigh line also shows a maximum value of T/T^*. The Mach number associated with the maximum value of the temperature ratio can be found by setting the derivative of the T/T^* expression in Table 7-6b to zero. Hence,

$$\frac{d}{dM}\left(\frac{T}{T^*}\right) = \frac{(\gamma + 1)^2 2M}{(1 + \gamma M^2)^2} - \frac{(\gamma + 1)^2 M^2}{(1 + \gamma M^2)^3}(2\gamma M) = 0. \tag{7-19}$$

This reduces to

$$M^2 = 1/\gamma. \tag{7-20}$$

Information about the Rayleigh process is schematically shown on Figure 7-28. Movement is to the right on an h–s diagram for heat addition and to the left for heat rejection. The sonic state occurs at the point of maximum entropy, and the Mach number for the maximum value of T/T^* is $1/\sqrt{\gamma}$. Subsonic Mach numbers are associated with the large temperature or enthalpy ratios, and supersonic Mach numbers with the smaller values of the ratios. Thus, the upper portion of the Rayleigh line corresponds to subsonic flow and the lower portion to supersonic flow.

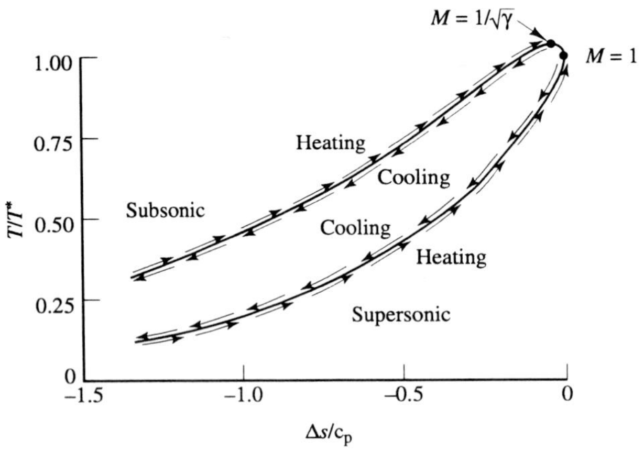

Figure 7-28 Dimensionless Rayleigh Line Representation ($\gamma = 1.4$)

The static temperature ratio decreases as heat is added in the region between $M = 1/\sqrt{\gamma}$ and the sonic point. The stagnation temperature increases during the heat addition process, but the flow accelerates in this region faster than heat is added, and thus additional thermal energy is required for conversion into kinetic energy. Moreover, since the sonic point cannot be traversed without moving into a region of decreasing entropy unless the sense of the heat transfer is changed, a Rayleigh flow that is initially subsonic must remain subsonic, and a Rayleigh flow that is initially supersonic must remain supersonic. For heat addition both supersonic and subsonic Rayleigh flows move toward the Mach number 1 state as energy is added. Both subsonic and supersonic Rayleigh flows move away from the sonic point as heat is rejected. The meaning of T_o^* is now clear; it is the stagnation temperature required in a Rayleigh flow for the Mach number to become 1. As such it represents the limit of heat addition for subsonic or supersonic flow. Any further heat addition will result in a decrease in mass flow rate and a jump to another dimensional Rayleigh line. As with the other simple flows, conditions at the location of Mach 1 in a Rayleigh flow are unique for that given flow and can be used as reference conditions to aid in calculations.

One interesting aspect of a Rayleigh flow is that it provides, at least in theory, an alternative to the converging–diverging nozzle for achieving supersonic flow. If sufficient heat is added to a subsonic flow from a reservoir in a Rayleigh flow situation, Mach 1 can be obtained. If heat is rejected after Mach 1 is attained, the flow is then expanded supersonically. Thus, with exactly the right amount of heat addition followed by heat rejection, supersonic flow in a constant-area duct can be achieved. Because of effects—such as wall friction and lack of flow uniformity—not considered in Rayleigh flow, the process is not really feasible, and to the best of our knowledge, no such device has ever been successful in obtaining steady supersonic flow.

Duct length does not enter into any of the Rayleigh flow expressions given in Table 7-6b. For a given Rayleigh flow process, the heat addition (or rejection) and the initial state determine the final state. Friction, which is the only driving potential concerned with length, is neglected in Rayleigh flows. Duct length can enter into the property distributions (P versus x, for example) for a Rayleigh flow in which energy is added or rejected at some specified rate in the one-dimensional duct.

Figure 7-28 is the Rayleigh line representation for all Rayleigh flows with $\gamma = 1.4$ as expressed in the dimensionless coordinates $\Delta s/c_p$ and T/T^* (or h/h^*). If the Rayleigh lines for flows with different initial stagnation conditions and/or different mass flow rates were represented in dimensional coordinates s and T (or h), then the shapes would be similar to that of Figure 7-28, but the Rayleigh lines for the individual mass flow rates would be distinct. Generally, for our purposes, the dimensionless Rayleigh line shown in Figure 7-28 will be used to explain various phenomena associated with Rayleigh flows.

Example 7-10

Plot the dimensional Rayleigh line for a flow with initial stagnation conditions of 1089 R and 100 psia. The initial Mach number is 0.7 for the Rayleigh flow.

Solution. The initial state occurs at $M = 0.7$ and is

$$M_i = 0.7;$$

$$T_i = (T_i/T_{o_i})T_{o_i} = 0.9107(1089) = 991.8\ R;$$

$$P_i = (P_i/P_{o_i})P_{o_i} = 0.7209(100) = 72.09 \text{ psia}$$

The entropy at the initial state was arbitrarily taken as 0.35 Btu/lbm. Because only the change in entropy between two states is meaningful, the entropy datum could have been taken as any number; however, the choice of this number yields a Rayleigh plot with positive entropy at each state considered. A computer program was written to generate the enthalpy–entropy values needed for the program. Using the static temperature at $M = 0.7$, the value of T^* at the sonic point becomes

$$T^* = \frac{T^*}{T}T = \frac{T}{T/T^*} = 991.8/0.993 = 998.8\ R.$$

With the Mach number as a parameter, the computation sequence was to compute the static temperature using (from Table 7-6b)

$$\frac{T}{T^*} = \frac{(\gamma + 1)^2 M^2}{(1 + \gamma M^2)^2}.$$

The dimensionless entropy becomes (from Table 7-6b)

$$\frac{\Delta s}{c_p} = \ln\left\{M^2\left(\frac{\gamma + 1}{1 + \gamma M^2}\right)^{(\gamma+1)/\gamma}\right\}.$$

The dimensional entropy was then generated from

$$s = s^* + c_p(\Delta s/c_p).$$

Plotting the preceding results yields the Rayleigh line in Figure 7-29. The initial state is indicated on the figure.

As heat is added, the process moves to the right in Figure 7-29. Heat can be added with no change in the mass flow rate until the sonic point, also indicated on the figure, is reached. Further addition of heat beyond that associated with sonic flow of the initial

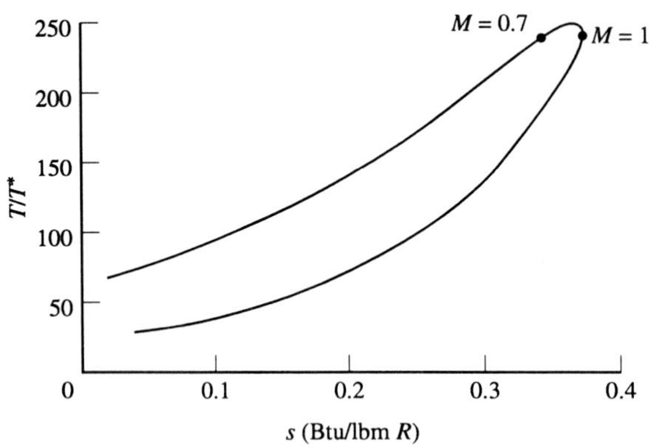

Figure 7-29 Dimensional Rayleigh Line for Example 7-10

Rayleigh line conditions will result in a decrease in mass flow rate and a shift to another dimensional Rayleigh line with a smaller initial Mach number.

Computational aids are available for Rayleigh flow. The expressions from Table 7-6b give the values of various property ratios as functions of Mach number. Tabulations of these relations for a specific heat ratio of 1.4 are presented in Appendix F. Figure 7-30 presents plots of the various property ratios as a function of Mach number. The property ratios are all referenced to sonic conditions.

Rayleigh flow computational aids are available under option **RA** in the COMPQ program. The menu for option **RA** is given in Figure 7-31a. Three cases are available. Cases **M** and **T** effectively replace the Rayleigh tables of Appendix F. The input for these cases is either the Mach number for case **M** or the static temperature ratio, T/T^*, for case **T**. The output for these cases are the Mach number and the static and stagnation pressure and temperature ratios. For a specified Mach number of 0.7, the output for case **M** is given in Figure 7-31b. Both stagnation temperature and pressure change in a Rayleigh flow. The third case, **HT**, is used for the process involved in heat addition or rejection and will be examined in Example 7-11. COMPQ provides added versatility since the value of the specific heat ratio can be specified.

Heat addition (or rejection) is a recurring circumstance involved in Rayleigh flow situations. In such a process, information on the initial state and the change in stagnation temperature or the heat transfer rate are given; the final state after the heat addition process is desired. The heat addition process is schematically illustrated in Figure 7-32. Conditions at station 1 are known as either ΔT_o or q_{1-2}. The two are related by

$$q_{1-2} = c_p(T_{o_2} - T_{o_1}) = c_p \Delta T_o. \tag{7-21}$$

The solution technique is based on the concept that the stagnation temperature T_o^* at the sonic point and the conditions at the sonic point in a given Rayleigh flow are unique. Thus, for state 1 the temperature ratio T_{o_1}/T_o^* relates the stagnation temperature to that at the sonic point, and for state 2, T_{o_2}/T_o^* represents the ratio. For both states T_o^* is the same. The two states are then related as

$$T_{o_2}/T_o^* = (T_{o_2}/T_{o_1})(T_{o_1}/T_o^*). \tag{7-22}$$

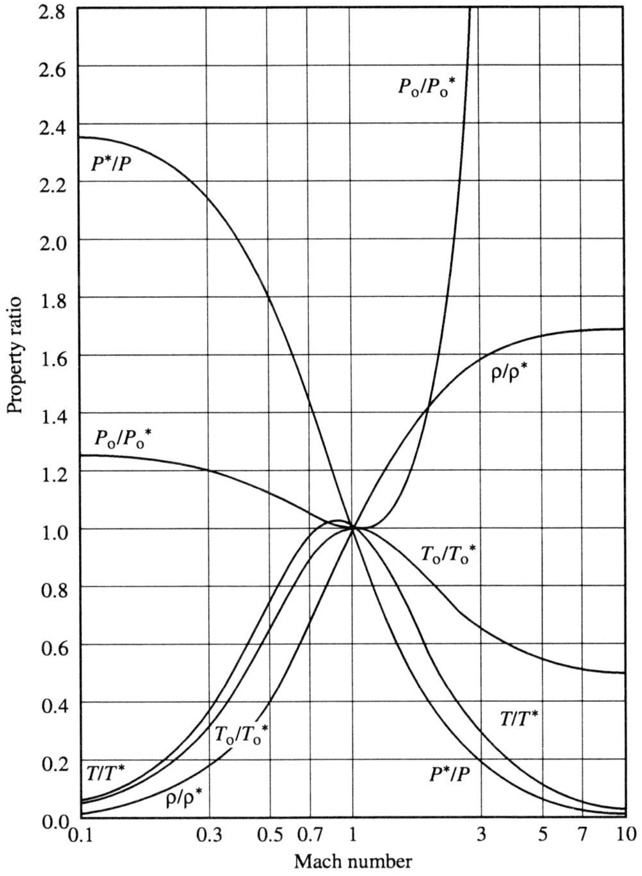

Figure 7-30 Property Ratios for Rayleigh Flow ($\gamma = 1.4$)

```
RAYLEIGH FLOW SELECTED
INPUT PARAMETERS ARE:
     MACH NUMBER            M
     TO/TO*                 T
     HEAT TRANSFER          HT
ENTER THE DESIRED LETTER(S)?
```

(a) Option **RA** menu

```
ENTER THE MACH NUMBER? 0.7

THE MACH NUMBER IS    0.700
   TO/TO*        T/T*        PO/PO*       P/P*       RHO/RHO*
   0.9085       0.9929       1.0431      1.4235       1.4337
```

(b) Case **M** sample output

Figure 7-31 Option **RA** Menu and Sample Output

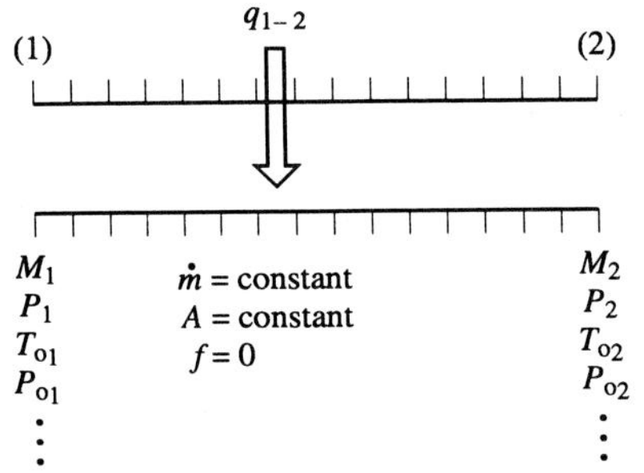

Figure 7-32 Heat Addition Process Schematic

Given the temperature ratio at the inlet, the temperature ratio at the exit can be found using Eq. 7-22. The Mach number and property ratios at state 2 can then be found using the values from the Rayleigh flow tables in Appendix F or from case **T** of option **RA** in COMPQ. Given the Mach number at state 2, the property ratios at state 2 can be computed by using the following ratios:

$$T_2/T_1 = (T_2/T^*)/(T_1/T^*);$$
$$P_{o_2}/P_{o_1} = (P_{o_2}/P_o^*)/(P_{o_1}/P_o^*); \qquad (7\text{-}23)$$
$$P_2/P_1 = (P_2/P^*)/(P_1/P^*).$$

Case **HT** of option **RA** accomplishes the heat addition process by going through the steps outlined earlier in this chapter. Consider the following example.

Example 7-11

Air with initial stagnation conditions of 1200 R and 120 psia is flowing at a Mach number of 0.4 in a constant-area duct. What heat transfer rate is required to raise the stagnation temperature to 1800 R? Find the static and stagnation properties after the simple heat addition, and sketch the process on the Rayleigh line.

Solution. The physical situation is illustrated in Figure 7-33a. The problem will be worked two ways:

1. using cases **M** and **T** of option **RA**
2. using heat transfer process case **HT** of option **RA**.

The heat transfer, sometimes called the *heat interaction*, required for the process is

$$q_{1-2} = c_p(T_{o_2} - T_{o_1})$$
$$= 0.24(1800 - 1200) = 144 \text{ Btu/lbm}.$$

Using case **M** of option **RA** with an input Mach number of 0.4 yields at station 1

$$T_{o_1}/T_o^* = 0.529 \qquad T_1/T^* = 0.615$$
$$P_{o_1}/P_o^* = 1.157 \qquad P_1/P^* = 1.961.$$

After the heat addition, $T_{o_2} = 1800\ R$, therefore

$$T_{o_2}/T_o^* = (T_{o_2}/T_{o_1})(T_{o_1}/T_o^*)$$
$$= (1800/1200)(0.529) = 0.7935.$$

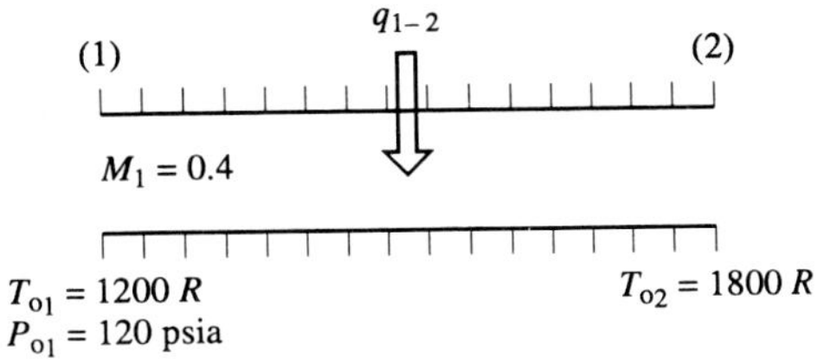

(a) Problem statement

```
THE MACH NUMBER IS    0.577
   TO/TO*        T/T*          PO/PO*         P/P*        RHO/RHO*
   0.7935      0.8927        1.0835        1.6362       1.8330

THE MACH NUMBER IS    1.999
   TO/TO*        T/T*          PO/PO*         P/P*        RHO/RHO*
   0.7935      0.5292        1.5024        0.3638       0.6876
```

(b) Case **T** output

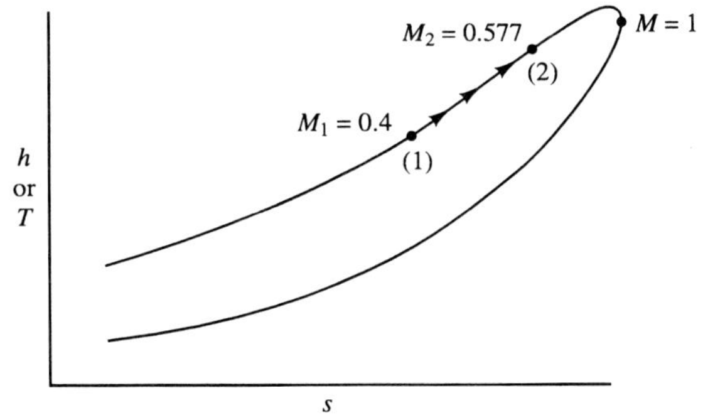

(c) Rayleigh line path

```
ENTER THE INITIAL MACH NUMBER AND THE STAGNATION TEMP (K OR R)? .4,1200.
ENTER THE CHANGE IN STAGNATION TEMP (+ ADD, - REJ)? 600

HEAT TRANSFER PROCESS COMPUTATIONS

     M1         M2          TO1         DELTO
   0.4000     0.5775     1200.00      600.00

   TO2/TO1    T2/T1      PO2/PO1     P2/P1     RHO2/RHO1
   1.5000     1.4512     0.9368     0.8344     0.5750
```

(d) Case **HT** output

Figure 7-33 Example Problem 7-11

With the value of T_{o_2}/T_o^* known, case **T** of option **RA** can be used to determine conditions at M_2. The output of case T for the preceding conditions is presented in Figure 7-33b. Since T_o/T_o^* is double-valued in Mach number, the output of case **T** contains entries for both supersonic and subsonic flows. Earlier we established that with the same sense (addition or rejection) of heat transfer, a process that was initially subsonic must remain subsonic and a process that was originally supersonic must remain supersonic. Because at station 1 the process in this problem started subsonically, subsonic flow must also exist at station 2. After the heat addition, the Mach number is 0.577. The static and stagnation properties at station 2 can be computed from the following various property ratios, using Eqs. 7-23:

$$T_{o_2}/T_{o_1} = (T_{o_2}/T_o^*)/(T_{o_1}/T_o^*)$$

$$= 0.793/0.529 = 1.500;$$

$$T_2/T_1 = (T_2/T^*)/(T_1/T^*)$$

$$= 0.893/0.615 = 1.452;$$

$$P_{o_2}/P_{o_1} = (P_{o_2}/P_o^*)/(P_{o_1}/P_o^*)$$

$$= 1.084/1.157 = 0.937;$$

$$P_2/P_1 = (P_2/P^*)/(P_1/P^*)$$

$$= 1.636/1.961 = 0.834.$$

At the initial Mach number of 0.4 the static properties are

$$P_1 = (P_1/P_{o_1})P_{o_1} = (0.8956)(120) = 107.5 \text{ psia};$$

$$T_1 = (T_1/T_{o_1})T_{o_1} = (0.9690)(1200) = 1162.8\ R.$$

The properties at the end of the heat addition process (station 2) then become

$$M_2 = 0.577;$$

$$T_{o_2} = 1800.0\ R;$$

$$T_2 = (1.452)(1162.8) = 1688.4\ R;$$

$$P_{o_2} = (0.937)(120) = 112.4 \text{ psia};$$

$$P_2 = (0.834)(107.5) = 89.7 \text{ psia}.$$

Because the change in entropy from state 1 to state 2 is positive, the stagnation pressure decreases from state 1 to state 2. Thus heat addition in a constant-area duct results in a loss in stagnation pressure even though the stagnation temperature has been increased by 600 R. The Rayleigh line path for this process is shown in Figure 7-33c.

This problem can also be worked using process case **HT** of option **RA**. The input and output for case **HT** for this problem are shown in Figure 7-33d. The input parameters required are the initial Mach number, the initial stagnation temperature, and the change in stagnation temperature for the process. The T_o change is positive for heat addition and negative for heat rejection. For this problem the heat addition is 600 R. The various output property ratios and the Mach number agree well with the results for the previous solution using cases **M** and **T**. Case **HT** thus provides a single-input-sequence solution, whereas the previous solution technique for this problem requires application of cases **M** and **T** in addition to some arithmetic.

The heat addition (or rejection) process is as fundamental to Rayleigh flow as the duct addition process is to Fanno flow. The use of case **HT** is generally the preferred technique. Case **HT** can be run "in reverse" in the same sense as case **DA** of option **FA**. Care must be taken to identify states 1 and 2 properly when moving upstream in a Rayleigh flow, and the sign on the change in T_o must be taken as negative with respect to the downstream process. An example involving heat rejection is presented next.

Example 7-12

Air, initially with a stagnation temperature of 800 K and a Mach number of 0.4, traverses a constant-area cold-wall duct, where its stagnation enthalpy is reduced by 300 kJ/kg. Determine the conditions at the duct exit.

Solution. Figure 7-34a schematically illustrates the duct arrangement. The energy transfer is related to the stagnation temperatures by

$$q_{1-2} = c_p(T_{o_2} - T_{o_1})$$
$$-300 \text{ kJ/kg} = 1.0035 \text{ kJ/kg K } (T_{o_2} - 800 \text{ K})$$
$$T_{o_2} = 501.0 \text{ K},$$

and

$$\Delta T_o = -299.0 \text{ K}.$$

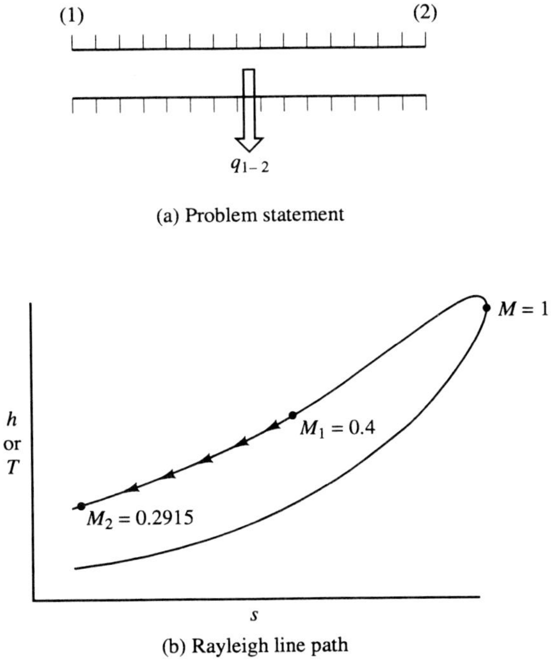

Figure 7-34 Example Problem 7-12

Using case **HT** of option **RA** with $M_1 = 0.4$, $T_{o_1} = 800$ K, and $\Delta T_o = -299.0$ K yields

$$M_2 = 0.2915 \qquad P_{o_2}/P_{o_1} = 1.0392$$

$$T_2/T_1 = 0.6354 \qquad P_2/P_1 = 1.0939.$$

Conditions at state 2 are

$$M_2 = 0.2915;$$

$$T_{o_2} = 501.0 \text{ K};$$

$$T_2 = \frac{T_2}{T_1}\frac{T_1}{T_{o_1}}T_{o_1} = (0.6354)(0.969)(800.0) = 492.56 \text{ K}.$$

The Rayleigh line path is indicated in Figure 7-34b.

7.6 RAYLEIGH FLOW APPLICATIONS

Combustors in gas turbine engines typically involve little change in area or mass flow rate and, because of the low Mach numbers involved, do not have significant frictional losses. They are dominated by the energy addition process. Hence, Rayleigh flow is a good model for the energy addition process in gas turbine combustion systems. Wilson (1984) gives the following typical ranges as those over which gas turbine combustors operate:

fuel/air mass ratio: 0.002–0.020
inlet pressure: 0.2–40 atm
inlet temperature: 245–945 K (440–1770 *R*)

The assumption of no change in mass flow rate is good for most gas turbine combustors since the maximum fuel/air mass ratio is 0.02, which is only a 2 percent change in mass flow rate due to the fuel addition.

Rayleigh flow assumes a calorically perfect gas. Obviously, when fuel is injected and burned, the chemical composition of the gas changes. In situations involving large changes in stagnation temperature and enthalpy, the assumption of a perfect gas may produce significant errors. However, for many gas turbine combustors the fuel/air mass ratio is small enough that the chemical changes involved are small compared to the heat addition effects. Table 7-7 lists densities and heating values for various common fuels.

TABLE 7-7 DENSITIES AND HEATING VALUES OF SELECTED FUELS

	Density heating value			
Fuel	(kg/l)	(lbm/gal)	(MJ/kg)	(Btu/lbm)
Jet A/A-1	0.775–0.830	6.47–6.93	42.8	18400
JP4/Jet B	0.751–0.802	6.27–6.69	42.8	18400
JP5	0.787–0.845	6.57–7.05	42.6	18315

Example 7-13

Air enters a Rayleigh flow region with a stagnation temperature of 800 K and a Mach number of 0.4. How much energy can be added to the flow without affecting the mass flow rate? If Jet A fuel is used, what fuel/air mass ratio is required?

Solution. The situation is depicted in Figure 7-35a, with the Rayleigh path sketched in Figure 7-35b. Starting at state 1, energy can be added until the Mach number at state 2 is 1. If more energy is added, the flow shifts Rayleigh lines and the initial Mach number will be altered. Using case **M** of option **RA** yields a stagnation temperature ratio of

$$\frac{T_{o_1}}{T_o^*} = 0.529.$$

Since $M_2 = 1$ represents the limit of energy addition without altering the mass flow rate,

$$T_{o_2} = T_o^* = 1512.3 \text{ K}.$$

The stagnation temperature difference is

$$\Delta T_o = T_{o_2} - T_{o_1} = T_o^* - T_{o_1} = 712.3 \text{ K}.$$

Assuming $\dot{m}_{\text{fuel}} \ll \dot{m}_{\text{air}}$, and letting H_v be the fuel heating value, results in the energy release by the fuel being equal to the energy gain by the air:

$$\dot{m}_{\text{fuel}} H_v = \dot{m}_{\text{air}} c_p (T_{o_2} - T_{o_1}).$$

The fuel/air mass ratio then becomes

$$\frac{\dot{m}_{\text{fuel}}}{\dot{m}_{\text{air}}} = \frac{c_p (T_{o_2} - T_{o_1})}{H_v}.$$

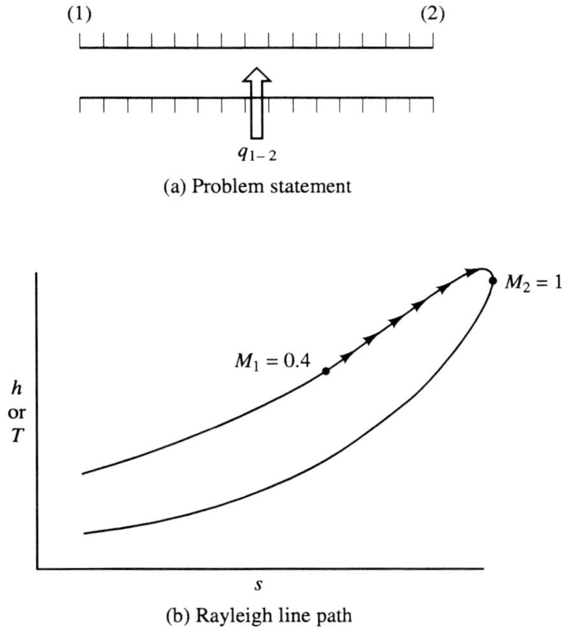

Figure 7-35 Example Problem 7-13

For Jet A in this example, this ratio reduces to

$$\frac{\dot{m}_{\text{fuel}}}{\dot{m}_{\text{air}}} = \frac{(1.0035)(712.3)}{42{,}800} = 0.0167.$$

The assumption of a small fuel/air mass ratio is reasonable for this problem. The stagnation pressure ratio for the process is

$$\frac{P_{o_2}}{P_{o_1}} = \frac{P_o^*}{P_{o_1}} = \frac{1}{1.157} = 0.8643,$$

producing a stagnation pressure loss of

$$\frac{\Delta P_o}{P_{o_1}} = \frac{(0.8643 - 1)P_{o_1}}{P_{o_1}} \times 100 = -13.57\%.$$

Throughout our discussion of Rayleigh flow, we have used the words *heat addition or rejection*. We have treated Rayleigh flow from a phenomenological standpoint and have not been concerned with the details of the mechanism that provides the heat addition or rejection. Heat addition can occur either by heat transfer from the duct surface to the flow or by injection and combustion of fuel. Heat rejection is usually associated with heat transfer from the flow to the duct surface. Additional considerations for combustion are presented later in this section. The average convective heat transfer is often written for incompressible fluids as

$$\frac{q}{A} = h_c(T_s - T_m), \tag{7-24}$$

where T_s is the surface temperature, T_m is the local mean temperature, and h_c is the average convective heat transfer coefficient. The driving potential for convective heat transfer is then $T_s - T_m$.

For a compressible fluid, however, the concept of adiabatic wall temperature is usually used in conjunction with convective heat transfer relations. The adiabatic wall temperature is defined as the surface temperature required for the local convective heat flux to vanish. It is the temperature that an insulated wall would attain in a compressible flow. Surface temperatures higher than the adiabatic wall temperature result in heat transfer *to* the fluid; surface temperatures lower than the adiabatic wall temperature result in heat transfer *from* the fluid. The adiabatic wall temperature is the natural reference for convective heat transfer processes in compressible flows. Its usual representation is

$$\frac{q}{A} = h_c(T_s - T_{aw}), \tag{7-25}$$

where T_{aw} is the adiabatic wall temperature. When $T_s = T_{aw}$, the surface is adiabatic. Since the no-slip conditions on a bounding surface require the fluid to be at rest with respect to the surface, the temptation is to declare the stagnation temperature and the adiabatic wall temperature to be the same. However, because of irreversible dissipation effects within the boundary layer, the stagnation temperature is not recovered at a bounding surface; the adiabatic wall temperature is recovered. The relationship between the adiabatic wall temperature, the Mach number, and the static temperature is given by

$$T_{aw} = T\left(1 + r\frac{\gamma - 1}{2}M^2\right), \tag{7-26}$$

where r is the recovery factor. The recovery factor is a function of Prandtl number; the usually accepted values are

$$\text{laminar} \quad r_{lam} = Pr^{1/2}; \tag{7-27}$$

$$\text{turbulent} \quad r_{turb} = Pr^{1/3}. \tag{7-28}$$

The Prandtl number for air is about 0.71, thus

$$\begin{aligned} r_{lam} &= 0.8426; \\ r_{turb} &= 0.8921. \end{aligned} \tag{7-29}$$

For subsonic flows the difference between the stagnation and adiabatic wall temperatures is very small, and the stagnation temperature may be used with little error.

Even though the exit state of a Rayleigh flow depends only on the total heat addition or rejection, the local heat flux is given by Eq. 7-25. For constant-property, fully developed flow, the convective transfer coefficient is usually taken as constant. Incropera and DeWitt (1990) provide an excellent discussion. Internal heat transfer is often discussed in terms of either constant heat flux or constant wall temperature conditions. The fully developed, constant heat flux condition requires a constant temperature difference between the fluid and the surface. Linear temperature distributions for both the surface temperature and the mean fluid temperature result from the constant heat flux condition.

The constant surface temperature condition produces a varying local heat flux and nonlinear adiabatic wall and stagnation temperature distributions. Temperature distributions for the constant surface temperature condition appear in Figure 7-36a. At a given position along the duct, an energy balance on a differential segment (as illustrated in Figure 7-36b) yields

$$dQ = -\dot{m}c_p\,dT_o = h_c\pi D\,dx(T_{aw} - T_s). \tag{7-30}$$

Approximating the adiabatic wall temperature by the stagnation temperature and integrating from the Rayleigh flow entrance to a distance x units down the duct gives

$$-\dot{m}c_p\int_{T_{o_1}}^{T_o(x)}\frac{dT_o}{T_o - T_s} = h_c\pi D\int_0^x dx. \tag{7-31}$$

The distribution of stagnation temperature then becomes

$$T_o(x) = T_s + (T_{o_1} - T_s)e^{-(h_c\pi D\,dx)/\dot{m}c_p}. \tag{7-32}$$

Although the local value of stagnation temperature for this condition varies, keep in mind that the Rayleigh flow exit conditions depend only on total energy transfer, not on the local distribution of that transfer.

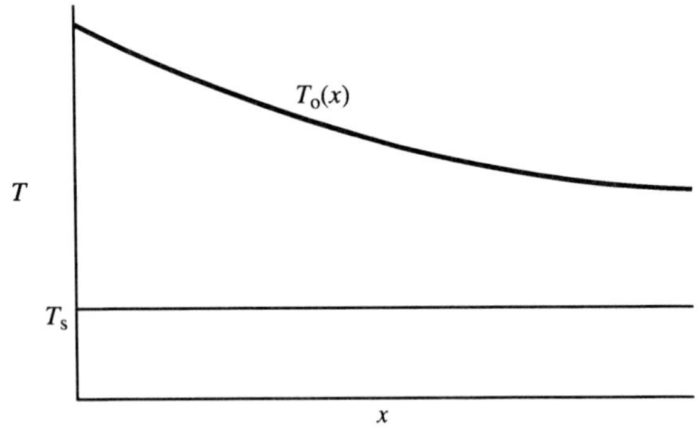

(a) Constant surface temperature

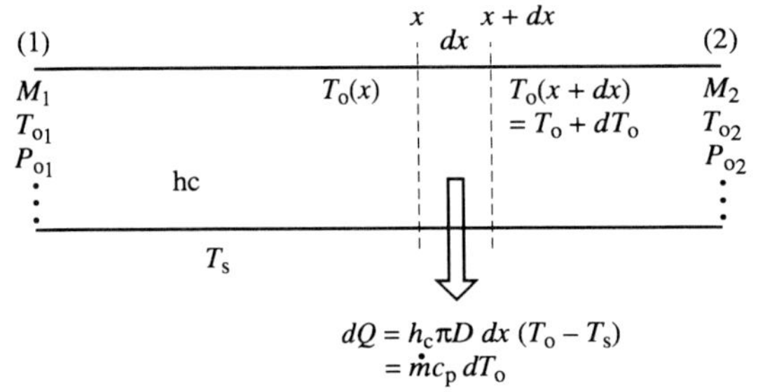

(b) Energy balance

Figure 7-36 Constant Surface Temperature $T_0(x)$ Derivation

7.7 RAYLEIGH FLOW APPLICATIONS INVOLVING NOZZLES

Rayleigh flow regions can be preceded by converging or converging–diverging nozzles. Such systems must obey the principles established for both simple area change and simple T_o change. Additionally, as with any other compressible flow, the exit pressure/back pressure relationship must be consistent with the exit Mach number. Consider the following example involving a converging nozzle/Rayleigh flow region.

Example 7-14

A converging nozzle/constant-area duct system is connected to a reservoir containing air at 500 R and 25 psia. Heat is added in the constant area duct.

(a) For an initial duct Mach number of 0.3, find the back pressure and q_{1-2} for the flow to be choked.

(b) For the same q_{1-2} as in part (a) and a back pressure of 15 psia, find the duct entrance and exit Mach numbers.

(c) For the back pressure of part (a) and a heat addition of 100 Btu/lbm, find the exit pressure and exit Mach number.

(d) For the back pressure of part (a) and a heat addition of 300 Btu/lbm, find the exit pressure and the exit Mach number.

Solution

(a) The system is schematically represented in Figure 7-37a. From case **M** of option **RA** we get the following values:

$$M_2 = 1$$

$$\frac{T_{o1}}{T_o^*} = 0.347$$

$$P_1/P^* = 2.131,$$

from which

$$T_o^* = \frac{500}{0.347} = 1441.5\ R,$$

and

$$\Delta T_o = 1441.5 - 500 = 941.5\ R.$$

The maximum back pressure to choke the system will occur when the exit Mach number is 1 and the exit and back pressures are equal. Hence,

$$P_e = P_b = P_2 = P^* = \frac{P^*}{P_1}\frac{P_1}{P_{o_1}}P_{o1};$$

$$P_e = P_b = (0.470)(0.9395)(25) = 11.039 \text{ psia}.$$

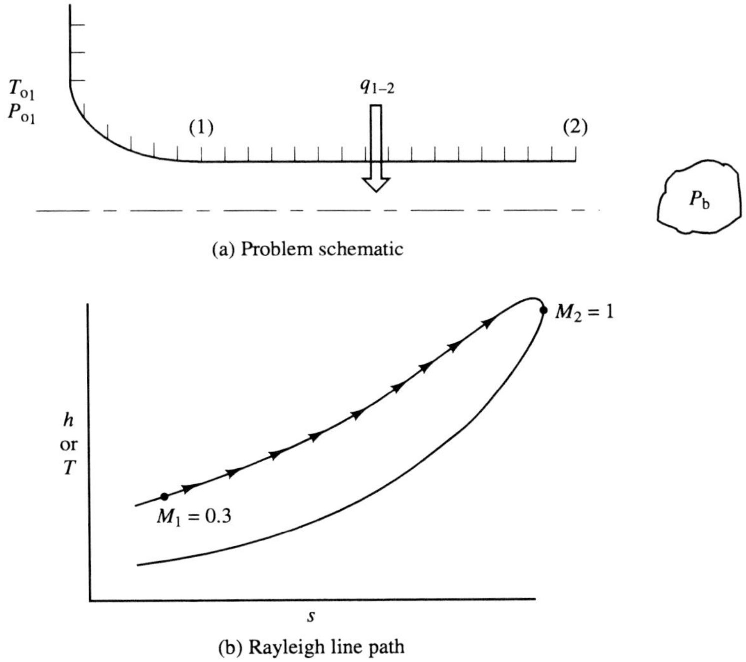

Figure 7-37 Example Problem 7-14

The nozzle/duct system will be choked for any back pressure less than or equal to 11.039 psia. The heat interaction is

$$\begin{aligned} q_{1-2} &= c_p(T_{o_2} - T_{o_1}) \\ &= 0.24(1441.5 - 500) \\ &= 225.8 \text{ Btu/lbm}. \end{aligned}$$

The Rayleigh line path is shown in Figure 7-37b.

(b) The energy addition is still 225.8 Btu/lbm, but the back pressure is now 15 psia. Since the back pressure is greater than 11.039 psia, the flow is not choked. Neither the entrance nor the exit Mach number is known, but the duct exit pressure must be equal to the back pressure. Two solution techniques are possible. An iterative procedure using case **HT** of option **RA** will be examined first. The procedure is to assume an entrance Mach number M_1, use case **M** of option **IS** to find P_1/P_o, and apply case **HT** of option **RA** to find P_2/P_1. The exit pressure is then computed as

$$P_e = P_2 = \frac{P_2}{P_1}\frac{P_1}{P_{o_1}}P_{o_1}.$$

When the assumed Mach number results in an exit pressure of 15 psia, convergence is attained. Table 7-8 delineates the sequence of the iterations. The entrance and exit Mach numbers are 0.287 and 0.735, respectively.

TABLE 7-8 ITERATIVE SEQUENCE FOR EXAMPLE 7-14b

M_1	P_1/P_{o_1}	M_2	P_2/P_1	P_2 (psia)
0.20	0.9725	0.384	0.875	19.09
0.25	0.9575	0.535	0.776	17.58
0.275	0.9488	0.650	0.695	16.49
0.287	0.9449	0.735	0.635	14.99

The alternate solution technique involves using the exit-to-entrance stagnation pressure ratio to devise an equation in terms of M_1 and M_2. The exit-to-entrance stagnation temperature is also used to devise another independent equation in terms of M_1 and M_2. The nonlinear algebraic system of two equations in the unknowns M_1 and M_2 is then solved. The ratio P_e/P_{o_1} can be expressed as

$$\frac{P_e}{P_{o_1}} = \frac{P_2}{P_{o_1}} = \frac{P_2}{P^*}\frac{P^*}{P_1}\frac{P_1}{P_{o_1}}.$$

Using the results of Table 7-6b for the P/P^* terms and the isentropic relation for P_1/P_o, the preceding equation becomes

$$\frac{P_e}{P_{o_1}} = \frac{15}{25} = \frac{\gamma+1}{1+\gamma M_2^2}\,\frac{1+\gamma M_1^2}{\gamma+1}\left(1+\frac{\gamma-1}{2}M_1^2\right)^{-\gamma/(\gamma-1)}.$$

In the same fashion, the stagnation temperature ratio is

$$\frac{T_{o_2}}{T_{o_1}} = \frac{T_{o_e}}{T_{o_1}} = \frac{T_{o_2}}{T_o^*}\frac{T_o^*}{T_{o_1}}.$$

Use of the relations in Table 7-6b for T_o/T_o^* leads to

$$\frac{1441.5}{500} = \left(\frac{M_2}{M_1}\right)^2\left(\frac{1 + \gamma M_1^2}{1 + \gamma M_2^2}\right)\left(\frac{1 + (\gamma - 1)/2M_2^2}{1 + (\gamma - 1)/2M_1^2}\right).$$

The nonlinear system represented by the two equations is then solved. With case **HT** of option **RA** available, the iterative process is usually easier and involves less arithmetic than solving the system of equations.

(c) The heat interaction is 100 Btu/lbm and the back pressure is 11.039 psia. The key to this part of the problem is deciding whether or not the flow is choked. The heat interaction has been reduced, so the loss in stagnation pressure has been reduced. This suggests that the exit pressure is greater than 11.039 psia. Thus, we shall initially assume that the system is choked. If this is the wrong assumption, then an inconsistency between the exit/back pressures and the exit Mach number will emerge.

The change in stagnation temperature can be calculated from consideration of

$$q_{1-2} = c_p(T_{o_2} - T_{o_1})$$

$$100 = 0.24(T_{o_2} - T_{o_1})$$

$$T_{o_2} = 916.7\ R,$$

so that

$$\Delta T_o = T_{o_2} - T_{o_1} = 416.7\ R.$$

Using case **HT** of option **RA** in the reverse sense yields

$$M_1 = 0.409;$$

$$\frac{P_1}{P_2} = 1.944.$$

The exit pressure is then

$$P_e = P_2 = \frac{P_2}{P_1}\frac{P_1}{P_{o_1}}P_{o_1}$$

$$= \frac{1}{1.944}(0.8912)(25) = 11.46 \text{ psia}.$$

Since the assumed exit Mach number of 1 results in an exit pressure greater than the back pressure, the assumption of choked flow is correct. Although this choking behavior is identified here for a specific problem, it can be generalized to conclude that if all other parameters, including the reservoir and back pressures, are equal, a decrease in the heat interaction, q_{1-2}, will result in the system remaining choked. The limit for this problem is for $q_{1-2} = 0$, for which the exit pressure is 13.21 psia. Thus, the effect of heat addition is to reduce the maximum back pressure required for choking when compared with a converging nozzle.

(d) If the energy transfer is increased to 300 Btu/lbm while the back pressure is maintained at 11.039 psia, the system will unchoke! This conclusion is logical based on the discussion of part (c). With an unchoked nozzle, the same procedure used in part (b) is applicable. If the initial assumption of unchoked flow is not correct, a contradiction between the exit/back pressures and the exit Mach number will arise. The iterative sequence is given in Table 7-9. The solution shows that an entrance Mach number of 0.266305 leads to an exit Mach number of 0.989 with an exit pressure of 11.04 psia. This confirms that increasing the heat interaction does indeed result in the unchoked flow. The iteration is very sensitive to the entrance Mach number.

TABLE 7-9 ITERATIVE SEQUENCE FOR EXAMPLE 7-14d

M_1	P_1/P_{o_1}	M_2	P_2/P_1	P_2 (psia)
0.300	0.9395		No Rayleigh flow solution	
0.250	0.9897	0.687	0.655	16.17
0.270	0.9506		No Rayleigh flow solution	
0.265	0.9524	0.898	0.516	12.29
⋮	⋮	⋮	⋮	⋮
finally				
⋮				
0.266305	0.9519	0.989	0.464	11.04

The four parts of Example 7-14 reasonably establish the behavior of converging nozzle/Rayleigh flow systems. For such systems, the key to many problem solutions is to determine the maximum back pressure required for choked flow.

Converging–diverging nozzles in conjunction with Rayleigh flows can be handled in essentially the same fashion, although the possibility of either a subsonic or supersonic entrance Mach number provides more of a variety of possible flow field configurations. Actually, supersonic Rayleigh flow conditions are difficult to produce and maintain because of the significant effect of friction at elevated Mach numbers and because of problems associated with maintaining uniform (one-dimensional) temperature profiles. Because of these difficulties, Rayleigh flows in conjunction with converging–diverging nozzles are of more academic than practical interest. We shall not pursue them in detail; however, we will examine one interesting aspect in the next paragraphs.

Section 7.4 provided extensive discussion of the effects of back pressure on the normal shock wave location in nozzle/constant-area duct systems. Changing the location of the shock wave in a duct for a Fanno flow results in a different exit pressure since the limiting duct length is different for supersonic and subsonic flows. A peculiar aspect of Rayleigh flow is that the exit pressure is independent of the shock wave location when the shock wave is in the duct, as in Figure 7-38, for the same overall change in stagnation temperature. Although Emanuel (1986) presents a formal proof of the independence of back pressure on shock wave location for Rayleigh flow in duct, we shall pursue a more heuristic argument. The limiting stagnation temperature condition, T_o/T_o^*, for the duct is set by the entrance duct Mach number. Even if a normal shock wave is present in the duct, that limiting temperature condition is not changed, since stagnation temperature is constant across a normal shock wave. Hence, since the Rayleigh flow limiting condition is unchanged, the duct exit conditions are not affected by the location of the normal shock; in addition, as long as the change in total temperature for the Rayleigh flow remains constant, duct shock wave location has no effect on back pressure.

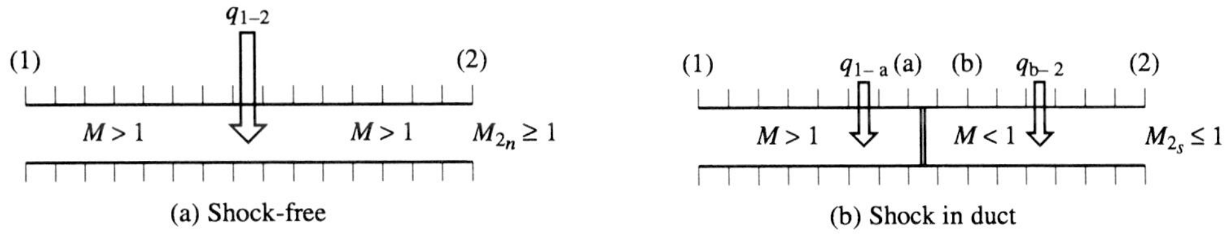

Figure 7-38 Shock Effects in Rayleigh Flow

One ramification of the preceding attributes of Rayleigh flow converging–diverging nozzle/duct systems is that for a given change in T_o the shock wave location cannot be used, as it can be in Fanno flow, to meet varying back pressure requirements. Generally in such Rayleigh flow situations, the normal shock must be placed in the diverging segment of the nozzle in order to have an effect on the exit plane back pressure.

7.8 NORMAL SHOCKS AND RAYLEIGH AND FANNO LINES

Do Rayleigh and Fanno lines intersect? If they do intersect, is there any physical interpretation of the meaning of the intersection(s)? The answer to both questions is yes. This section discusses the physical interpretation of their intersections and presents an example.

The Rayleigh (Figure 7-29) and Fanno (Figure 7-6) representations on the Mollier diagram intersect at points where the entropy and the enthalpy are the same for both processes. Tables 7-1b and 7-6b contain expressions for the entropy change referenced to the sonic states for the individual flows. The sonic states for the Rayleigh flow and the Fanno flow for given conditions are not the same, so the s in the two tables represents different entropy values. Consider both flows to have entropy s_r *at* M_r. For Rayleigh flow (from Table 7-6b),

$$\frac{s - s^*}{c_p} = \frac{\Delta s}{c_p} = \ln\left\{M^2\left(\frac{\gamma + 1}{1 + \gamma M^2}\right)^{(\gamma+1)/\gamma}\right\}. \tag{7-33}$$

Again from Table 7-6b, the entropy difference between M_r and the sonic point for Rayleigh flow is

$$\frac{s_r - s^*}{c_p} = \ln\left\{M_r^2\left(\frac{\gamma + 1}{1 + \gamma M_r^2}\right)^{(\gamma+1)/\gamma}\right\}. \tag{7-34}$$

Subtracting the two equations yields the following expression for the change in entropy referenced to s_r:

$$\frac{s - s_r}{c_p} = \ln\left\{\left(\frac{M}{M_r}\right)^2\left(\frac{1 + \gamma M_r^2}{1 + \gamma M^2}\right)^{(\gamma+1)/\gamma}\right\}. \qquad (7\text{-}35)$$

For Fanno flow (from Table 7-1b),

$$\frac{s - s^*}{c_p} = \ln\left\{M^{(\gamma-1)/\gamma}\left[\left(\frac{2}{\gamma + 1}\right)\psi\right]^{-(\gamma+1)/2\gamma}\right\}. \qquad (7\text{-}36)$$

Again using s_r as the entropy at M_r, the Fanno equivalent to Eq. 7-35 becomes

$$\frac{s - s_r}{c_p} = \ln\left\{\left(\frac{M}{M_r}\right)^{(\gamma-1)/\gamma}\left(\frac{1 + (\gamma - 1/2)M_r^2}{1 + (\gamma - 1/2)M^2}\right)^{(\gamma+1)/2\gamma}\right\}. \qquad (7\text{-}37)$$

If a Rayleigh line and a Fanno line are to intersect, then Eqs. 7-35 and 7-37 must be equal.

With the same reference for entropy, equating the expressions is meaningful to determine the point(s) of intersection. This results in

$$\left(\frac{M}{M_r}\right)^2\left(\frac{1 + \gamma M_r^2}{1 + \gamma M^2}\right)^{(\gamma+1)/\gamma} = \left(\frac{M}{M_r}\right)^{(\gamma-1)/\gamma}\left(\frac{1 + (\gamma - 1)/2M_r^2}{1 + (\gamma - 1)/2M^2}\right)^{(\gamma+1)/2\gamma}. \qquad (7\text{-}38)$$

After some simplification Eq. 7-38 can be expressed as

$$(1 + (\gamma - 1)/2M_r^2)\frac{M_r^2}{(1 + \gamma M_r^2)^2} = (1 + (\gamma - 1)/2M^2)\frac{M^2}{(1 + \gamma M^2)^2}. \qquad (7\text{-}39)$$

But we have seen this equation before; it is Eq. 2-14! The intersection of the Rayleigh and Fanno lines defines a normal shock wave occurring between M_r and M. If M_r is supersonic, then M is the subsonic Mach number across a normal shock wave. If M_r is subsonic, then M is the supersonic Mach number upstream of a normal shock wave.

At first this relationship looks strange because normal shock waves appear to have nothing to do with Rayleigh or Fanno flows. But consider the principles behind the entries in Tables 7-1 and 7-6. The entries in both of these tables represent the conservation equations, the equation of state, and various definitions applied to the generalized control volume of Figure 5-1 with only single driving potentials. The entries in Table 7-1 are associated with friction as the driving potential, while the entries in Table 7-6 are associated with heat addition or rejection as the driving potential. Friction plots as the Fanno line; heat addition or rejection plots as the Rayleigh line. For given initial flow conditions, movement along the Fanno line is determined by friction while movement along the Rayleigh line is determined by heat transfer. Friction is involved only with the momentum equation, and T_o change is involved only with the energy equation. Thus, movement along the Fanno line is controlled by the momentum equation, and movement along the Rayleigh line is controlled by the energy equation. The intersection points of the Rayleigh and Fanno lines simultaneously satisfy, for a constant-area duct, all conservation expressions as well as the equation of state—which are the basis of the derivation of the normal shock relations. Hence, while a normal shock may not be formally associated with either a Fanno or a Rayleigh flow, the two states where both Fanno and Rayleigh lines intersect are the functional equivalent of the conservation expressions applied across a shock wave.

Example 7-15

Flow at the entrance of a constant-area duct has a Mach number of 2.0 and a static temperature of 500 R. Using these initial conditions, plot the Rayleigh and Fanno lines. Show that the intersection points represent a normal shock wave with an upstream Mach number of 2.0.

Solution. The initial stagnation temperature is

$$T_{o_1} = 899.9\ R.$$

The computer programs previously used to generate the dimensional Fanno line (Figure 7-6) and the dimensional Rayleigh line (Figure 7-29) generated values of s, M, and h for the two lines. The entropy reference for both the Rayleigh and Fanno lines was taken to be zero at $M = 2$. The results are plotted in Figure 7-39a for the Mollier diagram and in Figure 7-39b for the Mach-number-versus-entropy plot.

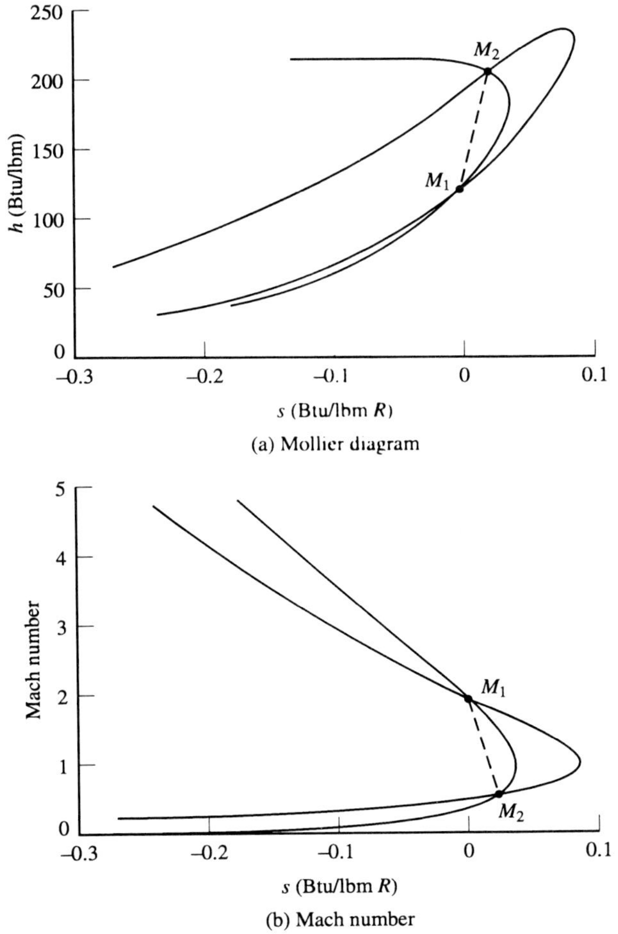

Figure 7-39 Normal Shock Definition Using Rayleigh and Fanno Flows

The Rayleigh and Fanno lines for these initial conditions do indeed intersect at two states. The $M = 2$ intersection was enabled by taking the entropy datum for both flows to be at that state. The other interaction in this case yields, from Figure 7-39b, $M = 0.57$. The normal shock tables or case **U** of option **NS** yields for $M_1 = 2$,

$$M_2 = 0.5774,$$

which agrees well with the value read from Figure 7-39b. As Eq. 7-38 suggests, the intersection points for this example define conditions across a normal shock at $M_1 = 2$. The intersections of Rayleigh and Fanno lines cast in terms of other properties would likewise define the states across a normal shock wave.

7.9 SIMPLE MASS ADDITION OR REMOVAL

If the only driving potential is mass flow rate change, then *simple mass addition or removal* is defined. This flow will be labeled simple mass addition even though mass removal can be treated. For convenience Table 7-10a reproduces the influence coefficient entries from Table 5-1 for simple mass addition. The independent variable is given as $d\dot{m}/\dot{m}$. A positive value of $d\dot{m}/\dot{m}$ signifies mass addition, and a negative value of $d\dot{m}/\dot{m}$ signifies mass removal. The closed-form expressions for simple mass addition from Table 5-2 are restated in Table 7-10b. The entries in Table 7-10b assume that

1. The stagnation enthalpy of the fluid added to (or removed from) the duct is equal to the stagnation enthalpy of the fluid within the duct.
2. The fluid is injected (or removed) with no axial velocity component.

The entropy change is referenced to the sonic state, so $\Delta s = s - s^*$. As with the other simple flows, the starred variables denote conditions at the sonic location. Simple mass addition is thus flow in an adiabatic (T_o = constant), frictionless ($f = dF_D = 0$), constant-area duct (A = constant) with mass flow rate change. If the conservation equations and various definitions are applied to an elemental control volume with constant area, no heat transfer, no friction, and the preceding assumptions, the expressions in Table 7-10a result. Integration of the entries in Table 7-10a with conditions referenced to the sonic point produce the entries of Table 7-10b.

The only expression with a new functional form appearing in Table 7-10b is the one that relates $\dot{m}/\dot{m}^*$ to the Mach number. For simple mass addition the increase in mass flow rate is the important parameter; no distinction is made involving the distance over which the fluid injection occurs. The influence coefficients for entropy and stagnation pressure show that mass addition produces an increase in entropy and a decrease in stagnation pressure; mass removal produces a decrease in entropy and an increase in stagnation pressure. Figure 7-40 schematically represents the conditions and nomenclature for simple mass addition.

Since the mass flow rate is not constant in this simple flow, the equivalents of the Rayleigh and Fanno lines, which are for constant mass flow rates, are slightly different. However, the process can be represented on an h–s or T–s diagram. Such

TABLE 7-10 SUMMARY OF RELATIONS FOR SIMPLE MASS FLOW RATE CHANGE (NORMAL INJECTION)

(a) Differential forms	(b) Property ratios
$\dfrac{dM^2}{M^2} = \dfrac{2(1+\gamma M^2)\psi}{1-M^2}\dfrac{d\dot{m}}{\dot{m}}$ $\quad \psi = 1 + \dfrac{\gamma - 1}{2}M^2$	$\dfrac{\dot{m}}{\dot{m}^*} = \dfrac{M\sqrt{2(\gamma+1)\psi}}{1+\gamma M^2}$
$\dfrac{dV}{V} = \dfrac{1+\gamma M^2}{1-M^2}\dfrac{d\dot{m}}{\dot{m}}$	$\dfrac{P}{P^*} = \dfrac{\gamma+1}{1+\gamma M^2}$
$\dfrac{dT}{T} = \dfrac{(\gamma-1)M^2(1+\gamma M^2)}{1-M^2}\dfrac{d\dot{m}}{\dot{m}}$	$\dfrac{\rho}{\rho^*} = \dfrac{2\psi}{1+\gamma M^2}$
$\dfrac{d\rho}{\rho} = -\dfrac{(\gamma+1)M^2}{1-M^2}\dfrac{d\dot{m}}{\dot{m}}$	$\dfrac{T}{T^*} = \left[\left(\dfrac{2}{\gamma+1}\right)\psi\right]^{-1}$
$\dfrac{dP}{P} = -\dfrac{2\gamma M^2\psi}{1-M^2}\dfrac{d\dot{m}}{\dot{m}}$	$\dfrac{V}{V^*} = M\left[\left(\dfrac{2}{\gamma+1}\right)\psi\right]^{-1/2}$
$\dfrac{dP_o}{P_o} = -\gamma M^2\dfrac{d\dot{m}}{\dot{m}}$	$\dfrac{P_o}{P_o^*} = \dfrac{\gamma+1}{1+\gamma M^2}\left[\left(\dfrac{2}{\gamma+1}\right)\psi\right]^{\gamma/(\gamma-1)}$
$\dfrac{dI}{I} = 0$	$\dfrac{\Delta s}{c_p} = \ln\left\{\left[\left(\dfrac{2}{\gamma+1}\right)\psi\right]^{-1}\left(\dfrac{1+\gamma M^2}{\gamma+1}\right)^{(\gamma-1)/\gamma}\right\}$
$\dfrac{ds}{c_p} = (\gamma-1)M^2\dfrac{d\dot{m}}{\dot{m}}$	$A = A^* = \text{constant}$
	$T_o = T_o^* = \text{constant}$
	$f = 0$
	$I = I^*$

a representation is a convenient mechanism for exploring this simple flow. The Mollier diagram representation for simple mass addition represents the locus of states for that particular process. Because of its relatively infrequent use, the process line is unnamed. Table 7-10b gives the change in entropy (referenced to the sonic state) as

$$\frac{\Delta s}{c_p} = \ln\left\{\left[\left(\frac{2}{\gamma+1}\right)\psi\right]^{-1}\left(\frac{1+\gamma M^2}{\gamma+1}\right)^{(\gamma-1)/\gamma}\right\}. \tag{7-40}$$

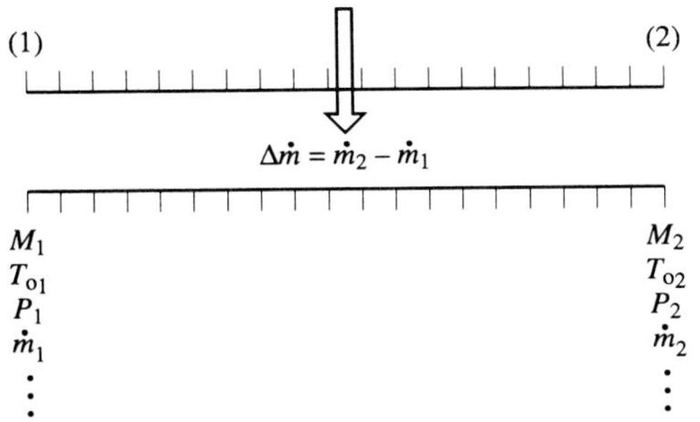

Figure 7-40 Simple Mass Addition Nomenclature

The static temperature ratio, T/T^*, as a function of Mach number is

$$\frac{T}{T^*} = \left[\left(\frac{2}{\gamma + 1}\right)\psi\right]^{-1}. \tag{7-41}$$

The dimensionless simple mass addition representation can be generated by using the Mach number as the parameter in conjunction with Eqs. 7-40 and 7-41. For a calorically perfect gas the temperature ratio is equal to the enthalpy ratio, since

$$\frac{T}{T^*} = \frac{c_p T}{c_p T^*} = \frac{h}{h^*}. \tag{7-17}$$

Equations 7-42 and 7-43 represent the simple mass addition line on a dimensionless h–s or T–s diagram, where the abscissa is the dimensionless entropy ($\Delta s/c_p$), and the ordinate is the dimensionless temperature or the enthalpy ratio (T/T^* or h/h^*). The dimensionless plot on a T–s diagram for a specific heat ratio of 1.4 is given in Figure 7-41a. Mach number as a function of $\Delta s/c_p$ for $\gamma = 1.4$ is plotted in Figure 7-41b.

By observation, Figure 7-41b suggests that the maximum entropy state is the

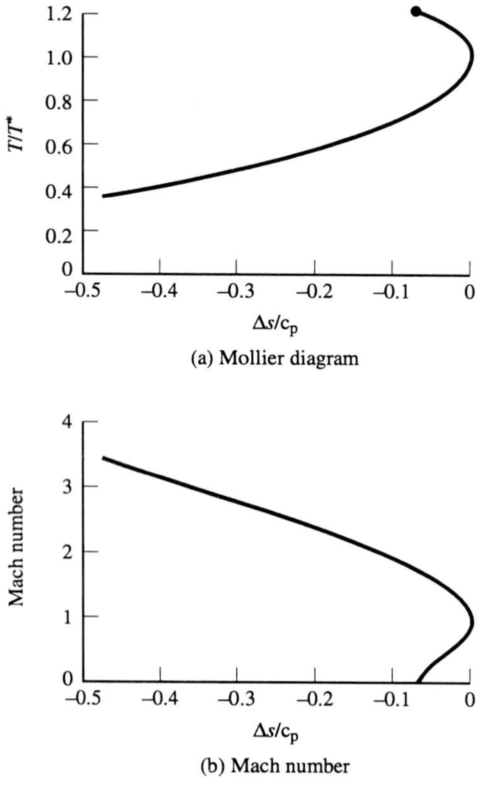

Figure 7-41 Simple Mass Addition Process Characteristics ($\gamma = 1.4$)

sonic state. This can be verified by finding the Mach number corresponding to the maximum entropy. The change in entropy for a simple mass addition flow, cast in terms of the Mach number, is given in Eq. 7-40. To find the Mach number at the maximum entropy state, the derivative of $\Delta s/c_p$ or $\exp(\Delta s/c_p)$ with respect to Mach number can be taken and set equal to zero. The result is

$$M = 1. \tag{7-42}$$

For a calorically perfect gas undergoing a simple mass addition, the state of maximum entropy is thus the sonic state, $M = 1$.

Information about the mass addition process is schematically shown in Figure 7-42. Movement is to the right (in the direction of increasing entropy) on an h–s diagram for mass addition and to the left (in the region of decreasing entropy) on that for mass removal. The sonic state occurs at the point of maximum entropy. Subsonic Mach numbers are associated with the large temperature or enthalpy ratios, and supersonic Mach numbers with the smaller values of the ratios. Thus, the upper portion of the process line corresponds to subsonic flow and the lower portion to supersonic flow. Unlike the Rayleigh or Fanno lines that possess unbounded (negative) values of entropy for low Mach numbers, the stagnation state for simple mass addition occurs at a finite value of entropy. Moreover, since the sonic point cannot be traversed without moving into a region of decreasing entropy unless the sense of the mass flow rate change (addition or removal) is reversed, a simple mass addition flow that is initially subsonic must remain subsonic, and one that is initially supersonic must remain supersonic. For mass addition both supersonic and subsonic flows move toward the $M = 1$ state as the mass flow rate is increased. Both subsonic and supersonic flows move away from the sonic point as the mass flow rate is reduced. The meaning of $\dot{m}^*$ is now clear: it is the mass flow rate required for M to become 1, and it represents the limit of mass addition for subsonic or supersonic flow. As with the other simple flows, conditions at the location of Mach 1 in a simple mass addition flow are unique for that given flow and can be used as reference conditions to aid in calculations.

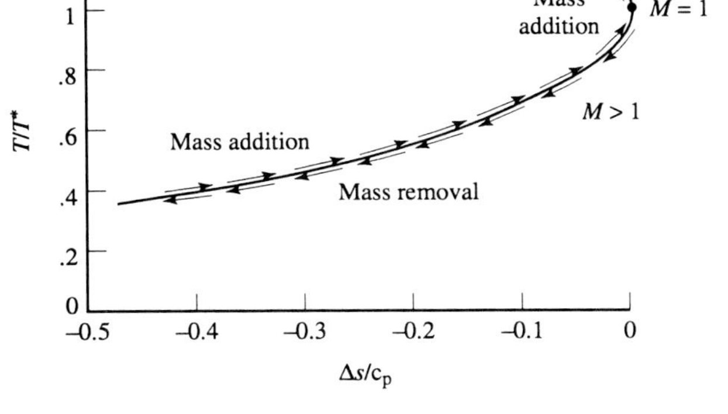

Figure 7-42 Dimensionless Simple Mass Addition Process Characteristics ($\gamma = 1.4$)

Simple mass addition provides, at least in theory, an alternative to the converging–diverging nozzle for achieving supersonic flow. If mass is added to a subsonic flow from a reservoir in a simple area change situation, Mach 1 can be obtained. If mass is removed after Mach 1 is attained, the flow is then expanded supersonically. Thus, with exactly the right amount of mass addition followed by mass removal, supersonic flow in a constant-area duct can be achieved. Because of effects such as wall friction and lack of flow uniformity that are not considered in simple mass addition, the process is not really feasible, and to the best of our knowledge, no such device has ever been successful in obtaining steady supersonic flow.

Duct length does not enter into any of the simple mass addition flow expressions given in Table 7-10b. For a given process, the mass addition (or removal) and the initial state determine the final state. Friction, which is the only driving potential concerned with length, is neglected in simple mass addition flows. Duct length can enter into the property distributions, P versus x for example, for a simple mass addition flow in which mass is added or removed at some specified rate in the one-dimensional duct.

Computational aids are available for simple mass addition. The expressions from Table 7-10b give the values of various property ratios as functions of Mach number. Tabulations of these relations for a specific heats ratio of 1.4 are presented in Appendix G. Figure 7-43 presents plots of the various property ratios as a function of Mach number. The property ratios are all referenced to sonic conditions. The property ratio $\dot{m}/\dot{m}^*$ is double-valued in Mach number, indicating that both a subsonic Mach number and a supersonic Mach number are possible for a given value of $\dot{m}/\dot{m}^*$. As the Mach number increases without bound, the asymptotic value of $\dot{m}/\dot{m}^*$ becomes $\sqrt{\gamma^2 - 1}/\gamma$. Thus, the ratio is double-valued only when $\dot{m}/\dot{m}^* > \sqrt{\gamma^2 - 1}/\gamma$.

Simple mass addition computational aids are available through option **MA** in COMPQ. The menu for option **MA** is given in Figure 7-44a. Three cases are available. Cases **M** and **R** effectively replace the simple mass addition tables of Appendix G. The input for these cases is either the Mach number for case **M** or the mass flow rate ratio, $\dot{m}/\dot{m}^*$, for case **R**. The output for these cases are the Mach number and the static and stagnation pressure and temperature ratios. For a specified Mach number of 0.3, the output for case **M** is given in Figure 7-44b. Although stagnation temperature is constant for this simple flow, the stagnation pressure is not. Case **FR**, which is used for the process involved in mass addition or removal, will be examined in Example 7-16. COMPQ provides added versatility since the value of the specific heats ratio can be specified.

Mass addition (or removal) is a recurring circumstance involved in many simple mass addition flow situations. In such a process, information on the initial state and the change in mass flow rate are given; the final state after the mass addition process is desired. The mass addition process is schematically illustrated in Figure 7-40. Conditions at station 1 and either the change in mass flow rate or the exit mass flow rate must be known. The solution technique is based on the concept that the mass flow rate at the sonic point and the state at the sonic point in a simple mass addition

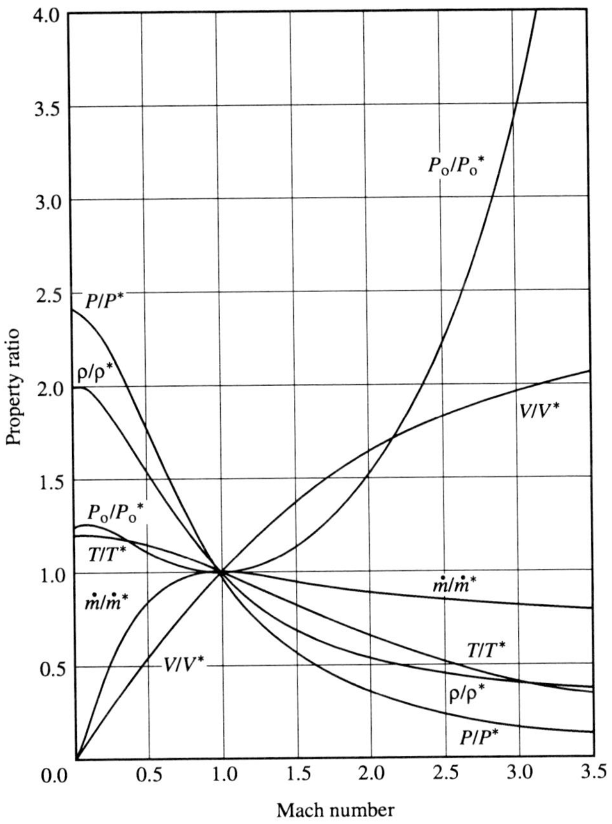

Figure 7-43 Property Ratios for Simple Mass Addition (γ = 1.4)

```
FLOW WITH MASS ADDITION OR REMOVAL SELECTED.
INPUT PARAMETERS ARE:
      MACH NUMBER           M
      MDOT/MDOT*            R
      FLOW RATE CHANGE      FR

SELECT THE DESIRED LETTER(S)?
```

(a) Option **MA** menu

```
ENTER THE MACH NUMBER? 0.3

      M1        MDOT/MDOT*       P/P*         PO/PO*       RHO/RHO*        T/T*
    0.3000        0.5889        2.1314        1.1985        1.8082        1.1788

STAGNATION TEMPERATURE IS CONSTANT FOR SIMPLE MASS ADDITION.
```

(b) Sample output for case **M**

Figure 7-44 Option **MA** Menu and Sample Output

flow are unique. Thus, for state 1 the mass flow rate ratio $\dot{m}_1/\dot{m}^*$ relates the mass flow rate to that at the sonic point, and for state 2, $\dot{m}_2/\dot{m}^*$ represents the ratio. For both states $\dot{m}^*$ is the same. The two states are then related as

$$\dot{m}_2/\dot{m}^* = (\dot{m}_2/\dot{m}_1)(\dot{m}_1/\dot{m}^*). \quad (7\text{-}45)$$

Given the ratio of the entrance and exit mass flow rates and the entrance Mach number, we can use Eq. 7-45 to find the mass flow rate ratio at the exit. The Mach number and property ratios at state 2 can then be found using the values from the simple mass addition tables in Appendix G or case **R** of option **MA** in COMPQ. Given the Mach number at state 2, the property ratios at state 2 can be computed by using the following ratios:

$$T_2/T_1 = (T_2/T^*)/(T_1/T^*);$$
$$P_{o_2}/P_{o_1} = (P_{o_2}/P_o^*)/(P_{o_1}/P_o^*); \quad (7\text{-}46)$$
$$P_2/P_1 = (P_2/P^*)/(P_1/P^*).$$

The procedure outlined from Eq. 7-45 to Eq. 7-46 could be done using **FR** in **MA**. Consider the following example.

Example 7-16

Air with initial stagnation conditions of 600 K and 1 MPa flows at a Mach number of 0.3 at the entrance to a constant-area, porous-walled duct. During passage through the duct, the mass flow rate is increased by 50 percent. Find the exit conditions, and sketch the process on the simple mass addition Mollier diagram.

Solution. The physical situation is illustrated in Figure 7-45a. The problem will be worked two ways: (1) using cases **M** and **R** of option **MA** and (2) using mass addition process case **FR** of option **MA**.

The mass flow rate at duct exit is

$$\dot{m}_2 = 1.5\,\dot{m}_1.$$

Using case **M** of option **MA** with an input Mach number of 0.3 yields at station 1

$$\dot{m}_1/\dot{m}^* = 0.5889 \qquad T_1/T^* = 1.1788$$
$$P_{o_1}/P_o^* = 1.1985 \qquad P_1/P^* = 2.1314.$$

After the mass addition, $\dot{m}_2 = 1.5\,\dot{m}_1$, and so

$$\dot{m}_2/\dot{m}^* = (\dot{m}_2/\dot{m}_1)(\dot{m}_1/\dot{m}^*)$$
$$= (1.5)(0.5889) = 0.88335.$$

With the value of $\dot{m}_2/\dot{m}^*$ known, case **R** of option **MA** can be used to determine conditions at M_2. The output of case **R** for the preceding conditions is presented in Figure 7-45b. Since $\dot{m}/\dot{m}^*$ is double-valued in Mach number, the output of case **R** contains entries for both supersonic and subsonic flows. Earlier we had established that with the same sense (addition or removal) of mass flow rate change, a process that was initially subsonic must remain subsonic and a process that was initially supersonic must remain supersonic. The process under consideration in this problem started subsoni-

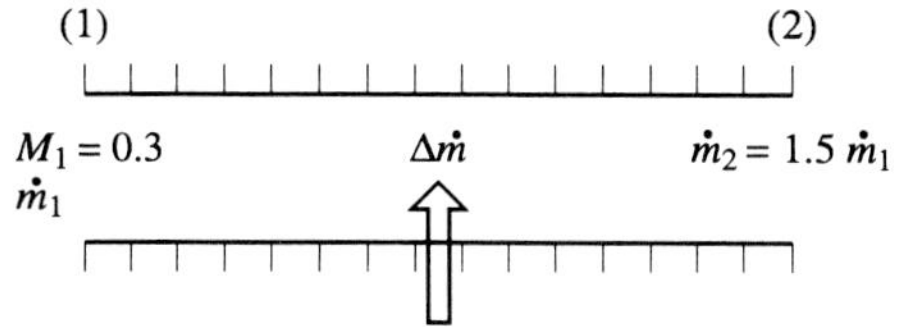

(a) Problem statement

```
ENTER THE RATIO MDOT/MDOT*? 0.88335

      M1        MDOT/MDOT*      P/P*        PO/PO*      RHO/RHO*      T/T*
    0.5663        0.8833       1.6563       1.0877       1.4688      1.1277

      M1        MDOT/MDOT*      P/P*        PO/PO*      RHO/RHO*      T/T*
    2.0668        0.8834       0.3438       1.5771       0.5313      0.6471

STAGNATION TEMPERATURE IS CONSTANT FOR SIMPLE MASS ADDITION.
```

(b) Case **R** output

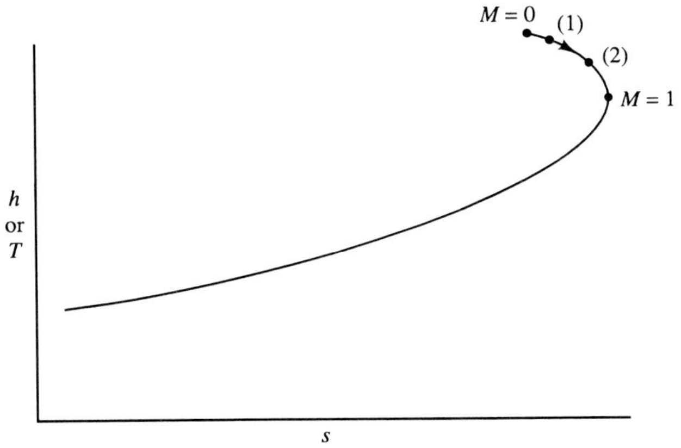

(c) Process path

```
ENTER THE INITIAL MACH NUMBER AND MDOT2/MDOT1? 0.3,1.5

  M1       m2/m1     (m/m*)1    (m/m*)2
  0.3000   1.5000    0.5889     0.8834

  M2       P2/P1     PO2/PO1    T2/T1     RHO2/RHO2
  0.5665   0.7770    0.9075     0.9566    0.8122

STAGNATION TEMPERATURE IS CONSTANT FOR SIMPLE MASS ADDITION.
```

(d) Case **FR** output

Figure 7-45 Example Problem 7-16

cally, meaning that the subsonic flow must also exist at station 2. After the mass addition the Mach number is 0.5663. The static and stagnation properties at station 2 can be computed from the following property ratios using Eqs. 7-46:

$$T_2/T_1 = (T_2/T^*)/(T_1/T^*)$$
$$= 1.1277/1.1788 = 0.9567;$$
$$P_{o_2}/P_{o_1} = (P_{o_2}/P_o^*)/(P_{o_1}/P_o^*)$$
$$= 1.0877/1.1985 = 0.9076;$$
$$P_2/P_1 = (P_2/P^*)/(P_1/P^*)$$
$$= 1.6563/2.1314 = 0.7771.$$

At the initial Mach number of 0.3 the static properties are

$$P_1 = (P_1/P_{o_1})P_{o_1} = (0.9395)(1) = 939.5 \text{ kPa};$$
$$T_1 = (T_1/T_{o_1})T_{o_1} = (0.9823)(600) = 589.4 \text{ K}.$$

The properties at the end of the mass addition process (station 2) then become

$$M_2 = 0.5663;$$
$$T_{o_1} = T_{o_2} = 600 \text{ K};$$
$$T_2 = (0.9567)(589.4) = 563.9 \text{ K};$$
$$P_{o_2} = (0.9076)(1) = 0.9076 \text{ MPa};$$
$$P_2 = (0.7771)(.9395) = 0.7301 \text{ MPa}.$$

Because the change in entropy from state 1 to state 2 is positive, the stagnation pressure decreases from state 1 to state 2. Thus mass addition in a constant-area duct results in a loss in stagnation pressure. The simple mass addition process path is shown in Figure 7-45c.

This problem can also be worked using the process case **FR** of option **MA**. The input and output for case **FR** for this problem are shown in Figure 7-45d. The input parameters required are the initial Mach number and the ratio $\dot{m}_2/\dot{m}_1$. The various output property ratios and the Mach number agree well with the results for the previous solution using cases **M** and **R**. Thus, in a single-input sequence, case **FR** accomplishes this problem's previous solution technique's requirements of application of cases **M** and **R** *and* some arithmetic.

Example Problem 7-16 is typical of many mass addition situations. As is common with the other simple flows, conditions at the sonic point represent both reference conditions and limits for the particular simple flow. For example, simple mass addition is limited to a mass flow rate ratio, $\dot{m}/\dot{m}^*$, less than 1. If the added mass results in a mass flow rate ratio greater than 1, then the entrance conditions must change to accommodate the added mass flow. Consider the following examples.

Example 7-17

With the same entrance conditions as in Example 7-16, find the exit conditions if $\dot{m}_2 = 3\ \dot{m}_1$.

Solution. The problem schematic is similar to that given in Figure 7-45a, except that the final mass flow rate is 3 $\dot{m}_1$. Case **FR** of option **MA** will be used. The inputs are the entrance Mach number, 0.3, and the mass flow rate ratio, 3.0. The output is presented in Figure 7-46. No solution is possible since $\dot{m}_2/\dot{m}^*$ is greater than 1. If the mass flow rate is to be increased to 3 $\dot{m}_1$, then conditions at the duct entrance must be altered.

```
ENTER THE INITIAL MACH NUMBER AND MDOT2/MDOT1? 0.3,3.0

     M1          m2/m1         (m/m*)1     (m/m*)2
   0.3000       3.0000        0.5889      1.7668

THE VALUE OF MDOT/MDOT* AT STATE 2 DOES NOT YIELD A MEANINGFUL SOLUTION.
MDOT/MDOT* MUST BE > 0.0 AND < 1.0 AND/OR >  .6998542  (SUPERSONIC).
```

Figure 7-46 Example Problem 7-17

Example 7-18

With the same entrance conditions as in Example 7-16, find the exit conditions if $\dot{m}_2 = 0.25\ \dot{m}_1$.

Solution. The problem schematic is similar to that given in Figure 7-45a, except that the final mass flow rate is 0.25 $\dot{m}_1$. Case **FR** of option **MA** will be used. The inputs are the entrance Mach number, 0.3, and the mass flow rate ratio, 0.25. The output is contained in Figure 7-47a. Using the static property values from Example 7-16, the properties at the end of the mass addition process (station 2) then become

$$M_2 = 0.068;$$

$$T_{o_2} = T_{o_1} = 600 \text{ K};$$

$$T_2 = (1.017)(589.4) = 599.4 \text{ K};$$

$$P_{o_2} = (1.054)(1) = 1.054 \text{ MPa};$$

$$P_2 = (1.119)(0.9395) = 1.051 \text{ MPa}.$$

The process path on the Mollier diagram is illustrated in Figure 7-47b.

Many problems that occur in propulsion applications can be modeled using simple mass addition. Example Problem 7-19 illustrates how simple mass addition and simple isentropic flow can be combined to analyze a solid propellant rocket motor.

Example 7-19

A solid rocket configuration is sketched in Figure 7-48a. The characteristics of the gas from the burning propellant are given in Table 7-11. With the configuration as given and with a mass flow rate of 5 kg/s, find the Mach number at station e, the head-end stagnation pressure P_{o_h}, and the change in static pressure from the head-end to the end of the propellant grain (station e).

```
ENTER THE INITIAL MACH NUMBER AND MDOT2/MDOT1? 0.3,.25

    M1          m2/m1       (m/m*)1     (m/m*)2
  0.3000      0.2500      0.5889      0.1472

    M2          P2/P1       PO2/PO1     T2/T1       RHO2/RHO2
  0.0676      1.1188      1.0545      1.0171       1.1001

STAGNATION TEMPERATURE IS CONSTANT FOR SIMPLE MASS ADDITION.
```

(a) Case **FR** output

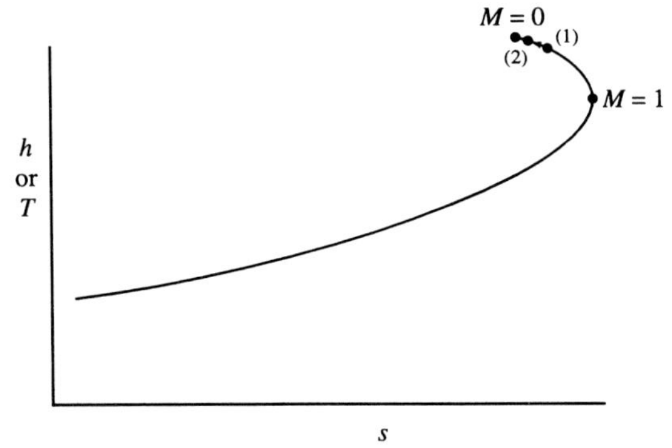

(b) Process path

Figure 7-47 Example Problem 7-18

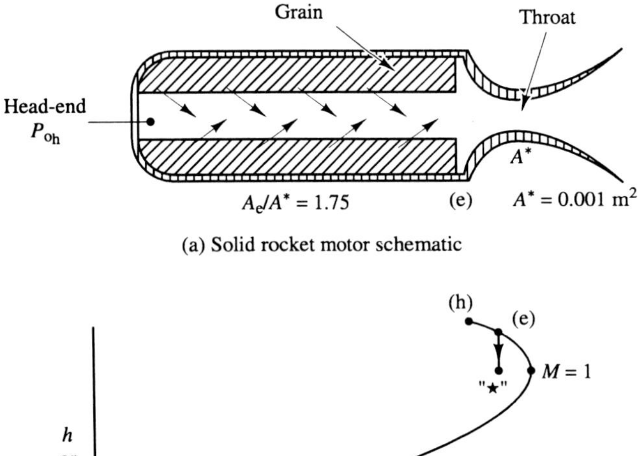

(a) Solid rocket motor schematic

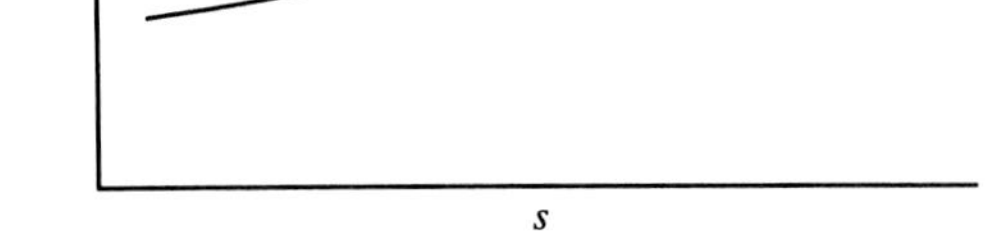

(b) Process path

Figure 7-48 Example Problem 7-19

TABLE 7-11 SOLID PROPELLANT GAS PROPERTIES

Property	Value units
γ	1.26
Molecular weight	22.0 kg/kg-mol
Stagnation temp.	4000.0 K
Gas constant	377.9 J/kg K

Solution. The solid propellant within the motor is called the *grain*. As the propellant burns, it is heated and undergoes pyrolysis from solid to liquid to gas phases. Combustion occurs in the gaseous phase, and the combustion product gases must be accelerated as they are added to the flowing stream. The head-end of the motor is indicated on the sketch. The head-end pressure must be high enough to permit the throat to pass the desired mass flow rate, 5 kg/s for this problem. The flow can be modeled as simple mass addition from the head-end to station e and as simple area change from station e to the throat of the motor. The process path is represented in Figure 7-48b. Stagnation conditions are indicated at the head-end with an isentropic expansion through the nozzle. The throat is choked for usual operation of a solid propellant rocket motor. The mass flow rate for a choked throat is (from Eq. 6-12)

$$\dot{m} = \sqrt{\frac{\gamma}{R}}\left(\frac{2}{\gamma + 1}\right)^{(\gamma+1)/2(\gamma-1)} \frac{A^* P_{o_e}}{\sqrt{T_o}}.$$

Solving for P_{o_e}, the stagnation pressure in the simple area change segment of the flow, gives

$$P_{o_e} = \dot{m}\sqrt{\frac{R}{\gamma}}\left(\frac{\gamma + 1}{2}\right)^{(\gamma+1)/2(\gamma-1)} \frac{\sqrt{T_o}}{A^*}.$$

Using the values specified,

$$P_{o_e} = 5\sqrt{\frac{377.9}{1.26}}\left(\frac{2.26}{2}\right)^{2.26/.52} \frac{\sqrt{4000}}{0.001},$$

from which

$$P_{o_e} = 9.5801 \text{ MPa}.$$

Using case **A** of option **IS** with $\gamma = 1.26$ and $A_e/A^* = 1.75$ yields

$$M_e = 0.3615.$$

Simple mass addition can be used to calculate property values from the head-end of the motor to station e. For simple mass addition, with case **M** of option **MA**, the reference stagnation pressure at the sonic point, P_o^*, can be calculated using conditions at station e:

$$\frac{P_{o_e}}{P_o^*} = 1.1645;$$

$$P_o^* = 8.2268 \text{ MPa}.$$

Since $P_o^* =$ constant for simple mass addition, the stagnation pressure ratio at the head-end of the motor, P_{o_h}/P_o^*, can be evaluated using case **R** of option **MA**. At the head-end of the motor the mass flow rate is zero; thus from case **R** with $\dot{m}_h/\dot{m}^* = 0$ as the input,

$$P_{o_h}/P_o^* = 1.2499,$$

and then

$$P_{o_h} = 10.2826 \text{ MPa}.$$

The static and stagnation pressures are equal at the head-end of the motor since the velocity there is zero. The static-to-stagnation pressure ratio at station e is, from case **M** of option **IS**,

$$P_e/P_{o_e} = 0.9216.$$

The static pressure becomes

$$P_e = 8.8296 \text{ MPa}.$$

The change in static pressure down the grain is

$$\Delta P = 8.8296 - 10.2826 = -1.4537 \text{ MPa}.$$

Hence, the loss in static pressure down the grain becomes

$$\frac{\Delta P}{P_{o_h}} \times 100 = \frac{-1.4537}{10.2826} \times 100 = -14.14\%.$$

The static pressure decreases by approximately 14.1 percent during the mass addition process.

The solid propellant rocket motor used in this example is a good illustration of the combination of simple mass addition and simple area change in a single situation. This problem is somewhat simplified as the mass flow rate in a sôlid propellant rocket motor is normally a function of pressure. For more information on solid propellant rocket motors, see the classic textbook by Sutton (1992).

Simple mass addition possesses two additional interesting features, both of which are concerned with normal shock waves in conjunction with mass injection. The double-valuedness of $\dot{m}/\dot{m}^*$ was mentioned earlier. Since both a subsonic Mach number and a supersonic Mach number exist for a given value of the mass flow rate ratio, the expression for $\dot{m}/\dot{m}^*$ from Table 7-10b yields

$$M_1\frac{\sqrt{1+\dfrac{\gamma-1}{2}M_1^2}}{1+\gamma M_1^2} = M_2\frac{\sqrt{1+\dfrac{\gamma-1}{2}M_2^2}}{1+\gamma M_2^2}. \tag{7-48}$$

But this is the equation expressing the relationship between the upstream and downstream Mach numbers for a normal shock wave; hence, on the process path for simple mass addition, the locations of subsonic and supersonic Mach numbers with the same mass flow rate ratio $\dot{m}/\dot{m}^*$ define a normal shock process. This is an important observation since it means that the process path for simple mass addition containing a normal shock wave can be represented on a single mass addition process line in the Mollier diagram. This is the same for Fanno and Rayleigh flows, except that those process paths are for constant mass flow rates (or mass velocities).

The second situation of interest involving shock waves and simple mass addition concerns the effect of simple mass addition on the back pressure in a duct segment containing a normal shock wave. In a converging–diverging nozzle feeding a constant-area duct with simple mass addition, the exit conditions at the duct will be the same regardless of the shock wave location within the duct. Only when the shock wave is in the diverging nozzle will location affect the duct exit conditions. This is similar to the behavior exhibited by a normal shock wave in a Rayleigh flow situation.

This section developed and examined simple mass addition processes in which the mass injection was normal (perpendicular) to the axis of the flow. Hodge and Young (1993) investigated mass addition with the normal injection constraint removed. As shown in Figure 5-1, α is identified as the angle the injectant mass makes with respect to the mainstream flow. The dimensionless parameter, $r_i = V_i \cos\alpha/V$, is zero for normal ($\alpha = 90°$) injection. For nonnormal injection, $\alpha \neq 90°$, r_i does not vanish, and the value of r_i governs the behavior of the general case of simple mass addition. Hodge and Young use influence coefficients as the starting point for their study. They found that for $r_i > 1$ simple mass addition exhibits a rich solution set including limiting conditions for critical values of r_i and Mach number.

7-10 CONCLUSION

In this and the preceding chapter we examined in some detail one-dimensional simple flows of calorically perfect gases. The sequence of presentation was essentially the same for all four of the simple flows examined. First, the differential forms and then the integrated forms of the property relations were listed (from Chapter 5) for each simple flow. These expressions were examined and their attributes analyzed. The conditions required for each simple flow were examined, and inferences were then made about the behavior of each simple flow for a number of situations.

One reason for approaching compressible flow in this fashion is to reinforce the concept that the four simple flows are each special cases involving only a single driving potential of generalized one-dimensional flow. Generalized one-dimensional flows can be much better understood if the behavior of each of the simple flows has been explored. In the next chapter we shall be concerned with nonsimple, one-dimensional compressible flows, and we will find that many of the concepts developed in this chapter and in Chapter 6 are needed if generalized one-dimensional flows are to be understood.

Many of the review questions and the chapter questions ask that a process path be sketched on the appropriate Mollier diagram representation (Fanno line for Fanno flow, Rayleigh line for Rayleigh flow, or mass addition line for simple mass addition). Figure 7-49 provides of Mollier diagrams to be copied and used for sketching the appropriate process paths for various problems. Indeed, even for problems that do not explicitly request process path sketches, considerable insight into these processes can be obtained by routinely showing process paths.

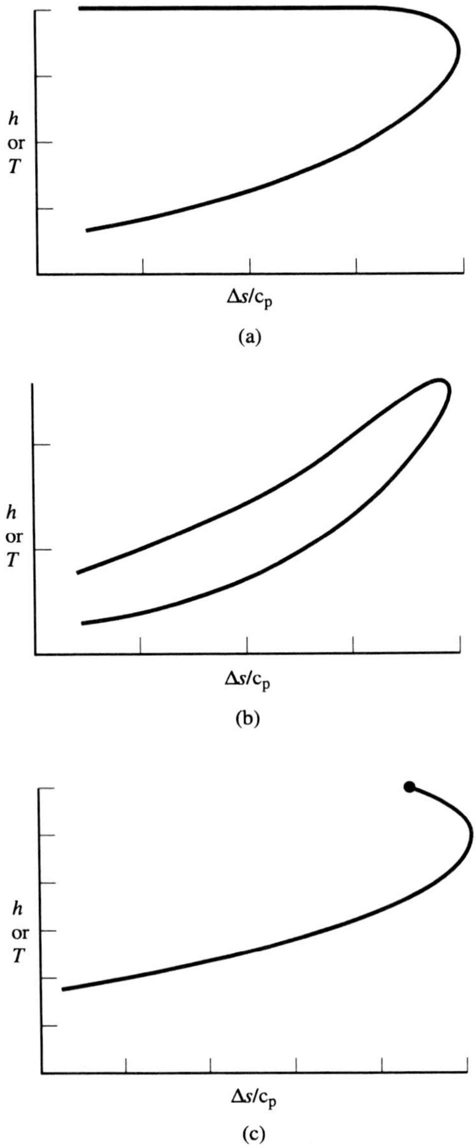

Figure 7-49 Mollier Diagrams for Use in Review Questions and Chapter Problems

CHAPTER 7 REVIEW QUESTIONS

1. In which configuration of Figure 7-50—a or b—will the high-pressure tank empty faster? Explain.

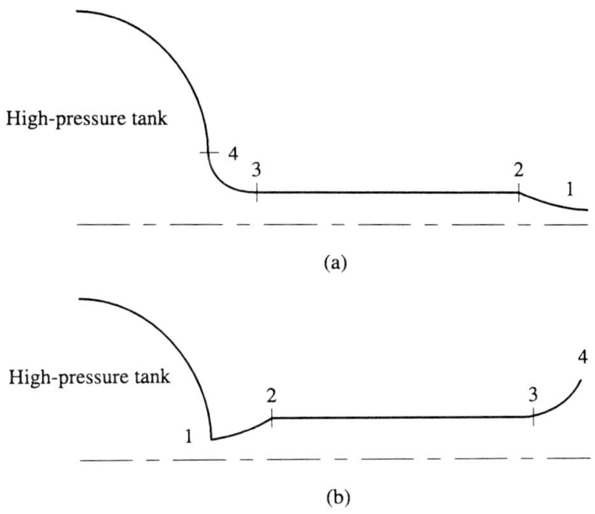

Figure 7-50

2. Sketch the Fanno line and pressure distribution for the situation in Figure 7-51.

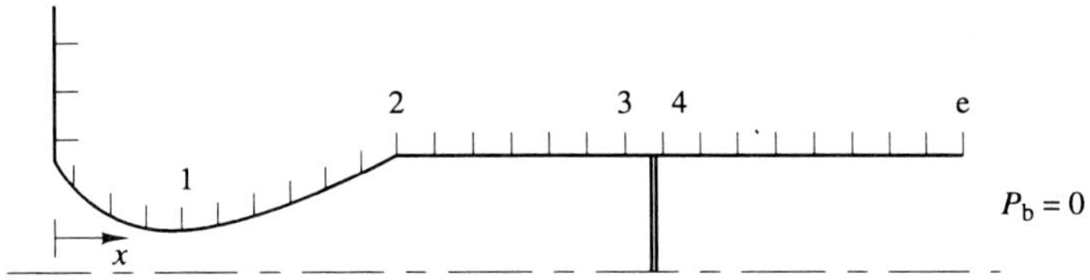

Figure 7-51

3. For the conditions indicated in Figure 7-52, sketch the process path on the Fanno line.

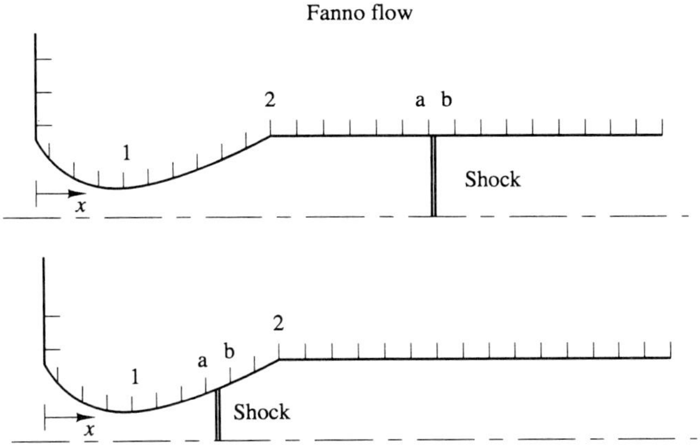

Figure 7-52

4. For the system shown in Figure 7-53, sketch the variations of Mach number and static pressure and the process path on the Fanno line.

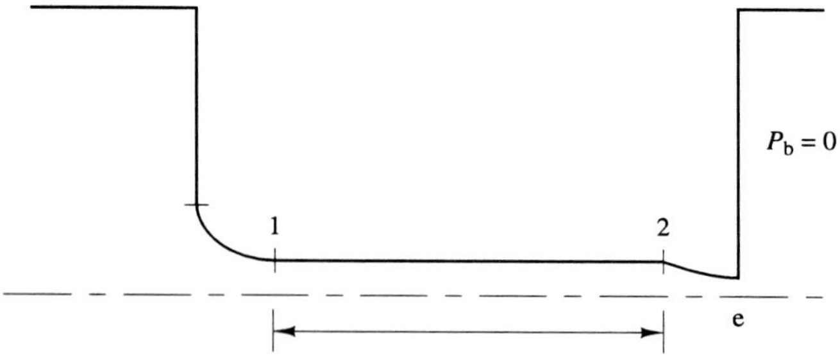

Figure 7-53

5. As illustrated in Figure 7-54, a constant-area duct is fed by a converging nozzle. Heat is lost from the constant-area duct at a constant rate. Discuss the variation of Mach number with x in the constant-area duct. Explain your answer.

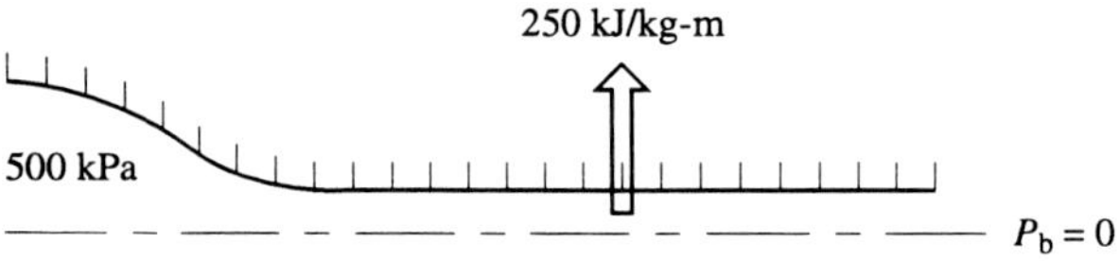

Figure 7-54

6. As shown in Figure 7-55, a converging nozzle feeds a constant-area duct in which heat is transferred at a uniform rate either to or from the flowing gas. For each process, sketch the Rayleigh line path and indicate the point at which choking occurs and the locations of stations (1) and (e) on the Rayleigh line.

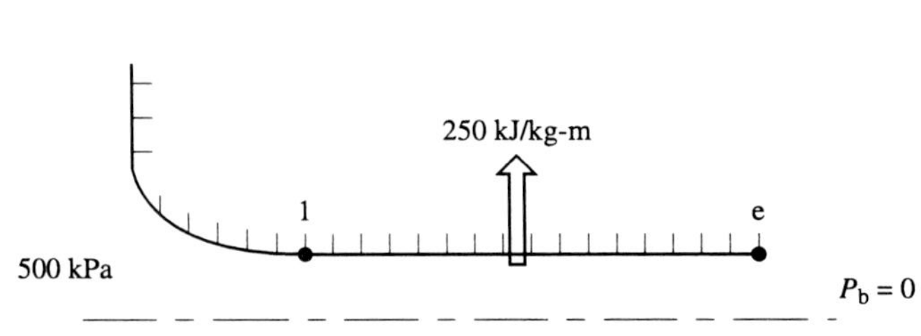

Figure 7-55

7. At what Mach number does simple mass addition flow choke?
8. Is choking by mass addition possible? Why?
9. Explain the meaning of L^* in Fanno flow.
10. Why must the Fanno line always be traversed from left to right?
11. Describe the important features of the Moody diagram.
12. How does the stagnation temperature vary in Fanno flow?
13. **(a)** In a subsonic Fanno flow what happens if the duct length exceeds L^* for a given inlet Mach number?
    (b) In a supersonic Fanno flow what happens if the duct length exceeds L^* for a given inlet Mach number?
14. Explain the meaning of T_o^* in Rayleigh flow.
15. What are the directions of movement on a Rayleigh line for heat addition and for heat rejection?
16. Why does T_o^* represent the limit of heat addition in a given Rayleigh flow?
17. Describe the behavior of the static temperature in a Rayleigh flow as heat is added until the flow is choked.
18. Explain the meaning of $\dot{m}^*$ in simple mass addition.
19. How does the simple mass addition process path representation differ from the Fanno line?
20. As shown in Figure 7-56, a converging nozzle feeds a constant-area duct in which mass addition takes place at a uniform rate. Sketch the path on the mass addition line and indicate the point at which choking occurs. Indicate stations (1) and (e) on the process path on the mass addition line.

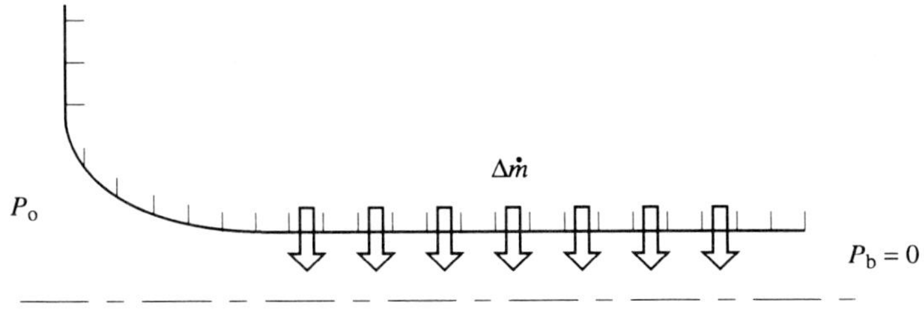

Figure 7-56

CHAPTER PROBLEMS

1. An air flow, initially at a Mach number of 0.25, enters a constant-area insulated duct with a 2-cm diameter. If the friction factor (Fanning) is 0.0042, determine the duct length required to reach Mach 0.75. What duct length is required to reach Mach 1? What happens if an additional 100 cm of duct length is added? The same initial stagnation conditions are maintained. Contrast this problem with Problem 6-1.
2. Hydrogen flows in a constant-area insulated duct with a 3-cm diameter and a length of 25 cm. If the initial velocity is 2500 m/s and the Fanning friction factor is 0.006, find the

exit Mach number and velocity. If the initial static conditions are 250 K and 500 kPa, calculate the exit static pressure and static temperature.

3. Air enters a constant-area insulated duct at a Mach number of 0.25 and stagnation conditions of 200 psia and 500 R. If the length and diameter of the duct are 2 feet and 0.5 inches, respectively, determine the shearing force on the wall of the duct. The friction factor (Fanning) is 0.006.
4. The tank shown in Figure 7-57 has a volume of 10 m^3 and is vented by an insulated constant-area duct that is 4 m long and 1.5 cm in diameter. The Fanning friction factor is estimated to be 0.005. If the initial tank pressure is 1 MPa and the back pressure is 101.3 kPa, estimate the time required for the pressure in the tank to decrease to 500 kPa. Be sure to state all assumptions.

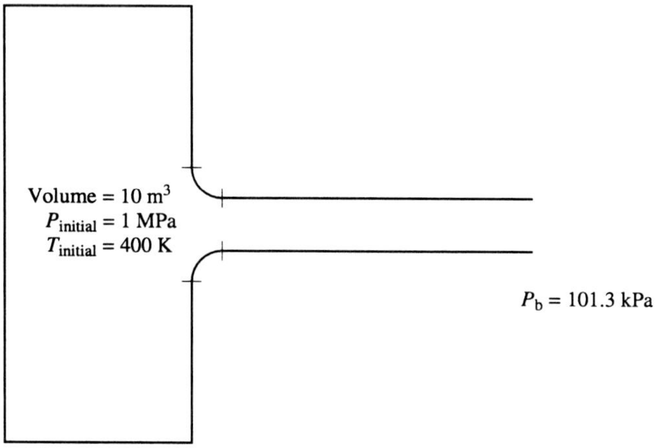

Figure 7-57

5. As illustrated in Figure 7-58, a 30-cm long duct that is 1.5 cm in diameter is attached to a reservoir containing air at 1 MPa and 500 K. If the back pressure is 101.3 kPa and the Fanning friction factor is 0.007, find the flow rate through the duct.

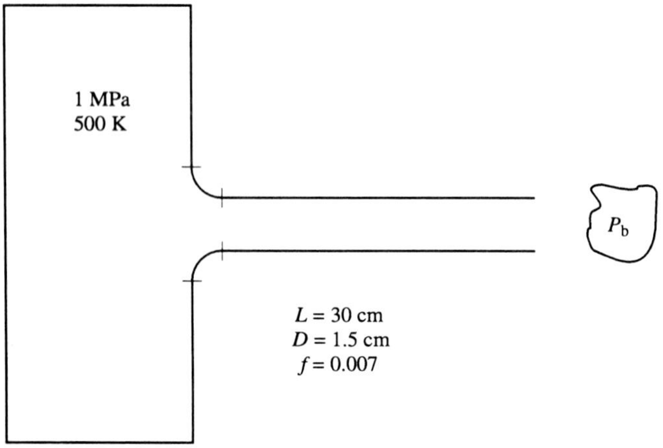

Figure 7-58

6. A converging–diverging nozzle with an area ratio of 3.3 is connected to a tank containing air at 200 psia and 600 R. The two configurations shown in Figure 7-59 are to be evaluated. The throat area for both configurations is 8 in^2. The bottom configuration is the same as the top configuration except for the addition of a 7-ft long insulated duct with a Fanning friction factor estimated as 0.005. Find the maximum flow rates through the nozzles and the ranges of back pressures for which the mass flow rates can be attained.

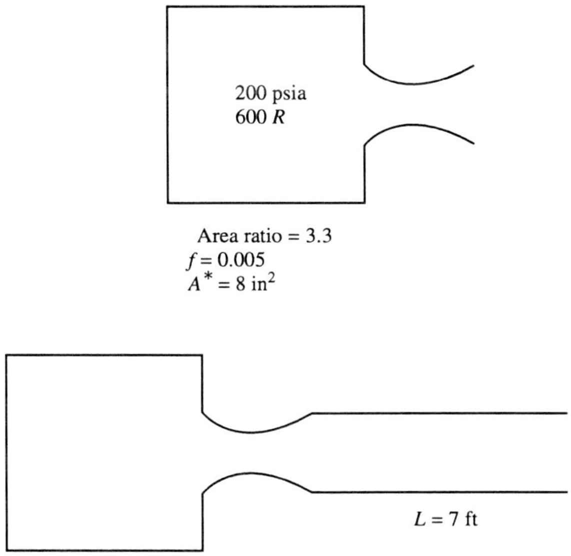

Figure 7-59

7. Two tanks are depicted in Figure 7-60. Each has a volume of 3 m^3 and is connected by a constant-area insulated duct that is 25-m long and 2.25 cm in diameter. Tank 2, shown on the right side of the figure, is initially at 0 kPa and is to be pressurized to 200 kPa. Tank 1, the left-hand side tank, is initially at 2.5 MPa and 40 C. If the Fanning friction factor is 0.006, find the time required to pressurize tank 2.

Figure 7-60

8. As shown in Figure 7-61, a converging–diverging nozzle followed by a constant-area insulated duct is connected to a reservoir containing air at 700 K and 750 kPa.

(a) Find the nozzle/duct mass flow rate for a back pressure of 0 kPa.

(b) Find the back pressure required for a normal shock to occur in the nozzle exit plane.

(c) Find the maximum back pressure for choked flow in the nozzle/duct system.

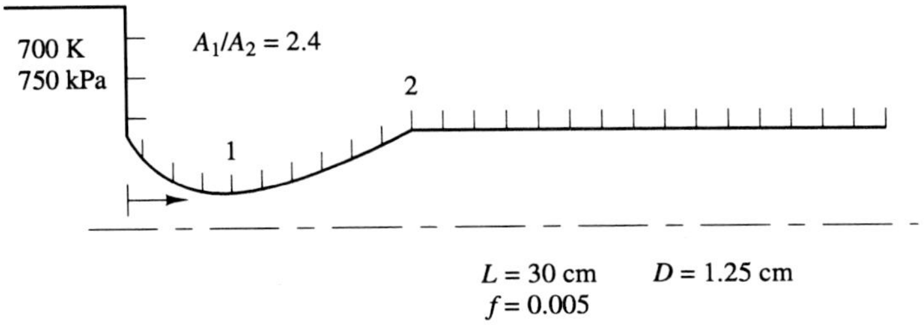

Figure 7-61

9. For the system and conditions illustrated in Figure 7-62, determine the mass flow rate through the system if $f = 0.003$.

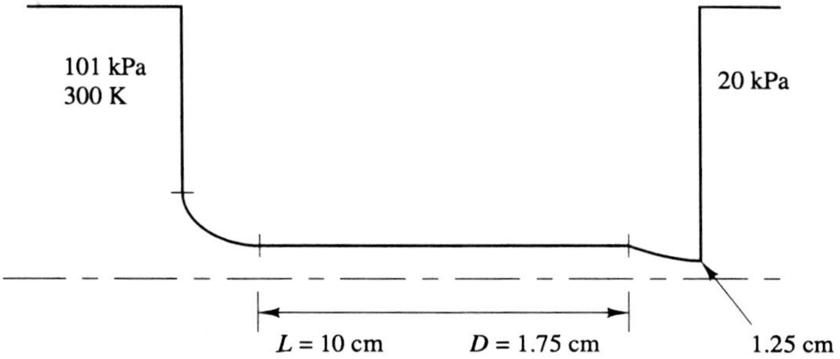

Figure 7-62

10. A converging–diverging nozzle with an area ratio of 2.5 (Figure 7-63) exhausts into an insulated duct that is 100 cm long and 1.5 cm in diameter, where $f = 0.004$. If the system back pressure is 45 kPa, find the range of reservoir pressures over which a normal shock will appear in the duct.

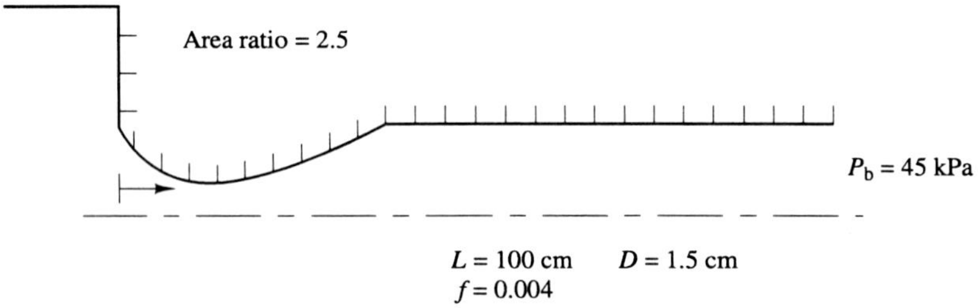

Figure 7-63

11. Some early rocket nozzles utilized a blast tube—a stretched-out throat arrangement—as illustrated in Figure 7-64. If stagnation conditions at the inlet to the blast tube are 1.5 MPa and 1500 K, find the blast tube exit velocity and the mass flow rate if the back pressure is 101.3 kPa. The nozzle exit-to-throat area ratio is 5. The exhaust gases have a ratio of specific heats of 1.33 and a gas constant of $R = 0.462$ kJ/kg K.

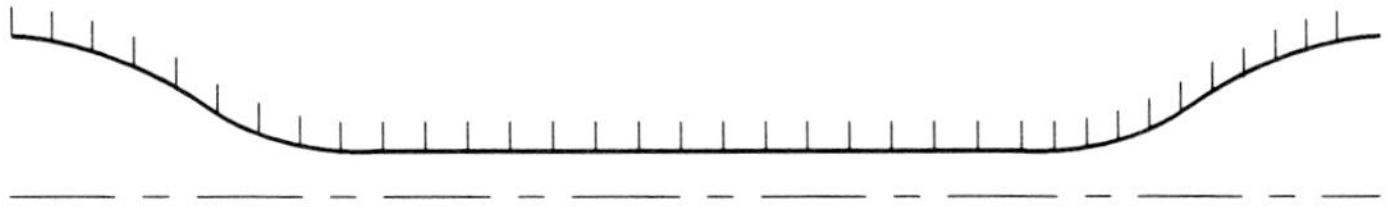

$L = 2$ m $D = 10$ cm
$f = 0.007$

Figure 7-64

12. As shown in Figure 7-65, a converging–diverging nozzle/constant-area insulated duct is connected to a reservoir with a pressure of 100 psia. If the nozzle area ratio is 3.2 and the duct L/D ratio is 60, determine the range of back pressures over which a normal shock wave will appear in the duct. The Fanning friction factor is 0.005.

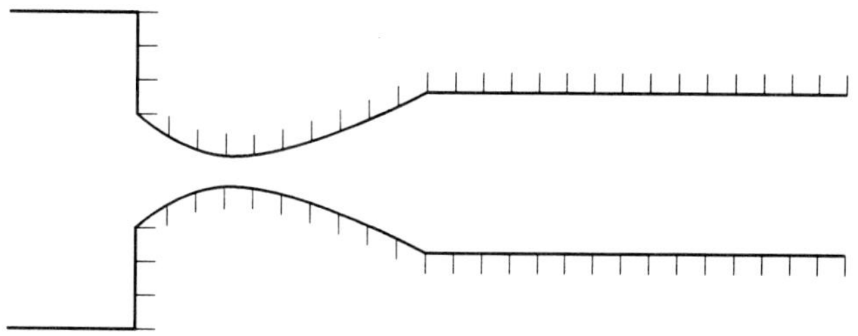

Figure 7-65

13. For the converging nozzle/constant-area duct arrangement illustrated in Figure 7-66, determine the range of back pressures for which the flow will be choked. Indicate the process path on the Fanno line. Determine the mass flow rate for the choked flow condition.

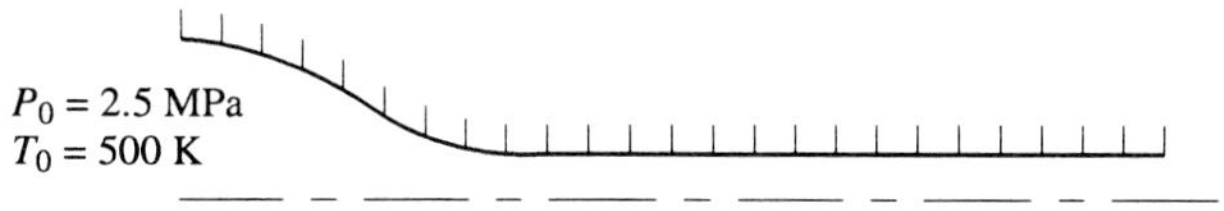

$L = 1$ m $D = .04$ m
$f = 0.006$

Figure 7-66

14. The converging nozzle depicted in Figure 7-67 feeds a constant-area duct in Fanno flow. The stagnation conditions in the reservoir are 500 kPa and 300 K. If $(4\,fL/D)_{\text{actual}} = 3.5$ and if the back pressure is 101 kPa, find the mass velocity, ρV, of the system and sketch the process on the Fanno line.

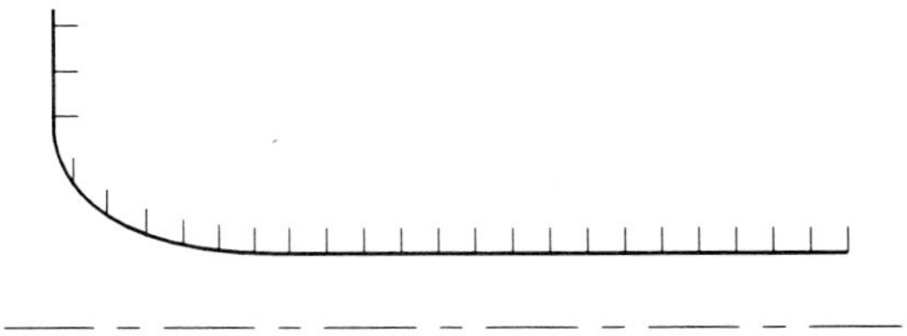

Figure 7-67

15. Find the back pressure for the conditions shown in Figure 7-68. Assume isentropic flow in the converging–diverging nozzle and Fanno flow in the duct. Sketch the process on a Fanno line, and sketch the pressure distribution. Air is the fluid.

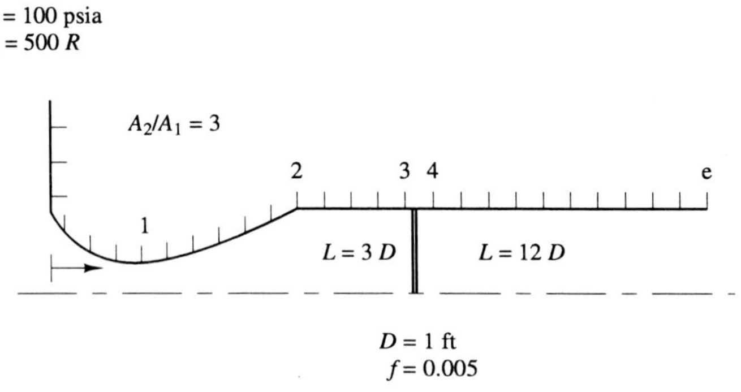

Figure 7-68

16. For the situation shown in Figure 7-69, assume Fanno flow in the constant-area duct and isentropic flow in the converging–diverging nozzle. Sketch the process on a Fanno line, and find the back pressure required for the conditions shown.

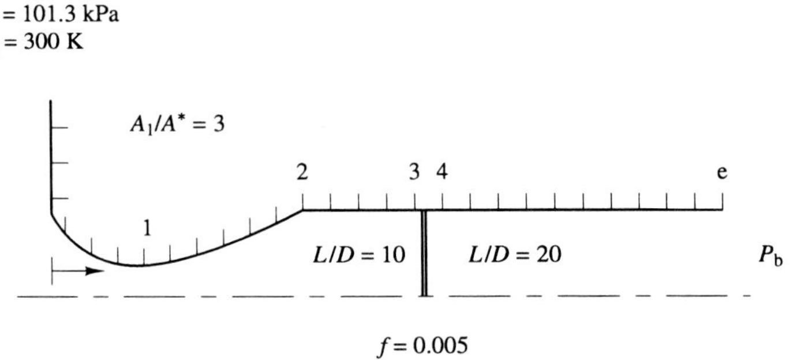

Figure 7-69

17. Devise a computer program to generate the Fanno table for an arbitrary ratio of specific heats. Verify using COMPQ.
18. Plot the Rayleigh line for air flow in a 5-cm diameter duct with a flow rate of 2 kg/s. Inlet stagnation conditions are 700 kPa and 300 K. Plot the Fanno line for the same inlet conditions. Plot both lines on the same graph, and demonstrate that the conditions at the intersection points define a normal shock wave.
19. Heat is rejected from air flowing in a constant-area duct. At the duct entrance the air is moving at 200 m/s and possesses static conditions of 300 K and 100 kPa. If 50,000 J/kg is rejected along the duct, find the Mach number, stagnation temperature change, and stagnation pressure change. Sketch the process on the Rayleigh line.
20. If air flows in a 2-cm diameter constant-area duct at a velocity of 125 m/s, a static temperature of 300 K, and a static pressure of 300 kPa, find the rate of heat input required to choke the duct.

21. Air flows at 0.75 kg/s in a constant-area duct of 12.5 cm diameter. If the inlet stagnation state is 150 kPa and 310 K, determine the two possible values of the inlet Mach number. For each Mach number, find the heat addition rate required to choke the duct.

22. At the inlet of a 1-in constant-diameter duct, the Mach number is 2 and the stagnation conditions are $P_o = 150$ psia and $T_o = 1000\ R$. If heat is added by a chemical reaction, determine the energy addition rate required to choke the duct. Changes in the composition of the gas stream due to the chemical reaction can be neglected.

23. Air enters a gas turbine combustor with an initial Mach number of 0.3 and stagnation conditions of 1500 F and 300 psia. If the exit stagnation temperature is to be 2500 F, find the exit Mach number and static temperature and the loss in stagnation pressure. Sketch the process on the Rayleigh line.

24. Air enters a converging–diverging nozzle with stagnation conditions of 400 kPa and 400 K. The area ratio of the nozzle is 4.0. After passing through the nozzle, the flow enters a duct where heat is added. At the end of the duct there is a normal shock wave. The stagnation temperature upstream (and downstream) of the shock wave is 500 K. Assume isentropic flow in the converging–diverging nozzle and Rayleigh flow in the duct. (See Figure 7-70.)

(a) Sketch the process on a T–s diagram.
(b) Sketch the pressure distribution.
(c) Determine the heat added (in kJ/kg).
(d) Determine the stagnation pressure loss (including the shock).

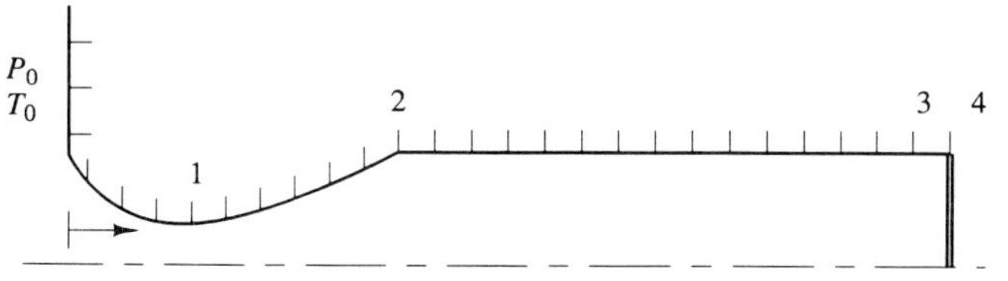

Figure 7-70

25. Air enters a converging–diverging nozzle with stagnation conditions of 1000 psia and 700 R. The area ratio of the nozzle is 4.0. After passing through the nozzle, the flow enters a constant-area duct where heat is added. At the end of the duct there is a normal shock wave after which the stagnation temperature is 900 R. Assume isentropic flow in the converging–diverging nozzle and Rayleigh flow in the duct. (See Figure 7-71.)

(a) Sketch the process on a T–s diagram.
(b) Sketch the pressure distribution.
(c) Determine the heat added (in kJ/kg).
(d) Determine the stagnation pressure loss (including the shock).

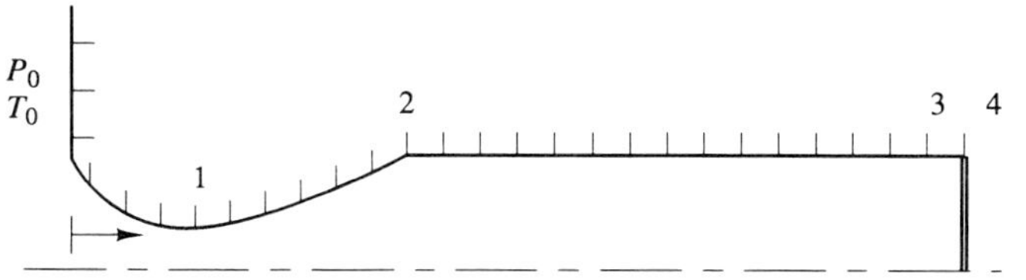

Figure 7-71

26. Flow enters a constant-area duct at a Mach number of 0.25 and static conditions of 250 K and 80 kPa. If heat is added such that the stagnation enthalpy is increased by 100.4 kJ/kg m, plot M versus x, P versus x, T versus x, and P_o versus x.

27. High-pressure steam condenses at 400 K on the outer surface of a thin-walled tube that is 5 cm in diameter. The overall heat transfer coefficient, U, between the steam and the air on the inside of the tube is 200 W/m^2 K. If the air enters the tube at 25 m/s, 150 kPa, and 280 K, find the tube length required for the air to be heated to 350 K.

28. At the inlet of the combustion chamber of a jet engine, air enters at 140 m/s with a stagnation state of 300 kPa and 650 K. After the combustion process is complete, the exit stagnation temperature is required to be 1200 K. If the fuel is Jet A, determine the required fuel/air ratio (on a mass basis). What fuel/air ratio would be required to choke the combustion chamber by heat addition? Be sure to state and defend all assumptions.

29. Air enters a jet engine combustor at a Mach number of 0.3 and stagnation conditions of 5 MPa and 800 K.

(a) If a fuel with a heating value of 40 MJ/kg is injected, find the maximum fuel addition per kg of air. Use the properties of air for the entire process and assume that the fuel flow rate is negligible compared to the air flow rate.

(b) Sketch the process on the Rayleigh line.

(c) What would happen if the fuel flow rate were larger than that specified in part (a)?

30. As illustrated in Figure 7-72, a constant-area duct is fed by a converging nozzle. Heat is added at the rate of 300 kJ/kg to the duct. If the duct diameter is 2 cm, what is the mass flow rate of the system if the back pressure is zero? What is the mass flow rate if the back pressure is 101.3 kPa? What happens if the heat is rejected from the flow?

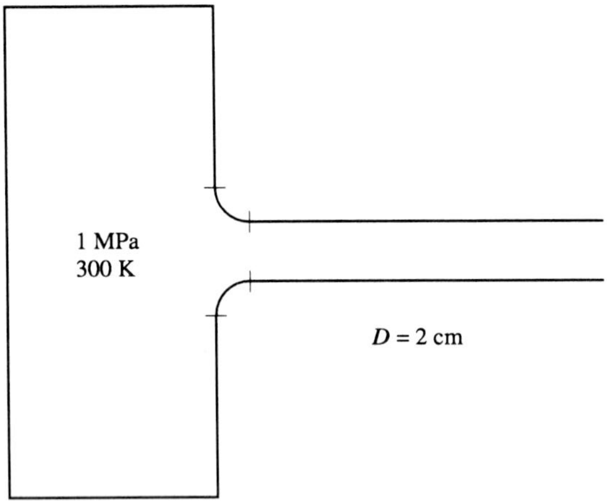

Figure 7-72

31. A converging–diverging nozzle feeds an insulated, constant-area duct with a length-to-diameter (L/D) ratio of 60. The nozzle is supplied with air at a stagnation pressure of 1.3 MPa, and the system exhausts into a back pressure of 500 kPa. The nozzle throat area is one-third of the duct area, and the duct Fanning friction factor is estimated to be 0.005. Calculate the Mach number at the duct's exit. For the conditions described, over what back pressure range would a normal shock wave stand in the constant-area duct?

32. Devise a computer program to calculate pressure as a function of time in Problem 7-4 if the duct is made of stainless steel with an absolute roughness of 0.005 cm and the friction factor is unknown.

33. Redo Problem 7-9 assuming that the constant-area segment is made of drawn copper tubing with a relative roughness of 0.00015 cm.

34. Air is flowing through an insulated pipe 1 in in diameter and 50-ft long. The absolute roughness of the pipe is 0.00015 ft. At the pipe inlet the stagnation conditions are 25 psia and 800 R. Find the inlet and exit Mach numbers and the mass flow rate for a back pressure of 8 psia. Repeat for a back pressure of 18 psia.

35. A rectangular duct six inches by 8 inches is 60-ft long. Find the exit Mach number and static properties if the inlet conditions are $M = 0.2$, $P = 300$ kPa, and $T = 500$ K. The average friction factor is 0.005.

36. At section 1 of an insulated, constant-area duct, conditions are $M = 0.2$, $P = 100$ kPa, and $T = 350$ K. If the duct diameter is 4 cm and the friction factor is 0.008, determine how long the duct must be in order to produce an exit Mach number of 0.75. Compute the loss in stagnation pressure.

37. Plot the variations in Mach number, static pressure, stagnation pressure, and static temperature down the Fanno segment in Problem 7-8.

38. Initial conditions for a Rayleigh flow are $T = 600\ R$, $P = 23$ psia, and $M = 0.5$. Can the stagnation temperature be increased to 2500 R? If not, what could be changed to permit such an increase?

39. Verify that for a Rayleigh flow

$$\left(\frac{T}{T^*}\right)_{\max} = \frac{(1+\gamma)^2}{4\gamma}.$$

40. In a Rayleigh flow the fuel/air ratio is 0.02, and Jet A is the fuel. At the inlet the air velocity is 120 m/s with static conditions of 500 kPa and 600 K. Estimate the Mach number and static and stagnation properties after combustion is complete.

41. Nitrogen at the inlet of a Rayleigh flow situation has a velocity of 250 m/s and a stagnation temperature of 400 K. If the exit Mach number is 0.25, determine whether heat is added or rejected and the rate of energy transfer.

42. If the inlet flow is sonic ($M = 1$), how much heat must be transferred to reach the two limiting Mach numbers, zero and infinity?

43. A converging–diverging nozzle with an area ratio of 2.5 feeds a constant-area duct where heat is added. The stagnation pressure upstream of the constant-area duct is 300 kPa, and the back pressure is 135 kPa. The ratio of the stagnation temperatures after and before the heat addition is greater than 1. Is the flow supersonic or subsonic in the duct? Determine the static pressure and Mach number at the inlet and exit of the constant-area duct.

44. An air flow, initially at a Mach number of 0.25 and a stagnation temperature of 500 K, enters a constant-area duct with heat addition. What is the heat interaction required to reach Mach 0.75? What is the heat interaction required to reach Mach 1.0? What happens if additional heat interaction takes place? Contrast this problem with Problems 6-1 and 7-1.

45. Air at 200 kPa enters a constant-area duct and exits at 125 kPa and a Mach number of 0.6 with a stagnation temperature of 700 K. Find the inlet Mach number and the heat transfer rate.

46. As shown in Figure 7-73, air initially at $M = 2$, $P = 10$ psia, and $T = 500\ R$ enters a constant-area duct that is followed by a converging segment with area ratio 1.5. At the minimum area, the Mach number is 1. Find the amount and direction (heat addition or rejection) of the heat interaction. If a normal shock wave stands in the duct, are the previous answers correct?

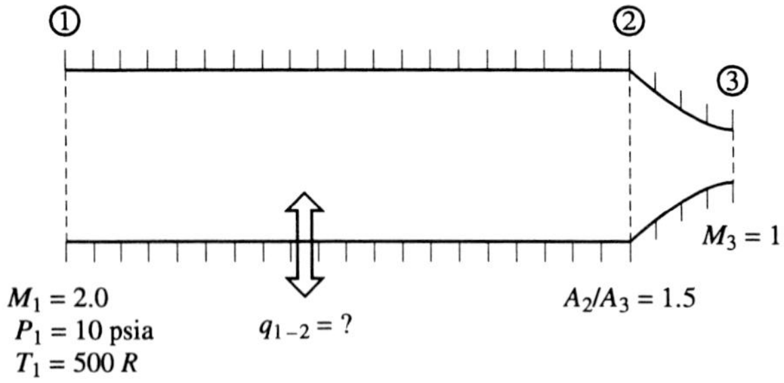

Figure 7-73

47. Air at stagnation conditions of 100 psia and 500 R flows through a converging–diverging nozzle of area ratio 8 into a Rayleigh flow arrangement. The system exhausts into a back pressure of 14.7 psia. If a normal shock wave occurs at the nozzle exit, find the exit Mach number and the stagnation temperature.

48. A duct of length L and constant diameter D is attached to a converging–diverging nozzle of area ratio 2. The system is supplied with air at stagnation conditions of 250 psia and 550 R. The back pressure is 100 psia. Find the $4fL/D$ of the nozzle-pipe arrangement for the following situations:
(1) A normal shock wave stands just downstream of the nozzle throat.
(2) A normal shock wave stands in the nozzle exit plane.
(3) A shock wave stands in the duct exit plane.

49. An air flow, initially at a Mach number of 0.25 and a stagnation temperature of 500 K, enters a constant-area duct with mass addition. What is the percentage increase in mass flow rate required to reach Mach 0.75? What is the percentage increase in mass flow rate required to reach Mach 1.0? What happens if the mass flow rate is increased beyond this? Contrast this problem with Problems 6-1, 7-1, and 7-44.

50. Verify that for simple mass addition

$$P_o = P + \rho V^2,$$

where P_o is the pressure for zero mass flow rate.

51. A tank of volume 10 m^3 is vented by an insulated constant-area duct that is 4 m long and 1.5 cm in diameter. The Fanning friction factor is estimated to be 0.005. If the initial tank pressure is 1 MPa and the back pressure is 101.3 kPa, estimate the time required for the pressure in the tank to decrease to 125 kPa. Be sure to state all assumptions. The system schematic is similar to that in Problem 7-4.

52. Air at a Mach number of 0.25 and stagnation conditions of 100 psia and 700 R enters a constant-area insulated duct with a 3-in diameter. Air at 700 R is injected normally into the duct such that the flow rate is increased by 30 percent. Find the conditions at the duct's exit. What is the maximum mass addition possible for the given initial conditions?

53. Air at a Mach number of 0.25 and stagnation conditions of 500 kPa and 700 K enters a constant-area insulated duct with an 8-cm diameter. Air at 700 K (stagnation temperature) is injected normally into the duct such that the flow rate is increased by 30 percent.

Find the conditions at the duct's exit. What is the maximum mass addition possible for the given initial conditions? Sketch the process on the mass addition process line.

54. As shown in Figure 7-74, mass is uniformly injected into a constant-area duct fed by a converging nozzle. If the inlet stagnation conditions are 300 kPa and 400 K, determine the duct exit pressure as a function of the total mass injection rate if the back pressure into which the duct exhausts is 101 kPa.

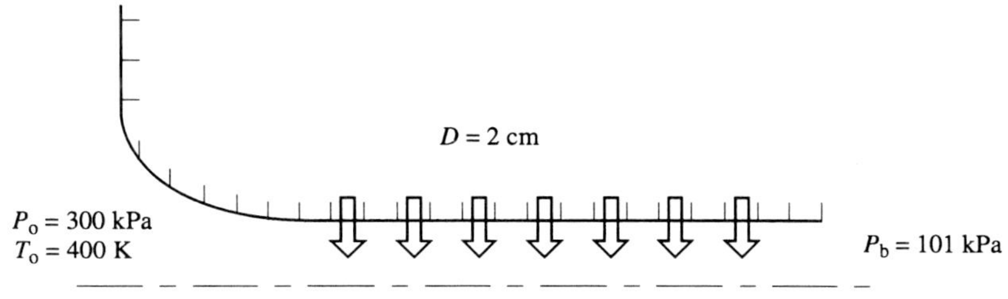

Figure 7-74

55. For the same physical arrangement and initial stagnation conditions as in Problem 7-54, 0.09 kg/s is injected in the constant-area duct. If the back pressure is very low, find the total mass flow rate for the system. How does this compare with the mass flow rate for a choked arrangement without any mass injection?

56. Redo Problem 7-55 assuming that the back pressure is 200 kPa.

57. Air at a Mach number of 2.00 and stagnation conditions of 500 kPa and 700 K enters a constant-area insulated duct with an 8-cm diameter. Air with a stagnation temperature of 700 K is injected normally into the duct in order to increase the flow rate by 30 percent. Find the conditions at the duct's exit. What is the maximum mass addition possible for the given initial conditions? Sketch the process on the mass addition process line.

58. Two methane streams, both with stagnation temperatures of 300 K, are to be mixed in a constant-area duct. The main stream has a flow rate of 0.25 kg/s with an inlet stagnation pressure of 200 kPa. The other stream, flowing at 0.15 kg/s, is to be injected with zero longitudinal momentum into the constant-area duct. If the duct's diameter is 4 cm, what back pressure or back pressure range is required?

59. Compute the thrust and estimate the throat and exit areas of a solid propellant rocket motor operating with a head-end pressure of 1000 psia and a nozzle expansion ratio (A_{exit}/A^*) of 5 at sea level. The propellant mass flow rate is 8 lbm/s. Properties of the propellant are estimated to be $\gamma = 1.25$, $T_o = 3000\ R$, and molecular mass = 22 lbm/lbm-mol.

60. The rate at which a solid propellant burns is sensitive to the pressure. An approximation is that the mass flow rate from a burning propellant grain is proportional to the head-end pressure raised to some power n. For the propellant and grain of Problem 7-59, what head-end pressure would be developed to expel the burned propellant through a throat area of 2 in^2 if $n = 0.4$?

Nonsimple and Generalized Flows

8

8.1 INTRODUCTION

In Chapters 6 and 7 the four simple flows resulting from the one-dimensional generalized analysis developed in Chapter 5 were examined, their behaviors determined, and various applications presented. This chapter is concerned with devising techniques suitable for nonsimple flows. In a *nonsimple flow* two or more of the driving potentials introduced in Chapter 5 are present. Generally, closed-form expressions cannot be developed for the nonsimple flows, and numerical techniques must be used to obtain solutions. Young (1993) has, however, recently greatly extended the class of problems for which closed-form solutions can be generated. The first topic discussed in this chapter, isothermal flow, is followed by an examination of other nonsimple situations. The last two topics of this chapter are devoted to numerical solutions of generalized one-dimensional compressible flow problems without and with sonic points present.

8.2 ISOTHERMAL FLOW

Isothermal flow is a nonsimple flow that has friction and heat transfer as driving potentials but that occurs in a constant-area duct with both the mass flow rate and the static temperature constant. The name "isothermal" is adopted from the attribute of constant static temperature. Isothermal flow situations arise when a compressible fluid flows at relatively low Mach numbers at a constant mass flow rate in a constant-area duct in close contact with a constant temperature environment. The flows of gases in long, buried pipelines are often modeled as isothermal flows.

The conditions for isothermal flow are schematically illustrated in Figure 8-1. Unlike most nonsimple flows, isothermal flow relations can be cast into closed form. The basic differential expressions for isothermal flow can be obtained from Table 5-2, the table of influence coefficients, by including only the influence coefficients for friction and heat transfer. However, the differential expressions obtained in this fashion contain both duct length ($4\,f dx/D_h$) and stagnation temperature change (dT_o/T_o) as driving potentials and cannot be integrated without further algebraic substitution. The usual substitution procedure is to express dT_o/T_o in terms of the Mach number and dM^2. In an isothermal flow at a point, the static and stagnation temperatures are related in the usual fashion as

$$T_o = T\left(1 + \frac{\gamma - 1}{2} M^2\right). \tag{1-66}$$

Taking the differential of the preceding equation yields

$$dT_o = dT\left(1 + \frac{\gamma - 1}{2} M^2\right) + T\frac{\gamma - 1}{2}\, dM^2, \tag{8-1}$$

but by definition $dT = 0$ since the static temperature is constant. Hence,

$$\frac{dT_o}{T_o} = \frac{\dfrac{\gamma - 1}{2}\, dM^2}{\psi}, \qquad \psi \equiv 1 + \frac{\gamma - 1}{2} M^2. \tag{8-2}$$

Table 5-2 gives the differential expression for dT/T, which must vanish for an isothermal flow, as

$$\frac{dT}{T} = 0 = -\frac{\gamma(\gamma - 1)M^4}{2(1 - M^2)} \frac{4f}{D_h} dx + \frac{(1 - \gamma M^2)\psi}{1 - M^2} \frac{dT_o}{T_o}. \tag{8-3}$$

Using Eq. 8-2, we obtain the following expression with x as the independent variable and M as the dependent variable:

$$M^4 \gamma \frac{4f}{D_h} dx = (1 - \gamma M^2)\, dM^2. \tag{8-4}$$

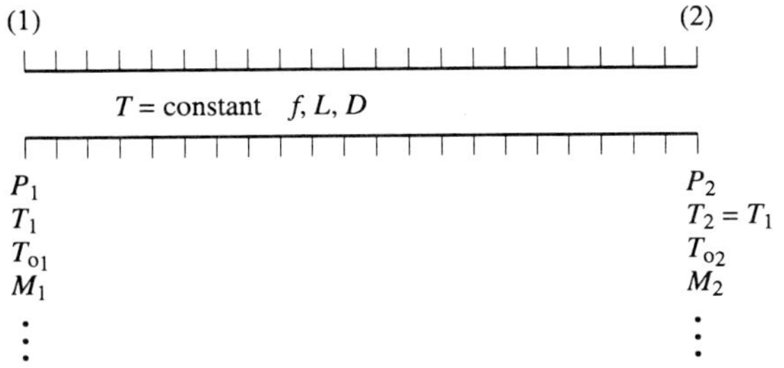

Figure 8-1 Isothermal Flow Nomenclature

Some useful information can be obtained from Eq. 8-4 by rearranging it in the following form:

$$\frac{dM^2}{M^2} = \frac{\gamma M^2}{1 - \gamma M^2} \frac{4f}{D_h} dx. \tag{8-5}$$

For $M < 1/\sqrt{\gamma}$, the Mach number increases with increasing x while for $M > 1/\sqrt{\gamma}$, the Mach number decreases with increasing x. $M = 1/\sqrt{\gamma}$ is the limiting Mach number for the case of isothermal flow and plays the same role in isothermal flow as $M = 1$ at the minimum area for simple area change. Isothermal flow thus chokes at $M = 1/\sqrt{\gamma}$, not at $M = 1$! This is certainly an unanticipated circumstance.

Moreover, the energy equation yields

$$dq = c_p \, dT_o. \tag{8-6}$$

Using Eq. 8-2 produces

$$dq = c_p \frac{\gamma - 1}{2} \frac{T_o}{\psi} dM^2. \tag{8-7a}$$

Finally, substituting for dM^2 from Eq. 8-5 results in

$$dq = c_p \frac{(\gamma - 1)M^4 \gamma/2}{\psi(1 - \gamma M^2)} \frac{4f}{D_h} dx. \tag{8-7b}$$

As the Mach number approaches its limiting value, $1/\sqrt{\gamma}$, the local heat addition required increases without bound. The limiting Mach number in an isothermal flow cannot be reached because of the unbounded heat transfer requirement. The generally low convective heat transfer coefficients associated with gas flows preclude the high heat transfer rates required to maintain high Mach number isothermal flows. Isothermal flow assumptions are usually viable only for low Mach number flows because the required heat addition increases dramatically as the limiting Mach number is approached.

Integrating Eq. 8-4 from station 1 to station 2 for the circumstance illustrated in Figure 8-1 gives

$$\int_{M_1^2}^{M_2^2} \frac{1 - \gamma M^2}{\gamma M^4} dM^2 = \int_0^L \frac{4f}{D_h} dx. \tag{8-8}$$

The result is

$$\frac{4fL}{D_h} = \frac{1}{\gamma M_1^2} - \frac{1}{\gamma M_2^2} + \ln(M_1/M_2)^2, \tag{8-9}$$

which is usually expressed as

$$\frac{4fL}{D_h} = \frac{1 - \gamma M_1^2}{\gamma M_1^2} - \frac{1 - \gamma M_2^2}{\gamma M_2^2} + \ln\left(\frac{M_1}{M_2}\right)^2. \tag{8-10}$$

The various property relations can be obtained by using the influence coefficients from Table 5-1, substituting dM^2 for dT_o from Eq. 8-2, and integrating from station

1 to station 2. However, a more direct approach is to use conservation of mass and the equation of state (remember T = constant). Then

$$\dot{m}_1 = \dot{m}_2. \tag{8-11}$$

Using the definition of Mach number and the equation of state yields

$$\frac{P_1}{RT_1} A_1 M_1 \sqrt{\gamma R T_1} = \frac{P_2}{RT_2} A_1 M_1 \sqrt{\gamma R T_2}. \tag{8-12}$$

Under the assumptions of isothermal flow, Eq. 8-12 reduces to

$$P_1 M_1 = P_2 M_2. \tag{8-13}$$

Equation 8-13 is one of the more unique expressions in compressible flow: the product of the static pressure and Mach number is constant in an isothermal flow. With the pressure and temperature relations defined, the remaining static and stagnation property ratios can be obtained. The stagnation temperature ratio is, from Eq. 1-66,

$$\frac{T_{o_2}}{T_{o_1}} = \frac{1 + (\gamma - 1)/2M_2^2}{1 + (\gamma - 1)/2M_1^2}. \tag{8-14}$$

The stagnation pressure ratio becomes

$$\frac{P_1}{P_{o_1}} \frac{P_{o_2}}{P_2} = \left(\frac{T_1}{T_{o_1}} \frac{T_{o_2}}{T_2}\right)^{\gamma/(\gamma-1)}. \tag{8-15}$$

Using Eqs. 8-13 and 1-66 allows the stagnation pressure ratio expressed in terms of the Mach number to become

$$\frac{P_{o_2}}{P_{o_1}} = \frac{M_1}{M_2}\left(\frac{1 + (\gamma - 1)/2M_2^2}{1 + (\gamma - 1)/2M_1^2}\right)^{\gamma/(\gamma - 1)}. \tag{8-16}$$

The limiting Mach number, $1\sqrt{\gamma}$, is the natural reference condition for the property ratios. Setting $M_1 = M$ and $M_2 = 1/\sqrt{\gamma}$, the property ratios become

$$\frac{P}{P_T^*} = \frac{1}{\sqrt{\gamma} M}; \tag{8-17a}$$

$$\frac{T_o}{T_{oT}^*} = \left(\frac{2\gamma}{3\gamma - 1}\right)\left(1 + \frac{\gamma - 1}{2} M^2\right); \tag{8-17b}$$

$$\frac{P_o}{P_{oT}^*} = \frac{1}{\sqrt{\gamma} M}\left[\left(\frac{2\gamma}{3\gamma - 1}\right)\left(1 + \frac{\gamma - 1}{2} M^2\right)\right]^{\gamma/(\gamma-1)}; \tag{8-17c}$$

$$\frac{4fL_T^*}{D_h} = \frac{1 - \gamma M^2}{\gamma M^2} + \ln(\gamma M^2). \tag{8-17d}$$

The symbol L_T^* is used to represent the limiting length for an isothermal flow and corresponds to the location where the Mach number is the limiting Mach number, $1/\sqrt{\gamma}$.

Computational aids are available for isothermal flow. The expressions from Eqs. 8-17 give the values of various property ratios as functions of Mach number. Tabulations of these relations for a specific heats ratio of 1.4 are presented in Appendix H. Because isothermal flow is normally associated with low Mach numbers, Appendix H contains isothermal flow tabular values only for $0 < M < 1/\sqrt{\gamma}$. The property ratios are all referenced to the limiting Mach number for isothermal flow.

Isothermal flow computational aids are available under option **PL** in COMPQ. The menu for option **PL** is given in Figure 8-2a. Three cases are available. Cases **M** and **F** effectively replace the isothermal tables in Appendix H. The input for these cases is either the Mach number for case **M** or $4fL_T^*/D_h$ for case **F**. The output for these cases are the Mach number and the various property ratios as well as $4fL_T^*/D_h$. For a specified Mach number of 0.3, the output for case **M** is given in Figure 8-2b. Although the static temperature is constant for isothermal flow, the stagnation pressure and temperature are not. Case **DA** is used for the process involved in duct length addition and will be demonstrated in Example 8-1. As stated earlier, COMPQ adds versatility since the value of the specific heats ratio can be specified.

```
ISOTHERMAL FLOW SELECTED.  ISOTHERMAL FLOW CHOKES AT 1/SQR(GAM).
ONLY THE MACH NUMBER RANGE < 1/SQR(GAM) IS COVERED.
THE FANNING FRICTION FACTOR IS USED IN ALL ISOTHERMAL CALCULATIONS.

INPUT PARAMETERS ARE:
      MACH NUMBER           M
      4 fL/D                F
      DUCT ADDITION         DA

ENTER THE DESIRED LETTER(S)?
```

(a) Option **PL** menu

```
      ENTER THE MACH NUMBER? .3

          M          P/P*       TO/TO*      PO/PO*     4 fL/D       T/T*
       0.3000      2.8172      0.8907      1.8791      4.8650      1.0000
```

(b) Sample output

Figure 8-2 Option **PL** Menu and Sample Output

Duct length addition is a recurring circumstance that is involved in many isothermal flow situations. In such a process, information on the initial state and the increase in $4fL/D_h$ are given; the final state after the duct length addition is desired. The isothermal duct length addition process is schematically illustrated in Figure 8-3. Conditions at station 1 are known, as is the value of $4fL/D_h$ for the duct addition. The solution technique is based on the concept that conditions at the limiting Mach number are unique for a given isothermal flow. Thus, for state 1,

$4fL^*_{T_1}/D_h$ is required to reach the limiting Mach number, and at state 2, $4fL^*_{T_2}/D_h$ is required. As in Fanno flow,

$$\frac{4fL^*_{T_1}}{D_h} = \left(4\frac{fL}{D_h}\right)_{\text{act}} + \frac{4fL^*_{T_2}}{D_h}. \tag{8-18}$$

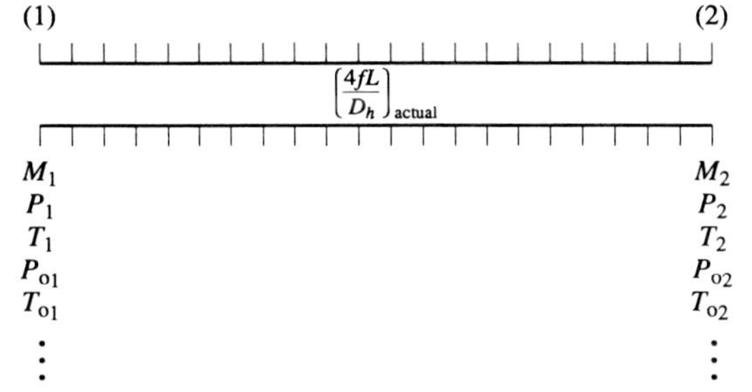

Figure 8-3 Duct Addition Schematic

Given the entrance conditions and the actual $4fL/D_h$ for the duct addition, the exit state can be found using Eq. 8-18. The Mach number and property ratios at state 2 are found using the values from the isothermal flow tables in Appendix H or from case **F** of option **PL** in COMPQ. Given the Mach number at state 2, the property ratios at state 2 can be computed by using the following ratios:

$$\begin{aligned} T_{o_2}/T_{o_1} &= (T_{o_2}/T^*_{o_T})/(T_{o_1}/T^*_{o_T}); \\ T_2/T_1 &= 1.0; \\ P_{o_2}/P_{o_1} &= (P_{o_2}/P^*_{o_T})/(P_{o_1}/P^*_{o_T}); \\ P_2/P_1 &= M_1/M_2. \end{aligned} \tag{8-19}$$

Case **DA** of option **PL** accomplishes the duct addition process by going through the steps just outlined. Consider the following example.

Example 8-1

Methane enters a pipeline with a Mach number of 0.05 and static conditions of 30 psia and 520 R. Find the exit conditions after a duct addition in which $4fL/D = 250.0$. What is the change in stagnation temperature and pressure?

Solution. The physical situation is illustrated in Figure 8-4a. Methane has a molecular mass of 16.043 lbm/lbm = mol, a gas constant of 96.322 ft-lbf/lbm R (518.25 J/kg K), and a ratio of specific heats (γ) of 1.32. The problem will be worked two ways: (1) using cases **M** and **F** of option **PL** and (2) using mass addition process case **DA** of option **PL**.

Using case **M** of option **PL** with an input Mach number of 0.05 yields at station 1

$$4fL^*_{T_1}/D = 296.3164 \qquad P_{o_1}/P^*_{o_T} = 10.8767$$

$$P_1/P^*_T = 17.4078$$

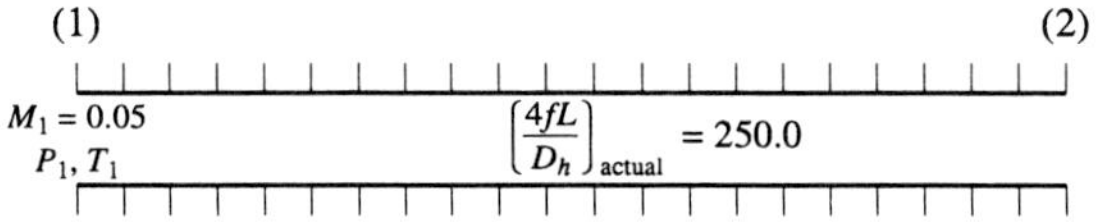

(a) Problem statement

```
ISOTHERMAL FLOW SELECTED.  ISOTHERMAL FLOW CHOKES AT 1/SQR(GAM).
ONLY THE MACH NUMBER RANGE < 1/SQR(GAM) IS COVERED.
THE FANNING FRICTION FACTOR IS USED IN ALL ISOTHERMAL CALCULATIONS.

INPUT PARAMETERS ARE:
      MACH NUMBER              M
      4 fL/D                   F
      DUCT ADDITION            DA

ENTER THE DESIRED LETTER(S)? f

ENTER THE VALUE OF 4 fL/D? 46.3164

       M         P/P*       TO/TO*     PO/PO*    4 fL/D      T/T*
    0.1216      7.1593     0.8940     4.5096    46.3192     1.0000

ARE MORE ISOTHEMAL COMPUTATIONS DESIRED?
```

(b) Case **F** output

```
ENTER THE MACH NUMBER AND THE 4 fL/D INCREMENT? 0.05,250

    M1       (4fL/D)1   D(4fL/D)    (4fL/D)2
  0.0500    296.3164   250.0000    46.3164

AT STATE 2 THE MACH NUMBER IS   0.1216

  T2/T1       TO2/TO1    P2/P1      PO2/PO1   RHO2/RHO1
  1.0000      1.0020     0.4113     0.4146     0.4113
```

(c) Case **DA** output

Figure 8-4 Example 8-1

After the duct addition,

$$\frac{4fL^*_{T_2}}{D_h} = \frac{4fL^*_{T_1}}{D_h} - \left(\frac{4fL}{D_h}\right)_{\text{act}} = 296.3164 - 250 = 46.3164.$$

With the value of $4fL^*_{T_2}/D_h$ known, case **F** of option **PL** can be used to determine conditions at M_2. The output of case **F** for the preceding conditions is presented in Figure 8-4b. After the duct addition the Mach number is 0.1216. The static and stagnation properties at station 2 can be computed from the various property ratios using Eqs. 8-19:

$$T_2 = T_1 = 520\ R;$$

$$T_{o_2}/T_{o_1} = (T_{o_1}/T^*_{oT})/(T_{o_1}/T^*_{oT})$$
$$= 0.8940/0.8922 = 1.0020;$$

$$P_{o_2}/P_{o_1} = (P_{o_2}/P^*_{oT})/(P_{o_1}/P^*_{oT})$$
$$= 4.5087/10.8767 = 0.4145;$$

$$P_2/P_1 = M_1/M_2$$
$$= 0.05/0.1216 = 0.4111.$$

At the initial Mach number of 0.05 the stagnation properties are

$$P_{o_1} = (P_{o_1}/P_1)P_1 = P_1/(P_1/P_{o_1}) = 30/0.9984 = 30.048 \text{ psia};$$
$$T_{o_1} = (T_{o_1}/T_1)T_1 = T_1/(T_1/T_{o_1}) = 520/0.9996 = 520.21\ R.$$

The properties at the end of the duct addition process (station 2) then become

$$M_2 = 0.1216;$$
$$T_{o_2} = 521.25\ R;$$
$$T_2 = 520.0\ R;$$
$$P_{o_2} = 12.455 \text{ psia};$$
$$P_2 = 12.333 \text{ psia}.$$

This problem can also be worked using the process case **DA** of option **PL**. The output for case **DA** for this problem is shown in Figure 8-4c. The input parameters required are the initial Mach number and the actual duct addition increment $4fL/D$. The various output property ratios and the Mach number agree well with the results for the previous solution using cases **M** and **F**. Case **DA** thus accomplished in a single input sequence what required, in the previous solution technique for this problem, application of cases **M** and **F** and some arithmetic.

Example 8-1 is typical of many duct addition situations. Conditions at the limiting value of $M = 1/\sqrt{\gamma}$ represent both reference conditions and limits for an isothermal flow process. A duct addition greater than $4fL^*_T/D$ would require changes in upstream conditions.

Example 8-2

Methane flows at 55° F through a 25,000-ft long buried pipeline that is 1 ft in diameter. At the entrance the average velocity is 50 ft/s and the static pressure is 25 psia. If the pipeline is considered smooth (relative roughness = 0), find the exit conditions assuming isothermal flow.

Solution. Thermodynamic constants for methane are given in Example 8-1. The physical situation for this problem is schematically illustrated in Figure 8-5a. At 55° F the viscosity of methane is 8.2314×10^{-6} lbm/ft-s. The density at the entrance is

$$\rho = P/RT = [(25)(144)]/[(96.322)(515)]$$
$$= 0.07257 \text{ lbm/ft}^3.$$

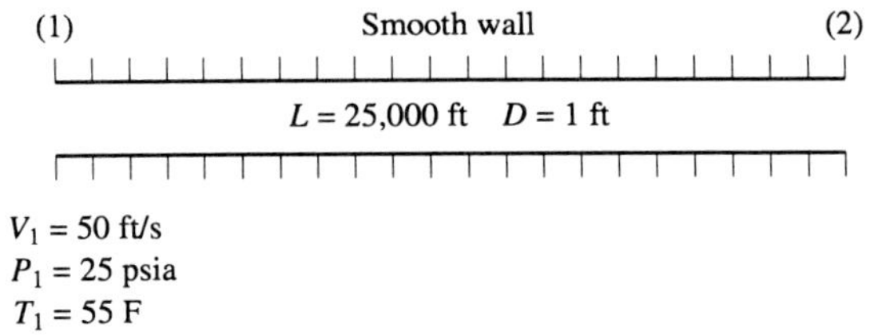

(a) Problem statement

```
ENTER THE MACH NUMBER AND THE 4 fL/D INCREMENT? .03446,340.0

    M1       (4fL/D)1   D(4fL/D)    (4fL/D)2
  0.0345    630.5047   340.0000    290.5047

AT STATE 2 THE MACH NUMBER IS   0.0505

  T2/T1       TO2/TO1     P2/P1       PO2/PO1    RHO2/RHO1
  1.0000      1.0002      0.6826      0.6832      0.6826
```

(b) Case **DA** output

Figure 8-5 Example 8-2

The Reynolds number becomes

$$Re_D = \rho VD/\mu = (0.07257)(50)(1)/8.2314 \times 10^{-6}$$
$$= 440{,}812.$$

From Appendix I, the Fanning friction factor is 0.0034. The duct addition then possesses a $4\,fL/D$ of

$$4\,fL/D = (4)(0.0034)(25000)/1 = 340.$$

Since ρV is a constant in steady-state duct flow and since the flow is isothermal (implying no change in viscosity), Re_D, and hence the friction factor, is invariant with length down the duct when fully developed flow has been established. The Mach number at the entrance is

$$M_1 = \frac{V_1}{a_1} = \frac{V_1}{\sqrt{\gamma R T_1}}$$
$$= \frac{50}{1451} = 0.03446.$$

Inlet stagnation conditions are

$$P_{o_1} = (P_{o_1}/P_1)P_1 = P_1/(P_1/P_{o_1}) = 25/0.9992 = 25.02 \text{ psia};$$
$$T_{o_1} = (T_{o_1}/T_1)T_1 = T_1/(T_1/T_{o_1}) = 520/0.9998 = 515.10\ R.$$

Conditions at state 2 can be calculated using case **DA** of option **PL** with an initial Mach number of 0.03446 and a $4fL/D$ increment of 340.0. The output is presented in Figure 8-5b. The exit conditions are

$$M_2 = 0.0505;$$

$$T_2 = 515\ R;$$

$$T_{o_2} \approx T_{o_1} = 515.10\ R;$$

$$P_2 = 17.07\ \text{psia};$$

$$P_{o_2} = 17.09\ \text{psia}.$$

Because of the low Mach numbers involved, the static and stagnation conditions at the exit are nearly equal. To three decimal places the entrance and exit stagnation temperatures are the same.

In the Mach number range covered in this example, Fanno flow and isothermal flow calculations result in nearly identical exit conditions. This should not be surprising as temperature is determined basically by Mach number while pressure drop is determined basically by duct length. The low Mach numbers and nearly constant stagnation temperature of the isothermal formulation for this example differ very little from the constant stagnation temperature and nearly constant static temperature associated with the equivalent Fanno flow formulation. The long duct length leads to a large static pressure drop and stagnation pressure loss even though the temperature changes are small. For many low Mach number applications, Fanno flow or isothermal flow calculations yield essentially the same results.

Isothermal flow is a good starting point for the generalized one-dimensional compressible flows. The nonsimple nature of isothermal flow demonstrates how more than one driving potential could be incorporated into a problem formulation. More importantly, the isothermal flow analysis showed that the limiting Mach number could be something other than 1.

8.3 ONE-DIMENSIONAL GENERALIZED FLOW CONCEPTS

In Chapter 5 we derived a system of differential equations describing the steady, one-dimensional flow of a calorically perfect gas with area change, friction, stagnation temperature change, and mass addition (or rejection) as the driving potentials. The results were expressed in terms of influence coefficients and were presented in Table 5-1. The four simple flows were treated by considering the effects of each of the driving potentials separately. Isothermal flow was treated as a special case of nonsimple flow.

With a few exceptions, such as isothermal flow and the special cases of Young (1993), generalized one-dimensional compressible flows require numerical solutions of nonlinear, ordinary differential equations. Differential equations for Mach number and various properties as function of the driving potentials can be obtained from Table 5-1. One possible procedure is to integrate the eight differential equations given in Table 5-1 for the Mach number and property distributions; however, if the

Mach number distribution is known, the property distributions can be obtained without integration of additional differential equations. We shall thus pursue this procedure in detail in the next section.

The differential equation with Mach number as the dependent variable is (from Table 5-1)

$$\frac{dM^2}{M^2} = -\frac{2\psi}{1-M^2}\frac{dA}{A} + \frac{\gamma M^2 \psi}{1-M^2}\left(\frac{4f}{D_h}dx + \frac{2dF_D}{\gamma M^2 PA}\right) + \frac{(1+\gamma M^2)\psi}{1-M^2}\frac{dT_o}{T_o} + \frac{2\psi(1+\gamma M^2)}{1-M^2}\frac{d\dot{m}}{\dot{m}}. \tag{8-20}$$

Expressed in terms of x-derivatives, Eq. 8-20 becomes

$$\frac{1}{M^2}\frac{dM^2}{dx} = -\frac{2\psi}{1-M^2}\frac{1}{A}\frac{dA}{dx} + \frac{\gamma M^2 \psi}{1-M^2}\left(\frac{4f}{D_h} + \frac{2dF_D/dx}{\gamma M^2 PA}\right) + \frac{(1+\gamma M^2)\psi}{1-M^2}\frac{1}{T_o}\frac{dT_o}{dx} + \frac{2\psi(1+\gamma M^2)}{1-M^2}\frac{1}{\dot{m}}\frac{d\dot{m}}{dx}. \tag{8-21}$$

For this first-order, nonlinear ordinary differential equation, the dependent variable is M, and the independent variable is x. If the variations of the four driving potentials with respect to x were known, then Eq. 8-21 could be integrated to yield the Mach number distribution as a function of x.

Even without integration, this equation can be used to infer characteristics of generalized one-dimensional solutions. A singular point of the differential equations occurs at the sonic point since the denominator vanishes for $M = 1$. Equation 8-21 can be expressed as

$$\frac{1}{M^2}\frac{dM^2}{dx} = \frac{\psi G(x,\gamma,M)}{1-M^2}, \tag{8-22}$$

where

$$G(x,\gamma,M) = -2\frac{1}{A}\frac{dA}{dx} + \gamma M^2\left(\frac{4f}{D_h} + \frac{2dF_D/dx}{\gamma M^2 PA}\right) + (1+\gamma M^2)\frac{1}{T_o}\frac{dT_o}{dx} + 2(1+\gamma M^2)\frac{1}{\dot{m}}\frac{d\dot{m}}{dx}. \tag{8-23}$$

When the flow is subsonic, a positive value of $G(x)$ results in an increase in Mach number, while a negative value of $G(x)$ yields a decrease in Mach number. If the flow is supersonic, just the opposite occurs. The sign of $G(x)$ is controlled by the sum of the products of the various influence coefficients and the dimensionless driving potentials. Hence, the Mach number response depends on the problem described by the driving potentials.

Consider a converging–diverging nozzle with friction. Equation 8-21 reduces to

$$\frac{1}{M^2}\frac{dM^2}{dx} = -\frac{2\psi}{1-M^2}\frac{1}{A}\frac{dA}{dx} + \frac{\gamma M^2 \psi}{1-M^2}\frac{4f}{D_h}, \tag{8-24}$$

which can be expressed as

$$(1 - M^2)\frac{dM^2}{dx} = M^2\psi\left(-\frac{2}{A}\frac{dA}{dx} + \gamma M^2\frac{4f}{D_h}\right). \tag{8-25}$$

At the sonic point, $M = 1$ and Eq. 8-25 reduces to

$$\frac{2}{A}\frac{dA}{dx} = \gamma\frac{4f}{D_h}. \tag{8-26}$$

The right-hand side of Eq. 8-26 is positive, so $dA/dx > 0$ at the sonic location. The sonic point in a converging–diverging nozzle with friction, but no heat transfer or mass addition, thus occurs in the diverging portion. Figure 8-6a illustrates the sonic point location for area change plus friction. The appearance of the sonic point in the diverging section is in contrast to simple area change, in which we concluded that a sonic point, if it is to exist, must always be at the minimum area, the nozzle throat. In a converging–diverging nozzle with friction, the location of the sonic point becomes, from Eq. 8-26, a function of the friction factor and area variation.

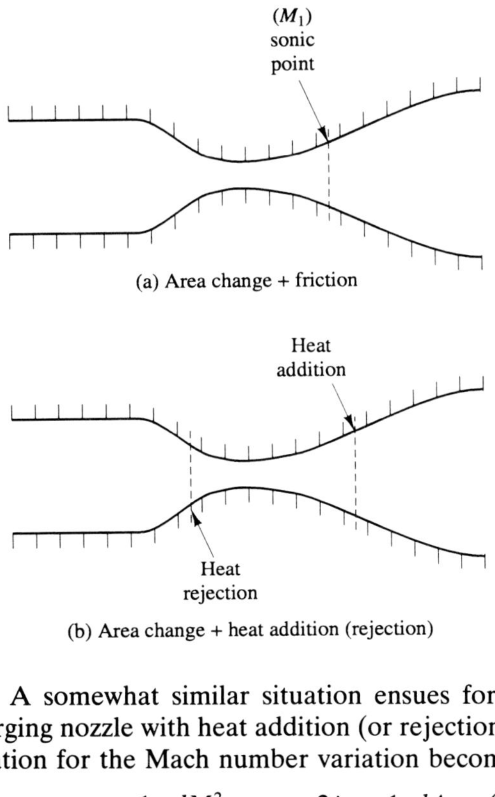

Figure 8-6 Sonic State Locations

A somewhat similar situation ensues for nonsimple flow in a converging–diverging nozzle with heat addition (or rejection). For this situation the differential equation for the Mach number variation becomes

$$\frac{1}{M^2}\frac{dM^2}{dx} = -\frac{2\psi}{1 - M^2}\frac{1}{A}\frac{dA}{dx} + \frac{(1 + \gamma M^2)\psi}{1 - M^2}\frac{1}{T_o}\frac{dT_o}{dx}, \tag{8-28a}$$

which becomes at the sonic location

$$\frac{2}{A}\frac{dA}{dx} = \frac{(1+\gamma)}{T_o}\frac{dT_o}{dx}. \tag{8-28b}$$

If heat is added to the flow, then dT_o/dx is positive and the location of the sonic point is in the diverging portion of the nozzle since $dA/A > 0$. If heat is rejected, then dT_o/dx is negative and the sonic point is located in the converging section since $dA/A < 0$. These are schematically illustrated in Figure 8-6b.

Combining Eqs. 8-24 and 8-28a yields an expression valid for friction plus heat addition in a converging–diverging nozzle flow. The results are

$$\frac{1}{M^2}\frac{dM^2}{dx} = \frac{\psi}{1-M^2}\left[-\frac{2}{A}\frac{dA}{dx} + \gamma M^2\frac{4f}{D_h} + (1+\gamma M^2)\frac{1}{T_o}\frac{dT_o}{dx}\right]. \tag{8-29}$$

For heat addition, dT_o/dx and f are positive and the sonic point is located in the diverging portion of the nozzle; its exact location depends on the distributions of $T_o(x)$, $A(x)$, and $f(x)$. If heat is rejected, then dT_o/dx is negative and f is positive and the location of the sonic point can be in either the diverging or converging segment of the nozzle, depending on driving potential distributions. The point behind all of this is to emphasize that only in simple area change converging–diverging nozzle flows do we know in advance the location of the sonic point. For nonsimple area change flows involving friction and/or heat addition, the sonic point location is prescribed by Eq. 8-29 evaluated at the sonic point.

Conditions for one-dimensional compressible flows in which a single parameter is held constant can also be obtained from Table 5-1. For example, if a constant Mach number flow considering friction and area change is desired, Eq. 8-22 reduces to

$$\frac{1}{M^2}\frac{dM^2}{dx} = -\frac{2\psi}{1-M^2}\frac{1}{A}\frac{dA}{dx} + \frac{\gamma M^2\psi}{1-M^2}\frac{4f}{D_h} = 0. \tag{8-30}$$

Then for a circular duct segment

$$\frac{1}{A}\frac{dA}{dx} = \frac{4}{\pi D^2}\frac{d}{dx}\left(\frac{\pi}{4}D^2\right) = \frac{2}{D}\frac{dD}{dx}, \tag{8-31}$$

so that

$$\frac{dD}{dx} = \frac{\gamma M^2}{4}4f. \tag{8-32}$$

Equation 8-32 describes the variation of the duct diameter with the specified constant Mach number and the friction factor.

Example 8-3

Air flows into a circular duct with an initial Mach number of 0.23. If the entrance diameter is 0.01 m and the friction factor is 0.002, find the shape of the duct that will permit constant Mach number flow over a 2-m long segment of duct.

Solution. The problem situation is illustrated in Figure 8-7a. This is a nonsimple constant Mach number flow with area change and friction. Thus, from Eq. 8-32,

$$\int_{D_1}^{D(x)} dD = \frac{\gamma M^2}{4} 4f \int_0^x dx.$$

Integrating yields

$$D(x) = D_1 + \frac{\gamma M^2}{4} 4fx,$$

and with the specific values for this problem,

$$D(x) = 0.01 + \frac{(1.4)(0.23)^2}{4} 4(.002)x = 0.01 + 0.00015x \quad (0 < x < 2).$$

Thus the desired constant Mach number duct segment is conical in shape, as illustrated in Figure 8-7b. Since the process is adiabatic with a constant Mach number, and thereby a constant static temperature, it actually describes a special kind of isothermal flow: constant Mach number with friction.

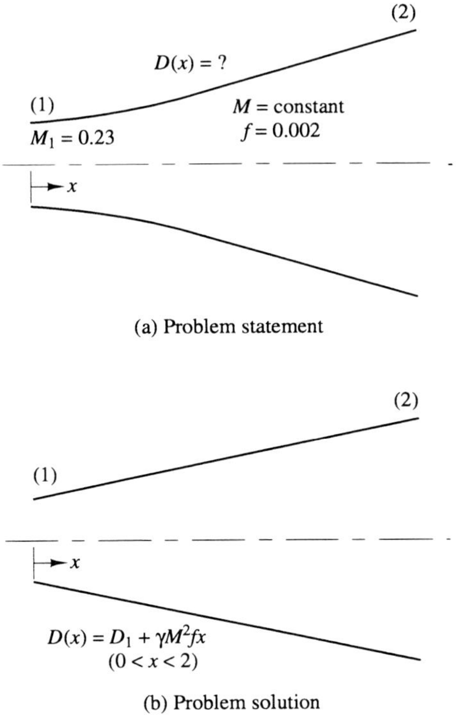

Figure 8-7 Example 8-3

Because of the singular point at the sonic location of Eq. 8-21, it is convenient to develop numerical techniques for integrating the equation in two parts: (1) flow fields with no sonic point present and (2) flow fields with sonic points present. We shall examine the first case next.

8.4 GENERALIZED FLOW ANALYSIS WITH NO SONIC POINT PRESENT

When no sonic points are present, Eq. 8-21 possesses no singular points and can be integrated once the four driving potentials are defined. The integration of Eq. 8-21 for a set of specified driving potentials yields the Mach number distribution as a function of x for one-dimensional flow. Situations involving sonic points are more complex and are examined in the next section.

Virtually any accurate numerical procedure can be used to integrate Eq. 8-21. The equation is well behaved and possesses no hidden problems. Because of its simplicity and accuracy, we shall use a fourth-order Runge-Kutta numerical procedure in integrating Eq. 8-21, which is of the functional form

$$\frac{dM}{dx} = f(x, \gamma, M). \tag{8-33}$$

Given the solution at some location x, the solution at location $x + \Delta x$ is

$$M(x + \Delta x) = M(x) + \frac{\Delta x}{6}(k_1 + 2k_2 + 2k_3 + k_4), \tag{8-34a}$$

where

$$\begin{aligned} k_1 &= f(x, \gamma, M(x)); \\ k_2 &= f\left(x + \frac{\Delta x}{2}, \gamma, M(x) + \frac{\Delta x}{2} k_1\right); \\ k_3 &= f\left(x + \frac{\Delta x}{2}, \gamma, M(x) + \frac{\Delta x}{2} k_2\right); \\ k_4 &= f(x + \Delta x, \gamma, M(x) + \Delta x\, k_3). \end{aligned} \tag{8-34b}$$

Equations 8-34 are sometimes called a *4–5 Runge-Kutta* since the numerical scheme possesses fourth-order accuracy with local truncation errors of order 5, usually written O(5). Additional details on Runge-Kutta procedures are provided in many numerical analysis textbooks; Conte and de Boor (1972) contains a particularly readable section on such methods.

If the driving potentials are known as functions of the independent variable x, then application of Eqs. 8-34 will yield the x-distribution of the Mach number. How are the remaining properties to be obtained? As previously mentioned, one procedure is to form differential equations for the remaining properties from Table 5-1 and then to integrate each of the equations. A much simpler procedure is available.

Consider an internal flow segment as illustrated in Figure 8-8. If the physical properties and Mach number are known at station 1 and the Mach number is known at station 2, then the properties at station 2 can be calculated by forming various ratios. The static temperature ratio T_1/T_2 is obtained from evaluating

$$T_o = T\left(1 + \frac{\gamma - 1}{2} M^2\right) \tag{1-66}$$

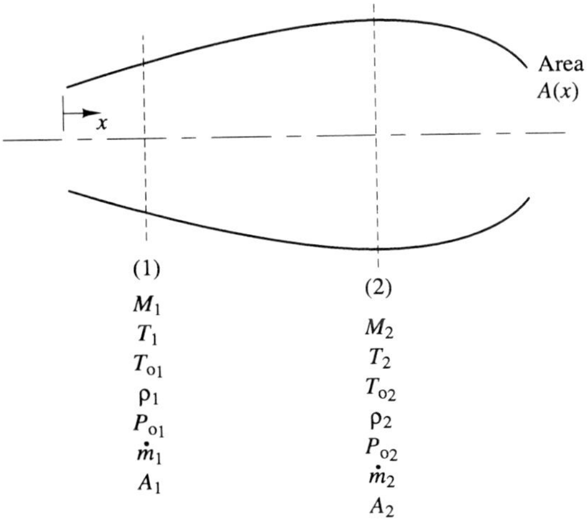

Figure 8-8 Integral Relations Development

at station 1 and station 2 and then dividing the results. This process gives

$$\frac{T_{o_2}}{T_{o_1}} = \frac{T_2}{T_1}\frac{(1 + (\gamma - 1)/2M_2^2)}{(1 + (\gamma - 1)/2M_1^2)}, \tag{8-35}$$

which can be solved for T_1/T_2 to yield

$$\frac{T_2}{T_1} = \frac{T_{o_2}}{T_{o_1}}\frac{(1 + (\gamma - 1)/2M_1^2)}{(1 + (\gamma - 1)/2M_2^2)}. \tag{8-36}$$

With T_{o_1} and T_{o_2} given by the driving potential, and T_1, M_1, and M_2 known, T_2 can then be evaluated.

The mass flow rate at a point can be written as

$$\dot{m} = \rho AV = \frac{P}{RT}AM\sqrt{\gamma RT} = \sqrt{\frac{\gamma}{RT}}PAM. \tag{8-37}$$

Evaluating Eq. 8-37 at stations 1 and 2 and then dividing yields

$$\frac{\dot{m}_1}{\dot{m}_2} = \frac{P_1 A_1 M_1}{P_2 A_2 M_2}\sqrt{\frac{T_2}{T_1}}, \tag{8-38}$$

which can be solved for P_2/P_1 to produce

$$\frac{P_2}{P_1} = \frac{\dot{m}_2 A_1 M_1}{\dot{m}_1 A_2 M_2}\sqrt{\frac{T_2}{T_1}}. \tag{8-39}$$

The mass flow rate ratio and the area distribution are known from the driving potentials. Using T_2/T_1 from Eq. 8-36 and the known Mach number and properties, we can evaluate P_2/P_1.

In a similar fashion, the definition of Mach number evaluated at stations 1 and 2 and then ratioed results in

$$\frac{V_2}{V_1} = \frac{M_2\sqrt{\gamma RT_2}}{M_1\sqrt{\gamma RT_1}}. \tag{8-40}$$

The velocity ratio V_2/V_1 then becomes

$$\frac{V_2}{V_1} = \frac{M_2}{M_1}\sqrt{\frac{T_2}{T_1}}, \tag{8-41}$$

where Eq. 8-36 is used for T_2/T_1. The equation of state for stations 1 and 2 can be arranged to give

$$\frac{\rho_2}{\rho_1} = \frac{P_2}{P_1}\frac{T_1}{T_2}, \tag{8-42}$$

where T_2/T_1 and P_2/P_1 are from Eqs. 8-36 and 8-41, respectively.

The stagnation and static pressures are related by

$$\frac{P}{P_o} = \left(\frac{T}{T_o}\right)^{\gamma/(\gamma-1)}. \tag{8-43}$$

Evaluating at stations 1 and 2 and dividing produces

$$\frac{P_{o_2}}{P_{o_1}} = \frac{P_2}{P_1}\left(\frac{T_{o_2}}{T_{o_1}}\frac{T_1}{T_2}\right)^{\gamma/(\gamma-1)}. \tag{8-44}$$

With T_{o_2}/T_{o_1} available from the driving potential and the static pressure and temperature ratios from previous expressions, the stagnation pressure ratio can be evaluated. The entropy can be calculated directly from the definition given in Chapter 1:

$$\frac{\Delta s}{c_p} = \ln\frac{T_2}{T_1} - \frac{\gamma - 1}{\gamma}\ln\frac{P_2}{P_1}. \tag{1-25}$$

The preceding property ratios are called *integral equations* since they permit the evaluation of quantities at two cross sections in a generalized flow. The integral relations are reproduced and grouped here in the usual order of their application.

$$\frac{T_2}{T_1} = \frac{T_{o_2}}{T_{o_1}}\frac{1 + (\gamma - 1)/2M_1^2}{1 + (\gamma - 1)/2M_2^2}; \tag{8-45a}$$

$$\frac{P_2}{P_1} = \frac{\dot{m}_2 A_1 M_1}{\dot{m}_1 A_2 M_2}\sqrt{\frac{T_2}{T_1}}; \tag{8-45b}$$

$$\frac{V_2}{V_1} = \frac{M_2}{M_1}\sqrt{\frac{T_2}{T_1}}; \tag{8-45c}$$

$$\frac{\rho_2}{\rho_1} = \frac{P_2}{P_1}\frac{T_1}{T_2}; \tag{8-45d}$$

$$\frac{P_{o_2}}{P_{o_1}} = \frac{P_2}{P_1}\left(\frac{T_{o_2}}{T_{o_1}}\frac{T_1}{T_2}\right)^{\gamma/(\gamma-1)}; \tag{8-45e}$$

$$\frac{\Delta s}{c_p} = \ln\frac{T_2}{T_1} - \frac{\gamma - 1}{\gamma}\ln\frac{P_2}{P_1}. \tag{8-45f}$$

Zucrow and Hoffman (1976) provide an outline for obtaining numerical solutions of nonsimple flows. An expanded version of their four-step procedure is:

1. Establish the initial conditions. Generally, the Mach number, the velocity, the density, and the static and stagnation pressures and temperatures are required at the entrance. Develop models for the driving potentials present. These models must be functions of the independent variable x.
2. Use Eqs. 8-34 to integrate Eq. 8-21 for the specified x-step of Δx.
3. Use the integral relations, Eqs. 8-45, to obtain the remaining properties.
4. Repeat steps 2 and 3 over the x-range of interest.

The entire procedure is best illustrated by an example.

Example 8-4

Air enters a combustor at $T_o = 1400\ R$, $P_o = 335$ psia, and $M = 0.2$. Energy is added to the air such that the exit stagnation temperature is 2200 R. The combustor is 2-ft long and tapers from a diameter of 1 ft to a diameter of 10 in. The friction factor is estimated to be 0.005. Neglect the effects of fuel mass addition and assume the properties are those of air. Solve for M, T, T_o, P, and P_o as functions of x.

Solution. The problem statement does not represent a simple flow, and so the influence coefficient method will be used. A sketch of the combustor is provided in Figure 8-9. Since the entrance stagnation conditions and Mach number are known, the static conditions at the entrance can be computed in the usual way. These evaluations will be accomplished in the computer program and will be discussed later.

The driving potentials are evaluated next. The diameter changes by 2 in in a 24-in length, so that

$$D(x) = 1.0 - x/12,$$

where x is expressed in feet. The area is

$$A(x) = \frac{\pi}{4}\left(1 - \frac{x}{12}\right)^2.$$

The area driving potential in the form $1/A\ dA/dx$ becomes for the above area distribution

$$\frac{1}{A}\frac{dA}{dx} = \frac{4}{\pi(1 - x/12)^2}\frac{\pi}{4}2\left(1 - \frac{x}{12}\right)\left(-\frac{1}{12}\right) = -\frac{1}{6(1 - x/12)}.$$

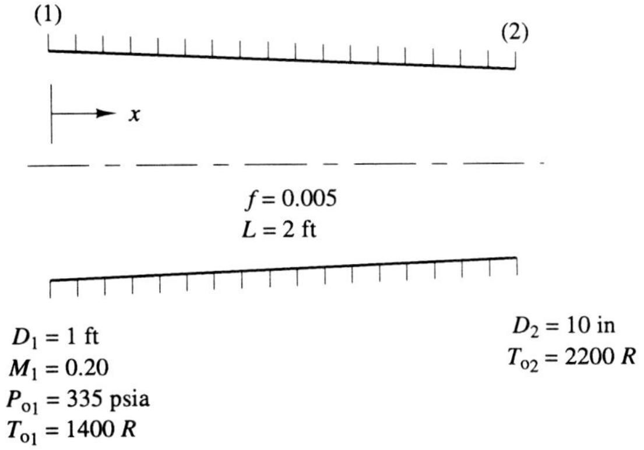

Figure 8-9 Example 8-4

The stagnation temperature is to vary from 1400 R to 2200 R. If the variation is assumed linear with distance

$$\frac{dT_o}{dx} = 400\,\frac{R}{\text{ft}},$$

so that

$$T_o(x) = 1400 + 400x.$$

The driving potential for heat addition, $1/T_o\, dT_o/dx$ is then cast as

$$\frac{1}{T_o}\frac{dT_o}{dx} = \frac{400}{1400 + 400x}.$$

The friction driving potential is $4\,f/D$. In terms of the diameter variation, this driving potential is

$$4\frac{f}{D} = 4\frac{f}{1 - x/12}.$$

Since there is no mass flow rate change, the driving potential for mass flow rate, $1/\dot{m}\, d\dot{m}/dx$, vanishes. Hence

$$\frac{1}{\dot{m}}\frac{d\dot{m}}{dx} = 0.$$

The differential equation for Mach number then becomes

$$\frac{dM}{dx} = M\frac{\psi}{1 - M^2}\left[\frac{1}{6(1 - x/12)} + \frac{\gamma M^2}{2}\frac{4f}{1 - x/12} + \frac{(1 + \gamma M^2)}{2}\frac{400}{1400 + 400x}\right].$$

The driving potentials have been modeled in terms of functions of x. The differential equation involves only γ, the independent variable x, and the dependent variable M. With the initial conditions known and the differential equation placed in the form

$$\frac{dM}{dx} = f(x, \gamma, M),$$

the Runge-Kutta procedure can be used to evaluate $M(x)$, and the integral relations can be used to obtain the various static and stagnation properties.

The computer program COMPINT1.BAS was developed to carry out the influence coefficient procedure discussed earlier. The computer program will be examined in detail later in this section. The results of the Runge-Kutta numerical integration and the integral relations property evaluations are presented in Figure 8-10. The exit Mach number is 0.398, the exit stagnation pressure is 325.71 psia, and the exit stagnation temperature is 2200 R. A loss of nearly 10 psia in stagnation pressure occurred in the system. If the program were rerun with the friction factor taken as zero, the net loss in stagnation pressure would be 8.4 psia. Hence, most of the stagnation pressure loss in the combustor is due to the heat addition effect. The earlier discussion of Rayleigh flow included the loss of stagnation pressure associated with heat addition.

This is a relatively complex problem involving area change, friction, and heat addition; nonetheless, the influence coefficient solution procedure was straightforward to develop and apply. Details of the computer program COMPINT1 are discussed next.

INPUT PARAMETERS:	GAMMA	F	DX
	1.400	0.005	0.100

I	X(FT)	MACH	PSTAG(PSI)	PS(PSI)	TSTAG(R)	TS(R)
1	0.000	0.200	335.00	325.79	1400.00	1388.89
2	0.100	0.207	334.71	324.88	1440.00	1427.79
3	0.200	0.214	334.40	323.92	1480.00	1466.60
4	0.300	0.221	334.09	322.91	1520.00	1505.30
5	0.400	0.228	333.75	321.85	1560.00	1543.89
6	0.500	0.236	333.41	320.72	1600.00	1582.36
7	0.600	0.244	333.05	319.53	1640.00	1620.69
8	0.700	0.252	332.67	318.26	1680.00	1658.88
9	0.800	0.261	332.28	316.92	1720.00	1696.90
10	0.900	0.270	331.87	315.50	1760.00	1734.75
11	1.000	0.279	331.44	314.00	1800.00	1772.41
12	1.100	0.289	330.99	312.39	1840.00	1809.86
13	1.200	0.299	330.51	310.69	1880.00	1847.07
14	1.300	0.309	330.02	308.87	1920.00	1884.02
15	1.400	0.320	329.50	306.93	1960.00	1920.68
16	1.500	0.331	328.95	304.86	2000.00	1957.02
17	1.600	0.343	328.37	302.65	2040.00	1993.01
18	1.700	0.356	327.76	300.27	2080.00	2028.59
19	1.800	0.369	327.11	297.72	2120.00	2063.73
20	1.900	0.383	326.43	294.97	2160.00	2098.35
21	2.000	0.398	325.71	292.00	2200.00	2132.40

Figure 8-10 COMPINT1 Output for Example 8-4

Example 8-4 was solved using Runge-Kutta integration for the differential equation resulting from the influence coefficient formulation. Since COMPINT1 contains virtually all of the logic needed for numerically solving problems based on influence coefficients, it is appropriate to examine the structure of the program. A block diagram of the program is presented in Figure 8-11, with a listing of the program in Figure 8-12.

As the listing shows, the program is structured about function definitions. Three groups of function definitions are used:

1. the driving potentials for a specific problem,
2. the differential equation used for all problems,
3. useful expressions used for all problems.

Lines 120–150 of the code in Figure 8-12 contain the function definition statements for the driving potential for Example 8-4. The variable **A** is a dummy variable and corresponds to the variable x in the problem discussion. Lines 160 and 170 contain the stagnation temperature distribution and the area distribution, respectively. The differential equation is defined in line 210. The driving potentials are identified in line 210 as **DPA**, **DPF**, **DPH**, and **DPM** for area, friction, heat addition, and mass

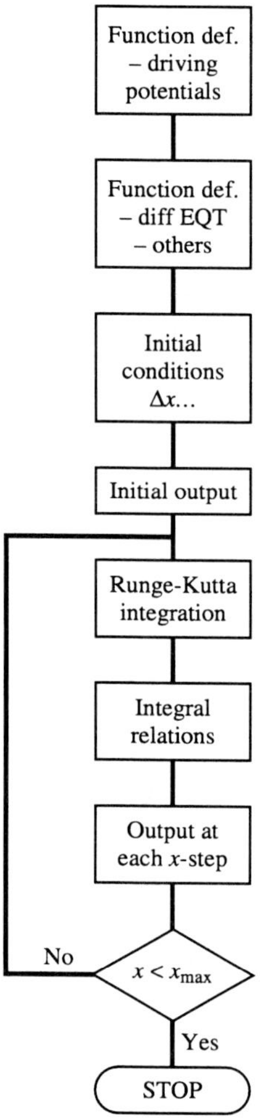

Figure 8-11 COMPINT1 Block Diagram

addition, respectively. Line 210 thus corresponds to the differential equation in the form

$$\frac{dM}{dx} = \frac{M\left(1 + \frac{\gamma - 1}{2}M^2\right)}{1 - M^2}\left(-\frac{1}{A}\frac{dA}{dx} + \frac{\gamma M^2}{2}\frac{4f}{D} + \frac{1 + \gamma M^2}{2}\frac{1}{T_o}\frac{dT_o}{dx} + (1 + \gamma M^2)\frac{1}{\dot{m}}\frac{d\dot{m}}{dx}\right). \tag{8-46}$$

Line 210 is the general representation of Eq. 8-21 or Eq. 8-46 and does not need to be altered for specific problems. The problem-specific driving potentials are defined in lines 120–150 and would be changed for different problems.

```
10 REM
20 REM INFLUENCE COEFFICIENT COMBUSTOR EXAMPLE
30 REM DEVELOPED FOR USE IN COMPRESSIBLE FLOW COURSES BY:
40 REM              MECHANICAL AND AEROSPACE ENGINEERING DEPARTMENTS
50 REM              MISSISSIPPI STATE UNIVERSITY
60 REM              MISSISSIPPI STATE, MS  39762
70 REM
80 PI = 3.1415926#
90  REM
100 REM DRIVING POTENTIAL DEFINITIONS
110 REM
120 DEF FN AREADP(A) = -1!/(6!*(1! - A/12!))  : REM AREA DRIVING POTENTIAL
130 DEF FN HEATDP(A) = 400!/(1400! + 400!*A)  : REM TO DRIVING POTENTIAL
140 DEF FN FRICDP(A) = 4!*F/(1! - A/12!)      : REM FRICTION DRIVING POTENTIAL
150 DEF FN MASSDP(A) = 0!                      : REM MDOT DRIVING POTENTIAL
160 DEF FN TSTAG(A)  = 1400! + 400!*A          : REM STAG. TEMP DISTRIBUTION
170 DEF FN AT(A) = PI/4!*(1! - A/12!)^2        : REM AREA VARIATION
180 REM
190 REM DIFFERENTIAL EQUATION DEFINITION
200 REM
210  DEF  FN  FM(AET,AMT,DPA,DPF,DPH,DPM,A,CM,GAM)  =  CM*AET/AMT*(-DPA  +
.5*GAM*CM*CM*DPF + .5*(1! + GAM*CM*CM)*DPH + (1! + GAM*CM*CM)*DPM)
220 REM
230 REM USEFUL FUNCTION DEFINITIONS
240 REM
250 DEF FN MT(CM) = 1! - CM*CM
260 DEF FN ET(CM) = 1! + .5*(GAM - 1!)*CM*CM
270 REM
280 REM INITIAL CONDITIONS
290 REM
300 XMI = .2: XM = XMI
310 F = .005
320 X = 0!: I = 1
330 GAM = 1.4
340 PEX = GAM/(GAM - 1!)
350 DX = .1
360 TSI = FN TSTAG(0!)/FN ET(XMI)
```

Figure 8-12 COMPINT1 Listing

```
370 PSTAGI = 335!: TSTAGI = FN TSTAG(0!)
380 PSI = PSTAGI/(FN ET(XMI)^PEX)
390 REM
400 REM INITIAL CONDITIONS OUTPUT
410 REM
420 INPUT "ENTER THE TITLE";TT$
430 CLS
440 PRINT: PRINT TT$: PRINT
450 PRINT "INPUT PARAMETERS:    GAMMA     F         DX"
460 PRINT "                        "; USING "  #.###   ";GAM,F,DX
470 PRINT: PRINT "   I      X(FT)      MACH     PSTAG(PSI)   PS(PSI)    TSTAG(R)
 TS(R)": PRINT
480 IS$ = STR$(I)
490 PRINT "  ";IS$;"  "; USING "  ###.###  ";X,XM;
500 PRINT USING "   ####.## ";PSTAGI,PSI,TSTAGI,TSI
510 REM
520 REM RUNGE-KUTTA INTEGRATION ROUTINE
530 REM
540 I = I + 1
550 REM FIRST RUNGE-KUTTA PASS
560 K1 = FN FM(FN ET(XM),FN MT(XM),FN AREADP(X),FN FRICDP(X),FN HEATDP(X),FN
MASSDP(X),X,XM,GAM)
570 REM SECOND RUNGE-KUTTA PASS
580 XA = X + .5*DX
590 XMA = XM + .5*DX*K1
600 K2 = FN FM(FN ET(XMA),FN MT(XMA),FN AREADP(XA),FN FRICDP(XA),FN HEATDP(XA),FN
MASSDP(XA),XA,XMA,GAM)
```

```
610 REM THIRD RUNGE-KUTTA PASS
620 XMA = XM + .5*DX*K2
630 K3 = FN FM(FN ET(XMA),FN MT(XMA),FN AREADP(XA),FN FRICDP(XA),FN HEATDP(XA),FN
MASSDP(XA),XA,XMA,GAM)
640 REM FOURTH RUNGE-KUTTA PASS
650 XA = X + DX
660 XMA = XM + DX*K3
670 K4 = FN FM(FN ET(XMA),FN MT(XMA),FN AREADP(XA),FN FRICDP(XA),FN HEATDP(XA),FN
MASSDP(XA),XA,XMA,GAM)
680 REM MACH NUMBER AT X + DX
690 XM = XM + DX*(K1 + 2!*(K2 + K3) + K4)/6!
700 X = XA
710 REM
720 REM PROPERTY CALCULATIONS
730 REM
740 T2OT1 = FN TSTAG(X)*FN ET(XMI)/(FN TSTAG(0!)*FN ET(XM))
750 TS = T2OT1*TSI
760 TSTAG = FN TSTAG(X)
770 P2OP1 = FN AT(0!)*XMI*SQR(T2OT1)/(FN AT(X)*XM)
780 PS = P2OP1*PSI
790 PSTAG = P2OP1*PSTAGI*(FN ET(XM)/FN ET(XMI))^PEX
800 IS$ = STR$(I)
810 PRINT "  ";IS$;"  "; USING "  ###.###  ";X,XM;
820 PRINT USING "   ####.## ";PSTAG,PS,TSTAG,TS
830 IF X < 2! THEN GOTO 540
840 END
```

Figure 8-12 (continued)

The initial conditions are specified in lines 300–380. The usual expressions relating static and stagnation temperatures and pressures are used in the order needed to compute the initial values of all properties. If an initial stagnation condition is taken as 1, then the program output for that property is in the form of the ratio of the local property value divided by the initial property value at the entrance. For different problems, the order of these lines could need rearranging. The step size for the independent variable x is defined in line 350.

Lines 560–700 contain the implementation of the Runge-Kutta procedure. The differential equation function definition is used, with that function's arguments being other previously defined function definitions. Since the driving potentials and other problem-specific functions can be defined for each problem in lines 120–170, no coding changes are required in lines 560–700 for different problems. Lines 740–790 evaluate other properties that use the integral relations. As with the integration procedure, these lines normally do not need to be changed for different problems.

COMPINT1 is an example of the implementation of the Runge-Kutta integration for the influence-coefficient–based differential equation. More complex problems might require some changes to the program, but the overall logic, in the absence of sonic points, would need little alteration.

Normal shock waves can be handled easily in one-dimensional generalized flows. The usual Runge-Kutta integration procedure is continued up to the shock wave location. At the shock wave location, conditions across the shock wave are calculated using the expressions, derived in Chapter 2, that relate conditions across a normal shock wave. For completeness, the so-called "jump conditions" are repeated here as follows:

$$M_2^2 = \frac{M_1^2 + 2/(\gamma - 1)}{2\gamma/(\gamma - 1)M_1^2 - 1}; \tag{8-47a}$$

$$\frac{P_2}{P_1} = \frac{2\gamma}{\gamma + 1}M_1^2 - \frac{\gamma - 1}{\gamma + 1}; \tag{8-47b}$$

$$\frac{\rho_2}{\rho_1} = \frac{(\gamma + 1)M_1^2}{2 + (\gamma - 1)M_1^2}; \tag{8-47c}$$

$$\frac{T_2}{T_1} = \frac{1 + (\gamma - 1)/2M_1^2}{1 + (\gamma - 1)/2M_2^2}, \tag{8-47d}$$

where subscript 1 represents upstream and subscript 2 downstream conditions. After the normal shock wave has been crossed, the Runge-Kutta procedure is continued throughout the remainder of the x-range.

Generalized flow problems are often composed of two different functions describing a driving potential. For example, the area change driving potential might be composed of a constant-area duct followed by a diverging or contracting segment. When a driving potential is different over various ranges of the independent variable

x, COMPINT1 needs to be modified slightly. The easiest way to accommodate the need for different driving potential descriptions with x is to place the different descriptions in a subroutine and to modify the Runge-Kutta function statements to reflect that change. Program COMPINT2.BAS illustrates the modifications needed to handle both a normal shock wave and different driving potential functions. The next example considers a situation involving two area functions and a normal shock wave at a specified location.

Example 8-5

Air enters a duct with an initial Mach number of 2.0 and stagnation conditions of 1000 K and 200 kPa. The 2-m long duct is composed of a constant diameter section for 1 m followed by a "sine-wave" shaped diameter distribution for 1 m. The initial diameter is 0.2 m, and the final diameter is 0.4 m. The friction factor is taken to be 0.005. The stagnation temperature varies in a linear fashion from 1000 K at the entrance to 600 K at the exit. The mass flow rate at the exit is 1.15 times the entrance mass flow rate; the injection is uniform and perpendicular to the axis. A normal shock wave is located 1.5 m from the entrance. The duct shape is sketched in Figure 8-13a. Determine the distributions of Mach number and static and stagnation pressure and temperature for this arrangement.

Solution. The problem possesses the driving potentials of area change, friction, heat rejection, and mass addition. Additionally, a shock wave is specified and the duct shape involves two different diameter functions. COMPINT2.BAS will be used to solve the problem. The driving potentials will be evaluated first. Over the first portion of the duct the diameter is

$$D(x) = 0.2; \qquad 0 \le x \le 1.$$

The diameter function for the second duct segment is

$$D(x) = 0.3 - 0.1 \cos[\pi(x - 1)]; \qquad 1 \le x \le 2.$$

The x-variable for all functions used in modeling the driving potentials must have its origin at the duct entrance. The cosine function was used since the sketch shows dD/dx to be zero at the beginning and ending of the last duct segment.

With the diameter functions specified in the preceding paragraph, the area functions become

$$A_1(x) = \pi/4(0.2)^2; \qquad 0 \le x \le 1;$$

$$A_2(x) = \pi/4[0.3 - 0.1 \cos[\pi(x - 1)]]^2; \qquad 1 \le x \le 2.$$

The two area change driving potentials, $1/A\, dA/dx$, then take the form

$$\frac{1}{A_1}\frac{dA_1}{dx} = 0; \qquad 0 \le x \le 1;$$

$$\frac{1}{A_2}\frac{dA_2}{dx} = \frac{0.2\pi \sin[\pi(x - 1)]}{0.3 - 0.1 \cos[\pi(x - 1)]}; \qquad 1 \le x \le 2.$$

The friction driving potentials, $4f/D$, are

$$\frac{4f}{D_1} = \frac{4f}{0.2}; \qquad 0 \le x \le 1;$$

$$\frac{4f}{D_2} = \frac{4f}{0.3 - 0.1\cos[\pi(x-1)]}; \qquad 1 \le x \le 2.$$

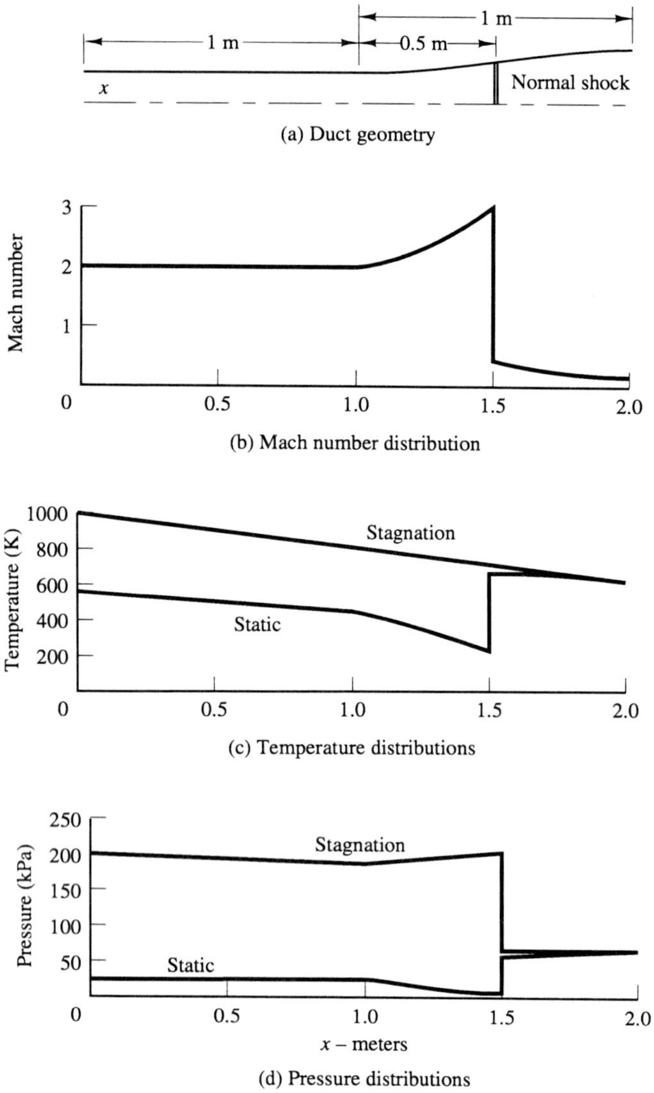

Figure 8-13 Example 8-5

The mass addition driving potential is not a function of diameter, so only one is needed. It is

$$\frac{1}{\dot{m}}\frac{d\dot{m}}{dx} = \frac{0.075}{1 + 0.075x}; \qquad 0 \leq x \leq 2.$$

With a uniform change in stagnation temperature, the distribution is

$$T_o(x) = 1000 - 200x; \qquad 0 \leq x \leq 2.$$

The heat rejection driving potential, $1/T_o\, dT_o/dx$, can then be cast as

$$\frac{1}{T_o}\frac{dT_o}{dx} = \frac{-1}{5 - x}; \qquad 0 \leq x \leq 2.$$

The shock wave is located at $x = 1.5$, so XSHOCK becomes 1.5. The solution to this generalized problem was obtained using COMPINT2. The output is presented in Figure 8-14. Plots of the Mach number and property distributions are given in Figures 8-13b, c, and d. The significant effect area change has on the Mach number and various properties is evident as is the dominant effect of the normal shock wave. As a result of the low exit Mach number (0.23), the static and stagnation properties values at the exit differ only slightly. This is a complex problem, but its solution is relatively straightforward using the generalized one-dimensional methodology.

The code for COMPINT2 is given in Figure 8-15. The overall logic is very similar to COMPINT1, except for COMPINT2's capability to handle normal shock waves and different diameter functions for various x-ranges. All problem-dependent functions, including the diameter functions (**FN DIA1** and **FN DIA2**), the area functions (**FN AREA1** and **FN AREA2**), and the area change driving potentials functions (**FN AREADP1** and **FN AREADP2**), are defined in lines 120–180. The initial conditions are evaluated as usual in lines 310–400. The variable **XBREAK** is the x-location where the constant diameter segment changes to the sine-wave shaped segment. The four Runge-Kutta passes each branch to the subroutine where the appropriate diameter, area, and area change driving potential are evaluated for the current value of x, **XDP**. Since the area change and friction driving potentials are functions of the diameter, they are evaluated in the subroutine and are simply referred to in the calling function statement as **DPA** and **DPF**. After application of the integral relations, the current value of x is compared to the shock wave location, and the branch is made to the subroutine to compute the conditions across the shock wave if the test is satisfied. The two subroutines are in lines 1000–1330. An examination of COMPINT1 and COMPINT2 shows that the overall logic for the two programs is essentially the same; COMPINT2 is more involved because it handles a larger array of circumstances.

This section has explored procedures for solving one-dimensional generalized compressible flow problems that do not possess local Mach numbers near 1. The two computer programs introduced are structured in a fashion to permit the solution of various problems by changing only the driving potential definitions and the initial conditions. The next section examines techniques for handling generalized flows with sonic points.

INPUT PARAMETERS:	GAMMA	F	DX
	1.400	0.005	0.050

X(m)	MACH	PSTAG(kPa)	PS(kPa)	TSTAG(K)	TS(K)	DELS/cp
0.000	2.000	200.00	25.56	1000.00	555.56	0.00
0.050	1.993	198.65	25.65	990.00	551.62	-0.01
0.100	1.987	197.38	25.73	980.00	547.53	-0.02
0.150	1.982	196.20	25.80	970.00	543.27	-0.02
0.200	1.977	195.11	25.85	960.00	538.86	-0.03
0.250	1.972	194.09	25.89	950.00	534.30	-0.04
0.300	1.968	193.16	25.93	940.00	529.58	-0.05
0.350	1.965	192.32	25.94	930.00	524.71	-0.06
0.400	1.962	191.55	25.95	920.00	519.70	-0.07
0.450	1.960	190.86	25.94	910.00	514.55	-0.08
0.500	1.959	190.25	25.93	900.00	509.25	-0.09
0.550	1.958	189.72	25.89	890.00	503.82	-0.10
0.600	1.957	189.26	25.85	880.00	498.24	-0.11
0.650	1.957	188.89	25.79	870.00	492.54	-0.12
0.700	1.958	188.60	25.72	860.00	486.70	-0.13
0.750	1.960	188.39	25.63	850.00	480.74	-0.15
0.800	1.962	188.26	25.53	840.00	474.65	-0.16
0.850	1.964	188.21	25.42	830.00	468.44	-0.17
0.900	1.968	188.26	25.29	820.00	462.10	-0.18
0.950	1.972	188.39	25.15	810.00	455.65	-0.19
1.000	1.977	188.61	25.00	800.00	449.08	-0.21
1.050	1.997	188.94	24.26	790.00	439.49	-0.22
1.100	2.046	189.41	22.52	780.00	424.51	-0.23
1.150	2.121	190.08	20.11	770.00	405.28	-0.25
1.200	2.217	191.04	17.39	760.00	383.23	-0.26
1.250	2.329	192.37	14.70	750.00	359.70	-0.28
1.300	2.453	194.16	12.22	740.00	335.81	-0.29
1.350	2.585	196.55	10.08	730.00	312.39	-0.31
1.400	2.723	199.65	8.28	720.00	290.00	-0.33
1.450	2.863	203.62	6.81	710.00	268.98	-0.35
1.500	3.005	208.60	5.64	700.00	249.47	-0.37
	NORMAL SHOCK WAVE ENCOUNTERED					
1.500	0.475	68.20	58.44	700.00	669.79	-0.05
1.550	0.414	68.24	60.65	690.00	667.14	-0.06
1.600	0.367	68.28	62.21	680.00	662.16	-0.08
1.650	0.330	68.31	63.33	670.00	655.68	-0.09
1.700	0.302	68.33	64.16	660.00	648.21	-0.11
1.750	0.279	68.36	64.76	650.00	640.04	-0.12
1.800	0.262	68.38	65.20	640.00	631.36	-0.14
1.850	0.248	68.40	65.52	630.00	622.32	-0.16
1.900	0.239	68.42	65.75	620.00	612.99	-0.17
1.950	0.233	68.44	65.90	610.00	603.44	-0.19
2.000	0.230	68.46	65.98	600.00	593.70	-0.20

Figure 8-14 Output for Example 8-5

```
10 REM
20 REM INFLUENCE COEFFICIENT EXAMPLE WITH MULTIPLE AREA DEFINITIONS
30 REM DEVELOPED FOR USE IN COMPRESSIBLE FLOW COURSES BY:
40 REM                 MECHANICAL AND AEROSPACE ENGINEERING DEPARTMENTS
50 REM                 MISSISSIPPI STATE UNIVERSITY
60 REM                 MISSISSIPPI STATE, MS  39762
70 REM
80 PI = 3.1415926#
90  REM
100 REM DRIVING POTENTIAL DEFINITIONS
110 REM
120 DEF FNAREADP1 (A) = 0!: DEF FNAREADP2 (A) = .2 * PI * SIN(PI * (A - 1!)) /
(.3 - .1 * COS(PI * (A - 1!)))
130 DEF FNHEATDP (A) = -1! / (5! - A)
140 DEF FNDIA1 (A) = .2: DEF FNDIA2 (A) = .3 - .1 * COS(PI * (A - 1!))
150 DEF FNMASSDP (A) = .075 / (1! + .075 * A)
160 DEF FNTSTAG (A) = 1000! - 200! * A
170 DEF FNAREA1 (A) = .25 * PI * .2 * .2: DEF FNAREA2 (A) = .25 * PI * (.3 - .1
* COS(PI * (A - 1!))) ^ 2
180 DEF FNMDOT (A) = 1! + .075 * A
190 REM
200 REM DIFFERENTIAL EQUATION DEFINITION
210 REM
220 DEF FNFM (AET, AMT, DPA, DPF, DPH, DPM, A, CM, GAM) = CM * AET / AMT * (-DPA
+ .5 * GAM * CM * CM * DPF + .5 * (1! + GAM * CM * CM) * DPH + (1! + GAM * CM *
CM) * DPM)
230 REM
240 REM USEFUL FUNCTION DEFINITIONS
250 REM
260 DEF FNMT (CM) = 1! - CM * CM
270 DEF FNET (CM) = 1! + .5 * (GAM - 1!) * CM * CM
280 REM
290 REM INITIAL CONDITIONS
300 REM
310 XMI = 2!: XM = XMI: DELS = 0!
320 F = .005
330 X = 0!: : XBREAK = 1!: I = 1
340 XSHOCK = 1.5: LOS = 0
```

Figure 8-15 COMPINT2 Listing

```
350 GAM = 1.4
360 PEX = GAM / (GAM - 1!)
370 DX = .05
380 TSI = FNTSTAG(0!) / FNET(XMI)
390 PSTAGI = 200!: TSTAGI = FNTSTAG(0!)
400 PSI = PSTAGI / (FNET(XMI) ^ PEX)
410 REM
420 REM INITIAL CONDITIONS OUTPUT
430 REM
440 INPUT "ENTER THE TITLE"; TT$
450 CLS
460 LPRINT : LPRINT TT$: LPRINT
470 LPRINT "INPUT PARAMETERS:    GAMMA      F          DX"
480 LPRINT "                        "; USING "  #.###   "; GAM; F; DX
490 LPRINT : LPRINT "    X(m)        MACH      PSTAG(kPa)    PS(kPa)    TSTAG(K)
TS(K)       DELS/cp": LPRINT
500 IS$ = STR$(I)
510 LPRINT USING "  ###.###  "; X; XM;
520 LPRINT USING "   ####.## "; PSTAGI; PSI; TSTAGI; TSI; DELS
530 REM
540 REM RUNGE-KUTTA INTEGRATION ROUTINE
550 REM
560 I = I + 1
570 REM FIRST RUNGE-KUTTA PASS
580 XDP = X
590 GOSUB 1000
600 K1 = FNFM(FNET(XM), FNMT(XM), DPA, DPF, FNHEATDP(X), FNMASSDP(X), X, XM, GAM)
610 REM SECOND RUNGE-KUTTA PASS
620 XA = X + .5 * DX: XDP = XA
630 XMA = XM + .5 * DX * K1
640 GOSUB 1000
650 K2 = FNFM(FNET(XMA), FNMT(XMA), DPA, DPF, FNHEATDP(XA), FNMASSDP(XA), XA,
XMA, GAM)
660 REM THIRD RUNGE-KUTTA PASS
670 XMA = XM + .5 * DX * K2
680 GOSUB 1000
```

```
690 K3 = FNFM(FNET(XMA), FNMT(XMA), DPA, DPF, FNHEATDP(XA), FNMASSDP(XA), XA,
XMA, GAM)
700 REM FOURTH RUNGE-KUTTA PASS
710 XA = X + DX: XDP = XA
720 XMA = XM + DX * K3
730 GOSUB 1000
740 K4 = FNFM(FNET(XMA), FNMT(XMA), DPA, DPF, FNHEATDP(XA), FNMASSDP(XA), XA,
XMA, GAM)
750 REM MACH NUMBER AT X + DX
760 XM = XM + DX * (K1 + 2! * (K2 + K3) + K4) / 6!
770 X = XA
780 REM
790 REM PROPERTY CALCULATIONS
800 REM
810 T2OT1 = FNTSTAG(X) * FNET(XMI) / (FNTSTAG(0!) * FNET(XM))
820 TS = T2OT1 * TSI
830 TSTAG = FNTSTAG(X)
840 P2OP1 = FNMDOT(X) * FNAREA1(0!) * XMI * SQR(T2OT1) / (FNMDOT(0!) * AREA * XM)
850 PS = P2OP1 * PSI
860 PSTAG = P2OP1 * PSTAGI * (FNET(XM) / FNET(XMI)) ^ PEX
865 DELS = LOG(TS / TSI) - (GAM - 1!) * LOG(PS / PSI) / GAM
870 IS$ = STR$(I)
880 LPRINT USING "  ###.###  "; X; XM;
890 LPRINT USING "   ####.## "; PSTAG; PS; TSTAG; TS; DELS
900 REM CHECK FOR SHOCK WAVE LOCATION
910 IF X > (XSHOCK - .5 * DX) AND LOS = 0 THEN GOSUB 1200
920 IF (X + .001) < 2! THEN GOTO 560
930 END
1000 REM
1010 REM SUBROUTINE FOR DRIVING POTENTIAL CHANGE WITH X
1020 REM
1030 IF XDP > XBREAK THEN GOTO 1080
```

Figure 8-15 (continued)

```
1040 DPA = FNAREADP1(XDP)
1050 DPF = 4! * F / FNDIA1(XDP)
1060 AREA = FNAREA1(XDP)
1070 RETURN
1080 DPA = FNAREADP2(XDP)
1090 DPF = 4! * F / FNDIA2(XDP)
1100 AREA = FNAREA2(XDP) : RETURN
1200 REM
1210 REM SHOCK WAVE JUMP CONDITIONS
1220 REM
1230 LOS = 1
1240 MO = XM
1250 XM = SQR((XM * XM + 2! / (GAM - 1!)) / (2! * GAM / (GAM - 1!) * XM * XM -
1!))
1260 PS = PS * (2! * GAM / (GAM + 1!) * MO * MO - (GAM - 1!) / (GAM + 1!))
1270 TS = TS * FNET(MO) / FNET(XM)
1280 PSTAG = PS * FNET(XM) ^ PEX
1290 DELS = LOG(TS / TSI) - (GAM - 1!) * LOG(PS / PSI) / GAM
1300 LPRINT "             NORMAL SHOCK WAVE ENCOUNTERED"
1310 LPRINT USING "  ###.###   "; X; XM;
1320 LPRINT USING "   ####.## "; PSTAG; PS; FNTSTAG(X); TS; DELS
1330 RETURN
```

Figure 8-15 (continued)

8.5 GENERALIZED FLOW ANALYSIS WITH A SONIC POINT LOCATION

In the previous section we established that the sonic point is a singular point of the generalized differential equation relating Mach number and the driving potentials. Consider Eqs. 8-22 and 8-23, which are reproduced here for convenience:

$$\frac{1}{M^2}\frac{dM^2}{dx} = \frac{\psi G(x,\gamma,M)}{1 - M^2}, \tag{8-22}$$

where

$$G(x,\gamma,M) = -2\frac{1}{A}\frac{dA}{dx} + \gamma M^2\left(4\frac{f}{D_h} + \frac{2dF_D/dx}{\gamma M^2 PA}\right) + (1 + \gamma M^2)\frac{1}{T_o}\frac{dT_o}{dx} + 2(1 + \gamma M^2)\frac{1}{\dot{m}}\frac{d\dot{m}}{dx}. \tag{8-23}$$

The denominator of Eq. 8-22 becomes unbounded in the neighborhood of $M = 1$, the sonic point. Thus, the procedure discussed in the previous section cannot be used near the sonic point, because "overflows" will be generated should the Mach number be sufficiently close to 1. Beans (1970) and Zucrow and Hoffman (1976) discuss procedures available to transit the sonic point location. We shall follow the technique proposed by Beans.

Equation 8-22 can be written as

$$\frac{1 - M^2}{M^2}\frac{dM^2}{dx} = \psi G(x,\gamma,M). \tag{8-48}$$

Evaluation at the sonic point leads to

$$G(x,\gamma,M)|_{M=1} = 0, \tag{8-49}$$

since ψ is not zero at the sonic point. The sonic point is thus the root of Eq. 8-49. Beans suggests that at the sonic point

$$\lim_{M\to 1}\frac{G(x,\gamma,M)}{1 - M^2} = \frac{0}{0} \tag{8-50}$$

is indeterminate since both the numerator and the denominator vanish. The value of dM/dx at the sonic point can then be evaluated by using l'Hospital's rule. Thus,

$$\lim_{M\to 1}\frac{dM}{dx} = \lim_{M\to 1}\frac{M\psi G(x,\gamma,M)}{2(1 - M^2)} = \frac{\gamma + 1}{4}\left(\frac{dG/dx}{-2MdM/dx}\right)_{M=1} \tag{8-51}$$

since $M = M(x)$.

The term $dG(x, \gamma, M)/dx$ is involved and becomes, neglecting dF_D/dx,

$$\begin{aligned}\frac{dG}{dx} &= \left[-2\frac{d}{dx}\left(\frac{1}{A}\frac{dA}{dx}\right) + \gamma M^2\frac{d}{dx}\left(4\frac{f}{D_h}\right)\right. \\ &\quad \left. + (1+\gamma M^2)\frac{d}{dx}\left(\frac{1}{T_o}\frac{dT_o}{dx}\right) + 2(1+\gamma M^2)\frac{d}{dx}\left(\frac{1}{\dot{m}}\frac{d\dot{m}}{dx}\right)\right] \\ &\quad + \left[2\gamma M\frac{dM}{dx}\left(\frac{4f}{D_h}\right) + 2\gamma M\frac{dM}{dx}\left(\frac{1}{T_o}\frac{dT_o}{dx}\right) + 4\gamma M\frac{dM}{dx}\left(\frac{1}{\dot{m}}\frac{d\dot{m}}{dx}\right)\right].\end{aligned} \tag{8-52}$$

Evaluation of Eq. 8-52 at the sonic point yields

$$\begin{aligned}\frac{dG}{dx} &= \left[-2\frac{d}{dx}\left(\frac{1}{A}\frac{dA}{dx}\right) + \gamma\frac{d}{dx}\left(\frac{4f}{D_h}\right) + (1+\gamma)\frac{d}{dx}\left(\frac{1}{T_o}\frac{dT_o}{dx}\right)\right. \\ &\quad \left. + 2(1+\gamma)\frac{d}{dx}\left(\frac{1}{\dot{m}}\frac{d\dot{m}}{dx}\right)\right] + \left.\frac{dM}{dx}\right|_{M=1}\left[2\gamma\left(\frac{4f}{D_h}\right)\right. \\ &\quad \left. + 2\gamma\left(\frac{1}{T_o}\frac{dT_o}{dx}\right) + 4\gamma\left(\frac{1}{\dot{m}}\frac{d\dot{m}}{dx}\right)\right].\end{aligned} \tag{8-53}$$

Equation 8-51 then becomes

$$\begin{aligned}\left(\frac{dM}{dx}\right)^2_{M=1} &= -\frac{\gamma+1}{8}\left[-2\frac{d}{dx}\left(\frac{1}{A}\frac{dA}{dx}\right)_{M=1} + \gamma\frac{d}{dx}\left(\frac{4f}{D_h}\right)_{M=1}\right. \\ &\quad \left. + (1+\gamma)\frac{d}{dx}\left(\frac{1}{T_o}\frac{dT_o}{dx}\right)_{M=1} + 2(1+\gamma)\frac{d}{dx}\left(\frac{1}{\dot{m}}\frac{d\dot{m}}{dx}\right)_{M=1}\right] \\ &\quad - \frac{\gamma+1}{8}\left(\frac{dM}{dx}\right)_{M=1}\left[2\gamma\left(\frac{4f}{D_h}\right) + 2\gamma\left(\frac{1}{T_o}\frac{dT_o}{dx}\right) + 4\gamma\left(\frac{1}{\dot{m}}\frac{d\dot{m}}{dx}\right)\right]_{M=1}.\end{aligned} \tag{8-54}$$

This expression is a quadratic equation in terms of dM/dx at the sonic point. The solution for Eq. 8-54 is double-valued; the negative value corresponds to subsonic flow downstream of the sonic point, and the positive value corresponds to supersonic flow downstream of the sonic location.

Beans's procedure is simple: in the neighborhood of the sonic point, the appropriate limiting value of dM/dx from Eq. 8-54 is used instead of Eq. 8-22. Singular behavior is thus avoided, and the solution proceeds smoothly through the sonic point.

The other remaining problem for nonsimple duct flows possessing sonic points is that the entrance Mach number is unknown. This situation did not arise in simple area change flow since the sonic point was known (at the minimum area location) and the area ratio determined the entrance Mach number. For the case of nonsimple flow with a sonic point, the Mach number (which is 1) and the sonic point location are known, but the entrance Mach number is unknown. The Runge-Kutta method is used to integrate *backward* in the negative x-direction, from the sonic point to the duct entrance. The Mach number at the duct entrance can thus be established. Once the entrance Mach number is known, the Runge-Kutta method is integrated forward from the entrance, $x = 0.0$, and the properties evaluated using the integral relations. For both integrations, forward and backward, the appropriate limiting value of dM/dx from Eq. 8-54 is used in the neighborhood of the sonic point.

The procedure for applying the generalized one-dimensional flow analysis technique to ducts with sonic points then becomes

1. Locate the sonic point, $x =$ **XSP**, using Eq. 8-49.
2. Evaluate dM/dx, **DMDXMAX**, at the sonic point using Eq. 8-54.
3. Starting at the sonic location ($x =$ **XSP**), integrate backward to establish the entrance Mach number, **XMI**.
4. Starting at the entrance ($x = 0.0$), integrate forward to find the distributions of Mach number and properties.

The COMPINT3.BAS program accomplishes these four steps for specified driving potentials. Details of this program will be considered after the next example.

Example 8-6

The flow field in a converging–diverging nozzle of diameter distribution

$$D/D^* = 1.0 + 0.5\cos(\pi x); \qquad 0 \le x \le 2,$$

where D^* is the minimum diameter is to be analyzed. The friction factor is 0.1, and the flow is adiabatic with a constant mass flow rate. A normal shock wave is located at the $x = 1.5$ position.

Solution. This problem is similar to that proposed by Beans (1970) and discussed by Zucrow and Hoffman (1976) and is schematically illustrated in Figure 8-16a. The value of 0.1 for the Fanning friction factor is unrealistically large, but it does serve to illustrate significant friction effects on both the Mach number and pressure distributions.

COMPINT3 will be used to solve this problem. With the diameter as just defined, the area change driving potential is

$$\frac{1}{A}\frac{dA}{dx} = \frac{4}{\pi[1 + .5\cos(\pi x)]^2}\frac{d}{dx}\frac{\pi}{4}[1 + .5\cos(\pi x)]^2 = \frac{-\pi\sin(\pi x)}{1 + 0.5\cos(\pi x)},$$

where D^* is taken as unity. The friction factor driving potential becomes

$$\frac{4f}{D_h} = \frac{4f}{1 + .5\cos(\pi x)}.$$

The remaining driving potential for heat and mass addition show no variation with x, so that

$$\frac{1}{T_o}\frac{dT_o}{dx} = 0;$$

$$\frac{1}{\dot{m}}\frac{d\dot{m}}{dx} = 0.$$

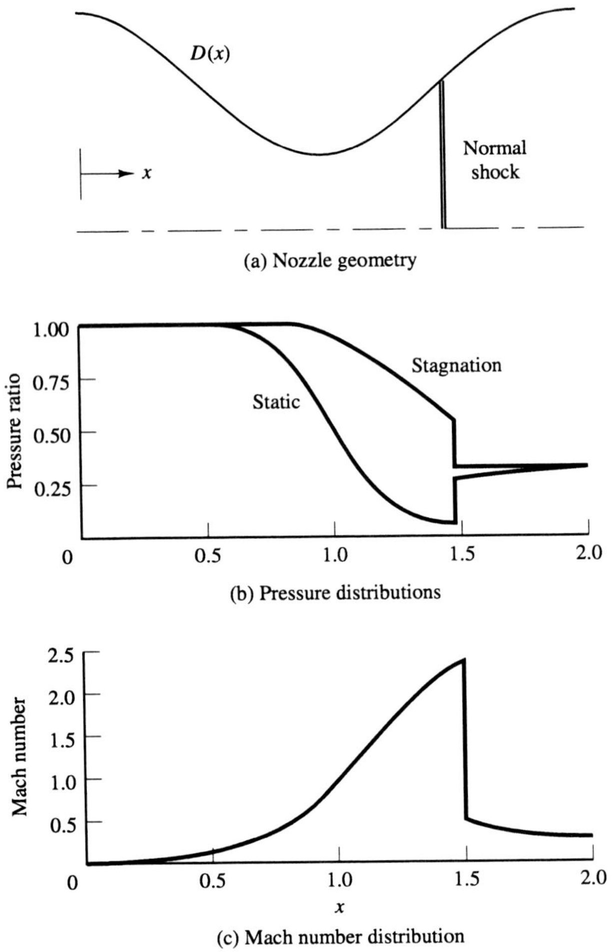

Figure 8-16 Example 8-6

The solution output for this problem is given in Figure 8-17. The sonic point is located at $x = 1.028$. As discussed previously, for nonsimple flows the sonic point location need not be at the minimum area. The backward integration from the sonic point yields an entrance Mach number of 0.0605; were the flow to be isentropic (simple area change only), the entrance Mach number would be 0.0645. The Mach number distribution confirms that sonic flow is reached downstream of the minimum area. Unlike the isentropic converging–diverging nozzle flows previously considered, this flow field shows a continuous degradation of the stagnation pressure from the entrance to the exit. This loss in stagnation pressure is caused by the presence of the large friction factor as well as the normal shock wave.

The stagnation and static pressure and Mach number distributions are presented in Figures 8-16b and c. Both the stagnation and static pressure ratios are nondimensionalized by the entrance stagnation pressure, P_{o_1}. The expected discontinuities at the normal shock wave are present. Although the friction factor is unrealistically large, the

```
INPUT PARAMETERS:        GAMMA      F          DX
                         1.400      0.100      0.050
SONIC POINT INFORMATION:            XSP        DMDXMAX
                                    1.028      3.109
MACH NUMBER AT ENTRANCE ESTABLISHED TO BE  0.0605
```

I	X	MACH	PSTAG	PSTATIC	TSTAG	TSTATIC
1	0.000	0.060	1.00000	0.99744	1.00000	0.99927
2	0.050	0.061	0.99997	0.99737	1.00000	0.99926
3	0.100	0.063	0.99993	0.99720	1.00000	0.99922
4	0.150	0.065	0.99989	0.99693	1.00000	0.99915
5	0.200	0.069	0.99985	0.99652	1.00000	0.99905
6	0.250	0.074	0.99979	0.99593	1.00000	0.99890
7	0.300	0.081	0.99973	0.99510	1.00000	0.99868
8	0.350	0.091	0.99965	0.99392	1.00000	0.99836
9	0.400	0.103	0.99954	0.99222	1.00000	0.99790
10	0.450	0.118	0.99939	0.98974	1.00000	0.99723
11	0.500	0.137	0.99917	0.98607	1.00000	0.99624
12	0.550	0.163	0.99884	0.98060	1.00000	0.99475
13	0.600	0.195	0.99834	0.97233	1.00000	0.99248
14	0.650	0.235	0.99755	0.95979	1.00000	0.98904
15	0.700	0.287	0.99626	0.94087	1.00000	0.98379
16	0.750	0.352	0.99416	0.91270	1.00000	0.97587
17	0.800	0.431	0.99073	0.87190	1.00000	0.96416
18	0.850	0.527	0.98520	0.81530	1.00000	0.94735
19	0.900	0.640	0.97651	0.74121	1.00000	0.92425
20	0.950	0.770	0.96339	0.65111	1.00000	0.89410
21	1.000	0.913	0.94456	0.55043	1.00000	0.85702
22	1.050	1.068	0.91910	0.44770	1.00000	0.81424
23	1.100	1.229	0.88666	0.35177	1.00000	0.76787
24	1.150	1.393	0.84788	0.26897	1.00000	0.72033
25	1.200	1.555	0.80409	0.20209	1.00000	0.67397
26	1.250	1.712	0.75706	0.15072	1.00000	0.63056
27	1.300	1.859	0.70862	0.11261	1.00000	0.59123
28	1.350	1.996	0.66041	0.08491	1.00000	0.55650
29	1.400	2.121	0.61366	0.06498	1.00000	0.52649
30	1.450	2.231	0.56924	0.05069	1.00000	0.50105
31	1.500	2.328	0.52768	0.04040	1.00000	0.47990
	NORMAL SHOCK WAVE ENCOUNTERED					
	1.500	0.531	0.30135	0.24868	1.00000	0.94659
32	1.550	0.434	0.30042	0.26389	1.00000	0.96363
33	1.600	0.368	0.29982	0.27304	1.00000	0.97362
34	1.650	0.320	0.29941	0.27889	1.00000	0.97992
35	1.700	0.284	0.29911	0.28276	1.00000	0.98407
36	1.750	0.258	0.29887	0.28536	1.00000	0.98687
37	1.800	0.238	0.29869	0.28711	1.00000	0.98877
38	1.850	0.224	0.29853	0.28825	1.00000	0.99004
39	1.900	0.215	0.29839	0.28895	1.00000	0.99086
40	1.950	0.209	0.29827	0.28929	1.00000	0.99131
41	2.000	0.208	0.29815	0.28932	1.00000	0.99145

Figure 8-17 COMPINT3 Output for Example 8-6

stagnation pressure shows little degradation until the local Mach number becomes sonic. As the Mach number increases, the friction effect, which is essentially proportional to M^2, becomes increasingly more significant. Just upstream of the shock wave, the stagnation pressure has been reduced by almost one-half of the entrance value. Downstream of the shock wave, when subsonic Mach numbers are again encountered, the loss in stagnation pressure due to friction is again small.

Example 8-6 illustrates the inputs required to use COMPINT3 to solve a one-dimensional generalized flow problem containing a sonic point. The listing of the program in the form used to solve the problem is presented in Figure 8-18, with the block diagram given in Figure 8-19. Although both the listing and block diagram for COMPINT3 appear markedly different from those of COMPINT1 and COMPINT2, all three programs are structured about the Runge-Kutta integration technique and the integral relations evaluations.

COMPINT3 utilizes subroutine calls (**GOSUB. . . .**) for most tasks. The subroutines start on line 1000 and are commented for ease of identification. The main program contains the various function definitions; they belong to the same three categories as discussed in COMPINT1 and COMPINT2. In implementing Beans' method, calls are made at line 400 to locate the sonic point, **XSP**; at line 410 to evaluate dM/dx, **DMDXMAX**, at the sonic point; and at line 460 to integrate in the negative x-direction to find the entrance Mach number, **XMI**. The forward integration and integral relations evaluations call is made on line 700.

The sonic point is located using a Newton-Raphson technique (lines 1250–1340). The value of dM/dx at the sonic point, **DMDXMAX**, is evaluated using Eq. 8-54. The value of the first grouping of terms in Eq. 8-53 at the sonic point is computed using a central finite difference approximation for dG/dx. This first group corresponds to the derivative of the defined function G evaluated at $M = 1$ before the derivative is taken. The positive root, corresponding to supersonic flow downstream of the sonic point is selected (line 1100). If subsonic flow were to occur downstream of the sonic location, then the negative root should be selected. The backward integration uses the specified value of Δx until the last step, when Δx is chosen in order to yield $x = 0.0$. The calculated value of the Mach number at $x = 0.0$ becomes the entrance value of the Mach number for the forward integration from $x = 0$ to $x =$ **XMAX**. The forward integration uses the specified value of Δx. Although the Mach number distribution from the earlier backward integration could be used in the range $0 \leq x \leq$ **XSP**, consistent spacing and simplicity dictate that the best procedure is to use just the entrance Mach number and to integrate from $x = 0$ with the desired x-spacing.

The integration and property evaluation subroutine (lines 2000–2490) is virtually identical to the Runge-Kutta/integral relations evaluation coding in COMPINT1 and COMPINT2. The only significant differences are the allowance made for integration near the sonic point and the logic to distinguish between forward and backward integration. Each of the four-step evaluations for the Runge-Kutta procedure must have the logic to use the value of dM/dx in the sonic limit (Eq. 8-54) when near the sonic point. If the absolute value of $(M - 1.0)$ is less than **MSONIC**, taken

```
10 REM
20 REM INFLUENCE COEFFICIENT EXAMPLE WITH SONIC LOCATION
30 REM DEVELOPED FOR USE IN COMPRESSIBLE FLOW COURSES BY:
40 REM                MECHANICAL AND AEROSPACE ENGINEERING DEPARTMENTS
50 REM                MISSISSIPPI STATE UNIVERSITY
60 REM                MISSISSIPPI STATE, MS  39762
70 REM
80 PI = 3.1415926#
90  REM
100 REM DRIVING POTENTIAL FUNCTION DEFINITIONS
110 REM
120 DEF FNAREADP (A) = -PI * SIN(PI * A) / (1! + .5 * COS(PI * A))
130 DEF FNHEATDP (A) = 0!
140 DEF FNDIA1 (A) = 1! + .5 * COS(PI * A)
150 DEF FNMASSDP (A) = 0!: DEF FNFRICDP (A, F) = 4! * F / (1! + .5 * COS(PI * A))
160 DEF FNTSTAG (A) = 1!
170 DEF FNAREA (A) = .25 * PI * (1! + .5 * COS(PI * A)) ^ 2
180 DEF FNMDOT (A) = 1!
190 DEF FNG (DPA, DPF, DPH, DPM, A, CM, GAM) = 2! * (-DPA + .5 * GAM * CM * CM
* DPF + .5 * (1! + GAM * CM * CM) * DPH + (1! + GAM * CM * CM) * DPM)
200 REM
210 REM DIFFERENTIAL EQUATION DEFINITION
220 REM
230 DEF FNFM (AET, AMT, DPA, DPF, DPH, DPM, A, CM, GAM) = CM * AET / AMT * (-DPA
+ .5 * GAM * CM * CM * DPF + .5 * (1! + GAM * CM * CM) * DPH + (1! + GAM * CM *
CM) * DPM)
240 REM
250 REM USEFUL FUNCTION DEFINITIONS
260 REM
270 DEF FNMT (CM) = 1! - CM * CM
280 DEF FNET (CM) = 1! + .5 * (GAM - 1!) * CM * CM
290 REM
300 REM INITIAL PARAMETERS
310 REM
320 F = .1
330 XMAX = 2!: XSHOCK = 1.5: LOS = 0
340 GAM = 1.4
350 PEX = GAM / (GAM - 1!)
```

Figure 8-18 COMPINT3 Listing

```
360 DX = .05: DXSAVE = DX
370 REM
380 REM BRANCH TO SUBROUTINES TO LOCATE SONIC POINT AND EVALUATE dM/dX AT M = 1
390 REM
400 GOSUB 1250: REM LOCATION OF SONIC POINT (XSP DETERMINED)
410 GOSUB 1000: REM EVALUATION OF dM/dX AT M = 1
420 REM
430 REM BACKWARDS INTEGRATION TO ESTABLISH ENTRANCE MACH NUMBER
440 REM
450 X = XSP: DX = -DXSAVE: XM = 1!: I = 0
460 GOSUB 2000: REM XMI DETERMINED
470 REM
480 REM INITIAL CONDITIONS OUTPUT
490 REM
500 XMI = XM: DX = DXSAVE: I = 1
510 TSI = FNTSTAG(0!) / FNET(XMI)
520 PSTAGI = 1!: TSTAGI = FNTSTAG(0!)
530 PSI = PSTAGI / (FNET(XMI) ^ PEX)
540 INPUT "ENTER THE TITLE"; TT$
550 CLS
560 LPRINT : LPRINT TT$: LPRINT
570 LPRINT "INPUT PARAMETERS:    GAMMA      F          DX"
580 LPRINT "                   "; USING "  #.###    "; GAM; F; DX
590 LPRINT "SONIC POINT INFORMATION:        XSP        DMDXMAX"
600 LPRINT "                             "; USING "  ##.###   "; XSP; DMDXMAX
610 LPRINT "MACH NUMBER AT ENTRANCE ESTABLISHED TO BE "; USING " #.####"; XMI
620 LPRINT : LPRINT "   I        X        MACH       PSTAG     PSTATIC     TSTAG
   TSTATIC": LPRINT
630 IS$ = STR$(I)
640 LPRINT "  "; IS$; "  "; USING "  ###.###  "; X; XM;
650 LPRINT USING "   #.##### "; PSTAGI; PSI; TSTAGI; TSI
660 REM
670 REM FORWARD RUNGE-KUTTA INTEGRATION AND PROPERTY EVALUATION CALL
680 REM
690 X = 0!: DX = DXSAVE
700 GOSUB 2000
710 END
1000 REM
```

```
1010 REM EVALUATION OF DMDXMAX FOR SONIC POINT EVALUATION
1020 REM
1030 XSPP = XSP + .001: XSPM = XSP - .001
1040  DGDX  =  (FNG(FNAREADP(XSPP),  FNFRICDP(XSPP,  F),  FNHEATDP(XSPP),
FNMASSDP(XSPP),  XSPP,  1!,  GAM)  -  FNG(FNAREADP(XSPM),  FNFRICDP(XSPM,  F),
FNHEATDP(XSPM), FNMASSDP(XSPM), XSPM, 1!, GAM)) / (XSPP - XSPM)
1050 CTERM = (GAM + 1!) * DGDX / 8!
1060 BTERM = (GAM + 1!) * (2! * GAM / 8!) * (FNFRICDP(XSP, F) + FNHEATDP(XSP) +
2! * FNMASSDP(XSP))
1070 DMDX1 = -.5 * BTERM + .5 * SQR(BTERM * BTERM - 4! * CTERM): DMDX2 = -.5 *
BTERM - SQR(BTERM * BTERM - 4! * CTERM)
1080 REM PRINT "B";"C";"DMDX1";"DMDX2"
1090 REM PRINT BTERM,CTERM,DMDX1,DMDX2
1100 DMDXMAX = DMDX1: REM DMDXMAX FOR SUPERSONIC FLOW CHOSEN
1110 RETURN
1250 REM
1260 REM NEWTON-RAPHSON TO LOCATE SONIC POINT
1270 REM
1280 XINT = 1!
1290 XPLUS = XINT + .001: XMIN = XINT - .001
1300  DGDX  =  (FNG(FNAREADP(XPLUS),  FNFRICDP(XPLUS,  F),  FNHEATDP(XPLUS),
FNMASSDP(XPLUS),  XPLUS,  1!,  GAM)  -  FNG(FNAREADP(XMIN),  FNFRICDP(XMIN,  F),
FNHEATDP(XMIN), FNMASSDP(XMIN), XMIN, 1!, GAM)) / (XPLUS - XMIN)
1310  XNEW  =  XINT  -  FNG(FNAREADP(XINT),  FNFRICDP(XINT,  F),  FNHEATDP(XINT),
FNMASSDP(XINT), XINT, 1!, GAM) / DGDX
1320 IF ABS(XNEW - XINT) < .0001 THEN GOTO 1340
1330 XINT = XNEW: GOTO 1290
1340 XSP = XINT
1350 RETURN
1500 REM
1510 REM SHOCK WAVE JUMP CONDITIONS
1520 REM
1530 LOS = 1
1540 MO = XM
1550 XM = SQR((XM * XM + 2! / (GAM - 1!)) / (2! * GAM / (GAM - 1!) * XM * XM -
1!))
1560 PS = PS * (2! * GAM / (GAM + 1!) * MO * MO - (GAM - 1!) / (GAM + 1!))
1570 TS = TS * FNET(MO) / FNET(XM)
```

Figure 8-18 (continued)

```
1580 PSTAG = PS * FNET(XM) ^ PEX
1590 DELS = LOG(TS / TSI) - (GAM - 1!) * LOG(PS / PSI) / GAM
1600 LPRINT "           NORMAL SHOCK WAVE ENCOUNTERED"
1610 LPRINT "       "; USING "  ###.###  "; X; XM;
1620 LPRINT USING "   #.##### "; PSTAG; PS; FNTSTAG(X); TS
1630 RETURN
2000 REM
2010 REM RUNGE-KUTTA INTEGRATION ROUTINE
2020 REM
2030 I = I + 1: MSONIC = .05
2040 REM FIRST RUNGE-KUTTA PASS
2050 IF ABS(XM - 1!) < MSONIC THEN GOTO 2070
2060 K1 = FNFM(FNET(XM), FNMT(XM), FNAREADP(X), FNFRICDP(X, F), FNHEATDP(X),
FNMASSDP(X), X, XM, GAM): GOTO 2090
2070 K1 = DMDXMAX
2080 REM SECOND RUNGE-KUTTA PASS
2090 XA = X + .5 * DX
2100 XMA = XM + .5 * DX * K1
2110 IF ABS(XMA - 1!) < MSONIC THEN GOTO 2130
2120 K2 = FNFM(FNET(XMA), FNMT(XMA), FNAREADP(XA), FNFRICDP(XA, F), FNHEATDP(XA),
FNMASSDP(XA), XA, XMA, GAM): GOTO 2150
2130 K2 = DMDXMAX
2140 REM THIRD RUNGE-KUTTA PASS
2150 XMA = XM + .5 * DX * K2
2160 IF ABS(XMA - 1!) < MSONIC THEN GOTO 2180
2170 K3 = FNFM(FNET(XMA), FNMT(XMA), FNAREADP(XA), FNFRICDP(XA, F), FNHEATDP(XA),
FNMASSDP(XA), XA, XMA, GAM): GOTO 2200
2180 K3 = DMDXMAX
2190 REM FOURTH RUNGE-KUTTA PASS
2200 XA = X + DX
2210 XMA = XM + DX * K3
2220 IF ABS(XMA - 1!) < MSONIC THEN GOTO 2240
2230 K4 = FNFM(FNET(XMA), FNMT(XMA), FNAREADP(XA), FNFRICDP(XA, F), FNHEATDP(XA),
FNMASSDP(XA), XA, XMA, GAM): GOTO 2260
2240 K4 = DMDXMAX
2250 REM MACH NUMBER AT X + DX
2260 XM = XM + DX * (K1 + 2! * (K2 + K3) + K4) / 6!
2270 X = XA
2280 IF DX < 0! THEN GOTO 2440
```

```
2290 REM
2300 REM PROPERTY CALCULATIONS
2310 REM
2320 T2OT1 = FNTSTAG(X) * FNET(XMI) / (FNTSTAG(0!) * FNET(XM))
2330 TS = T2OT1 * TSI
2340 TSTAG = FNTSTAG(X)
2350 P2OP1 = FNMDOT(X) * FNAREA(0!) * XMI * SQR(T2OT1) / (FNMDOT(0!) * FNAREA(X)
* XM)
2360 PS = P2OP1 * PSI
2370 PSTAG = P2OP1 * PSTAGI * (FNET(XM) / FNET(XMI)) ^ PEX
2380 IS$ = STR$(I)
2390 LPRINT "  "; IS$; "  "; USING "  ###.###  "; X; XM;
2400 LPRINT USING "   #.##### "; PSTAG; PS; TSTAG; TS
2410 REM CHECK FOR SHOCK WAVE LOCATION
2420 IF X > (XSHOCK - .5 * DX) AND LOS = 0 THEN GOSUB 1500
2430 REM CHECK ON END OF INTEGRATION SEQUENCES
2440 IF DX < 0! THEN GOTO 2470
2450 IF (X + .001) < XMAX THEN GOTO 2030
2460 RETURN
2470 IF ABS(DX) > X THEN DX = -X
2480 IF X > .0001 THEN GOTO 2030
2490 RETURN
```

Figure 8-18 (continued)

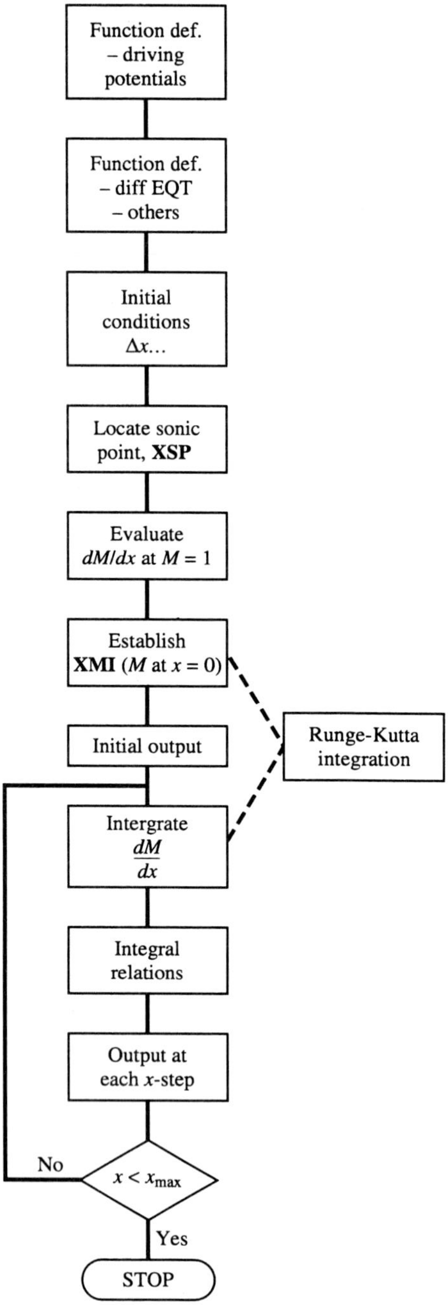

Figure 8-19 COMPINT3 Block Diagram

as 0.05 in COMPINT3, then dM/dx from Eq. 8-54 is used and the function call is avoided by branching. Calculations in the neighborhood of the sonic point for situations involving subsonic flow everywhere except the sonic point are made by using the positive root of dM/dx upstream of the sonic point and the negative root of dM/dx downstream of the sonic point.

Although Example 8-6 is relatively simple, it does illustrate Beans' procedure for handling sonic points in generalized one-dimensional flow. For more complex situations, COMPINT3 can be modified as needed. The generalized flow analysis can be used in much the same fashion as various options of COMPQ in solving nonsimple nozzle flow problems. Consider the following example.

Example 8-7

A converging–diverging nozzle with the hyperbolic diameter distribution, where $D(x)$ and x are measured in feet,

$$D(x) = \sqrt{1 + 0.25(x - 3)^2}$$

is connected to a reservoir with stagnation conditions of 100 psia and 1000 R. The nozzle exhausts into a variable back pressure arrangement. Determine the nozzle behavior with respect to back pressure if the stagnation temperature increases by 20 percent and the mass flow rate increases by 10 percent. The increase in both temperature and mass flow rate is linear with respect to x, the nozzle axis coordinate. The friction factor is 0.01.

Solution. The nozzle geometry is sketched in Figure 8-20a. COMPINT3 will be used to determine the back pressures corresponding to a normal shock at the sonic point and in the exit plane. These back pressures are analogous to key pressures 1 and 2 in the discussion of nozzle flow in Chapter 6. Key pressure 3 from Chapter 6 is analogous to the back pressure required for supersonic shock-free flow in the nozzle.

The driving potentials must be evaluated if COMPINT3 is to be used. The driving potential for area change for this configuration becomes

$$\frac{1}{A}\frac{dA}{dx} = \frac{4}{\pi[1 + 0.25(x - 3)^2]}\frac{d}{dx}\left[\frac{\pi}{4}(1 + 0.25(x - 3)^2)\right] = \frac{x - 3}{2(1 + 0.25(x - 3)^2)}.$$

The driving potential for friction can be expressed as

$$\frac{4f}{D_h} = \frac{4f}{\sqrt{1 + 0.25(x - 3)^2}}.$$

The stagnation temperature change for a 20-percent linear increase in temperature is expressed as

$$T_o(x) = 1000 + 20x,$$

which yields the driving potential for heat addition in the form of

$$\frac{1}{T_o}\frac{dT_o}{dx} = \frac{1}{50 + x}.$$

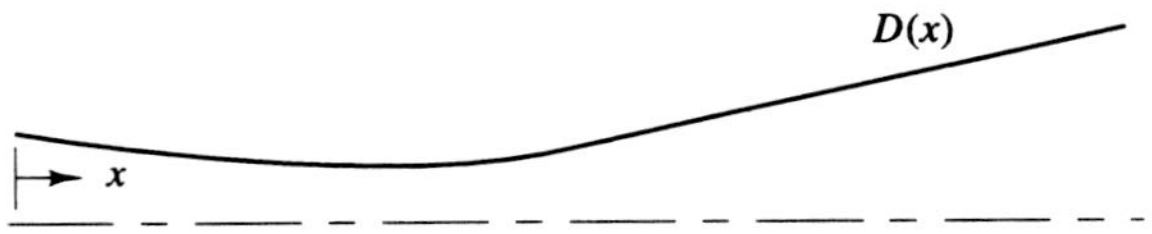

$D(x) = \sqrt{1 + 0.25\,(x-3)^2}$

$T_o(x) = 1000 + 20x$

$\dot{m}(x) = \dot{m}(0)\,[1 + 0.1x]$

$f = 0.01$

(a) Problem statement

```
160 REM
170 REM DRIVING POTENTIAL FUNCTION DEFINITIONS
180 REM
190 DEF FNAREADP (A) = (A - 3!) / (2! * (1! + .25 * (A - 3!) ^ 2))
200 DEF FNHEATDP (A) = 1! / (50! + A)
210 DEF FNDIA1 (A) = SQR(1! + .25 * (A - 3!) ^ 2)
220 DEF FNMASSDP (A) = 1! / (100! + A): DEF FNFRICDP (A, F) = 4! * F / SQR(1! +
.25 * (A - 3!) ^ 2)
230 DEF FNTSTAG (A) = 1000! + 20! * A
240 DEF FNAREA (A) = .25 * PI * (1! + .25 * (A - 3!) ^ 2)
250 DEF FNMDOT (A) = 1! + .01 * A
260 DEF FNG (DPA, DPF, DPH, DPM, A, CM, GAM) = 2! * (-DPA + .5 * GAM * CM * CM
* DPF + .5 * (1! + GAM * CM * CM) * DPH + (1! + GAM * CM * CM) * DPM)
```

(b) Function definitions

Figure 8-20 Example 8-7

The stagnation temperature distribution nondimensionalized by the entrance stagnation temperature can be cast as

$$\frac{T_o(x)}{T_o(0)} = 1 + 0.02x.$$

For simplicity of output the nondimensionalized version is used. In the same vein, the driving potential for mass addition becomes

$$\frac{1}{\dot{m}}\frac{d\dot{m}}{dx} = \frac{1}{100 + x}.$$

The initial conditions for the COMPINT3 executions are

$$f = 0.01; \qquad \Delta x = 0.25; \qquad x_{\text{max}} = 10.0;$$

$$\textbf{XSHOCK} = 20,\ 10,\ 3.25$$

$$P_o/P_{o_i} = 1.0; \qquad T_o/T_{o_i} = 1.0.$$

Since the minimum area occurs at $x = 3$, the initial guess for the sonic point location iteration is taken as 3.

With the aforementioned values coded into COMPINT3, the various required back pressures were found. Figure 8-20b presents the driving potentials required for this problem. The program output for shock-free nozzle flow is provided in Figure 8-21a. For the conditions of this problem, the back pressure required for this situation is 1.136 psia. The back pressure required for the shock to stand in the nozzle exit plane was generated by using the variable **XSHOCK** to require a shock to be positioned at $x = 10$. Since none of the flow conditions upstream of the shock are changed, only the last few stations of the output are given in Figure 8-21c.

The conditions for a normal shock wave of vanishing strength to stand exactly at the sonic point cannot be precisely attained because of the finite value of Δx used in the integration. The conditions are approximated by placing the shock at the nearest grid point location, **XSHOCK** = 3.25. The actual sonic point is $x = 3.148$, and thus the difference is slight. The subsonic flow distributions upstream of the sonic point are unchanged from the output given in Figure 8-21a, and so they are not repeated in Figure 8-21b, the output for the shock wave located near the sonic point. The back pressure required to place the shock wave at $x = 3.25$ is 91.8 psia.

The behavior of the nozzle for the nonsimple flow is thus established. The back pressures, expressed in psia, for the various ranges are as follows:

subsonic throughout	$91.8 < P_b < 100;$
shock in nozzle	$9.34 < P_b < 91.8;$
shock at nozzle exit	$P_b = 9.34;$
oblique shock at exit	$1.14 < P_b < 9.34;$
shock-free flow	$P_b = 1.14;$
outside expansion wave	$P_b < 1.14.$

```
INPUT PARAMETERS:      GAMMA       F           DX
                       1.400       0.010       0.250
SONIC POINT INFORMATION:           XSP         DMDXMAX
                                   3.148       0.512
MACH NUMBER AT ENTRANCE ESTABLISHED TO BE  0.1636
```

I	X	MACH	PSTAG	PSTATIC	TSTAG	TSTATIC
1	0.000	0.164	1.00000	0.98150	1.00000	0.99468
2	0.250	0.186	0.99967	0.97590	1.00500	0.99811
3	0.500	0.212	0.99922	0.96842	1.01000	1.00100
4	0.750	0.243	0.99863	0.95839	1.01500	1.00314
5	1.000	0.280	0.99784	0.94488	1.02000	1.00423
6	1.250	0.324	0.99674	0.92669	1.02500	1.00388
7	1.500	0.377	0.99524	0.90228	1.03000	1.00154
8	1.750	0.439	0.99316	0.86989	1.03500	0.99654
9	2.000	0.513	0.99026	0.82773	1.04000	0.98807
10	2.250	0.598	0.98625	0.77442	1.04500	0.97524
11	2.500	0.696	0.98072	0.70970	1.05000	0.95731
12	2.750	0.805	0.97323	0.63507	1.05500	0.93386
13	3.000	0.925	0.96331	0.55410	1.06000	0.90507
14	3.250	1.053	0.95053	0.47176	1.06500	0.87181
15	3.500	1.184	0.93456	0.39336	1.07000	0.83562
16	3.750	1.317	0.91528	0.32284	1.07500	0.79818
17	4.000	1.447	0.89277	0.26241	1.08000	0.76119
18	4.250	1.573	0.86728	0.21248	1.08500	0.72595
19	4.500	1.691	0.83922	0.17224	1.09000	0.69331
20	4.750	1.802	0.80906	0.14028	1.09500	0.66373
21	5.000	1.905	0.77733	0.11509	1.10000	0.63734
22	5.250	1.999	0.74453	0.09525	1.10500	0.61407
23	5.500	2.085	0.71115	0.07959	1.11000	0.59371
24	5.750	2.163	0.67761	0.06716	1.11500	0.57604
25	6.000	2.233	0.64428	0.05722	1.12000	0.56076
26	6.250	2.296	0.61146	0.04921	1.12500	0.54763
27	6.500	2.352	0.57941	0.04270	1.13000	0.53641
28	6.750	2.402	0.54832	0.03737	1.13500	0.52686
29	7.000	2.447	0.51833	0.03296	1.14000	0.51879
30	7.250	2.486	0.48956	0.02928	1.14500	0.51204
31	7.500	2.521	0.46206	0.02619	1.15000	0.50645
32	7.750	2.551	0.43589	0.02357	1.15500	0.50188
33	8.000	2.577	0.41104	0.02135	1.16000	0.49823
34	8.250	2.600	0.38752	0.01943	1.16500	0.49540
35	8.500	2.619	0.36531	0.01778	1.17000	0.49329
36	8.750	2.635	0.34436	0.01634	1.17500	0.49183
37	9.000	2.649	0.32465	0.01508	1.18000	0.49096
38	9.250	2.660	0.30612	0.01398	1.18500	0.49061
39	9.500	2.669	0.28871	0.01300	1.19000	0.49074
40	9.750	2.676	0.27238	0.01214	1.19500	0.49130
41	10.000	2.681	0.25707	0.01136	1.20000	0.49225

(a) Shock-free output

Figure 8-21 COMPINT3 Outputs for Example 8-7

I	X	MACH	PSTAG	PSTATIC	TSTAG	TSTATIC
12	2.750	0.805	0.97323	0.63507	1.05500	0.93386
13	3.000	0.925	0.96331	0.55410	1.06000	0.90507
14	3.250	1.053	0.95053	0.47176	1.06500	0.87181
	NORMAL SHOCK WAVE ENCOUNTERED					
	3.250	0.951	0.95037	0.53118	1.06500	0.90191
15	3.500	0.814	0.94055	0.60826	1.07000	0.94471
16	3.750	0.699	0.93342	0.67334	1.07500	0.97923
17	4.000	0.600	0.92830	0.72753	1.08000	1.00736
18	4.250	0.516	0.92464	0.77095	1.08500	1.03008
19	4.500	0.445	0.92201	0.80473	1.09000	1.04844
20	4.750	0.385	0.92011	0.83052	1.09500	1.06342
21	5.000	0.335	0.91873	0.85000	1.10000	1.07583
22	5.250	0.293	0.91771	0.86464	1.10500	1.08635
23	5.500	0.257	0.91694	0.87564	1.11000	1.09548
24	5.750	0.227	0.91637	0.88394	1.11500	1.10358
25	6.000	0.202	0.91593	0.89024	1.12000	1.11093
26	6.250	0.180	0.91559	0.89504	1.12500	1.11773
27	6.500	0.162	0.91532	0.89874	1.13000	1.12412
28	6.750	0.146	0.91511	0.90162	1.13500	1.13019
29	7.000	0.132	0.91494	0.90387	1.14000	1.13604
30	7.250	0.120	0.91480	0.90564	1.14500	1.14171
31	7.500	0.110	0.91469	0.90705	1.15000	1.14725
32	7.750	0.100	0.91460	0.90818	1.15500	1.15268
33	8.000	0.092	0.91453	0.90910	1.16000	1.15803
34	8.250	0.085	0.91446	0.90984	1.16500	1.16332
35	8.500	0.079	0.91441	0.91046	1.17000	1.16855
36	8.750	0.073	0.91436	0.91096	1.17500	1.17375
37	9.000	0.068	0.91433	0.91138	1.18000	1.17891
38	9.250	0.063	0.91429	0.91173	1.18500	1.18405
39	9.500	0.059	0.91426	0.91202	1.19000	1.18917
40	9.750	0.056	0.91424	0.91227	1.19500	1.19426
41	10.000	0.052	0.91422	0.91248	1.20000	1.19935

(b) Shock near sonic point output

I	X	MACH	PSTAG	PSTATIC	TSTAG	TSTATIC
35	8.500	2.619	0.36531	0.01778	1.17000	0.49329
36	8.750	2.635	0.34436	0.01634	1.17500	0.49183
37	9.000	2.649	0.32465	0.01508	1.18000	0.49096
38	9.250	2.660	0.30612	0.01398	1.18500	0.49061
39	9.500	2.669	0.28871	0.01300	1.19000	0.49074
40	9.750	2.676	0.27238	0.01214	1.19500	0.49130
41	10.000	2.681	0.25707	0.01136	1.20000	0.49225
	NORMAL SHOCK WAVE ENCOUNTERED					
	10.000	0.497	0.11061	0.09342	1.20000	1.14348

(c) Shock at exit output

Figure 8-21 (continued)

These are indicated in Figure 8-22a, which was generated from the outputs of the program executions. For this same nozzle but with simple area change only, the corresponding back pressure ranges are as follows:

subsonic throughout	$99.9 < P_b < 100$;
shock in nozzle	$9.99 < P_b < 99.9$;
shock at nozzle exit	$P_b = 9.99$;
oblique shock at exit	$0.48 < P_b < 9.99$;
shock-free flow	$P_b = 0.48$;
outside expansion wave	$P_b < 0.48$.

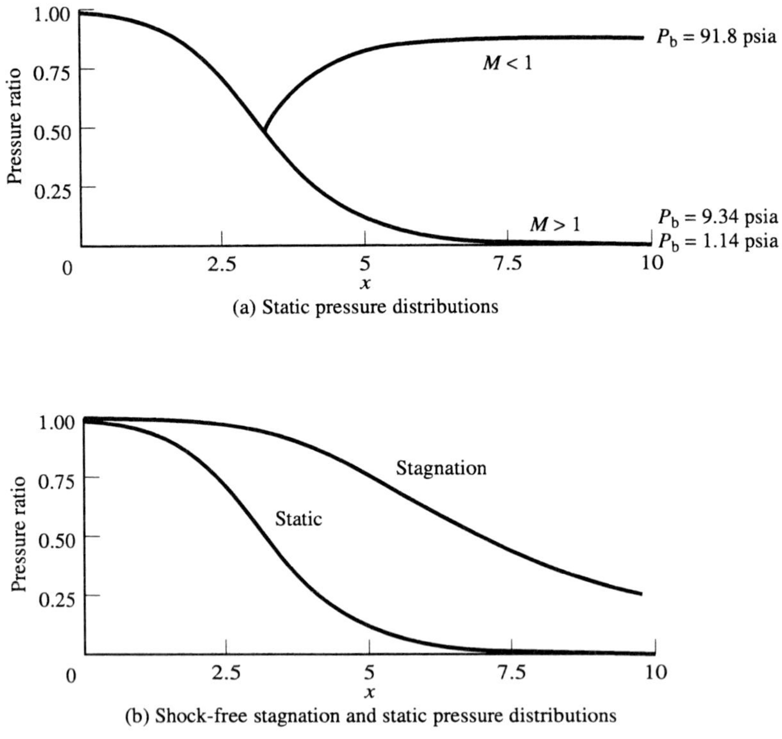

Figure 8-22 Pressure Distributions for Example 8-7

The driving potentials other than area change thus have a noticeable effect on the nozzle behavior with respect to back pressure. One of the most significant effects is the degradation of the stagnation pressure. Figure 8-22b shows the pressure distribution for supersonic shock-free nozzle flow for this problem. The loss of stagnation pressure is pronounced. An important point to note is that once the nozzle behavior is characterized and the nonsimple analysis is available, the added complexity of multiple driving potentials can be readily handled.

8.6 CONCLUSION

This chapter has examined many aspects of nonsimple flows and has provided and discussed three software elements useful in solving nonsimple flows. Additionally, isothermal flow was developed and examined. The three software elements—COMPINT1, COMPINT2, and COMPINT3—illustrate virtually every procedure needed for the solution of generalized one-dimensional flow problems. All three elements were developed with some generality, so that different problems can be worked without significant changes in logic.

This chapter ends the coverage of one-dimensional and quasi–one-dimensional flows. In succeeding chapters we shall be developing topics applicable for multidimensional flows. Nonetheless, the basic knowledge established in the first eight chapters of this book is a necessary prerequisite if multidimensional compressible flow techniques are to be understood and mastered.

CHAPTER 8 REVIEW QUESTIONS

1. What is a simple flow?
2. Why is isothermal flow in a constant-area duct not a simple flow?
3. What is the relationship between static pressure and Mach number in an isothermal flow?
4. What are some unusual behavioral characteristics about isothermal flow (static T = constant)?
5. Select the correct answer for isothermal flow in a long, constant-area duct.
   (a) The flow chokes at $M = 1$, and it is possible.
   (b) The flow chokes at $M = 1/\sqrt{\gamma}$, and it is possible.
   (c) The flow chokes at $M = 1$, and it is not possible.
   (d) The flow chokes at $M = 1/\sqrt{\gamma}$, and it is not possible.
6. For nonsimple flows, must $M = 1$ always occur at the throat? Explain your answer.
7. What are the integral relations?
8. What simplification does the use of the integral relations permit in generalized flow solutions?
9. Why cannot the generalized flow differential equation for Mach number be integrated through a sonic point?
10. Why is $G(x)$ taken to be zero at a sonic point?
11. Explain the strategy involved in determining the one-dimensional Mach number and properties distributions for a converging–diverging nozzle possessing a sonic point.
12. How can the results of the example problems in this chapter be used to justify the concept implicit in Chapter 6 that area change is more important than the other driving potentials in nozzle problems?

CHAPTER PROBLEMS

1. Write a differential equation for Mach number if only area change and mass flow rate change are driving potentials.
2. In a converging–diverging nozzle flow with heat rejection will the sonic point be upstream of, at, or downstream of the minimum area? Justify your answer.
3. For choked flow in a converging–diverging nozzle with friction, does the location of $M = 1$ occur upstream of, at, or downstream of the minimum area? Why?
4. In a converging–diverging nozzle flow with heat addition will the sonic point be upstream of, at, or downstream of the minimum area? Justify your answer.
5. Air initially at 25° C and 1,000,000 Pa is piped through an underground pipeline. The initial Mach number is 0.05. What is the exit Mach number if the actual duct $4fL/D$ is 23?
6. To maintain a constant Mach number (subsonic) in a constant-area duct with friction and heat transfer, should heat be added or removed? What if the flow is supersonic?
7. A piping arrangement for natural gas is illustrated in Figure 8-23. The distance between pumping stations is specified so as to reduce the pressure in the 50-cm diameter pipe from 500 kPa at the exit of a compressor to 150 kPa at the inlet of the next compressor. If the natural gas is assumed to flow isothermally at 300 K and if the pipe possesses a Fanning friction factor of 0.004, estimate the distance between compressors. If the compression process is isentropic, what must be the power output of the compressors? If the flow is cooled after the compression process, what is the required heat rejection rate?

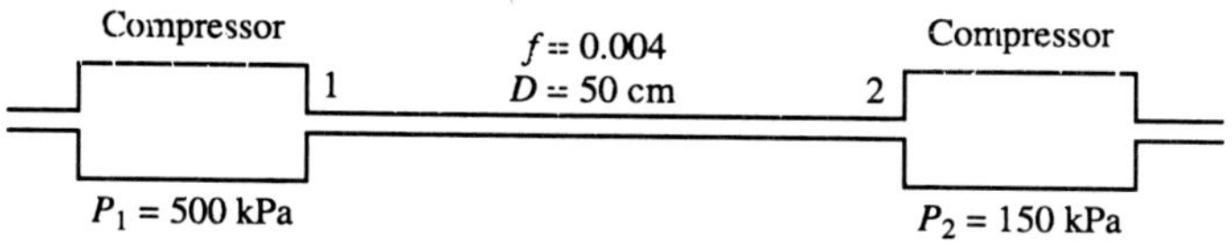

Figure 8-23

8. The gaseous oxygen feed system illustrated in Figure 8-24 is composed of a compressor and an uninsulated 3-cm diameter pipe that is 1000-m long. At the inlet to the pipe segment, the oxygen is at 750 kPa and 300 K. If the pressure at the exit of the pipe is 101.3 kPa, find the system mass flow rate and the power that the compressor delivers to the oxygen.

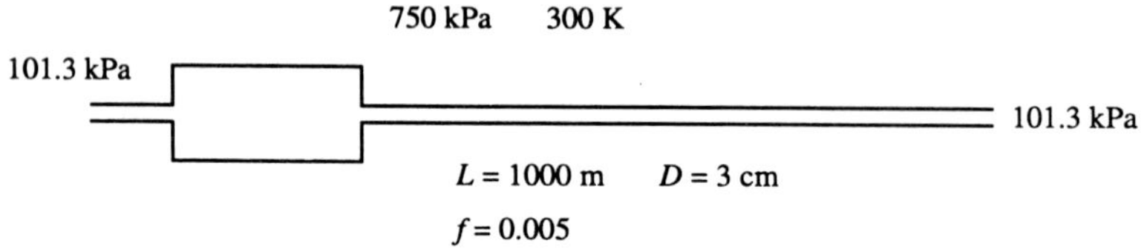

Figure 8-24

9. Nitrogen flows in an insulated tube with a circular cross section. The conditions at the inlet to the tube are $M = 0.25$, $T_1 = 300$ K, and $P_1 = 200$ kPa. The tube cross-sectional area is to be changed so that the static temperature of the stream remains constant. The tube L/D_1 ratio is 125, and the Fanning friction factor can be treated as a constant

with a value of 0.005. Specify the exit Mach number, D_2/D_1, and the exit pressure. What is $D(x)$?

10. Methane flows isothermally at 300 K in a smooth pipeline of diameter 0.125 m. The initial Mach number is 0.075, and the initial pressure is 500 kPa. Find the Mach number and pressure at the exit of the 1000-m line. How long could the line be and isothermal flow still be maintained? What would happen if the line were as long as the limiting case?

11. Prove the following: in isothermal flow the maximum heat transfer possible is $RT/2$.

12. Air is transported isothermally at 300 K in a buried pipe. At the inlet to the pipe the pressure is 200 kPa and the Mach number is 0.1. At the exit of the system the Mach number is 0.5. For this arrangement, find the change in stagnation pressure, the change in stagnation temperature, the change in entropy, and the heat transfer per unit mass of air. If the pipe is commercial steel of nominal diameter 0.15 m, estimate its length.

13. Air flows isothermally at 310 K through a pipe 200 m in length with a 0.25-m diameter. The inlet pressure is 400 kPa, and over the length of the pipe the pressure loss is 200 kPa. If the friction factor is estimated to be 0.003, determine the flow rate of the air through the pipe.

14. A converging–diverging nozzle of exit-to-throat area ratio 3 is followed by a constant-diameter duct. The nozzle is fed from a reservoir with stagnation conditions of 750 kPa and 400 K. The converging–diverging nozzle segment can be considered isentropic, but the constant-diameter duct segment is taken to be isothermal. The duct discharges into a back pressure of 101 kPa, and a normal shock is located at the inlet to the constant-area segment. For these conditions, find the exit Mach number and the heat transfer rate per unit mass required to sustain the isothermal flow.

15. Methane enters a constant-area duct at $M = 0.15$, $P = 30$ psia, and $T = 550\,R$. The duct diameter is 3 inches, and the friction factor is estimated to be 0.005. If the duct is 100 ft long, estimate the exit conditions if (a) the flow is isothermal and (b) the flow is adiabatic (Fanno). What does this say about the distinction between Fanno flow and isothermal flow at low subsonic Mach numbers?

16. A constant-area duct with a friction factor of f possesses the capability to be heated or cooled. If constant Mach number flow is to be maintained in subsonic flow, will heat need to be added or rejected? What about for supersonic flow?

17. Air flows at 1.5 lbm/s in a variable area duct with uniform heat addition of 1,000,000 Btu/hr. At the inlet to the duct, the cross-sectional area is 0.1 ft^2 and the air has static properties of 15 psia and 500 R. Devise a diameter variation, $D(x)$, that will result in a constant static pressure heat addition. Friction may be neglected.

18. Heat is added to a gas flowing frictionlessly through a pipe of varying cross-sectional area. The static pressure is to be maintained at a constant rate.

(a) Demonstrate that the relationship between the area and the stagnation temperature is

$$\frac{A}{A^*} = \frac{\gamma + 1}{2}\frac{T_o}{T_o^*} - \frac{\gamma - 1}{2}.$$

What are the meanings of A^* and T_o^*?

(b) Prove that the velocity along the pipe is constant.

(c) Verify that the total heat interaction is

$$\frac{\gamma}{\gamma - 1}PV(A_2 - A_1).$$

19. Air enters a linearly converging circular duct at an initial Mach number of 0.4. If the initial diameter is 1 ft and the Fanning friction factor is 0.006, determine the Mach number distribution as the Mach number increases to unity.

20. What must be the discharge (output) pressure of a pipeline compressor to deliver 1 kg/s over a 3-km length of a 0.1-m diameter line with an average friction factor of 0.005? The pressure at the end of the segment is 100 kPa. The temperature of the soil surrounding the buried pipeline is 285 K.

PROJECTS

1. Specify the dimensions of a supersonic inlet that must handle 10 kg/s of air at a speed of Mach 3 at an altitude of 5000 m. The air must be delivered to the engine at a Mach number of 0.4. Length, width, and height are needed if you choose a two-dimensional inlet, and length and maximum radius are needed if you choose an axisymmetric inlet. Desirable characteristics are low stagnation pressure loss and low cross-sectional area. State carefully the assumptions made in the design process, and justify your final design.

2. Air at $M = 0.8$, $T_o = 1000\ R$, and $P_o = 140$ psia enters a circular duct 6 in in diameter. Air at high pressure and 500 R is available from another source. A design is required to provide air at $M = 0.2$ and $T_o = 800\ R$. Using the generalized one-dimensional compressible flow program COMPINT1, devise a system that will accomplish the requirements. Be sure to state and defend all assumptions.

3. Use the generalized flow analysis discussed in this chapter to obtain preliminary dimensions of a combustor for a gas turbine engine. A sketch of the combustor configuration is shown in Figure 8-25. Air from the compressor enters the combustor; part of the air is diverted for dilution and the remainder is mixed with fuel for combustion. The diverted air is injected back into the mainstream air to protect the surfaces of the combustor and to reduce to an allowable level the temperature of the air entering the turbine. For this problem we are interested only in modeling the segment between station (1) and station (2). Design specifications and input values are as follows:

   Station 1 (combustion chamber inlet):
   stagnation pressure = 5 bar
   stagnation temperature = 488 K
   mass flow rate = 50 kg/s
   Mach number = 0.3
   Station 2 (combustion chamber outlet) specifications:
   stagnation temperature = 1200 K
   Mach number < 0.35.

   The heating value of the fuel is 42,000 kJ/kg. Use the properties of calorically perfect air ($\gamma = 1.4$) for all calculations. All of the diverted air must eventually be mixed with the mainstream air. The rate of heat addition and the rate of mass addition distributions will both determine the rate of change of T_o.

4. Solve the following simple problems by modifying COMPINT2 or developing a generalized one-dimensional compressible flow computer program using the influence-coefficient method. If a new program is developed, devise the coding in such a fashion that only the initial conditions and the driving potentials need to be changed in order to

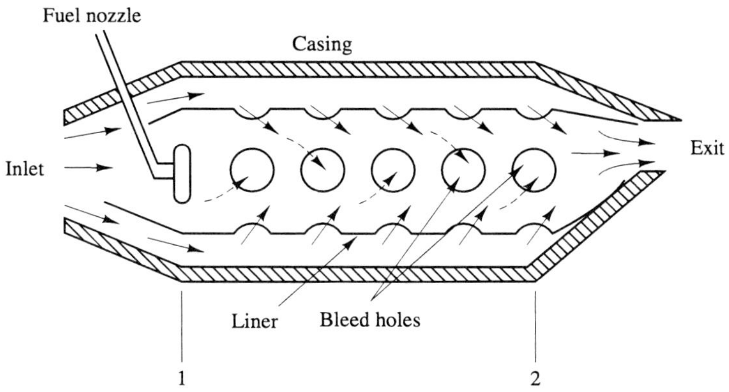

Figure 8-25

work different problems, and use the integral relations to compute the other properties and parameters once the Mach number distribution is known. The program should also possess the capabilities to handle normal shock waves and at least two different area distribution functions in a given problem. For each of the simple flows, plot the Mach number and pressure distributions as functions of length and indicate the COMPQ results at important locations.

(1) Simple area change (See Figure 8-26.)
(2) Simple frictional flow (A Fanno flow problem from Chapter 7)
(3) Simple stagnation temperature change (A Rayleigh flow problem from Chapter 7).

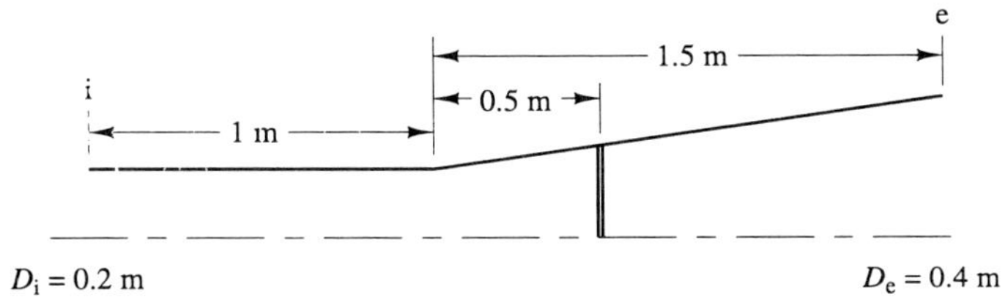

Figure 8-26

5. (a) Solve the following simple problems by modifying COMPINT2 or developing a generalized one-dimensional compressible flow computer program using the influence-coefficient method. If a new program is developed, devise the coding in such a fashion that only the initial conditions and the driving potentials need to be changed in order to work different problems, and use the integral relations to compute the other properties and parameters once the Mach number distribution is known. The program should also possess the capabilities to handle normal shock waves and at least two different area distribution functions in a given problem. For each of the simple flows, plot the Mach number and pressure distributions as functions of length and indicate the COMPQ results at important locations.

(1) Simple area change as shown in Figure 8-27a
(2) Simple frictional flow as in Figure 8-27b
(3) Simple heat addition as in Figure 8-27c.

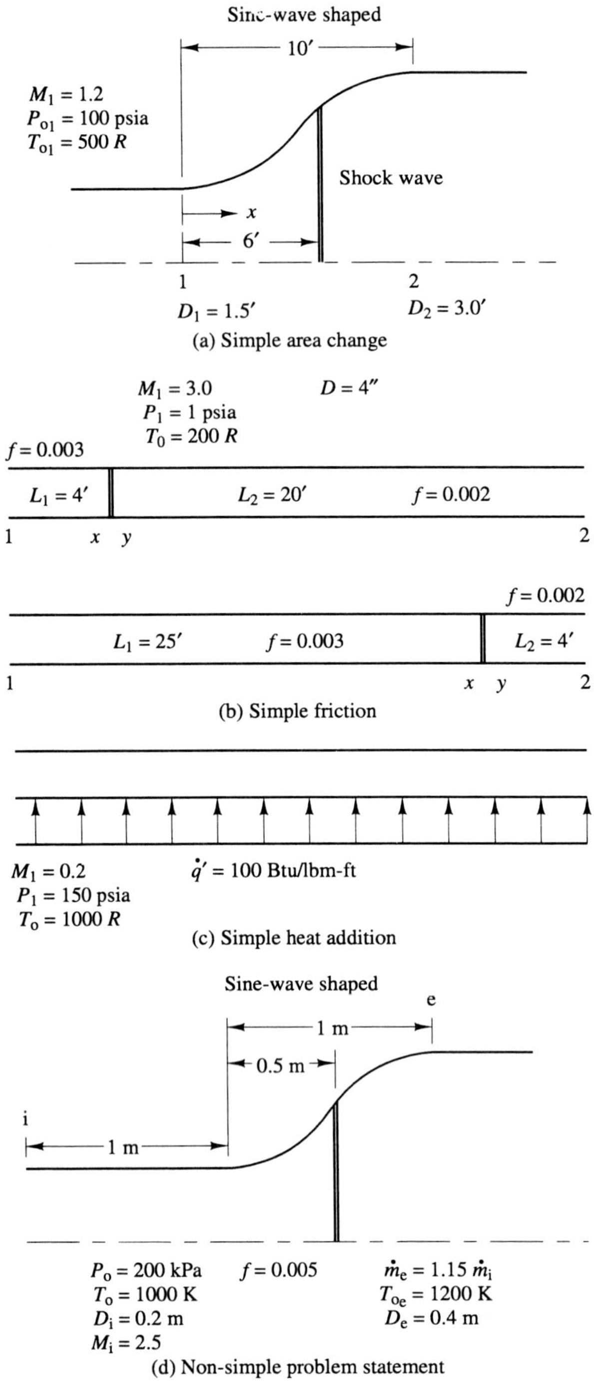

Sine-wave shaped
10′
$M_1 = 1.2$
$P_{o1} = 100$ psia
$T_{o1} = 500\ R$
Shock wave
x
6′
1
2
$D_1 = 1.5'$
$D_2 = 3.0'$
(a) Simple area change
$M_1 = 3.0$
$P_1 = 1$ psia
$T_0 = 200\ R$
$D = 4''$
$f = 0.003$
$L_1 = 4'$
$L_2 = 20'$
$f = 0.002$
1
x y
2
$f = 0.002$
$L_1 = 25'$
$f = 0.003$
$L_2 = 4'$
1
x y
2
(b) Simple friction
$M_1 = 0.2$
$P_1 = 150$ psia
$T_o = 1000\ R$
$\dot{q}' = 100$ Btu/lbm-ft
(c) Simple heat addition
Sine-wave shaped
e
1 m
0.5 m
i
1 m
$P_o = 200$ kPa
$T_o = 1000$ K
$D_i = 0.2$ m
$M_i = 2.5$
$f = 0.005$
$\dot{m}_e = 1.15\ \dot{m}_i$
$T_{oe} = 1200$ K
$D_e = 0.4$ m
(d) Non-simple problem statement

Figure 8-27

(b) Use the program from part (a) to solve the nonsimple problem in Figure 8-27d. Plot the Mach number and pressure distributions as a function of x. Comment on the general utility of the influence-coefficient method for solving nonsimple one-dimensional compressible flow problems.

6. Modify COMPINT2 to solve the nonsimple problem shown in Figure 8-28. Plot the Mach number and pressure distributions as functions of the length. Comment on the general usefulness of the generalized method for handling nonsimple compressible flows.

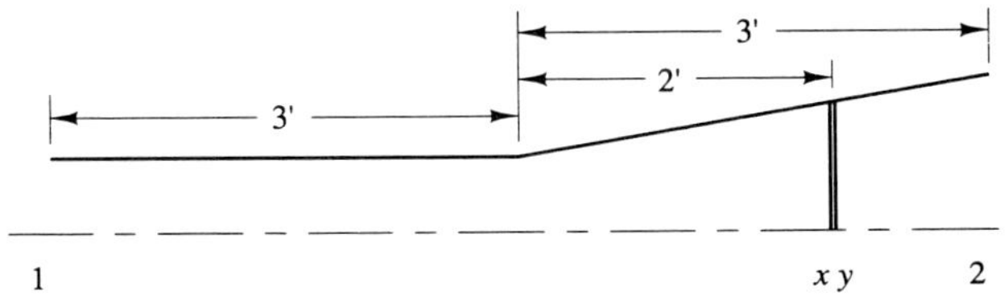

$M_1 = 2.5$ $f = 0.002$ $\Delta T_o = -500\ R$ (uniform change)
$P_1 = 2$ psia $\dot{m}_2 = 1.15\ \dot{m}_1$ (uniform change)
$T_o = 1500\ R$ $D_2 = 9''$
$D_i = 4''$

T_o and $\dot{m}$ changes are uniform along the length of the duct. **Figure 8-28**

7. Develop expressions that relate the characteristics of steady, frictionless, one-dimensional isothermal flow in a variable area passage. Start with the influence coefficients for this flow condition and determine first that the variation in area required is of the form

$$\frac{A_2}{A_1} = \frac{M_1}{M_2} e^{\gamma(M_2^2 - M_1^2)/2}.$$

Develop expressions for the variations in static and stagnation pressure and stagnation temperature. Investigate choking for this flow. Zucrow and Hoffman (1976) present a detailed discussion for this flow.

8. Use COMPINT1, COMPINT2, or COMPINT3, as appropriate, to solve some of the simple flow problems from Chapters 6 and 7. Compare the numerical results of the generalized solutions with the results obtained from using COMPQ.

Part IV

MULTIDIMENSIONAL COMPRESSIBLE FLOWS

The Equations of Inviscid Multidimensional Compressible Flow

9

9.1 INTRODUCTION

One of the most intriguing aspects of compressible flow is the overall utility of the combined assumptions of a perfect gas and an inviscid, one-dimensional flow. Indeed, as we have seen thus far in this textbook, these assumptions permit the quantitative behavior of many internal and external flows to be understood and assessed. In many instances the perfect gas and one-dimensional flow assumptions yield answers that are accurate for engineering purposes. However, logic dictates that many compressible flow applications cannot be one-dimensional and involve a calorically perfect gas. The remainder of our study will be concerned with circumstances in which the one-dimensional assumption is inadequate or in which the behavior of the gas under consideration can no longer be considered calorically perfect. The treatment of viscous, compressible flows is generally beyond the scope of this work. The reader is referred to the excellent texts by Schlichting (1979), Sherman (1990), and White (1991) for details concerning viscous flows.

The purpose of this chapter is to examine the derivation of and understand the principles behind the equations describing the multidimensional flow of an inviscid compressible fluid. Generally we shall be interested in expressing the conservation concepts (mass, momentum, and energy) in the form of differential equations that are valid at a point in a continuum. The integral conservation expressions that we obtained in Chapter 1 are appropriate for multidimensional flows and will be the starting point for these derivations. The differential equations describing the inviscid flow of a compressible fluid can be obtained by applications of the integral conservation expressions to a differential control volume or by applications of the divergence theorem to the integral conservation expressions. Of the two approaches, the former is usually thought to result in a deeper appreciation of the meaning of the individual terms in the differential equations. We shall present both approaches, although not necessarily for the same conservation statements.

9.2 DIFFERENTIAL FORMS OF THE CONSERVATION EQUATIONS

The general approach followed in deriving differential equations for conservation of mass, momentum, and energy will be to apply the integral conservation equations to a differential control volume. Each conservation statement will be examined, and various forms of the associated differential equations considered.

Conservation of Mass

The differential control volume for the conservation of mass derivation is illustrated in Figure 9-1. The density and velocity components are specified at the center of the control volume, and Taylor series expansions are used to estimate the products of the density and velocity components at the faces of the control volume. For example, the product ρV_x at the x-plane control volume face located at $x + dx/2$ is $(\rho V_x + \partial(\rho V_x)/\partial x\, dx/2)$. The integral form of conservation of mass is reproduced here for convenience:

$$\frac{\partial}{\partial t}\int_{\text{cv}} \rho\, d\text{Vol} + \int_{\text{cs}} \rho\overline{V} \cdot \overline{dA} = 0. \tag{1-33}$$

The mass contained within the control volume of Figure 9-1 is $\rho\, dx\, dy\, dz$; hence

$$\frac{\partial}{\partial t}\int_{\text{cv}} \rho\, d\text{Vol} = \frac{\partial}{\partial t}(\rho\, dx\, dy\, dz). \tag{9-1}$$

Adhering to the sign convention of a positive unit outward normal, the mass efflux at the two faces located at $x - dx/2$ and $x + dx/2$ becomes

$$-\left(\rho V_x - \frac{\partial \rho V_x}{\partial x}\frac{dx}{2}\right) dy\, dz + \left(\rho V_x + \frac{\partial \rho V_x}{\partial x}\frac{dx}{2}\right) dy\, dz.$$

Following the same procedure for the y- and z-faces, Eq. 1-33 for the control volume of Figure 9-1 takes the form

$$\frac{\partial}{\partial t}(\rho\, dx\, dy\, dz) + \frac{\partial \rho V_x}{\partial x} dx\, dy\, dz + \frac{\partial \rho V_y}{\partial y} dx\, dy\, dz + \frac{\partial \rho V_z}{\partial z} dx\, dy\, dz = 0. \tag{9-2}$$

Simplifying and noting that $dx\, dy\, dz$ are independent of the time t allow Eq. (9-2) to be written as

$$\frac{\partial \rho}{\partial t} + \frac{\partial}{\partial x}(\rho V_x) + \frac{\partial}{\partial y}(\rho V_y) + \frac{\partial}{\partial z}(\rho V_z) = 0. \tag{9-3}$$

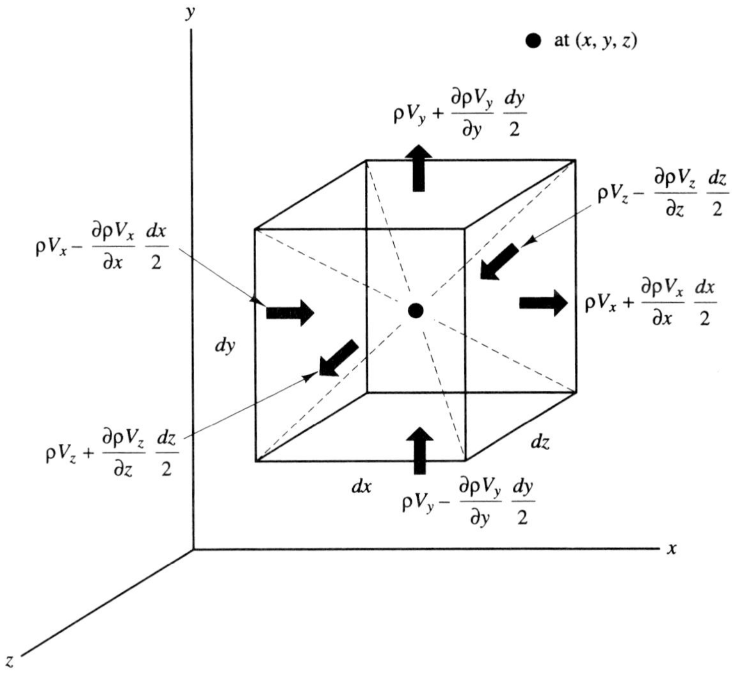

Figure 9-1 Differential Control Volume for Mass Conservation Derivation

Equation 9-3 is the differential equation representing conservation of mass for a compressible fluid. The del operator is

$$\nabla = \frac{\partial}{\partial x}i + \frac{\partial}{\partial y}j + \frac{\partial}{\partial z}k. \tag{9-4}$$

Recognizing that $\nabla \cdot \rho\overline{V} = \partial\rho V_x/\partial x + \partial\rho V_y/\partial y + \partial\rho V_z/\partial z$ lets conservation of mass be expressed as

$$\frac{\partial\rho}{\partial t} + \nabla \cdot \rho\overline{V} = 0. \tag{9-5}$$

This form of conservation of mass is often called the *divergence form* since the formulation contains the divergence of a vector. Alternately, Eq. 9-5 is also referred to as the *conservation form* since it was obtained directly from the integral conservation expression.

If the flow is steady, then $\partial\rho/\partial t = 0$, and the equation reduces to

$$\nabla \cdot \rho\overline{V} = 0. \tag{9-6a}$$

An incompressible fluid possesses constant density, and Eq. 9-6a takes the form

$$\nabla \cdot \overline{V} = 0. \tag{9-6b}$$

Expressions such as Eq. 9-6 are often called *continuity equations*.

Conservation of Momentum

The integral form of conservation of linear momentum is repeated for convenience:

$$\sum \bar{F} = \frac{\partial}{\partial t}\int_{\text{cv}} \rho \bar{V}\, d\text{Vol} + \int_{\text{cs}} \rho \bar{V}\bar{V} \cdot \overline{dA}. \tag{1-35}$$

Equation 1-35 is a vector equation with the x-component of the equation appearing as

$$\sum F_x = \frac{\partial}{\partial t}\int_{\text{cv}} \rho V_x\, d\text{Vol} + \int_{\text{cs}} \rho V_x \bar{V} \cdot \overline{dA}. \tag{9-7}$$

Figure 9-2 represents schematics of a differential control volume. Part (a) illustrates the pressures associated with the control volume and part (b) shows the x-momentum fluxes associated with the control volume. Since we are dealing with an inviscid fluid, pressure represents the only stress component acting on the control volume faces. The pressure is P at the center of the control volume, and its values on the faces are developed from Taylor series expansions, as in Figure 9-1. The body force per unit mass, $\bar{f}$, is broken into its components (f_x, f_y, and f_z). The summation of forces in the x-direction acting on the mass within the control volume then becomes

$$\left(P - \frac{\partial P}{\partial x}\frac{dx}{2}\right) dy\, dz - \left(P + \frac{\partial P}{\partial x}\frac{dx}{2}\right) dy\, dz + \rho f_x\, dx\, dy\, dz. \tag{9-8}$$

The x-direction momentum flux needs some examination. At the faces of the control volume the mass flux is described by $\rho\bar{V} \cdot \overline{dA}$. The y-face at $y + dy/2$, for example, possesses a mass flux of $(\rho V_y + \partial(\rho V_y)/\partial y\, dy/2)\, dx\, dz$. This mass flux coming out of the control volume carries x-momentum across the boundary of the control volume at the $y + dy/2$ face. The flux of x-momentum crossing the control volume boundary of the y-face at $y + dy/2$ then becomes

$$\left(\rho V_y V_x + \frac{\partial(\rho V_y V_x)}{\partial y}\frac{dy}{2}\right) dx\, dz, \tag{9-9}$$

where a Taylor series has once again been employed to define a product value at the face of the differential control volume based on the values at the center. The total x-momentum flux can then be written as

$$\begin{aligned}
&-\left(\rho V_x V_x - \frac{\partial}{\partial x}(\rho V_x V_x)\frac{dx}{2}\right) dy\, dz + \left(\rho V_x V_x + \frac{\partial}{\partial x}(\rho V_x V_x)\frac{dx}{2}\right) dy\, dz \\
&-\left(\rho V_y V_x - \frac{\partial}{\partial y}(\rho V_y V_x)\frac{dy}{2}\right) dx\, dz + \left(\rho V_y V_x + \frac{\partial}{\partial y}(\rho V_y V_x)\frac{dy}{2}\right) dx\, dz \\
&-\left(\rho V_z V_x - \frac{\partial}{\partial z}(\rho V_z V_x)\frac{dz}{2}\right) dx\, dy + \left(\rho V_z V_x + \frac{\partial}{\partial z}(\rho V_z V_x)\frac{dz}{2}\right) dx\, dy,
\end{aligned} \tag{9-10}$$

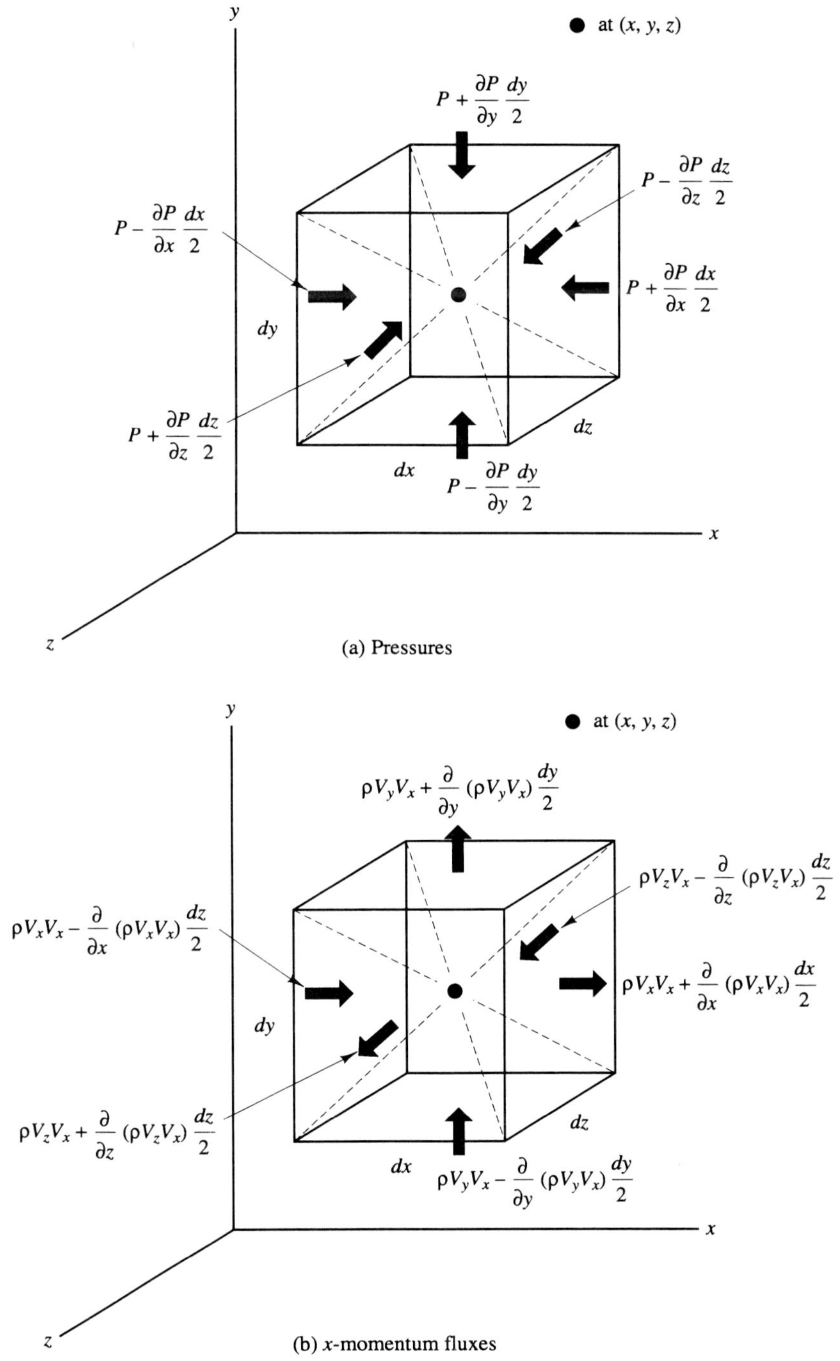

Figure 9-2 Differential Control Volume for Momentum Conservation Derivation

which reduces to

$$\frac{\partial}{\partial x}(\rho V_x V_x)\,dx\,dy\,dz + \frac{\partial}{\partial y}(\rho V_y V_x)\,dx\,dy\,dz + \frac{\partial}{\partial z}(\rho V_z V_x)\,dx\,dy\,dz. \qquad (9\text{-}11)$$

The time rate of change of x-momentum in the control volume is expressed as

$$\frac{\partial}{\partial t}\int_{\text{cv}} \rho V_x\,d\text{Vol} = \frac{\partial}{\partial t}(\rho V_x\,dx\,dy\,dz) = \frac{\partial \rho V_x}{\partial t}\,dx\,dy\,dz. \qquad (9\text{-}12)$$

Conservation of linear momentum in the x-direction for a compressible, inviscid fluid then has the form

$$-\frac{\partial P}{\partial x} + \rho f_x = \frac{\partial}{\partial x}(\rho V_x V_x) + \frac{\partial}{\partial y}(\rho V_y V_x) + \frac{\partial}{\partial z}(\rho V_z V_x) + \frac{\partial \rho V_x}{\partial t}. \qquad (9\text{-}13)$$

Using the ∇ (del) operator, Eq. 9-4, the right-hand side terms of Eq. 9-13 can be written as

$$\nabla \cdot \rho V_x \overline{V} = \frac{\partial}{\partial x}(\rho V_x V_x) + \frac{\partial}{\partial y}(\rho V_x V_y) + \frac{\partial}{\partial z}(\rho V_x V_z). \qquad (9\text{-}14)$$

Thus, the divergence form of the x-momentum equation becomes

$$\frac{\partial \rho V_x}{\partial t} + \nabla \cdot \rho V_x \overline{V} = -\frac{\partial P}{\partial x} + \rho f_x. \qquad (9\text{-}15a)$$

In similar fashion, the divergence forms of the y- and z-momentum equations become

$$\frac{\partial \rho V_y}{\partial t} + \nabla \cdot \rho V_y \overline{V} = -\frac{\partial P}{\partial y} + \rho f_y; \qquad (9\text{-}15b)$$

$$\frac{\partial \rho V_z}{\partial t} + \nabla \cdot \rho V_z \overline{V} = -\frac{\partial P}{\partial z} + \rho f_z. \qquad (9\text{-}15c)$$

Conservation of Energy

Recall from Chapter 1 that the first law of thermodynamics is the basis of the integral form of the conservation of energy statement, which is reproduced here for completeness:

$$\frac{dq}{dt} - \frac{dw_s}{dt} = \frac{\partial}{\partial t}\int_{\text{cv}} \rho\left(u + \frac{V^2}{2}\right) d\text{Vol} + \int_{\text{cs}} \rho\left(h + \frac{V^2}{2}\right)\overline{V} \cdot \overline{dA}. \qquad (1\text{-}40c)$$

The work term, dw_s/dt, in Eq. 1-40c represents only shaft work since the so-called flow work is accounted for in the enthalpy flux term. Figure 9-3 is a representation of the differential control volume used for the energy equation derivation. As with the previous control volume representations, the fluxes at the faces are developed by use of a Taylor series about the center of the control volume. The mass flux at each face is transporting the sum of the static enthalpy and kinetic energy across each face. The rate of increase of the internal energy plus kinetic energy within the

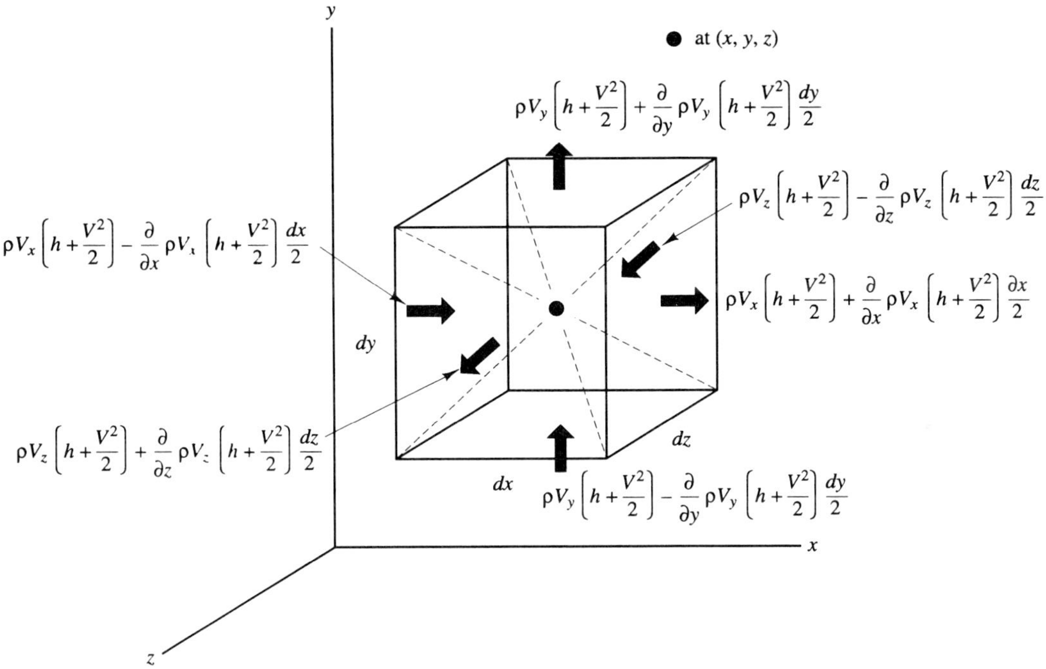

Figure 9-3 Differential Control Volume for Energy Conservation Derivation

differential control volume is $\partial(\rho(u + V^2/2)\,dx\,dy\,dz)/\partial t$. The energy production term per unit mass is $\dot{q}$; thus, $\rho\dot{q}\,dx\,dy\,dz$ represents the energy addition within the volume. Finally the only non-flow work associated with the volume is for $\bar{f}$ and is $\rho\bar{f}\cdot\bar{V}\,dx\,dy\,dz$ for the entire volume. Equation (1-40c) can then be written as

$$\begin{aligned}
&\rho\dot{q}\,dx\,dy\,dz + \rho\bar{f}\cdot\bar{V}\,dx\,dy\,dz \\
&= \frac{\partial}{\partial t}\left[\rho\left(u + \frac{V^2}{2}\right)\right]dx\,dy\,dz - \left[\rho V_x\left(h + \frac{V^2}{2}\right) - \frac{\partial}{\partial x}\rho V_x\left(h + \frac{V^2}{2}\right)\frac{dx}{2}\right]dy\,dz \\
&\quad + \left[\rho V_x\left(h + \frac{V^2}{2}\right) + \frac{\partial}{\partial x}\rho V_x\left(h + \frac{V^2}{2}\right)\frac{dx}{2}\right]dy\,dz \\
&\quad - \left[\rho V_y\left(h + \frac{V^2}{2}\right) - \frac{\partial}{\partial y}\rho V_y\left(h + \frac{V^2}{2}\right)\frac{dy}{2}\right]dx\,dz \\
&\quad + \left[\rho V_y\left(h + \frac{V^2}{2}\right) + \frac{\partial}{\partial y}\rho V_y\left(h + \frac{V^2}{2}\right)\frac{dy}{2}\right]dx\,dz \\
&\quad - \left[\rho V_z\left(h + \frac{V^2}{2}\right) - \frac{\partial}{\partial z}\rho V_z\left(h + \frac{V^2}{2}\right)\frac{dz}{2}\right]dx\,dy \\
&\quad + \left[\rho V_z\left(h + \frac{V^2}{2}\right) + \frac{\partial}{\partial z}\rho V_z\left(h + \frac{V^2}{2}\right)\frac{dz}{2}\right]dx\,dy.
\end{aligned} \tag{9-16}$$

After some algebra a form of the differential energy equation is

$$\rho\dot{q} + \rho\bar{f} \cdot \bar{V} = \frac{\partial}{\partial t}\left[\rho\left(u + \frac{V^2}{2}\right)\right] + \frac{\partial}{\partial x}\left[\rho V_x\left(h + \frac{V^2}{2}\right)\right] + \frac{\partial}{\partial y}\left[\rho V_y\left(h + \frac{V^2}{2}\right)\right] + \frac{\partial}{\partial z}\left[\rho V_z\left(h + \frac{V^2}{2}\right)\right]. \tag{9-17}$$

The del operator can again be used to advantage, and Eq. 9-17 reduces to

$$\frac{\partial}{\partial t}\left[\rho\left(u + \frac{V^2}{2}\right)\right] + \nabla \cdot \left[\rho\bar{V}\left(h + \frac{V^2}{2}\right)\right] = \rho\dot{q} + \rho\bar{f} \cdot \bar{V}. \tag{9-18}$$

Expressing the enthalpy as $u + P/\rho$, Eq. 9-18 can be expanded to yield

$$\frac{\partial}{\partial t}\left[\rho\left(u + \frac{V^2}{2}\right)\right] + \nabla \cdot \left[\rho\bar{V}\left(u + \frac{V^2}{2}\right)\right] = \rho\dot{q} - \nabla \cdot P\bar{V} + \rho\bar{f} \cdot \bar{V}. \tag{9-19}$$

The flow work term, $\nabla \cdot P\bar{V}$, explicitly appears with the result that the flux terms are expressed in terms of the internal energy. Equations 9-18 and 9-19 are often called divergence or conservation forms of the energy equation. None of the equations developed to this point in the chapter has been associated with any equation of state or any assumption of thermal or caloric perfection in the gas.

Divergence Theorem Approach

The divergence form of the differential equations for conservation of mass, momentum, and energy can be obtained directly from the integral forms of the equations by using the divergence theorem. Gauss's theorem states that the integral taken over the volume of the divergence of a vector field is equal to the integral of the vector taken over the closed surface surrounding the vector field; the mathematical statement of the theorem is

$$\int_{cv} \nabla \cdot \bar{B}\, d\text{Vol} = \int_{cs} \bar{B} \cdot \overline{dA}. \tag{9-20a}$$

Virtually any textbook on vector analysis or applied mathematics contains a discussion on Gauss's theorem; Pipes and Harvill (1970) is typical. If scalar rather than vector functions are involved, then

$$\int_{cv} \nabla B\, d\text{Vol} = \int_{cs} B\, \overline{dA}. \tag{9-20b}$$

Consider the integral form of the continuity equation (Eq. 1-33):

$$0 = \frac{\partial}{\partial t}\int_{cv} \rho\, d\text{Vol} + \int_{cs} \rho\bar{V} \cdot \overline{dA}. \tag{1-33}$$

Gauss's theorem, Eq. 9-20a, permits the surface integral to be cast in the terms of the volume integral such that

$$\int_{cs} \rho\bar{V} \cdot \overline{dA} = \int_{cv} \nabla \cdot \rho\bar{V}\, d\text{Vol}. \tag{9-21}$$

Hence,

$$0 = \frac{\partial}{\partial t}\int_{\text{cv}} \rho \, d\text{Vol} + \int_{\text{cv}} \nabla \cdot \rho \overline{V} \, d\text{Vol}, \tag{9-22a}$$

which can be written as

$$0 = \int_{\text{cv}} \left(\frac{\partial \rho}{\partial t} + \nabla \cdot \rho \overline{V} \right) d\text{Vol} \tag{9-22b}$$

for a fixed control volume. Since the fixed control volume shape is arbitrary, the only way the equality can be satisfied is if the integrand is identically equal to zero. Thus,

$$\frac{\partial \rho}{\partial t} + \nabla \cdot \rho \overline{V} = 0, \tag{9-23}$$

which was derived earlier in this section as Eq. 9-4.

The second and last divergence theorem example to be presented will be the x-momentum equation, Eq. 9-6. The forces acting in the x-direction are composed of surface forces, the pressure in an inviscid flow, and volume forces. Pressure is directed inward, and the unit outward normal is directed outward. Then

$$\sum F_x = \int_{\text{cv}} \rho f_x \, d\text{Vol} - \int_{\text{cs}} [P \overline{dA}]_x, \tag{9-24}$$

where $[\ldots]_x$ indicates the x-component. With the use of Eq. 9-24, the x-momentum integral conservation equation becomes

$$\int_{\text{cv}} \rho f_x \, d\text{Vol} - \int_{\text{cs}} [P \overline{dA}]_x = \frac{\partial}{\partial t}\int_{\text{cv}} \rho V_x \, d\text{Vol} + \int_{\text{cs}} \rho V_x \overline{V} \cdot \overline{dA}. \tag{9-25}$$

Applications of Eqs. 9-20 result in

$$\int_{\text{cv}} \rho f_x \, d\text{Vol} - \int_{\text{cv}} [\nabla P \, d\text{Vol}]_x = \frac{\partial}{\partial t}\int_{\text{cv}} \rho V_x \, d\text{Vol} + \int_{\text{cv}} \nabla \cdot \rho V_x \overline{V} \, d\text{Vol} \tag{9-26a}$$

or in

$$\int_{\text{cv}} \left[\rho f_x - \frac{\partial P}{\partial x} - \frac{\partial}{\partial t}(\rho V_x) - \nabla \cdot \rho V_x \overline{V} \right] d\text{Vol} = 0. \tag{9-26b}$$

An arbitrary control volume specification requires that

$$\frac{\partial}{\partial t}(\rho V_x) + \nabla \cdot \rho V_x \overline{V} = -\frac{\partial P}{\partial x} + \rho f_x, \tag{9-27}$$

which is the same as the previously derived Eq. 9-15a.

This same procedure could obviously be applied to the remaining integral conservation expressions, with the same results as in the previous derivations. Use of Gauss's theorem is perhaps a shorter approach to obtaining the divergence forms of the differential equations, but it is a more abstract approach.

9.3 ALTERNATE FORMS OF THE DIFFERENTIAL EQUATIONS

The divergence forms of the differential equations governing an inviscid, compressible flow are Eqs. 9-5, 9-15, and 9-19 for mass, momentum, and energy, respectively. These equations can be expressed in a convenient, compact vector form as

$$\frac{\partial \overline{U}}{\partial t} + \frac{\partial \overline{E}}{\partial x} + \frac{\partial \overline{F}}{\partial y} + \frac{\partial \overline{G}}{\partial z} + \overline{S} = 0, \tag{9-28}$$

where

$$\overline{U} = \begin{bmatrix} \rho \\ \rho V_x \\ \rho V_y \\ \rho V_z \\ \rho(u + V^2/2) \end{bmatrix}; \quad \overline{E} = \begin{bmatrix} \rho V_x \\ \rho V_x V_x + P \\ \rho V_x V_y \\ \rho V_x V_z \\ [\rho(u + V^2/2) + P]V_x \end{bmatrix}; \quad \overline{F} = \begin{bmatrix} \rho V_y \\ \rho V_x V_y \\ \rho V_y V_y + P \\ \rho V_y V_z \\ [\rho(u + V^2/2) + P]V_y \end{bmatrix};$$

$$\overline{G} = \begin{bmatrix} \rho V_z \\ \rho V_x V_z \\ \rho V_y V_z \\ \rho V_z V_z + P \\ [\rho(u + V^2/2) + P]V_z \end{bmatrix}; \quad \overline{S} = \begin{bmatrix} 0 \\ -\rho f_x \\ -\rho f_y \\ -\rho f_z \\ -\rho(\dot{q} + \bar{f} \cdot \overline{V}) \end{bmatrix}. \tag{9-29}$$

In the Eulerian description we consider that any property B is a function of position as well as time; that is

$$B = B(x, y, z, t). \tag{9-30}$$

The differential can then be expressed as

$$dB = \frac{\partial B}{\partial x}dx + \frac{\partial B}{\partial y}dy + \frac{\partial B}{\partial z}dz + \frac{\partial B}{\partial t}dt, \tag{9-31}$$

such that

$$\frac{dB}{dt} = \frac{\partial B}{\partial x}\frac{dx}{dt} + \frac{\partial B}{\partial y}\frac{dy}{dt} + \frac{\partial B}{\partial z}\frac{dz}{dt} + \frac{\partial B}{\partial t}. \tag{9-32}$$

A *streamline* is normally defined as the loci of positions of a fluid particle, and along a streamline dx/dt, dy/dt, and dz/dt become, respectively, V_x, V_y, and V_z. Hence, the time rate of change of the property B along a streamline becomes

$$\frac{DB}{Dt} = \frac{dB}{dt} = \underbrace{V_x\frac{\partial B}{\partial x} + V_y\frac{\partial B}{\partial y} + V_z\frac{\partial B}{\partial z}}_{\text{convective}} + \underbrace{\frac{\partial B}{\partial t}}_{\text{local}}. \tag{9-33}$$

Equation 9-33 is called the *substantial*, *material*, or *particle derivative*, each connoting the special relationship that the motion of a fluid element along a streamline implies. In order to emphasize the special nature of the substantial derivative, it is often written in operator form as D/Dt. As is indicated at the bottom of Eq. 9-33, the substantial derivative is composed of two parts: (1) the *local*, due to unsteadiness at the point in question, and (2) the *convective*, due to motion of the fluid element along a streamline. The most common form of Eq. 9-33 utilizes the del operator and appears as

$$\frac{DB}{Dt} = \frac{\partial B}{\partial t} + (\overline{V} \cdot \nabla)B. \tag{9-34}$$

If B is the velocity, $\overline{V}$, for example, the acceleration can be expressed as

$$\overline{a} = \frac{D\overline{V}}{Dt} = \underbrace{\frac{\partial \overline{V}}{\partial t}}_{\text{local}} + \underbrace{(\overline{V} \cdot \nabla)\overline{V}}_{\text{convective}}, \tag{9-35}$$

with the convective and local accelerations as indicated.

The divergence or conservative form of the differential equations was listed previously. An alternative form of the equations, using the substantial derivative and called the nonconservative form of the differential equations, appears frequently in the literature. Consider the continuity equation, Eq. 9-4, with the argument of the divergence expanded to yield

$$\frac{\partial \rho}{\partial t} + (\overline{V} \cdot \nabla)\rho + \rho\nabla \cdot \overline{V} = 0. \tag{9-36}$$

Since $D\rho/Dt = \partial\rho/\partial t + (\overline{V} \cdot \nabla)\rho$, however, Eq. 9-36 can be written as

$$\frac{D\rho}{Dt} + \rho\nabla \cdot \overline{V} = 0. \tag{9-37}$$

The x-momentum equation, Eq. 9-15a, can be expanded in a similar fashion as continuity with the following result:

$$\rho\frac{\partial V_x}{\partial t} + V_x\frac{\partial \rho}{\partial t} + V_x\nabla \cdot \rho\overline{V} + (\rho\overline{V} \cdot \nabla)V_x = -\frac{\partial P}{\partial x} + \rho f_x. \tag{9-38}$$

Grouping results in

$$V_x\left(\frac{\partial \rho}{\partial t} + \nabla \cdot \rho\overline{V}\right) + \rho\left(\frac{\partial V_x}{\partial t} + (\overline{V} \cdot \nabla)_x\right) = -\frac{\partial P}{\partial x} + \rho f_x. \tag{9-39}$$

But the first term on the left-hand side is identically zero from the continuity equation, and the remaining term is simply the substantial derivative of V_x. Thus,

$$\rho\frac{DV_x}{Dt} = -\frac{\partial P}{\partial x} + \rho f_x. \tag{9-40a}$$

The y- and z-component equations follow:

$$\rho\frac{DV_y}{Dt} = -\frac{\partial P}{\partial y} + \rho f_y; \tag{9-40b}$$

$$\rho\frac{DV_z}{Dt} = -\frac{\partial P}{\partial z} + \rho f_z. \tag{9-40c}$$

Equations 9-40 can be combined using vector notation to yield

$$\rho\frac{D\bar{V}}{Dt} = -\nabla P + \rho\bar{f}. \tag{9-41}$$

Perhaps the most important equation for an inviscid, compressible flow, Eq. 9-41 is known as *Euler's equation*. We shall use it on numerous occasions.

If performed on the energy equation, the same sequence leads to the nonconservative form of the energy equation:

$$\rho\frac{D(u + V^2/2)}{Dt} = -\nabla \cdot P\bar{V} + \rho\dot{q} + \rho\bar{f} \cdot \bar{V}. \tag{9-42}$$

Equation 9-42 is cast in the form of the internal energy plus the kinetic energy. If this form is expanded and combined with Euler's equation, an expression involving only the substantial derivative of the internal energy results:

$$\rho\frac{Du}{Dt} = -P\nabla \cdot \bar{V} + \rho\dot{q}. \tag{9-43}$$

Since $h = u + P/\rho$,

$$\rho\frac{Dh}{Dt} = \frac{DP}{Dt} + \rho\dot{q}. \tag{9-44}$$

Adding Eq. 9-44 to and subtracting Eq. 9-43 from Eq. 9-42 lead to the following energy equation involving stagnation enthalpy:

$$\rho\frac{Dh_o}{Dt} = \frac{\partial P}{\partial t} + \rho\dot{q} + \rho(\bar{f} \cdot \bar{V}). \tag{9-45}$$

Equation 9-45 is one of the more revealing forms of the energy equation since it tells us that the enthalpy of a fluid element in an inviscid flow is constant only if

1. No body forces are present.
2. No heat transfer is present.
3. The flow is steady.

For such conditions, $Dh_o/Dt = 0$ and h_o = constant throughout the flow field.

A classic equation from Crocco (1937) provides an excellent platform from which to discuss classification of compressible flows. We shall derive Crocco's equation and infer from it some important features of compressible flows. Euler's equation in the absence of body forces can be written as

$$\rho\frac{D\bar{V}}{Dt} = \rho\frac{\partial\bar{V}}{\partial t} + \rho(\bar{V} \cdot \nabla)\bar{V} = -\nabla P. \tag{9-46}$$

Recall from Chapter 1 the combined first- and second-laws expression

$$Tds = dh - \frac{dP}{\rho}. \tag{1-23b}$$

For multidimensional flows an appropriate form for this statement is

$$T\nabla s = \nabla h - \frac{\nabla P}{\rho}. \tag{9-47}$$

Stagnation enthalpy is defined as $h_o = h + V^2/2$, so that $h = h_o - V^2/2$. Then the gradient of the static enthalpy becomes

$$\nabla h = \nabla h_o - \nabla\left(\frac{V^2}{2}\right). \tag{9-48}$$

When Eq. 9-48 is substituted for enthalpy in Eq. 9-47, and when the gradient of the pressure, ∇P, is substituted from Eq. 9-46, the following results:

$$T\nabla s = \nabla h_o - \nabla\left(\frac{V^2}{2}\right) + \frac{\partial \overline{V}}{\partial t} + (\overline{V} \cdot \nabla)\overline{V}. \tag{9-49}$$

This doesn't appear to be useful, but the vector identity

$$\nabla\left(\frac{V^2}{2}\right) = (\overline{V} \cdot \nabla)\overline{V} + \overline{V} \times (\nabla \times \overline{V}) \tag{9-50}$$

allows Eq. 9-50 to be cast as

$$T\nabla s = \nabla h_o + \frac{\partial \overline{V}}{\partial t} - \overline{V} \times \nabla \times \overline{V}. \tag{9-51}$$

Since Euler's equation was derived under the assumption of inviscid flow, Eq. 9-51 is restricted to inviscid flows. If the condition of steady flow is invoked for Eq. 9-51, then the following, which is Crocco's equation, is obtained:

$$T\nabla s = \nabla h_o - \overline{V} \times \nabla \times \overline{V}. \tag{9-52}$$

The curl of the velocity, $\nabla \times \overline{V}$, is called the *vorticity* and plays a significant role in fluid mechanics. Flows for which $\nabla \times \overline{V} = 0$ are called *irrotational*; flows for which $\nabla \times \overline{V} \neq 0$ are called *rotational*. In physical terms, the vorticity is equal to twice the angular velocity of a fluid element, hence the involvement of the word "rotational." Virtually any fluid mechanics textbook presents a derivation and an elaboration of the concept; Fox and McDonald (1992) and White (1986) are good examples. Crocco's equation relates the gradients of entropy and stagnation enthalpy to the vorticity. It is pertinent to consider further the implications of irrotational and rotational flows.

In Chapters 2 and 3 we concluded that the stagnation enthalpy was constant across a stationary normal or oblique shock wave; hence $\nabla h_o = 0$. Moreover, we found that the change in entropy across a shock wave depended only on the ratio of specific heats, the upstream Mach number, and the shock wave angle (with respect to the freestream). Consider, as in Figure 9-4, an oblique attached shock wave and a curved detached shock wave; both might be in conjunction with a wedge or a cone. Since the oblique shock wave is not curved, the shock angle is constant and the entropy change across the shock wave is the same at any location on the shock wave. The value of the entropy behind the shock wave is thus uniform, although larger than

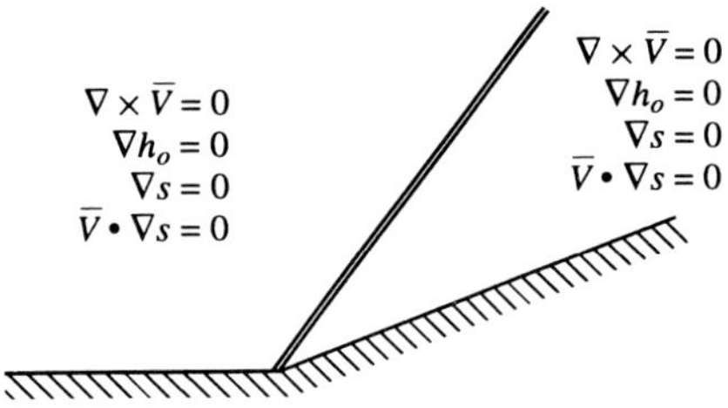

(a) Oblique shock wave

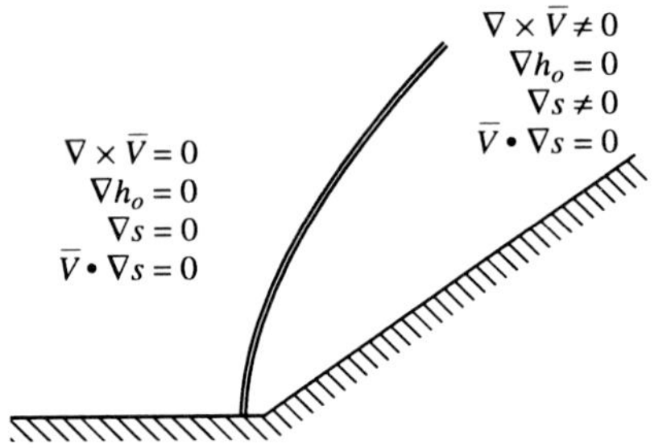

(b) Detached shock wave

Figure 9-4 Flow Field Classifications

the entropy upstream of the shock wave. Across the detached, curved shock wave, the value of the entropy is dependent on the local shock angle and is different at various locations behind the shock wave. For the attached shock wave the gradient of the entropy is zero. From Crocco's equation, the flow field behind the shock wave is irrotational. However, the gradient of the entropy behind the curved, detached shock wave is not zero; therefore, the flow behind a curved shock wave must be rotational according to Crocco's equation. A generalization of this is that inviscid flows involving curved shock waves are rotational.

Another significant class of rotational flows is viscous flows. Boundary-layer flows, for example, are rotational. Unlike inviscid flows, viscous flows possess shearing stresses, and we can envision these shearing stresses as imparting rotation to the fluid elements.

A version of Crocco's equation valid along a streamline can be obtained by taking the dot product of Crocco's equation. Since $\bar{V} \cdot \bar{V} \times (\nabla \times \bar{V}) = 0$, this version becomes

$$T\bar{V} \cdot \nabla s = \bar{V} \cdot \nabla h_o. \tag{9-53}$$

Many general flow classifications are made from Eqs. 9-52 and 9-53, two versions of Crocco's equation. The following are defined on the basis of Crocco's two relations:

irrotational flow	$\nabla \times \bar{V} = 0$
isoenergetic flow	$\nabla h_o = 0$
homentropic flow	$\nabla s = 0$
isentropic flow	$\bar{V} \cdot \nabla s = 0$

We have already briefly examined irrotational flow. *Isoenergetic flow* is adiabatic in the steady state. In *homentropic flow* the gradient of the entropy vanishes, and thus entropy is uniform throughout the flow. Isentropic flow is defined as $\overline{V} \cdot \nabla s = 0$, which can be interpreted as entropy being constant along any given streamline. From Eqs. 9-52 and 9-53 we see that flow behind a curved shock wave is rotational and isentropic, but not homentropic. Flow behind an oblique shock wave is irrotational, isentropic, and homentropic.

Euler's equation in the absence of body forces (Eq. 9-46) for steady flow can be written as

$$\rho(\overline{V} \cdot \nabla)\overline{V} = -\nabla P. \tag{9-54}$$

The convective terms, $(\overline{V} \cdot \nabla)\overline{V}$, can be replaced by using the vector identity given in Eq. 9-50 so that Eq. 9-54 appears as

$$\nabla\left(\frac{V^2}{2}\right) - \overline{V} \times \nabla \times \overline{V} = -\frac{\nabla P}{\rho}. \tag{9-55}$$

If the flow is irrotational, $\overline{V} \times \nabla \times \overline{V} = 0$ and Eq. 9-55 simplifies to

$$\nabla\left(\frac{V^2}{2}\right) + \frac{\nabla P}{\rho} = 0. \tag{9-56}$$

The value of the terms in Eq. 9-56 in any arbitrary direction can be found by taking the dot product of the equation with an arbitrary differential vector $d\bar{r} = dx\,\mathbf{i} + dy\,\mathbf{j} + dz\,\mathbf{k}$, where $\mathbf{i}$, $\mathbf{j}$, and $\mathbf{k}$ are unit vectors in the x, y, and z directions. Then

$$\nabla\left(\frac{V^2}{2}\right) \cdot d\bar{r} + \frac{\nabla P}{\rho} \cdot d\bar{r} = 0. \tag{9-57}$$

But

$$\begin{aligned} \nabla\left(\frac{V^2}{2}\right) \cdot d\bar{r} &= \frac{\partial V^2/2}{\partial x}dx + \frac{\partial V^2/2}{\partial y}dy + \frac{\partial V^2/2}{\partial z}dz \\ &= d\left(\frac{V^2}{2}\right) \\ &= V\,dV, \end{aligned} \tag{9-58}$$

and

$$\begin{aligned} \frac{\nabla P}{\rho} \cdot d\bar{r} &= \frac{1}{\rho}\frac{\partial P}{\partial x}dx + \frac{1}{\rho}\frac{\partial P}{\partial y}dy + \frac{1}{\rho}\frac{\partial P}{\partial z}dz \\ &= \frac{dP}{\rho}. \end{aligned} \tag{9-59}$$

The final forms of Eqs. 9-58 and 9-59 are achieved by recognizing the exact differential in both expressions. Equation 9-57 then reduces to

$$V\,dV + \frac{dP}{\rho} = 0. \tag{9-60}$$

Equation 9-60 is Euler's equation for an irrotational flow and is valid throughout a steady-state, irrotational flow field with no body forces.

9.4 THE POTENTIAL EQUATION

Irrotationality—the absence of vorticity—has another very important consequence. Recall from vector analysis that if $\nabla \times B = 0$, then B can always be represented as the gradient of a potential, $\nabla\phi$. Thus, for irrotational flow a velocity potential must exist and takes the form

$$\bar{V} = \nabla\phi. \tag{9-61a}$$

The velocity components are expressed as

$$V_x = \frac{\partial\phi}{\partial x} = \phi_x, \qquad V_y = \frac{\partial\phi}{\partial y} = \phi_y, \qquad V_z = \frac{\partial\phi}{\partial z} = \phi_z \tag{9-61b}$$

in Cartesian coordinates, and as

$$V_r = \phi_r, \qquad V_\theta = \frac{1}{r}\phi_\theta, \qquad V_z = \phi_z \tag{9-61c}$$

in cylindrical coordinates, where partial differentiation is indicated by a subscript; for example, $\phi_x = \partial\phi/\partial x$. Continuity appears for steady-state flow as

$$\frac{\partial}{\partial x}(\rho V_x) + \frac{\partial}{\partial y}(\rho V_y) = \frac{\partial}{\partial z}(\rho V_z) = 0. \tag{9-62}$$

Substituting the results of Eq. 9-61b yields

$$\frac{\partial}{\partial x}(\rho\phi_x) + \frac{\partial}{\partial y}(\rho\phi_y) + \frac{\partial}{\partial z}(\rho\phi_z) = 0. \tag{9-63}$$

Carrying out the indicated differentiations and collecting terms results in

$$\rho(\phi_{xx} + \phi_{yy} + \phi_{zz}) + \phi_x\frac{\partial\rho}{\partial x} + \phi_y\frac{\partial\rho}{\partial y} + \phi_z\frac{\partial\rho}{\partial z} = 0. \tag{9-64}$$

Since a potential flow is both irrotational and isoenergetic, it is also homentropic and throughout the flow field $dP/d\rho$ must be equal to the square of local speed of sound, a^2, since $(\partial P/\partial\rho)_s = a^2$. Then Euler's equation for an irrotational flow requires that

$$a^2\,d\rho = dP = -\rho V\,dV = -\frac{\rho}{2}d(V^2), \tag{9-65a}$$

which can be expressed as

$$d\rho = -\frac{\rho}{a^2}d\left(\frac{\phi_x^2 + \phi_y^2 + \phi_z^2}{2}\right), \tag{9-65b}$$

since $V^2 = V_x^2 + V_y^2 + V_z^2 = \phi_x^2 + \phi_y^2 + \phi_z^2$. Effectively, this couples the momentum equation to the continuity equation in the derivation and thus implicitly satisfies conservation of momentum. Each of the spatial density derivative terms in Eq. 9-64 can be cast in terms of the velocity potential derivatives as follows:

$$\begin{aligned}
\frac{\partial\rho}{\partial x} &= -\frac{\rho}{a^2}(\phi_x\,\phi_{xx} + \phi_y\,\phi_{yx} + \phi_x\,\phi_{zx});\\
\frac{\partial\rho}{\partial y} &= -\frac{\rho}{a^2}(\phi_x\,\phi_{xy} + \phi_y\,\phi_{yy} + \phi_z\,\phi_{zy});\\
\frac{\partial\rho}{\partial z} &= -\frac{\rho}{a^2}(\phi_x\,\phi_{xz} + \phi_y\,\phi_{yz} + \phi_z\,\phi_{zz}).
\end{aligned} \tag{9-66}$$

An intermediate form of the continuity equation is obtained by using Eqs. 9-66 in Eq. 9-64:

$$\begin{aligned}
&\left(1 - \frac{\phi_x^2}{a^2}\right)\phi_{xx} + \left(1 - \frac{\phi_y^2}{a^2}\right)\phi_{yy} + \left(1 - \frac{\phi_z^2}{a^2}\right)\phi_{zz}\\
&\qquad - \frac{2\phi_x\,\phi_y}{a^2}\phi_{xy} - \frac{2\phi_x\,\phi_z}{a^2}\phi_{xz} - \frac{2\phi_y\,\phi_z}{a^2}\phi_{yz} = 0.
\end{aligned} \tag{9-67}$$

But the utility of this form is limited because the square of the local speed of sound, a^2, appears in the equation. The local speed of sound depends on the local Mach number, and our goal is to express the continuity equation completely in terms of the velocity potential and its derivatives. To accomplish this, recall the definition of stagnation enthalpy,

$$c_p T + \frac{V^2}{2} = c_p T_o, \tag{1-64}$$

which is equivalent to

$$\frac{\gamma RT}{\gamma - 1} + \frac{V^2}{2} = \frac{\gamma RT_o}{\gamma - 1}. \tag{9-68}$$

Recognizing that $a^2 = \gamma RT$ and $a_o^2 = \gamma RT_o$, a convenient form of Eq. 9-68 is

$$\frac{a^2}{\gamma - 1} + \frac{V^2}{2} = \frac{a_o^2}{\gamma - 1}. \tag{9-69}$$

The square of the local speed of sound takes the following form:

$$a^2 = a_o^2 - \frac{\gamma - 1}{2}V^2 = a_o^2 - \frac{\gamma - 1}{2}(\phi_x^2 + \phi_y^2 + \phi_z^2). \tag{9-70}$$

Equation 9-70 provides an expression for the local speed of sound in terms of the stagnation speed of sound and derivatives of the velocity potential. It also couples the energy equation into the derivation via the constant stagnation enthalpy statement. The combination of Eqs. 9-67 and 9-70 allows conservation of mass, momentum, and energy to be expressed completely in terms of a single, nonlinear, partial differential equation involving the derivatives of the velocity potential and a constant (the stagnation speed of sound). The usual boundary conditions associated with the potential equation are maintenance of a finite velocity far away from the surface configuration under consideration and tangency of the flow on the surface of the configuration. Once the velocity potential is known, the whole flow field is specified since all local velocities and properties can be obtained.

The full potential equation, Eq. 9-67 combined with Eq. 9-70, is nonlinear and not generally amenable to closed-form general solutions. Indeed, except for some special cases of fairly limited utility, the most effective utilization of the full potential equation has been made via numerical solutions. An examination of the computational fluid dynamics literature shows both an impressive number of papers and articles dealing with the full potential equation and an unmistakable trend toward more sophisticated applications. Anderson, Tannehill, and Pletcher (1984) present a good introduction to numerical solution of the full potential equation.

The system composed of Eqs. 9-67 and 9-70 is a nonlinear partial differential equation that is valid for any irrotational, isentropic flow. For an incompressible flow, the local speed of sound is unbounded and Eq. 9-67 reduces to Laplace's well-known linear partial differential equation:

$$\phi_{xx} + \phi_{yy} + \phi_{zz} = 0. \tag{9-71}$$

Laplace's equation has a long and rich history of study, and many solutions and solution techniques are available for it, many of which are directly applicable to problems in fluid mechanics and aerodynamics. Anderson (1990), Milne-Thomson (1968), and Robertson (1965) contain excellent discussions of this equation and its use in flow problems.

9.5 THE SMALL DISTURBANCE EQUATION

Historically, difficulties associated with obtaining solutions of the nonlinear full potential equation provided the impetus for seeking linear approximations to the full potential equation. In this section we shall derive and examine the *small disturbance potential equation* since it represents the most frequently used approximation of the full potential equation. A uniform flow in the x-direction has velocity components $V_x = V_\infty$, $V_y = 0$, and $V_z = 0$. The uniform motion is disturbed if a body is placed in the uniform flow; thus, $V_x = V_\infty + u'(x,y,z)$, $V_y = v'(x,y,z)$, and $V_z = w'(x,y,z)$. If the deviations—u', v', and w'—from the uniform flow are small in comparison with the freestream velocity, then

$$\frac{u'}{V_\infty}, \frac{v'}{V_\infty}, \frac{w'}{V_\infty} \ll 1. \tag{9-72}$$

We wish to ascertain what terms in the full potential equation might be eliminated by neglecting higher order terms if the disturbances are small.

Cast in the form of small disturbances, the velocity can be written as

$$\bar{V} = (V_\infty + u')i + v'j + w'k = \nabla\phi, \tag{9-73}$$

with the relationship as indicated to the velocity potential. The full potential, as expressed in Eq. 9-73, can be viewed as being composed of that due to the uniform flow plus that due to deviations from the uniform flow; hence

$$\phi = V_\infty x + \phi'(x, y, z), \tag{9-74}$$

where ϕ' is called the *disturbance potential.* The velocity components then become

$$V_x = V_\infty + u' = V_\infty + \phi'_x, \qquad V_y = v' = \phi'_y, \qquad V_z = w' = \phi'_z, \tag{9-75}$$

with the subscript on ϕ' indicating partial differentiation. Liepmann and Roshko (1957) and Anderson (1990) point out that physical insight is enhanced in this development if the terms in the full potential equation are expressed as velocities and derivatives of velocities rather than as partial derivatives of the potential. Use of Eq. 9-74 then leads to

$$\begin{aligned}[a^2 - (V_\infty + u')^2]\frac{\partial u'}{\partial x} &+ (a^2 - v'^2)\frac{\partial v'}{\partial y} + (a^2 - w'^2)\frac{\partial w'}{\partial z} \\ &- 2(V_\infty + u')v'\frac{\partial u'}{\partial y} - 2(V_\infty + u')w'\frac{\partial u'}{\partial z} - 2v'w'\frac{\partial v'}{\partial z}.\end{aligned} \tag{9-76}$$

Like Eq. 9-67, Eq. 9-76 contains the square of the local speed of sound. However, the most useful formulation of this term for the small disturbance development involves use of the freestream speed of sound rather than the stagnation speed of sound as for the full potential equation. For an isoenergetic flow,

$$h_o = \text{constant} = c_p T_\infty + \frac{V_\infty^2}{2} = c_p T + \frac{V^2}{2}. \tag{9-77}$$

Using Eq. 9-75 for the velocity components and using the relationship between the specific heat, gas constant, and ratio of specific heats permits Eq. 9-77 to be expressed as

$$\frac{\gamma R T_\infty}{\gamma - 1} + \frac{V_\infty^2}{2} = \frac{\gamma R T}{\gamma - 1} + \frac{[(V_\infty + u')^2 + v'^2 + w'^2]}{2}. \tag{9-78}$$

With γRT identified as the square of the speed of sound, Eq. 9-78 can be solved for a^2 as

$$a^2 = a_\infty^2 - \frac{\gamma - 1}{2}(2u'V_\infty + u'^2 + v'^2 + w'^2). \tag{9-79}$$

The next steps are to substitute Eq. 9-79 into Eq. 9-76, collect coefficients of the derivatives of the velocities, and place the linear forms on the left-hand side and the nonlinear forms on the right-hand side to yield

$$\begin{aligned}(1 - M_\infty^2)\frac{\partial u'}{\partial x} + \frac{\partial v'}{\partial y} + \frac{\partial w'}{\partial z} &= \frac{\partial u'}{\partial x}\left[\frac{\gamma + 1}{2}(2u'V_\infty + u'^2) + \frac{\gamma - 1}{2}(v'^2 + w'^2)\right]\frac{1}{a_\infty^2} \\ &+ \frac{\partial v'}{\partial y}\left[\frac{\gamma + 1}{2}v'^2 + \frac{\gamma - 1}{2}(2u'V_\infty + u'^2 + w'^2)\right]\frac{1}{a_\infty^2} \\ &+ \frac{\partial w'}{\partial z}\left[\frac{\gamma + 1}{2}w'^2 + \frac{\gamma - 1}{2}(2u'V_\infty + u'^2 + v'^2)\right]\frac{1}{a_\infty^2} \\ &+ \left[2V_\infty v'\frac{\partial u'}{\partial y} + 2u'v'\frac{\partial u'}{\partial y} + 2V_\infty w'\frac{\partial u'}{\partial z} + 2u'w'\frac{\partial u'}{\partial z} + 2u'w'\frac{\partial v'}{\partial z}\right]\frac{1}{a_\infty^2}.\end{aligned} \tag{9-80}$$

A more useful version of Eq. 9-80 can be obtained by dividing by the square of the freestream velocity, V_∞^2, and realizing that the mixed partial derivative terms can take alternate, equivalent forms; for example, $2\phi_{xy} = \phi_{xy} + \phi_{xy} = \phi_{xy} + \phi_{yx} = \partial u'/\partial y + \partial v'/\partial x$. Then

$$\begin{aligned}(1 - M_\infty^2)\frac{\partial u'}{\partial x} + \frac{\partial v'}{\partial y} + \frac{\partial w'}{\partial z} \\ = M_\infty^2\left[(\gamma + 1)\frac{u'}{V_\infty} + \frac{\gamma + 1}{2}\frac{u'^2}{V_\infty^2} + \frac{\gamma - 1}{2}\frac{v'^2 + w'^2}{V_\infty^2}\right]\frac{\partial u'}{\partial x} \\ + M_\infty^2\left[(\gamma - 1)\frac{u'}{V_\infty} + \frac{\gamma - 1}{2}\frac{w'^2 + u'^2}{V_\infty^2} + \frac{\gamma + 1}{2}\frac{v'^2}{V_\infty^2}\right]\frac{\partial v'}{\partial y} \\ + M_\infty^2\left[(\gamma - 1)\frac{u'}{V_\infty} + \frac{\gamma + 1}{2}\frac{w'^2}{V_\infty^2} + \frac{\gamma - 1}{2}\frac{u'^2 + v'^2}{V_\infty^2}\right]\frac{\partial w'}{\partial z} \\ + M_\infty^2\left[\frac{v'}{V_\infty}\left(1 + \frac{u'}{V_\infty}\right)\left(\frac{\partial u'}{\partial y} + \frac{\partial v'}{\partial x}\right) + \frac{w'}{V_\infty}\left(1 + \frac{u'}{V_\infty}\right)\left(\frac{\partial u'}{\partial z} + \frac{\partial w'}{\partial x}\right)\right. \\ \left. + \frac{v'w'}{V_\infty^2}\left(\frac{\partial v'}{\partial z} + \frac{\partial w'}{\partial y}\right)\right].\end{aligned} \tag{9-81}$$

Remember that in spite of all the manipulations, Eq. 9-81 is still a version of the full potential equation; everything is just expressed in terms of disturbance velocities and their derivatives. The assumption of small disturbances is now invoked by comparing each of the first three groups of terms on the right-hand side with the corresponding terms on the left-hand side of Eq. 9-81. Except for $M_\infty \simeq 1$,

$$|1 - M_\infty^2| > M_\infty^2\left[(\gamma + 1)\frac{u'}{V_\infty} + \frac{\gamma + 1}{2}\frac{u'^2}{V_\infty^2} + \frac{\gamma - 1}{2}\frac{v'^2 + w'^2}{V_\infty^2}\right]. \tag{9-82}$$

Terms 2 and 3 on the right-hand side are negligible compared to their counterparts on the left-hand side as long as $M_\infty < 5$. Mach number 5 for the upper limit is rather arbitrary, but the point is that as the Mach number increases, the terms on the right-hand side of the equation become increasingly important. The last term on the right-hand side of Eq. 9-81 is usually taken as negligibly small since it involves products of velocity components and their cross derivatives. Thus, for subsonic Mach numbers in the range $0 < M_\infty < 0.8$ and for supersonic Mach numbers in the range $1.2 < M_\infty < 5$, the small disturbance potential equation becomes

$$(1 - M_\infty^2)\phi'_{xx} + \phi'_{yy} + \phi'_{zz} = 0. \tag{9-83}$$

The usually accepted small disturbance potential equation valid for subsonic, transonic, and supersonic flow, $0 < M_\infty < 5$, is

$$(1 - M_\infty^2)\phi'_{xx} + \phi'_{yy} + \phi'_{zz} = M_\infty^2(\gamma + 1)\frac{\phi'_x}{V_\infty}\phi'_{xx}, \tag{9-84}$$

which is still nonlinear. Hence, the small disturbance potential equation remains nonlinear for the transonic regime.

Many applications, especially those involving external flows, require information about the pressure on a surface. The pressure coefficient, C_p, is the usually preferred dimensionless parameter associated with the pressure and is defined as

$$C_p = \frac{P - P_\infty}{(\rho_\infty V_\infty^2)/2}. \tag{9-85a}$$

Recall that $\rho V_\infty^2 = \gamma P M_\infty^2$; the pressure coefficient therefore becomes

$$C_p = \frac{2(P/P_\infty - 1)}{\gamma M_\infty^2}. \tag{9-85b}$$

The isentropic relation, $P/P_\infty = (T/T_\infty)^{\gamma/(\gamma-1)}$, can be used to relate pressure and temperature at any location in a homentropic flow. Equation 9-77 can be written as

$$T_o = T + \frac{V^2}{2c_p} = T_\infty + \frac{V_\infty^2}{2c_p} \tag{9-86a}$$

or as

$$\frac{T}{T_\infty} - 1 = \frac{\gamma - 1}{2}\frac{V_\infty^2 - V^2}{\gamma R T_\infty} = \frac{\gamma - 1}{2}\frac{V_\infty^2 - V^2}{a_\infty^2}. \tag{9-86b}$$

Since $V_\infty^2 = (V_\infty + u')^2 + v'^2 + w'^2$, Eq. 9-86b can be solved for the ratio T/T_∞ with the form

$$\frac{T}{T_\infty} = 1 - \frac{\gamma - 1}{2a_\infty^2}(2V_\infty u' + u'^2 + v'^2 + w'^2). \tag{9-87}$$

Then

$$\frac{P}{P_\infty} = \left[1 - \frac{\gamma - 1}{2a_\infty^2}(2V_\infty u' + u'^2 + v'^2 + w'^2)\right]^{\gamma/(\gamma-1)}. \tag{9-88}$$

With the use of the symbolic manipulation language DERIVE, the Taylor series expansion for the pressure ratio becomes

$$\frac{P}{P_\infty} = 1 - \frac{\gamma}{2a_\infty^2}(2V_\infty u' + u'^2 + v'^2 + w'^2) + \frac{\gamma}{8a_\infty^2}(2V_\infty u' + u'^2 + v'^2 + w'^2)^2 + \cdots. \tag{9-89}$$

Including up to quadratic velocity disturbance terms, the pressure coefficient then takes the form

$$C_p = -\left[2\frac{u'}{V_\infty} + (1 - M_\infty^2)\frac{u'^2}{V_\infty^2} + \frac{v'^2}{V_\infty^2} + \frac{w'^2}{V_\infty^2} + \cdots\right], \tag{9-90}$$

which for small disturbances reduces to

$$C_p = -\frac{2u'}{V_\infty}. \tag{9-91}$$

Hence, in terms of small disturbance theory the pressure coefficient depends only on the deviation of the x-component velocity from the freestream.

Results from the subsonic–supersonic small disturbance equation, Eq. 9-83, not only are useful but also provide a mechanism for understanding much of the physics of multidimensional compressible flows. The classical example used to elicit this understanding is the two-dimensional *wavy wall*. We shall devote the remainder of this chapter to the subsonic and supersonic wavy wall solutions that fall within the small disturbance assumption. Consider, as is illustrated in Figure 9-5, a wavy wall over which a compressible fluid is flowing. The wall surface is cosine shaped with a wavelength l and a half-height h and has the functional form

$$y_w = h\cos\left(\frac{2\pi x}{l}\right). \tag{9-92}$$

If h, the half-height, is small in comparison to the wavelength, l, then the small disturbance assumption will be fulfilled and Eq. 9-83 will provide an adequate solution. Subsonic flow will be examined first.

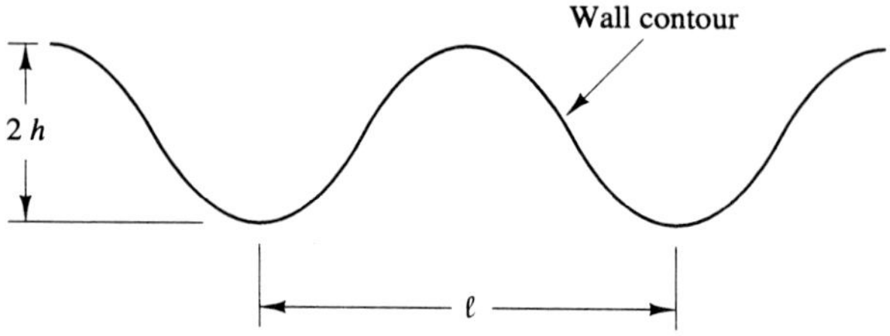

Figure 9-5 Wavy Wall Geometry

Subsonic Wavy Wall

This development will be for subsonic flow, which for the small disturbance equation is usually constrained to be less than a Mach number of 0.8. The two-dimensional version of the small disturbance equation is

$$(1 - M_\infty^2)\phi'_{xx} + \phi'_{yy} = 0, \tag{9-93}$$

with the boundary conditions of bounded disturbance velocities as y becomes unbounded and tangency of flow at the surface of the wall. Thus

$$\begin{aligned} y = y_w \quad & \frac{dy_w}{dx} = \frac{v'_w}{V_\infty + u'_w} \\ y \to \infty \quad & \nabla\phi' \text{ remains finite.} \end{aligned} \tag{9-94}$$

Subsonic flow yields $(1 - M_\infty^2) > 1$; hence, the result is an elliptic equation similar to Laplace's equation. The transformation of Eqs. 9-93 and 9-94 from (x, y) coordinates to (ξ, η) coordinates with $\xi = x$ and $\eta = (1 - M_\infty^2)^{1/2}$ will result in Laplace's equation in the form of $\phi_{\xi\xi} + \phi_{\eta\eta} = 0$, with the appropriately transformed boundary conditions. The classical separation-of-variables technique suitable for solutions to Laplace's equation is then a valid approach for the small disturbance equation solution. Any textbook concerned with mathematical methods of physics will contain an extensive treatment of Laplace's equation solutions. Typical classic references are Margenau and Murphy (1956) and Pipes and Harvill (1970).

The usual starting point for a separation-of-variables solution is to assume the multivariable solution to be composed of products of single variable functions such that

$$\phi'(x, y) = X(x)Y(y). \tag{9-95a}$$

Then

$$(1 - M_\infty^2)X''Y + Y''X = 0, \tag{9-95b}$$

and Eq. 9-93 can be expressed as

$$\frac{X''}{X} = -\frac{1}{1 - M_\infty^2}\frac{Y''}{Y} = -K^2. \tag{9-95c}$$

Since the right-hand side of Eq. 9-95c is a function of x only and since the left-hand side is a function of y only, each side must be a constant and independent of the other. The most general condition that will satisfy this is that each side be equal to a constant, usually called the *separation constant*. The sign of the separation constant is taken in order to give a periodic—in this case sine and cosine—general solution in the bounded direction, in this case the x-direction. The $-K^2$ for the separation constant then results in

$$X'' + K^2X = 0, \tag{9-96a}$$

with the general solution

$$X(x) = A \sin(Kx) + B \cos(Kx); \tag{9-96b}$$

and

$$Y'' - (1 - M_\infty^2)K^2 Y = 0, \tag{9-96c}$$

with the general solution

$$Y(y) = Ce^{K\sqrt{1-M_\infty^2}y} + De^{-K\sqrt{1-M_\infty^2}y}. \tag{9-96d}$$

The general disturbance potential solution can be written as

$$\phi'(x, y) = [A \sin(Kx) + B \cos(Kx)][Ce^{K\sqrt{1-M_\infty^2}y} + De^{-K\sqrt{1-M_\infty^2}y}]. \tag{9-96e}$$

The boundary conditions will be applied next. As y increases without bound, $C = 0$ if the disturbance is to remain bounded. The boundary condition on the surface requires some further examination. At the wall $y = y_w$, and the tangency condition requires that

$$\left(\frac{dy}{dx}\right)_w = \frac{v_w'}{V_\infty + u_w'}. \tag{9-97a}$$

But with small disturbances, $u' \ll V_\infty$, and Eq. 9-97a becomes

$$\left(\frac{dy}{dx}\right)_w = \frac{v_w'}{V_\infty} = \frac{1}{V_\infty}\left(\frac{\partial\phi'}{\partial y}\right)_w. \tag{9-97b}$$

The wall surface description, Eq. 9-92, provides the slope of the surface,

$$\frac{dy_w}{dx} = -h \sin\left(\frac{2\pi x}{l}\right)\frac{2\pi}{l}. \tag{9-97c}$$

It follows that

$$\left(\frac{\partial\phi'}{\partial y}\right)_w = -V_\infty h \sin\left(\frac{2\pi x}{l}\right)\frac{2\pi}{l}. \tag{9-97d}$$

Since $y_w \simeq 0$ within the constraints of the small disturbance development, $(\partial\phi'/\partial y)_{y=0} \simeq (\partial\phi'/\partial y)_w$ and

$$\left(\frac{\partial\phi'}{\partial y}\right)_{y=0} \simeq \left(\frac{\partial\phi'}{\partial y}\right)_w = -V_\infty h \sin\left(\frac{2\pi x}{l}\right)\frac{2\pi}{l}. \tag{9-97e}$$

The expression for $\partial\phi'/\partial y$ from Eq. 9-96e with C taken as zero has the form

$$\frac{\partial\phi'}{\partial y} = [AD \sin(Kx) + BD \cos(Kx)](-K\sqrt{1 - M_\infty^2})e^{-K\sqrt{1-M_\infty^2}y}, \tag{9-98a}$$

and has the following form when evaluated at $y = 0$:

$$\left(\frac{\partial\phi'}{\partial y}\right)_{y=0} = [AD \sin(Kx) + BD \cos(Kx)](-K\sqrt{1 - M_\infty^2}). \tag{9-98b}$$

Equating Eqs. 9-97d and 9-98b produces

$$-V_\infty h\frac{2\pi}{l}\sin\left(\frac{2\pi x}{l}\right) = [AD\sin(Kx) + BD\cos(Kx)]\left(-K\sqrt{1-M_\infty^2}\right). \qquad (9\text{-}99)$$

Comparing coefficients of like terms reveals that

$$BD = 0;$$

$$K = \frac{2\pi}{l}; \qquad (9\text{-}100)$$

$$AD = \frac{2\pi}{l}\frac{V_\infty h}{K\sqrt{1-M_\infty^2}},$$

so that the disturbance potential for the subsonic wavy wall becomes

$$\phi'(x,y) = \frac{V_\infty h}{\sqrt{1-M_\infty^2}}\sin\left(\frac{2\pi x}{l}\right)e^{-2\pi\sqrt{1-M_\infty^2}y/l} \qquad (9\text{-}101)$$

and is independent of the ratio of specific heats, γ.

The disturbance potential given in Eq. 9-101 results in a u' value of

$$u' = \frac{V_\infty h}{\sqrt{1-M_\infty^2}}\frac{2\pi}{l}\cos\left(\frac{2\pi x}{l}\right)e^{-2\pi\sqrt{1-M_\infty^2}y/l}. \qquad (9\text{-}102)$$

The pressure coefficient then appears as

$$C_p = \frac{-4\pi}{\sqrt{1-M_\infty^2}}\frac{h}{l}\cos\left(\frac{2\pi x}{l}\right)e^{-2\pi\sqrt{1-M_\infty^2}y/l}. \qquad (9\text{-}103a)$$

Approximating the surface at the $y = 0$ position, the surface pressure coefficient takes the form

$$C_{p_w} = \frac{-4\pi}{\sqrt{1-M_\infty^2}}\frac{h}{l}\cos\left(\frac{2\pi x}{l}\right). \qquad (9\text{-}103b)$$

The subsonic wavy wall flow field is described completely by Eqs. 9-101 to 9-103. The nature of the solution is illustrated in Figure 9-6, which provides comparison of the streamlines for low and high subsonic Mach numbers. The lack of attenuation with distance above the wall for the high subsonic case is quite visible in the illustration. The figure also presents the variation of the surface pressure coefficient from which the symmetry is again apparent. The solution is completely symmetrical in terms of all parameters. The symmetry of the surface pressure coefficient distribution precludes a net drag force on the wavy wall since positive and negative contributions exactly counter each other for a wall with complete cyclic shape. This is consistent with our understanding of inviscid, incompressible flows. The streamlines maintain their approximately cosine shape and are in phase with the wall shape, but the amplitude of the streamlines decreases with increasing distance from the wall. The pressure coefficient is maximum at the troughs of the wall and minimum at the crests; this variation is in agreement with one-dimensional behavior with respect to area change.

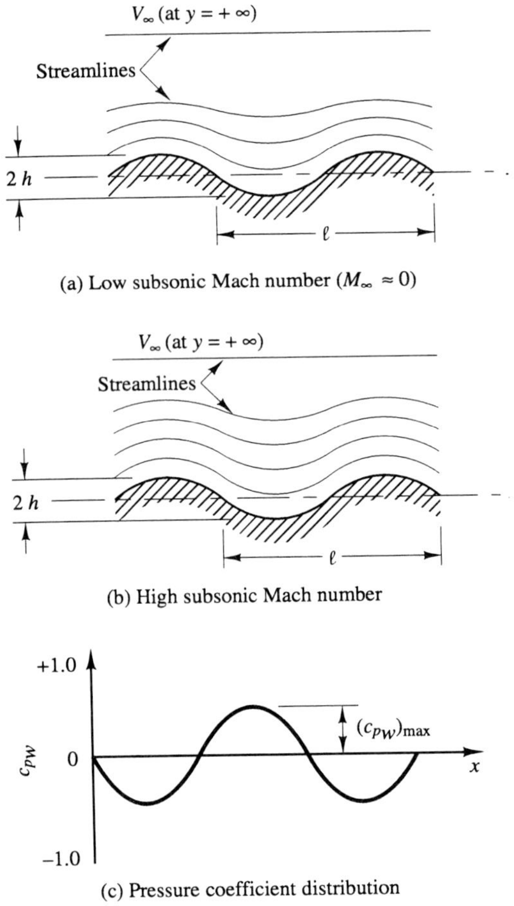

Figure 9-6 Streamlines and Pressure Coefficient Distribution for Subsonic Wavy Wall. *From Shapiro (1953).*

The x-component of the disturbance velocity, u', in addition to being symmetrical about the surface possesses a $1/(1 - M_\infty^2)^{1/2}$ dependence on magnitude and is damped with respect to y as an exponential with $(1 - M_\infty^2)^{1/2}$ in the argument. Thus, as the Mach number increases subsonically, the magnitude of the disturbance increases and the rate at which the disturbance is damped in the y-direction decreases exponentially. The behavior with Mach number has a number of ramifications. One of the more important is that wind tunnel testing in the near-transonic regime is severely hampered by the presence of the wind tunnel walls, which constrain the y-behavior to something different than an ambient environment with no constraining walls.

One wavy wall result that is generalizable to external subsonic flows can be inferred from the pressure coefficient relation, Eq. 9-103b. The pressure coefficient for an incompressible flow occurs when M_∞ is essentially zero; thus

$$C_{p_w}\big|_{M_\infty=0} = -4\pi\frac{h}{l}\cos\left(\frac{2\pi x}{l}\right). \tag{9-104a}$$

The ratio of the pressure coefficient for the incompressible and compressible subsonic cases, Eqs. 9-103b and 9-104a, respectively, yields

$$C_{p_w} = \frac{C_{p_w}|_{M_\infty=0}}{\sqrt{1 - M_\infty^2}}, \tag{9-104b}$$

which is universally known as the *Prandtl-Glauert rule* and is widely used in understanding and estimating the behavior of external flows with respect to Mach number variations. Since the lift and moment coefficients represent integrations of the pressure coefficient distribution, the relation also carries over to lift and moments as

$$C_L = \frac{C_L|_{M_\infty=0}}{\sqrt{1 - M_\infty^2}};$$

$$C_M = \frac{C_M|_{M_\infty=0}}{\sqrt{1 - M_\infty^2}}. \tag{9-105}$$

This simple wavy wall example has illustrated much about the behavior of compressible subsonic flows. The supersonic wavy wall solution is just as revealing and will be examined next.

Supersonic Wavy Wall

The same wall geometry, Figure 9-5, used for the subsonic wavy wall discussion will be used for the supersonic wavy wall. For supersonic flow, $M_\infty^2 - 1 = \lambda^2$ will be a negative number and the small disturbance potential equation becomes

$$\lambda^2 \phi'_{xx} - \phi'_{yy} = 0. \tag{9-106}$$

A comparison of Eq. 9-106 with Eq. 9-93 reveals that fundamental differences exist between the two equations. In particular, Eq. 9-106, valid for supersonic flow, exhibits the characteristics of a wave equation with the form $\phi_{\xi\xi} - \phi_{\eta\eta} = 0$, whereas Eq. 9-93 exhibits the characteristics of Laplace's equation, with the form $\phi_{\xi\xi} + \phi_{\eta\eta} = 0$. The general solution for Eq. 9-106 is

$$\phi' = f(x + \lambda y) + g(x - \lambda y), \tag{9-107}$$

where f and g are functions to be determined by the boundary conditions. That Eq. 9-107 represents a general solution to supersonic small disturbance potential equation can be verified by taking the indicated partial derivatives and substituting these derivatives into Eq. 9-106.

The most interesting feature of the general solution is that the value of the functions are constants along lines for which

$$x \pm \lambda y = \text{constant} \tag{9-108}$$

and which have the slopes

$$\frac{dy}{dx} = \pm\frac{1}{\lambda} = \pm\frac{1}{\sqrt{M_\infty^2 - 1}}. \tag{9-109}$$

But we have seen a form of the slope expressions before! Turn back and examine Eq. 4-7 and Figure 4-3 for the interpretation of the slope expression; the slopes are simply the Mach lines associated with M_∞. Hence, the general small disturbance potential equation solutions for supersonic flow are constant along Mach lines. Along a given Mach line, the disturbance potential is not diminished with distance, but remains constant. The boundary condition as y becomes unbounded is then automatically satisfied. Cast into the form familiar to us from previous discussions, the lines of f = constant and g = constant emanate from a disturbance as in Figure 9-7. When Eqs. 9-108 and 9-109 are used in the wavy wall supersonic solution, disturbances cannot propagate upstream; hence, for the problem as stated, only the g family of Mach lines is needed. These appear schematically in Figure 9-8, but the specific form of the function g will depend on the surface geometry.

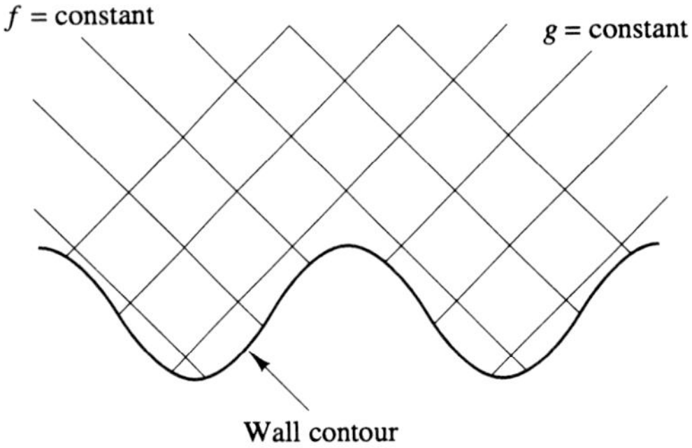

Figure 9-7 Lines of g = constant and f = constant for Supersonic Flow

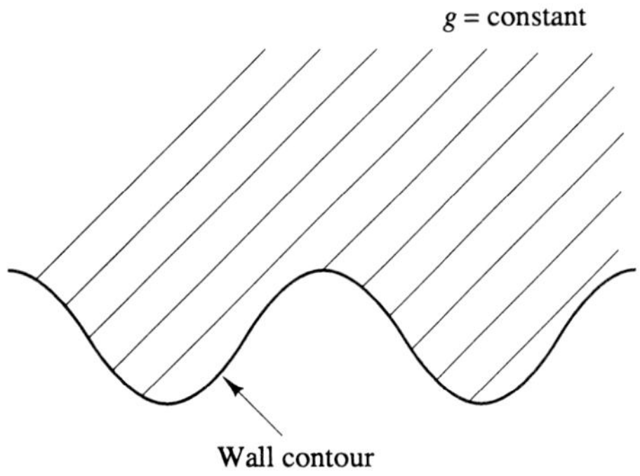

Figure 9-8 g-family of Mach Lines

For the wavy wall geometry of Figure 9-5, the disturbance potential solution is

$$\phi'(x, y) = g(x - \lambda y). \tag{9-110}$$

The surface boundary condition is the same as for the subsonic version and is available from Eq. 9-97b as

$$\frac{\partial \phi'}{\partial y} = V_\infty \frac{dy_w}{dx}. \tag{9-111}$$

The small disturbance assumption permits dy_w/dx to be expressed as $(dy/dx)_{y=0}$; see Eq. 9-97c. Finally,

$$\frac{\partial \phi'}{\partial y} = \frac{dg}{d(x - \sqrt{M_\infty^2 - 1}y)}\left(-\sqrt{M_\infty^2 - 1}\right). \tag{9-112}$$

Combining Eqs. 9-92, 9-111, and 9-112 produces

$$-\sqrt{M_\infty^2 - 1}\left[\frac{dg}{d(x - \sqrt{M_\infty^2 - 1}y)}\right]_{y=0} = -V_\infty h \frac{2\pi}{l} \sin\left(\frac{2\pi x}{l}\right), \tag{9-113a}$$

from which

$$\left[\frac{dg}{d(x - \sqrt{M_\infty^2 - 1}y)}\right]_{y=0} = g_w' = \frac{-V_\infty h}{\sqrt{M_\infty^2 - 1}} \frac{2\pi}{l} \sin\left(\frac{2\pi x}{l}\right). \tag{9-113b}$$

Generalizing this expression for any y results in

$$\frac{dg}{d(x - \sqrt{M_\infty^2 - 1}y)} = -\frac{V_\infty h}{\sqrt{M_\infty^2 - 1}} \frac{2\pi}{l} \sin\left[\frac{2\pi}{l}(x - \sqrt{M_\infty^2 - 1}y)\right]. \tag{9-113c}$$

Integrating leads to the functional form for g, which is in fact the disturbance potential

$$g = \phi'(x,y) = -\frac{V_\infty h}{\sqrt{M_\infty^2 - 1}} \cos\left[\frac{2\pi}{l}(x - \sqrt{M_\infty^2 - 1}y)\right] + C. \tag{9-114}$$

Proceeding as in the subsonic case leads to

$$u' = +\frac{V_\infty h}{\sqrt{M_\infty^2 - 1}} \frac{2\pi}{l} \sin\left[\frac{2\pi}{l}(x - \sqrt{M_\infty^2 - 1}y)\right], \tag{9-115a}$$

where the pressure coefficient becomes

$$C_p = -\frac{4\pi}{\sqrt{M_\infty^2 - 1}} \frac{h}{l} \sin\left[\frac{2\pi}{l}(x - \sqrt{M_\infty^2 - 1}y)\right], \tag{9-115b}$$

with the surface value of the pressure coefficient being

$$C_{p_w} = -\frac{4\pi}{\sqrt{M_\infty^2 - 1}} \frac{h}{l} \sin\left(\frac{2\pi x}{l}\right). \tag{9-115c}$$

Figure 9-9 illustrates the results of the supersonic wavy wall small disturbance potential equation solution. The streamlines are shown as shifted in order to stay aligned with Mach lines. The surface pressure coefficient is not symmetrical with respect to the wall shape; thus, unlike the subsonic case, there will be a net drag force on the wavy wall even with an inviscid flow assumption. Since the parameters are transmitted undiminished along Mach lines, supersonic disturbances do not damp

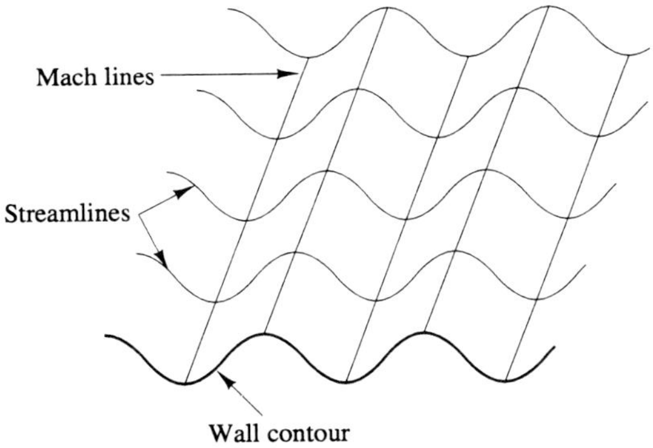

Figure 9-9 Streamlines for Supersonic Wavy Wall

out with increasing y-distance as in the subsonic case. The contrast between the subsonic and supersonic solutions should not have been a great surprise; we saw in the first part of this textbook that subsonic and supersonic behaviors can be quite different. The wavy wall solutions to the subsonic and supersonic small disturbance potential equations demonstrate that in two dimensions distinctions between the behavior of a subsonic and a supersonic flow will be evident. The linking of the two-dimensional supersonic behavior to the Mach line is very significant and will carry over beyond the small disturbance considerations.

9.6 CONCLUSION

This chapter, in contrast to the previous chapters, has emphasized derivations instead of applications. If we are to continue beyond the one-dimensional and quasi–one-dimensional considerations, the appropriate governing equations in multidimensional forms are required. Moreover, various versions of these equations permit us to infer some important ideas associated with more complex compressible flows.

The only solutions presented in this chapter were for the subsonic and supersonic wavy wall viewed in terms of the small disturbance potential equation. The small disturbance analyses pointed out many salient features of the behavior of subsonic and supersonic flows. Indeed, the connection between Mach lines and supersonic flow will be used extensively in the next chapter. It is important that we understand and can distinguish between various forms of the multidimensional equations for an inviscid compressible flow.

CHAPTER 9 REVIEW QUESTIONS

1. How do the divergence (or conservation) and nonconservation forms of mass, momentum, and energy differ?
2. What is the substantial (or material or particle) derivative? What form does it take? What are the two different terms called?
3. What is Euler's equation?
4. What three conditions must be met if the total enthalpy of a fluid element in an inviscid flow is to be constant?
5. What is Crocco's equation? What is the defining equation for vorticity?
6. Is the flow field downstream of a curved shock wave rotational or irrotational? Why?
7. Define
   (1) irrotational flow
   (2) isoenergetic flow
   (3) homentropic flow
   (4) isentropic flow.

8. How are the velocity components and the velocity potential related?
9. Verify that superposition holds for

$$\phi_{xx} + \phi_{yy} + \phi_{zz} = 0.$$

10. What is Laplace's equation?
11. What is the small disturbance potential equation?
12. Why is the small disturbance potential equation not valid for transonic flow?
13. How do solutions to the small disturbance potential equation differ for subsonic and supersonic flows?
14. Sketch the streamlines for low and high subsonic Mach numbers for flow over a wavy wall. Sketch the pressure distributions on the surface.
15. Sketch the streamlines for a supersonic flow over a wavy wall. Sketch the pressure distribution on the surface.
16. What constraints on equations of state or thermal or caloric perfection of gases are placed on Eq. 9-19?
17. Why are wind tunnel walls a problem in high subsonic Mach number testing?
18. How are Mach lines and the solution to the supersonic small disturbance equation related?

CHAPTER PROBLEMS

1. Derive Eq. 9-15b. Start with a differential control volume (similar to Figure 9-2) showing the forces and momentum fluxes for the derivation.
2. Verify that Eq. 9-17 can be expressed as Eq. 9-18 or 9-19.
3. Verify Eq. 9-28.
4. Verify Eq. 9-45.
5. Verify Eqs. 9-67 and 9-70.
6. Verify Eq. 9-81.
7. Verify Eqs. 9-83 and 9-91.
8. Verify Eq. 9-101.
9. Verify Eq. 9-115a.
10. Given the velocity components

$$V_x = 4 + 2xy + t^2 \quad \text{and} \quad V_y = 7 - xy^2 + 5t:$$

   (a) Find the acceleration at $x = 2$, $y = 3$, and $t = 2$.
   (b) Compute the vorticity at the same time and spatial location.
11. As illustrated in Figure 9-10, subsonic flow takes place between two wavy walls. Find the disturbance potential and the pressure distribution along the centerline.

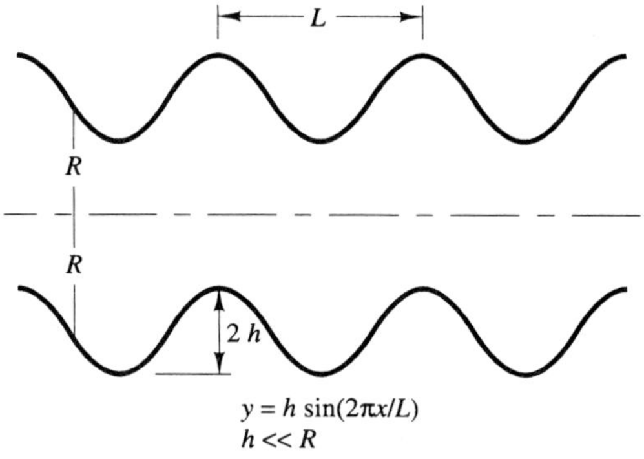

Figure 9-10

12. A planar channel is composed of a wavy wall (with waves of amplitude h and wavelength l) and a straight wall. The straight wall and the centerline of the wavy wall are separated by a distance H. Verify that the disturbance potential is

$$\phi' = \frac{V_\infty}{\beta} \frac{h}{1 - e^{-4\pi\beta H/l}} \sin\left(\frac{2\pi x}{l}\right) e^{(-2\pi\beta y)/l}(1 + e^{4\pi\beta(y-H)})$$

where $\beta = \sqrt{1 - M^2}$.

13. A wing has the shape of a repeating sine wave, as shown in Figure 9-11. Neglecting any leading and trailing edge effects, solve for the disturbance potential on the upper and lower sides of the wing. Develop an expression for the pressure coefficient on the upper and lower surfaces of the wing. Consider that $h \ll L$.

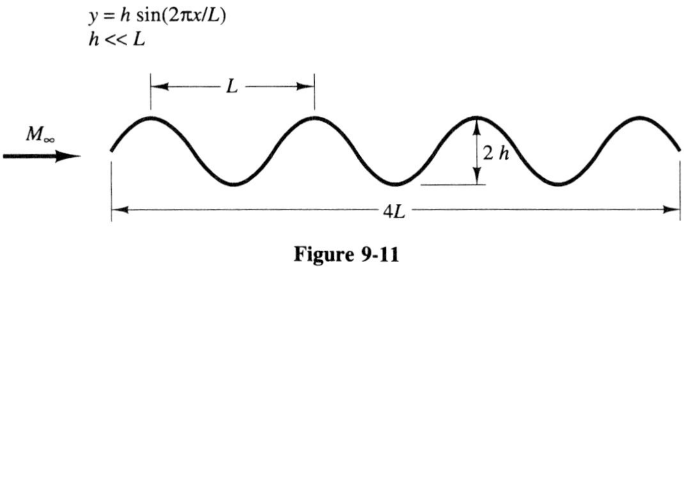

Figure 9-11

The Method of Characteristics

10

10.1 INTRODUCTION

The oblique shock (Chapter 3) and the Prandtl-Meyer expansion fan (Chapter 4) are examples of steady supersonic flows in two dimensions. Upon encountering one of these wave systems, an initially one-dimensional flow turns and gains a velocity component in a second direction, thus becoming two-dimensional. The oblique shock relations and the Prandtl-Meyer function provide us with the means to analyze simple two-dimensional, steady, supersonic flows. In this chapter we will develop a technique for analyzing more complicated two-dimensional supersonic flows. To demonstrate the usefulness of this technique, we will then apply it to the design of supersonic nozzles.

The technique we will develop here is known as the *method of characteristics*. To help understand this technique examine Figure 10-1, which shows a uniform supersonic stream encountering two expansion corners in series. A fan of expansion Mach waves emanates from each corner, and as the flow crosses these waves it turns and accelerates (steadily and isentropically). Between the corners, where the Mach waves are all parallel, no changes occur. We can think of the Mach waves as carrying information into the flow concerning the shape of the wall. Where the wall shape changes, such as at the corners, the spreading fans of Mach waves "inform" the flow of these changes. This concept is essentially the basis of the method of characteristics. Information about boundary conditions is propagated into the flow along certain lines, or *characteristics*. We will find that the characteristics for our problems are Mach waves. The flow equations (continuity, momentum, and energy) will be written in order to follow these lines or characteristics. We will also find that the equations of motion written in this way are greatly simplified and can be readily solved (at least for two-dimensional planar motion) for the entire flow field.

The method of characteristics we will develop here will be restricted to steady supersonic flow in two dimensions (planar and axisymmetric). We will also assume that the flow is homenergetic ($\nabla h_o = 0$) and homentropic ($\nabla s = 0$) and hence irrotational by Crocco's equation, Eq. 9-52. Extensions of the method to less restricted flows will be briefly discussed at the end of this chapter.

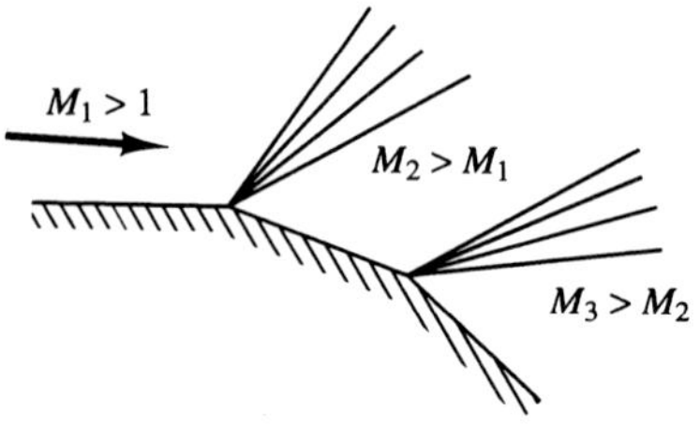

Figure 10-1 Two Expansion Corners in Tandem Illustrate the Propagation of Boundary Conditions by Mach Waves into a Supersonic Flow

10.2 EQUATION DEVELOPMENT

There are several ways by which to develop the method of characteristics. The approach we will take here relies as much as possible on physical concepts. This approach follows closely the development in Thompson (1972), which, in turn, is based on the derivation in Liepmann and Roshko (1957).

As we mentioned, the basis for our derivation is the notion that Mach waves carry information from the boundaries into the flow. Now recall the derivation in Chapter 4 of the Prandtl-Meyer function ν. That derivation began by considering the infinitesimal change in flow velocity (magnitude) across a Mach wave created by an infinitesimal change in flow direction. This suggests that the proper dependent variables to first consider are flow velocity magnitude, V, and direction θ. We will take this suggestion and write the equations of motion so that V and θ are the unknowns.

For the independent variables we need two spatial coordinates. The coordinate system most consistent with the dependent variables is what is known as the *natural*, or *streamline*, coordinate system. As sketched in Figure 10-2, one coordinate, s, lies along the streamlines of the flow and the other coordinate, n, follows curves normal to the streamlines. The velocity vector is everywhere tangent to the s-coordinate lines and is everywhere normal to the n-coordinate lines. The inclination of the s-lines with respect to the horizontal direction is the flow direction θ, which will be measured positive counterclockwise from the positive x-axis. For axisymmetric flow we also have the radial position, r, of each point from the axis of symmetry, which is aligned with $\theta = 0$. For planar problems this radius goes to infinity.

Figure 10-2 shows a cell formed by two adjacent streamlines and two adjacent normal lines. To help us express the equations of motion in terms of V and θ it is convenient to introduce the radii of curvature, R_s and R_n, of the coordinate lines in this cell. An explanation of the radius of curvature of a general three-dimensional curve can be found in most good texts on multivariable calculus or vector analysis.

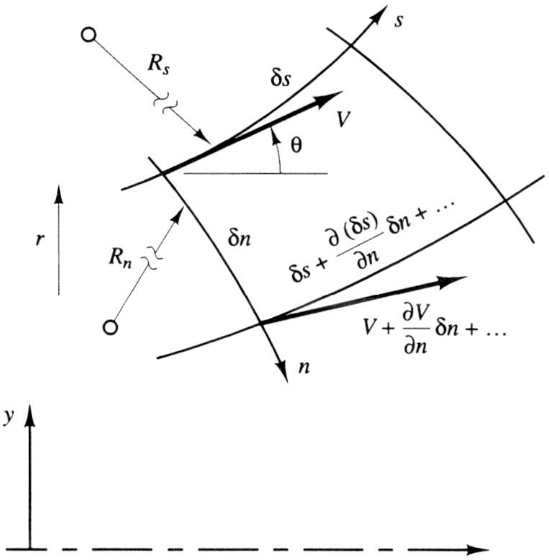

Figure 10-2 The Natural or Streamline Coordinate System

For example, Section 6.6 of Hildebrand (1962) states that the inverse of the radius of curvature "is a measure of the rate at which the tangent vector changes its direction with distance along the curve." For the s-lines the direction of the tangent vector is θ, which increases as the distance along the lines, s, increases; hence

$$\frac{1}{R_s} = \frac{\partial\theta}{\partial s}. \tag{10-1a}$$

For the n-lines the tangent vector direction is $\theta + 3\pi/2$. This angle decreases, however, as the distance, n, along the lines increases, and thus

$$\frac{1}{R_n} = -\frac{\partial\theta}{\partial n}. \tag{10-1b}$$

Now examine Figure 10-3a. The two triangles shown are similar (assuming δn is small) so that the sides of each are proportional. That is,

$$\frac{\delta s}{R_s} = \frac{\partial(\delta s)}{\partial n} \tag{10-2a}$$

In the same manner, from Figure 10-3b we have

$$\frac{\delta n}{R_n} = \frac{\partial(\delta n)}{\partial s}. \tag{10-2b}$$

From Eqs. 10-1 and 10-2 we conclude

$$\frac{1}{\delta s}\frac{\partial(\delta s)}{\partial n} = \frac{\partial\theta}{\partial s} \tag{10-3a}$$

and

$$\frac{1}{\delta n}\frac{\partial(\delta n)}{\partial s} = -\frac{\partial\theta}{\partial n}. \tag{10-3b}$$

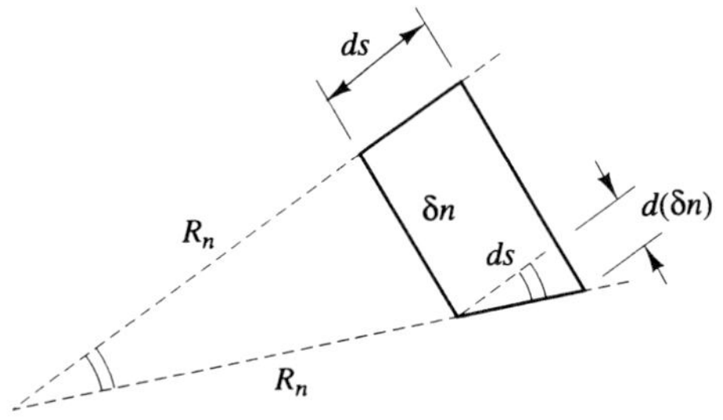

Figure 10-3a Differential Geometry along Adjacent Normal Coordinate Lines

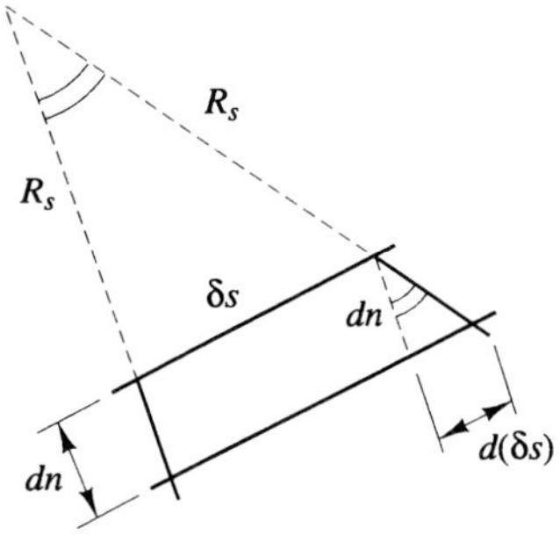

Figure 10-3b Differential Geometry along Adjacent Streamlines

So far we have only been concerned with the geometry of the steady streamline pattern in Figure 10-2. Now let us consider the physics of the flow. In addition to being steady the flow is supersonic with uniform enthalpy and entropy throughout. These conditions also mean that the flow is irrotational. Consequently, the circulation around any contour is zero. Let us evaluate the circulation around the cell in Figure 10-2 using the definition, Anderson (1991). Realizing that on the n lines $\overline{V} \cdot \overline{dl}$ is zero, gives

$$V\delta s - \left(V + \frac{\partial V}{\partial n}\delta n + \cdots\right)\left(\delta s + \frac{\partial(\delta s)}{\partial n}\delta n + \cdots\right) = 0. \tag{10-4}$$

Simplifying and dropping terms which are second order or higher in the arc lengths δs and δn we have

$$V\frac{\partial(\delta s)}{\partial n} + \delta s\frac{\partial V}{\partial s} = 0$$

Using Eq. 10-3a the irrotationality condition then becomes

$$\frac{\partial V}{\partial n} + V\frac{\partial \theta}{\partial s} = 0. \tag{10-5}$$

The continuity equation, Eq. 1-33, applied to a unit depth of the cell in Figure 10-2 is

$$\rho V r \delta n = \text{constant}. \tag{10-6}$$

Next, we differentiate Eq. 10-6 with respect to s, divide through by the left-hand side of Eq. 10-6 (a process equivalent to logarithmically differentiating), and then use Eq. 10-3b to give

$$\frac{1}{\rho}\frac{\partial \rho}{\partial s} + \frac{1}{V}\frac{\partial V}{\partial s} - \frac{\partial \theta}{\partial n} = -\frac{1}{r}\frac{\partial r}{\partial s}.$$

A differential length along a streamline, as depicted in Figure 10-4a, shows that

$$\sin\theta = \frac{\partial r}{\partial s},$$

so that the continuity equation becomes

$$\frac{1}{\rho}\frac{\partial \rho}{\partial s} + \frac{1}{V}\frac{\partial V}{\partial s} - \frac{\partial \theta}{\partial n} = -\frac{\sin\theta}{r}. \tag{10-7}$$

The momentum equation along a streamline is Euler's equation, Eq. 9-60. Written in terms of derivatives with respect to s, momentum is

$$V\frac{\partial V}{\partial s} + \frac{1}{\rho}\frac{\partial P}{\partial s} = 0. \tag{10-8}$$

Since the flow is homentropic we can replace derivatives of P with derivatives of ρ by Eq. 1-47:

$$dP = a^2\,d\rho. \tag{1-47}$$

Combining Eqs. 10-7, 10-8, and 1-47 finally gives

$$(M^2 - 1)\frac{1}{V}\frac{\partial V}{\partial s} + \frac{\partial \theta}{\partial n} = \frac{\sin\theta}{r}. \tag{10-9}$$

Equations 10-5 and 10-9 form a system of equations for the unknowns V and θ. (Actually, M is also an unknown. To have a complete system we must include Eq. 1-47 and an equation of state, such as the perfect gas law.) These equations have been developed using all of the fundamental conservation laws (mass, momentum, and energy) and therefore incorporate all of the basic physics of the flow. We now wish to solve this system for V and θ, taking advantage of the notion that Mach waves carry information through the flow field.

We use Eq. 4-7 to introduce Mach waves explicitly into the problem:

$$\tan\mu = \frac{1}{\sqrt{M^2 - 1}}. \tag{4-7}$$

Equations 10-5 and 10-9 then become, respectively,

$$\tan\mu\frac{\sqrt{M^2 - 1}}{V}\frac{\partial V}{\partial n} + \frac{\partial \theta}{\partial s} = 0 \tag{10-10}$$

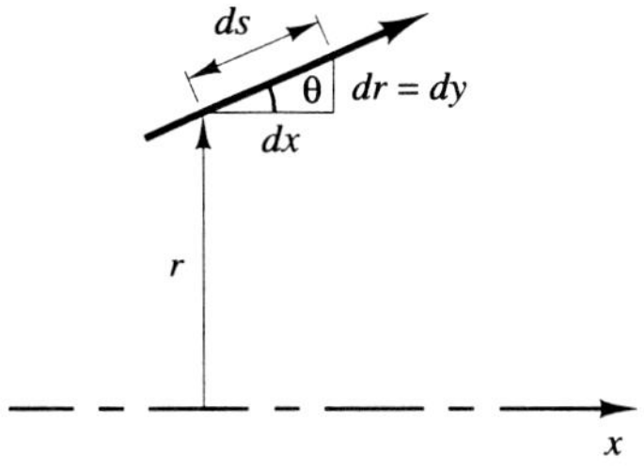

Figure 10-4a Differential Geometry along a *Normal Coordinate Line*

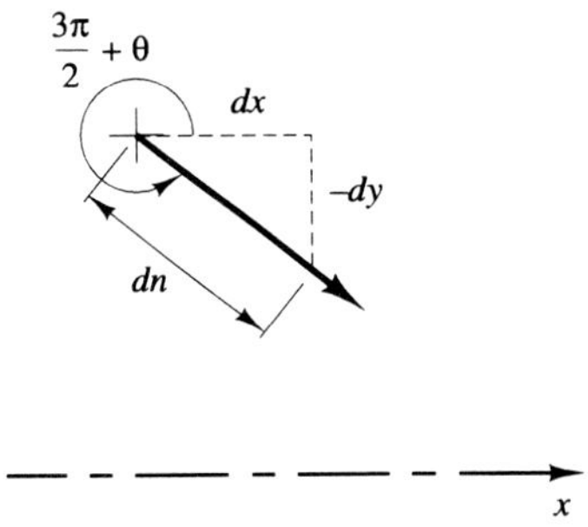

Figure 10-4b Differential Geometry along a *Streamline*

and

$$\frac{\sqrt{M^2-1}}{V}\frac{\partial V}{\partial s} + \tan\mu\frac{\partial\theta}{\partial n} = \frac{\tan\mu\sin\theta}{r}. \tag{10-11}$$

This system can be further simplified by using Eq. 4-8 to replace derivatives of V with derivatives of ν:

$$\frac{dV}{V} = \frac{d\nu}{\sqrt{M^2-1}}. \tag{4-8}$$

We then have

$$\tan\mu\frac{\partial\nu}{\partial n} + \frac{\partial\theta}{\partial s} = 0 \tag{10-12}$$

and

$$\frac{\partial\nu}{\partial s} + \tan\mu\frac{\partial\theta}{\partial n} = \frac{\tan\mu\sin\theta}{r}. \tag{10-13}$$

Equations 10-12 and 10-13 form a complete system of equations for ν and θ (since μ can be immediately found if ν is known). The velocity magnitude is embodied in the Prandtl-Meyer function. Mach waves are also embodied in the system through ν and μ. One last step is needed, however, to take full advantage of the Mach waves. This step follows.

If we add Eqs. 10-12 and 10-13 together we obtain

$$\tan\mu\frac{\partial\nu}{\partial n} + \frac{\partial\theta}{\partial s} + \frac{\partial\nu}{\partial s} + \tan\mu\frac{\partial\theta}{\partial n} = \frac{\tan\mu\sin\theta}{r},$$

which can be rearranged into the form

$$\left(\frac{\partial}{\partial s} + \tan\mu\frac{\partial}{\partial n}\right)(\theta+\nu) = +\frac{\tan\mu\sin\theta}{r}. \tag{10-14}$$

Likewise, subtracting Eq. 10-13 from Eq. 10-12 gives

$$\left(\frac{\partial}{\partial s} - \tan\mu\frac{\partial}{\partial n}\right)(\theta-\nu) = -\frac{\tan\mu\sin\theta}{r}. \tag{10-15}$$

Equations 10-14 and 10-15 are now in a form that can be interpreted in a meaningful way and readily solved. (Indeed, Thompson (1972) does just that.) However, we find it more convenient to convert from natural coordinates to rectangular Cartesian (x, y) coordinates. Referring to Figure 10-4, we see that along an s-line

$$dx = \cos\theta\,ds \quad\text{and}\quad dy = \sin\theta\,ds,$$

while along an n-line

$$dx = \cos\left(\frac{3}{2}\pi + \theta\right)dn = \sin\theta\,dn,$$

and

$$dy = \sin\left(\frac{3}{2}\pi + \theta\right) dn = -\cos\theta\, dn.$$

Using the *chain rule of differentiation* gives us

$$\frac{\partial}{\partial s} = \frac{dx}{ds}\frac{\partial}{\partial x} + \frac{dy}{ds}\frac{\partial}{\partial y} = \cos\theta\frac{\partial}{\partial x} + \sin\theta\frac{\partial}{\partial y},$$

and

$$\frac{\partial}{\partial n} = \frac{dx}{dn}\frac{\partial}{\partial x} + \frac{dy}{dn}\frac{\partial}{\partial y} = \sin\theta\frac{\partial}{\partial x} - \cos\theta\frac{\partial}{\partial y}.$$

The derivative operator on the left-hand side of Eq. 10-14 becomes, under this transformation,

$$\begin{aligned}\frac{\partial}{\partial s} + \tan\mu\frac{\partial}{\partial n} &= \cos\theta\frac{\partial}{\partial x} + \sin\theta\frac{\partial}{\partial y} + \tan\mu\left(\sin\theta\frac{\partial}{\partial x} - \cos\theta\frac{\partial}{\partial y}\right)\\ &= (\cos\theta + \tan\mu\,\sin\theta)\frac{\partial}{\partial x} + (\sin\theta - \tan\mu\,\cos\theta)\frac{\partial}{\partial y}\\ &= \sec\mu(\cos\theta\,\cos\mu + \sin\theta\,\sin\mu)\frac{\partial}{\partial x}\\ &\quad + \sec\mu\,(\sin\theta\,\cos\mu - \cos\theta\,\sin\mu)\frac{\partial}{\partial y}.\end{aligned}$$

Using the trigonometric angle-sum and angle-difference relations, we then have

$$\begin{aligned}\frac{\partial}{\partial s} + \tan\mu\frac{\partial}{\partial n} &= \sec\mu\left(\cos(\theta - \mu)\frac{\partial}{\partial x} + \sin(\theta - \mu)\frac{\partial}{\partial y}\right)\\ &= \sec\mu\,\cos(\theta - \mu)\left(\frac{\partial}{\partial x} + \tan(\theta - \mu)\frac{\partial}{\partial y}\right).\end{aligned}$$

Equation 10-14 can then be written as

$$\begin{aligned}\cos(\theta - \mu)\left(\frac{\partial}{\partial x} + \tan(\theta - \mu)\frac{\partial}{\partial y}\right)(\theta + \nu) &= \frac{1}{r}\frac{\tan\mu\,\sin\theta}{\sec\mu}\\ &= \frac{\sin\theta\,\sin\mu}{r}.\end{aligned} \tag{10-16}$$

A similar procedure transforms the derivative operator on the left-hand side of Eq. 10-15 into

$$\frac{\partial}{\partial s} - \tan\mu\frac{\partial}{\partial n} = \sec\mu\,\cos(\theta + \mu)\left(\frac{\partial}{\partial x} + \tan(\theta + \mu)\frac{\partial}{\partial y}\right)$$

and transforms Eq. 10-15 itself into

$$\cos(\theta + \mu)\left(\frac{\partial}{\partial x} + \tan(\theta + \mu)\frac{\partial}{\partial y}\right)(\theta - \nu) = -\frac{\sin\theta\,\sin\mu}{r}. \tag{10-17}$$

Equations 10-16 and 10-17 are nearly in their final form for our purposes. To help us understand these equations let us consider a point in space through which a streamline and two Mach waves pass, as depicted in Figure 10-5. The upper and lower Mach waves are labeled mw^+ and mw^-, respectively. At the point in question the streamline makes an angle of θ with respect to the horizontal direction. By definition, Mach waves make an angle of magnitude μ with respect to the local flow direction. Consequently, the waves mw^+ and mw^- make angles of $\theta + \mu$ and $\theta - \mu$, respectively, with the x-axis. Distance along the waves is measured by l^+ and l^-. Let us determine the derivative with respect to these distances of any function, say $F(x, y)$. By the chain rule, we have

$$\frac{dF}{dl^\pm} = \frac{dx}{dl^\pm}\frac{\partial F}{\partial x} + \frac{dy}{dl^\pm}\frac{\partial F}{\partial y}.$$

Figure 10-5b shows that along mw^+

$$dx = \cos(\theta + \mu)\, dl^+ \quad \text{and} \quad dy = \sin(\theta + \mu)\, dl^+,$$

while along mw^-

$$dx = \cos(2\pi - (\mu - \theta))\, dl^- = \cos(2\pi + (\theta - \mu))\, dl^- = \cos(\theta - \mu)\, dl^-,$$

and

$$dy = \sin(2\pi - (\mu - \theta))\, dl^- = \sin(2\pi + (\theta - \mu))\, dl^- = \sin(\theta - \mu)\, dl^-.$$

The derivative along each Mach wave is then

$$\begin{aligned} \frac{dF}{dl^\pm} &= \cos(\theta \pm \mu)\frac{\partial F}{\partial x} + \sin(\theta \pm \mu)\frac{\partial F}{\partial y} \\ &= \cos(\theta \pm \mu)\left(\frac{\partial}{\partial x} + \tan(\theta \pm \mu)\frac{\partial}{\partial y}\right)F. \end{aligned} \tag{10-18}$$

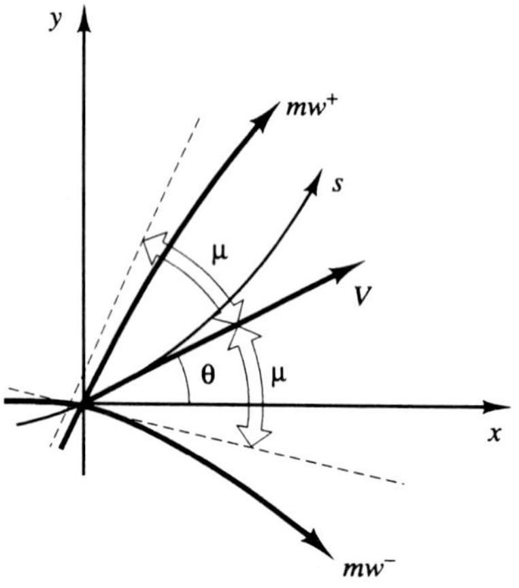

Figure 10-5a The Relationship among the Plus and Minus Mach Waves and the Streamline through a Point in Space

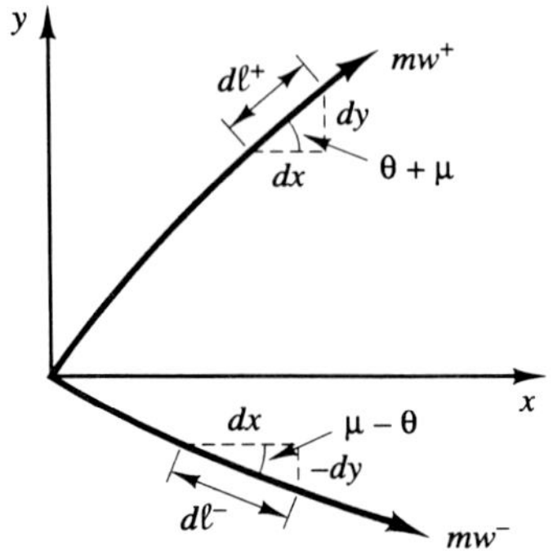

Figure 10-5b Differential Geometry along Mach Waves

Equation 10-18 provides us with the rate of change of the function F as we move along either of the two Mach waves. Notice, considering just the derivative operator in Eq. 10-18,

$$\frac{\partial}{\partial x} + \tan(\theta \pm \mu)\frac{\partial}{\partial y},$$

that the coefficients of $\partial/\partial y$ are the slopes of the Mach waves. That is,

$$\frac{dy}{dx} = \tan(\theta \pm \mu) \quad \text{along } mw^{\pm}. \tag{10-19}$$

Now compare Eq. 10-18 to Eqs. 10-16 and 10-17. The left-hand sides of Eqs. 10-16 and 10-17 are identical to the right-hand side of Eq. 10-18 if we take the function F to be $\theta - \nu$ or $\theta + \nu$. Consequently, the interpretation of Eqs. 10-18 and 10-19 can be applied to Eqs. 10-16 and 10-17. Thus we conclude that

$$\frac{d}{dl^+}(\theta - \nu) = -\frac{\sin\theta \sin\mu}{r} \tag{10-20a}$$

along curves given by

$$\frac{dy}{dx} = \tan(\theta + \mu),$$

that is, along mw^+. We also conclude that

$$\frac{d}{dl^-}(\theta + \nu) = \frac{\sin\theta \sin\mu}{r} \tag{10-20b}$$

along curves given by

$$\frac{dy}{dx} = \tan(\theta - \mu),$$

that is, along mw^-. One further manipulation of the equations is useful. Since, from Figure 10-5b,

$$dy = dr = \sin(\theta + \mu)dl^+ \quad \text{on } mw^+$$

and

$$dy = dr = \sin(\theta - \mu)dl^- \quad \text{on } mw^-,$$

we can finally write that along mw^+

$$d(\theta - \nu) = -\frac{\sin\theta \sin\mu}{\sin(\theta + \mu)}\frac{dr}{r} \tag{10-21a}$$

and that along mw^-

$$d(\theta + \nu) = \frac{\sin\theta \sin\mu}{\sin(\theta - \mu)}\frac{dr}{r}. \tag{10-21b}$$

Equations 10-19 and 10-20 or 10-21 are the working equations for the method of characteristics as applied to two-dimensional (planar or axisymmetric), steady, supersonic, homentropic, homenergetic flow. They are a combination of the continuity, momentum, and energy equations and thus provide a complete physical description of the flow. In the formal theory of the method of characteristics Eqs. 10-20 and 10-21 are known as *compatibility relations*. The equations describing the curves along which the compatibility relations apply, that is Eqs. 10-19, are the *characteristic equations* and the curves themselves are the *characteristics*. Here the characteristics are the Mach waves in the flow. Thus, our original goal of explicitly incorporating Mach waves into the flow solution has been achieved.

The compatibility equations are solved for their arguments, $\theta \pm \nu$, by integrating along the characteristics. The details of this procedure for planar flow are presented in the next section.

10.3 THE METHOD OF CHARACTERISTICS FOR PLANAR FLOW

Planar Flow Equations

When the flow is planar, $r \to \infty$ and Eqs. 10-21 reduce to

$$d(\theta - \nu) = 0 \quad \text{along } mw^{+} \tag{10-22a}$$

and

$$d(\theta + \nu) = 0 \quad \text{along } mw^{-}. \tag{10-22b}$$

We can immediately integrate Eqs. 10-22 to give the solution

$$\theta - \nu = \text{constant} \tag{10-23a}$$

along curves given by

$$\frac{dy}{dx} = \tan(\theta + \mu) \tag{10-19a}$$

and the solution

$$\theta + \nu = \text{constant} \tag{10-23b}$$

along curves given by

$$\frac{dy}{dx} = \tan(\theta - \mu). \tag{10-19b}$$

These remarkably simple expressions are the solutions to the combined continuity, momentum, and energy equations for planar, steady, supersonic, homentropic, homenergetic flow. In other words, Eq. 10-23a tells us that along any given positive Mach wave the difference between the flow angle and the Prandtl-Meyer function is a constant. Similarly, Eq. 10-23b says that along any given negative Mach wave the sum of the flow angle and the Prandtl-Meyer function is a constant. The quantities $\theta - \nu$ and $\theta + \nu$ are called the *Riemann invariants*, K^+ and K^-, of the solution. The Mach waves, mw^+ and mw^-, will be referred to as the C^+ and C^- characteristics, respectively. These two types of characteristics are also referred to as *families*. With this notation we can write the solution in the extremely compact form:

$$K^{\pm} = \text{constant on } C^{\pm}. \tag{10-24}$$

As we indicated in the introduction to this chapter, Mach waves play a key role in this type of flow. Mach waves originating at the boundary of a flow region carry information into the interior of that region. For planar flow, the information carried by a Mach wave is that the corresponding Riemann invariant is constant.

Mach wave systems in steady, supersonic flow can form four different types of regions. Figure 10-6, which shows a uniform, supersonic flow approaching an expansion section in a duct, illustrates these regions. Ahead of the expansion corner the flow is *uniform* and is designated as such. In a uniform region both families of characteristics are straight and parallel. The expansion fan emanating from the corner initially consists of straight, but nonparallel, C^- characteristics. This type of field is a *simple minus region*. In a simple region one family of characteristics is straight but nonparallel. (We will see that the characteristics of the other family in a simple region are curved.) The initial expansion fan reflects upon encountering the lower wall. The flow zone where the reflected characteristics cross the incoming, or incident, characteristics is called a *nonsimple region*. In a nonsimple region both families of characteristics are curved. Finally, the reflected characteristics emerge from the nonsimple region to form a *simple plus region*, with the C^+ characteristics straight but nonparallel. Notice that the simple regions are bounded on two sides by uniform regions and that the nonsimple region is bounded by simple regions. All steady, supersonic flows can be regarded as a combination of these regions.

In any region, implementation of the method of characteristics occurs through a set of basic computational procedures known as *unit processes*. A unit process considers the interaction of a single characteristic of one family with one of four types of entities: a characteristic of the opposite family, a solid surface, a free surface, or a shock wave. We will now discuss each of these interactions.

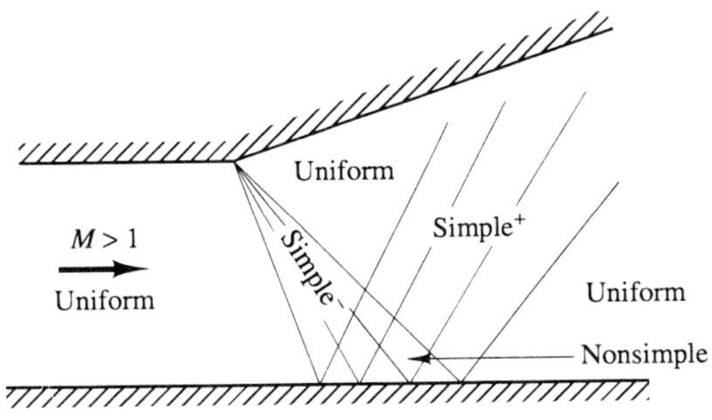

Figure 10-6 The Four Types of Characteristic Regions

Field Point Unit Process

Figure 10-7 shows C^+ and C^- characteristics that cross somewhere in the midst of a flow. The C^- and C^+ characteristics pass through points *a* and *b*, respectively, prior to the intersection that occurs at point *c*. Let us assume that all flow properties and coordinates are known at points *a* and *b* but that nothing is known at point *c*. We wish to determine the flow properties and coordinates at *c*.

Because points *a* and *c* lie on the same C^-, then by Eq. 10-24 we can say that

$$K_a^- = K_c^- = \theta_c + \nu_c. \tag{10-25a}$$

Similarly, *b* and *c* lie on the same C^+ and so again by Eq. 10-24 we have

$$K_b^+ = K_c^+ = \theta_c - \nu_c. \tag{10-25b}$$

Because all of the properties are known at *a* and *b*, the Riemann invariants K_a^-, K_c^-, K_a^+, and K_c^+ are also known. Thus, Eqs. 10-25 are a pair of algebraic equations with unknowns θ_c and ν_c. Solving this simple system gives

$$\theta_c = \frac{1}{2}(K_c^- + K_c^+) \tag{10-26a}$$

and

$$\nu_c = \frac{1}{2}(K_c^- - K_c^+). \tag{10-26b}$$

With ν_c now known, we can immediately find the Mach number at *c* from the Prandtl-Meyer relation, Eq. 4-18 (or, equivalently, from either Appendix D or option **PM** in COMPQ). The isentropic relations, Eqs. 1-66, 1-70, and 1-71 (or either Appendix A or option **IS** in COMPQ), provide us with the static-to-stagnation ratios of all the properties at *c*. Finally, since the flow is homentropic, all stagnation properties are constant through the flow. Assuming that the upstream stagnation conditions are known, then the local static properties can be readily determined. Thus, we have found all of the flow properties at point *c*. It only remains now to find the location of point *c*.

To locate point *c* we will make the approximation that the characteristics connecting *c* to each of its neighbors *a* and *b* are straight lines as depicted in Figure 10-8. The angle each of these characteristics makes with the *x*-axis will be approximated by the average of the characteristic angle at each point. The two characteristics C_{ac}^-

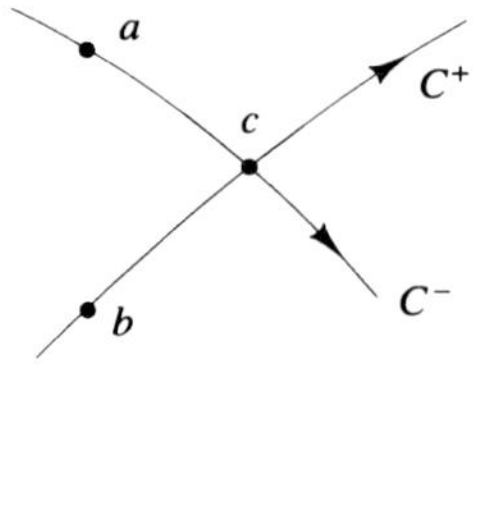

Figure 10-7 The General Schematic for a Field Point Unit Process

and C_{bc}^+ have slopes s_{ac} and s_{bc}, respectively, given by the tangent of this average angle. That is,

$$s_{ac} = \tan\left(\frac{1}{2}((\theta - \mu)_a + (\theta - \mu)_c)\right) \tag{10-27a}$$

and

$$s_{bc} = \tan\left(\frac{1}{2}((\theta + \mu)_b + (\theta + \mu)_c)\right). \tag{10-27b}$$

The equations for C_{ac}^- and C_{bc}^+ have the form

$$y = c_0 + c_1 x.$$

Since we know the coordinates of a and b and the slopes of the characteristics, the constants in this linear equation can be readily found for each characteristic. Thus we have for C_{ac}^-,

$$y = y_a + s_{ac}(x - x_a), \tag{10-28a}$$

and for C_{bc}^+

$$y = y_b + s_{bc}(x - x_b). \tag{10-28b}$$

Equations 10-28 are both valid at point c since C_{ac}^- and C_{bc}^+ intersect there. Consequently, for C_{ac}^-,

$$y_c = y_a + s_{ac}(x_c - x_a), \tag{10-29a}$$

and for C_{bc}^+,

$$y_c = y_b + s_{bc}(x_c - x_b). \tag{10-29b}$$

Equations 10-29 constitute a pair of linear algebraic equations for the two unknowns x_c and y_c. Solving this simple system provides us with the coordinates of point c. Thus,

$$x_c = \frac{y_a - y_b + s_{bc}x_b - s_{ac}x_a}{s_{bc} - s_{ac}}, \tag{10-30}$$

and y_c is given by either of Eqs. 10-29.

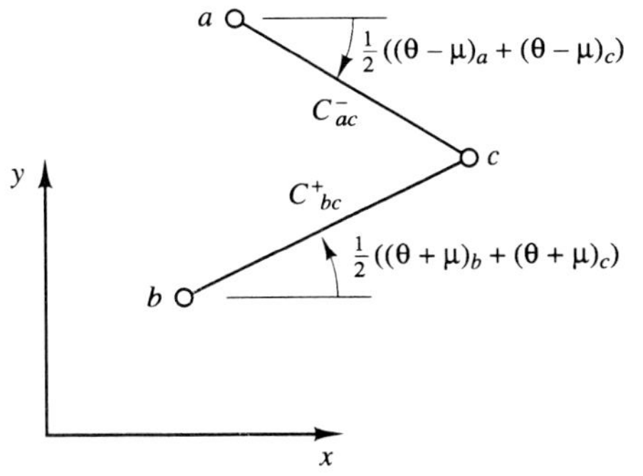

Figure 10-8 Locating the Unknown Point in a Field Point Unit Process

We now have the complete solution for the flow properties at, and the location of, point c. This procedure is valid anywhere away from a boundary.

Wall Point Unit Process

Figure 10-9 depicts the interaction of characteristics with a wall. In part a of this figure a C^- characteristic is incident on a lower wall, while in part b a C^+ characteristic is incident on an upper wall. In each interaction the characteristic reflects from the wall. Let us first determine the flow properties at the wall and the value of the Riemann invariant on the reflected characteristic.

We will assume that the value of K (K_i^- or K_i^+) on each incident characteristic and the wall shape are known. If we assume that the walls are solid, then at the wall the flow must be tangent to the wall so that the flow angle equals the wall angle. Thus, for Figure 10-9a,

$$K_i^- = K_w^- = \theta_w + \nu_w,$$

and for Figure 10-9b,

$$K_i^+ = K_w^+ = \theta_w - \nu_w.$$

The only unknowns in these expressions are the values of ν_w. Solving gives

$$\nu_w = K_i^- - \theta_w \tag{10-31a}$$

for the lower wall and

$$\nu_w = \theta_w - K_i^+ \tag{10-31b}$$

for the upper wall. With ν_w known, all of the wall flow properties can be found exactly as described for point c in the field point unit process.

The reflected Riemann invariants for Figures 10-9a and b are

$$K_r^+ = K_w^+ = \theta_w - \nu_w$$

and

$$K_r^- = K_w^- = \theta_w + \nu_w,$$

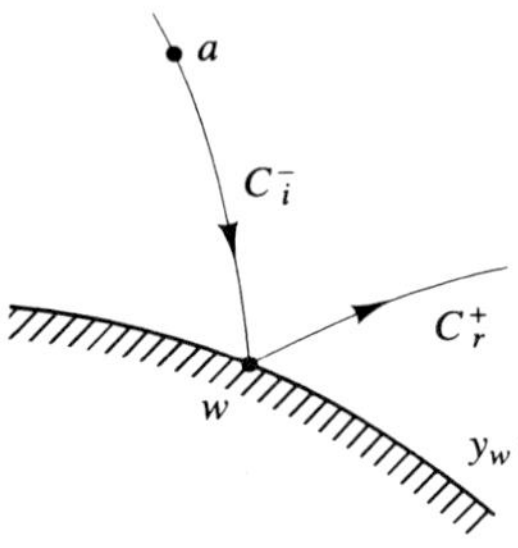

Figure 10-9a Reflection of a C^- Characteristic from a Lower Wall

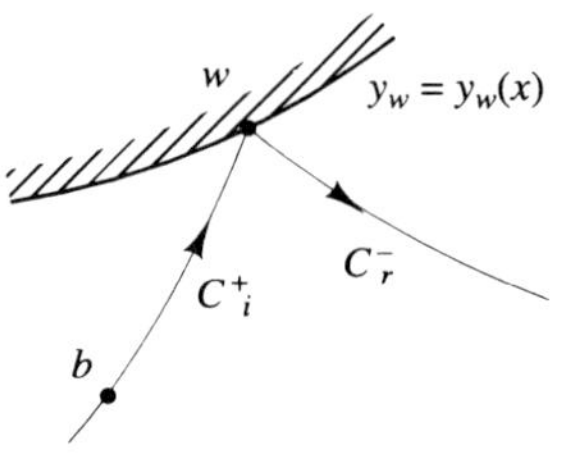

Figure 10-9b Reflection of a C^+ Characteristic from an Upper Wall

respectively. These can be put into a more useful form with the aid of Eqs. 10-31:

$$K_r^+ = 2\theta_w - K_i^-, \tag{10-32a}$$

$$K_r^- = 2\theta_w - K_i^+. \tag{10-32b}$$

Equations 10-32 express the reflected Riemann invariants in terms of the known incident Riemann invariants and the wall boundary condition.

The location of the intersection of the wall and the incident characteristic, (x_w, y_w) is found by a procedure that is essentially the same as that followed for the field point c. The only difference here is that if the wall shape is known, that is, if y_w is a known function of x,

$$y_w = y_w(x), \tag{10-33}$$

then this expression should be used instead of one of Eqs. 10-28 and 10-29. For example, for the lower wall assume that $y_w(x)$ is known and that the (known) incident characteristic passes through a point a whose location is also known. At the intersection we can then write

$$y_w = y_a + s_{aw}(x_w - x_a) \tag{10-34a}$$

for C_i^- and

$$y_w = y_w(x_w) \tag{10-34b}$$

for the wall. Similarly, for the upper wall, with the known incident characteristic passing through the known point b, we have

$$y_w = y_b + s_{bw}(x_w - x_b) \tag{10-35a}$$

for C_i^+ and

$$y_w = y_w(x_w) \tag{10-35b}$$

for the wall. The slopes s_{aw} and s_{bw} are obtained from expressions similar to Eqs. 10-27 but with the subscript c replaced by w. As before, Eqs. 10-34 and 10-35 are algebraic simultaneous systems for the unknowns x_w and y_w of the lower and upper wall intersections, respectively. These systems can be solved for the wall coordinates although the algebra may be more involved if the wall function $y_w(x)$ is complicated.

Finally, the angle of the reflected characteristic at the wall is $\theta_w \pm \mu_w$ for $C_r^{\pm}$, where

$$\mu_w = \sin^{-1}\left(\frac{1}{M_w}\right).$$

The wall Mach number, M_w, is found from ν_w and the Prandtl-Meyer relation, Eq. 4-18.

We now have the complete solution for the flow properties at, and the location of, point w. This procedure is valid anywhere along a solid boundary.

Free Surface Point Unit Process

The boundary of a supersonic jet issuing from a nozzle, such as that depicted in Figure 10-10, is an example of a free surface. Across a free surface the static pressure is constant. For the case of the exhaust flow in Figure 10-10 this says that

$$P_{\text{jet}} = P_{\text{b}}, \tag{10-36}$$

where P_{b} is the back or ambient pressure in the volume into which the nozzle is emptying. If the back pressure is uniform in space, then the static pressure along the free surface is constant, both inside and outside of the jet. For isentropic flow in the jet the stagnation pressure will also be constant so that,

$$\frac{P}{P_{\text{o}}} = \text{constant along the free surface.} \tag{10-37}$$

Hence, the Mach number, the Prandtl-Meyer function, and all other properties (except flow direction) will also be constant along the free surface. This gives us the information we need to determine the interaction of a characteristic with a free surface.

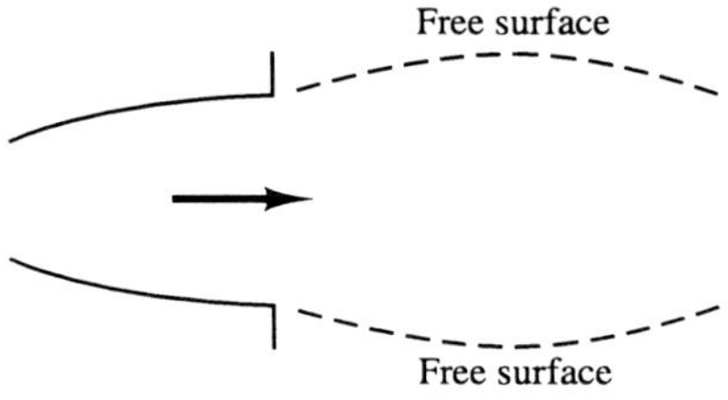

Figure 10-10 Free Surface Formed by a Jet Entering a Quiescent Fluid

Figure 10-11 shows a C^- characteristic incident upon a lower free surface in part and a C^+ characteristic incident upon an upper free surface in part b. For both cases we will assume that the ratio in Eq. 10-37 is given so that ν_{fs} is known. We will also assume that all flow properties and coordinates at points a and b are known so that K_i^- and K_i^+ are known. Then, for Figure 10-11a,

$$K_i^- = K_f^- = \theta_f + \nu_f,$$

and for Figure 10-11b,

$$K_i^+ = K_f^+ = \theta_f - \nu_f.$$

The only unknowns in these expressions are the values of θ_f. Solving gives

$$\theta_f = K_i^- - \nu_f \tag{10-38a}$$

for the lower free surface and

$$\theta_f = \nu_f - K_i^+ \tag{10-38b}$$

for the upper free surface. With ν_f known, all of the free surface flow properties can be found exactly as described for point c in the field point unit process.

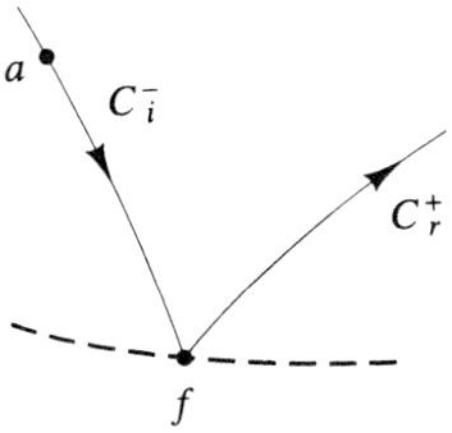

Figure 10-11a Reflection of a C^- Characteristic from a Lower Free Surface

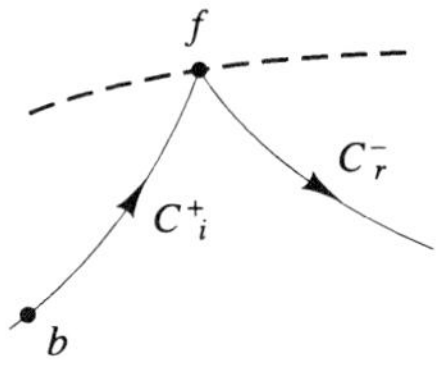

Figure 10-11b Reflection of a C^+ Characteristic from an Upper Free Surface

The reflected Riemann invariants for Figures 10-11a and b are

$$K_r^+ = K_f^+ = \theta_f - \nu_f$$

and

$$K_r^- = K_f^- = \theta_f + \nu_f,$$

respectively. These can be put into a more useful form with the aid of Eqs. 10-31:

$$K_r^+ = K_i^- - 2\nu_f, \tag{10-39a}$$

$$K_r^- = 2\nu_f - K_i^+. \tag{10-39b}$$

Equations 10-39 express the reflected Riemann invariants in terms of the known incident Riemann invariants and the free surface boundary condition.

To locate point f we will make the approximation that the characteristic and free surface segments connecting f to the neighboring points a and b are straight lines as depicted in Figure 10-12. The angle the characteristic makes with the x-axis will be approximated by the average of the characteristic angle at each point. The angle the free surface makes with the x-axis will be approximated by the average of the flow angle at each point. The slopes s_{af} and s_{bf} are given by the tangent of these average angles.

For Figure 10-12a, where a C^- hits a lower free surface, the slopes are

$$s_{af} = \tan\left(\frac{1}{2}((\theta - \mu)_a + (\theta - \mu)_f)\right) \tag{10-40a}$$

and

$$s_{bf} = \tan\left(\frac{1}{2}(\theta_b + \theta_f)\right). \tag{10-40b}$$

The equations for C_{af}^- and the free surface segment bf have the form

$$y = c_0 + c_1 x.$$

Assuming that we know the coordinates and slopes at a and b, the constants in this linear equation can be readily found for each segment. Thus we have for C_{af}^-,

$$y = y_a + s_{af}(x - x_a), \tag{10-41a}$$

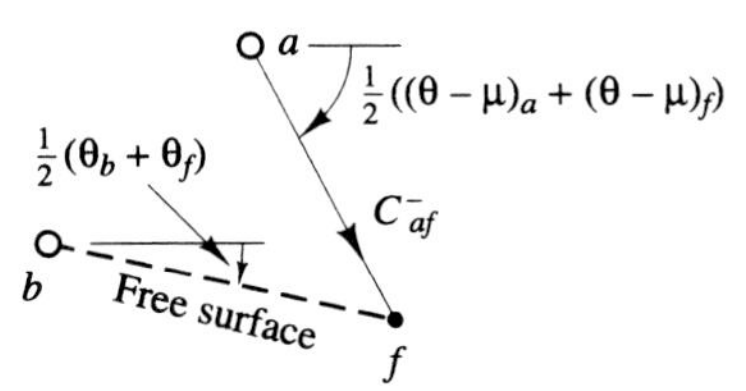

Figure 10-12a Locating the Unknown Point at a Lower Free Surface

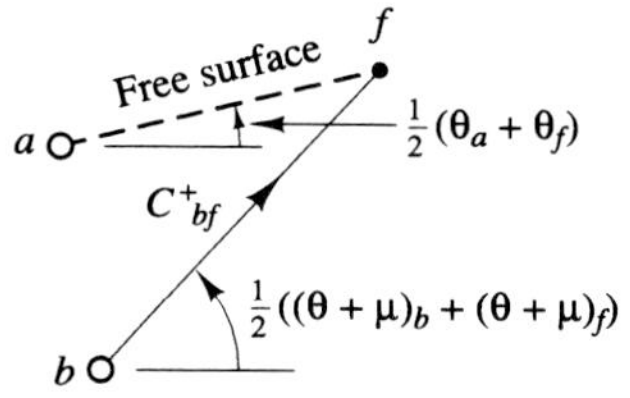

Figure 10-12b Locating the Unknown Point at an Upper Free Surface

and for segment bf

$$y = y_b + s_{bf}(x - x_b). \tag{10-41b}$$

Equations 10-41 are both valid at point f since C^-_{af} and segment bf both pass through point f. Consequently, for C^-_{bf},

$$y_f = y_a + s_{af}(x_f - x_a), \tag{10-42a}$$

and for segment bf,

$$y_f = y_b + s_{bf}(x_f - x_b). \tag{10-42b}$$

Equations 10-42 constitute a pair of linear algebraic equations for the two unknowns x_f and y_f. Solving this simple system provides us with the coordinates of point f. Thus,

$$x_f = \frac{y_a - y_b + s_{bf}x_b - s_{af}x_a}{s_{bf} - s_{af}}, \tag{10-43}$$

and y_f is given by either of Eqs. 10-42.

For the interaction in Figure 10-12b the characteristic and free surface segment subscripts switch. The slope of the free surface segment is now

$$s_{af} = \tan\left(\frac{1}{2}(\theta_a + \theta_f)\right), \tag{10-44a}$$

while the slope of the characteristic C^+_{bf} becomes

$$s_{bf} = \tan\left(\frac{1}{2}((\theta + \mu)_b + (\theta + \mu)_f)\right). \tag{10-44b}$$

The rest of the analysis proceeds as shown earlier with the result that the coordinates of point f are given by Eqs. 10-43 and 10-42.

Finally, the angle of the reflected characteristic at the free surface is $\theta_f \pm \mu_f$ for $C^\pm_r$, where

$$\mu_f = \sin^{-1}\left(\frac{1}{M_f}\right).$$

The free surface Mach number, M_f, is found from ν_f and the Prandtl-Meyer relation, Eq. 4-18.

We now have the complete solution for the flow properties at, and the location of, point f. This procedure is valid anywhere along a free surface.

Shock Point Unit Process

The procedure for determining the interaction of a characteristic with a shock wave is similar to the wall and free surface processes but with one additional complication. At both the wall and the free surface points, one of the dependent variables is known—θ_w for the wall and ν_f for the free surface. For a curved shock, however, the shock angle is, in general, not known. Consequently, neither θ nor ν is known on the downstream side of the shock. (We can assume that the values of θ and M (and thus ν) are known on the upstream side of the shock.) The solution requires searching for the shock angle such that the value of the Riemann invariant on the downstream side of the shock is equal to the value of the Riemann invariant on the incident characteristic. Once this shock angle is found, the rest of the solution is straightforward.

Interactions of characteristics with shocks are depicted in Figure 10-13. We assume that the flow properties upstream of the shock and the flow properties and coordinates of points a and b are known. The known shock points, b in Figure 10-13a and a in Figure 10-13b, are immediately downstream of the shock. We wish to determine the conditions at point s, which is at the intersection of the incident characteristic and the downstream side of the shock.

For Figure 10-13a,

$$K_i^- = K_s^- = \theta_s + \nu_s,$$

and for Figure 10-13b,

$$K_i^+ = K_s^+ = \theta_s - \nu_s.$$

The Riemann invariants in these two expressions are known, but neither θ_s nor ν_s is known individually. We must now use the known upstream conditions, M_1 and θ_1, and the oblique shock relations (Eqs. 3-10 or option **OS** of COMPQ) to find the shock that gives downstream conditions, M_2 and θ_2, that are compatible with the known downstream Riemann invariants. That is, for Figure 10-13a we must have

$$\theta_2 + \nu_2(M_2) = K_s^-, \tag{10-45a}$$

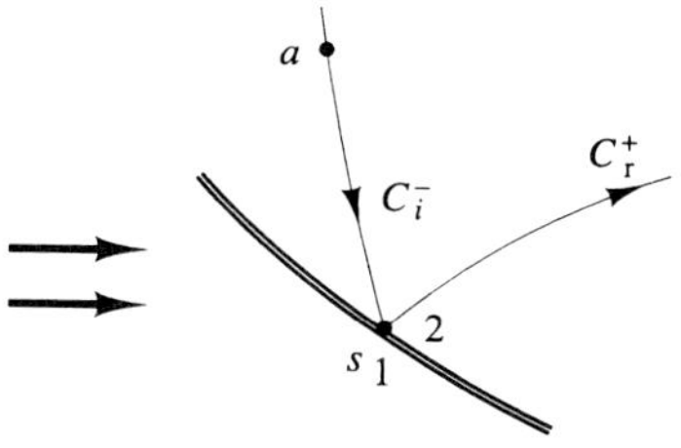

Figure 10-13a Reflection of a C^- Characteristic from a Lower Shock Wave

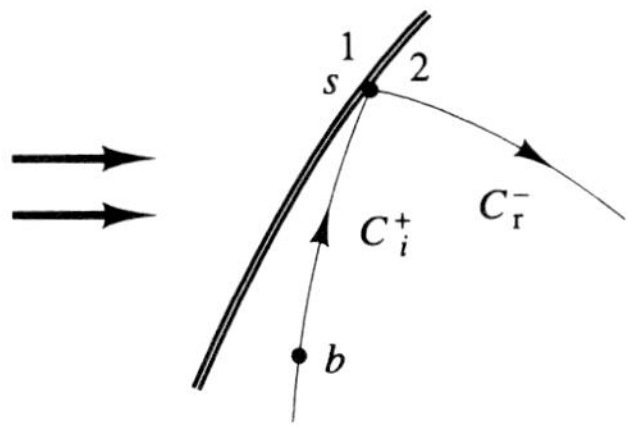

Figure 10-13b Reflection of a C^+ Characteristic from an Upper Shock Wave

and for Figure 10-13b,

$$\theta_2 - \nu_2(M_2) = K_s^+ . \qquad \text{(10-45b)}$$

The search for the proper shock can be done by first guessing a local shock angle σ. With σ and the (known) upstream Mach number M_1, we use the oblique shock solution to find the flow turn angle δ and the downstream Mach number M_2. The downstream Prandtl-Meyer angle is found from M_2 and the Prandtl-Meyer relation, Eq. 4-18. The downstream flow angle is determined by

$$\theta_2 = \theta_1 - \delta$$

for Figure 10-13a and by

$$\theta_2 = \theta_1 + \delta$$

for Figure 10-13b. We next check to see if Eq. 10-45 is satisfied. If it is satisfied, we have the correct shock solution. If it is not satisfied, we must select a new value for σ and repeat the preceding procedure.

With the correct shock solution we can then say that the final σ is σ_s and that all of the downstream shock properties (with subscript 2) are actually the properties we labeled with subscript s. That is,

$$\theta_s = \theta_2, \qquad M_s = M_2, \qquad \nu_s = \nu_2,$$

and so forth. Thus the flow properties at point s are completely known, including the reflected Riemann invariants. For Figure 10-13a,

$$K_r^+ = \theta_s - \nu_s, \qquad \text{(10-46a)}$$

while for Figure 10-13b,

$$K_r^- = \theta_s + \nu_s. \qquad \text{(10-46b)}$$

The procedure to locate point s is identical to the procedure to find point f on a free surface if the coordinates and flow properties at points a and b in Figure 10-14 are known and the segments as and bs are approximated by straight lines. Then Eqs. 10-40 through 10-44 and the associated discussion are directly applicable after minor changes in notation—the subscript f in Eqs. 10-40 through 10-44 is now an s;

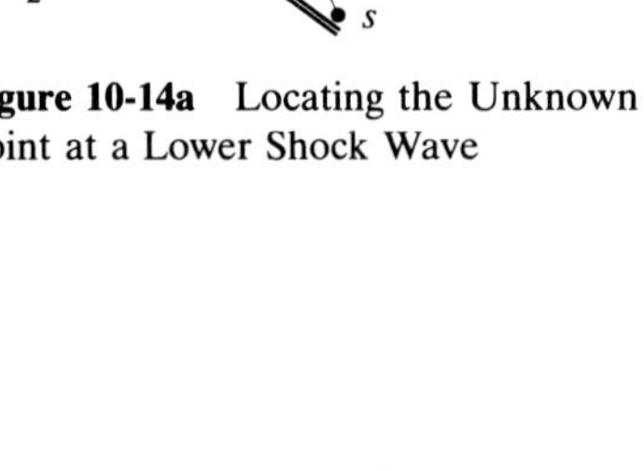

Figure 10-14a Locating the Unknown Point at a Lower Shock Wave

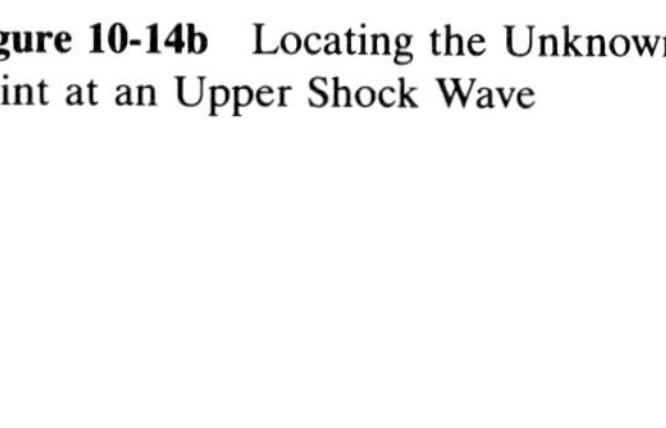

Figure 10-14b Locating the Unknown Point at an Upper Shock Wave

the angle θ in Eqs. 10-40b and 10-44a is now the shock angle σ; references to the free surface are now references to the shock; and references to Figure 10-12 become references to Figure 10-14. With these changes point s can be readily located.

Finally, the angle of the reflected characteristic at the shock is $\theta_s \pm \mu_s$ for $C_r^{\pm}$, where

$$\mu_s = \sin^{-1}\left(\frac{1}{M_s}\right).$$

We now have the complete solution for the flow properties at, and the location of, point s. This procedure is valid anywhere along a shock.

Initial Value Line

Each of the unit processes described so far requires that we know the coordinates and flow conditions at points a and b. But how do we obtain these properties? The answer is that they are determined through unit process calculations based on previous points. Conditions at these previous points are likewise found from unit processes based on even more previous points. Continuing this line of reasoning, we eventually must ask where do these calculations actually begin? To answer this question let us consider a particular example for which we might use the method of characteristics.

A choked converging–diverging nozzle with steady flow is depicted in Figure 10-15a. The sonic line, which separates the subsonic and supersonic flows, is indicated. (Notice the shape and location of the sonic line; we will discuss these in more detail later in this section.) Because the method of characteristics as we have developed it is applicable only to supersonic flow, it can be used only in the region downstream of the sonic line. The flow upstream of the sonic line and the details of the sonic line must be determined in some other way. We conclude from this that the sonic line serves as a sort of "boundary" for the method of characteristics (MOC) and might perhaps be used as the starting point for the MOC procedure. Actually,

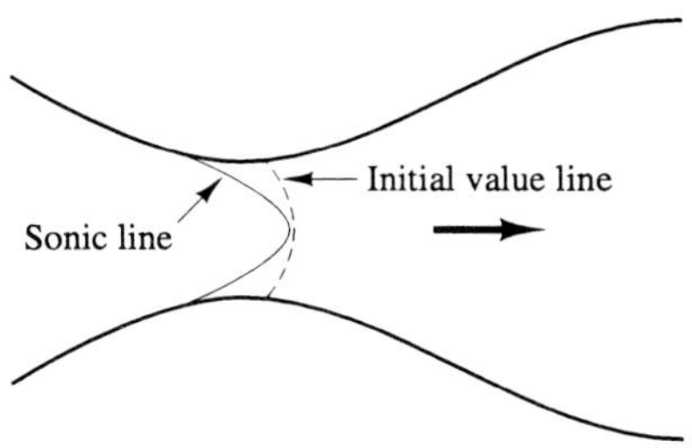

Figure 10-15a The Sonic Line and Initial Value Line in the Throat Region of a Supersonic Converging–Diverging Duct

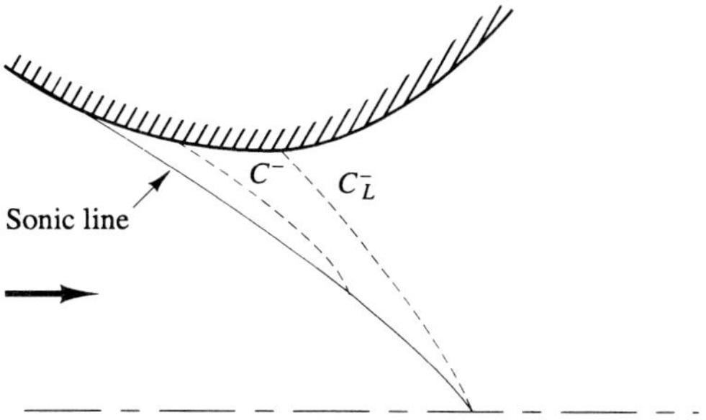

Figure 10-15b The Limiting Characteristic in the Throat Region of a Supersonic Converging–Diverging Duct

certain problems arise if we try to start the MOC calculations at the sonic line unless some additional approximations are made. There is, in fact, another line slightly downstream of the sonic line on which the procedure must generally be started. This line, also shown in Figure 10-15a, is known as the *initial value* (or *data*) *line* (IVL). Like the sonic line, conditions on the IVL must be determined by some method other than the method of characteristics. However, once the IVL conditions are known, the IVL may then be used as the starting point for the unit processes of the method of characteristics.

Let us examine the nozzle throat region a little more closely. Figure 10-15b presents an enlargement of the upper half of the throat. The sonic line is shown. Also shown is a C^- characteristic emanating from the upper wall a small distance downstream from where the sonic line intersects the wall. This characteristic intercepts the sonic line. Consequently, the information carried by this characteristic concerning the shape of the wall will have an influence on the sonic line and subsequently on the upstream subsonic region. This is a situation in which the method of characteristics is not valid. Hence, the MOC calculations must start downstream of any wall characteristics that intercept the sonic line. The characteristic in Figure 10-15b labeled C_L^- is the most downstream wall characteristic to intercept the sonic line. It is known as the *limiting characteristic*. The IVL must lie at, or downstream of, the limiting characteristic. Various methods have been developed for locating the IVL in nozzle flows. Sauer (1947) and Zucrow and Hoffman (1976) provide detailed discussions of some of these methods.

Figure 10-16 shows uniform, steady supersonic flow approaching a sharp-nosed airfoil. The airfoil has an attached shock with entirely supersonic flow downstream. Hence the downstream flow is amenable to analysis by the method of characteristics. Furthermore, we do not have to worry here about a sonic line nor a subsonic region. However, there is also not a clear starting place for our unit processes. One common approach to obtaining an IVL is sketched in Figure 10-16b. A short section of the airfoil leading-edge surface is approximated by a wedge with an angle equal to the airfoil leading-edge angle. The conditions between the shock and wedge are found from the oblique shock relations. The line marking the terminus of this wedge flow region (and connecting the shock and body) serves as the IVL. Of course, since the flow behind an attached shock produced by a wedge is uniform, the IVL conditions

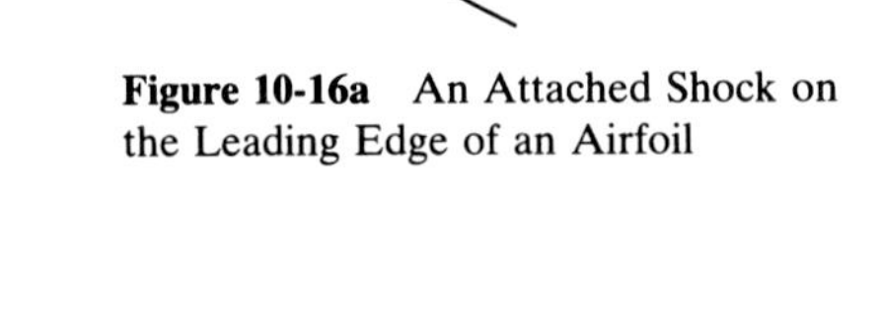

Figure 10-16a An Attached Shock on the Leading Edge of an Airfoil

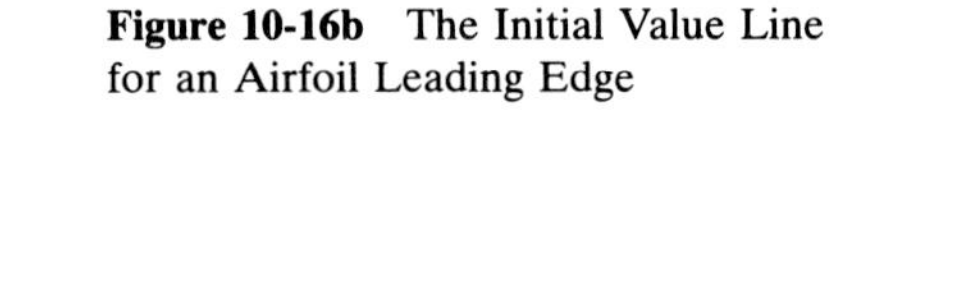

Figure 10-16b The Initial Value Line for an Airfoil Leading Edge

are constant along its length. This is quite satisfactory, however, and an MOC solution can be constructed from this starting line.

A blunt body in supersonic flow, such as the cylinder shown in Figure 10-17, produces a situation similar to that in the throat region of a nozzle. The sonic line and characteristics lie in such a way that some of the characteristics intercept the sonic line. This permits disturbances that originate in the supersonic region to be transmitted into (and throughout) the subsonic flow. The method of characteristics can be applied only downstream of the last characteristic to intercept the sonic line. That is, the IVL for blunt body flow must lie downstream of the limiting characteristic.

The approach Mach number in Figure 10-17a is slightly greater than 1. Here, the limiting characteristic on the upper half of the body is the C^- characteristic that originates at the body and intercepts the shock–sonic line junction. C^- characteristics leaving the body upstream of this limiting characteristic intersect the sonic line and carry information about the body shape in the supersonic region into the subsonic flow. As the freestream Mach number becomes large relative to 1, the sonic line moves toward the stagnation point and becomes more nearly parallel to the freestream. This situation is shown in Figure 10-17b. Now the limiting characteristic on the upper surface is the C^+ that originates at the shock and intercepts the body–sonic line junction. C^+ characteristics leaving the shock upstream of this limiting characteristic intersect the sonic line and carry information about the shock shape in the supersonic region into the subsonic flow. In both cases the IVL must lie downstream of the limiting characteristics. The texts by Hayes and Probstein (1966) and Zucrow and Hoffman (1976) provide detailed discussions on using the method of characteristics to solve blunt body flows.

Zones of Influence and Dependence

Recall our discussion concerning Figures 1-12, 1-13, and 1-14. This discussion pointed out an extremely important concept: a disturbance in steady subsonic flow is felt everywhere throughout the flow field, while a disturbance in steady supersonic flow is felt only inside the Mach cone (or wedge in a planar flow) trailing downstream from the disturbance. This observation is equivalent to saying that information, or disturbances, cannot propagate upstream in a supersonic flow although they can in a subsonic flow. We have encountered these ideas several times previously in this text.

The downstream Mach cone or wedge is sometimes referred to as the *region of influence* of a supersonic disturbance. We should note that in a general multidimensional flow the characteristics curve so that the Mach cone somewhat distorts in shape. Despite any distortion, we shall retain the nomenclature "downstream Mach cone"; the meaning should usually be obvious. Figure 10-18a shows several tiny disturbances in a planar supersonic flow. Although they are slightly distorted, the downstream Mach wedges are still clearly identifiable.

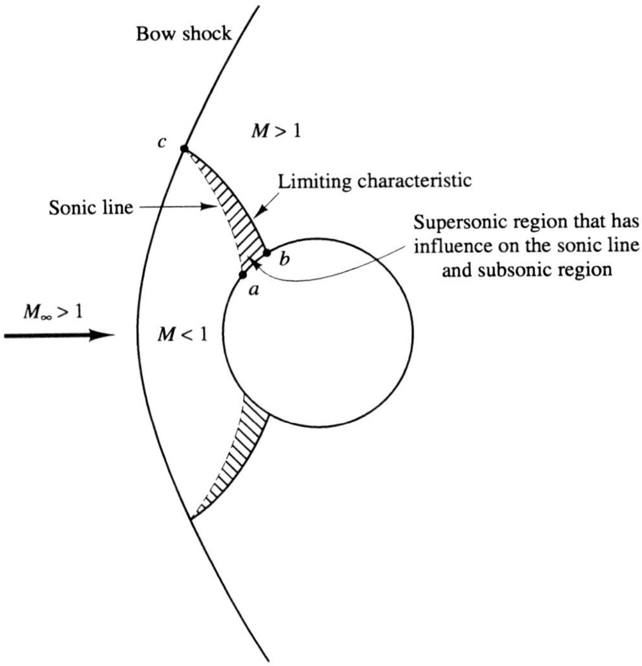

Figure 10-17a The Limiting Characteristic for Low Supersonic Flow Past a Cylinder. *From Anderson (1990).*

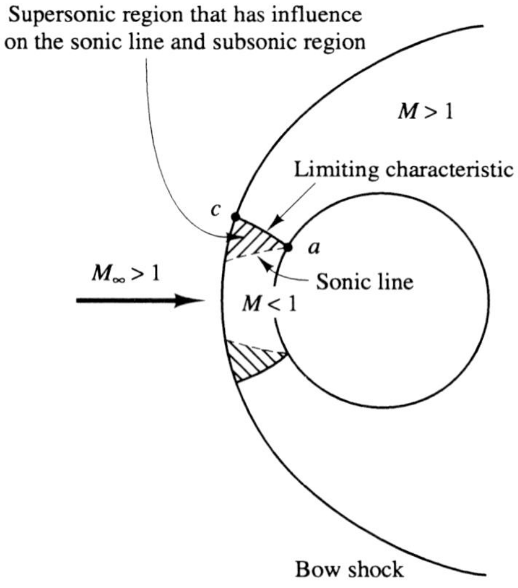

Figure 10-17b The Limiting Characteristic for High Supersonic Flow Past a Cylinder. *From Anderson (1990).*

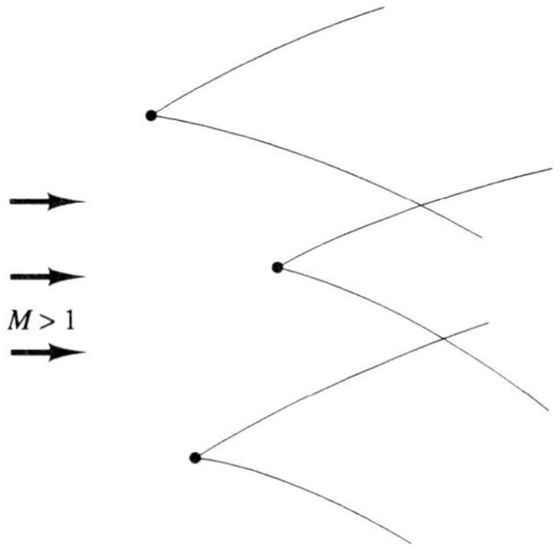

Figure 10-18a Mach Cones Propagating Downstream from Small Disturbances in a Supersonic Flow

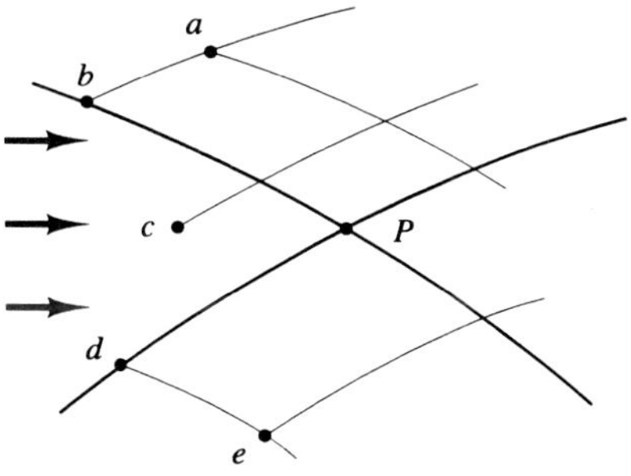

Figure 10-18b Regions of Influence and Dependence in Supersonic Flow

In Figure 10-18b several more disturbances have been added to the flow. In this particular situation point P lies outside the regions of influence of points a and e, and so they have no effect at P. Conversely, P lies inside the Mach wedge of point c and thus experiences an effect from c. P lies on an edge of the Mach wedges from points b and d, and so these points also influence P.

Examine Figure 10-18b more closely. Points b and P lie on a common C^- characteristic, call it C_P^-, while d and P lie on a common C^+ characteristic, C_P^+. Downstream of point P these two characteristics form the boundaries of the downstream Mach wedge of P and thus are also the boundaries of the region of influence of P. Upstream of P the characteristics C_P^- and C_P^+ delimit another important flow zone. Any disturbance that exists in this upstream zone, such as the disturbance at point c, exerts an effect on point P. We might say that conditions at point P depend on disturbances within this zone. This upstream zone is the upstream Mach wedge, or cone for three-dimensional flow, of point P. It is also sometimes referred to as the *domain of dependence* of a point in a steady supersonic flow.

The fact that there are limited zones of influence and dependence for any given point in a steady supersonic flow is quite useful and could, in fact, be regarded as the physical basis of the method of characteristics.

Properties of Characteristic Regions

In Figure 10-6 we observed and defined four regions that can be formed by Mach wave systems. Since Mach waves are the characteristics in planar supersonic flow, we can regard the characteristics as forming these regions. Let us now examine each of these regions more closely.

The uniform region is quite easy to describe. By definition, Mach number, flow direction, and all other properties are constant throughout the uniform region. Thus the Riemann invariants and the angles of the characteristics are also constant throughout the region. Hence, we conclude that all C^- characteristics are straight and parallel and that all C^+ characteristics are straight and parallel. If the uniform region is designated by the subscript 1, then

$$K^- = \theta_1 + \nu_1 \quad \text{and} \quad K^+ = \theta_1 - \nu_1 \quad \text{everywhere,}$$

and the angles of the characteristics are

$$\theta_1 - \mu_1 \quad \text{on all } C^- \quad \text{and} \quad \theta_1 + \mu_1 \quad \text{on all } C^+.$$

A simple$^+$ region is illustrated by supersonic flow over the rounded expansion corner depicted in Figure 10-19. Chapter 4 discussed this flow in some detail; the Prandtl-Meyer function was an important and highly useful result of that discussion. Now we can ask, according to the characteristic theory we have developed in this chapter, what are the properties of this simple$^+$ region?

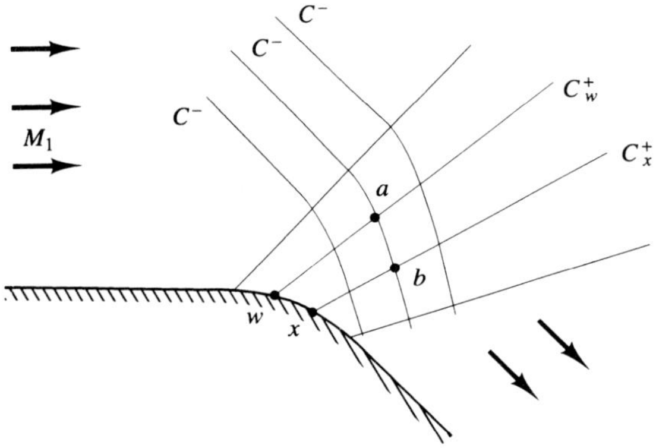

Figure 10-19 Characteristics in a Simple$^+$ Region Formed by a Gradual Expansion Turn

A characteristic leaves from each point along the wall and carries with it the Riemann invariant associated with the conditions at the originating wall point. For example, assuming that the uniform approach flow is known then at wall point w, we have

$$\theta_w = \text{the given wall angle at } w;$$

$$\nu_w = \nu_1 + (\theta_1 - \theta_w);$$

$$M_w = M(\nu_w);$$

$$\mu_w = \mu(M_w);$$

$$K_w^+ = \theta_w - \nu_w = \text{constant along } C_w^+;$$

$$\text{angle of characteristic } C_w^+ \text{ at } w = \theta_w + \mu_w.$$

The C^- characteristics entering the simple region originate in the uniform region. Consequently, they all have the same value of $K^- = K_1^-$. Prior to entering the simple region, they are also straight and parallel.

Now consider any point, say a, on C_w^+. At this point we have

$$K_a^- = \theta_a + \nu_a = K_1^- = \theta_1 + \nu_1 = \text{known}$$

and

$$K_a^+ = \theta_a - \nu_a = K_w^+ = \theta_w - \nu_w = \text{known}.$$

Solving first for θ_a, we have

$$\begin{aligned}\theta_a &= \frac{1}{2}(K_1^- + K_w^+) = \frac{1}{2}(\theta_1 + \nu_1 + \theta_w - \nu_w) \\ &= \frac{1}{2}(\theta_1 + \nu_1 + \theta_w - (\nu_1 + \theta_1 - \theta_w)) \\ &= \theta_w.\end{aligned}$$

Next solving for ν_a, we have

$$\begin{aligned}\nu_a &= \frac{1}{2}(K_1^- - K_w^+) = \frac{1}{2}(\theta_1 + \nu_1 - \theta_w + \nu_w) \\ &= \frac{1}{2}(\nu_w + \nu_w) \\ &= \nu_w.\end{aligned}$$

Since point a is arbitrarily located along C_w^+, we conclude that θ and ν remain constant along C_w^+. With ν constant all other properties, including μ, are constant. Finally, with both θ and μ constant then the angle of C_w^+ is constant. Thus, we have proven that C_w^+ is straight and has constant flow properties along it.

Now consider an adjacent wall point x. C_x^+ is straight and carries on it the constant values θ_x, ν_x, and so forth, which are different than the values carried by C_w^+. From the shape of the wall and the behavior of the Prandtl-Meyer function and the Mach angle, we can easily show that

$$\theta_w + \mu_w > \theta_x + \mu_x,$$

and so the two characteristics diverge, as we always depict for an expansion.

Behavior along the C^- characteristics is our next concern. Consider point b which lies at the intersection of C_x^+ and the C^- through point a. The conditions at b are the same as the conditions at x; these conditions are not the same as those at point a. It is important to realize that even though a and b are connected by a common C^- characteristic, only the Riemann invariant is necessarily constant along this C^-. The individual properties (θ, ν, M, and so on) may vary. Such is the case here. However, since the C^- characteristics originate in a uniform region, they all have the same value of K^-.

Finally, the angle of the C^- characteristics is of interest. At point a the angle is $\theta_w - \mu_w$ and at point b it is $\theta_x - \mu_x$. Although it is not readily apparent, these two angles will usually not be equal. (This inequality can be easily demonstrated, but not proved, with some simple Prandtl-Meyer calculations.) Hence we reach the conclusion that the C^- characteristics are generally curved in a simple$^+$ region.

To summarize, in a simple$^+$ region the C^+ characteristics are straight but not parallel. All properties are constant along any given C^+, but they vary between the C^+ characteristics. All C^- characteristics carry the same constant value of K^-, but the individual flow properties vary along any given C^-. The C^- characteristics are curved.

The preceding discussion can be immediately applied to a simple$^-$ region by appropriate changes in signs.

In a nonsimple region each individual characteristic carries on it a constant value for the Riemann invariant. This value varies between adjacent characteristics, however. Along any given characteristic the individual flow properties also vary. The characteristics are all curved.

Centered Expansions

The abrupt turn in Figure 10-20 produces a centered Prandtl-Meyer expansion fan. This is, of course, one of the basic flows that we discussed in Chapter 4. The downstream conditions are easily determined with the aid of the Prandtl-Meyer function and the isentropic relations. The fan is a simple region (a simple$^+$ region in this case). Conditions within the fan can also be determined quite easily if we treat the sharp turn as the limit of a gradual turn as the radius of curvature of the turn goes to zero. Equivalently, imagine viewing a gradual turn from far away. The turn will then shrink to a point with a centered expansion fan. Zooming in on the turn reveals details similar to those shown in Figure 10-19.

To analyze the sharp turn, simply choose a set of flow angles, θ_w, with values between the initial and final flow angles of the turn,

$$\theta_1 \geq \theta_w \geq \theta_2.$$

Each angle corresponds to a unique characteristic emanating from the corner. The

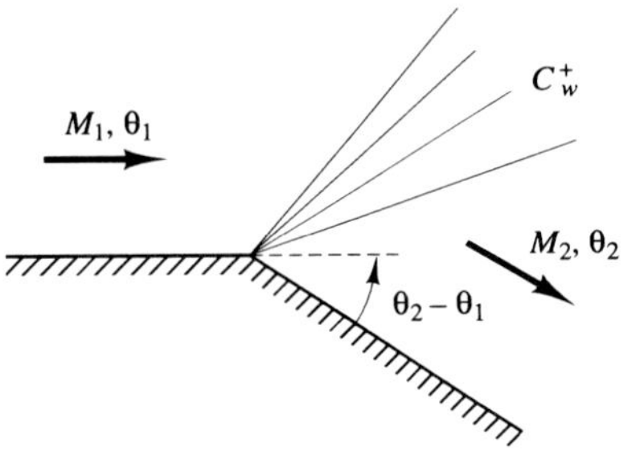

Figure 10-20 The Simple$^+$ Region of a Centered Prandtl-Meyer Expansion Fan

properties on each characteristic are constant and are found using the procedure described in the previous section for the simple$^+$ region. That is,

$$\begin{aligned}\theta_w &= \text{the given wall angle at } w;\\ \nu_w &= \nu_1 + (\theta_1 - \theta_w);\\ M_w &= M(\nu_w);\\ \mu_w &= \mu(M_w);\\ K_w^+ &= \theta_w - \nu_w = \text{constant along } C_w^+;\\ \text{angle of characteristic } C_w^+ \text{ at } w &= \theta_w + \mu_w.\end{aligned}$$

For a sharp corner producing a simple$^-$ region, the same procedure is used except for appropriate changes in signs.

Compression Turns

A gradual compression turn, such as that shown in Figure 10-21, produces an isentropic wave system near the wall. For the situation shown in Figure 10-21, the wave system forms a simple$^+$ region. Because the flow is isentropic, the Prandtl-Meyer function is valid. Consequently, the analysis for the gradual expansion applies here also. Observe that through the expansion in Figure 10-19, θ_w continually becomes more negative so that ν_w continually increases; however, through the compression, θ_w increases so that ν_w decreases. In addition, whereas the angle of successive characteristics in the expansion decreases so that the characteristics diverge, the angle of characteristics in the compression increases so that the characteristics converge. Figure 10-22 presents the experimentally measured pressure distribution on a curved compression surface with an approach Mach number of 2.7. Also shown is the pressure distribution found by the isentropic method we have developed here. The agreement is quite good and demonstrates the validity of our analysis.

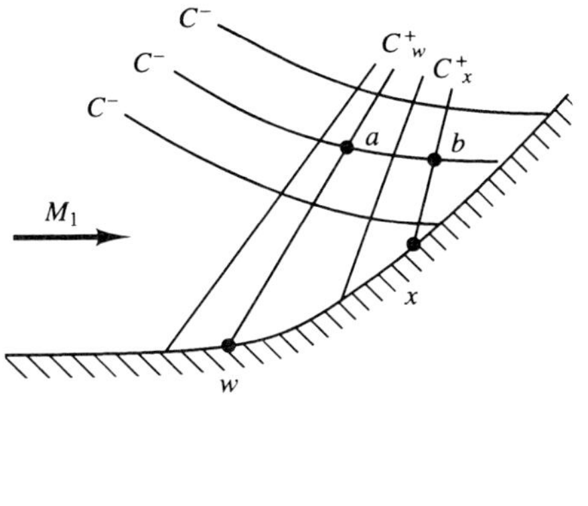

Figure 10-21 Characteristics in a Simple$^+$ Region Formed by a Gradual Compression Turn

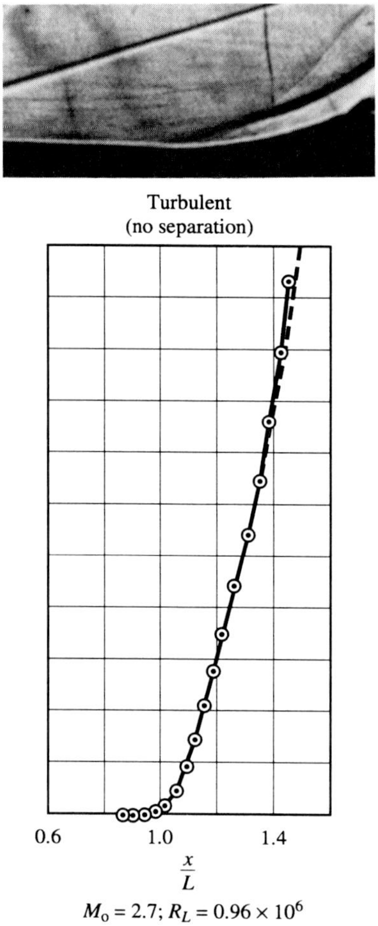

Figure 10-22 Comparison of the Experimental and Theoretical (Inviscid) Pressure Distributions on a Curved Compression Surface in Supersonic Flow *(from Chapman et al., 1957)*. The Mach number approaching the turn is 2.7.

Convergence, or coalescence, of the compression characteristics poses an important problem. Suppose that two characteristics from a gradual compression ultimately intersect, as shown in Figure 10-23. Let us determine the conditions at the intersection point. This has the appearance of a field point unit process and so we will apply that approach. Assume that conditions at a point on each of the incident characteristics, say points a and b, are known. Then K_a^+ and K_c^+ are known and have distinctly different values. Since point c lies on the characteristic through a, then

$$K_a^+ = K_c^+ = \theta_c - \nu_c.$$

But, point c also lies on the characteristic through b, and so

$$K_b^+ = K_c^+ = \theta_c - \nu_c.$$

This is an impossible situation—the Riemann invariant K_c^+ cannot be equal to two

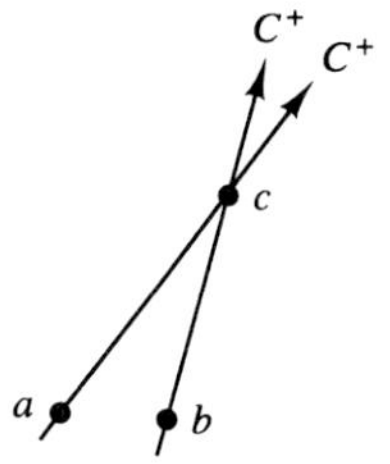

Figure 10-23a Intersection of C^+ Characteristics at a Field Point in a Simple$^+$ Compression Region

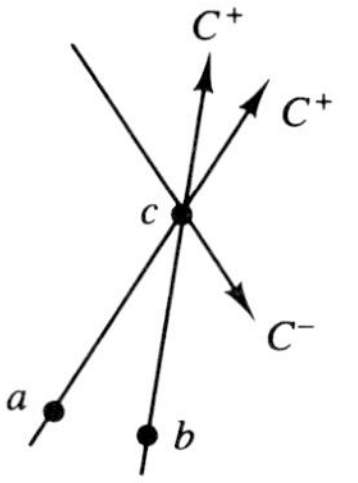

Figure 10-23b Intersection of C^+ and C^- Characteristics at a Field Point in a Simple$^+$ Compression Region

distinctly different numbers. To complicate matters further, we have not accounted for the C^- that also passes through point c. Doing so requires

$$K_1^- = K_c^- = \theta_c + \nu_c,$$

where K_1^- is the value in the uniform approach flow. The net result here is that we now have three equations for the two unknowns, θ_c and ν_c. This is an overconstrained algebraic system, and there is not a unique solution for θ_c and ν_c.

This difficulty arises not from an error in our solution procedure but in a failure of one of the fundamental assumptions we used in developing the method of characteristics. Recall from the beginning of this chapter that we are assuming here that the flow has uniform enthalpy and entropy throughout. One requirement in order for this assumption to be true is that the flow is inviscid. In practice this means that we are assuming that viscous effects are negligibly small. However, in the vicinity of converging compression characteristics gradients in the velocity field ($\partial u/\partial x$, $\partial u/\partial y$, and so on) become extremely large. In regions of large velocity gradients viscous effects are significant and must be included in any analysis. Since the method of characteristics does not account for viscous effects, it is not valid in regions of large gradients and is therefore not valid in the vicinity of converging characteristics. The method of characteristics can be used for gradual compression turns as long as we stay away from regions where the characteristics begin to cross. Realize that we are talking about characteristics of the same family crossing. It is quite satisfactory for characteristics of the opposite family to cross. Indeed, that is an intrinsic feature of the characteristic method. But characteristics of the same family may not cross. In an actual flow situation coalescing compression Mach waves ultimately merge to form an oblique shock wave. This is illustrated in Figure 10-24.

Figure 10-24 Formation of an Oblique Shock from Coalescing Mach Waves Produced by a Gradual Compression Turn in Supersonic Flow *(from Johannsen 1952).* The Mach number approaching the turn is 1.96.

10.4 SUPERSONIC NOZZLES

Introduction

In Chapter 6 we learned that a converging–diverging duct or nozzle is required in order to accelerate a gas that is initially at rest to supersonic speeds. An example appears in Figure 10-25. We also developed the notion of the area ratio A/A^*, which allows us to determine the flow conditions at any axial station along a duct. There is an important limitation to the area ratio analysis, however. A key assumption in Chapter 6 is that the flow is quasi-one-dimensional, meaning that we consider only variations in the axial, or longitudinal, direction and neglect variations across the duct, in the transverse direction. The true flow is multidimensional so that there are transverse variations. To determine the flow more accurately within a duct, a more complete analysis technique is needed. The method of characteristics provides us with such a technique.

There is an additional concern for which the quasi-one-dimensional approach is inadequate. To achieve a particular nozzle exit Mach number, the quasi-one-dimensional method can tell us what the exit area ratio must be. However, it

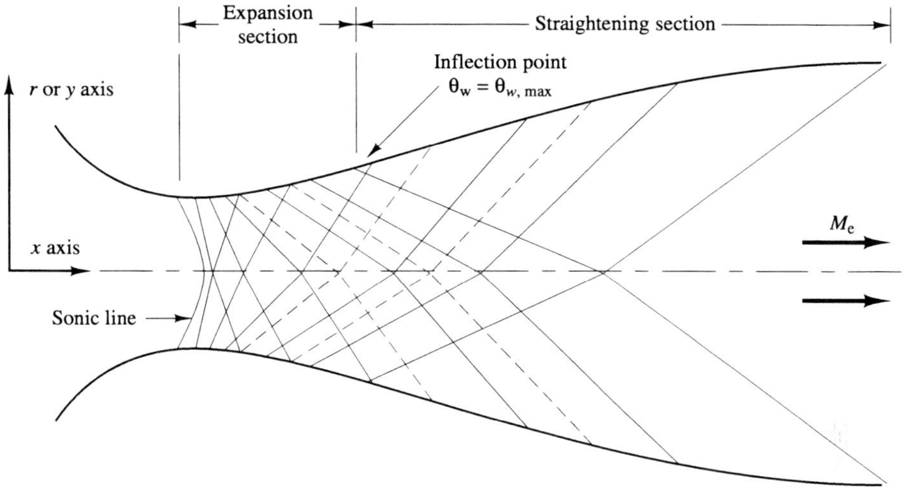

Figure 10-25 A Converging–Diverging Duct with Subsonic Flow Upstream of the Throat and Supersonic Flow Downstream of the Throat

does not tell us what the nozzle shape should be to achieve this design condition. Because of the multidimensionality of the flow, shock waves can form in the duct if it is not shaped correctly. The proper shape cannot be determined from the quasi–one-dimensional approach. Again, the method of characteristics can provide this information.

Terminology

Let us now examine the nozzle in Figure 10-25 more closely. The flow passes from subsonic to supersonic speeds in the throat region. As our discussion in Section 10.3 on initial value lines noted, the sonic line in the throat is actually curved. The sonic line starts at the wall slightly upstream of the minimum area and crosses the nozzle centerline downstream of the minimum area. Downstream of the throat the duct walls diverge, forming what is known as the *expansion section*. In many applications the expansion section eventually terminates and is followed by a section in which the duct wall angle decreases. This section with decreasing wall angle is variously known as the *straightening*, *converging*, or *compression* section. We will adopt the first term, straightening, to avoid confusion with the subsonic converging section of a nozzle. The point at which the nozzle transitions from expansion to straightening is an inflection point in the wall shape and is also the location of the maximum wall angle $\theta_{w,\max}$ (with respect to the centerline). The straightening section often ends with the walls parallel, or nearly parallel, to the centerline.

As shown in Figure 10-25, Mach waves, or characteristics, are created and reflected all along the walls of the expansion section. These waves form a nonsimple region in the duct. For many nozzle applications uniform, or nearly uniform, flow is required at the nozzle exit. In these situations the nonsimple region must end prior to the nozzle exit. Terminating the nonsimple region can be achieved by preventing the creation and reflection of waves at the downstream walls. This is the purpose of the straightening section. The waves are eliminated in the straightening section by matching the straightening section wall angle at each point to the value of θ on the incident characteristic at that point. No reflected or created wave is then necessary, because the wall boundary condition is automatically satisfied.

Nozzle applications requiring uniform exit conditions usually have long expansion and straightening sections, such as shown in Figure 10-25, to provide the high-quality flow needed at the exit. The nozzles of supersonic wind tunnels are an example of this. Other applications require the nozzle to be as short as possible. Examples of this include rocket nozzles where minimum weight is desired and nozzles for gas-dynamic lasers where very rapid expansion is needed to produce a certain type of condition known as *nonequilibrium flow* (see Anderson (1976) and Anderson (1990) for good introductions to gas-dynamic lasers and nonequilibrium flow). In a *minimum-length nozzle*, Figure 10-26, the initial expansion is produced by a sharp corner at the throat. The throat turn angle is $\theta_{w,\,max,\,ml}$. The straightening section of a minimum-length nozzle is often prematurely terminated to yield a *truncated* nozzle. Rocket nozzles on interplanetary and deep-space vehicles often are *conical*, having no straightening section, in order to keep the nozzle as light in weight as possible. An example of a conical rocket nozzle appears in Figure 10-27.

Let us assume that the nozzles in Figures 10-25 and 10-26 have the same exit

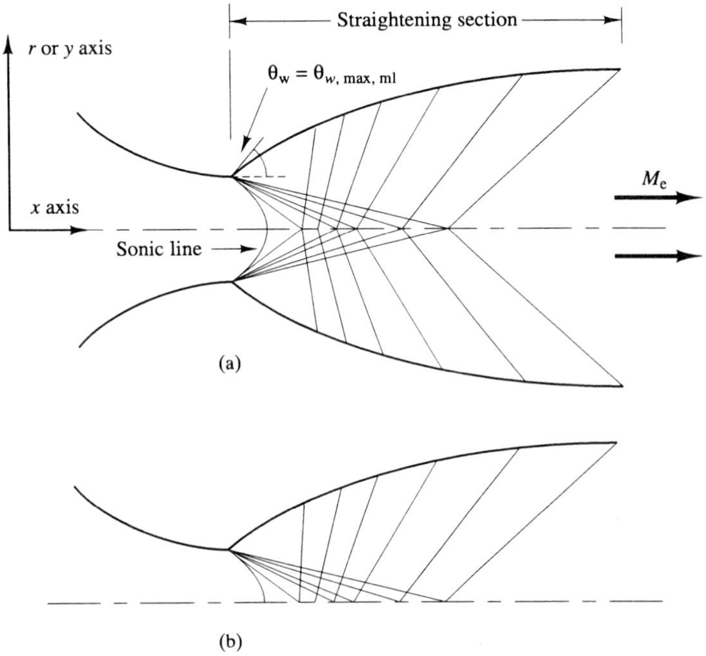

Figure 10-26 A Minimum-Length Nozzle

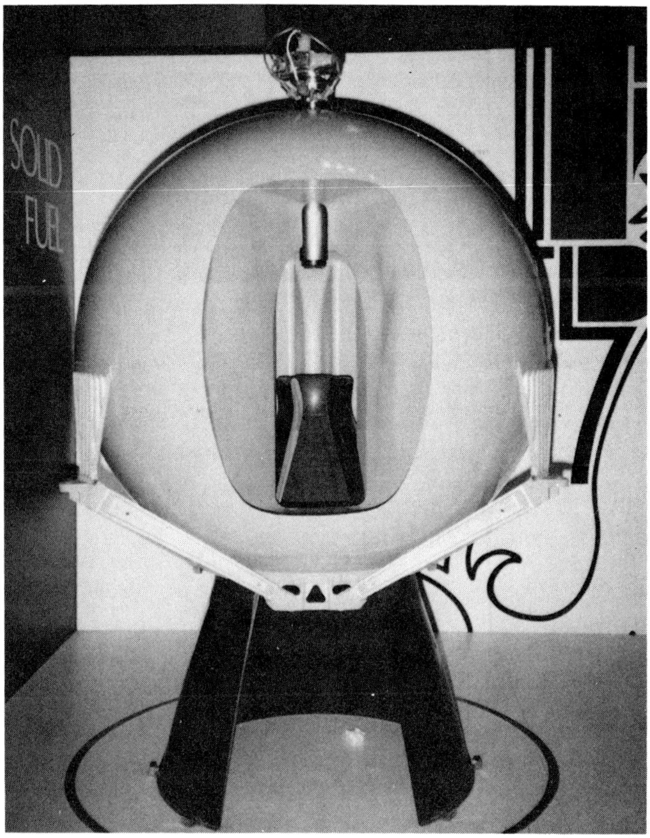

Figure 10-27 A Rocket Engine Used for Insertion of Satellites into Orbit. The diverging portion of the rocket nozzle is conical.

Mach number and let us compare their wave patterns. For the gradual expansion nozzle in Figure 10-25, fluid moving through the expansion section encounters Mach waves that originate at both the upper and lower walls, and it also encounters reflections of these waves from the walls. All of these are expansion waves, and the fluid experiences a continual acceleration. In contrast, for the minimum-length nozzle in Figure 10-26, the fluid encounters only the waves originating at the walls; there are no reflected waves. As a consequence, in order to obtain the same net acceleration (and the same exit Mach number), $\theta_{w,\max,\mathrm{ml}}$ must be larger than $\theta_{w,\max}$.

There is one additional observation to make from the wave patterns in Figures 10-25 and 10-26. In both figures the x-axis is an axis of symmetry and the wave patterns are symmetrical top and bottom. Because of this symmetry we can treat the centerline as if it is a horizontal wall and the C^+ characteristics as if they are all the result of reflections of the C^- characteristics from the centerline "wall." This is sketched in Figure 10-26b. Then we only need to analyze the upper half of the nozzle; the lower half is symmetrically the same.

Minimum-Length Nozzle Expansion Conditions

Here we will establish a general relation for $\theta_{w,\max,\mathrm{ml}}$, the throat turn angle of a minimum-length planar nozzle as shown in Figure 10-28. In this analysis the sonic line is approximated by a straight line normal to the flow at the throat. Centered expansions emanate from the two corners. These expansion fans form a nonsimple region in the middle of the duct. The nozzle walls in the straightening section are shaped in such a way (which we will find) that no reflections occur. Thus the expansions emerge from the nonsimple region to form simple regions approaching the nozzle walls. Downstream of these simple regions is the uniform region in which the flow is parallel to the x-axis and is at the design Mach number M_e.

The flow conditions at point c, the last point in the nonsimple region on the x-axis, are the design conditions,

$$\theta = 0, \qquad M = M_e, \qquad \text{and} \qquad \nu = \nu(M_e). \tag{10-47}$$

The last C^- in the upper expansion is C_{ac}^- and Eq. 10-47 provides known conditions at one point, point c, on this characteristic. At the corner the conditions on C_{ac}^- are

$$\theta = \theta_{w,\max,\mathrm{ml}} \qquad \text{and} \qquad M = M_a,$$

where both quantities are to be determined. Now, since the upper expansion begins at the throat where $M = 1$, then by the definition of the Prandtl-Meyer function we can say that on C_{ac}^- at the corner,

$$\nu_a = \theta_{w,\max,\mathrm{ml}}.$$

Then,

$$K_a^- = \theta_a + \nu_a = \theta_{w,\max,\mathrm{ml}} + \theta_{w,\max,\mathrm{ml}} = 2\theta_{w,\max,\mathrm{ml}}.$$

Similarly at point c we have

$$K_c^- = \theta_c + \nu_c = 0 + \nu(M_e) = \nu(M_e).$$

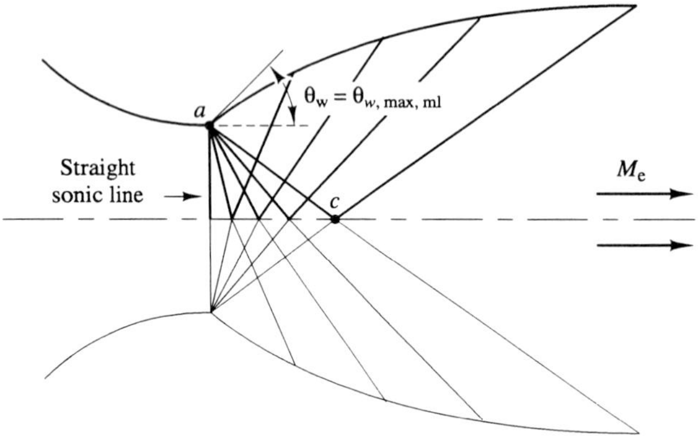

Figure 10-28 Details for Determining the Initial Turn Angle of a Minimum-Length Nozzle

Since K^- is constant along C^-_{ac}, then

$$K^-_a = 2\theta_{w,\,\max,\,\mathrm{ml}} = K^-_c = \nu(M_e). \tag{10-48}$$

From Eq. 10-48 we finally conclude that

$$\theta_{w,\,\max,\,\mathrm{ml}} = \frac{1}{2}\nu(M_e). \tag{10-49}$$

The required turn angle at the throat of a minimum-length planar nozzle is equal to one-half of the design Prandtl-Meyer function.

Design of a Planar Minimum-Length Nozzle with Uniform Exit Flow

To illustrate the implementation of the method of characteristics we will now design the supersonic portion of a converging–diverging nozzle. This will be a planar (two-dimensional) minimum-length nozzle, symmetric with respect to the x-axis and with uniform flow at the exit. The design exit Mach number is 1.92, for which $\theta_{w,\,\max,\,\mathrm{ml}} = 12.075°$ by Eq. 10-49. Because the nozzle is symmetric we will only analyze the upper half. A schematic of this portion of the nozzle appears in Figure 10-29.

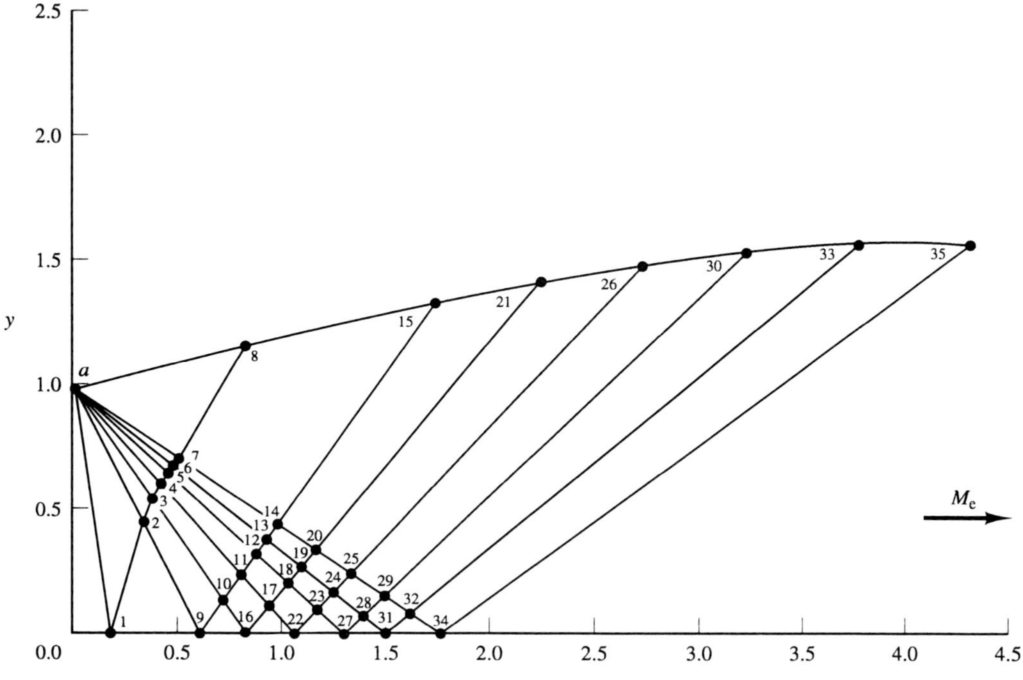

Figure 10-29 The Nozzle Designed for an Exit Mach Number of 1.92 Using Seven Initial Characteristics at the Throat and the Program CHARQ

When we use the method of characteristics, certain engineering judgments must be made. The first judgment involves the number of characteristics to be analyzed. The expansion process at the throat produces an infinite number of Mach waves or characteristics. We obviously cannot analyze every possible characteristic. Thus we must represent the expansion by a finite number of waves, the number of which we must decide. In principle, the more characteristics we use, the better we will represent the flow. Numerical errors become important, however, if we use too many characteristics. For this example we will represent the initial expansion by seven C^- characteristics, as depicted in Figure 10-29. This number will give us reasonable accuracy without requiring an excessive number of unit process calculations and without encountering numerical problems.

The second judgment or decision we must make involves the conditions on the initial characteristics. The design Mach number prescribes the value of $\theta_{w,\max,\mathrm{ml}}$ and the value of K^- (by Eq. 10-48) on the last characteristic in the initial expansion. In the interior of the initial expansion we are free to specify one of the conditions on each of the C^- characteristics, as long as the specified condition lies within the physical bounds of the problem. The physical bounds for a minimum-length planar nozzle with a sonic throat are

$$\nu(M) = \theta; \qquad 0 \leq \theta \leq \theta_{w,\max,\mathrm{ml}} = \nu_a; \qquad 1 \leq M \leq M(\nu_a). \tag{10-50}$$

On each of the C^- characteristics representing the initial expansion we may specify one of the variables θ, ν, or M. (Actually, since the expansion is isentropic another property such as P or T could be specified. However, this is usually not done.) It makes the most sense to use the same variable for all of the characteristics. It is also usually best to use θ for the specified variable with equal increments in θ between the characteristics. For our specific design we will specify θ on each of the seven initial C^- characteristics and the increment between characteristics will be 2° (the reason for which will be made clear shortly).

For this design there is a third (and final) engineering decision we must make. A condition on the first characteristic of the expansion must be specified. This is somewhat problematic. Recall that we are assuming that the sonic line is straight and vertical. With this assumption the sonic line corresponds to a C^- characteristic and, in fact, is the first C^- characteristic in the expansion. If we choose the sonic line as the first characteristic in our solution, which certainly seems to make sense since it is the beginning of the expansion process, a difficulty arises. This difficulty is due to the fact that the vertical sonic line "reflects" from the centerline back onto itself. Consequently, the information it is carrying is not propagated downstream. The unit processes to determine the local conditions on the next C^- characteristic are thus stymied and, in effect, we cannot get the calculation procedure started. To avoid this problem we choose the first C^- characteristic to be displaced slightly downstream from the sonic line. This can be done by specifying a small positive value for θ on the first characteristic. In the present example, since $\theta_{w,\max,\mathrm{ml}} = 12.075°$, a logical starting value is $\theta = 0.075°$. With this as the value of θ on the first characteristic, we need the remaining six initial C^- characteristics to accomplish an additional 12° of turning. The selection of 2° increments between characteristics is then apparent (12° ÷ 6 = 2°).

We may now begin the unit processes. Table 10-1 presents the conditions at all intersections in the characteristic mesh. Some of these values have already been specified. Others are fixed by boundary conditions. Most, however, come from the unit process calculations.

The first characteristic propagates undistorted all the way to the centerline, where it reflects at point 1. The wall unit process determines K^+ on C_r^+ at point 1.

TABLE 10-1 MINIMUM-LENGTH NOZZLE WITH UNIFORM EXIT FLOW AT $M = 1.92$

Point	K^+	K^-	θ	ν	M	μ	x	y
a	—	—	—	—	—	—	0.000	1.000
1	0.151	0.000	0.075	0.075	1.014	80.409	0.170	0.000
2	4.151	0.000	2.075	2.075	1.136	61.667	0.318	0.458
3	8.151	0.000	4.075	4.075	1.221	55.004	0.367	0.548
4	12.151	0.000	6.075	6.075	1.297	50.470	0.403	0.605
5	16.151	0.000	8.075	8.075	1.368	46.963	0.434	0.650
6	20.151	0.000	10.075	10.075	1.438	44.077	0.461	0.689
7	24.151	0.000	12.075	12.075	1.506	41.614	0.487	0.724
8	24.151	0.000	12.075	12.075	1.506	41.614	0.818	1.175
9	4.151	−4.151	0.000	4.151	1.224	54.806	0.613	0.000
10	8.151	−4.151	2.000	6.151	1.299	50.323	0.715	0.138
11	12.151	−4.151	4.000	8.151	1.371	46.844	0.792	0.235
12	16.151	−4.151	6.000	10.151	1.440	43.977	0.857	0.314
13	20.151	−4.151	8.000	12.151	1.508	41.528	0.917	0.384
14	24.151	−4.151	10.000	14.151	1.576	39.384	0.973	0.450
15	24.151	−4.151	10.000	14.151	1.576	39.384	1.751	1.357
16	8.151	−8.151	0.000	8.151	1.371	46.844	0.842	0.000
17	12.151	−8.151	2.000	10.151	1.440	43.977	0.938	0.101
18	16.151	−8.151	4.000	12.151	1.508	41.528	1.022	0.187
19	20.151	−8.151	6.000	14.151	1.576	39.384	1.098	0.265
20	24.151	−8.151	8.000	16.151	1.644	37.474	1.170	0.338
21	24.151	−8.151	8.000	16.151	1.644	37.474	2.250	1.436
22	12.151	−12.151	0.000	12.151	1.508	41.528	1.052	0.000
23	16.151	−12.151	2.000	14.151	1.576	39.384	1.151	0.088
24	20.151	−12.151	4.000	16.151	1.644	37.474	1.243	0.169
25	24.151	−12.151	6.000	18.151	1.712	35.750	1.331	0.247
26	24.151	−12.151	6.000	18.151	1.712	35.750	2.729	1.495
27	16.151	−16.151	0.000	16.151	1.644	37.474	1.266	0.000
28	20.151	−16.151	2.000	18.151	1.712	35.750	1.373	0.083
29	24.151	−16.151	4.000	20.151	1.780	34.177	1.476	0.163
30	24.151	−16.151	4.000	20.151	1.780	34.177	3.225	1.538
31	20.151	−20.151	0.000	20.151	1.780	34.177	1.495	0.000
32	24.151	−20.151	2.000	22.151	1.850	32.729	1.615	0.082
33	24.151	−20.151	2.000	22.151	1.850	32.729	3.756	1.566
34	24.151	−24.151	0.000	24.151	1.920	31.388	1.750	0.000
35	24.151	−24.151	0.000	24.151	1.920	31.388	4.334	1.576

The second characteristic propagates undistorted until it intersects at point 2 with the reflected characteristic from point 1. The field point unit process determines the conditions at point 2. The C^- emerging from point 2 continues on to the centerline, where it reflects at point 9. The wall unit process, with $\theta_w = 0$, is used. The third initial characteristic encounters, at point 3, the C^+ emerging from point 2 and then, at point 10, the C^+ from point 9. Field point unit processes determine the conditions at these intersections. This procedure is continued, "marching" downstream through the flow. Indeed, the method of characteristics is often referred to as a *marching technique.*

The initial expansion ends with C_{a7}^-, which passes through points 7, 14, 20, 25, 29, 32, and 34. These points form the downstream boundary of the nonsimple region. (Points 1, 2, . . . , 7 form the upstream boundary of this region.) Emerging from these boundary points are C^+ characteristics that form a simple$^+$ region. These C^+ characteristics propagate to the upper nozzle wall, carrying with them constant flow properties. In particular, the flow angle, θ, is constant along each C^+.

Now we must consider the interaction of the waves with the upper wall. In the typical wall point unit process the wall angle, θ_w, is given. The incident characteristic usually is carrying a different value of θ, and thus a reflection occurs in order to satisfy the wall boundary condition. Here, however, the wall angle is not known. It is, in fact, our goal to determine the wall shape subject to the constraint that no characteristic reflects from the wall. We can satisfy this constraint by matching the value of θ_w to the value of θ on the incident characteristic. Consequently, for our design we set θ_8 equal to θ_7, θ_{15} equal to θ_{14}, and so on. The last C^+, the one emerging from point 34, carries $\theta = 0°$. At the wall (point 35) we then also have $\theta = 0°$. The flow emerging from the simple$^+$ region is thus all in the x-direction. Furthermore, since the Mach number is also constant along each C^+ in the simple$^+$ region, and since $\theta_{34} = 1.92 = M_e$, then the flow emerging from the simple$^+$ region is also at the design Mach number. Point 35 thus terminates the nozzle, and the design is finished. The nozzle and the characteristic mesh for this design are shown in Figure 10-29.

Notice in Figure 10-29 that the wall from a to 8 is straight. This is a consequence of the fact that the first characteristic used to represent the expansion, C_{a1}^-, is displaced by a small, but finite, angle from the sonic line. The straight wall segment from a to 8 shortens as the angle of C_{a1}^- is made smaller. Also notice in Table 10-1 that the flow angle at point 1, θ_1, is not zero. Elsewhere along the centerline the angle is zero, as it should be. The problem at point 1 is also caused by the starting characteristic, C_{a1}^-. In our solution we have specified the sonic line and C_{a1}^- to be straight, which means, in effect, that we are assuming the immediate vicinity of the throat is a simple region. In reality, the throat flow is nonuniform and the sonic line and C_{a1}^- are curved. In the real flow θ varies along C_{a1}^- so that at the centerline θ_1 is actually zero. Because our present method of characteristics has no way to predict the nonuniform transonic throat flow, we must make an assumption that introduces an error, the nonzero flow angle at point 1. As long as we specify a small angle for C_{a1}^-, this error will not be serious and our overall solution will be acceptable.

The Programs CHARQ and CHARP

The software element CHARQ (Al-Habbash (1993)) is an implementation of the method of characteristics to design the supersonic section of planar minimum-length nozzles. CHARQ is based directly on the ideas and equations developed in this section and in Section 10.3. The program CHARP uses the output of CHARQ to produce profiles of flow properties across the nozzle. Both programs are quite simple to use.

CHARQ is run simply by typing its name, CHARQ. The dialogue necessary to run CHARQ for the example in Figure 10-29 and Table 10-1 follows:

```
INPUT 1 FOR TERMINAL OUTPUT OR 2 FOR PRINTER OUTPUT
? 1
WHAT IS THE NOZZLE EXIT MACH NUMBER
? 1.92
WHAT IS GAMMA
? 1.4

THE MAXIMUM WALL DEFLECTION ANGLE (THETAM) = 12.0753

WHAT DO YOU WANT TO SPECIFY:

  1- NUMBER OF CHARACTERISTICS WITH MINIMUM INCLINATION
                                  OR
  2- NUMBER OF CHARACTERISTICS, EQUALLY SPACED
                                  OR
  3- ANGLE FOR SPACING BETWEEN CHARACTERISTICS

INPUT 1 OR 2 FOR # OF CHARAC. OR 3 FOR SPACING
? 1
HOW MANY CHARACTERISTICS DO YOU WANT (MINIMUM
INCLINATION)
? 7
WHAT DO YOU WANT TO PRINT OUT:
  1- COMPLETE TABLE OF NODES & PROPERTIES
  2- NOZZLE CONTOUR COORDINATES
  3- CHARACTERISTIC NET (CELL-NODE) COORDINATES & PROPERTIES

INPUT YOUR CHOICE 1, 2 OR 3?
1

                                  [Table]


DO YOU WANT TO CREATE AN OUTPUT FILE
INPUT Y FOR YES
? Y
WHAT DO YOU WANT TO NAME YOUR OUTPUT FILE
  EXAMPLE: MACH#4.DAT
  A NAME WITH MAX. 6 CHARACTERS WITH EXT. DAT
? M192_7.DAT
```

The table produced in this dialogue and the output file M192_7.DAT contain the same information that is in Table 10-1. In addition to the output file M192_7.DAT, two other files, GR.DAT and GP.DAT, are produced. These files are for use by CHARP.

Typically, when we design or analyze a nozzle we would like to know the condition of the flow across the nozzle at particular axial, or streamwise, stations. The method of characteristics used in CHARQ directly computes flow properties at the grid points in the characteristic mesh. These grid points are not known prior to running CHARQ, and they do not lie on lines of constant x. CHARP (Al-Habbash (1993)) takes output from CHARQ and produces profiles of flow properties across the nozzle at user-specified x-locations. The number of points in each profile is also user-specified.

CHARP is run by simply typing its name, CHARP. The dialogue to run CHARP for the example in Figure 10-29 and Table 10-1 follows:

```
INPUT 1 FOR TERMINAL OUTPUT OR 2 FOR PRINTER OUTPUT
? 1
HOW MANY NODES DO YOU HAVE (MAX=1500):
? 35
HOW MANY INITIAL CHARACTERISTICS DO YOU HAVE:
? 7
WHAT IS GAMMA
? 1.4

THE TOTAL LENGTH OF THE NOZZLE IS 4.334

INPUT THE LOCATION WHERE YOU WANT TO FIND THE PROFILE,

CHOOSE (X/ TOTAL LENGTH) LOCATION BETWEEN 0.0 AND 1.0:
? .2

THE TOTAL HEIGHT OF THE NOZZLE AT THE CROSS SECTION IS 1.185

HOW MANY STATIONS DO YOU WANT (MAX=100)
? 10
NOZZLE LENGTH          L        = 4.334
THE CROSS SECTION IS AT X/L       = 0.2
NOZZLE SECTION LENGTH             = 0.8667
CROSS SECTION HEIGHT    Yw/Y* = 1.185
11 EQUALLY SPACED PROFILE POINTS

CELL   X/Y*    Y/Y*    K-       K+        THETA    NU       M

21     0.867   0.000    8.629   -8.629    0.000    8.629   1.387
:
14     0.867   1.185   24.151   -3.917   10.117   14.034   1.572

DO YOU WANT TO CREATE AN OUTPUT FILE
INPUT Y FOR YES
? Y
WHAT DO YOU WANT TO NAME YOUR OUTPUT FILE
  EXAMPLE: MACH#4.DAT
  A NAME WITH MAX. 6 CHARACTERS WITH EXT. DAT
? M192_P.DAT
```

Figure 10-30 presents two of the profiles produced by this run of CHARP.

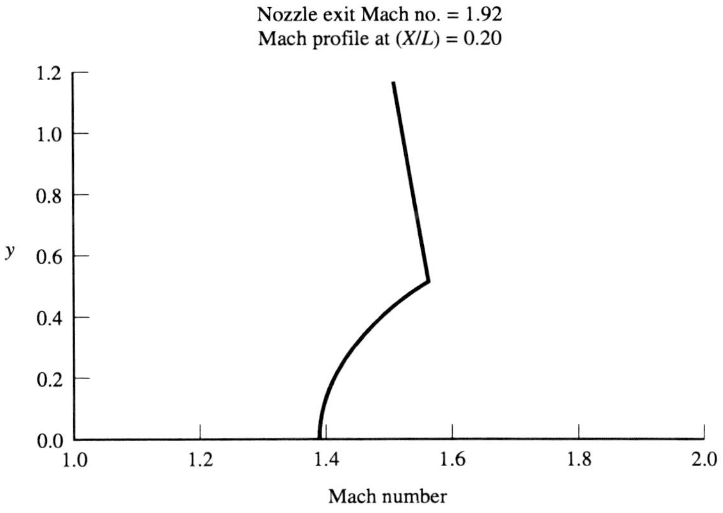

(a)

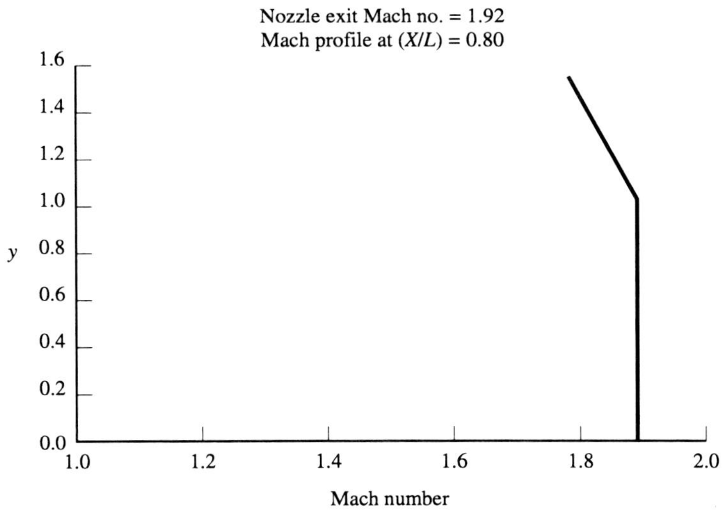

(b)

Figure 10-30 Profiles of Mach Number in the Nozzle of Figure 10-29, Produced by the Program CHARP

10.5 GENERALIZATIONS OF THE METHOD OF CHARACTERISTICS

Introduction

In Section 10.2 we developed the differential equations for the method of characteristics for steady, supersonic, homenergetic, and homentropic (hence irrotational) flow in two dimensions. These equations are the characteristic equations, Eqs. 10-19, and the compatibility equations, Eqs. 10-20 or 10-21. In Section 10.3 we assumed planar flow for which the equations of the method of characteristics reduced to the remarkably simple solution

$$K^{\pm} = \text{constant on } C^{\pm}. \tag{10-24}$$

Section 10.3 detailed the nature of this solution, and Section 10.4 described its use to design a supersonic planar nozzle with a straight sonic line. In these discussions the downstream propagation of information along characteristics, the satisfaction of simultaneous equations at intersections, and the conditions imposed at boundaries were stressed.

Obviously many flows do not satisfy some, or even any, of the assumptions on which Eq. 10-24 is based. In many instances, however, the method of characteristics is still valid. In this section we will consider less restricted flow problems in which the method of characteristics can be used. In these less restricted flows the method of characteristics becomes considerably more complicated and generally requires elaborate numerical solution techniques. Nevertheless, in all of these flows the same concepts inherent in Eq. 10-24—downstream propagation of information along characteristics, satisfaction of simultaneous equations at intersections, and conditions imposed at boundaries—are present.

Our discussions here will be limited to very brief outlines of the essential features of the method of characteristics for more general flows. The reader is urged to see Anderson (1990) for additional insight. The text by Zucrow and Hoffman (1976) is an excellent source for insight and computational details, including computer code, regarding most of the flows we will mention here.

Axisymmetric Flow

We begin by repeating the compatibility equations and characteristic equations for general two-dimensional, steady, supersonic, homenergetic, and homentropic flow.

$$d(\theta - \nu) = -\frac{\sin\theta \sin\mu}{\sin(\theta + \mu)} \frac{dr}{r}, \tag{10-21a}$$

on

$$\frac{dy}{dx} = \tan(\theta + \mu), \tag{10-19a}$$

and

$$d(\theta + \nu) = \frac{\sin\theta \sin\mu}{\sin(\theta - \mu)} \frac{dr}{r} \tag{10-21b}$$

on

$$\frac{dy}{dx} = \tan(\theta - \mu). \tag{10-19b}$$

By letting r go to infinity in Eqs. 10-21 we obtain planar flow, which is described by Eq. 10-24. For finite r the preceding equations describe axisymmetric flow. The spatial coordinate y is the same as r and can be replaced by it so that the equations are in terms of the independent variables x and r.

Equations 10-19 show that for axisymmetric flow the characteristics are still Mach lines. However, from Eqs. 10-21 we see that the quantities $\theta \pm \nu$ are no longer constant along the characteristics. In the planar case the compatibility equations are simple algebraic equations for θ and ν and are independent of the spatial coordinates; they are *uncoupled* from the characteristic equations. For axisymmetric flow the compatibility equations are differential equations that are dependent on r. Thus, in axisymmetric flow the compatibility and characteristic equations are coupled and must be solved simultaneously. This is a problem of numerical integration of a system of ordinary differential equations. Zucrow and Hoffman (1976) detail a procedure for performing this integration. Martensen (1990) describes an implementation of the Zucrow and Hoffman procedure for the design and analysis of supersonic axisymmetric nozzles.

Rotational Flow

By Crocco's theorem, Eq. 9-52, a flow that is nonisentropic and/or nonadiabatic is rotational. An important instance of inviscid but nonisentropic fluid motion is the flow downstream of a curved shock, which we discussed in Chapter 9. An example of inviscid but nonadiabatic flow is one in which the gas static temperature, say downstream of a shock, becomes high enough for radiation to be important. Rotation adds an important complication to the method of characteristics. Shapiro (1953) and Zucrow and Hoffman (1976) present derivations of the method of characteristics for rotational flow.

The most important alteration to the method of characteristics for two-dimensional rotational flow is the addition of a third characteristic direction. Plus and minus Mach lines still form two of the characteristic families. In addition, the streamlines in the flow now form a third family of characteristics. The nature of the characteristic formulation for rotational flow is such that an iterative numerical solution is required, even for planar motions. Again, Zucrow and Hoffman (1976) detail a procedure for performing the integration, while Martensen (1990) implements this procedure for supersonic nozzles.

We should note that in Section 10.3 we considered the shock point unit process.

Our discussion of this process implied that it is valid for curved shocks (see Figure 10-13, for example). Since the discussion in Section 10.3 was in the context of irrotational flow, we see now that, strictly speaking, that analysis is not quite correct. However, in the immediate vicinity of a shock point, the curvature of the shock is small and the entropy gradient is also small. The assumption of locally homentropic flow is sometimes acceptable so that the shock point unit process we described is then approximately valid.

Three-Dimensional Flow

The addition of a third independent spatial coordinate further complicates the method of characteristics. The principal consequence of the third dimension is that the characteristics are no longer simply lines but are now three-dimensional surfaces. Allowing for the possibility of rotation in the flow, the characteristic surfaces are Mach surfaces, "warped" Mach cones, and stream surfaces, surfaces everywhere tangent to the local velocity vector. In addition to these characteristic surfaces, another set of curves known as *bicharacteristics* appears. Certain aspects of the solution must be carried out along the bicharacteristics instead of along the characteristic surfaces.

As may be apparent, solution procedures for three-dimensional flows are quite complex. Nevertheless, practical solutions have been obtained. An example of a particularly successful solution method can be found in Rakich (1969) and in Rakich and Cleary (1970).

Improved Throat Flow Prediction

In Section 10.3 we discussed the importance of the initial value line (IVL) from which a method of characteristics solution is begun. For flows with subsonic zones the IVL generally must lie somewhat downstream of the sonic line. In order to properly locate the IVL the transonic flow in the vicinity of the sonic line must be determined. For the planar nozzle design in Section 10.4 we took the most simple approach and simply assumed that the sonic line and the first characteristic in the expansion were straight, with the first characteristic serving as the IVL. These assumptions simplified the subsequent analysis but introduced error, such as the nonzero centerline flow angle at point 1, into the solution. It is generally preferable to analyze the sonic flow region more carefully.

Supersonic flow past blunt bodies and converging–diverging supersonic nozzles are two practical situations for which the sonic line region is an important concern. The method of characteristics has perhaps seen more (successful) use in nozzle problems; we mention here a few of the methods that have been developed for prediction of nozzle throat flows.

Sauer's method. The analysis of Sauer (1947) (repeated in detail in Zucrow and Hoffman (1976)) stands out as an early, reasonably successful description of throat flow. Sauer's method is based on a power series solution to the small disturbance transonic potential equation, Eq. 9-86, for axisymmetric or planar flow. The power series is truncated and the final equation set is closed in form, which makes it especially easy to apply. The analysis assumes that the sonic line is parabolic in shape and that the generators of the wall in the vicinity of the throat lie on circles of constant radius R_t. Thus, R_t is the wall radius of curvature. The throat height (if planar) or radius (if axisymmetric) is r_t. A very important assumption made in Sauer's analysis is that the throat wall radius of curvature is considerably larger than the throat height or radius. In practice, acceptable results from this analysis require that

$$\frac{R_t}{r_t} \geq 2.$$

The initial value line of Sauer's method is chosen as the line on which the transverse (vertical if planar or radial if axisymmetric) velocity component is zero and which passes through the sonic line/centerline intersection. The geometry and nomenclature of Sauer's method are shown in Figure 10-31.

The final equation set for Sauer's method is presented next. In these equations

$$\delta = 0 \text{ for planar flow} \quad \text{and} = 1 \quad \text{for axisymmetric flow.} \tag{10-51}$$

The parameter α is the *coefficient of linear nondimensional axial velocity perturbation* and is given by

$$\alpha = \sqrt{\frac{1}{\gamma + 1}\frac{1 + \delta}{R_t r_t}}. \tag{10-52}$$

Notice that the dimension of α is length^{-1}. The dimensional equation of the sonic line is

$$x^* = -\frac{\gamma + 1}{2}\frac{\alpha y^2}{1 + \delta}, \tag{10-53}$$

where $x^* = 0$ at the centerline. The $x^* = 0$ location relative to the minimum area cross-sectional plane is

$$\epsilon = -\frac{\gamma + 1}{2}\frac{\alpha r_t^2}{3 + \delta}. \tag{10-54}$$

The dimensional equation of the $v = 0$ line (which is the IVL for this analysis) is

$$x_{\mathrm{IVL}} = -\frac{\gamma + 1}{2}\frac{\alpha y^2}{3 + \delta}. \tag{10-55}$$

The speed of sound on the sonic line is

$$a^* = \sqrt{\gamma R T^*} = \sqrt{\gamma R \frac{T^*}{T_0} T_o} = \sqrt{2\frac{\gamma}{\gamma + 1} R T_0}. \tag{10-56}$$

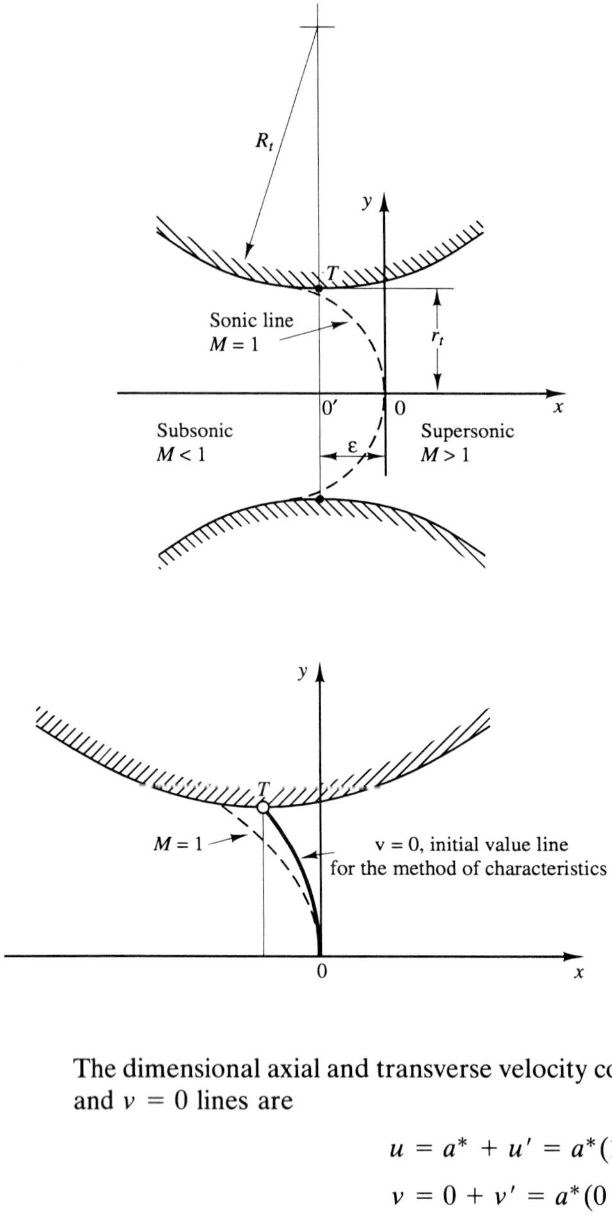

Figure 10-31 Geometry and Nomenclature for Sauer's Method of Analyzing the Throat Region of a Supersonic Converging–Diverging Nozzle. *From Zucrow and Hoffman (1976).*

The dimensional axial and transverse velocity components in the region of the sonic and $v = 0$ lines are

$$u = a^* + u' = a^*(1 + u''),$$
$$v = 0 + v' = a^*(0 + v''), \tag{10-57}$$

where u' and v' are the dimensional perturbation velocities, and u'' and v'' are the nondimensional perturbation velocities. These latter velocities are respectively given by

$$u'' = \alpha x + \frac{\gamma + 1}{2}\frac{\alpha^2 y^2}{1 + \delta} \tag{10-58a}$$

and

$$v'' = (\gamma + 1)\frac{\alpha^2 xy}{1 + \delta} + \frac{(\gamma + 1)^2}{2}\frac{\alpha^3 y^3}{(1 + \delta)(3 + \delta)}. \tag{10-58b}$$

The most important purpose of this analysis is to provide the coordinates of the initial value line and the values of all of the fluid properties on this line. Equation 10-55 gives the IVL coordinates. Substitution of Eqs. 10-55 and 10-58 into Eqs. 10-57 gives the dimensional velocity components on the IVL:

$$u_{\text{IVL}} = a^*\left(1 + \frac{(\gamma + 1)\alpha^2 y^2}{(1 + \delta)(3 + \delta)}\right) \tag{10-59a}$$

and

$$v_{\text{IVL}} = 0 \text{ (as it should!).} \tag{10-59b}$$

This analysis assumes that the flow in the vicinity of the throat is isentropic. The most convenient way to obtain the IVL speed of sound is to use the adiabatic energy equation in the form of Eq. 9-70:

$$a_{\text{IVL}}^2 = a_0^2 - \frac{\gamma - 1}{2}(u_{\text{IVL}}^2 + v_{\text{IVL}}^2), \tag{10-60}$$

where

$$a_0 = \sqrt{\gamma R T_0}. \tag{10-61}$$

The Mach number, Mach angle, and Prandtl-Meyer function are

$$M_{\text{IVL}} = \frac{u_{\text{IVL}}}{a_{\text{IVL}}}, \tag{10-62}$$

$$\mu_{\text{IVL}} = \sin^{-1}\left(\frac{1}{M_{\text{IVL}}}\right), \tag{10-63}$$

and

$$\nu_{\text{IVL}} = \nu(M_{\text{IVL}}). \tag{10-64}$$

Since the flow is isentropic, we can use M_{IVL}, the stagnation properties, and the isentropic relations to find the static properties. At this point we have everything necessary to use the $v = 0$ line as the initial value line for a method of characteristics solution.

As a final note regarding Sauer's method, it is important to observe that because u_{IVL} is a function of y (by Eq. 10-59a), then all properties vary along the IVL except the transverse velocity. However, the simple algebraic closed form of these equations makes determination of the IVL properties by this method an easy task.

Other throat flow prediction techniques. Sauer's method does provide a better description of the throat region than, say, the simple assumptions we used in Section 10.4. However, it is based on several limiting assumptions, an important one of which is that R_t is much larger than r_t. This assumption means, for instance, that Sauer's method is not appropriate for minimum-length nozzles. There have been a number of other methods developed for predicting throat flow which are applicable to smaller ratios of throat radius of curvature to throat height. Some of these methods also are more accurate than Sauer's method at larger values of R_t/r_t. However, as is usually the case, these improved techniques generally increase in complexity as their range of validity and/or accuracy increases.

Hall (1962) developed a power series solution to the full nonlinear potential equation, Eq. 9-83, for axisymmetric or planar flow. This solution is restricted to $R_t/r_t > 1$. Kliegel and Levine (1969) extended Hall's analysis to $R_t/r_t < 1$. The analyses of Hall and of Kliegel and Levine provide algebraic expressions for the velocity components, as does Sauer's method. As Zucrow and Hoffman observe, for large values of R_t/r_t the three methods are nearly the same, while for small values of R_t/r_t the method of Kliegel and Levine appears to be the most accurate.

The three methods mentioned so far involve power series solutions to steady-flow potential equations. Cline (1974) developed a numerical solution to the full unsteady nonlinear equations of motion. For a flow with steady boundary conditions, the unsteady equations yield a steady solution as time becomes large. Cline's solution has proven to be accurate, computationally efficient, and capable of analyzing a variety of nozzle flows.

One of the best, readily available, experimental data sets of nozzle throat flow appears in Back, Massier, and Grier (1965) and in Cuffel, Back, and Massier (1969). This experiment involved two axisymmetric nozzles with conical converging and diverging sections and circular arc throat walls. For both nozzles the converging and diverging wall angles were 45° and 15°, respectively. The two nozzles differed in throat radius ratios, R_t/r_t, 2 and 0.625 being examined. Figure 10-32, from Zucrow and Hoffman (1976), shows comparisons among the experimental results and two of the prediction methods of Mach number contours for the $R_t/r_t = 0.625$ nozzle. In order of decreasing accuracy are the methods of Kliegel and Levine, and Sauer, as we would expect. Notice that the flow is clearly nonuniform.

Unsteady Flow

Unsteady, adiabatic, and isentropic flow can also be analyzed by the method of characteristics. The most simple case is that of unsteady flow in one spatial dimension. There is a remarkable qualitative analogy between the cases of one-dimensional, unsteady, isentropic flow and steady, planar, supersonic flow (which we developed in Section 10.3). Thorough and very lucid developments of the unsteady, one-dimensional method appear in Thompson (1972) and Anderson (1990). Zucrow and Hoffman (1976) discuss in detail the extension of the method to unsteady, two-dimensional, isentropic flow.

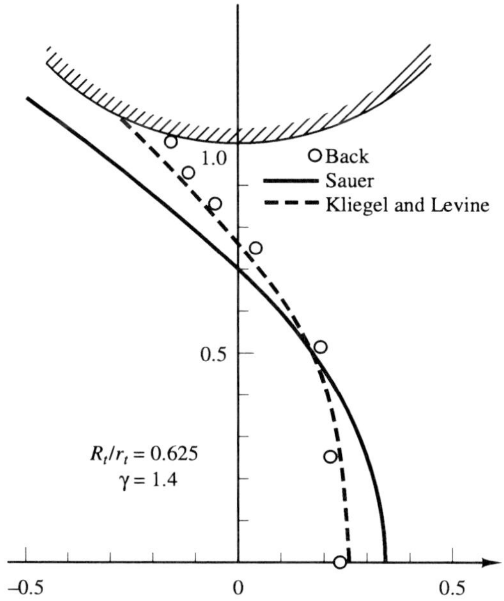

Figure 10-32 Experimental and Computed Sonic Line Locations in the Throat Region of a Supersonic Converging–Diverging Nozzle. *From Zucrow and Hoffman (1976).*

Chemically Reacting Flows

A variety of flow situations exist for which the method of characteristics might seem appropriate, but in which chemical reactions occur. Examples of these include hypersonic flight vehicles, gas turbine and rocket engine exhaust nozzles, and nozzles used in gas-dynamic and chemical lasers. (See Anderson (1976) and Anderson (1991) for good introductions to these situations.) The method of characteristics can be extended to include chemical reactions; again, Zucrow and Hoffman (1976) show how this is done. An example of a widely used method-of-characteristics code for rotational and chemically reacting flow in axisymmetric nozzles is the program TDK, for "Two-Dimensional Kinetics" (Nickerson et al., 1989). This code also accounts for wall friction and heat transfer. Table 10-2 presents area-averaged flow properties at three axial stations in the Space Shuttle Main Engine, which burns hydrogen and oxygen, as computed by TDK. Notice the large combustion chamber temperature and pressure, the large exit velocity, and the changes in the gas chemical composition as the gas progresses through the nozzle.

TABLE 10-2 AREA-AVERAGED SPACE SHUTTLE MAIN ENGINE FLOW PROPERTIES COMPUTED BY TDK WITH BOUNDARY-LAYER CORRECTIONS

Total propellant mass flow rate = 1138 lbm/s
Vacuum thrust = 517,745 lbf
Vacuum specific impulse = 455 s

Station	M	Velocity ft/s	Static temperature °R	Static pressure psia	Molecular mass lbm/lbm-mol	γ	C_p Btu/lbm R
Combustion chamber	0.199	1066	6592	3208	13.67	1.196	0.8853
Throat	0.979	5069	6217	1890	13.80	1.195	0.8812
Nozzle exit $A/A^* = 76.5$	4.661	14,550	2222	2.97	14.23	1.254	0.6901

Total propellant mass flow rate = 516 kg/s
Vacuum thrust = 2.303E+06 N
Vacuum specific impulse = 455 s

Station	M	Velocity m/s	Static temperature °K	Static pressure N/m^2	Molecular mass kg/kg-mol	γ	C_p kJ/kg K
Combustion chamber	0.199	325	3662	2.21E+07	13.67	1.196	3.707
Throat	0.979	1545	3454	1.30E+07	13.80	1.195	3.689
Nozzle exit $A/A^* = 76.5$	4.661	4435	1234	2.05E+04	14.23	1.254	2.889

MASS FRACTIONS OF REACTION PRODUCTS

Mixture ratio = (mass of O_2)/(mass of H_2) = 6.05

Station	H_2O	H_2	O_2	H	O	OH
Combustion chamber	9.02E−01	3.58E−02	7.00E−03	2.05E−03	3.14E−03	4.98E−02
Throat	9.19E−01	3.50E−02	4.67E−03	1.67E−03	2.00E−03	3.75E−02
Nozzle exit	9.66E−01	3.35E−02	2.08E−06	7.78E−05	1.80E−07	3.92E−05

10.6 CONCLUSION

The method of characteristics is an older computational technique for analyzing certain types of flow situations. Although numerical solutions of the Euler or Navier-Stokes equations have superseded it in many instances, it still has considerable usefulness as a design and analysis tool. A particular advantage it has is that method of characteristics solutions for two-dimensional (planar and axisymmetric) rotational flow can be easily executed on personal computers, for example, as described by Martensen (1990).

Perhaps more importantly, many of the recent methods developed to solve the Euler and Navier-Stokes equations incorporate characteristic concepts. An understanding of the ideas presented in this chapter better prepares the reader to work with more precise (and complicated!) flow analysis techniques.

CHAPTER 10 REVIEW QUESTIONS

1. What is the physical basis for the method of characteristics as developed here?
2. What physical principles are embodied in Eqs. 10-14 and 10-15?
3. What are the characteristic equations and the compatibility equations for steady, irrotational, two-dimensional supersonic flow?
4. What are the boundary conditions at a wall, a free surface, and a shock wave?
5. What problems arise in starting a method of characteristics solution?
6. What complications arise in the method of characteristics as we generalize to axisymmetric flow, to rotational flow, or to three-dimensional flow?
7. What are the physical conditions (in terms of velocity components and/or Mach number) on the initial value line in Sauer's method?

PROBLEMS

1. Consider flow over a horizontal lower wall with an abrupt expansion corner. Determine the angle with respect to horizontal of the leading and trailing Mach waves in the expansion fan for the following approach Mach numbers, M_1, and turn angles, θ.
   (a) $M_1 = 2$, $\theta = 2°$
   (b) $M_1 = 2$, $\theta = 32°$
   (c) $M_1 = 5$, $\theta = 2°$
   (d) $M_1 = 5$, $\theta = 16°$
2. **(a)** By using the Prandtl-Meyer function, determine the Mach number and static pressure distributions along a continuously curved expansion turn where the wall shape is part of a circular arc. The approach Mach number is 3 and the net turn is 30°.
   (b) Compare the answer in part a to the final flow produced by an abrupt expansion turn with the same conditions.
3. **(a)** By using the Prandtl-Meyer function, determine the Mach number and static pressure distributions along a continuously curved compression turn where the wall shape is part of a circular arc. The approach Mach number is 3 and the net turn is 30°.
   (b) Compare the answer in part a to the final flow produced by an abrupt compression turn with the same conditions.
4. Prove that $\theta_w + \mu_w > \theta_x + \mu_x$ in Figure 10-19.

5. Consider a lower wall abrupt expansion corner that accelerates a flow from $M_1 = 1$ to $M_2 = 1.4$. Determine the inclination $(\theta - \mu)$ of several C^- characteristics in the simple$^+$ region. Repeat for a turn from $M_1 = 1.4$ to $M_2 = 1.8$, and then again for a turn from $M_1 = 2$ to $M_2 = 3$. What conclusion can you draw from a comparison of these turns?
6. Repeat Problem 2a using the MOC.
7. Repeat Problem 3a using the MOC.
8. Use seven initial characteristics to create a table like Table 10.1 for a planar nozzle with $M_e = 2.4$. Use equal increments in θ between the initial characteristics. Try performing this solution by hand. Check your results using CHARQ.
9. Repeat the design in Problem 8 by using CHARQ with 19 initial characteristics. Compare the results with those of Problem 8 by using CHARP to obtain the wall shape and Mach number profiles at common values of x/h^*, where h^* is the throat half height.
10. **(a)** Using Sauer's method, determine the dimensional shape and flow properties on the sonic line and initial value line of a planar nozzle. The gas is standard air with $T_o = 500$ K. The throat half height is 1 cm and the throat wall radius of curvature is 10 cm.
    (b) Repeat for an axisymmetric nozzle with the same conditions.

Part V

HIGH-TEMPERATURE GAS DYNAMICS

High-Temperature Gas Dynamics

11

11.1 INTRODUCTION

The calorically perfect gas assumption, which results in a mathematically simple model of compressible fluids, has provided the basis of the gas behavior for our study thus far in this text. Recall that a calorically perfect gas satisfies the thermal equation of state (also called the perfect gas law or the ideal gas law)

$$P = \rho R T \tag{11-1}$$

and the specific heats, c_p and c_v, for it, and thereby the ratio of specific heats are constant. However, as Table 1-4 indicates, the specific heats and their ratio do vary as the temperature is increased. This chapter examines other gas models and develops some computational techniques for accounting for deviations from the calorically perfect assumption. An introduction to the gas dynamics of noncalorically perfect gases is provided. This chapter covers three primary topics:

1. the thermodynamics of nonperfect gases
2. computations involving thermally perfect gases with nonconstant specific heats
3. gases with chemical reactions.

With the recent revived interest in hypersonic flow, the importance of gases with thermal and caloric imperfections as well as chemical reactions is increasing. Anderson (1989) provides an excellent contemporary treatment of high-temperature gas dynamics. He lists atmospheric entry, rocket engines, high-enthalpy wind tunnels, high-power lasers, and ramjet and SCRAMjet engines as typical contemporary topics for which the calorically perfect gas assumption is not adequate. Anderson's extensive treatment of high-temperature gas dynamics is perhaps the most lucid development available. Pai and Luo (1991) present a comprehensive survey of compressible flow analysis techniques, including classical and computational methods. They consider in detail procedures available for calorically imperfect and reacting gas situations, including radiative gas dynamics models.

11.2 BEHAVIORAL CLASSIFICATION OF GASES

As suggested by Anderson (1989), the behavior of gases can be classified, in order of increasing complexity, as calorically perfect, thermally perfect, chemically reacting mixtures of perfect gases, and real gases. We shall follow this hierarchy in our examination, but first an examination of the thermal equation of state is appropriate.

The Thermal Equation of State

The thermal equation of state, Eq. 11-1, is a good starting point for our examination of gas behavior. Macroscopic thermodynamics accepts this equation as an experimental fact for gases at low-to-moderate pressures and temperatures. It can be derived from kinetic theory of gases considerations under the assumptions of negligible intermolecular forces and negligible molecular volumes. As the pressure increases, the number of molecules per unit volume increases, causing the molecular spacing to decrease and both the intermolecular forces and the volume occupied by the molecules to become increasingly important. The assertion that a gas is thermally perfect, $P = \rho RT$, implies much beyond just the form of the equation of state. Consider the combined first and second laws expression

$$T\,ds = du + P\,dv \tag{11-2}$$

where $u = u(v, T)$ and $s = s(v, T)$. Then differential expressions for the internal energy and entropy become

$$ds = \frac{\partial s}{\partial v}dv + \frac{\partial s}{\partial T}dT \tag{11-3a}$$

and

$$du = \frac{\partial u}{\partial v}dv + \frac{\partial u}{\partial T}dT. \tag{11-3b}$$

By substituting Eqs. 11-3 for ds and du into Eq. 11-2 and equating the coefficients of dv and dT, the following equations result:

$$\frac{\partial s}{\partial T} = \frac{1}{T}\frac{\partial u}{\partial T} \tag{11-4a}$$

and

$$\frac{\partial s}{\partial v} = \frac{1}{T}\left(\frac{\partial u}{\partial v} + P\right). \tag{11-4b}$$

Taking the partial derivative of the first expression with respect to v and the second with respect to T yields

$$\frac{\partial^2 s}{\partial v\, \partial T} = \frac{1}{T}\frac{\partial^2 u}{\partial v\, \partial T} \tag{11-5a}$$

and

$$\frac{\partial^2 s}{\partial T\, \partial v} = -\frac{1}{T^2}\left(\frac{\partial u}{\partial v} + P\right) + \frac{1}{T}\left(\frac{\partial^2 u}{\partial T\, \partial v} + \frac{\partial P}{\partial T}\right). \tag{11-5b}$$

When the two expressions for $\partial^2 s/(\partial T\, \partial v)$ are equated and simplified, the following is obtained:

$$\frac{\partial P}{\partial T} = \frac{1}{T}\left(\frac{\partial u}{\partial v} + P\right). \tag{11-6}$$

Since $\partial P/\partial T$ for a thermally perfect gas becomes R/v, Eq. 11-6 for such a gas reduces to

$$\frac{\partial u}{\partial v} = 0. \tag{11-7}$$

However, if $\partial u/\partial v$ is zero, then for a thermally perfect gas $u(v, T)$ cannot be a function of v and T as the original assertion claimed. Thus, $u = u(T)$, and we have established the (very important) principle that for a thermally perfect gas the internal energy is only a function of the temperature. Since $h = u + Pv = u + P/\rho$, it follows that

$$h = u(T) + P/\rho = u(T) + RT = h(T); \tag{11-8}$$

hence, enthalpy is also a function of only temperature for a thermally perfect gas.

That internal energy and enthalpy are functions of only temperature for a thermally perfect gas are great simplifications over the initial assertions that both might be functions of more than a single thermodynamic property. However, classical thermodynamics is unable to establish the actual functional relationships of internal energy and enthalpy with temperature. Results from the kinetic theory of gases and from statistical thermodynamics can provide the actual functional forms of these relationships and can enhance our understanding of the physics of gases.

Calorically Perfect Gas

A calorically perfect gas obeys the thermal equation of state and possesses constant specific heats. Thus, the internal energy and the enthalpy become

$$u = c_v T; \tag{11-9a}$$

$$h = c_p T. \tag{11-9b}$$

Since the specific heats are constant, it also follows that the ratio of specific heats, γ, is constant.

A calorically perfect gas provides a convenient and realistic framework within which to develop and understand the behavior and the phenomena associated with compressible fluid flow. Derivations of the relatively uncomplicated expressions for the simple flows were possible because of the assumption of a calorically perfect gas. Other equations of state and variable specific heats generally lead to very complex expressions or to differential equations that cannot be integrated in closed form. To the accuracy required in many engineering computations, air and many other common gases behave as perfect gases. Hence, even though they provide mathematical simplicity, the calorically perfect gas assumptions also yield results that are widely used for many compressible flow applications.

Thermally Perfect Gas

A thermally perfect gas obeys the thermal equation of state but is calorically imperfect since the specific heats are variables (functions of temperature), not constants. Thus, as derived previously, the internal energy and enthalpy are functions of only the temperature, and

$$du = c_v\, dT; \tag{11-10a}$$

$$dh = c_p\, dT. \tag{11-10b}$$

Since differences, rather than absolute values, of internal energy and enthalpy are typically of interest, the usual expressions become

$$u - u_{\text{ref}} = \int_{T_{\text{ref}}}^{T} c_v(T)\, dT; \tag{11-11a}$$

$$h - h_{\text{ref}} = \int_{T_{\text{ref}}}^{T} c_p(T)\, dT. \tag{11-11b}$$

In order to deal quantitatively with thermally perfect gases, functional relationships must be obtained for the specific heats. These functional relationships will be examined later in this chapter.

Chemically Reacting Mixtures of Perfect Gases

Our previous examinations of calorically and thermally perfect gases implicitly assume that the gas compositions, and hence the molecular weights, do not change. As we shall later see, this is a viable assumption for air if the temperature is below about 2500 K. Above 2500 K, chemical reactions occur, and thus the air composition, and hence the molecular weight, varies. Chemical reactions for air in this context mean that the relative amounts of molecular and atomic oxygen and nitrogen (O_2, O, N_2, and N) vary as functions of pressure and temperature. The latter portions of this chapter will develop techniques for determining the composition of gases as functions of temperature and pressure.

When we consider chemically reacting mixtures of perfect gases, the distinction between equilibrium and nonequilibrium reactions is important. If a chemically reacting mixture of gases is maintained at constant pressure and temperature for an extended period of time, then the composition of the mixture becomes invariant and equilibrium is said to be attained. If, on the other hand, the pressure and temperature of a mixture were suddenly changed, the relative amounts of the species would initially change as functions of time. The chemical reaction during this period of change is said to be one of nonequilibrium. In the typical gas flow of interest, the time required to attain chemical equilibrium is of the order of a few milliseconds. Although chemical nonequilibrium is certainly possible, this discussion is confined to mixtures in chemical equilibrium. Park (1990) provides an extensive contemporary treatment of nonequilibrium gas flows. He also cites and discusses a number of applications, especially in the hypersonic flow regime, where chemical nonequilibrium effects are important.

Chemical equilibrium thus implies that in a mixture of gases the number of particles of each species is not changing with time. Moreover, since each constituent is a perfect gas, each species i must obey the thermal equation of state in the form

$$P_i = \rho_i R_i T, \tag{11-12}$$

where $R_i = \overline{R}/MW_i$. Since the composition of a mixture depends on the pressure and temperature, the mixture molecular weight and the internal energy and the enthalpy are all functions of both pressure and temperature. Thus

$$u = u(T, P), \tag{11-13a}$$

$$h = h(T, P); \tag{11-13b}$$

and in general, the specific heats have the functional form

$$c_v = c_v(T, P), \tag{11-14a}$$

$$c_p = c_p(T, P). \tag{11-14b}$$

The gas dynamic processes of reacting mixtures of perfect gases are thus considerably more involved than those of single species perfect gases.

Real Gas

The three previous classifications of gas behavior have all had in common the thermal equation of state, Eq. 11-1. *Real gases* are gases for which intermolecular forces are not negligible and must be considered. The equation of state is then more involved than the thermal equation of state. Perhaps the best known of the real gas equations of state is Van der Waals' equation of state [Lee et al. (1973)].

$$\left(P + \frac{a}{v^2}\right)(v - b) = RT. \tag{11-15}$$

The constants a and b are dependent on the gas; a is associated with the intermolecular forces, and b is associated with the molecular volumes. As Anderson (1989) points out, the aerodynamic literature often refers to a real gas as one in which caloric imperfections are important—gases whose behavior we would characterize as thermally perfect or as a mixture of chemically reacting perfect gases. Traditional physical-chemistry nomenclature uses the phrase "real gas" only in conjunction with a gas whose behavior depends in part on the intermolecular forces. In this textbook, real gas will be used in the traditional physical chemistry sense.

11.3 THE THERMODYNAMICS OF HIGH-TEMPERATURE GASES

The calorically perfect gas assumption has served us well, but we need to examine the causes and effects of departures from that assumption. The results of the microscopic approach, as introduced in Chapter 1, provide a quantitative explanation of the behavior of gases over a wide temperature range. Although the kinetic theory of gases, statistical thermodynamics, and quantum statistics accurately and rigorously develop quantitative aspects of the behavior of gases over a wide range of pressures and temperatures, we shall adopt a more heuristic approach and seek simply to understand qualitatively what is happening.

Recall from the introductory discussion in Chapter 1 the relationship between macroscopic (classical) and microscopic thermodynamics. The microscopic description assumes that a gas consists of a large number of individual particles (molecules). The simplest molecules are those composed of a single atom and are called *monatomic*. Examples of monatomic gas molecules include helium and argon. The simplest molecular configuration containing more than one atom is a diatomic molecule composed of two atoms of the same element. Molecular oxygen, O_2, and molecular nitrogen, N_2, are examples of such *diatomic* molecules. Diatomic molecules can also be formed from atoms of two different elements. Air, composed primarily of oxygen and nitrogen, is also frequently modeled as a diatomic gas, although it is a mixture of two diatomic gases. The usual schematic representation of a diatomic molecule is as a simple dumbbell, illustrated in Figure 11-1a, composed of two masses (atoms) attached to a rigid rod. *Polyatomic* molecules are composed of three or more atoms of one or more different elements and are more complex and less fruitful for simple examination than diatomic molecules. Polyatomic molecules are called *linear* if all the atoms are modeled as attached to a single rigid rod or called *nonlinear* if more than one rigid rod is required to model the molecule. Our attention will be focused on diatomic molecules.

The simplest implementation of the microscopic approach to the behavior of gases is the kinetic theory of gases, which uses Newtonian mechanics to determine the properties of gases. The number of *degrees of freedom* of a molecule is a central concept in the microscopic description of a gas according to kinetic theory. Each energy storage mode of a molecule is called a degree of freedom. For example, the

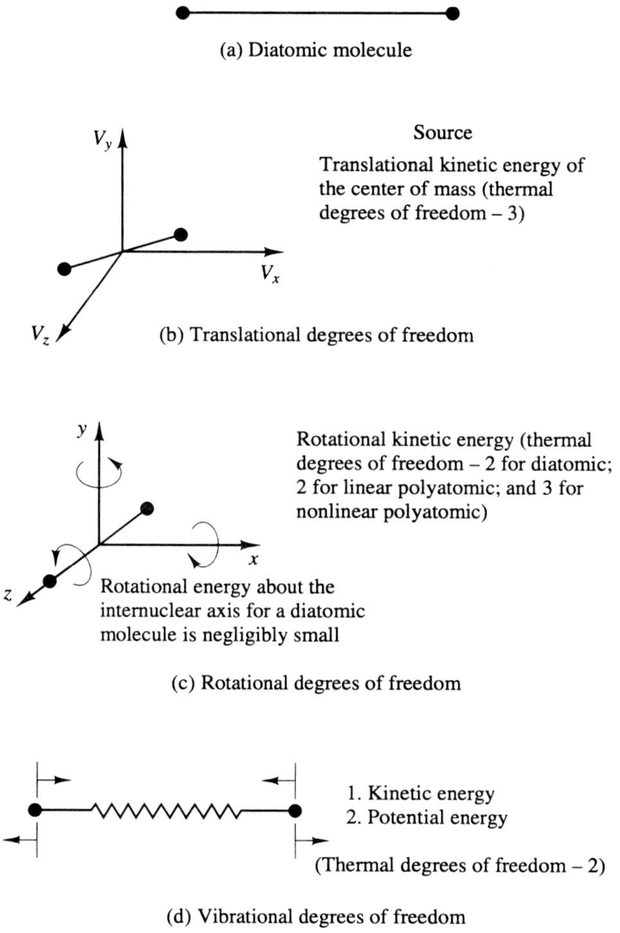

Figure 11-1 Representation of Modes of Molecular Energy Storage. *From Anderson (1989).*

translational kinetic energy of a molecule is determined by the three components of velocity, and so translation is said to have three degrees of freedom. Kinetic theory computes the ratio of specific heats for a thermally perfect (ideal) gas to be

$$\gamma = \frac{f + 2}{f}, \tag{11-16}$$

where f is the number of degrees of freedom of energy storage available to the gas molecule. The derivation of the above expression, which is cited without proof, is available in virtually any kinetic theory of gases or statistical thermodynamics textbook (Lee, Sears, and Turcotte (1973), for example). For a diatomic molecule, we can envision both translation and rotation as contributing possible degrees of freedom of motion. Figures 11-1b and 11-1c are schematic representations of the degrees of freedom available in translation and rotation. The molecule can translate in the three coordinate directions for three degrees of freedom. Since the molecule can also rotate, an additional three degrees of freedom for rotation seems reasonable. But the angular momentum associated with rotation about the longitudinal (connecting) axis is much smaller than the angular momentum associated with rotation about the other two coordinate axes. Hence, only two degrees of freedom are available from rotation since longitudinal rotation adds little energy to the molecule. For a diatomic molecule this would then suggest that the ratio of specific heats, γ, should become $(5 + 2)/5 = 1.4$. Figure 11-2 illustrates the variations of the specific heat at constant volume and the ratio of specific heats for nitrogen. The assumptions of constant specific heats and a constant value of the ratio of specific

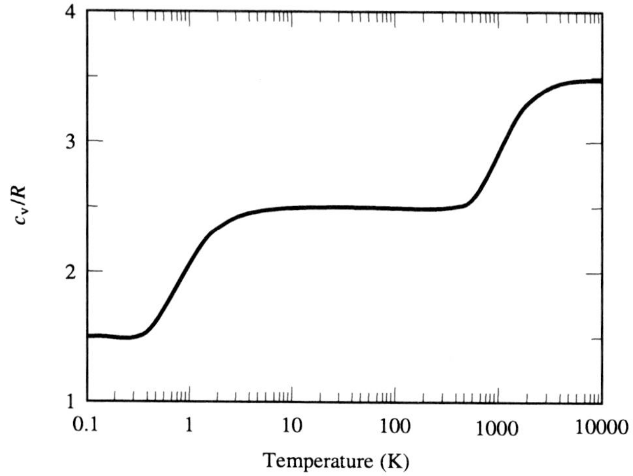

(a) Specific heat at constant volume

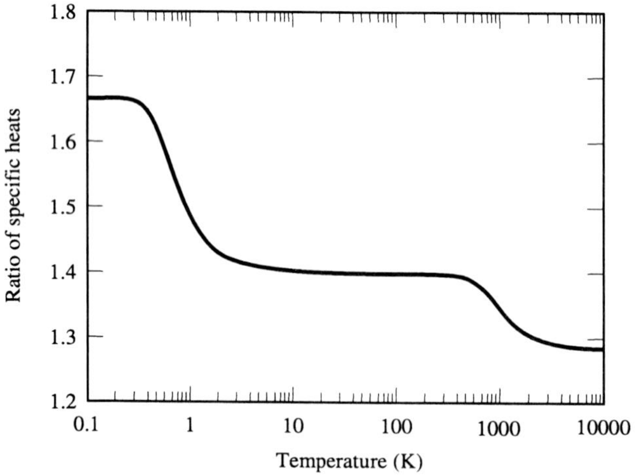

(b) Ratio of specific heats

Figure 11-2 Temperature Variation of c_v/R and γ for N_2

heats are quite reasonable estimates over a significant temperature range. But the figure also shows that as the temperature increases, the value of the constant volume specific heat starts to increase and the value of the ratio of specific heats ratio starts to decrease. Why?

Equation 11-16 demands that if the ratio of specific heats is to decrease, then the number of degrees of freedom must increase. As the temperature is increased, the next degrees of freedom are said to result from vibrational excitation. This is schematically indicated in Figure 11-1d, where the vibration mode is indicated as a spring instead of as a rigid rod connecting the two constituent atoms of the diatomic molecule. Since the "spring" is compressing and the atoms have relative motion, two new degrees of freedom appear associated with the kinetic and potential energy of the added motion. The ratio of specific heats should then be $\gamma = (2 + 7)/7 = 1.286$. But, as Figure 11-2b shows, only at elevated temperature does the value of the ratio of specific heats approach this number. Kinetic theory and the degrees of freedom interpretation predict only discrete values of the ratio of specific heats as various degrees of freedom are excited. The kinetic theory of gases is not able to explain the gradual variation of the specific heats with temperature.

Further refinements in predicting the behavior of the specific heats as a function of temperature must come from statistical thermodynamics and quantum mechanics. Quantum mechanics predicts that the energy of a molecule is restricted to certain discrete values (or levels). Spectroscopic data provide information on the energy levels and on the distribution of molecules in the various energy levels. From the spectroscopic data it is possible to develop characteristic temperatures expressing the degree of excitation of the rotational and vibrational degrees of freedom. For temperatures much greater than the characteristic temperature, the degrees of freedom ascribable to that mode are fully excited. Examples of characteristic rotational temperatures, θ_{rot}, and of characteristic vibrational temperatures, θ_{vib}, are presented in Table 11-1 for some common gases. The characteristic vibrational temperature for air is suggested by Technical Report 1135 of the National Advisory Committee for Aeronautics (NACA) (1953). Thus, both nitrogen and oxygen—and hence air—possess fully excited rotational modes well before room temperature is reached. Any variations in specific heats or in the ratio of specific heats with temperature are due to vibrational excitation. Quantum mechanics provides an expression for the temperature variations of the specific heats of a diatomic molecule undergoing vibrational excitation by modeling the molecule as a simple harmonic oscillator. For a thermally perfect, but calorically imperfect, gas the specific heats and the ratio of specific heats become

$$c_v = \frac{5}{2}R + R\left(\frac{\theta_{\text{vib}}}{T}\right)^2 \frac{e^{\theta_{\text{vib}}/T}}{(e^{\theta_{\text{vib}}/T} - 1)^2}; \tag{11-17}$$

$$c_p = \frac{7}{2}R + R\left(\frac{\theta_{\text{vib}}}{T}\right)^2 \frac{e^{\theta_{\text{vib}}/T}}{(e^{\theta_{\text{vib}}/T} - 1)^2}; \tag{11-18}$$

$$\gamma = \frac{c_p}{c_v} = \frac{7 + 2(\theta_{\text{vib}}/T)^2(e^{\theta_{\text{vib}}/T}/(e^{\theta_{\text{vib}}/T} - 1)^2)}{5 + 2(\theta_{\text{vib}}/T)^2(e^{\theta_{\text{vib}}/T}/(e^{\theta_{\text{vib}}/T} - 1)^2)}. \tag{11-19}$$

TABLE 11-1 CHARACTERISTIC ROTATIONAL AND VIBRATIONAL TEMPERATURES

Molecule	θ_{rot} (K)	θ_{rot} (R)	θ_{vib} (K)	θ_{vib} (R)
H_2	87.20	157.00	6340	11,412
N_2	2.86	5.15	3390	6102
O_2	2.09	3.76	2270	4086
Air			3056	5500

Adapted from John (1984).

The preceding expressions assume that the rotational mode is fully excited and that the vibrational mode is not fully excited. They are generally accepted to be valid for temperatures up to 2500 K (~5000 R).

If the temperature continues to increase, the composition of the gas can change as chemical activity and then ionization becomes possible. Figure 11-3 provides a very clear delineation of temperature ranges of air for various excitation levels and chemical activity. As Figure 11-2 illustrates, rotation in nitrogen and oxygen is essentially fully excited at a few degrees K. Vibrational excitation in nitrogen (and oxygen and, hence, air) starts around 800 K. At temperatures between fully excited

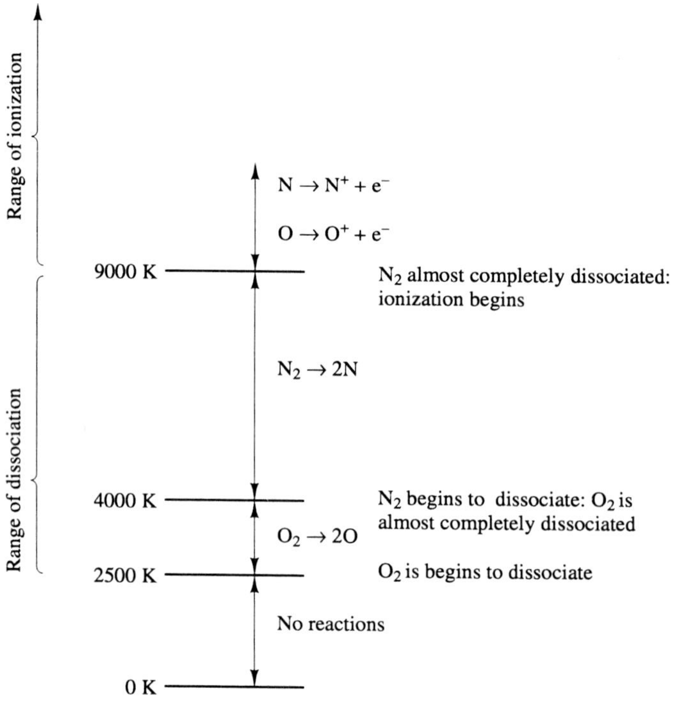

Figure 11-3 Ranges of Dissociation and Ionization for Air at 1-atm Pressure. *From Anderson (1989).*

rotation and the start of vibrational excitation, the specific heats and the ratio of specific heats are essentially constant. The first of the chemical reactions—*dissociation* of O_2 into O—starts at about 2500 K and is almost completed at 4000 K, when dissociation of N_2 is initiated. Dissociation of nitrogen is almost complete at 9000 K, at which point *ionization* of N and O begins. Dissociation and ionization are chemical reactions that change the chemical structure of the gas's molecules and/or atoms and that result in a change in the gas's molecular weight. We are perhaps used to thinking of a chemical reaction as something that occurs when reactive compounds are combined in a test tube. The dissociation of molecular oxygen and nitrogen and the ionization of the O and N atoms are just as much chemical reactions as the combination of two reactive compounds that produce light, heat, and/or smoke.

The mathematical development of expressions for the rotational and vibrational modes for polyatomic molecules is considerably more complex than for diatomic molecules such as O_2. Fortunately, since air is primarily a mixture of two diatomic molecules, we can continue our introductory study without consideration of the more complex molecular structures.

With Eqs. 11-17 through 11-19 used to describe the temperature variations of the specific heats and the ratio of specific heats, procedures for incorporating caloric imperfection can be considered. However, prior to launching into the development of these procedures it is useful to consider some general questions about the relationship of calorically perfect and thermally perfect gases. In a very effective pedagogical approach, Anderson (1989) casts similar considerations as questions.

Question: What forms do the inviscid conservation equations (Eqs. 9-28 and 9-29) have for gas behavior other than calorically perfect?

Answer: To answer these questions, consider the derivations of each of the conservation expressions. Equation 9-5 is simply a statement of conservation of mass. Nowhere in the derivation was any specific equation of state utilized nor were any references made to the functional forms of specific heats. We conclude then that conservation of mass as derived is acceptable for high-temperature flows. Much the same conclusion is reached in examining the derivation of conservation of momentum. Nothing in the derivation depends on the equation of state or specific heats. Thus, Eqs. 9-15 are acceptable for inviscid, high-temperature flows. Conservation of energy is the application of the first law of thermodynamics to the differential control volume in Figure 9-3. Equation 9-18 is also valid for inviscid, high-temperature flows. Anderson points out that this validity also includes chemically reacting flows.

The conservation expressions, valid for high-speed inviscid flows, are repeated here for convenience:

$$\frac{\partial \rho}{\partial t} + \nabla \cdot \rho \overline{V} = 0; \tag{11-20a}$$

$$\rho \frac{D\overline{V}}{Dt} = -\nabla P; \tag{11-20b}$$

$$\frac{D}{Dt}\left(u + \frac{V^2}{2}\right) = -\nabla \cdot (\rho \overline{V}) + \rho \dot{q}. \tag{11-20c}$$

Question: Can the flow of a thermally perfect gas be isentropic?

Answer: Consider the adiabatic flow of an inviscid, thermally perfect gas. For adiabatic, steady flow, the stagnation enthalpy, h_o, is constant, and

$$dh_o = dh + V\,dV = 0. \tag{11-21}$$

In Chapter 9 we obtained, independent of any equation of state or specific heats variation, the following version of Euler's equation:

$$dP = -\rho V\,dV, \tag{11-22a}$$

which can be cast in the following form:

$$V\,dV = -dP/\rho = -v\,dP. \tag{11-22b}$$

Then the differential of the enthalpy can be expressed as

$$dh - v\,dP = 0. \tag{11-23a}$$

But a combined first and second laws expression is $T\,ds = dh - v\,dP$, and so

$$T ds = 0 \tag{11-23b}$$

and the flow is isentropic. This makes sense even for equilibrium chemically reacting flows since in an equilibrium system the chemical composition depends on only the pressure and the temperature and any excursion that returns to the original pressure and temperature would leave all thermodynamic properties unchanged.

Question: For the one-dimensional flow of an inviscid, high-temperature gas, does $M = 1$ still occur at the minimum area (the throat)?

Answer: Consider a differential control volume with area variation, as shown in Figure 11-4. Conservation of mass as applied to this control volume yields

$$dA/A + d\rho/\rho + dV/V = 0. \tag{11-24a}$$

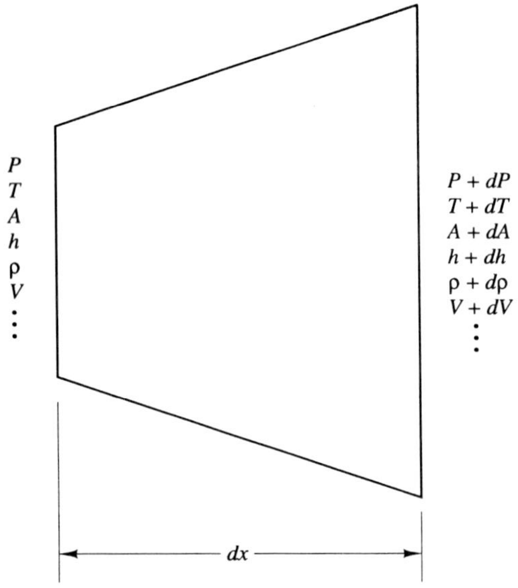

Figure 11-4 Differential Control Volume with Area Variation

Conservation of energy, in the absence of frictional forces, applied to the control volume becomes

$$dP + \rho V\, dV = 0. \tag{11-24b}$$

Recall from Chapter 1 that the general expression for the speed of sound is

$$a^2 = (\partial P/\partial \rho)_s. \tag{1-53}$$

Like many of the fundamental expressions we derived in Part I of this text, the local speed of sound was obtained without specifying an equation of state or any other auxiliary expressions. Since we established in the answer to the previous question that adiabatic, chemically reacting frictionless flow is isentropic, the speed of sound becomes simply $a^2 = dP/d\rho$. Using the speed of sound in the conservation of mass relationship produces

$$dP/(\rho a^2) + dA/A + dV/V = 0. \tag{11-25a}$$

Substituting dP from Euler's equation then yields

$$-V\,dV/a^2 + dA/A + dV/V = 0, \tag{11-25b}$$

which becomes

$$dA/A = dV/V(V^2/a^2 - 1). \tag{11-25c}$$

Letting $V^2/a^2 = M^2$ obtains the needed form:

$$dA/A = (M^2 - 1)\, dV/V. \tag{11-26}$$

An equivalent expression to Eq. 11-26 was obtained in Chapter 5 in conjunction with the simple isentropic flow of a calorically perfect gas. Equation 11-26, however, has no such restrictions; it is valid for gas characterizations up to chemically reacting flows. We reach the same conclusion about the relationship of the minimum area to the sonic point as was found for the calorically perfect gas—namely that sonic flow does exist in the throat for the isentropic flow of a chemically reacting gas.

Question: For a chemically reacting mixture is the speed of sound given by $a = \sqrt{\gamma R T}$?

Answer: The first and second laws expressions are

$$T\,ds = du + P\,dv; \tag{11-27a}$$

$$T\,ds = dh - v\,dP. \tag{11-27b}$$

Since a sound wave is isentropic, Eqs. 11-27 reduce to

$$0 = du + P\,dv; \tag{11-27c}$$

$$0 = dh - v\,dP. \tag{11-27d}$$

But we have seen that for an equilibrium chemically reacting gas, $u = u(v, T)$ and $h = h(P, T)$. For these functional relationships, the differentials become

$$\begin{aligned} du &= \left(\frac{\partial u}{\partial v}\right)_T dv + \left(\frac{\partial u}{\partial T}\right)_v dT \\ &= \left(\frac{\partial u}{\partial v}\right)_T dv + c_v\, dT, \end{aligned} \tag{11-28a}$$

$$dh = \left(\frac{\partial h}{\partial P}\right)_T dP + \left(\frac{\partial h}{\partial T}\right)_P dT$$
$$= \left(\frac{\partial h}{\partial P}\right)_T dP + c_p\, dT, \tag{11-28b}$$

where as usual $c_v = (\partial u/\partial T)_v$ and $c_p = (\partial h/\partial T)_p$. Combining Eqs. 11-28 with Eqs. 11-27 and collecting terms produce

$$c_v\, dT + \left[P + \left(\frac{\partial u}{\partial v}\right)_T\right] dv = 0 \tag{11-29a}$$

and

$$c_p\, dT + \left[\left(\frac{\partial h}{\partial P}\right)_T - v\right] dP = 0. \tag{11-29b}$$

Dividing yields

$$\frac{c_p}{c_v} = \frac{\left(\frac{\partial h}{\partial P}\right)_T - v}{\left(\frac{\partial u}{\partial v}\right)_T + P}\, \frac{dP}{dv}. \tag{11-30}$$

Recognizing that $dv = -d\rho/\rho^2$ obtains

$$\frac{c_p}{c_v} = \frac{\left(\frac{\partial h}{\partial P}\right)_T - v}{\left(\frac{\partial u}{\partial v}\right)_T + P}\,(-\rho^2)\,\frac{dP}{d\rho}. \tag{11-31}$$

However, since the sound wave is isentropic, $a^2 = dP/d\rho = (\partial P/\partial \rho)_s$ and

$$\left(\frac{\partial P}{\partial \rho}\right)_s = \frac{c_p}{c_v}\frac{1}{\rho^2}\, \frac{\left(\frac{\partial u}{\partial v}\right)_T + P}{v - \left(\frac{\partial h}{\partial P}\right)_T}. \tag{11-32a}$$

Some algebraic rearrangement and the use of the thermal equation of state produce

$$a^2 = \gamma R T\, \frac{1 + \left(\frac{\partial u}{\partial v}\right)_T \frac{1}{P}}{1 - \left(\frac{\partial h}{\partial P}\right)_T \rho}, \tag{11-32b}$$

where $\gamma = c_p/c_v$. Equation 11-32b gives us the answer to the question. The simple relation $a = \sqrt{\gamma R T}$ is *not* the speed of sound in a chemically reacting flow. However, if the flow is calorically perfect or thermally perfect, Eq. 11-32b reduces to the familiar form $a = \sqrt{\gamma R T}$.

11.4 COMPUTATIONAL PROCEDURES FOR THERMALLY PERFECT GASES

We have now developed sufficient background in the thermodynamics of calorically imperfect gases to initiate development of useful computational procedures. Such procedures will be inherently more complex than for calorically perfect gases; nonetheless, our goal is to attain computational utility for some processes comparable to that for calorically perfect gases. The sequence of development will be as follows:

1. Develop the necessary mathematical approaches.
2. Explore the numerical techniques needed to obtain the desired results.
3. Examine software that accomplishes the desired results.
4. Apply the software to situations of interest and/or utility.

The software element available for calorically imperfect gases is called CIGQ and is available on the diskette for this textbook.

Many of the ideas developed in Chapter 6 are applicable for calorically imperfect gases. Virtually all of the simple flow/calorically perfect gas development involved the ratio of specific heats, γ. Indeed, it appeared in strange combinations both as exponents in equations and as various terms in equations. The ratio of specific heats is a meaningful parameter for calorically imperfect gases, but it will not occupy the central position that it did for calorically perfect gas processes. The notion of a simple flow is also a useful concept from our earlier study. Isentropic flow of a calorically imperfect gas—a simple flow process—will be the first flow examined. Normal shock waves are an integral part of virtually any compressible flow study and will be examined within the calorically imperfect gas development. The following development of calorically imperfect gases computational capability is based on material from NACA TR-1135.

Calculations involving calorically perfect simple flows, shock waves, and expansion waves rely heavily on the relationships between static and stagnation conditions at a point in a flow field. Applications involving calorically imperfect gases will share that same involvement. Consider, from Chapter 1, the steady-state energy equation

$$c_p\,dT + d\left(\frac{V^2}{2}\right) = 0. \qquad (11\text{-}33)$$

Integration of this equation from an arbitrary velocity V and static temperature T to the stagnation state, $V = 0$ and T_o, yields

$$\int_o^V d\left(\frac{V^2}{2}\right) = -\int_{T_o}^{T} c_p\,dT, \qquad (11\text{-}34a)$$

which becomes

$$\frac{V^2}{2} = \int_T^{T_o} R\left(\frac{7}{2} + \left(\frac{\theta_{vib}}{T}\right)^2 \frac{e^{\theta_{vib}/T}}{(e^{\theta_{vib}/T} - 1)^2}\right) dT \tag{11-34b}$$

when the simple harmonic model specific heat relation is used for c_p. Carrying out the integration results in

$$\frac{V^2}{2} = R\left(\frac{7}{2}(T_o - T) + \theta_{vib}\left(\frac{1}{e^{\theta_{vib}/T_o} - 1} - \frac{1}{e^{\theta_{vib}/T} - 1}\right)\right). \tag{11-35}$$

In Section 11.3, the speed of sound for a thermally perfect gas was established as $a = \sqrt{\gamma RT}$ and the ratio of specific heats was found to be

$$\gamma = \frac{7 + 2(\theta_{vib}/T)^2(e^{\theta_{vib}/T}/(e^{\theta_{vib}/T} - 1)^2)}{5 + 2(\theta_{vib}/T)^2(e^{\theta_{vib}/T}/(e^{\theta_{vib}/T} - 1)^2)}. \tag{11-19}$$

From Eq. 11-35, the square of the Mach number can be cast as

$$M^2 = \frac{V^2}{a^2} = \frac{1}{\gamma}\left(7\left(\frac{T_o}{T} - 1\right) + \frac{2\theta_{vib}}{T}\left(\frac{1}{e^{\theta_{vib}/T_o} - 1} - \frac{1}{e^{\theta_{vib}/T} - 1}\right)\right), \tag{11-36}$$

where γ is defined as in Eq. 11-19. This expression is the calorically imperfect equivalent to Eq. 1-66, the calorically perfect steady-state energy equation. Equation 11-36 provides a relationship between M and T/T_o. For an imperfect gas, note that the stagnation temperature must be specified in order to obtain either the Mach number for the temperature or the temperature from the Mach number. With T/T_o the Mach number can be directly calculated, but obtaining T/T_o given the Mach number requires a numerical solution.

The combined first and second laws expression for a thermally perfect gas is

$$ds = c_p\, dT/T - R\, dP/P. \tag{11-37}$$

For an isentropic flow, $ds = 0$ and Eq. 11-37 becomes

$$c_p\, dT/T = R\, dP/P. \tag{11-38}$$

Integrating Eq. 11-38 from static state (T, P) to stagnation state (T_o, P_o), and using Eq. 11-18 for the specific heat, result in

$$\ln\frac{P_o}{P} = \int_T^{T_o}\left(\frac{7}{2} + \left(\frac{\theta_{vib}}{T}\right)^2 \frac{e^{\theta_{vib}/T}}{(e^{\theta_{vib}/T} - 1)^2}\right)\frac{dT}{T}, \tag{11-39a}$$

which can be cast as an expression for the stagnation-to-static pressure ratio, P_o/P, for a calorically imperfect gas:

$$\frac{P_o}{P} = \left(\frac{T_o}{T}\right)^{7/2}\left(\frac{e^{\theta_{vib}/T} - 1}{e^{\theta_{vib}/T_o} - 1}\right)\exp\left(\left(\frac{\theta_{vib}}{T_o}\frac{e^{\theta_{vib}/T_o}}{e^{\theta_{vib}/T_o} - 1}\right) - \left(\frac{\theta_{vib}}{T}\frac{e^{\theta_{vib}/T}}{e^{\theta_{vib}/T} - 1}\right)\right). \tag{11-39b}$$

This equation is the calorically imperfect version of the isentropic relationship for pressure and temperature. It is explicit in the pressure ratio given the temperature ratio, but it requires a numerical root extraction for the temperature ratio given the pressure ratio. The additional complexity that caloric imperfection introduces is evident when Eq. 11-39b is compared to the calorically perfect version. With Eqs. 11-36 and 11-39b available, applications involving calorically imperfect gases can be handled.

Even with the relatively simple expressions for simple flows and perfect gases, we found it convenient to use the software element COMPQ in most applications. The added complexity of the expressions for a calorically imperfect gas virtually demand that applications software be generated if mathematical drudgery is to be avoided. The Mach number equation is a monotonically decreasing function of temperature ratio, and the pressure ratio is a monotonic function of the temperature ratio. The software element CIGQ takes advantage of these functional relationships to use the *method of false position* (Conte and deBoor, 1972) to numerically solve for the roots of these equations.

Applications involving calorically imperfect gases will require pressure and temperature ratios, Mach numbers, stagnation pressures and temperatures, and/or velocities. The approaches developed through this section's examples will permit us to involve these parameters in typical applications easily. An examination of these capabilities is in order.

Example 11-1

For air at a stagnation temperature of 1000 K, find

a. the static temperature and velocity for a Mach number of 0.8;
b. the Mach number and velocity for a static temperature of 800 K;
c. the Mach number and static temperature for a velocity of 1000 m/s.
d. Compare these results with the calorically perfect solutions, which are given in Example 1-6.

Solution. The calorically imperfect gas software element CIGQ will be used for solution of this problem.

a. This problem illustrates quite well the additional complexity that a calorically imperfect gas assumption involves. Were the gas calorically perfect, the solution would be straightforward and similar to that of Example 1-6b. The Mach number and the stagnation temperature are known, which enables Eq. 11-36 to provide the solution to this part of the example. With the Mach number, stagnation temperature, and characteristic vibrational temperature θ_{vib} known, the equation can be solved for the static temperature. With the static temperature known, the speed of sound can be evaluated using Eq. 11-19 for the ratio of specific heats. The velocity then becomes simply $V = Ma$. The problem is that the static temperature is not explicit in Eq. 11-36; the solution is the root of the equation

$$.8^2 = .64 = \frac{1}{\gamma}\left(7\left(\frac{1000}{7} - 1\right) + \frac{2\theta_{vib}}{T}\left(\frac{1}{e^{\theta_{vib}/1000} - 1} - \frac{1}{e^{\theta_{vib}/T} - 1}\right)\right),$$

where $\theta_{vib} = 3056$ K and γ is as defined in Eq. 11-19. Thus, simply finding the static temperature corresponding to a given Mach number is not trivial for calorically imperfect gases. The software element CIGQ provides solutions for many of the same options and cases as CIGQ provides for computations involving calorically perfect gases. CIGQ is limited to diatomic gases at temperatures below the dissociation threshold and to pressures low enough for the thermal equation of state to be valid. Figure 11-5 illustrates the initial input sequence and the menu for CIGQ. Computations for this software element generally require the gas constant as well as the stagnation temperature, so that either the SI (kg/m/s) or EE (lbm/ft/s) units must be specified. For this problem, the isentropic option, **IS**, is required. The menu for the isentropic option is shown in Figure 11-6a. The Mach number is known, so case **M** is the correct selection. For an input Mach number of 0.8, the output of CIGQ is presented in Figure 11-6b. The velocity in the proper units, m/s or ft/s, is also calculated in this option and presented in the output. For a Mach number of 0.8 and a stagnation temperature of 1000 K, the velocity is 471.68 m/s and the static temperature is 901.88 K.

b. The given data for the second part of the problem is the static temperature, 800 K. The Mach number and velocity are required. This part of the problem is somewhat simpler than part (a) since Eq. 11-36 can be solved directly for the Mach number; nonetheless, CIGQ will be used. Option **IS** with case **T** is the appropriate choice. The output is given in Figure 11-6c. The solution yields a Mach number of 1.2025 and a velocity of 670.17 m/s.

c. The third requirement is to compute the Mach number and static temperature given a velocity of 1000 m/s. Again, CIGQ will be used; the output is in Figure 11-6d and shows a Mach number of 2.1559 and a static temperature of 542.25 K.

```
MISSISSIPPI STATE UNIVERSITY
DEPARTMENTS OF MECHANICAL AND AEROSPACE ENGINEERING
CALORICALLY IMPERFECT GAS EVALUATION FOR DIATOMIC GASES

SELECT SI OR EE UNITS? SI

SI UNITS SELECTED
ENTER THE CHARACTERISTIC VIBRATION TEMPERATURE (K)? 3056.
ENTER THE GAS CONSTANT IN J/kg-K? 287.
ENTER THE TOTAL TEMPERATURE IN K? 1000.


OPTIONS FOR CALORICALLY IMPERFECT GASES ARE AS FOLLOWS:
          PROCESS                    CODE
      ISENTROPIC FLOW                 IS
      NORMAL SHOCK WAVE               NS
      PRANDTL-MEYER FLOW              PM
      FINISHED                        FI

CHARACTERISTIC VIBRATION TEMPERATURE = 3056
TOTAL TEMPERATURE =    1000

ENTER THE CODE FOR THE DESIRED PROCESS?
```

Figure 11-5 CIGQ Startup and Options Menu

```
ISENTROPIC FLOW OPTION SELECTED

THE FOLLOWING INPUTS ARE AVAILABLE FOR ISENTROPIC FLOW OF A DIATOMIC
CALORICALLY IMPERFECT GAS
  INPUT VARIABLE      SYMBOL
    MACH NUMBER         M
    TTOTAL FROM V,T     TS
    TTOTAL FROM M,T     TT
       T/TO             T
       P/PO             P
       A/A*             A

ENTER THE SYMBOL FOR THE DESIRED INPUT VARIABLE? M

ENTER THE VALUE OF THE MACH NUMBER? .8
```

(a) Isentropic option menu

```
ISENTROPIC RELATIONS FOR CALORICALLY IMPERFECT DIATOMIC GAS

       R            VIB           TT
    287.00        3056.00      1000.00

     MACH          T/TO          P/PO        GAM        ARATIO        A         VEL
    0.8000       0.90188       0.66504     1.34303      1.0392     589.60     471.68

FOR ISENTROPIC FLOW THE STAGNATION PRESSURE AND TEMPERATURE ARE CONSTANT.

ARE ADDITIONAL ISENTROPIC COMPUTATIONS DESIRED (Y FOR YES)?
```

(b) Isentropic option output

Figure 11-6 CIGQ Input and Output for Example 11-1

```
ISENTROPIC RELATIONS FOR CALORICALLY IMPERFECT DIATOMIC GAS

      R           VIB          TT
   287.00       3056.00      1000.00

    MACH         T/TO         P/PO         GAM        ARATIO        A          VEL
   1.2025      0.80000      0.41796     1.35280      1.0323     557.32      670.17

FOR ISENTROPIC FLOW THE STAGNATION PRESSURE AND TEMPERATURE ARE CONSTANT.

ARE ADDITIONAL ISENTROPIC COMPUTATIONS DESIRED (Y FOR YES)?
```

(c) Output for Example 11-1(b)

```
ISENTROPIC RELATIONS FOR CALORICALLY IMPERFECT DIATOMIC GAS

      R           VIB          TT
   287.00       3056.00      1000.00

    MACH         T/TO         P/PO         GAM        ARATIO        A          VEL
   2.1559     5.422E-01    9.851E-02    1.38254      1.9896     463.85     1000.01

FOR ISENTROPIC FLOW THE STAGNATION PRESSURE AND TEMPERATURE ARE CONSTANT.

ARE ADDITIONAL ISENTROPIC COMPUTATIONS DESIRED (Y FOR YES)?
```

(d) Output for Example 11-1(c)

Figure 11-6 (continued)

d. Table 11-2 is a tabulation of the results from this example problem and from Example 1-6. Table 11-2 shows small differences between the calorically perfect example (1-6) and the calorically imperfect example (11-1) results and serve to illustrate the discrepancies involved in the calorically perfect gas assumption for moderate stagnation temperatures.

TABLE 11-2 COMPARISON OF EXAMPLES 1-6 AND 11-1

Part	Example	M	V (m/s)	T (K)
(a)	1-6	0.8	477.5	886.5
	11-1	0.8	471.7	901.9
(b)	1-6	1.118	633.9	800.0
	11-1	1.203	670.2	800.0
(c)	1-6	2.227	1000.0	502.6
	11-1	2.156	1000.0	542.3

Converging–diverging nozzle flows of calorically imperfect gases follow the basic principles explored in Chapter 6 for simple converging–diverging nozzle flows. With software elements developed for area change and normal shock waves for calorically imperfect gas flows, a treatment analogous to that in Chapter 6 is possible for this more complex gas flow.

Some capabilities of CIGQ for the isentropic option, **IS**, have already been developed and explored. Earlier in this chapter we established that if a sonic point occurs, it must occur at a minimum area for a calorically imperfect gas flow. With this premise, the area ratio, A/A^*, assumes much the same function in calorically imperfect gas flows as in calorically perfect gas flows. Indeed, CIGQ's isentropic option computes the area ratio as a standard output parameter and accepts the area ratio as an input case. The relative simplicity of the calorically perfect gas assumption permits a closed-form expression (given in Table 6-1b) to be derived for the area ratio. The added complexity of a calorically imperfect gas prohibits derivation of a simple closed-form expression for the area ratio, and thus we must resort to numerical solutions. A brief description of the numerics involved in the area ratio computations is in order. For a steady flow the area ratio can be expressed as

$$\frac{A}{A^*} = \frac{\rho^* a^*}{\rho M a}. \tag{11-40}$$

All isentropic input cases except A/A^* result in the Mach number being either input or computed. Given the Mach number or temperature, the various terms in Eq. 11-40 can be readily computed using Eqs. 11-32b, 11-35, 11-36, and 11-39b as well as the thermal equation of state. The case for A/A^* as the input is more complex. As with calorically perfect gases, both subsonic and supersonic Mach number solutions exist. The subsonic solution is calculated by iteration using the bisection procedure to find the Mach number that yields the specified area ratio. The supersonic solution is found by initially determining the two whole-number supersonic Mach numbers that bracket the specified area ratio and then using bisection iteration. Both the subsonic and supersonic solutions use Eq. 11-40 in the iteration processes. Consider the following example.

Example 11-2

A supersonic air flow with a stagnation temperature of 4000 R is expanded through an area change $A_2/A_1 = 2$ from an initial Mach number of 1.5. Find the final Mach number and the pressure and temperature ratios for the process. Compare the results with Example 6-1, which solved the problem using the calorically perfect gas assumption.

Solution. See Figure 6-7a for a schematic representation of this problem. Although not physically present, the throat can be imagined at either or both ends of the diverging passage. The solution starts with

$$\frac{A_2}{A^*} = \frac{A_2}{A_1}\frac{A_1}{A^*}.$$

For a Mach number of 1.5, the isentropic option of CIGQ yields

$$A_1/A^* = 1.190$$

$$T_1/T_o = 0.7458$$

$$P_1/P_o = 0.2824,$$

so that

$$\frac{A_2}{A^*} = 2(1.190) = 2.380,$$

from which

$$M_2 = 0.2554 \quad \text{and} \quad 2.2984.$$

Since the flow is initially supersonic, the supersonic root will be chosen. Then for $M_2 = 2.2984$,

$$T_2/T_o = 0.5496;$$

$$P_2/P_o = 0.07866.$$

With T_o and P_o constant for an isentropic flow,

$$\frac{T_2}{T_1} = \frac{T_2}{T_o}\frac{T_o}{T_1} = \frac{T_2/T_o}{T_1/T_o} = \frac{0.5496}{0.7458} = 0.7370;$$

$$\frac{P_2}{P_1} = \frac{P_2}{P_o}\frac{P_o}{P_1} = \frac{P_2/P_o}{P_1/P_o} = \frac{0.07866}{0.2824} = 0.2785.$$

Figure 11-7 presents the various CIGQ inputs and outputs required to solve this problem. Example 6-1 is essentially the same problem except that the solution was obtained under the assumption of a calorically perfect gas. Table 11-3 contains a tabulation of the results for Examples 11-2 and 6-1.

```
ISENTROPIC RELATIONS FOR CALORICALLY IMPERFECT DIATOMIC GAS

      R           VIB         TT
    53.35       5500.00     4000.00

    MACH        T/TO         P/PO        GAM        ARATIO       A        VEL
   1.5000     0.74581      0.28240     1.30683     1.1900    2586.86   3880.30

FOR ISENTROPIC FLOW THE STAGNATION PRESSURE AND TEMPERATURE ARE CONSTANT.


ISENTROPIC RELATIONS FOR CALORICALLY IMPERFECT DIATOMIC GAS

      R           VIB         TT
    53.35       5500.00     4000.00

    MACH        T/TO         P/PO        GAM        ARATIO       A        VEL
   0.2554     0.99037      0.95873     1.29817     2.3800    2971.08    758.91
   2.2984    5.496E-01    7.866E-02    1.32170     2.3800    2233.34   5133.12

FOR ISENTROPIC FLOW THE STAGNATION PRESSURE AND TEMPERATURE ARE CONSTANT.

ARE ADDITIONAL ISENTROPIC COMPUTATIONS DESIRED (Y FOR YES)?
```

Figure 11-7 CIGQ Output for Example 11-2

TABLE 11-3 COMPARISON OF EXAMPLES 6-1 AND 11-2

Example	M_1	A_2/A_1	M_2	T_2/T_1	P_2/P_1
6-1	1.5	2.0	2.3768	0.6807	0.2603
11-2	1.5	2.0	2.2984	0.7370	0.2785

Caloric imperfection thus causes a decrease in the exit Mach number and an increase in the temperature and pressure ratios. In this example the percentage errors are 3.4 percent in Mach number, 7.64 percent in pressure ratio, and 6.54 percent in temperature ratio. For many estimates the calorically perfect gas assumption results would be acceptable, but this example does illustrate the magnitude of errors that might be expected because of the calorically perfect gas assumption. It is instructive to compare the results, presented in Table 11-3, of the calorically perfect and calorically imperfect computational procedures. Expansion from the same initial Mach number through the same area ratio produces a higher temperature and a lower Mach number for the imperfect gas assumption than for the perfect gas assumption. Consider the reason for this. Because of the increase in specific heats with temperature in the calorically imperfect gas assumption, more energy is stored in a calorically imperfect gas than in a calorically perfect gas at the same temperature. As the velocity is increased during the expansion process, the conversion of thermal to directed kinetic energy of motion (gas velocity) is driven by the static temperature decrease during the expansion process. Since more energy is stored in the calorically imperfect gas, the required thermal energy can be obtained at a higher static temperature. Since the static temperature is higher, the speed of sound is higher and the Mach number is lower. The pressure simply tracks the temperature in an isentropic fashion. A similar analysis could be made for virtually every example problem in this and the next section and the explanations would be similar. The key to understanding the effects of caloric imperfection is the realization that vibrational excitation requires additional energy, which is stored as internal energy, as a gas is heated. As the gas is cooled, as in the expansion process, that additional energy is recovered.

As the last example illustrates, case **A** of option **IS** can be used to solve isentropic area change problems in calorically imperfect gas flows. The capability to handle normal shock waves in a calorically imperfect gas environment is needed if nozzle flows as well as a number of other compressible flow problems are to be effectively treated. Chapter 2 developed the conventional calorically perfect gas analysis of normal shock waves and examined a number of applications. The computational procedure required for normal shock waves in a calorically imperfect gas flow starts with the conservation equations from Chapter 2 (Figure 2-3 is the appropriate schematic representation). Conservation of mass, momentum, and energy across a normal shock wave becomes

$$\rho_1 V_1 = \rho_2 V_2; \tag{11-41a}$$

$$P_1 + \rho_1 V_1^2 = P_2 + \rho_2 V_2^2; \tag{11-41b}$$

$$h_1 + V_1^2/2 = h_2 + V_2^2/2. \tag{11-41c}$$

These conservation expressions, in conjunction with the thermal equation of state

$$P_1/\rho_1 T_1 = P_2/\rho_2 T_2, \tag{11-41d}$$

express the property relationships across a normal shock wave. Note that $T_{o_1} = T_{o_2}$ for a calorically imperfect shock wave. Equations 11-41 are identical to the starting point for the analysis of calorically perfect normal shock waves since no assumptions about specific heat variations have been made. Basically, what must be accomplished is solving the system of equations in the unknowns ρ_2, V_2, P_2, and T_2. For a calorically perfect gas such solution was accomplished algebraically; for a calorically imperfect gas, the solution must be numerical. All of the cases in CIGQ involve iteration in the solution. The iteration procedure starts by assuming a value of T_2 and then uses conservation of mass in the form $\rho_2/\rho_1 = V_1/V_2$ to express conservation of momentum as

$$T_2 = \frac{1}{R}\left(\frac{\rho_1}{\rho_2}(RT_1 + V_1^2) - \left(\frac{\rho_1}{\rho_2}V_1\right)^2\right). \tag{11-42}$$

The basic solution kernel for finding conditions at state 2 given conditions at state 1 is composed of the following sequence:

1. Given T_1, T_o, and M_1, use Eq. 11-35 to find V_1.
2. Guess a value of T_2.
3. Use T_2 and T_o to find M_2 and V_2, using Eqs. 11-36 and 11-35.
4. Compute the density ratio using $\rho_2/\rho_1 = V_1/V_2$.
5. Use Eq. 11-42 to compute T_2, which is compared with its estimated value in step 2.
6. If the two values of T_2 do not agree, then use bisection to make a new guess on T_2 and repeat steps 2–6 until convergence is reached.

Figure 11-8a presents the CIGQ menu for the normal shock option. The input choices are similar to those of the normal shock option in COMPQ. All of the cases in the normal shock option use the iteration procedure just given; in most cases the procedure is embedded in another iteration. As with other options in CIGQ, the software removes much of the arithmetical drudgery associated with calorically imperfect gas computations. Consider the following example.

Example 11-3

Air, moving at $M = 2.5$ and with a stagnation temperature of 4000 R, undergoes a normal shock. Find the downstream conditions and the loss in stagnation pressure for the process. Compare the results with the solution if the process were assumed to be calorically perfect.

Solution. The solution is illustrated in Figure 11-8. With the upstream Mach number, M_1, known, case **U** of option **NS** can be used. The output for this example is shown in Figure 11-8b, and the output for the same problem using COMPQ is presented in Figure 11-8c. A comparison of the two outputs reveals the differences between the calorically perfect and calorically imperfect gas assumptions. The loss in stagnation pressure is 53.7 percent of the initial stagnation pressure.

```
NORMAL SHOCK WAVE OPTION SELECTED

THE FOLLOWING INPUTS ARE AVAILABLE FOR NORMAL SHOCK WAVES IN A DIATOMIC
CALORICALLY IMPERFECT GAS
        INPUT VARIABLE                 SYMBOL
    UPSTREAM MACH NUMBER                 U
    DOWNSTREAM MACH NUMBER               D
    STATIC PRESSURE RATIO, P2/P1         P
    STAG. PRESSURE RATIO, PO2/PO1        S

ENTER THE SYMBOL FOR THE DESIRED INPUT VARIABLE?
```

(a) Normal shock CIGQ options

```
NORMAL SHOCK WAVE SOLUTION USING CALORICALLY IMPERFECT GAS

UPSTREAM CONDITIONS:
MACH NUMBER      P/PO1        T/TO1      GAMMA       A1          U1
    2.5000    5.604E-02   5.059E-01    1.3269    2146.8      5366.9

DOWNSTREAM CONDITIONS:
MACH NUMBER      P/PO2        T/TO2      GAMMA       A2          U2
   0.4950       0.8553       0.9647    1.2988    2933.1      1451.9

PROPERTY RATIOS ACROSS NORMAL SHOCK WAVE:
    P2/P1   PO2/PO1      T2/T1   RHO2/RHO1
     7.05   0.46192      1.907     3.697

CHARACTERISTIC VIBRATION TEMPERATURE = 5500
TOTAL TEMPERATURE =  4000

STAGNATION TEMPERATURE IS CONSTANT ACROSS A STATIONARY NORMAL SHOCK WAVE

ARE ADDITIONAL NORMAL SHOCK WAVE COMPUTATIONS DESIRED (Y FOR YES)?
```

(b) Calorically imperfect output (CIGQ)

Figure 11-8 CIGQ Output for Example 11-3

```
NORMAL SHOCK WAVE CALLED FOR
INPUT PARAMETERS ARE:
      UPSTREAM MACH NUMBER          U
      DOWNSTREAM MACH NUMBER        D
      STATIC PRESSURE RATIO         P
      STAGNATION PRESSURE RATIO     S

ENTER THE DESIRED LETTER? u

ENTER THE UPSTREAM MACH NUMBER? 2.5

    M1          M2          P2/P1       T2/T1       RHO2/1      PO2/PO1
  2.5000      0.5130      7.1250      2.1375      3.3333      0.4990

STAGNATION TEMPERATURE IS CONSTANT ACROSS A NORMAL SHOCK.

THE CHANGE IN ENTROPY IS DEL S/CP =     0.1986

ARE ADDITIONAL NORMAL SHOCK WAVE COMPUTATIONS DESIRED?
```

(c) Calorically perfect output (CIGQ)

Figure 11-8 (continued)

CIGQ also provides capability for a Prandtl-Meyer process for calorically imperfect gases. However, unlike the remainder of the processes in COMPQ and CIGQ, this implementation involves generation of a table that is used for input and output interpolation. The suggested basis, from NACA TR-1135, for a Prandtl-Meyer process for calorically imperfect gases is the integration of the equation

$$\nu = \int_{P^*}^{P} \frac{\sin 2\mu}{2\gamma P}\, dP. \tag{11-43}$$

The first call to the **PM** option of CIGQ results in the generation of a table containing the Mach number, Prandtl-Meyer angle, and Mach angle. Equation 11-43 is integrated to produce $\nu(M)$ using a fourth-order Runge-Kutta procedure (the same as for CONEQ from Chapter 3) with determination of the ratio of specific heats from Eq. 11-19, computation of the temperature ratio T/T_0 from Eq. 11-39b, and calculation of the Mach number from Eq. 11-36 (from which the Mach angle is computed). A table interpolation implementation was chosen since numerical integration of Eq. 11-43 for each call would be too time-consuming for convenient use. The menu for the first call to option **PM** is illustrated in Figure 11-9a. All three input cases utilize the table and a second-order Lagrangian interpolating polynomial to find the remaining two parameters. Additional calls to the **PM** option produce a menu similar to that in Figure 11-9a except that information relating to the generation of the table is not present. The typical output for this case is illustrated in Figure 11-9b. Consider the following example.

```
INITIAL CALL FOR PRANDTL-MEYER CALCULATION
PRANDTL-MEYER FUNCTION TABLE BEING GENERATED

PRANDTL-MEYER FUNCTION TABLE GENERATION COMPLETE

THE FOLLOWING INPUTS ARE AVAILABLE FOR PRANDTL-MEYER CALORICALLY
IMPERFECT GAS FLOW:
     INPUT VARIABLE             SYMBOL
   MACH NUMBER                    M
   PRANDT-MEYER ANGLE             N
   MACH ANGLE                     A

ENTER THE SYMBOL FOR THE DESIRED INPUT VARIABLE?
```

(a) Menu for first call to option **PM**

```
PRANDTL-MEYER FUNCTION FOR CALORICALLY IMPERFECT DIATOMIC GAS

       R            VIB          TT
    287.00        3056.00      1644.70

     MACH        NU(DEG)      T/TO        P/PO        GAM       MU(DEG)
    1.1547        2.605    8.278E-01   4.509E-01    1.3157      60.000

THE PRANDTL-MEYER PROCESS IS ISENTROPIC.

ARE ADDITIONAL PRANDTL-MEYER CALCULATIONS DESIRED (Y FOR YES)?
```

(b) **PM** sample output

Figure 11-9 Menu and Sample Output for **PM** Option

Example 11-4

A supersonic flow initially at $M = 2$ and with static properties of 1000 K and 10 kPa is expanded around a 10° sharp corner. Find the downstream Mach number and properties, and construct a schematic of the expansion fan. Compare the results with that of Example 4-1, which assumed a calorically perfect gas.

Solution. The situation is illustrated in Figure 11-10a. The stagnation temperature must be found first. Using case **TT** of the isentropic option in CIGQ, the stagnation temperature is computed as 1644.7 K. Figure 11-10b contains the output for the stagnation temperature calculation. The logic in CIGQ is such that calls to cases **TS** or **TT** of the isentropic option reset the value of the input stagnation temperature. The Prandtl-Meyer angle is then obtained by a call to option **PM**. The first call to **PM** results in the generation of a Mach number/Prandtl-Meyer angle/Mach angle table. Case **M** of the **PM** option gives the Prandtl-Meyer angle at state 1 as 28.438 degrees. For an expansion, the Prandtl-Meyer angle at state 2 is then 28.438 + 10 = 38.438 degrees. Case **NU** of the Prandtl-Meyer option then returns the Mach number at state 2: 2.3458.

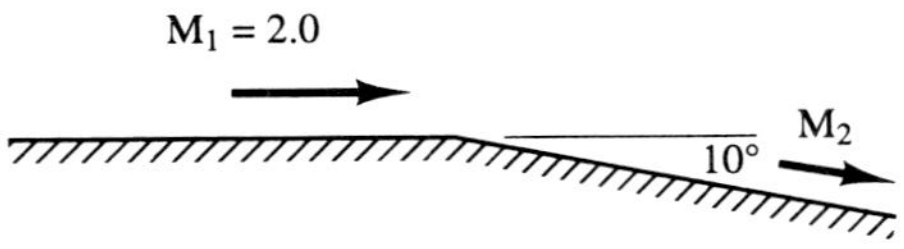

(a) Schematic for Example 11.4

```
ISENTROPIC RELATIONS FOR CALORICALLY IMPERFECT DIATOMIC GAS

      R            VIB          TT
   287.00        3056.00      1644.70

    MACH          T/TO         P/PO         GAM        ARATIO        A          VEL
   2.0000       0.60801      0.12807     1.33510      1.7630      619.01     1238.02

FOR ISENTROPIC FLOW THE STAGNATION PRESSURE AND TEMPERATURE ARE CONSTANT.

ARE ADDITIONAL ISENTROPIC COMPUTATIONS DESIRED (Y FOR YES)?
```

(b) Determination of stagnation temperature

Figure 11-10 CIGQ and COMPQ Outputs for Example 11-4

```
PRANDTL-MEYER FUNCTION FOR CALORICALLY IMPERFECT DIATOMIC GAS

        R              VIB            TT
     287.00         3056.00        1644.70

      MACH         NU(DEG)        T/TO          P/PO          GAM          MU(DEG)
     2.0000        28.438      6.080E-01    1.281E-01      1.3351        30.000

THE PRANDTL-MEYER PROCESS IS ISENTROPIC.

ARE ADDITIONAL PRANDTL-MEYER CALCULATIONS DESIRED (Y FOR YES)?


PRANDTL-MEYER FUNCTION FOR CALORICALLY IMPERFECT DIATOMIC GAS

        R              VIB            TT
     287.00         3056.00        1644.70

      MACH         NU(DEG)        T/TO          P/PO          GAM          MU(DEG)
     2.3458        38.438      5.257E-01    7.225E-02      1.3464        25.233

THE PRANDTL-MEYER PROCESS IS ISENTROPIC.
```

(c) Cases **M** and **NU** outputs

```
ENTER THE UPSTREAM MACH NUMBER AND THE TURNING ANGLE? 2.0,10.0

PRANDTL-MEYER TURNING SOLUTION
    M1              M2          ANGLE(DEG)          T2/T1            P2/P1
  2.0000          2.3849         10.0000           0.8421           0.5480

THE PROCESS IS ISENTROPIC; HENCE NO CHANGE IN STAGNATION PROPERTIES.
```

(d) COMPQ solution

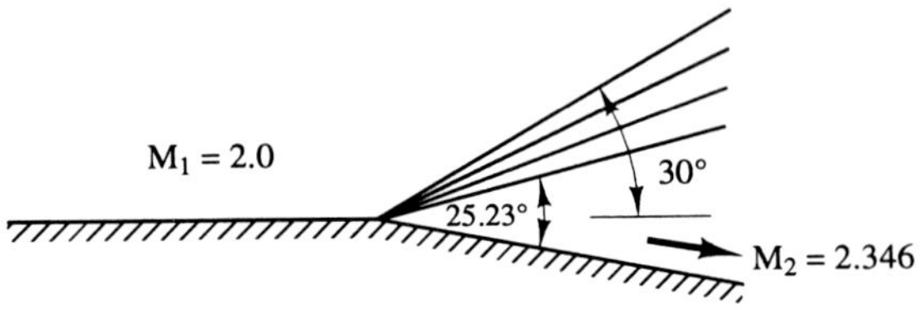

$$\frac{T_2}{T_1} = 0.8646 \quad \frac{P_2}{P_1} = 0.5644$$

(e) Solution including Mach wave bounds

Figure 11-10 (continued)

The output of CIGQ for these cases is shown in Figure 11-10c. Finally, the temperature and pressure ratios are computed as

$$\frac{T_2}{T_1} = \frac{T_2}{T_o}\frac{T_o}{T_1} = \frac{T_2/T_o}{T_1/T_o} = \frac{0.5257}{0.6080} = 0.8646;$$

$$\frac{P_2}{P_1} = \frac{P_2}{P_o}\frac{P_o}{P_1} = \frac{P_2/P_o}{P_1/P_o} = \frac{0.0723}{0.1281} = 0.5644.$$

The calorically perfect version of this example problem is obtained using COMPQ with the output given in Figure 11-10d. A comparison of the results of the two approaches can be made by considering the entries in Table 11-4. As with previous examples in this chapter, the effects of caloric imperfection are noticeable, although not severe.

TABLE 11-4 COMPARISON OF EXAMPLE 11-4 RESULTS

Caloric assumption	M_1	M_2	P_2/P_1	T_2/T_1
Imperfect	2.0	2.346	0.5644	0.8644
Perfect	2.0	2.385	0.5480	0.8420

11.5 APPLICATIONS INVOLVING CALORICALLY IMPERFECT GASES

Because the results of the examples in the previous section illustrate a range of errors caused by the assumption of a calorically perfect gas, they should permit quantitative assessment of that assumption. CIGQ allows many of the individual processes encountered in earlier chapters to be solved within the framework of the harmonic vibrator model of caloric imperfection. This section will consider some of the applications.

Example 11-5

A converging–diverging nozzle with an area ratio of 3.0 is attached to a reservoir with stagnation conditions of 500 kPa and 2000 K. Identify the key pressures and the back pressure ranges for each region.

Solution. Except for the stagnation temperature, this is the same problem as that in Example 6-3. The solution was obtained using CIGQ. Sketches of the various nozzle flow fields are similar to those in Figure 6-18. The examination of converging–diverging nozzles in Chapter 6 contains a discussion of the key pressures and regions. Key pressure (1) is the maximum back pressure at which the nozzle will choke. Key pressure (2) is the back pressure required for a normal shock wave to stand in the nozzle exit plane. Key pressure (3) is the back pressure required for shock-free supersonic flow through the diverging portion of the nozzle. The four regions are delineated by the three key pressures. A sketch of the problem is provided in Figure 11-11. Key pressures (1) and

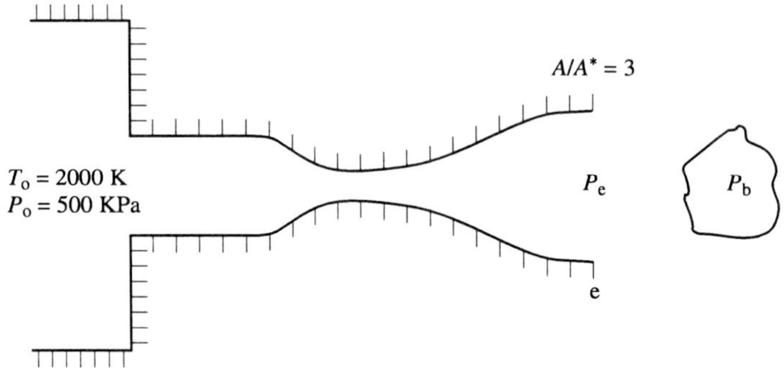

Figure 11-11 Converging–Diverging Schematic for Example 11-5

(3) correspond to the two isentropic conditions for an area ratio of 3. Using the isentropic option of CIGQ with the specified area ratio yields

$$A/A^* = 3.0$$

$$M_e = 0.1996 \qquad M_e = 2.5239$$

$$T_e/T_o = 0.9941 \qquad T_e/T_o = 0.4965$$

$$P_e/P_o = 0.9746 \qquad P_e/P_o = 0.0538,$$

from which the following key pressures are obtained:

$$P_1 = 487.3 \text{ kPa};$$

$$P_3 = 26.89 \text{ kPa}.$$

Key pressure (2) occurs when a normal shock wave stands in the exit plane of the converging–diverging nozzle. For this example the normal shock wave would occur at a Mach number of 2.5239. Case **U** of CIGQ option **NS** produces

$$M_1 = 2.5239 \qquad M_2 = 0.4937 \qquad P_2/P_1 = 7.220,$$

where subscript 1 means upstream of the shock wave and 2 indicates downstream. Key pressure (2), downstream of the normal shock wave, then becomes 194.11 kPa. This is also the back pressure since the Mach number downstream of the shock wave is subsonic. Table 11-5 presents the back pressure ranges associated with the various regions. For comparison purposes the back pressure ranges from Example 6-3 are also given in the table. Given any back pressure for the conditions of this problem, consult Table 11-5 to determine the nozzle flow region.

Example 11-6

For the conditions of Example 11-5, find the exit Mach number and pressure and describe the flow field for a back pressure of 250 kPa.

Solution. An examination of Table 11-5 reveals that a back pressure of 250 kPa places the flow in region 2—the region where a shock wave stands in the diverging section of the nozzle and the exit plane pressure is equal to the back pressure. For a calorically perfect gas such problems can be solved either by Eq. 6-38 or by iteration. (See Exam-

TABLE 11-5 BACK PRESSURE RANGES FOR EXAMPLE 11-5

	Back pressure range (kPa)	
Region	Calorically imperfect	Calorically perfect
1	$P_b > 487.3$	$P_b > 486.6$
2	$194.1 < P_b < 487.3$	$188.0 < P_b < 486.6$
3	$26.89 < P_b < 194.1$	$23.65 < P_b < 188.0$
4	$P_b < 26.89$	$P_b < 23.65$

ple 6-5 for details of such calorically perfect solution techniques.) However, for a calorically imperfect gas, only the iteration technique is possible. The iteration procedure used is similar to that of Example 6-5 except that CIGQ is used instead of COMPQ. The iteration sequence is shown in Table 11-6. The procedure is to first choose a shock wave location and then compute the exit pressure corresponding to that location. When the chosen location of the shock wave yields an exit plane pressure of 250 kPa, convergence has been attained. The solution to Example 6-5 places the shock wave in the diverging portion of the nozzle at an area ratio of 2.34.

The previous converging–diverging nozzle examples illustrated the use of CIGQ and should have reinforced the notion that the basic concepts of converging–diverging nozzle flow remain unchanged when caloric imperfections are considered. As the following example demonstrates, moving normal shock waves can also be readily handled using CIGQ.

Example 11-7

The December 1917 explosion of a fully loaded ammunition ship in the Halifax, Nova Scotia, harbor created a blast wave that destroyed much of that city. The wave was estimated to be moving at 2100 mph as it swept over the city. Considering a calorically imperfect gas, estimate the stagnation temperature and the velocity of the gas after passage of the wave.

Solution. The concepts involved with moving normal shock waves are covered in Chapter 2. The scenario for the problem statement is illustrated in Figure 11-12a. Since the normal shock relationships for both the calorically perfect and the calorically imperfect gases are for stationary normal shock waves, the problem must be cast into a stationary reference frame. The schematic for the stationary reference frame version

TABLE 11-6 ITERATION SEQUENCE FOR EXAMPLE 11-6

A_1/A_1^*	M_1	M_2	A_2/A_2^*	A_e/A_2^*	M_e	P_e/P_o	P_{o_2}/P_{o_1}	P_e (kPa)
2.2	2.2241	0.5289	1.2957	1.7669	0.3561	0.9215	0.5889	271.3
2.3	2.2681	0.5230	1.3058	1.7032	0.3719	0.9148	0.5678	259.7
2.4	2.3099	0.5176	1.3153	1.6441	0.3879	0.9077	0.5481	248.8
2.39	2.3058	0.5181	1.3144	1.6499	0.3863	0.9085	0.5500	249.8

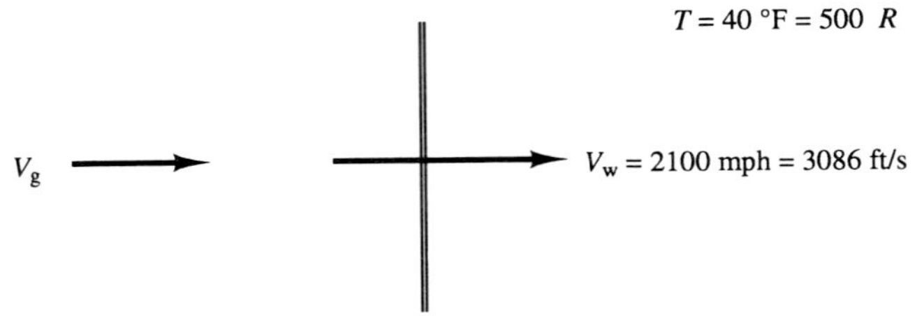

(a) Problem formulation in moving reference frame

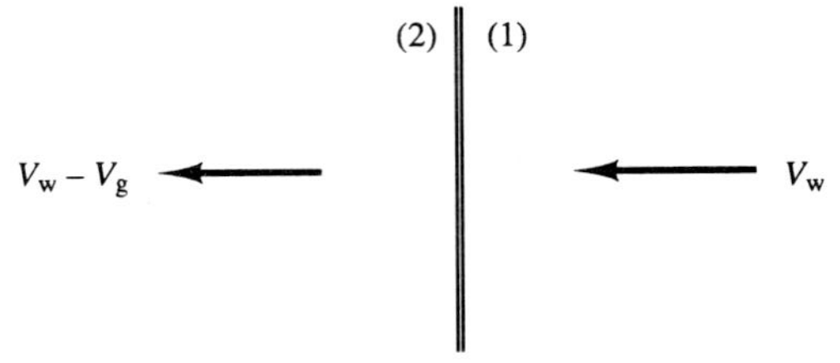

(b) Stationary equivalent

```
ISENTROPIC RELATIONS FOR CALORICALLY IMPERFECT DIATOMIC GAS

      R           VIB          TT
    53.35       5500.00      1271.56

STAGNATION TEMPERATURE HAS BEEN RESET

    MACH         T/TO         P/PO        GAM        ARATIO        A          VEL
   2.8156     3.932E-01    3.551E-02    1.39968     3.6033     1096.02    3086.00

FOR ISENTROPIC FLOW THE STAGNATION PRESSURE AND TEMPERATURE ARE CONSTANT.

ARE ADDITIONAL ISENTROPIC COMPUTATIONS DESIRED (Y FOR YES)?
```

(c) CIGQ output for evaluation of T_o

```
NORMAL SHOCK WAVE SOLUTION USING CALORICALLY IMPERFECT GAS

UPSTREAM CONDITIONS:
MACH NUMBER      P/PO1         T/TO1      GAMMA       A1          U1
    2.8156    3.551E-02   3.932E-01    1.3997    1096.0      3086.0

DOWNSTREAM CONDITIONS:
MACH NUMBER      P/PO2         T/TO2      GAMMA       A2          U2
   0.4870      0.8533        0.9585     1.3665    1690.8       823.4

PROPERTY RATIOS ACROSS NORMAL SHOCK WAVE:
    P2/P1    PO2/PO1     T2/T1    RHO2/RHO1
     9.14    0.38017     2.438      3.748

CHARACTERISTIC VIBRATION TEMPERATURE = 5500
TOTAL TEMPERATURE =  1271.561

STAGNATION TEMPERATURE IS CONSTANT ACROSS A STATIONARY NORMAL SHOCK WAVE
```

(d) CIGQ output for normal shock crossing

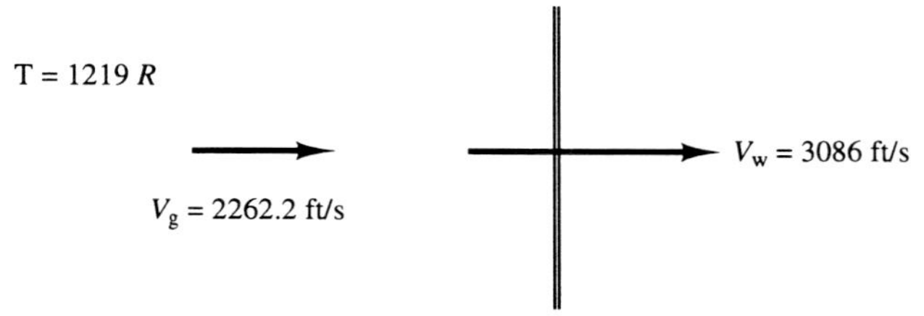

(e) Solution in moving reference frame

Figure 11-12 Schematics and Output of Example 11-7

```
ISENTROPIC RELATIONS FOR CALORICALLY IMPERFECT DIATOMIC GAS

       R          VIB        TT
     53.35      5500.00    1609.11

STAGNATION TEMPERATURE HAS BEEN RESET

     MACH       T/TO        P/PO       GAM      ARATIO      A        VEL
    1.3378    0.75756     0.34652   1.36651    1.0862   1690.94   2262.20

FOR ISENTROPIC FLOW THE STAGNATION PRESSURE AND TEMPERATURE ARE CONSTANT.

ARE ADDITIONAL ISENTROPIC COMPUTATIONS DESIRED (Y FOR YES)?
```

(f) CIGQ output for evaluation of T_o after shock wave passage

Figure 11-12 (continued)

is shown in Figure 11-12b. The first call to CIGQ is for case **TS** of option **IS**. Case **TS** requires as input the static temperature and the velocity and returns the stagnation temperature and Mach number as well as other parameters. The output for the conditions of this problem is contained in Figure 11-12c. The upstream Mach number is 2.8156 for the stationary reference frame. Case **U** of option **NS** yields the downstream conditions; the output for this case with an input Mach number of 2.8156 is given in Figure 11-12d. The static temperature downstream of the shock wave is 1219 R. Conservation of mass across the shock wave yields

$$\frac{\rho_2}{\rho_1} = \frac{V_2}{V_2 - V_g},$$

from which the gas velocity downstream of the shock wave becomes

$$V_g = 2262.2 \text{ ft/s}.$$

The output of Figure 11-12d is for a stationary normal shock wave, but the problem statement is for a moving shock wave. Casting the solution back to the moving reference frame results in the conditions shown in Figure 11-12e. With a static temperature of 1219 R and a velocity of 2262.2 ft/s, case **TS** of option **NS** can be used to find the stagnation temperature. The output of Figure 11-13f indicates a stagnation temperature of 1609.1 R. Were the same sequence to be performed for calorically perfect gases using COMPQ, the solution would yield a gas velocity of 2248 ft/s and a stagnation temperature of 1657.3 R. Table 11-7 contains a summary of calorically imperfect and calorically perfect gas results for this problem.

TABLE 11-7 CALORICALLY PERFECT AND IMPERFECT GAS RESULTS FOR EXAMPLE 11-7

Caloric assumption	M_1	M_g	V_g (ft/s)	T_o (R)
Imperfect	2.8156	1.338	2262.2	1609.1
Perfect	2.8200	1.304	2248.0	1657.3

11.6 GAS DYNAMICS WITH CHEMICAL ACTIVITY

The two previous sections have developed and explored techniques and software for routine computations using a calorically imperfect gas modeled as a simple harmonic vibrator. As Figure 11-3 illustrates, no chemical reactions occur in air at temperatures below about 2500 K. For temperatures in excess of 2500 K, dissociation of molecular oxygen, O_2, is initiated, and for temperatures in excess of about 4000 K, dissociation of molecular nitrogen, N_2, is initiated. Ionization is possible above 4000 K. Until now in this textbook, we have considered situations for which the gas composition—and hence the molecular weight and the gas constant—were invariant. Once chemical activity is initiated, the chemical composition of the gas is altered and the mixture molecular weight and gas constant are no longer invariant. Several questions are of importance for the present consideration:

1. For a given temperature and pressure, what is the equilibrium composition of a gas?
2. Given the equilibrium composition of a gas, how can the physical properties be computed?
3. What are the effects of chemically reacting perfect gas mixtures on various compressible flow processes?

The most convenient approach is to examine question 2 first. We have thus far treated air, a constant chemical composition mixture of nitrogen and oxygen, as a "gas" with its own property relationships; thus the thermodynamic properties of mixtures can obviously be computed. Earlier the concept that each constituent species in a mixture of perfect gases must obey the thermal equation of state was introduced as Eq. 11-12, $P_i = \rho_i R_i/T$. P_i is called the *partial pressure of the species* i and is defined as the pressure that would exist in the system if all other species were removed and the particles of species i occupied the whole system of volume V and temperature T. The pressure of the mixture is specified by Dalton's law of partial pressures to be the sum of the partial pressures of all the species of the mixture; that is,

$$P = \sum P_i. \tag{11-44}$$

The species density ρ_i can be written as M_i/V and the species gas constant R_i can be expressed as $\bar{R}/MW_i$. With the universal gas constant $\bar{R}$ divided by the molecular weight MW_i, Eq. 11-12 takes the form

$$P_i = (M_i/V)(\bar{R}/MW_i)T, \tag{11-45a}$$

which becomes

$$P_i V = (M_i/MW_i)\bar{R}T = N_i \bar{R}T, \tag{11-45b}$$

where N_i is the number of moles of species i. However, for the gas mixture, $P = \rho RT$, and the analogous substitutions yield

$$PV = N\bar{R}T. \tag{11-45c}$$

Dividing Eq. 11-45b by Eq. 11-45c results in

$$P_i/P = N_i/N = X_i, \tag{11-45d}$$

where X_i is the mole fraction of species i, and the ratio of the partial pressure of species i to the mixture pressure is equal to the mole fraction of species i.

Consider the relationship between the species gas constant, R_i, and the mixture gas constant, R. Dalton's law with the substitution of ρRT becomes

$$\rho RT = \sum_i \rho_i R_i T, \tag{11-46a}$$

which reduces to

$$R = \sum_i (\rho_i/\rho)R_i = \sum_i c_i R_i, \tag{11-46b}$$

where c_i is the mass fraction of species i. In terms of the universal gas constant, this expression takes the form

$$(\overline{R}/MW) = \sum_i c_i(\overline{R}/MW_i). \tag{11-47a}$$

Then

$$1/MW = \sum_i c_i/MW_i \quad \text{or} \quad MW = 1/\sum_i c_i/MW_i. \tag{11-47b}$$

An alternate form cast in terms of the mole fractions can be obtained by considering the total mass to be the sum of the masses of the species:

$$M = \sum_i M_i, \tag{11-48a}$$

which becomes

$$N\,MW = \sum_i N_i\,MW_i, \tag{11-48b}$$

and therefore,

$$MW = \sum_i (N_i/N)MW_i = \sum_i X_i\,MW_i. \tag{11-48c}$$

Consider the following example.

Example 11-8

Air is often considered to be a mixture of perfect gases, containing (in terms of mole fractions) 0.781 N_2, 0.21 O_2, and 0.009 argon. Determine the molecular weight of the mixture.

Solution. Equation 11-48c will be used. The calculation then becomes

$$MW_{\text{air}} = 0.781(28.0) + 0.21(32.0) + 0.009(40.0) = 28.95 \text{ mass/mass-mol.}$$

The "exact" result is 28.967 mass/mass-mol. The difference between this calculation and the accepted value is the presence of other gases in trace amounts. The problem statement is of interest. We have often seen the approximation that air is composed of 80 percent nitrogen and 20 percent oxygen. This is not a complete specification since it is not clear what is meant by the 80 percent and the 20 percent; are these the percentages by mass fraction, the percentages by mole fraction, or some other percentages? As this example problem illustrates, these commonly quoted numbers are the mole fraction percentages. From Eq. 11-45d, the 80 percent and the 20 percent can also be interpreted as the partial pressure ratios for the constituents. However, if the composition percentages had been interpreted as mass fractions, the results would have been similar since the molecular weights of molecular nitrogen and molecular oxygen are close to each other; hence, the mass fractions are approximately the mole fractions.

Given the temperature and composition of a gas, the partial pressure can be computed, but properties such as entropy and enthalpy are also of interest. Recall from Chapter 1 the expression for the entropy change between two states in a calorically perfect gas:

$$\Delta s = c_p \ln T_2/T_1 - R \ln P_2/P_1, \tag{1-25}$$

which can also be expressed as

$$s = c_p \ln T/T_{ref} - R \ln P/P_{ref} + (s)_{ref}, \tag{11-49}$$

with $(s)_{ref}$ typically taken as zero. For a thermally perfect gas, c_p is a function of temperature, and the equivalent expression becomes

$$s = \int_{T_{ref}}^{T} c_p \, dT/T - R \ln P/P_{ref} + (s)_{ref}. \tag{11-50}$$

A working definition of the entropy of a mixture of thermally perfect gases is

$$s = \sum_i c_i s_i, \tag{11-51}$$

where c_i is the species mass fraction and s_i the species entropy. For a given species Eq. 11-50 becomes

$$s_i = \int_{T_{ref}}^{T} c_{pi} \, dT/T - R_i \ln P_i/P_{ref} + (s_i)_{ref}. \tag{11-52a}$$

The mixture entropy, Eq. 11-51, can then be expressed as

$$s = \sum_i c_i \left(\int_{T_{ref}}^{T} c_{pi} \, dT/T - R_i \ln P_i/P_{ref} \right) + \sum_i c_i (s_i)_{ref}. \tag{11-52b}$$

Frequently $s_{o,i}$ is defined as

$$s_{o,i}(T) = \int_{T_{ref}}^{T} c_{pi} \, dT/T, \tag{11-53a}$$

with

$$s_o(T) = \sum_i c_i s_{o,i}(T) \tag{11-53b}$$

and

$$s_{ref} = \sum_i c_i (s_i)_{ref}, \tag{11-53c}$$

in which case the expression for entropy can then be cast as

$$s(T, P) = s_o - \sum_i c_i R_i \ln P_i/P_{ref} + s_{ref}. \tag{11-53d}$$

A very revealing expression for entropy can be deduced by considering

$$P_i/P_{\text{ref}} = (P_i/P)(P/P_{\text{ref}}), \tag{11-54a}$$

where P is the mixture pressure (the sum of the partial pressures). Taking the natural logarithm yields

$$\ln(P_i/P_{\text{ref}}) = \ln(P_i/P) + \ln(P/P_{\text{ref}}). \tag{11-54b}$$

Substituting Eq. 11-54b into Eq. 11-53d results in

$$s(T, P) = s_o - \sum_i c_i R_i \ln(P_i/P) - \sum_i c_i R_i \ln(P/P_{\text{ref}}) + s_{\text{ref}}. \tag{11-54c}$$

But Eq. 11-45d shows that $P_i/P = X_i$, and the mixture gas constant R is defined as $\sum c_i R_i$. Moreover, since $c_i = X_i(MW_i/MW)$,

$$c_i R_i = X_i R_i(MW_i/MW) = X_i R. \tag{11-54d}$$

The expression for entropy in a mixture reduces to

$$s(T, P) = s_o - R \ln(P/P_{\text{ref}}) - R \sum_i X_i \ln X_i + s_{\text{ref}}. \tag{11-54e}$$

Example 11-9

A container is partitioned into two volumes. Initially one of the volumes contains molecular nitrogen at a temperature T and pressure P and the other volume contains molecular oxygen at the same temperature T and pressure P. The partition is removed and the two species mix and finally reach equilibrium. Determine the change in entropy for the mixing process.

Solution. The situation is initially portrayed as in Figure 11-13a. After the partition is removed, the situation is as in Figure 11-13b. Since the initial pressure and temperature of the gases in both volumes are the same, no change in either temperature or pressure will be experienced during the mixing process. Thus, the final temperature and pressure are equal to the initial temperature and pressure. The initial mass of the nitrogen is taken as m_N, and the initial mass of the oxygen is taken as m_O. The final mass of the mixture must then be $(m_N + m_O)$. For convenience, the entropies at the reference conditions are all taken as zero.

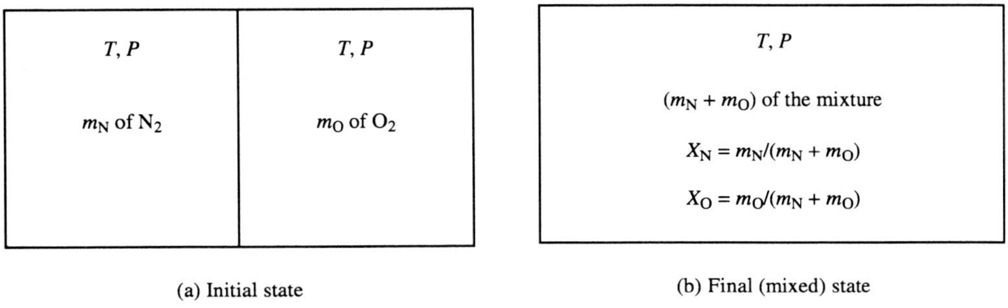

Figure 11-13 System Schematics for Example 11-9

The total entropy of the nitrogen at the initial state is simply the specific entropy as given by Eq. 11-49 multiplied by the mass of the nitrogen or

$$S_N = m_N s_N,$$

which can be expressed as

$$S_N = m_N(s_0(T)_N - R_N \ln(P/P_{ref})),$$

where the subscript N indicates nitrogen. Similarly, the entropy for oxygen is

$$S_O = m_O(s_0(T)_O - R_O \ln(P/P_{ref})).$$

The initial state total entropy is simply the sum of the two constituent parts:

$$S_1 = m_N s_0(T)_N + m_O s_0(T)_O - (m_N R_N + m_O R_O) \ln(P/P_{ref}).$$

After the mixing process is complete, the total entropy of the mixture is

$$S_2 = (m_N + m_O)s(T, P),$$

where $s(T, P)$ is defined by Eq. 11-54e. The final entropy then becomes

$$S_2 = (m_N + m_O)(s_0 - R \ln(P/P_{ref}) - R \sum_i X_i \ln X_i].$$

But using Eq. 11-53b, and realizing that the mass fractions are simply $m_N/(m_N + m_O)$ for nitrogen and $m_O/(m_N + m_O)$ for oxygen, let the expression for s_0 take the form

$$s_0 = m_N/(m_N + m_O)s_0(T)_N + m_O/(m_N + m_O)s_0(T)_O.$$

Similarly, the mixture gas constant becomes

$$R = m_N/(m_O + m_N)R_N + m_O/(m_N + m_O)R_O.$$

Assembling the preceding expressions for the final entropy state gives

$$S_2 = (m_N + m_O)[m_N/(m_N + m_O)s_0(T)_N + m_O/(m_N + m_O)s_0(T)_O$$
$$- (m_N/(m_N + m_O)R_N + m_O/(m_N + m_O)R_O) \ln(P/P_{ref}) - R \sum_i X_i \ln X_i],$$

which reduces to

$$S_2 = m_N s_0(T)_N + m_O s_0(T)_O - [(m_N R_N + m_O R_O) \ln(P/P_{ref})$$
$$+ (m_N + m_O)R \sum_i X_i \ln X_i].$$

Finally, the change in entropy for the initial state to the final state is $S_2 - S_1$:

$$S_2 - S_1 = m_N s_0(T)_N + m_O s_0(T)_O - [(m_N R_N + m_O R_O) \ln(P/P_{ref})$$
$$+ (m_N + m_O)R \sum_i X_i \ln X_i] - m_N s_0(T)_N + m_O s_0(T)_O$$
$$+ (m_N R_N + m_O R_O) \ln(P/P_{ref}).$$

Many of the terms in the previous equation add out with the simple result

$$S_2 - S_1 = -(m_N + m_O)R \sum_i X_i \ln X_i.$$

Since X_i, the mole fraction, must be less than 1, the logarithm of X_i will always be a negative number; hence $S_2 - S_1$ will be positive, and the change in entropy for the mixing process will always be positive. The result is very interesting, as it shows that a mixing process at constant temperature and pressure produces in an increase in entropy. We should not have been surprised since the mixing of two gases is certainly irreversible. Also, the entropy increase for this mixing process, on a unit mass basis, depends on only the mole fractions of the two gases involved.

The enthalpy of a mixture of thermally perfect gases will be examined next. The specific enthalpy of a mixture can be expressed as

$$h = \sum_i c_i h_i. \tag{11-55}$$

The usual way of evaluating enthalpy is to assign a reference value, e_o, at absolute zero temperature; thus

$$h_i = h_i + (e_o)_i. \tag{11-56}$$

This is a very convenient approach since the enthalpy description from statistical mechanics is based on the so-called zero-point value, e_o. For example, for a gas with translation and rotation fully excited, but with vibration not fully excited, the enthalpy can be expressed as

$$\underset{\text{}}{(h - e_0)_i} = \underset{\text{tran}}{3/2RT} + \underset{\text{rot}}{RT} + \underset{\text{vib}}{h\nu/kT/(e^{h\nu/hT} - 1)RT} + \underset{\text{P}/\rho}{RT}, \tag{11-57}$$

where the various contributions are identified. The use of a zero-point energy or enthalpy does not present any problems since we are normally interested only in the change in enthalpy. The enthalpy of a mixture of thermally perfect gases then can be expressed as

$$h = \sum_i c_i[(h - \theta_0)_i + (\theta_0)_i] = \sum_i c_i(h - \theta_0)_i + \sum_i c_i(\theta_0)_i, \tag{11-58}$$

where θ_0 is the zero-point energy, that is the energy of the species at absolute zero temperature. Frequently, the first term on the right-hand side is called the "sensible" enthalpy of the mixture and the last term is labeled the zero-point energy of the mixture; hence

$$h = h_{\text{sens}} + \theta_0, \tag{11-59a}$$

where

$$h_{\text{sens}} = \sum_i c_i(h - \theta_0)_i \tag{11-59b}$$

and

$$\theta_0 = \sum_i c_i(\theta_0)_i. \tag{11-59c}$$

If we denote the enthalpy at state 1 as h_1 and at state 2 as h_2, then the change in enthalpy from state 1 to state 2 becomes

$$h_2 - h_1 = (h_{\text{sens}} + \theta_0)_2 - (h_{\text{sens}} + \theta_0)_1 \tag{11-60a}$$

$$= \Delta h_{\text{sens}} + \Delta\theta_0. \tag{11-60b}$$

The sensible portion of the mixture change in enthalpy can be computed from a knowledge of the enthalpy and mass fraction of each species. The change in zero-point energy needs further examination. The change in zero-point energy is closely related to the heat of formation for a given species. Anderson (1989) verifies the following theorem:

Theorem. In a chemical reaction, the change in zero-point energy is equal to the difference between the heats of formation of the products at absolute zero and the heats of formation of the reactants at absolute zero.

For example, consider the combustion of hydrogen and oxygen. The chemical representation is

$$H_2 + 1/2\, O_2 \rightarrow H_2O.$$

By definition the heat of formation at absolute zero temperature, also called the *chemical enthalpy*, is

$$\Delta h_f = h^{\text{O}}_{H_2O} - h^{\text{O}}_{H_2} - 1/2\, h^{\text{O}}_{O_2}, \tag{11-61a}$$

but at absolute zero $h = \theta_{\text{O}}$ and

$$\Delta h_f = (\theta_0)_{H_2O} - (\theta_0)_{H_2} - 1/2\,(\theta_0)_{O_2}, \tag{11-61b}$$

or

$$\Delta h_f = (\theta_0)_2 - (\theta_0)_1. \tag{11-61c}$$

A consistent definition of enthalpy then becomes

$$h_i = (h - \theta_0)_i + (\Delta h_f)_i, \tag{11-62}$$

where it is understood that Δh_f is the heat of formation at absolute zero. The enthalpy for a mixture follows as simply

$$h = \sum_i c_i(h - \theta_0)_i + \sum_i c_i(\Delta h_f)_i. \tag{11-63}$$

Information on the heat of formation for a wide variety of species is available in a number of publications. The two most frequently cited sources are McBride et al. (1963) and Stull (1971).

Equation 11-63 has some interesting ramifications. As we shall see just ahead, the composition of a mixture depends on the pressure and temperature of the mixture. Thus, the mass fractions and the mole fractions are functions of temperature and pressure, and the enthalpy of a mixture of reacting gases will likewise be a function of temperature and pressure. That is $h = h(P, T)$ for a chemically reacting mixture of thermally perfect gases. This is in stark contrast to our earlier conclusion that a single species thermally perfect gas or a nonreacting mixture of thermally perfect gases possessed a functional relation for enthalpy of $h = h(T)$. For a nonreacting mixture of thermally perfect gases, Eq. 11-63 can be used to compute the enthalpy, but the contribution of the last term, $\sum_i c_i(\Delta h_f)_i$, will be invariant with temperature since the species mass fractions will be invariant with temperature. As a nonreacting gas undergoes a process from state 1 to state 2, the change in enthalpy will reduce to just the change in the sensible contribution, Δh_{sens}, since the change in the zero-point energies of states 1 and 2 will be identically zero. The zero-point energies or the chemical enthalpies simply add in and then subtract out for the nonreacting cases. Since they make no contribution to the change in enthalpy for nonreacting gases, they are simply taken as zero for convenience.

Figure 11-3 indicates the nature of chemical activity in air as a function of temperature. Dissociation of air, for example, possesses the chemical equation

$$1/2\,\mathrm{O}_2 \leftrightarrow \mathrm{O}. \tag{11-64}$$

The double-ended arrow symbol indicates that the reaction might not go to completion; that is, all of the molecular oxygen, O_2, may not dissociate into atomic oxygen, O. The question at hand is to determine, as a function of temperature and pressure, the species molecular and atomic composition of air. Once the species composition is known, mixture properties can be computed using procedures examined earlier in this section. The basis for determining the equilibrium composition of air is the *law of mass action*. Virtually any engineering thermodynamics textbook contains a derivation, explanation, and examples of use of the law of mass action; see, for instance, Howell and Buckius (1992) or Wark (1983). Anderson (1989), which has been so frequently referenced in this chapter, also presents a readable discussion of the development of the law of mass action.

The law of mass action is usually formulated for and applied to chemical equations. Consider an expression representing reactants A and B and products C and D:

$$\nu_{\mathrm{A}}\,\mathrm{A} + \nu_{\mathrm{B}}\,\mathrm{B} \leftrightarrow \nu_{\mathrm{C}}\,\mathrm{C} + \nu_{\mathrm{D}}\,\mathrm{D}, \tag{11-65}$$

where A, B, C, and D represent different chemical species (elements or compounds) and the ν_i are the stoichiometric coefficients (or stoichiometric mole numbers) for those species. The general form of the law of mass action for any chemical equation takes the form

$$\prod_i P_i^{\nu_i} = K_p(T), \tag{11-66}$$

where P_i are the partial pressures of the chemical species and $K_p(T)$ is the equilibrium constant for the reaction expressed by the chemical equation. The sign convention for the stoichiometric coefficients is that products are positive and reactants are negative. Thus, when applied to Eq. 11-65, the law of mass action becomes

$$\frac{P_{\mathrm{C}}^{\nu_{\mathrm{C}}}\,P_{\mathrm{D}}^{\nu_{\mathrm{D}}}}{P_{\mathrm{A}}^{\nu_{\mathrm{A}}}\,P_{\mathrm{B}}^{\nu_{\mathrm{B}}}} = K_p(T). \tag{11-67}$$

The equilibrium constant, $K_p(T)$, is needed if the computations are to proceed. Although pressures are involved on the left-hand side of Eq. 11-67, the equilibrium constant is not a function of pressure; it is a function of temperature only. Since $K_p(T)$ is a function of temperature only, a tabulation for a given reaction can be made. Equilibrium constants can be calculated using statistical thermodynamics or obtained from experiments. They are available in the thermodynamics literature for the reactions of interest to us. Although many engineering specialty books contain abbreviated tabulations of equilibrium constants, most cite either the JANAF tables (Stull, 1971), or McBride et al. (1963) as the sources of various tabulations. Many engineering thermodynamics textbooks contain abbreviated versions (for example, Howell and Buckius (1992) and Wark (1984). Strehlow (1984) presents a useful, abbreviated version of the JANAF tables, and Kuo (1986) provides an abbreviated version of the NBS equilibrium constants. From this point on in developing the procedure for computing equilibrium chemical composition, we shall presume that the equilibrium constants are available.

Since air is the primary fluid of interest, we will pursue the computation of its equilibrium chemical composition. The simplest viable model for air is composed of five species: O_2, O, NO, N_2, and N. The following three independent chemical reactions are possible:

$$1/2\, O_2 \leftrightarrow O; \tag{11-68a}$$

$$1/2\, N_2 \leftrightarrow N; \tag{11-68b}$$

$$1/2\, N_2 + 1/2\, O_2 \leftrightarrow NO. \tag{11-68c}$$

The law of mass action applied to the preceding chemical reactions yields

$$\frac{P_O}{P_{O_2}^{1/2}} = K_{O_2}(T); \tag{11-69a}$$

$$\frac{P_N}{P_{N_2}^{1/2}} = K_{N_2}(T); \tag{11-69b}$$

$$\frac{P_{NO}}{P_{N_2}^{1/2} P_{O_2}^{1/2}} = K_{O_2N_2}(T), \tag{11-69c}$$

where the subscripts indicate the species and where K_{O_2}, K_{N_2}, and $K_{O_2N_2}$ are the equilibrium reactions for oxygen dissociation (Eq. 11-68a), nitrogen dissociation (Eq. 11-68b), and the formation of NO (Eq. 11-68c), respectively. Equations 11-69 are a set of three independent expressions in five unknowns, the species partial pressures; hence, two additional independent relations are required to close the system. Since any additional chemical reactions expressions generally either add additional unknowns or are not independent, we must look elsewhere to complete the system of equations. The pressure of a mixture of thermally perfect gases is simply the sum of the partial pressures of the various species; see Eq. 11-44. Thus,

$$P_{O_2} + P_O + P_{N_2} + P_N + P_{NO} = P, \tag{11-70}$$

where P is the mixture pressure.

The last equation is based on the concept of conservation of nuclei; that is, the total number of nitrogen atoms remains constant whether in a free or combined state (NO, N_2, or N). N_N, the number of nitrogen nuclei per unit mass of the mixture, is given by

$$N_N = N_A(2\eta_{N_2} + \eta_{NO} + \eta_N), \tag{11-71a}$$

where N_A is Avogadro's number and the ηs are the mole–mass ratios of the various species. The mole–mass ratio is the number of moles of a species per unit mass of the mixture. Similarly, the number of oxygen nuclei per unit mass of the mixture becomes

$$N_O = N_A(2\eta_{O_2} + \eta_{NO} + \eta_O). \tag{11-71b}$$

The mole–mass ratio for a species can be obtained by starting with Eq. 11-45b, which is repeated here for convenience:

$$P_i V = N_i \bar{R} T. \tag{11-45b}$$

The mole–mass ratio for species i, η_i, is simply the number of moles of species i divided by the total mass of the system; that is,

$$P_i V/M = P_i v = N_i/M\bar{R}T = \eta_i \bar{R} T, \tag{11-72a}$$

from which

$$\eta_i = P_i v/\bar{R}T. \tag{11-72b}$$

Substituting Eq. 11-72b into the expressions for N_N and N_O yields

$$N_N = N_A\, v/RT(2P_{N_2} + P_{NO} + P_N); \tag{11-73a}$$

$$N_O = N_A\, v/RT(2P_{O_2} + P_{NO} + P_O). \tag{11-73b}$$

Dividing the two results in an expression for the ratio of nitrogen nuclei to oxygen nuclei:

$$\frac{2P_{N_2} + P_{NO} + P_N}{2P_{O_2} + P_{NO} + P_O} = \frac{N_N}{N_O}. \tag{11-74}$$

But the ratio of nitrogen to oxygen nuclei is a constant for the mixture! That is, the total number of nitrogen nuclei and the total number of oxygen nuclei do not change as the mixture species composition changes. In the five-equation model of air in chemical equilibrium we know at low temperatures that air consists of approximately 80 percent molecular nitrogen and approximately 20 percent molecular oxygen with negligible amounts of O, NO, and N. Thus,

$$N_N/N_O \approx 0.8/0.2 \approx 4.0 \tag{11-75}$$

for the entire temperature and pressure range of the five-equation model for air.

The system of equations is now complete and is made up of the three law of mass action expressions, the mixture pressure as the sum of the partial pressures of the species, and the conservation of nuclei statement. The complete system for the five-equation model consists of the following five equations:

$$\begin{aligned} \frac{P_O}{P_{O_2}^{1/2}} &= K_{O_2}(T); \\ \frac{P_N}{P_{N_2}^{1/2}} &= K_{N_2}(T); \\ \frac{P_{NO}}{P_{N_2}^{1/2} P_{O_2}^{1/2}} &= K_{O_2N_2}(T); \\ P_{O_2} + P_O + P_{N_2} + P_N + P_{NO} &= P; \\ 2P_{O_2} + P_O + 0.75P_{NO} - 0.50P_{N_2} - 0.25P_N &= 0. \end{aligned} \tag{11-76}$$

The unknowns are the species partial pressures P_{O_2}, P_O, P_{NO}, P_{N_2}, and P_N. The various equilibrium constants are available from the literature.

An examination of Eqs. 11-76 is in order. Equations 11-76 are a nonlinear system of algebraic equations since the exponents of the unknowns are different. If the system is to be solved for the partial pressures of the species, the temperature must be specified since the equilibrium constants are a function of temperature, and the mixture pressure, P, must be specified. Solution techniques are available for nonlinear algebraic systems. The multivariable Newton-Raphson procedure (Conte and deBoor, 1972) is widely used for the solution of such systems and will be adapted for use in solving the system represented by Eqs. 11-76.

The multivariable Newton-Raphson method is closely tied to the Newton-Raphson method for one variable. We wish to find the correct value x_c of x, such that

$$f(x_c) = 0, \tag{11-77}$$

since, in general,

$$f(x) = y(x) \neq 0. \tag{11-78a}$$

A truncated Taylor's series about x_c can be written as

$$y(x) = y(x_c) + y'(x_c)(x - x_c). \tag{11-78b}$$

If a trial guess x_t is made for x_c, then

$$y(x_t) = y(x_c) + y'(x_c)(x_t - x_c). \tag{11-78c}$$

But since $y(x_c) = 0$ and x_c is not known, Eq. 11-78c can be approximated by being solved to yield an estimate for x_c:

$$x_c \approx x_t - y(x_t)/y'(x_t). \tag{11-78d}$$

This equation is the classical Newton-Raphson equation for finding the root (or roots) of an equation in a single variable. Exercise care in using Eq. 11-78d for expressions with multiple roots, as convergence may be erratic or may not even occur, especially if the initial trial guess is far removed from the correct results.

The multivariable Newton-Raphson technique is developed, as is the single variable approach, by considering a Taylor series expansion. In this case, more than a single independent variable is present, and the more complicated series is

$$y(x_1, x_2, \ldots, x_n) = y(a_1, a_2, \ldots, a_n) + \sum_{j=1}^{n} \left(\frac{\partial y}{\partial x_j}\right)_a (x_j - a_j) + \frac{1}{2}\sum_{i=1}^{n}\sum_{j=1}^{n}\left(\frac{\partial^2 y}{\partial x_j\, \partial x_i}\right)_a (x_i - a_i)(x_j - a_j) + \cdots. \tag{11-79a}$$

Consider the following system of three equations in the independent variables x_1, x_2, and x_3:

$$\begin{aligned} f_1(x_1, x_2, x_3) &= 0; \\ f_2(x_1, x_2, x_3) &= 0; \\ f_3(x_1, x_2, x_3) &= 0. \end{aligned} \tag{11-79b}$$

The correct solution point is (x_{1c}, x_{2c}, x_{3c}), and the initially assumed solution point is (x_{1t}, x_{2t}, x_{3t}). The Taylor series for $f_1(x_{1t}, x_{2t}, x_{3t})$ can be written as

$$f_1(x_{1t}, x_{2t}, x_{3t}) = f_1(x_{1c}, x_{2c}, x_{3c}) + \left(\frac{\partial f_1}{\partial x_1}\right)_c (x_{1t} - x_{1c}) + \left(\frac{\partial f_1}{\partial x_2}\right)_c (x_{2t} - x_{2c}) + \left(\frac{\partial f_1}{\partial x_3}\right)_c (x_{3t} - x_{3c}) + \cdots, \tag{11-80}$$

and similarly for $f_2(x_{1t}, x_{2t}, x_{3t})$ and $f_3(x_{1t}, x_{2t}, x_{3t})$. Approximating $(\partial f_i/\partial x_j)_c$ by $(\partial f_i/\partial x_j)_t$, and realizing that $f_i(x_{1c}, x_{2c}, x_{3c}) = 0, i = 1, 2, 3$, we obtain

$$\begin{aligned} f_1(x_{1t}, x_{2t}, x_{3t}) &\approx \left(\frac{\partial f_1}{\partial x_1}\right)_t (x_{1t} - x_{1c}) + \left(\frac{\partial f_1}{\partial x_2}\right)_t (x_{2t} - x_{2c}) + \left(\frac{\partial f_1}{\partial x_3}\right)_t (x_{3t} - x_{3c}) + \cdots; \\ f_2(x_{1t}, x_{2t}, x_{3t}) &\approx \left(\frac{\partial f_2}{\partial x_1}\right)_t (x_{1t} - x_{1c}) + \left(\frac{\partial f_2}{\partial x_2}\right)_t (x_{2t} - x_{2c}) + \left(\frac{\partial f_2}{\partial x_3}\right)_t (x_{3t} - x_{3c}) + \cdots; \\ f_3(x_{1t}, x_{2t}, x_{3t}) &\approx \left(\frac{\partial f_3}{\partial x_1}\right)_t (x_{1t} - x_{1c}) + \left(\frac{\partial f_3}{\partial x_2}\right)_t (x_{2t} - x_{2c}) + \left(\frac{\partial f_3}{\partial x_3}\right)_t (x_{3t} - x_{3c}) + \cdots. \end{aligned} \tag{11-81}$$

Equations 11-81 constitute a *linear* system of algebraic equations for the unknowns $(x_{it} - x_{ic}), i = 1, 2, 3$, which can be cast in matrix form as

$$\begin{bmatrix} \frac{\partial f_1}{\partial x_1} & \frac{\partial f_1}{\partial x_2} & \frac{\partial f_1}{\partial x_3} \\ \frac{\partial f_2}{\partial x_1} & \frac{\partial f_2}{\partial x_2} & \frac{\partial f_2}{\partial x_3} \\ \frac{\partial f_3}{\partial x_1} & \frac{\partial f_3}{\partial x_2} & \frac{\partial f_3}{\partial x_3} \end{bmatrix} \begin{bmatrix} x_{1t} - x_{1c} \\ x_{2t} - x_{2c} \\ x_{3t} - x_{3c} \end{bmatrix} = \begin{bmatrix} f_1 \\ f_2 \\ f_3 \end{bmatrix}, \tag{11-82}$$

so that

$$\begin{bmatrix} x_{1t} - x_{1c} \\ x_{2t} - x_{2c} \\ x_{3t} - x_{3c} \end{bmatrix} = \begin{bmatrix} \dfrac{\partial f_i}{\partial x_j} \\ i = 1, 2, 3 \\ j = 1, 2, 3 \end{bmatrix}^{-1} \begin{bmatrix} f_1 \\ f_2 \\ f_3 \end{bmatrix} \tag{11-83}$$

and the new corrected values for the next iterative pass become

$$\begin{aligned} x_{1\text{cnew}} &= x_{1\text{told}} - (x_{1t} - x_{1c}); \\ x_{2\text{cnew}} &= x_{2\text{told}} - (x_{2t} - x_{2c}); \\ x_{3\text{cnew}} &= x_{3\text{told}} - (x_{3t} - x_{3c}). \end{aligned} \tag{11-84}$$

When $(x_{it} - x_{ic}), i = 1, 2, 3$, are sufficiently close to zero, the procedure has converged and x_{1c}, x_{2c}, and x_{3c} are the required solution of the nonlinear system of equations. Stoecker (1989) and Hodge (1990) give a seven-step procedure for implementing the multivariable Newton-Raphson technique:

1. Rewrite all the equations in the form $f_i(x_1, x_2, x_3) = 0, i = 1, 2, 3$.
2. Assume values for x_{1t}, x_{2t}, and x_{3t}.
3. Calculate $f_i(x_{1t}, x_{2t}, x_{3t})$ for $i = 1, 2, 3$.
4. Compute $\partial f_i/\partial x_j$ at (x_{1t}, x_{2t}, x_{3t}) for $i = 1, 2, 3$ and $j = 1, 2, 3$.
5. Solve the set of linear equations (Eqs. 11-82) for $(x_{it} - x_{ic}), i = 1, 2, 3$.
6. Compute the new values of $x_{i\text{new}}, i = 1, 2, 3$, from Eq. 11-84.
7. Substitute the values computed in step 6 into step 3, test for convergence; if convergence fails, repeat steps 4–7.

Consider this procedure applied to the five-equation chemistry model for air. Equations 11-76 are the five equations in the systems with the unknowns P_{O_2}, P_O, P_{NO}, P_{N_2}, and P_N. Following the seven steps just delineated, the system of equation is expressed as

$$\begin{aligned} f_1 &= P_O - K_{O_2}(T)P_{O_2}^{1/2}; \\ f_2 &= P_N - K_{N_2}(T)P_{N_2}^{1/2}; \\ f_3 &= P_{NO} - K_{O_2N_2}(T)P_{N_2}^{1/2}P_{O_2}^{1/2}; \\ f_4 &= P_{O_2} + P_O + P_{N_2} + P_N + P_{NO} - P; \\ f_5 &= 2P_{O_2} + P_O + 0.75P_{NO} - 0.50P_{N_2} - 0.25P_N. \end{aligned} \tag{11-85}$$

The mixture pressure and temperature as well as the various equilibrium constants must be specified prior to the start of the solution. The equilibrium constants can be found in the previously cited literature; however, the JANAF results as reported by Stull (1971) have been curve-fit for the reactions involved with the five-species air model. These curve fits are as follows:

$$\begin{aligned} K_{O_2} &= \exp[(-27500.45/T) + (-0.018628 - 459.7472/T)\log_e(T) \\ &\quad + 8.422563]; \\ K_{N_2} &= \exp[(-40830.43/T) + (0.1416532 - 432.0607/T)\log_e(T) \\ &\quad + 7.046769]; \\ K_{O_2N_2} &= \exp[(-9899.007/T) + (-0.147708 - 193.7042/T)\log_e(T) \\ &\quad + 2.888774]. \end{aligned} \tag{11-86}$$

The functional form of the curve fits was taken from Hodgson and Lamas (1990), with the coefficients evaluated using least-squares curve fits of the JANAF tabulations. This is a very convenient approach for obtaining the values of the equilibrium constants. The accuracy of the curve fits is quite good because Eqs. 11-86 and the tabular JANAF data are virtually indistinguishable when plotted.

The various partial derivatives, $\partial f_i/\partial x_j$, for Eqs. 11-84 with $x_1 = P_{O_2}$, $x_2 = P_O$, $x_3 = P_{N_2}$, $x_4 = P_{NO}$, and $x_5 = P_N$ are

$$\begin{aligned} &\frac{\partial f_1}{\partial x_1} = -\frac{1}{2}K_{O_2}P_{O_2}^{-1/2} \quad \frac{\partial f_1}{\partial x_2} = 1 \quad \frac{\partial f_1}{\partial x_3} = 0 \quad \frac{\partial f_1}{\partial x_4} = 0 \quad \frac{\partial f_1}{\partial x_5} = 0; \\ &\frac{\partial f_2}{\partial x_1} = 0 \quad \frac{\partial f_2}{\partial x_2} = 0 \quad \frac{\partial f_2}{\partial x_3} = -\frac{1}{2}K_{N_2}P_{N_2}^{-1/2} \quad \frac{\partial f_2}{\partial x_4} = 0 \quad \frac{\partial f_2}{\partial x_5} = 1; \\ &\frac{\partial f_3}{\partial x_1} = -\frac{1}{2}K_{O_2N_2}P_{N_2}^{1/2}P_{O_2}^{-1/2} \quad \frac{\partial f_3}{\partial x_2} = 0 \quad \frac{\partial f_3}{\partial x_3} = -\frac{1}{2}K_{O_2N_2}P_{N_2}^{-1/2}P_{O_2}^{1/2}; \\ &\frac{\partial f_3}{\partial x_4} = 1 \quad \frac{\partial f_3}{\partial x_5} = 0; \\ &\frac{\partial f_4}{\partial x_1} = \frac{\partial f_4}{\partial x_2} = \frac{\partial f_4}{\partial x_3} = \frac{\partial f_4}{\partial x_4} = \frac{\partial f_4}{\partial x_5} = 1; \\ &\frac{\partial f_5}{\partial x_1} = 2 \quad \frac{\partial f_5}{\partial x_2} = 1 \quad \frac{\partial f_5}{\partial x_3} = -0.5 \quad \frac{\partial f_5}{\partial x_4} = 0.75 \quad \frac{\partial f_5}{\partial x_5} = -0.25. \end{aligned} \tag{11-87}$$

The computer program that implements the five-species equilibrium air model is called CHEMQ and is listed in Figure 11-14. The logic of the program follows that of the numbered procedure delineated previously. Figure 11-15 presents a sample of the input and output screens of CHEMQ for a pressure of 10 atm and a temperature of 4000 K.

```
10 REM
20 REM EQUILIBRIUM CHEMISTRY USING NON-LINEAR ALGEBRAIC EQUATIONS
30 REM DEVELOPED FOR MECHANICAL AND AEROSPACE ENGINEERNG DEPARTMENTS
50 REM              MISSISSIPPI STATE UNIVERSITY
60 REM              MISSISSIPPI STATE, MS  39762
70 REM
80 DIM C(5, 5), B(5), V(5)
90 REM
100 REM FUNCTION DEFINITIONS
110 REM
120 COLOR 7, 1, 1: N = 5: CLS
130 DEF FNF1 (PO2, PO, KPO2) = PO - KPO2 * SQR(PO2)
140 DEF FNF2 (PN, PN2, KPN2) = PN - KPN2 * SQR(PN2)
150 DEF FNF3 (PO2, PN2, PNO, KPO2N2) = PNO - KPO2N2 * SQR(PN2 * PO2)
160 DEF FNF4 (PO2, PO, PNO, PN2, PN, P) = PO2 + PO + PNO + PN2 + PN - P
155 DEF FNF5 (PO2, PO, PNO, PN2, PN) = 2! * PO2 + PO + .75 * PNO - .5 * PN2 - .25
* PN
170 REM
180 REM INITIAL CONDITIONS
190 ITER = 0: ITERMX = 200
200 INPUT "ENTER THE TITLE"; TT$
210 INPUT "ENTER THE VALUE OF THE ITERATION TOLERANCE"; EPS
211 INPUT "ENTER THE VALUES OF THE MIXTURE PRESSURE (ATM) AND TEMPERATURE (K)";
P, T
223 REM
224 REM EVALUATION OF EQUILIBRIUM CONSTANTS
225 REM
231 KPO2 = EXP((-27500.45 / T) + (-.018628 - 459.7472 / T) * LOG(T) + 8.422563)
233 KPN2 = EXP((-40830.43 / T) + (.1416532# - 432.0607 / T) * LOG(T) + 7.046769)
234 KPO2N2 = EXP((-9899.007 / T) + (-.147708 - 193.7042 / T) * LOG(T) + 2.888774)
235 REM
236 REM INITIAL ESTIMATES OF PARTIAL PRESSURES
237 REM
240 PO2 = .2 * P
250 PO = .2 * P
260 PNO = .2 * P
270 PN2 = .2 * P
275 PN = .05 * P
```

```
280 REM INITIAL CONDITIONS PRINTOUT
285 REM PRINT : PRINT "CHEMICAL EQUILIBRIUM CALCULATION ": PRINT " PRESSURE = ";
P; "   TEMPERATURE = "; T
290 REM PRINT : PRINT "    PO2          PO          PN2          PNO          PN";
PRINT "      F1          F2          F3          F4          F5": PRINT
300 B(1) = FNF1(PO2, PO, KPO2)
310 B(2) = FNF2(PN, PN2, KPN2)
320 B(3) = FNF3(PO2, PN2, PNO, KPO2N2)
330 B(4) = FNF4(PO2, PO, PNO, PN2, PN, P)
335 B(5) = FNF5(PO2, PO, PNO, PN2, PN)
340 IT$ = STR$(ITER)
350 REM PRINT USING "  ####.###  "; PO2; PO; PN2; PNO; PN: PRINT "  "; USING "
####.###  "; B(1); B(2); B(3); B(4); B(5)
360 REM
370 REM ITERATION SEQUENCE
380 REM
390 ITER = ITER + 1
400 PO2O = PO2
410 POO = PO
420 PNOO = PNO
430 PN2O = PN2
435 PNOLD = PN
440 C(1, 1) = -.5 * KPO2 / SQR(PO2)
450 C(1, 2) = 1!
460 C(1, 3) = 0!
470 C(1, 4) = 0!
475 C(1, 5) = 0!
480 C(2, 1) = 0!
490 C(2, 2) = 0!
500 C(2, 3) = -.5 * KPN2 / SQR(PN2)
510 C(2, 4) = 0!
515 C(2, 5) = 1!
520 C(3, 1) = -.5 * KPO2N2 * SQR(PN2 / PO2)
530 C(3, 2) = 0!
540 C(3, 3) = -.5 * KPO2N2 * SQR(PO2 / PN2)
550 C(3, 4) = 1!
555 C(3, 5) = 0!
560 C(4, 1) = 1!
```

Figure 11-14 Listing of Program CHEMQ

```
570 C(4, 2) = 1!
580 C(4, 3) = 1!
590 C(4, 4) = 1!
591 C(4, 5) = 1!
592 C(5, 1) = 2!
593 C(5, 2) = 1!
594 C(5, 3) = -.5
595 C(5, 4) = .75
596 C(5, 5) = -.25
600 GOSUB 10000
605 SPLIM = 1E-08
610 PO2 = PO2O - V(1): IF PO2 < SPLIM THEN PO2 = SPLIM
620 PO = POO - V(2): IF PO < SPLIM THEN PO = SPLIM
630 PN2 = PN2O - V(3): IF PN2 < SPLIM THEN PN2 = SPLIM
640 PNO = PNOO - V(4): IF PNO < SPLIM THEN PNO = SPLIM
645 PN = PNOLD - V(5): IF PN < SPLIM THEN PN = SPLIM
650 REM
660 REM CHECK ON CONVERGENCE
670 REM
680 B(1) = FNF1(PO2, PO, KPO2)
690 B(2) = FNF2(PN, PN2, KPN2)
700 B(3) = FNF3(PO2, PN2, PNO, KPO2N2)
710 B(4) = FNF4(PO2, PO, PNO, PN2, PN, P)
715 B(5) = FNF5(PO2, PO, PNO, PN2, PN)
720 IT$ = STR$(ITER)
730 REM PRINT : PRINT IT$; USING "  ####.###  "; PO2; PO; PN2; PNO; PN: PRINT "
"; USING "  ####.###  "; B(1); B(2); B(3); B(4)
740 FOR I = 1 TO N
750 IF ABS(B(I)) > EPS THEN GOTO 828
760 NEXT I
761 REM
762 REM CONVERGED SOLUTION
763 REM
766 PRINT : PRINT TT$: PRINT "SOLUTION CONVERGED AT ITERATION "; USING "####";
ITER
768 PRINT : PRINT "PRESSURE (ATM) = "; P; "      TEMPERATURE (K) = "; T
```

```
772 MFO2 = PO2 / P: MFO = PO / P: MFN2 = PN2 / P: MFN = PN / P: MFNO = PNO / P
774 PRINT : PRINT "MOLE FRACTIONS"
776 PRINT "     O2          O          N2          N          NO"
778 PRINT USING "  ##.####  "; MFO2; MFO; MFN2; MFN; MFNO
780 PRINT : PRINT "PARTIAL PRESSURES (ATM)"
781 PRINT "    PO2         PO         PN2         PN         PNO"
782 PRINT USING "  ##.####  "; PO2; PO; PN2; PN; PNO
784 PRINT : PRINT "EQUILIBRIUM CONSTANT VALUES ARE: KPO2 = "; KPO2
786 PRINT "                                        KPN2 = "; KPN2
788 PRINT "                                        KPO2N2 = "; KPO2N2
790 PRINT : INPUT "ARE ADDITIONAL AIR EQUILIBRIUM COMPUTATIONS DESIRED"; AEQ$
792 AEQ$ = UCASE$(AEQ$): IF AEQ$ = "N" THEN END
794 CLS : GOTO 211
826 END
827 REM NEXT PASS
828 REM
829 IF ITER < ITERMX THEN GOTO 390
830 PRINT : PRINT "ITERATION COUNT EXCEEDED"
840 END
10000 REM
10010 REM GAUSSIAN ELIMINATION SUBROUTINE
10020 REM
10030 REM N      = ORDER OF COEFFICIENT MATRIX
10040 REM C(I,J) = COEFFICIENT MATRIX TERMS
10050 REM B(I)   = COLUMN MATRIX TERMS
10060 REM V(I)   = RETURNED SOLUTION
10070 REM SET ARRAY DIMENSIONS AND SPECIFY N IN MAIN PROGRAM
10080 L = N - 1
10090 FOR I = 1 TO L
10100 M = I
10110 IT = I + 1
10120 FOR J = IT TO N
10130 IF ABS(C(M, I)) > ABS(C(J, I)) GOTO 10150
10140 M = J
```

Figure 11-14 (continued)

```
10150 NEXT J
10160 IF M = 1 THEN GOTO 10260
10170 NI = I
10180 FOR J = NI TO N
10190 TEMP = C(M, J)
10200 C(M, J) = C(I, J)
10210 C(I, J) = TEMP
10220 NEXT J
10230 TEMP = B(M)
10240 B(M) = B(I)
10250 B(I) = TEMP
10260 FOR J = IT TO N
10270 TEMP = C(J, I) / C(I, I)
10280 C(J, I) = 0!
10290 FOR K = IT TO N
10300 C(J, K) = C(J, K) - TEMP * C(I, K)
10310 NEXT K
10320 B(J) = B(J) - TEMP * B(I)
10330 NEXT J
10340 NEXT I
10350 V(N) = B(N) / C(N, N)
10360 K = N - 1
10370 ITEMP = K + 1
10380 TEMP = 0!
10390 FOR I = ITEMP TO N
10400 TEMP = TEMP + C(K, I) * V(I)
10410 NEXT I
10420 V(K) = (B(K) - TEMP) / C(K, K)
10430 K = K - 1
10440 IF K <> 0 THEN GOTO 10370
10450 RETURN
```

Figure 11-14 (continued)

```
Equilibrium Air Composition
SOLUTION CONVERGED AT ITERATION    4

PRESSURE (ATM) =  10         TEMPERATURE (K) =  4000

MOLE FRACTIONS
     O2          O          N2          N          NO
   0.0801     0.1389      0.6960      0.0148     0.0702

PARTIAL PRESSURES (ATM)
    PO2         PO         PN2          PN         PNO
   0.8008     1.3890      6.9603      0.1478     0.7021

EQUILIBRIUM CONSTANT VALUES ARE: KPO2 =   1.552175
                                 KPN2 =   5.603141E-02
                                 KPO2N2 =   .2973863
```

Figure 11-15 Input and Output of Program CHEMQ for 10 atm Pressure and 4000 K

The species mole fraction distributions of air at atmospheric pressure as a function of temperature are shown in Figure 11-16, which was generated by Anderson (1989) from the compilation of Hilsenrath and Klein (1965). The mole fraction distributions presented in the figure conform to the general chemical behavior described in Figure 11-3. In particular, little chemical activity is observed for temperatures less than about 2500 K; O_2 dissociation is nearly complete by 4000–5000 K; N_2 dissociation starts at about 4000 K and is nearly complete at 9000 K; and NO is present in the temperature range 2000–6000 K. The behavior of some species, especially oxygen, may be deceptive because mole fractions are plotted and the number of moles increases as the degree of dissociation of N_2 and O_2 increases. Thus, although the mole fraction of O decreases for temperatures in excess of about 5000 K, the number of moles of O remains essentially constant after the number of moles of NO becomes small. The decrease the mole fraction of O at elevated temperatures reflects the increasing number of moles in the mixture, not a decreasing number of O moles.

Pressure also affects the equilibrium composition of air. Table 11-8 presents the computed species mole fractions for air at 4000 K and 0.1, 1.0, and 10.0 atm pressure. CHEMQ was used to generate the entries in the table. As the pressure increases, the mole fractions of molecular oxygen and nitrogen increase at 4000 K. The mole fractions of O and N decrease with increasing pressure, and the mole fraction of NO increases with increasing pressure. A comparison of Table 11-8 with Figure 11-16 leads us to conclude that increasing pressure would result in a general shift of the various mole fractions plots to the right in Figure 11-16; that is, higher pressure would increase the temperature required for various chemical reactions. Conversely, a reduction in pressure would decrease the temperature required for various chemical reactions and would generally shift the various plots in Figure 11-16 to the left.

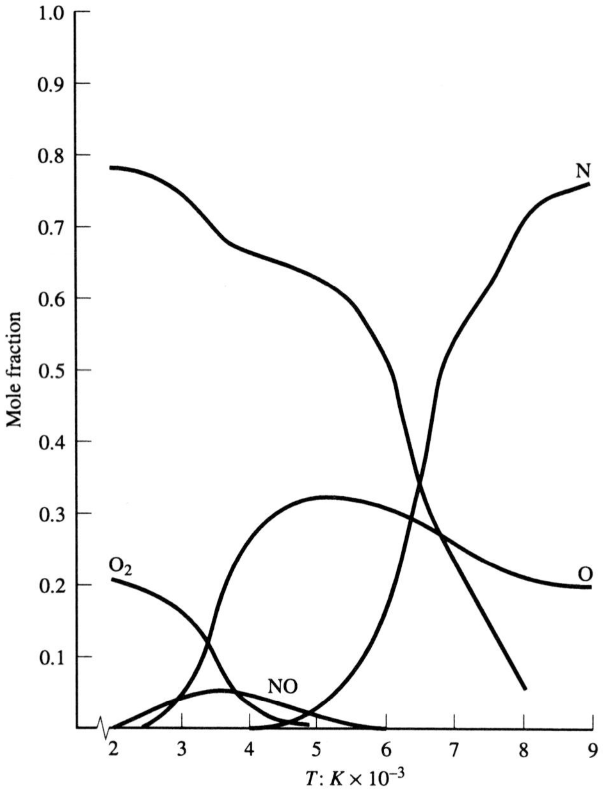

Figure 11-16 Species Distributions for Equilibrium Air. *From Anderson (1989).*

The five-species equilibrium air model that has been discussed here is perhaps the simplest model that captures most of the important aspects of the behavior of air with regards to temperature and pressure. However, an examination of Figure 11-3 shows that ionization begins as the gas mixture temperature approaches 9000 K. Since the five-species model of air does not treat ionization, a different model is needed if ionization effects are to be considered. The simplest equilibrium chemistry air model that accounts for ionization is the seven-species model. In

TABLE 11-8 SPECIES MOLE FRACTIONS AT 4000 K AND VARIOUS PRESSURES

	Mole fractions				
Pressure (atm)	O_2	O	N_2	N	NO
0.1	0.0036	0.2942	0.5567	0.1322	0.0133
1.0	0.0260	0.2505	0.6403	0.0448	0.0384
10.0	0.0801	0.1389	0.6960	0.0148	0.0702

addition to the three chemical equations for nitrogen and oxygen dissociation and NO formation, the seven-species model adds the ionization reaction

$$NO \leftrightarrow NO^+ + e^-, \tag{11-88a}$$

with the following law of mass action representation:

$$P_{NO^+} P_{e^-}/P_{NO} = K_{P_{NOe}}. \tag{11-88b}$$

The two additional species are ionized NO, NO^+, and free electrons, e^-. Each possesses a partial pressure. Equation 11-88b adds one additional equation but two unknowns; thus, an additional relation is required for closure. The additional equation is based on the conservation of electrical charge

$$\eta_{NO^+} = \eta_{e^-}, \tag{11-89a}$$

which can be cast in terms of the partial pressures as

$$P_{NO^+} = P_{e^-}. \tag{11-89b}$$

Additionally, the equation for conservation of nuclei, Eq. 11-74, must be changed to reflect the new species NO^+:

$$\frac{2P_{O_2} + P_O + P_{NO} + P_{NO^+}}{2P_{N_2} + P_N + P_{NO} + P_{NO^+}} = \frac{N_O}{N_N}. \tag{11-90}$$

The seven-species model results in a nonlinear system of seven equations that can be solved using the multivariable Newton-Raphson procedure.

Other, more complex, models for air can be generated by adding other ionization reactions. For example, the usual form of a nine-species model adds the reactions for ionization of *atomic* oxygen and nitrogen:

$$\begin{aligned} N &\leftrightarrow N^+ + e^-; \\ O &\leftrightarrow O^+ + e^-. \end{aligned} \tag{11-91}$$

An eleven-species model considers ionization of *molecular* oxygen and nitrogen:

$$\begin{aligned} N_2 &\leftrightarrow N_2^+ + e^-; \\ O_2 &\leftrightarrow O_2^+ + e^-. \end{aligned} \tag{11-92}$$

For models more complex than eleven species, additional trace species are added. A thirteen-species model usually adds the ionization of argon:

$$Ar \leftrightarrow Ar^+ + e^-, \tag{11-93}$$

which introduces the two additional species Ar and Ar^+. Beyond the thirteen-species model, additional species such as CO_2 and CO and their various ionization reactions are added. From the five-species model to the thirteen- and beyond-species model, a nonlinear system of equations similar to Eqs. 11-76 must be solved and the equilibrium constants must be supplied. So although the higher-species models are more complex, the system formulation and solution procedure are identical to those of the five-equation model.

We have examined various models for equilibrium air, starting with the five-equation formulation and going on to more complex models. Although applied to air, the procedure examined here can be used to find the equilibrium composition of any chemically reacting mixture of thermally perfect gases. The key step in applying the technique is to realize that if a noninert species is to appear in the composition it must be included in a chemical equation. The usual approach is to include all species that are likely to be important; the computations will identify those species that are present in only trace amounts. If an important species is omitted, then the equilibrium composition will be in error since the omitted species will not appear in the results.

The information presented in this section shows how properties of chemically reacting mixtures of thermally perfect gases can be computed. However, an alternative to computing properties of chemically reacting mixtures is to use existing tabulations and codes from the literature. In addition to the tabulations of McBride et al. (1963), the JANAF tables (Stull, 1971), and the tables of Hilsenrath and Klein (1965), many other sources of thermophysical properties exist. The equilibrium air curve fits of Tannehill and Mugge (1974) are widely used for computational fluid dynamics applications. Tannehill and Mugge provide polynomial curve fits for $P = P(u, \rho)$, $h = h(P, \rho)$, $T = T(u, \rho)$, and $T = T(P, \rho)$. These curve fits span three orders of magnitude in temperature, twelve orders of magnitude in pressure, and ten orders of magnitude in density. Liu and Vinokur (1989) contain a review of algorithms for determining properties of equilibrium gas mixtures. An interesting alternative approach is provided by Cox (1992), who developed a general equilibrium flow solver, called the "Black Box," for use in computational fluid dynamic (CFD) applications. The central idea was to devise a fast, robust equilibrium chemistry solver that could be used in a wide range of CFD applications. This approach offers an alternative to curve fits or to tabulations or local equilibrium calculations using techniques similar to the multivariable Newton-Raphson. The "Black Box" provides for CFD applications local equilibrium calculations without an undue penalty in computer time and is structured to permit additional chemistry models and chemical reactions to be easily incorporated.

11.7 EXAMPLES INVOLVING CHEMICALLY REACTING GASES IN EQUILIBRIUM

The purpose of this section is to provide an examination of some examples of the effects of chemically reacting gases (typically air) in equilibrium. The sequence of presentation will be to start with equilibrium chemistry qualitative and quantitative effects on individual processes such as shock waves and then to proceed to more complex situations such as nozzle flows and external flows. Some CFD-based results will also be utilized. One of the more interesting aspects of flows considering equilibrium chemistry is the time of appearance in the literature. Although some scattered early examples exist, the decade of the 1960s saw the first widespread efforts to use equilibrium chemistry for engineering applications. Indeed, many of the classic studies for individual processes were accomplished during this timeframe. During the mid-1970s through the late 1980s, interest waned, only to be revived during the late 1980s and the early 1990s. Much of the early interest was driven by the quest for solutions to the then-new reentry problem and for support of the Apollo–Saturn program. New interest has been stimulated by such research and development activities as the National Aerospace Plane (NASP).

At the start of Chapter 2, normal shock waves were the first individual compressible flow process to be examined, so it is fitting to ascertain the effects of equilibrium chemistry on that process. The results of applying the conservation equations to the control volume of Figure 2-3, which contains a normal shock wave, are listed below for convenience:

$$\rho_1 V_1 = \rho_2 V_2; \tag{2-1b}$$

$$P_1 + \rho_1 V_1^2 = P_2 + \rho_2 V_2^2; \tag{2-3}$$

$$h_1 + V_1^2/2 = h_2 + V_2^2/2. \tag{2-7}$$

These expressions were used in the development of the solution for normal shock waves in calorically perfect gases. However, an examination of the application of the conservation equations in Chapter 2 reveals that Eqs. 2-1b, 2-3, and 2-7 were obtained without invoking any assumptions involving the caloric perfection or equation of state of the gas involved. Hence, these equations are valid for both nonreacting and reacting gases. In Chapter 2 we eventually expressed the ratio of properties across a calorically perfect normal shock wave as functions of only the upstream Mach number, M_1. That is, the final forms were

$$P_2/P_1 = P_2/P_1(M_1) \quad \text{or} \quad \text{Eq. 2-18a};$$

$$T_2/T_1 = T_2/T_1(M_1) \quad \text{or} \quad \text{Eq. 2-18b};$$

$$\rho_2/\rho_1 = \rho_2/\rho_1(M_1) \quad \text{or} \quad \text{Eq. 2-18c};$$

$$M_2 = M_2(M_1) \quad \text{or} \quad \text{Eq. 2-18d}.$$

The normal shock wave in air in chemical equilibrium is more complex since the static temperature and pressure determine the chemical composition of the gas. Thus, the functional form of the normal shock relations for reacting air in chemical equilibrium becomes

$$\begin{aligned} P_2/P_1 &= P_2/P_1(V_1, P_1, T_1); \\ T_2/T_1 &= T_2/T_1(V_1, P_1, T_1); \\ \rho_2/\rho_1 &= \rho_2/\rho_1(V_1, P_1, T_1). \end{aligned} \tag{11-94}$$

In addition to clearly showing the dependence on the pressure and temperature, Eqs. 11-94 also indicate that the Mach number no longer occupies as central a position as it did for calorically perfect gases. The simple expression for the speed of sound in calorically perfect gases greatly simplified casting the Mach number as the central variable in the development of the normal shock wave relations for calorically perfect gases; the more complex form of the speed of sound in chemically reacting gases precludes such simple involvement with the Mach number. Relations for the flow of a chemically reacting gas are thus usually expressed in terms of the so-called primitive variables—velocity, temperature, and pressure.

The classic studies involving equilibrium air and normal shock waves are the Cornell Aeronautical Laboratory reports by Wittliff and Curtiss (1961) and Marrone (1962). Anderson (1989) presents one of the most revealing comparisons of calorically perfect and equilibrium normal shock waves. His results are reproduced as Figure 11-17. In this figure, the effects of pressure on the temperature ratio T_2/T_1 are presented as a function of upstream velocity in km/s. The initial temperature T_1 was taken as 225 K. Additionally, the results for a calorically perfect gas are shown as the dashed line. The figure illustrates that for a given upstream velocity, the higher the upstream pressure, the higher the temperature downstream of the shock wave. For a calorically perfect shock wave, much of the kinetic energy upstream of the shock wave is converted into internal energy, which results in the increase in temperature downstream of the shock wave. When a chemically reacting gas undergoes a shock wave, some of the upstream kinetic energy is required by the chemical reaction and the resulting temperature downstream of the shock wave is reduced from the calorically perfect value. The behavior with increasing pressure exhibited by Figure 11-17 is consistent with the observation made in the previous section that increasing pressure tends to shift the equilibrium composition plots from Figure 11-15 to the right, indicating decreasing chemical activity as the pressure increases. With decreasing chemical activity, more of the upstream kinetic energy is recovered as internal energy. Figure 11-17 also demonstrates graphically the significant effects that chemical reactions can have on the normal shock process. Upstream of a normal shock wave, the relatively high Mach number results in a relatively low static temperature. As the flow traverses the shock wave, the Mach number is dramatically reduced, resulting in a significant increase in the static temperature. The chemical activity is driven by the high static temperatures recovered downstream of the shock wave.

Oblique shock waves in air with equilibrium chemistry will be the next process examined. Consider, as in Figure 3-1, the schematic for the analysis of oblique shock waves. Conservation of mass, normal momentum, tangential momentum, and energy for the oblique shock situation as illustrated in Figure 3-1 yields, as in Chapter 3,

$$\rho_1 V_{1n} = \rho_2 V_{2n}; \tag{3-1}$$

$$P_1 + \rho_1 V_{1n}^2 = P_2 + \rho_2 V_{2n}^2; \tag{3-3}$$

$$V_{1t} = V_{2t}; \tag{3-6}$$

$$h_1 + V_{1n}^2/2 = h_2 + V_{2n}^2/2. \tag{3-9}$$

Hence, as with the normal shock relations, the application of the conservation

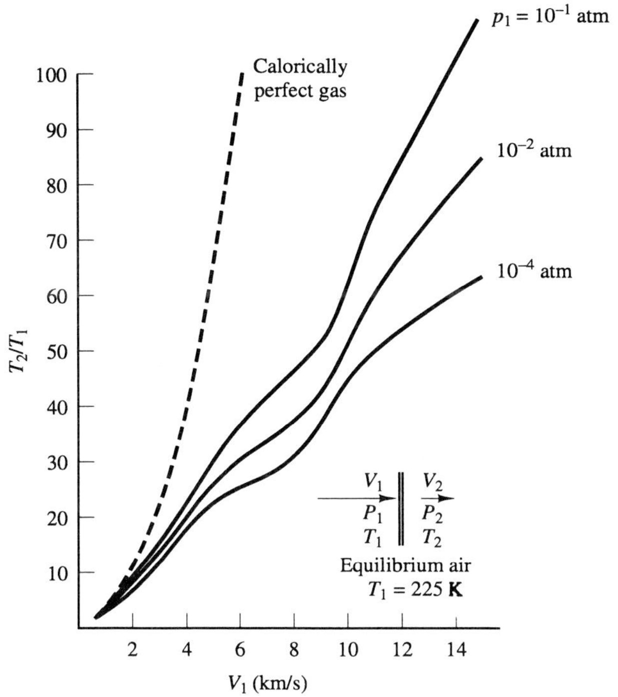

Figure 11-17 Equilibrium Air Normal Shock Results. *From Anderson (1989).*

expressions are valid for nonreacting as well as reacting flows. In fact, from geometrical considerations alone, Eq. 3-21 is also valid for chemically reacting flows. With the similarities that exist between the analysis for calorically perfect gas oblique shock waves and the analysis for equilibrium air, we should not be surprised that the formats of the presentations will be similar. The calorically perfect functional form of the relationship between the turning angle, the upstream Mach number, and shock wave angle is given in Chapter 3 as $\theta = \theta(M_1, \delta)$. When oblique shock waves with chemical reactions are considered, the relationship becomes more complex than for the calorically perfect form. Because the temperature and pressure change across an oblique shock wave (as they do across a normal shock wave), the equilibrium composition of the gas upstream and downstream of the oblique shock wave changes. Chemical reactions then add pressure and temperature dependencies, so that for oblique shock waves with chemical reaction the functional form becomes $\theta = \theta(V_1, \delta, P_1, T_1)$. As with the normal shock wave the upstream velocity V_1 is a more appropriate parameter than the upstream Mach number M_1. One of the early, classic results for the analysis of oblique shock waves in chemically reacting air in equilibrium is that of Moeckel (1957). Figure 11-18a, from Anderson (1989), presents Moeckel's results for an altitude of 100,000 ft for a fixed value of P_1 and T_1. The format of the plot is the familiar turning angle versus shock wave angle with the various cases parameterized with the upstream velocity. Calorically perfect gas results are presented for a $\gamma = 1.4$ fluid with an upstream velocity of 5000 ft/s and with an unbounded upstream Mach number. As illustrated in Figure 11-18b, the

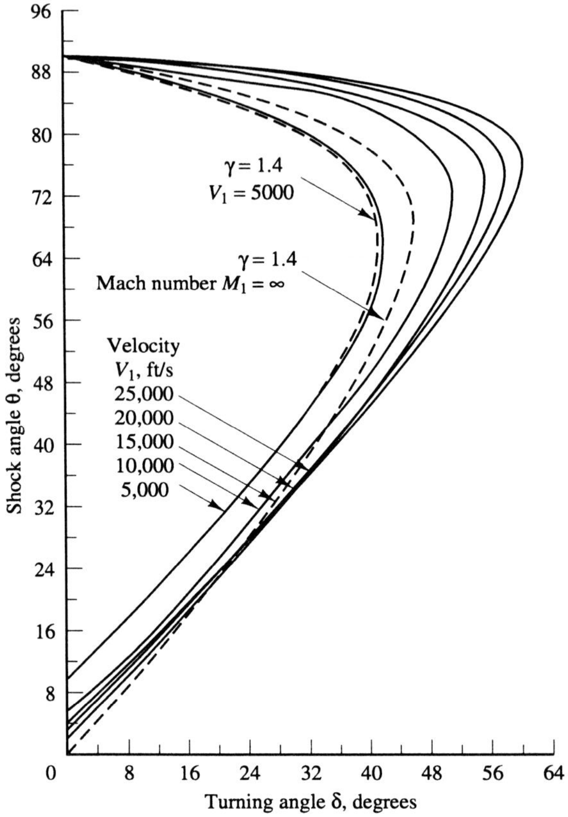

(a) Turning angle versus shock wave angle from Anderson, 1989, based on Moeckel, 1957

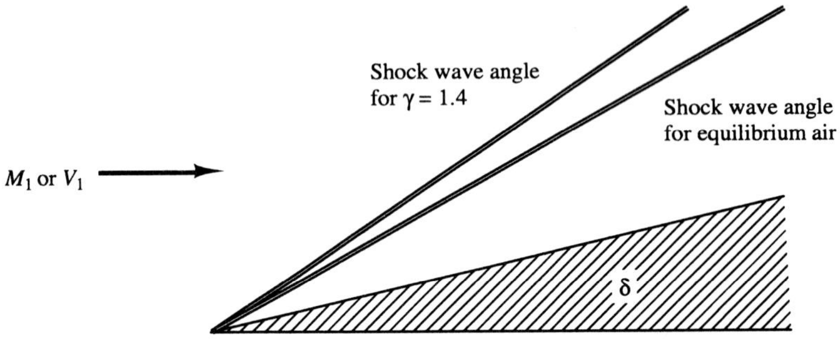

(b) Schematic for equilbrium and calorically perfect oblique shock waves

Figure 11-18 Equilibrium Air Oblique Shock Results

weak shock solution for the chemically reacting oblique shock lies closer to the surface than for the calorically perfect weak oblique shock. The opposite occurs for the chemically reacting strong shock solution, which lies closer to the normal than for the calorically perfect strong oblique shock. The maximum deflection angle for a given upstream velocity is also increased for the chemically reacting oblique shock when compared with the calorically perfect.

The flow of chemically reacting air over a right-circular cone presents some interesting contrasts to the calorically perfect analog. Much of the general discussion contained in Chapter 3 on the supersonic conical flow of calorically perfect gases is also appropriate for the supersonic flow of a gas in chemical equilibrium over a right-circular cone. The general supersonic conical flow schematic and nomenclature are provided in Figure 3-16. Indeed the salient features of chemically reacting flow over a right-circular cone are quite similar to those of calorically perfect flow over a right-circular cone. Classic results are by Romig (1960) and Hudgins (1964, 1965). Anderson (1989) used the tabulations of Hudgins (1965) to generate several interesting plots. Figure 11-19 shows the cone surface pressure to ambient pressure ratio

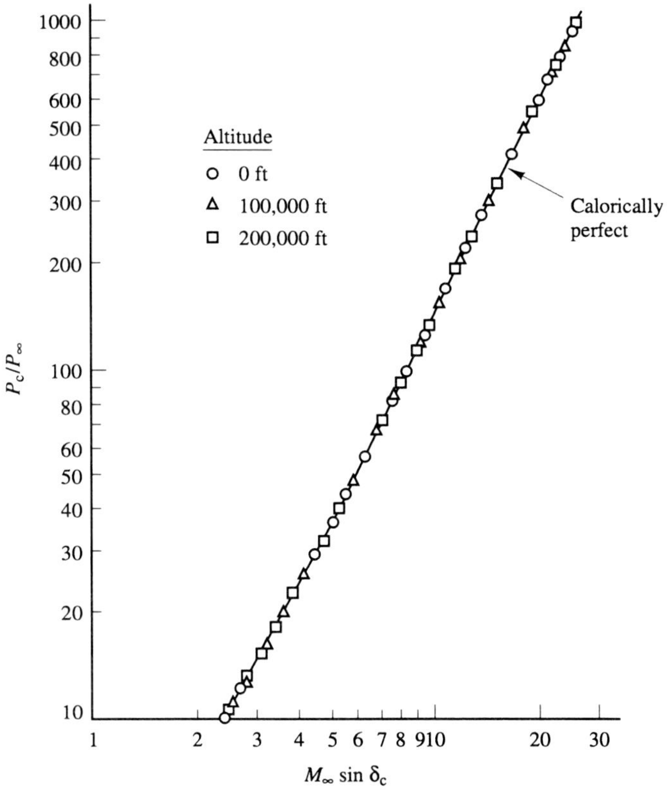

Figure 11-19 Conical Surface Pressure Ratios. *(From Anderson, 1989, based on Hudgins, 1965).*

as a function of the parameter $M_\infty \sin \delta_c$ for different altitudes, and Figure 11-20 provides the cone surface density to ambient density ratio for the same altitudes. The calorically perfect results are also indicated on the figures. A comparison of the two figures is very interesting; chemical reactions have little effect on the surface pressure but a noticeable effect on the surface density. As Anderson (1989) states, "The pressure is a mechanical variable, strongly dependent on the mechanical aspects of the flow, and essentially uninfluenced by chemically reacting effects." The density, however, is greatly influenced by chemical reactions. As the freestream Mach number increases, the chemical reaction effects become increasingly important. The distinctive effects of altitude occur since, as we saw in the previous section, the species composition is influenced by the pressure, which varies with altitude.

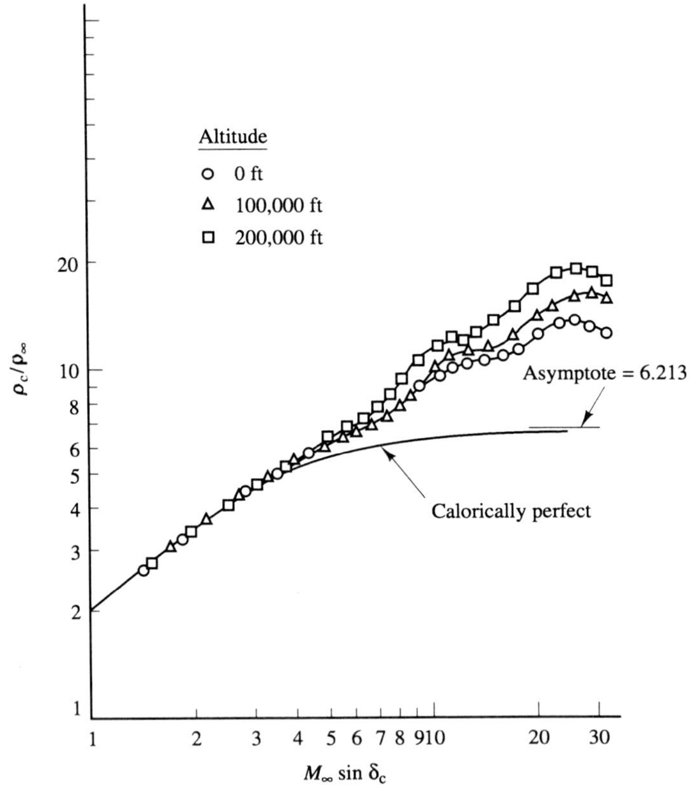

Figure 11-20 Conical Surface Density Ratios. *(From Anderson, 1989, based on Hudgins, 1965).*

These examples of equilibrium chemically reacting flows are for the most part considered classics and are frequently referenced. The maturation of computational fluid dynamics has resulted in a number of vehicle flow field studies involving chemically reacting gases. Consider, for example, the blunt-body computational study of Palmer (1987). He computed equilibrium chemically reacting flow about blunt bodies at hypersonic Mach numbers. Figure 11-21 illustrates the differences between calorically perfect and equilibrium chemically reacting flows for a blunt body in a Mach-20 flow. In addition to visibly affecting the density contours downstream of the bow shock wave, the entire structure of the flow field, including the shock wave stand-off distance, is altered.

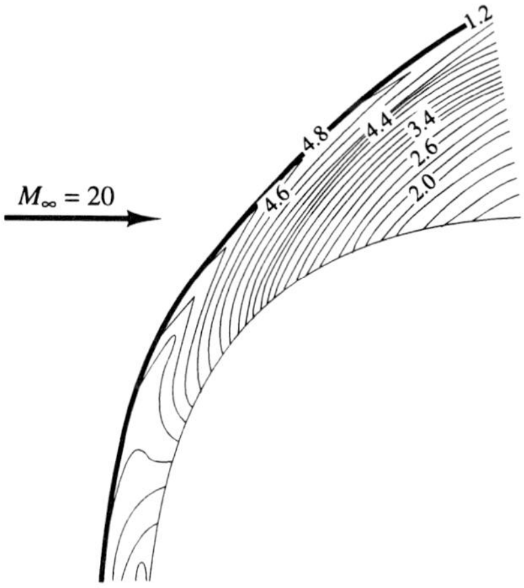

(a) Calorically perfect gas results

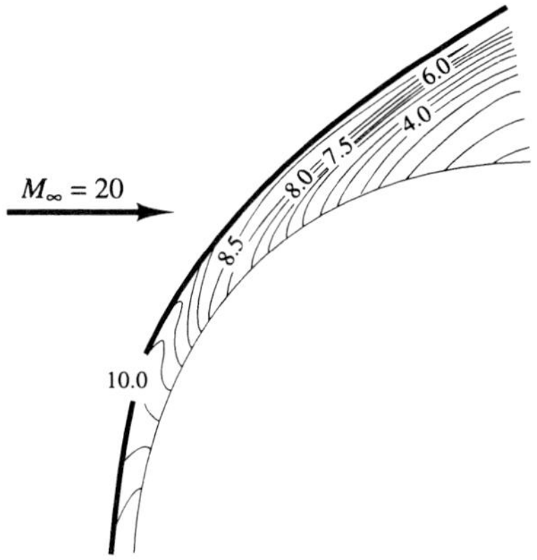

(b) Equilibrium air results

Figure 11-21 Equilibrium and Calorically Perfect Blunt-Body Results. *(From Anderson, 1989, from Palmer 1987).*

One of the most interesting speculations of the importance of chemistry on gas flows was put forth by Maus et al. (1984). Data from several of the early Space Shuttle flights, STS-1 through STS-4, indicated forward shifts of the center-of-pressure locations from those inferred from wind tunnel tests. The wind tunnel tests generally were made in facilities with fluid temperatures less than 1200 K so that caloric imperfections and chemistry effects were minimal. Figure 11-22, from Romere and Whitnah (1983), presents actual versus predicted center-of-pressure values for STS-2 for several Mach numbers. For all Mach numbers the actual center-of-pressure locations are forward of the predicted and the shifts increase with increasing Mach number. Woods et al. (1983) investigated the possibility that the shifts were caused by high-temperature effects on transport properties. They concluded that changes in transport properties would result in a backward, not a forward shift, so that transport property variations with temperature appear an unlikely cause of the observed shifts. Maus et al. (1984) made a series of computations assuming either a perfect gas with a ratio of specific heats of 1.4 or equilibrium air chemistry. Their results, presented in Figure 11-23, show a forward shift of the computed center-of-pressure locations for the equilibrium air computations. The computations of Maus

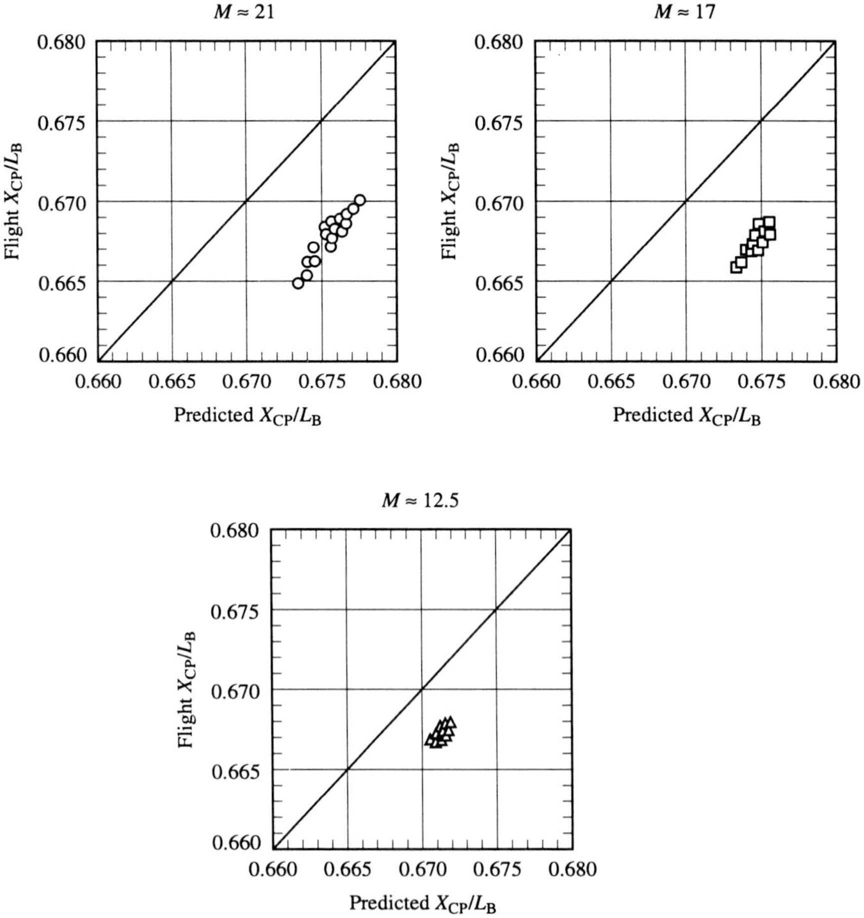

Figure 11-22 Space Shuttle STS-2 Center of Pressure. *(From Romere and Whitnah, 1983).*

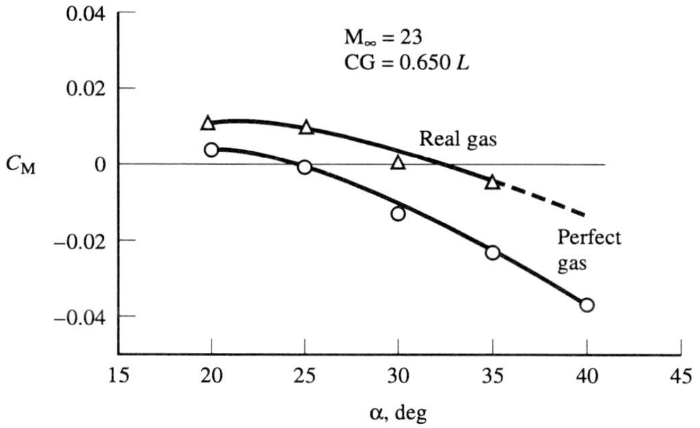

Figure 11-23 Computed Pitching Moment Coefficients for Equilibrium and Calorically Perfect Gas Assumptions. *(From Maus et al., 1983).*

et al. also show increasing forward shifts as the Mach number increases: hence, the speculation by Maus et al. that the shifts in centers-of-pressure locations over those predicted from wind tunnel tests are caused by chemistry effects. Park (1990) points out that chemical nonequilibrium effects on the center-of-pressure location for the Space Shuttle have not been investigated.

Cox (1992), whose "Black Box" chemical equilibrium solver was mentioned previously in this section, uses a five-species model of equilibrium air to compute the flow field over a blunt-cone geometry at a Mach number of 10. The geometry and computational grid used by Cox are shown in Figure 11-24. The results of his

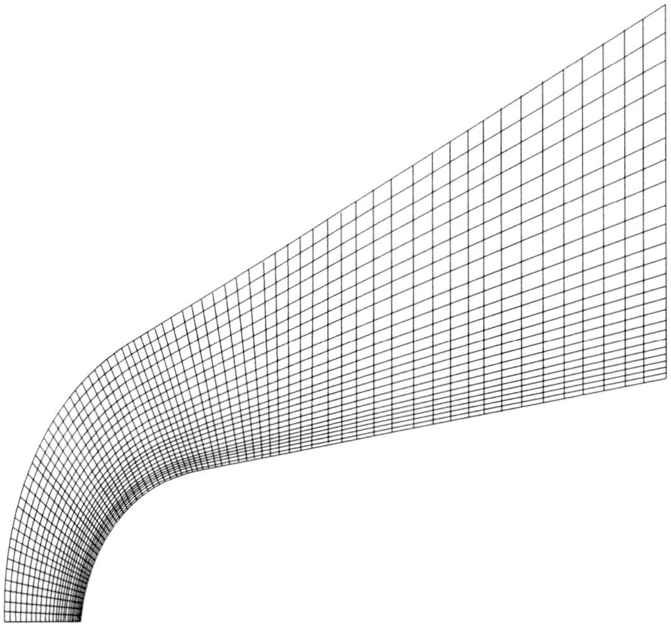

Figure 11-24 Blunt Cone Geometry and Grid. *(From Cox, 1992).*

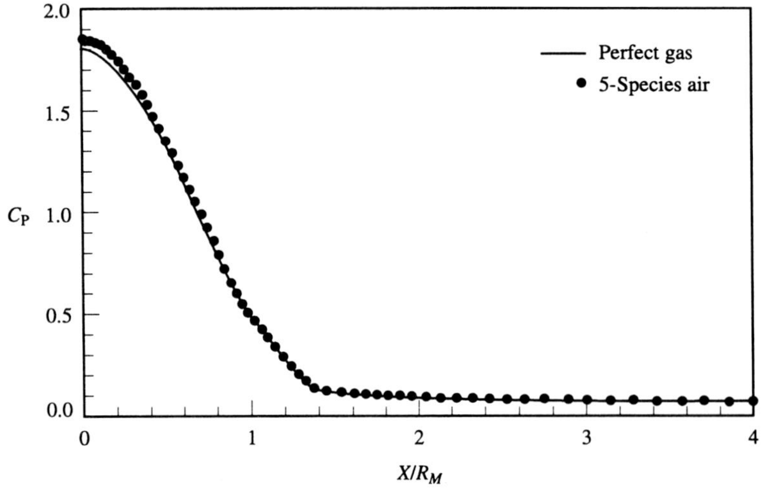

Figure 11-25 Blunt Cone Surface Pressure Distributions. *(From Cox, 1992).*

computations for the pressure coefficient on the surface are shown in Figure 11-25. The differences between the calorically perfect gas calculations and the five-species equilibrium air model are slight. The same, however, is not true of the surface temperature presented in Figure 11-26. Over the spherical portion of the body, the perfect gas computations greatly overestimate the surface temperature. The mole percentages for the calorically perfect gas computations and the five-species computations are presented in Figure 11-27.

Nozzle flows with chemical equilibrium will be considered next. The answers to several of the questions in Section 11-3 are germane to chemically reacting nozzle flows: (1) the flow of a thermally perfect as can be isentropic and (2) $M = 1$ occurs

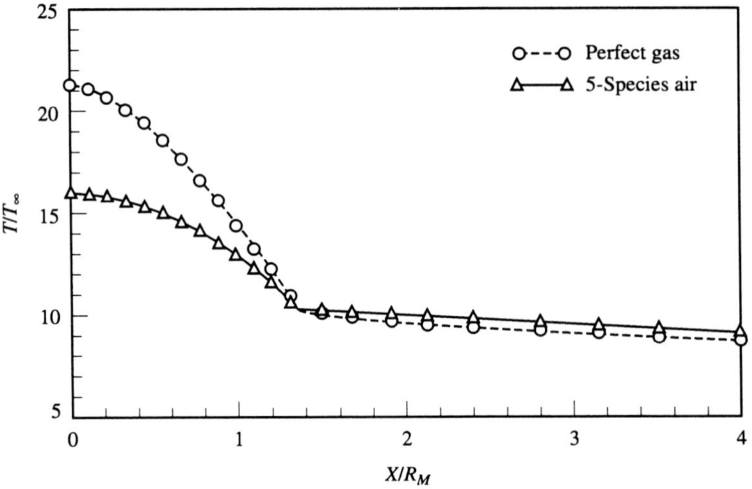

Figure 11-26 Blunt Cone Surface Temperature Distributions. *(From Cox, 1992).*

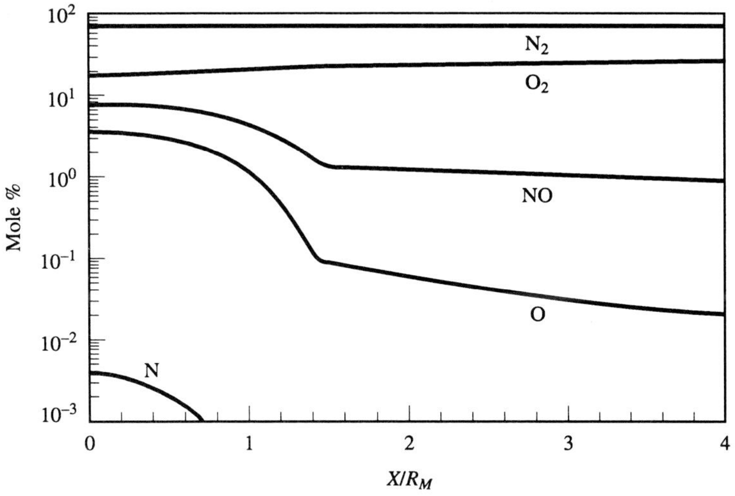

Figure 11-27 Blunt Cone Equilibrium Mole Percentages. *(From Cox, 1992).*

at the throat in a chemically reacting flow. In Chapter 6, the property ratios for isentropic calorically perfect gas nozzle flow are functions of only the local Mach number; that is from Table 6-1b,

$$A/A^* = A/A^*(M);$$
$$T/T_o = T/T_o(M);$$
$$P/P_o = P/P_o(M).$$

As with other processes examined in this section, the functional forms for a chemically reacting flow will differ from those of a calorically perfect flow. The chemical composition of the gas changes as it transits the nozzle since the static temperature and pressure change. The functional forms of the preceding equations for a chemically reacting gas in equilibrium then become

$$A/A^* = A/A^*(V, P_o, T_o),$$
$$T/T_o = T/T_o(V, P_o, T_o), \tag{11-95}$$
$$P/P_o = P/P_o(V, P_o, T_o),$$

where V is the local velocity. The velocity is used because it is a more convenient variable than the local Mach number. Consider the composition of an equilibrium chemically reacting gas in a nozzle with supersonic flow in the diverging segment of the nozzle. As the flow accelerates through and traverses the nozzle, the local static pressure and temperature decrease. A gas that is significantly dissociated in the nozzle reservoir will recombine as the temperature is decreased. For example, the atomic species will tend to recombine to form molecular species. The recombination of species is an exothermic reaction with the results that the static temperature of the equilibrium chemically reacting gas will be higher than for a calorically perfect gas flow. The temperature distribution through a converging–diverging nozzle with supersonic flow is shown schematically in Figure 11-28.

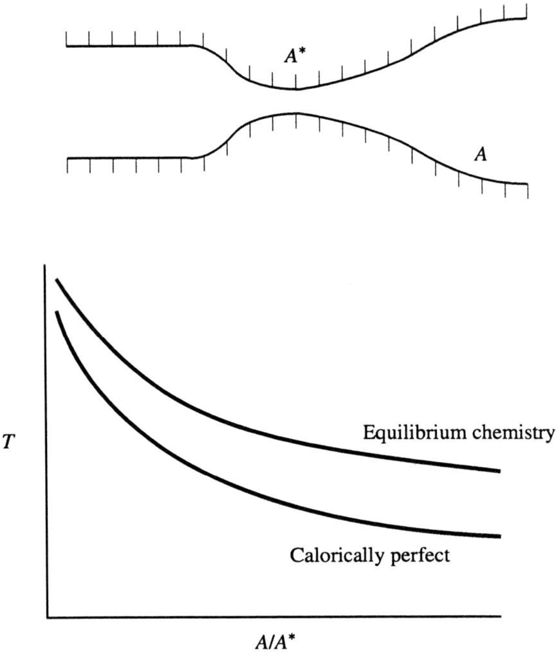

Figure 11-28 Equilibrium and Calorically Perfect Converging–Diverging Nozzle Performance Schematic

The Space Shuttle Main Engine (SSME) computations of Cox (1992) provide an excellent study in equilibrium chemistry effects in a supersonic nozzle flow. The SSME utilizes liquid hydrogen and liquid oxygen so that the combustion product is water vapor. The nozzle gas is thus composed of water vapor, oxygen, and hydrogen. Figure 11-29 shows the computational grid Cox used in the SSME computations. The

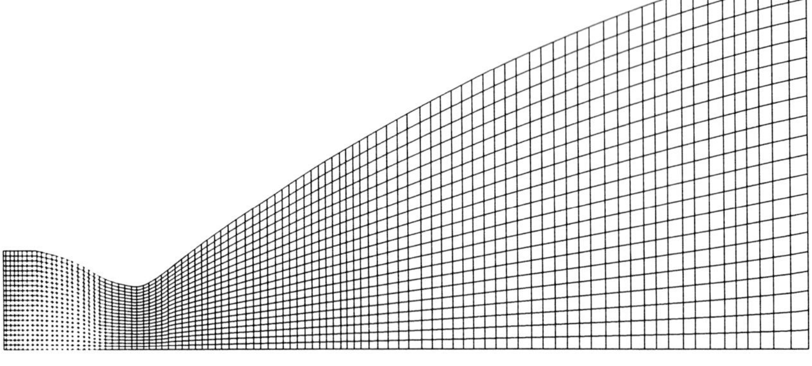

Figure 11-29 SSME Geometry and Grid. *(From Cox, 1992).*

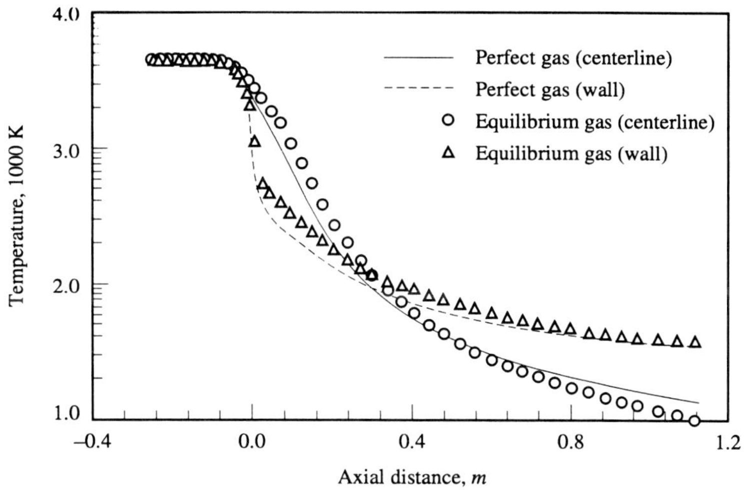

Figure 11-30 SSME Equilibrium and Calorically Perfect Centerline and Wall Temperature Distributions. *(From Cox, 1992).*

wall and centerline temperature distributions for both calorically perfect and equilibrium chemistry calculations are presented in Figure 11-30. As expected, over much of the nozzle wall the equilibrium chemistry temperature is higher than that of the calorically perfect gas. The species mole percentages along the SSME nozzle wall are given in Figure 11-31. The decrease in temperature of the exhaust products as they traverse the nozzle results in the recombination of species, which decreases the mole percentages of O and OH. As the flow is expanded supersonically in the

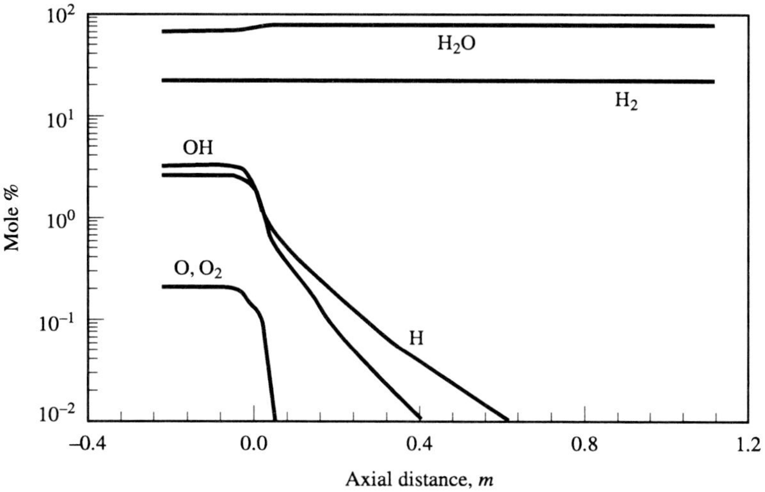

Figure 11-31 SSME Equilibrium Mole Percentages. *(From Cox, 1992).*

diverging portion of the nozzle, the temperature decreases and the mole percentages of the OH, H, O, and O_2 species decrease.

11.8 ADDITIONAL CONSIDERATIONS

The latter portions of this chapter have emphasized and examined equilibrium chemistry effects in high-temperature gas dynamics. Equilibrium chemistry essentially presumes that as the fluid traverses the flow field, the mixture species composition instantaneously attains the composition specified by the local temperature and pressure. The equilibrium chemistry assumption requires no postulates about the mechanism required to attain equilibrium because the rates of change associated with the species and the vibrational state are taken as being unbounded. In actuality, neither the chemical composition nor the vibrational energy level of the mixture instantaneously reaches the equilibrium value as specified by the local conditions (pressure and/or temperature). Flows not in chemical and vibrational equilibrium are broadly called nonequilibrium flows. Such flows constitute a topic of major interest in high-temperature gas dynamics. The book by Park (1990) is a contemporary assessment and study of nonequilibrium in hypersonics. Much of the latter chapters of Anderson (1989) are devoted to nonequilibrium effects. Both of these references are recommended reading for additional information on nonequilibrium effects. Additionally, the contemporary literature in the thermal and thermal/computational sciences, particularly that from the aerospace sector, contains contributions on reacting equilibrium and nonequilibrium flows.

A flow in which the chemical composition remains constant in space and time is called a *frozen flow*. If the vibrational state of the flow is also invariant with space and time, then the frozen flow possesses constant specific heats and is calorically perfect.

As we reach the end of this textbook, what has not been covered should be mentioned. To start with, this is an introductory textbook with sufficient material for perhaps a two-semester sequence at the advanced undergraduate or the beginning graduate level. Its purpose is to supply enough information and, we hope, insight to provide a basis for either initial engineering practice in or further study of compressible flow. Two important areas that have not received much treatment are computational fluid dynamics (CFD) and experimental techniques. Only a glance at the literature is required to ascertain how important CFD has become to compressible flow. However, CFD has matured to where it is perhaps an identifiable separate branch of fluid mechanics (see Graves, 1982), with its own literature and an increasing number of textbooks. Most of the authors' CFD colleagues seem to agree that a single chapter on CFD in an introductory textbook would not be very effective in conveying sufficient information for meaningful applications; hence the decision not to include such a chapter. Likewise, we do not include a chapter on experimental techniques, for much the same reason. A single chapter on such an important topic as experimental techniques would be inadequate to convey the needed information. At this point, however, the basis for further study involving compressible flow has been established.

CHAPTER 11 REVIEW QUESTIONS

1. Distinguish between a calorically perfect gas, a thermally perfect gas, a chemically reacting mixture of perfect gases, and a real gas.
2. What important characteristic of internal energy and enthalpy does the assumption of a thermally perfect gas infer?
3. What are the functional forms of the internal energy and enthalpy for a mixture of chemically reacting thermally perfect gases? Explain why these forms are as they are.
4. Sketch and explain the variations of c_p and c_v for a thermally perfect gas.
5. For the one-dimensional flow of an inviscid, high-temperature gas, does $M = 1$ occur at the minimum area?
6. Does $a = \sqrt{\gamma RT}$ in a thermally perfect but calorically imperfect gas?
7. Explain why (as in Example 11-2) the effects of caloric imperfection on a given area ratio are to reduce the Mach number.
8. For a given upstream Mach number, explain why the downstream Mach numbers are different for the case of calorically perfect and calorically imperfect gases?
9. What is the partial pressure of a component of a mixture of thermally perfect gases?
10. Why does the mixing of two diatomic gases at the same temperature and pressure result in an increase in entropy?
11. What is the law of mass action?
12. What type of system of equations results when the equilibrium composition of a mixture is obtained? What principles are used to formulate the system? How is the system solved?
13. When a chemically reacting gas undergoes a shock wave, why is the resulting downstream temperature reduced from the calorically perfect behavior?
14. Given the same converging–diverging nozzle, the downstream temperatures are higher for an equilibrium chemistry case than for a calorically perfect gas. Explain why.

CHAPTER PROBLEMS

1. Compute the speed of sound in air at 1000 R, 2000 R, 3000 R, and 4000 R. How do these values differ from the calorically perfect values?
2. Compute the speed of sound in air at 500 K, 1000 K, 1500 K, and 2000 K. How do these values differ from the calorically perfect values?
3. Generate a plot of the ratio of specific heats versus temperature for 300 $R < T <$ 4000 R. Indicate degrees Kelvin on the plot. What does this say about the validity of the calorically perfect assumption?
4. Generate a plot of Mach number versus T/T_o for T_o = 250 K, 500 K, 1000 K, 1500 K, and 2000 K. The fluid is air. What does this say about the calorically perfect gas assumption?

5. Generate a plot of pressure ratio, P/P_o, versus T/T_o for $T_o = 250$ K, 500 K, 1000 K, 1500 K, and 2000 K. The fluid is air. What does this say about the calorically perfect gas assumption?
6. For nitrogen at a stagnation temperature of 2000 R,
   (a) Find the static temperature and velocity for a Mach number of 0.75.
   (b) Find the Mach number and velocity for a static temperature of 1000 R.
   (c) Find the Mach number and static temperature for a velocity of 2500 ft/s.
   (d) Compare these results with the calorically perfect values.
7. For air at a stagnation temperature of 1500 K,
   (a) Find the static temperature and velocity for a Mach number of 0.75.
   (b) Find the Mach number and velocity for a static temperature of 1000 K.
   (c) Find the Mach number and static temperature for a velocity of 1000 m/s.
   (d) Compare these results with the calorically perfect values.
8. Find the stagnation conditions for $T = 1250$ K, $P = 100$ kPa, and $V = 400$ m/s.
9. Find the stagnation conditions for $T = 2500\ R$, $P = 50$ psia, and $V = 1000$ ft/s.
10. Redo Problem 6-2, assuming the reservoir temperature is 3000 K.
11. Redo Problem 6-3, assuming the static temperature at $M = 1.75$ is 1000 K.
12. Redo Problem 6-8, assuming the stagnation temperature is 1500 K.
13. Redo Problem 6-12 if the reservoir temperature is 1000 K.
14. Redo Problem 6-14 if the stagnation temperature is changed to 2500 R.
15. Redo Problem 6-18 using both calorically imperfect and calorically perfect assumptions.
16. Redo Problem 6-21 assuming a calorically imperfect gas.
17. Redo Problem 6-24 if the stagnation temperature is 4000 R.
18. Redo Problem 6-25 if the reservoir temperature is 2500 R.
19. Redo Problem 6-33 if the initial tank temperature is 2000 K and if the tank discharge is isothermal.
20. A shock wave is moving at a Mach number of 3 into air at $T = 300$ K and $P = 101.3$ kPa. What are the velocity and stagnation temperature (with respect to a stationary observer) in the air after passage of the shock wave?
21. A projectile moves down a gun barrel with a velocity of 2000 ft/s. Calculate the velocity of the normal shock wave that would move ahead of the projectile. The pressure and density of the undisturbed air ahead of shock wave are 70° F and 14.7 psia.
22. Air at Mach 2.5, with stagnation conditions of 5 MPa and 2000 K, expands around a 15° convex corner. Determine the Mach number, the stagnation conditions, and the static conditions downstream of the expansion.
23. Redo Problem 6-4, assuming the stagnation temperature is 3500 R.
24. Redo Problem 6-5 if the stagnation temperature is 2000 R.
25. Redo Problem 6-6, assuming the stagnation temperature is 3500 R.
26. Redo Problem 6-10 if the stagnation temperature is 2000 K.
27. A converging–diverging nozzle with exit and inlet areas 10 in^2 and 5 in^2, respectively, is used to vent a large chamber containing hot air. For a back pressure of 14.7 psia and a chamber temperature of 3550 R, compute the following:
   (a) minimum chamber pressure to choke the nozzle;
   (b) mass flow rate for a chamber pressure of 60 psia;
   (c) mass flow rate for a chamber pressure of 30 psia.

28. A converging–diverging nozzle, designed to operate isentropically with air at an exit Mach number of 1.75, is connected to a reservoir with stagnation conditions of 4 MPa and 1600 K. Compute the following:
(a) the maximum back pressure to choke the nozzle;
(b) the flow rate for a back pressure of 101 kPa;
(c) the flow rate for a back pressure of 1 MPa.
The nozzle exit area is 0.15 m^2.

29. Determine the thrust of a rocket nozzle designed to operate isentropically at 45,000 ft with a chamber pressure of 350 psia and a chamber temperature of 4000 *R*. The nozzle is supplied with heated air. The exit area of the converging–diverging nozzle is 1.25 ft^2.

30. Redo Problem 6-29, assuming the stagnation temperature is 2000 K.

31. Compare the results of CHEMQ for the equilibrium composition of air with Anderson's results as presented in Figure 11-16.

32. Modify CHEMQ to compute the equilibrium composition of a 50–50 mixture of nitrogen and air. The temperature range of interest is from 300 K to 2000 K, and the pressure is 2 atm.

33. Develop a computer subroutine to find the static temperature given the Mach number and the stagnation temperature. Use Eq. 11-36, and verify the results using CIGQ.

34. Develop a computer subroutine to find the static temperature given the velocity and the stagnation temperature. Use Eq. 11-35, and verify the results using CIGQ.

35. Develop a computer subroutine to find the static temperature given the pressure ratio, P/P_o, and the stagnation temperature. Use Eq. 11-39b, and verify the results using CIGQ.

36. Given T_1, T_o, and M_1, develop a computer program to find the conditions downstream of a normal shock wave using the iterative sequence given after Eq. 11-42. Verify the results using CIGQ.

APPENDICES

Isentropic Flow

APPENDIX A

TABLE A-1 ISENTROPIC FLOW (PERFECT GAS, $\gamma = 1.4$)

M	$\frac{P}{P_0}$	$\frac{\rho}{\rho_0}$	$\frac{T}{T_0}$	$\frac{A}{A*}$
0.00	1.0000	1.0000	1.0000	∞
0.02	0.9997	0.9998	0.9999	2.8942+01
0.04	0.9989	0.9992	0.9997	1.4481+01
0.06	0.9975	0.9982	0.9993	9.6659
0.08	0.9955	0.9968	0.9987	7.2616
0.10	0.9930	0.9950	0.9980	5.8218
0.12	0.9900	0.9928	0.9971	4.8643
0.14	0.9864	0.9903	0.9961	4.1824
0.16	0.9823	0.9873	0.9949	3.6727
0.18	0.9776	0.9840	0.9936	3.2779
0.20	0.9725	0.9803	0.9921	2.9635
0.22	0.9668	0.9762	0.9904	2.7076
0.24	0.9607	0.9718	0.9886	2.4956
0.26	0.9541	0.9670	0.9867	2.3173
0.28	0.9470	0.9618	0.9846	2.1656
0.30	0.9395	0.9564	0.9823	2.0351
0.32	0.9315	0.9506	0.9799	1.9219
0.34	0.9231	0.9445	0.9774	1.8229
0.36	0.9143	0.9380	0.9747	1.7358
0.38	0.9052	0.9313	0.9719	1.6587
0.40	0.8956	0.9243	0.9690	1.5901
0.42	0.8857	0.9170	0.9659	1.5289
0.44	0.8755	0.9094	0.9627	1.4740
0.46	0.8650	0.9016	0.9594	1.4246
0.48	0.8541	0.8935	0.9560	1.3801
0.50	0.8430	0.8852	0.9524	1.3398
0.52	0.8317	0.8766	0.9487	1.3034
0.54	0.8201	0.8679	0.9449	1.2703
0.56	0.8082	0.8589	0.9410	1.2403
0.58	0.7962	0.8498	0.9370	1.2130

TABLE A-1 (continued)

M	$\frac{P}{P_0}$	$\frac{\rho}{\rho_0}$	$\frac{T}{T_0}$	$\frac{A}{A*}$
0.60	0.7840	0.8404	0.9328	1.1882
0.62	0.7716	0.8310	0.9286	1.1656
0.64	0.7591	0.8213	0.9243	1.1451
0.66	0.7465	0.8115	0.9199	1.1265
0.68	0.7338	0.8016	0.9154	1.1097
0.70	0.7209	0.7916	0.9107	1.0944
0.72	0.7080	0.7814	0.9061	1.0806
0.74	0.6951	0.7712	0.9013	1.0681
0.76	0.6821	0.7609	0.8964	1.0570
0.78	0.6690	0.7505	0.8915	1.0471
0.80	0.6560	0.7400	0.8865	1.0382
0.82	0.6430	0.7295	0.8815	1.0305
0.84	0.6300	0.7189	0.8763	1.0237
0.86	0.6170	0.7083	0.8711	1.0179
0.88	0.6041	0.6977	0.8659	1.0129
0.90	0.5913	0.6870	0.8606	1.0089
0.92	0.5785	0.6764	0.8552	1.0056
0.94	0.5658	0.6658	0.8498	1.0031
0.96	0.5532	0.6551	0.8444	1.0014
0.98	0.5407	0.6445	0.8389	1.0003
1.00	0.5283	0.6339	0.8333	1.0000
1.02	0.5160	0.6234	0.8278	1.0003
1.04	0.5039	0.6129	0.8221	1.0013
1.06	0.4919	0.6024	0.8165	1.0029
1.08	0.4800	0.5920	0.8109	1.0051
1.10	0.4683	0.5817	0.8051	1.0079
1.12	0.4568	0.5714	0.7994	1.0113
1.14	0.4455	0.5612	0.7937	1.0153
1.16	0.4342	0.5511	0.7879	1.0198
1.18	0.4232	0.5411	0.7822	1.0248
1.20	0.4124	0.5311	0.7764	1.0304
1.22	0.4017	0.5213	0.7706	1.0366
1.24	0.3912	0.5115	0.7648	1.0432
1.26	0.3809	0.5019	0.7590	1.0504
1.28	0.3708	0.4923	0.7532	1.0581
1.30	0.3609	0.4829	0.7474	1.0663
1.32	0.3512	0.4736	0.7416	1.0750
1.34	0.3417	0.4644	0.7358	1.0842
1.36	0.3323	0.4553	0.7300	1.0940
1.38	0.3232	0.4463	0.7242	1.1042
1.40	0.3142	0.4374	0.7184	1.1149
1.42	0.3055	0.4287	0.7126	1.1262
1.44	0.2969	0.4201	0.7068	1.1379
1.46	0.2886	0.4116	0.7011	1.1501
1.48	0.2804	0.4032	0.6954	1.1629
1.50	0.2724	0.3950	0.6897	1.1762
1.52	0.2646	0.3869	0.6840	1.1899
1.54	0.2570	0.3789	0.6783	1.2042
1.56	0.2496	0.3711	0.6726	1.2190
1.58	0.2423	0.3633	0.6670	1.2344

TABLE A-1 (continued)

M	$\frac{P}{P_0}$	$\frac{\rho}{\rho_0}$	$\frac{T}{T_0}$	$\frac{A}{A*}$
1.60	0.2353	0.3557	0.6614	1.2502
1.62	0.2284	0.3483	0.6558	1.2666
1.64	0.2217	0.3409	0.6502	1.2836
1.66	0.2151	0.3337	0.6447	1.3010
1.68	0.2088	0.3266	0.6392	1.3190
1.70	0.2026	0.3197	0.6337	1.3376
1.72	0.1966	0.3129	0.6283	1.3567
1.74	0.1907	0.3062	0.6229	1.3764
1.76	0.1850	0.2996	0.6175	1.3967
1.78	0.1794	0.2932	0.6121	1.4175
1.80	0.1740	0.2868	0.6068	1.4390
1.82	0.1688	0.2806	0.6015	1.4610
1.84	0.1637	0.2745	0.5963	1.4836
1.86	0.1587	0.2686	0.5910	1.5069
1.88	0.1539	0.2627	0.5859	1.5308
1.90	0.1492	0.2570	0.5807	1.5553
1.92	0.1447	0.2514	0.5756	1.5804
1.94	0.1403	0.2459	0.5705	1.6062
1.96	0.1360	0.2405	0.5655	1.6326
1.98	0.1318	0.2352	0.5605	1.6597
2.00	0.1278	0.2300	0.5556	1.6875
2.05	0.1182	0.2176	0.5433	1.7600
2.10	0.1094	0.2058	0.5314	1.8369
2.15	0.1011	0.1946	0.5196	1.9185
2.20	9.3522-02	0.1840	0.5081	2.0050
2.25	8.6482-02	0.1740	0.4969	2.0964
2.30	7.9973-02	0.1646	0.4859	2.1931
2.35	7.3957-02	0.1556	0.4752	2.2953
2.40	6.8399-02	0.1472	0.4647	2.4031
2.45	6.3267-02	0.1392	0.4544	2.5168
2.50	5.8528-02	0.1317	0.4444	2.6367
2.55	5.4153-02	0.1246	0.4347	2.7630
2.60	5.0115-02	0.1179	0.4252	2.8960
2.65	4.6389-02	0.1115	0.4159	3.0359
2.70	4.2950-02	0.1056	0.4068	3.1830
2.75	3.9777-02	0.0999	0.3980	3.3377
2.80	3.6848-02	9.4626-02	0.3894	3.5001
2.85	3.4146-02	8.9616-02	0.3810	3.6707
2.90	3.1651-02	8.4889-02	0.3729	3.8498
2.95	2.9349-02	8.0431-02	0.3649	4.0376
3.00	2.7224-02	7.6226-02	0.3571	4.2346
3.05	2.5261-02	7.2260-02	0.3496	4.4410
3.10	2.3449-02	6.8517-02	0.3422	4.6573
3.15	2.1775-02	6.4986-02	0.3351	4.8838
3.20	2.0228-02	6.1654-02	0.3281	5.1210
3.25	1.8798-02	5.8510-02	0.3213	5.3691
3.30	1.7477-02	5.5541-02	0.3147	5.6286
3.35	1.6255-02	5.2739-02	0.3082	5.9000
3.40	1.5125-02	5.0093-02	0.3019	6.1837
3.45	1.4079-02	4.7594-02	0.2958	6.4801

TABLE A-1 ISENTROPIC FLOW (PERFECT GAS, $\gamma = 1.4$)

M	$\frac{P}{P_0}$	$\frac{\rho}{\rho_0}$	$\frac{T}{T_0}$	$\frac{A}{A*}$
3.50	1.3111-02	4.5233-02	0.2899	6.7896
3.55	1.2215-02	4.3002-02	0.2840	7.1128
3.60	1.1385-02	4.0894-02	0.2784	7.4501
3.65	1.0616-02	3.8901-02	0.2729	7.8020
3.70	9.9029-03	3.7017-02	0.2675	8.1691
3.75	9.2420-03	3.5235-02	0.2623	8.5517
3.80	8.6290-03	3.3549-02	0.2572	8.9506
3.85	8.0601-03	3.1954-02	0.2522	9.3661
3.90	7.5320-03	3.0445-02	0.2474	9.7990
3.95	7.0417-03	2.9015-02	0.2427	1.0250+01
4.00	6.5861-03	2.7662-02	0.2381	1.0719+01
4.05	6.1627-03	2.6379-02	0.2336	1.1207+01
4.10	5.7690-03	2.5164-02	0.2292	1.1715+01
4.15	5.4028-03	2.4013-02	0.2250	1.2243+01
4.20	5.0621-03	2.2921-02	0.2209	1.2792+01
4.25	4.7449-03	2.1886-02	0.2168	1.3362+01
4.30	4.4494-03	2.0903-02	0.2129	1.3955+01
4.35	4.1742-03	1.9971-02	0.2090	1.4571+01
4.40	3.9176-03	1.9087-02	0.2052	1.5210+01
4.45	3.6784-03	1.8247-02	0.2016	1.5873+01
4.50	3.4553-03	1.7449-02	0.1980	1.6562+01
4.55	3.2470-03	1.6691-02	0.1945	1.7277+01
4.60	3.0526-03	1.5971-02	0.1911	1.8018+01
4.65	2.8709-03	1.5286-02	0.1878	1.8786+01
4.70	2.7012-03	1.4635-02	0.1846	1.9583+01
4.75	2.5426-03	1.4016-02	0.1814	2.0408+01
4.80	2.3943-03	1.3427-02	0.1783	2.1264+01
4.85	2.2555-03	1.2866-02	0.1753	2.2150+01
4.90	2.1256-03	1.2333-02	0.1724	2.3067+01
4.95	2.0040-03	1.1824-02	0.1695	2.4017+01
5.00	1.8900-03	1.1340-02	0.1667	2.5000+01
5.10	1.6832-03	1.0439-02	0.1612	2.7070+01
5.20	1.5013-03	9.6204-03	0.1560	2.9283+01
5.30	1.3411-03	8.8753-03	0.1511	3.1649+01
5.40	1.1997-03	8.1965-03	0.1464	3.4175+01
5.50	1.0748-03	7.5775-03	0.1418	3.6869+01
5.60	9.6430-04	7.0124-03	0.1375	3.9740+01
5.70	8.6635-04	6.4959-03	0.1334	4.2797+01
5.80	7.7941-04	6.0233-03	0.1294	4.6050+01
5.90	7.0214-04	5.5904-03	0.1256	4.9507+01
6.00	6.3336-04	5.1936-03	0.1220	5.3180+01
6.10	5.7206-04	4.8293-03	0.1185	5.7077+01
6.20	5.1735-04	4.4947-03	0.1151	6.1210+01
6.30	4.6845-04	4.1870-03	0.1119	6.5590+01
6.40	4.2468-04	3.9037-03	0.1088	7.0227+01
6.50	3.8547-04	3.6427-03	0.1058	7.5134+01
6.60	3.5028-04	3.4020-03	0.1030	8.0323+01
6.70	3.1867-04	3.1797-03	0.1002	8.5805+01
6.80	2.9024-04	2.9744-03	9.7580-02	9.1594+01
6.90	2.6464-04	2.7845-03	9.5039-02	9.7702+01

TABLE A-1 (continued)

M	$\frac{P}{P_0}$	$\frac{\rho}{\rho_0}$	$\frac{T}{T_0}$	$\frac{A}{A*}$
7.00	2.4156-04	2.6088-03	9.2593-02	1.0414+02
7.10	2.2072-04	2.4460-03	9.0236-02	1.1093+02
7.20	2.0189-04	2.2950-03	8.7966-02	1.1808+02
7.30	1.8485-04	2.1550-03	8.5778-02	1.2560+02
7.40	1.6942-04	2.0249-03	8.3668-02	1.3352+02
7.50	1.5543-04	1.9040-03	8.1633-02	1.4184+02
7.60	1.4273-04	1.7915-03	7.9669-02	1.5058+02
7.70	1.3119-04	1.6868-03	7.7773-02	1.5977+02
7.80	1.2069-04	1.5893-03	7.5942-02	1.6940+02
7.90	1.1114-04	1.4984-03	7.4173-02	1.7951+02
8.00	1.0243-04	1.4135-03	7.2464-02	1.9011+02
8.10	9.4485-05	1.3343-03	7.0811-02	2.0121+02
8.20	8.7231-05	1.2603-03	6.9214-02	2.1285+02
8.30	8.0602-05	1.1911-03	6.7668-02	2.2502+02
8.40	7.4537-05	1.1264-03	6.6173-02	2.3776+02
8.50	6.8984-05	1.0658-03	6.4725-02	2.5109+02
8.60	6.3895-05	1.0090-03	6.3323-02	2.6501+02
8.70	5.9228-05	9.5582-04	6.1966-02	2.7957+02
8.80	5.4943-05	9.0590-04	6.0650-02	2.9477+02
8.90	5.1006-05	8.5904-04	5.9375-02	3.1063+02
9.00	4.7386-05	8.1504-04	5.8140-02	3.2719+02
9.10	4.4055-05	7.7369-04	5.6941-02	3.4446+02
9.20	4.0986-05	7.3480-04	5.5779-02	3.6246+02
9.30	3.8158-05	6.9822-04	5.4651-02	3.8123+02
9.40	3.5549-05	6.6378-04	5.3556-02	4.0078+02
9.50	3.3141-05	6.3134-04	5.2493-02	4.2113+02
9.60	3.0916-05	6.0077-04	5.1462-02	4.4232+02
9.70	2.8860-05	5.7194-04	5.0459-02	4.6437+02
9.80	2.6957-05	5.4474-04	4.9485-02	4.8730+02
9.90	2.5195-05	5.1907-04	4.8539-02	5.1115+02
10.00	2.3563-05	4.9483-04	4.7619-02	5.3594+02
11.00	1.2448-05	3.1369-04	3.9683-02	8.4191+02
12.00	6.9222-06	2.0628-04	3.3557-02	1.2762+03
13.00	4.0223-06	1.3998-04	2.8736-02	1.8761+03
14.00	2.4278-06	9.7597-05	2.4876-02	2.6854+03
15.00	1.5148-06	6.9680-05	2.1739-02	3.7552+03
16.00	9.7309-07	5.0795-05	1.9157-02	5.1446+03
17.00	6.4147-07	3.7719-05	1.7007-02	6.9205+03
18.00	4.3272-07	2.8473-05	1.5198-02	9.1593+03
19.00	2.9800-07	2.1813-05	1.3661-02	1.1946+04
20.00	2.0908-07	1.6935-05	1.2346-02	1.5377+04
21.00	1.4918-07	1.3307-05	1.1211-02	1.9558+04
22.00	1.0810-07	1.0572-05	1.0225-02	2.4607+04
23.00	7.9433-08	8.4834-06	9.3633-03	3.0651+04
24.00	5.9126-08	6.8704-06	8.6059-03	3.7832+04
25.00	4.4535-08	5.6114-06	7.9365-03	4.6305+04
26.00	3.3914-08	4.6191-06	7.3421-03	5.6236+04
27.00	2.6089-08	3.8299-06	6.8120-03	6.7806+04
28.00	2.0259-08	3.1969-06	6.3371-03	8.1212+04
29.00	1.5871-08	2.6853-06	5.9102-03	9.6663+04

TABLE A-1 (concluded)

M	$\frac{P}{P_0}$	$\frac{\rho}{\rho_0}$	$\frac{T}{T_0}$	$\frac{A}{A*}$
30.00	1.2535-08	2.2688-06	5.5249-03	1.1439+05
31.00	9.9764-09	1.9274-06	5.1760-03	1.3462+05
32.00	7.9973-09	1.6458-06	4.8591-03	1.5763+05
33.00	6.4541-09	1.4122-06	4.5704-03	1.8369+05
34.00	5.2418-09	1.2172-06	4.3066-03	2.1309+05
35.00	4.2828-09	1.0536-06	4.0650-03	2.4615+05
36.00	3.5190-09	9.1565-07	3.8432-03	2.8319+05
37.00	2.9070-09	7.9884-07	3.6390-03	3.2457+05
38.00	2.4135-09	6.9945-07	3.4507-03	3.7065+05
39.00	2.0135-09	6.1452-07	3.2765-03	4.2184+05
40.00	1.6875-09	5.4167-07	3.1153-03	4.7853+05
41.00	1.4203-09	4.7894-07	2.9656-03	5.4117+05
42.00	1.2005-09	4.2472-07	2.8265-03	6.1021+05
43.00	1.0186-09	3.7770-07	2.6969-03	6.8613+05
44.00	8.6757-10	3.3679-07	2.5760-03	7.6943+05
45.00	7.4158-10	3.0108-07	2.4631-03	8.6064+05
46.00	6.3607-10	2.6982-07	2.3574-03	9.6031+05
47.00	5.4736-10	2.4237-07	2.2584-03	1.0690+06
48.00	4.7251-10	2.1821-07	2.1654-03	1.1873+06
49.00	4.0913-10	1.9687-07	2.0781-03	1.3159+06
50.00	3.5528-10	1.7799-07	1.9960-03	1.4555+06

Normal Shock Waves

APPENDIX B

TABLE B-1 NORMAL SHOCK (PERFECT GAS, $\gamma = 1.4$)

M_1	M_2	$\frac{P_2}{P_1}$	$\frac{\rho_2}{\rho_1}$	$\frac{T_2}{T_1}$	$\frac{P_{02}}{P_{01}}$
1.00	1.0000	1.0000	1.0000	1.0000	1.0000
1.02	0.9805	1.0471	1.0334	1.0132	1.0000
1.04	0.9620	1.0952	1.0671	1.0263	0.9999
1.06	0.9444	1.1442	1.1009	1.0393	0.9998
1.08	0.9277	1.1941	1.1349	1.0522	0.9994
1.10	0.9118	1.2450	1.1691	1.0649	0.9989
1.12	0.8966	1.2968	1.2034	1.0776	0.9982
1.14	0.8820	1.3495	1.2378	1.0903	0.9973
1.16	0.8682	1.4032	1.2723	1.1029	0.9961
1.18	0.8549	1.4578	1.3069	1.1154	0.9946
1.20	0.8422	1.5133	1.3416	1.1280	0.9928
1.22	0.8300	1.5698	1.3764	1.1405	0.9907
1.24	0.8183	1.6272	1.4112	1.1531	0.9884
1.26	0.8071	1.6855	1.4460	1.1657	0.9857
1.28	0.7963	1.7448	1.4808	1.1782	0.9827
1.30	0.7860	1.8050	1.5157	1.1909	0.9794
1.32	0.7760	1.8661	1.5505	1.2035	0.9758
1.34	0.7664	1.9282	1.5854	1.2162	0.9718
1.36	0.7572	1.9912	1.6202	1.2290	0.9676
1.38	0.7483	2.0551	1.6549	1.2418	0.9630
1.40	0.7397	2.1200	1.6897	1.2547	0.9582
1.42	0.7314	2.1858	1.7243	1.2676	0.9531
1.44	0.7235	2.2525	1.7589	1.2807	0.9476
1.46	0.7157	2.3202	1.7934	1.2938	0.9420
1.48	0.7083	2.3888	1.8278	1.3069	0.9360
1.50	0.7011	2.4583	1.8621	1.3202	0.9298
1.52	0.6941	2.5288	1.8963	1.3336	0.9233
1.54	0.6874	2.6002	1.9303	1.3470	0.9166
1.56	0.6809	2.6725	1.9643	1.3606	0.9097
1.58	0.6746	2.7458	1.9981	1.3742	0.9026

TABLE B-1 (continued)

M_1	M_2	$\frac{P_2}{P_1}$	$\frac{\rho_2}{\rho_1}$	$\frac{T_2}{T_1}$	$\frac{P_{02}}{P_{01}}$
1.60	0.6684	2.8200	2.0317	1.3880	0.8952
1.62	0.6625	2.8951	2.0653	1.4018	0.8877
1.64	0.6568	2.9712	2.0986	1.4158	0.8799
1.66	0.6512	3.0482	2.1318	1.4298	0.8720
1.68	0.6458	3.1261	2.1649	1.4440	0.8639
1.70	0.6405	3.2050	2.1977	1.4583	0.8557
1.72	0.6355	3.2848	2.2304	1.4727	0.8474
1.74	0.6305	3.3655	2.2629	1.4873	0.8389
1.76	0.6257	3.4472	2.2952	1.5019	0.8302
1.78	0.6210	3.5298	2.3273	1.5167	0.8215
1.80	0.6165	3.6133	2.3592	1.5316	0.8127
1.82	0.6121	3.6978	2.3909	1.5466	0.8038
1.84	0.6078	3.7832	2.4224	1.5617	0.7948
1.86	0.6036	3.8695	2.4537	1.5770	0.7857
1.88	0.5996	3.9568	2.4848	1.5924	0.7765
1.90	0.5956	4.0450	2.5157	1.6079	0.7674
1.92	0.5918	4.1341	2.5463	1.6236	0.7581
1.94	0.5880	4.2242	2.5768	1.6394	0.7488
1.96	0.5844	4.3152	2.6070	1.6553	0.7395
1.98	0.5808	4.4071	2.6369	1.6713	0.7302
2.00	0.5774	4.5000	2.6667	1.6875	0.7209
2.05	0.5691	4.7362	2.7400	1.7285	0.6975
2.10	0.5613	4.9783	2.8119	1.7704	0.6742
2.15	0.5540	5.2263	2.8823	1.8132	0.6511
2.20	0.5471	5.4800	2.9512	1.8569	0.6281
2.25	0.5406	5.7396	3.0186	1.9014	0.6055
2.30	0.5344	6.0050	3.0845	1.9468	0.5833
2.35	0.5286	6.2762	3.1490	1.9931	0.5615
2.40	0.5231	6.5533	3.2119	2.0403	0.5401
2.45	0.5179	6.8362	3.2734	2.0885	0.5193
2.50	0.5130	7.1250	3.3333	2.1375	0.4990
2.55	0.5083	7.4196	3.3919	2.1875	0.4793
2.60	0.5039	7.7200	3.4490	2.2383	0.4601
2.65	0.4996	8.0262	3.5047	2.2901	0.4416
2.70	0.4956	8.3383	3.5590	2.3429	0.4236
2.75	0.4918	8.6563	3.6119	2.3966	0.4062
2.80	0.4882	8.9800	3.6635	2.4512	0.3895
2.85	0.4847	9.3096	3.7139	2.5067	0.3733
2.90	0.4814	9.6450	3.7629	2.5632	0.3577
2.95	0.4782	9.9862	3.8106	2.6206	0.3428
3.00	0.4752	1.0333+01	3.8571	2.6790	0.3283
3.05	0.4723	1.0686+01	3.9025	2.7383	0.3145
3.10	0.4695	1.1045+01	3.9466	2.7986	0.3012
3.15	0.4669	1.1410+01	3.9896	2.8598	0.2885
3.20	0.4643	1.1780+01	4.0315	2.9220	0.2762
3.25	0.4619	1.2156+01	4.0723	2.9851	0.2645
3.30	0.4596	1.2538+01	4.1120	3.0492	0.2533
3.35	0.4573	1.2926+01	4.1507	3.1142	0.2425
3.40	0.4552	1.3320+01	4.1884	3.1802	0.2322
3.45	0.4531	1.3720+01	4.2251	3.2471	0.2224
3.50	0.4512	1.4125+01	4.2609	3.3150	0.2129
3.55	0.4492	1.4536+01	4.2957	3.3839	0.2039
3.60	0.4474	1.4953+01	4.3296	3.4537	0.1953
3.65	0.4456	1.5376+01	4.3627	3.5245	0.1871
3.70	0.4439	1.5805+01	4.3949	3.5962	0.1792
3.75	0.4423	1.6240+01	4.4262	3.6689	0.1717
3.80	0.4407	1.6680+01	4.4568	3.7426	0.1645
3.85	0.4392	1.7126+01	4.4866	3.8172	0.1576
3.90	0.4377	1.7578+01	4.5156	3.8928	0.1510
3.95	0.4363	1.8036+01	4.5439	3.9694	0.1448

TABLE B-1 (continued)

M_1	M_2	$\frac{P_2}{P_1}$	$\frac{\rho_2}{\rho_1}$	$\frac{T_2}{T_1}$	$\frac{P_{02}}{P_{01}}$
4.00	0.4350	1.8500+01	4.5714	4.0469	0.1388
4.05	0.4336	1.8970+01	4.5983	4.1253	0.1330
4.10	0.4324	1.9445+01	4.6245	4.2048	0.1276
4.15	0.4311	1.9926+01	4.6500	4.2852	0.1223
4.20	0.4299	2.0413+01	4.6749	4.3666	0.1173
4.25	0.4288	2.0906+01	4.6992	4.4489	0.1126
4.30	0.4277	2.1405+01	4.7229	4.5322	0.1080
4.35	0.4266	2.1910+01	4.7460	4.6165	0.1036
4.40	0.4255	2.2420+01	4.7685	4.7017	9.9481-02
4.45	0.4245	2.2936+01	4.7904	4.7879	9.5501-02
4.50	0.4236	2.3458+01	4.8119	4.8751	9.1698-02
4.55	0.4226	2.3986+01	4.8328	4.9632	8.8062-02
4.60	0.4217	2.4520+01	4.8532	5.0523	8.4587-02
4.65	0.4208	2.5060+01	4.8731	5.1424	8.1263-02
4.70	0.4199	2.5605+01	4.8926	5.2334	7.8086-02
4.75	0.4191	2.6156+01	4.9116	5.3254	7.5047-02
4.80	0.4183	2.6713+01	4.9301	5.4184	7.2140-02
4.85	0.4175	2.7276+01	4.9482	5.5124	6.9359-02
4.90	0.4167	2.7845+01	4.9659	5.6073	6.6699-02
4.95	0.4160	2.8420+01	4.9831	5.7031	6.4153-02
5.00	0.4152	2.9000+01	5.0000	5.8000	6.1716-02
5.10	0.4138	3.0178+01	5.0326	5.9966	5.7151-02
5.20	0.4125	3.1380+01	5.0637	6.1971	5.2966-02
5.30	0.4113	3.2605+01	5.0934	6.4014	4.9126-02
5.40	0.4101	3.3853+01	5.1218	6.6097	4.5600-02
5.50	0.4090	3.5125+01	5.1489	6.8218	4.2361-02
5.60	0.4079	3.6420+01	5.1749	7.0378	3.9383-02
5.70	0.4069	3.7738+01	5.1998	7.2577	3.6643-02
5.80	0.4059	3.9080+01	5.2236	7.4814	3.4120-02
5.90	0.4050	4.0445+01	5.2464	7.7091	3.1795-02
6.00	0.4042	4.1833+01	5.2683	7.9406	2.9651-02
6.10	0.4033	4.3245+01	5.2893	8.1760	2.7672-02
6.20	0.4025	4.4680+01	5.3094	8.4153	2.5845-02
6.30	0.4018	4.6138+01	5.3287	8.6585	2.4156-02
6.40	0.4011	4.7620+01	5.3473	8.9055	2.2594-02
6.50	0.4004	4.9125+01	5.3651	9.1564	2.1148-02
6.60	0.3997	5.0653+01	5.3822	9.4113	1.9808-02
6.70	0.3991	5.2205+01	5.3987	9.6700	1.8566-02
6.80	0.3985	5.3780+01	5.4145	9.9326	1.7414-02
6.90	0.3979	5.5378+01	5.4298	1.0199+01	1.6345-02
7.00	0.3974	5.7000+01	5.4444	1.0469+01	1.5352-02
7.10	0.3968	5.8645+01	5.4586	1.0744+01	1.4428-02
7.20	0.3963	6.0313+01	5.4722	1.1022+01	1.3569-02
7.30	0.3958	6.2005+01	5.4853	1.1304+01	1.2769-02
7.40	0.3954	6.3720+01	5.4980	1.1590+01	1.2023-02
7.50	0.3949	6.5458+01	5.5102	1.1880+01	1.1329-02
7.60	0.3945	6.7220+01	5.5220	1.2173+01	1.0680-02
7.70	0.3941	6.9005+01	5.5334	1.2471+01	1.0075-02
7.80	0.3937	7.0813+01	5.5444	1.2772+01	9.5102-03
7.90	0.3933	7.2645+01	5.5550	1.3078+01	8.9819-03
8.00	0.3929	7.4500+01	5.5652	1.3387+01	8.4878-03
8.10	0.3925	7.6378+01	5.5751	1.3700+01	8.0254-03
8.20	0.3922	7.8280+01	5.5847	1.4017+01	7.5924-03
8.30	0.3918	8.0205+01	5.5940	1.4338+01	7.1866-03
8.40	0.3915	8.2153+01	5.6030	1.4663+01	6.8061-03
8.50	0.3912	8.4125+01	5.6116	1.4991+01	6.4492-03
8.60	0.3909	8.6120+01	5.6201	1.5324+01	6.1141-03
8.70	0.3906	8.8138+01	5.6282	1.5660+01	5.7994-03
8.80	0.3903	9.0180+01	5.6361	1.6000+01	5.5036-03
8.90	0.3901	9.2245+01	5.6438	1.6345+01	5.2255-03

TABLE B-1 (concluded)

M_1	M_2	$\frac{P_2}{P_1}$	$\frac{\rho_2}{\rho_1}$	$\frac{T_2}{T_1}$	$\frac{P_{02}}{P_{01}}$
9.00	0.3898	9.4333+01	5.6512	1.6693+01	4.9639-03
9.10	0.3895	9.6445+01	5.6583	1.7045+01	4.7175-03
9.20	0.3893	9.8580+01	5.6653	1.7401+01	4.4855-03
9.30	0.3891	1.0074+02	5.6721	1.7760+01	4.2669-03
9.40	0.3888	1.0292+02	5.6787	1.8124+01	4.0608-03
9.50	0.3886	1.0513+02	5.6850	1.8492+01	3.8664-03
9.60	0.3884	1.0735+02	5.6912	1.8863+01	3.6828-03
9.70	0.3882	1.0961+02	5.6972	1.9238+01	3.5095-03
9.80	0.3880	1.1188+02	5.7031	1.9617+01	3.3458-03
9.90	0.3878	1.1418+02	5.7088	2.0000+01	3.1911-03
10.00	0.3876	1.1650+02	5.7143	2.0388+01	3.0448-03
11.00	0.3859	1.4100+02	5.7619	2.4471+01	1.9451-03
12.00	0.3847	1.6783+02	5.7987	2.8944+01	1.2866-03
13.00	0.3837	1.9700+02	5.8276	3.3805+01	8.7709-04
14.00	0.3829	2.2850+02	5.8507	3.9055+01	6.1380-04
15.00	0.3823	2.6233+02	5.8696	4.4694+01	4.3953-04
16.00	0.3817	2.9850+02	5.8851	5.0722+01	3.2119-04
17.00	0.3813	3.3700+02	5.8980	5.7138+01	2.3899-04
18.00	0.3810	3.7783+02	5.9088	6.3944+01	1.8072-04
19.00	0.3806	4.2100+02	5.9180	7.1138+01	1.3865-04
20.00	0.3804	4.6650+02	5.9259	7.8722+01	1.0777-04
21.00	0.3802	5.1433+02	5.9327	8.6694+01	8.4778-05
22.00	0.3800	5.6450+02	5.9387	9.5055+01	6.7414-05
23.00	0.3798	6.1700+02	5.9438	1.0381+02	5.4140-05
24.00	0.3796	6.7183+02	5.9484	1.1294+02	4.3877-05
25.00	0.3795	7.2900+02	5.9524	1.2247+02	3.5859-05
26.00	0.3794	7.8850+02	5.9559	1.3239+02	2.9534-05
27.00	0.3793	8.5033+02	5.9591	1.4269+02	2.4500-05
28.00	0.3792	9.1450+02	5.9620	1.5339+02	2.0460-05
29.00	0.3791	9.8100+02	5.9645	1.6447+02	1.7193-05
30.00	0.3790	1.0498+03	5.9668	1.7594+02	1.4531-05
31.00	0.3790	1.1210+03	5.9689	1.8780+02	1.2349-05
32.00	0.3789	1.1945+03	5.9708	2.0005+02	1.0548-05
33.00	0.3789	1.2703+03	5.9726	2.1269+02	9.0526-06
34.00	0.3788	1.3485+03	5.9742	2.2572+02	7.8044-06
35.00	0.3788	1.4290+03	5.9756	2.3914+02	6.7570-06
36.00	0.3787	1.5118+03	5.9769	2.5294+02	5.8738-06
37.00	0.3787	1.5970+03	5.9782	2.6714+02	5.1254-06
38.00	0.3786	1.6845+03	5.9793	2.8172+02	4.4885-06
39.00	0.3786	1.7743+03	5.9803	2.9669+02	3.9441-06
40.00	0.3786	1.8665+03	5.9813	3.1205+02	3.4771-06
41.00	0.3785	1.9610+03	5.9822	3.2780+02	3.0748-06
42.00	0.3785	2.0578+03	5.9830	3.4394+02	2.7271-06
43.00	0.3785	2.1570+03	5.9838	3.6047+02	2.4255-06
44.00	0.3785	2.2585+03	5.9845	3.7739+02	2.1630-06
45.00	0.3784	2.3623+03	5.9852	3.9469+02	1.9339-06
46.00	0.3784	2.4685+03	5.9859	4.1239+02	1.7332-06
47.00	0.3784	2.5770+03	5.9865	4.3047+02	1.5571-06
48.00	0.3784	2.6878+03	5.9870	4.4894+02	1.4019-06
49.00	0.3784	2.8010+03	5.9875	4.6780+02	1.2650-06
50.00	0.3784	2.9165+03	5.9880	4.8705+02	1.1438-06

Oblique and Conical Shock Waves

APPENDIX C

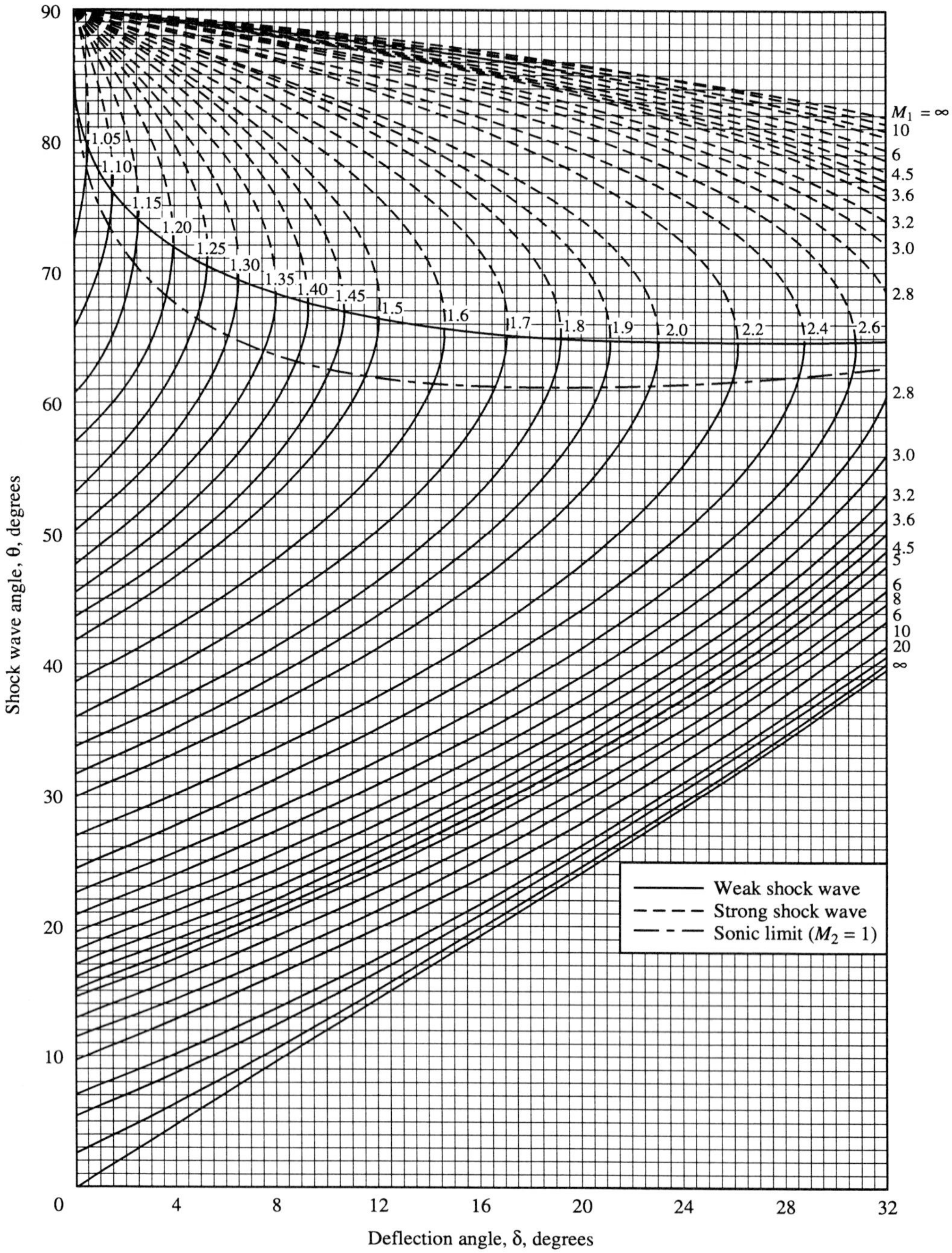

Figure C-1 Variation of Shock Wave Angle with Flow-Deflection Angle for Various Upstream Mach Numbers

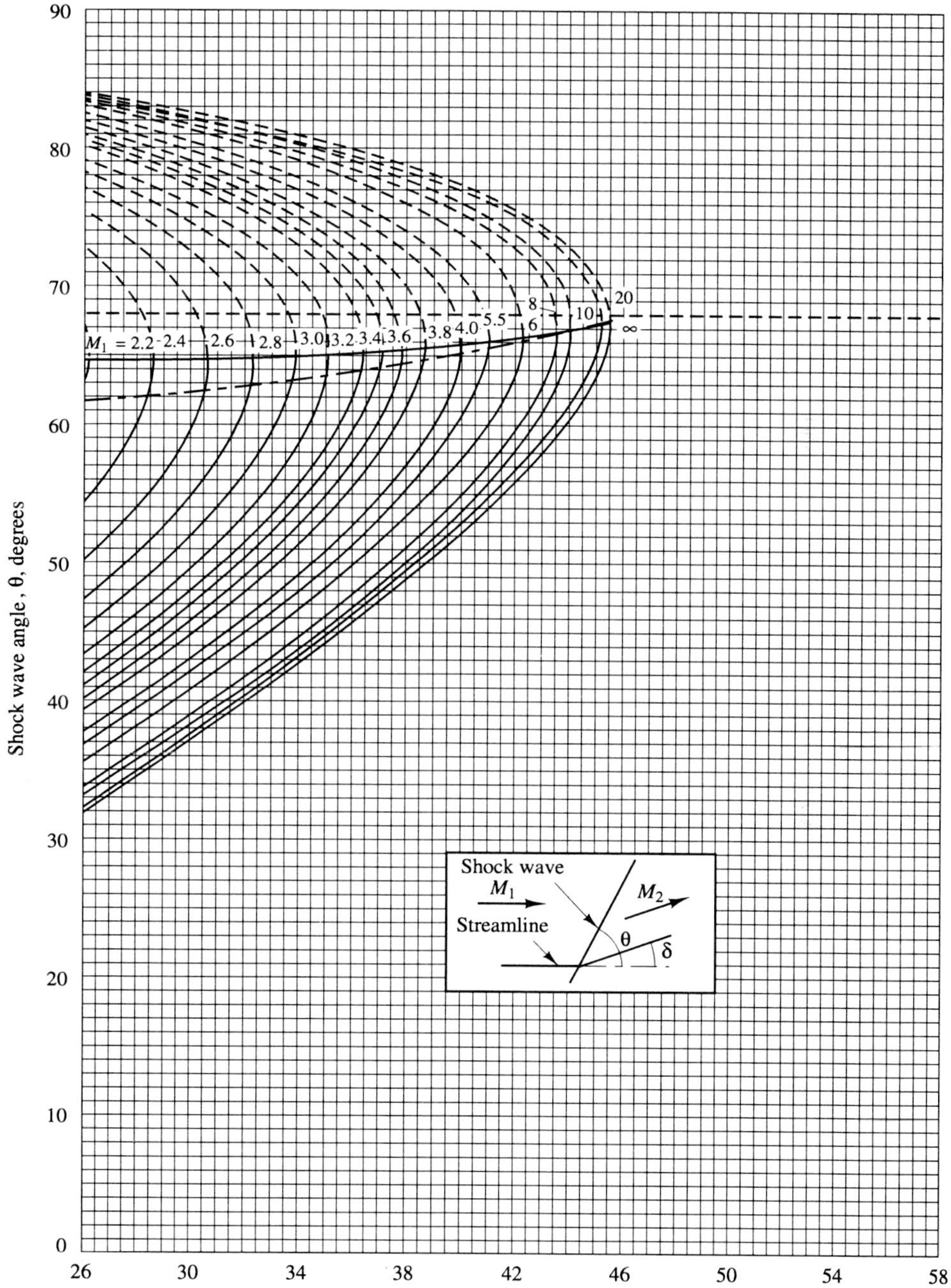

Figure C-1 (continued)

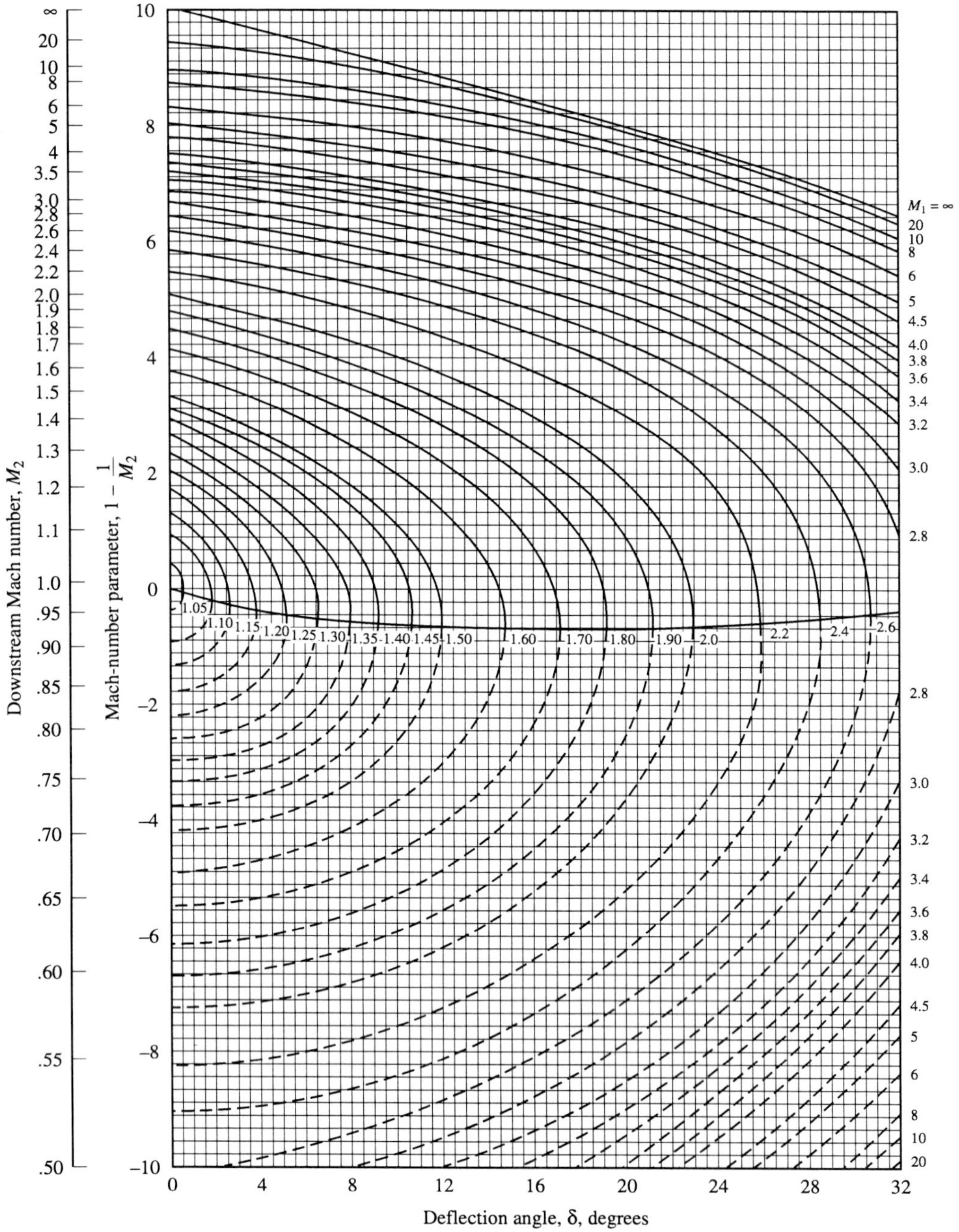

Figure C-2 Variation of Mach Number Downstream of a Shock Wave with Flow-Deflection Angle for Various Upstream Mach Numbers

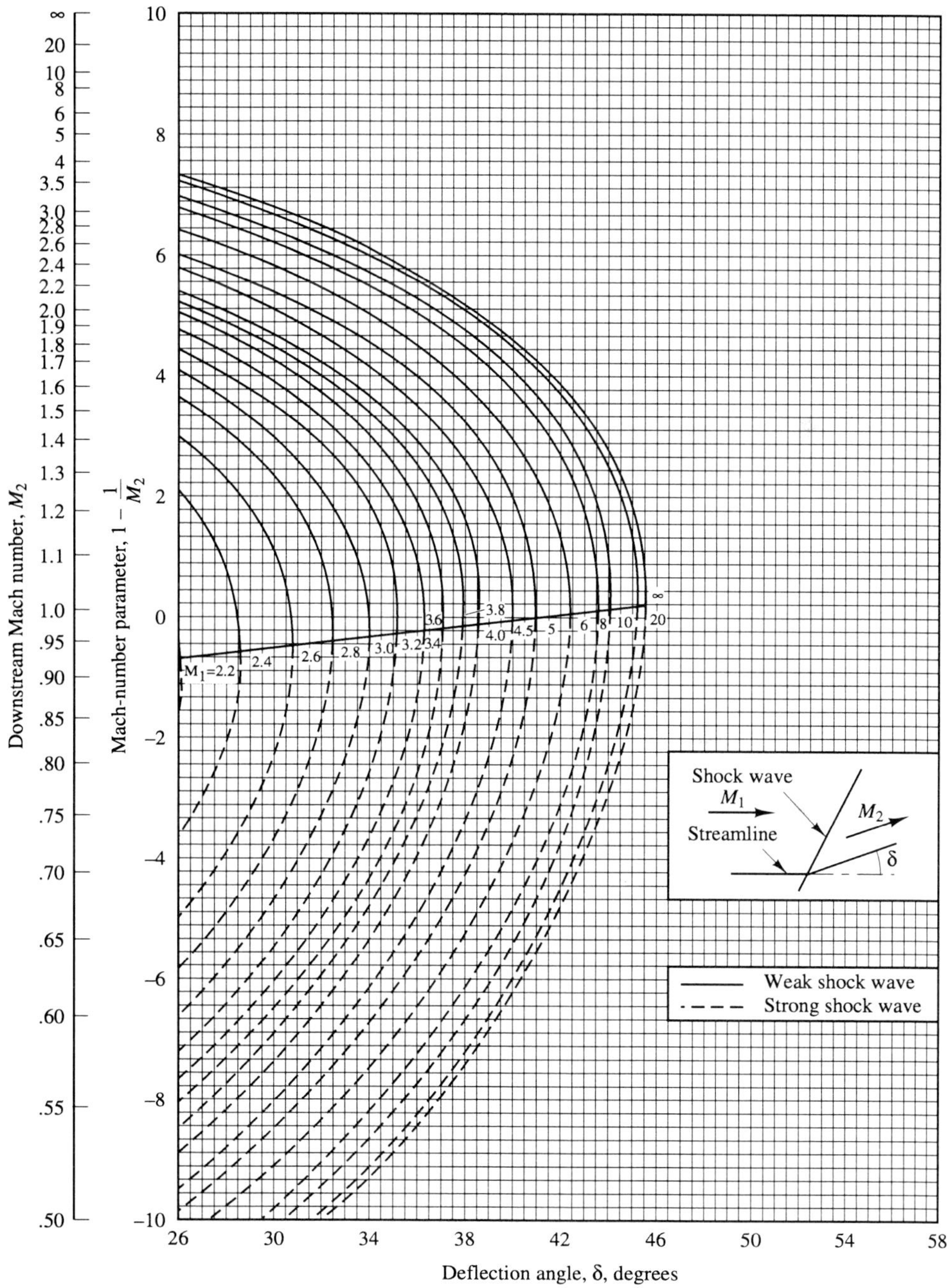

Figure C-2 (continued)

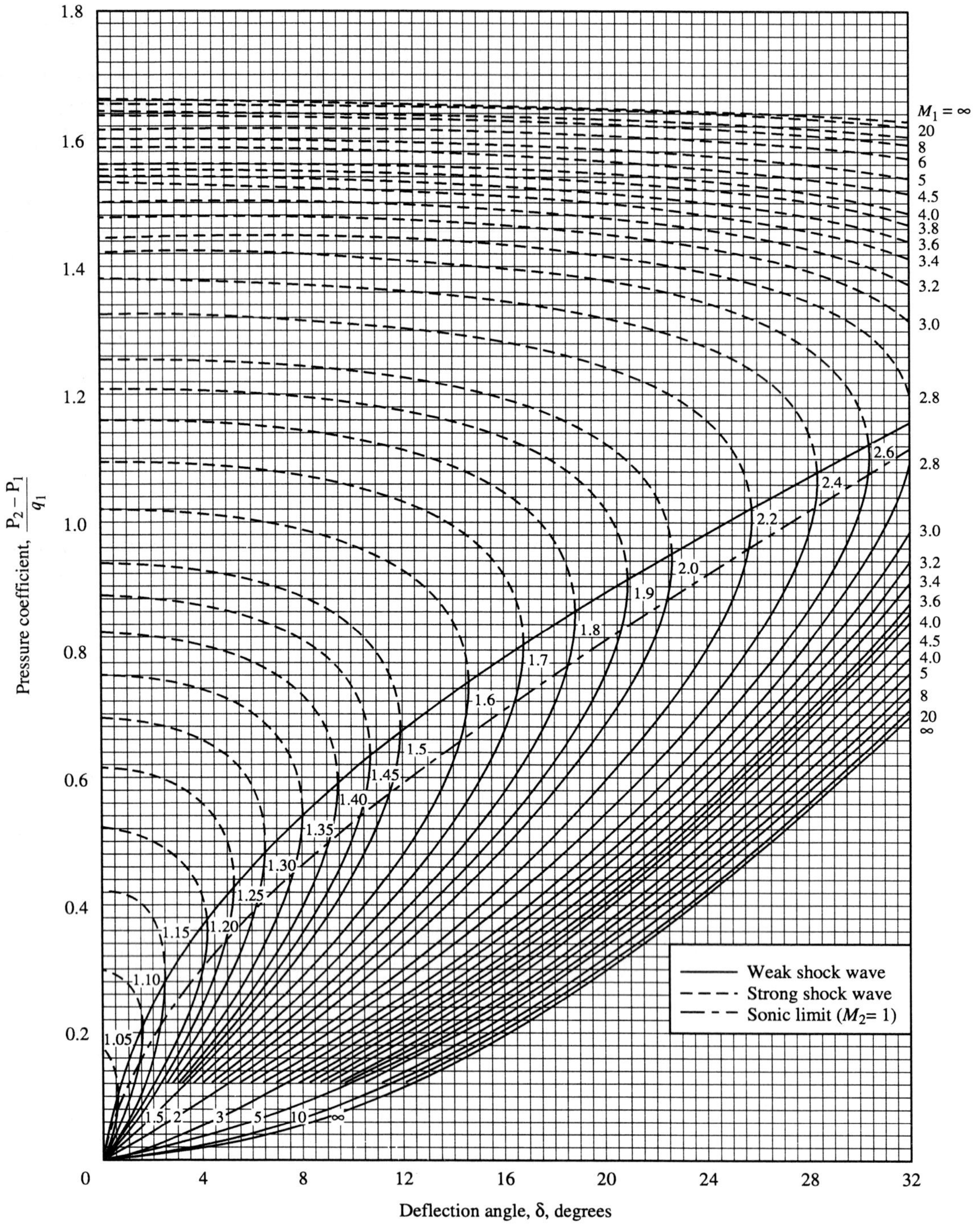

Figure C-3 Variation of Pressure Coefficient across Shock Waves with Flow-Deflection Angle for Various Upstream Mach Numbers. Perfect gas, $\gamma = 1.4$.

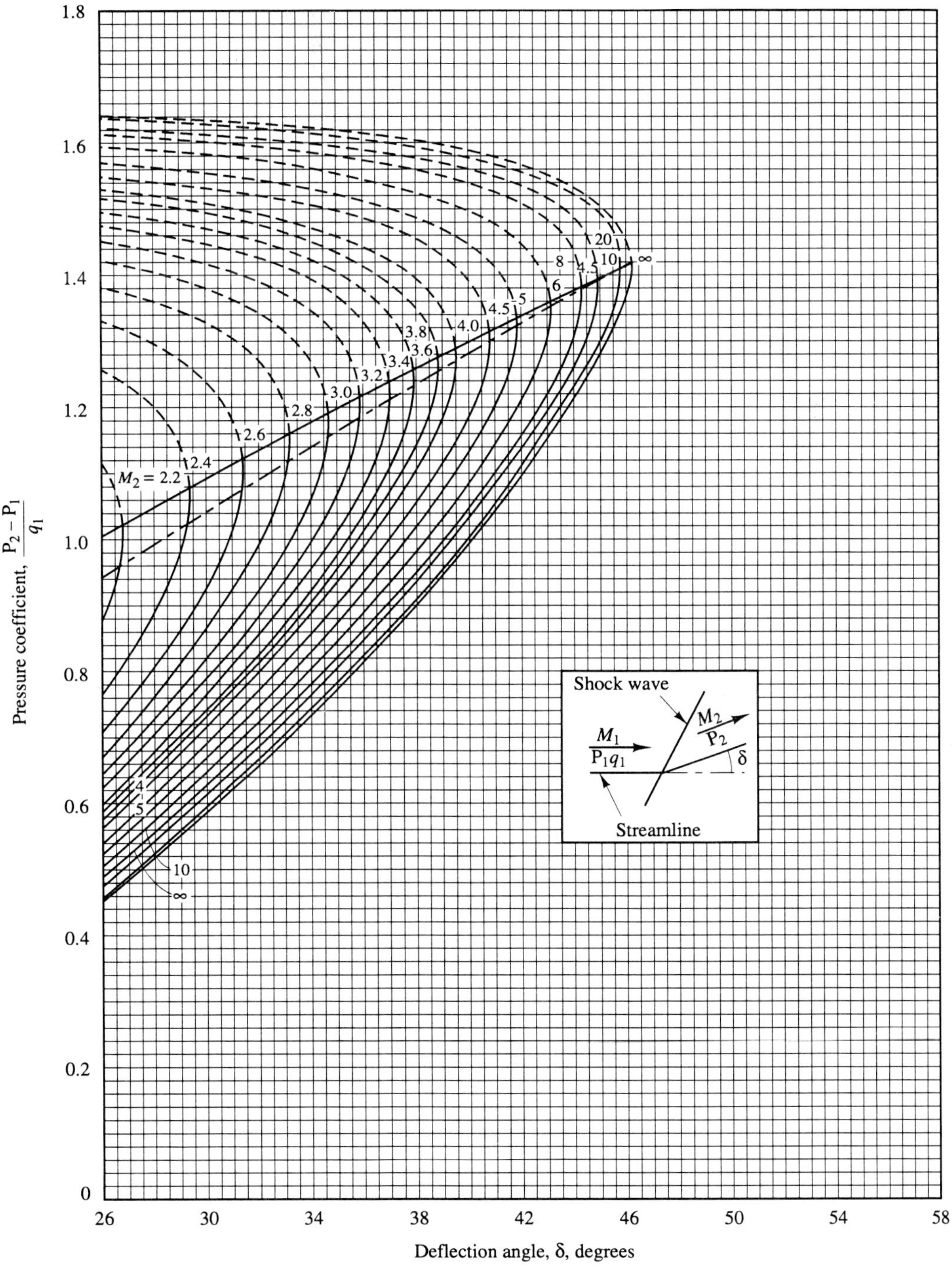

Figure C-3 (continued)

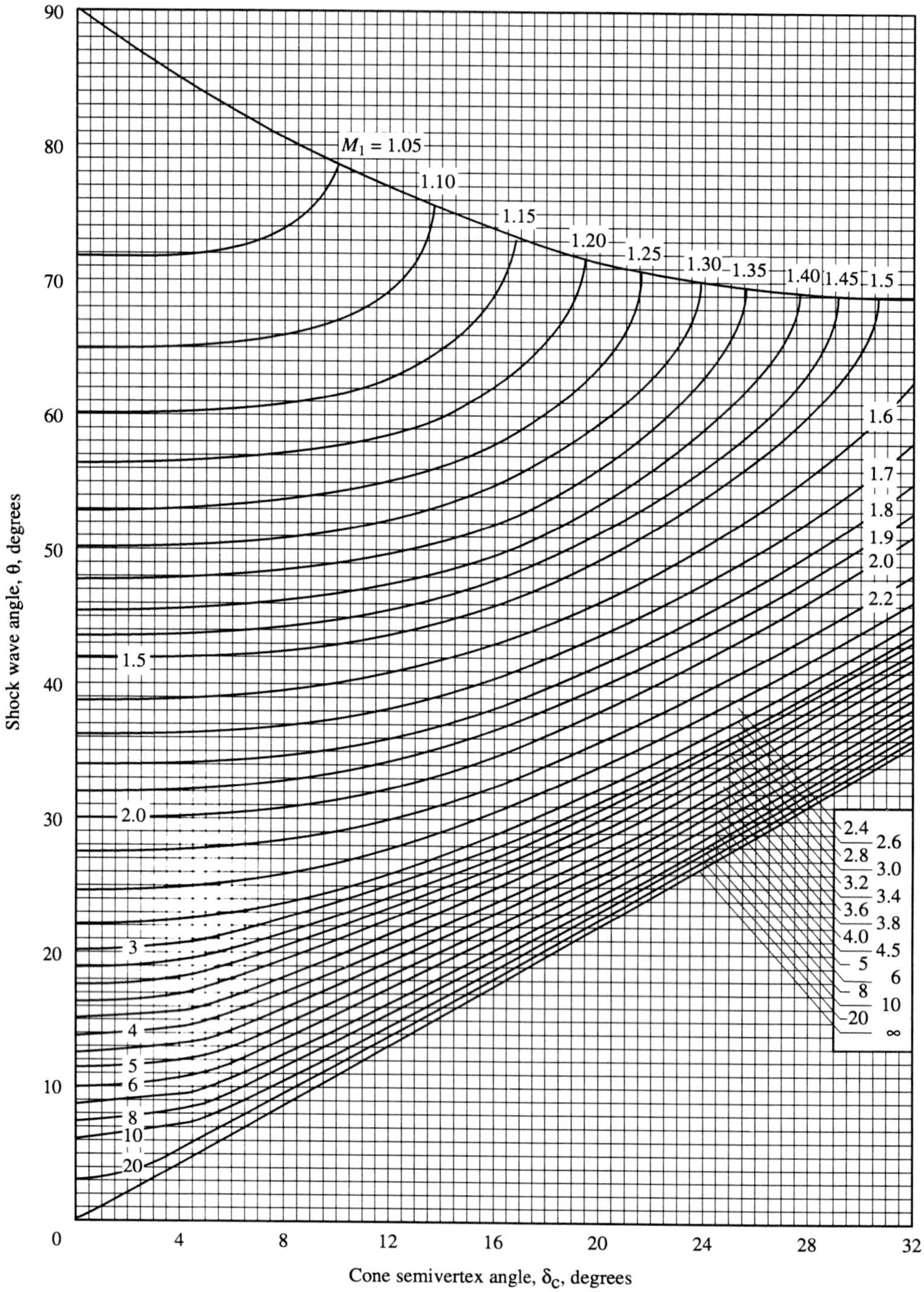

Figure C-4 Conical Shock Charts ($\gamma = 1.405$)

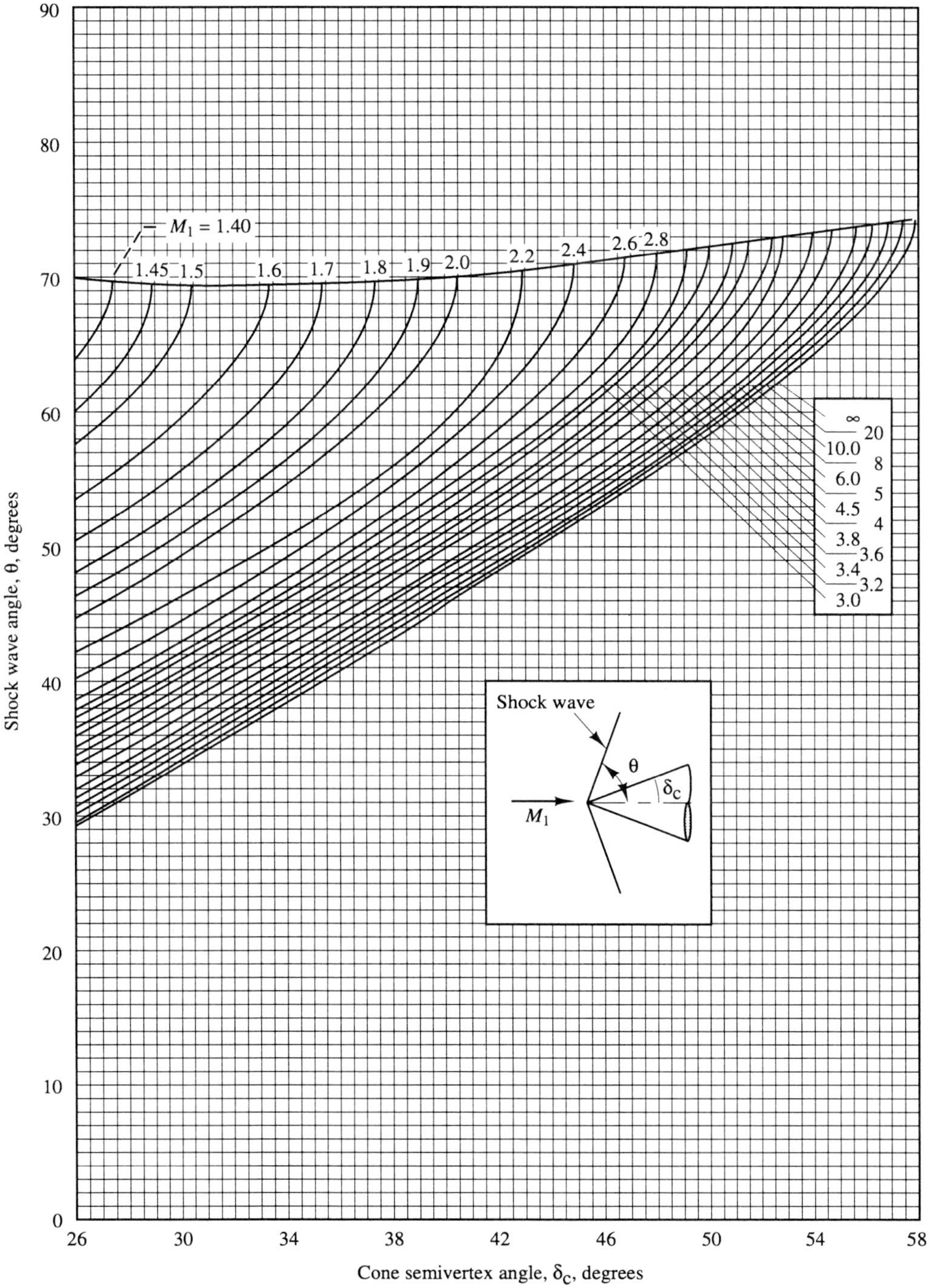

Figure C-4 (continued)

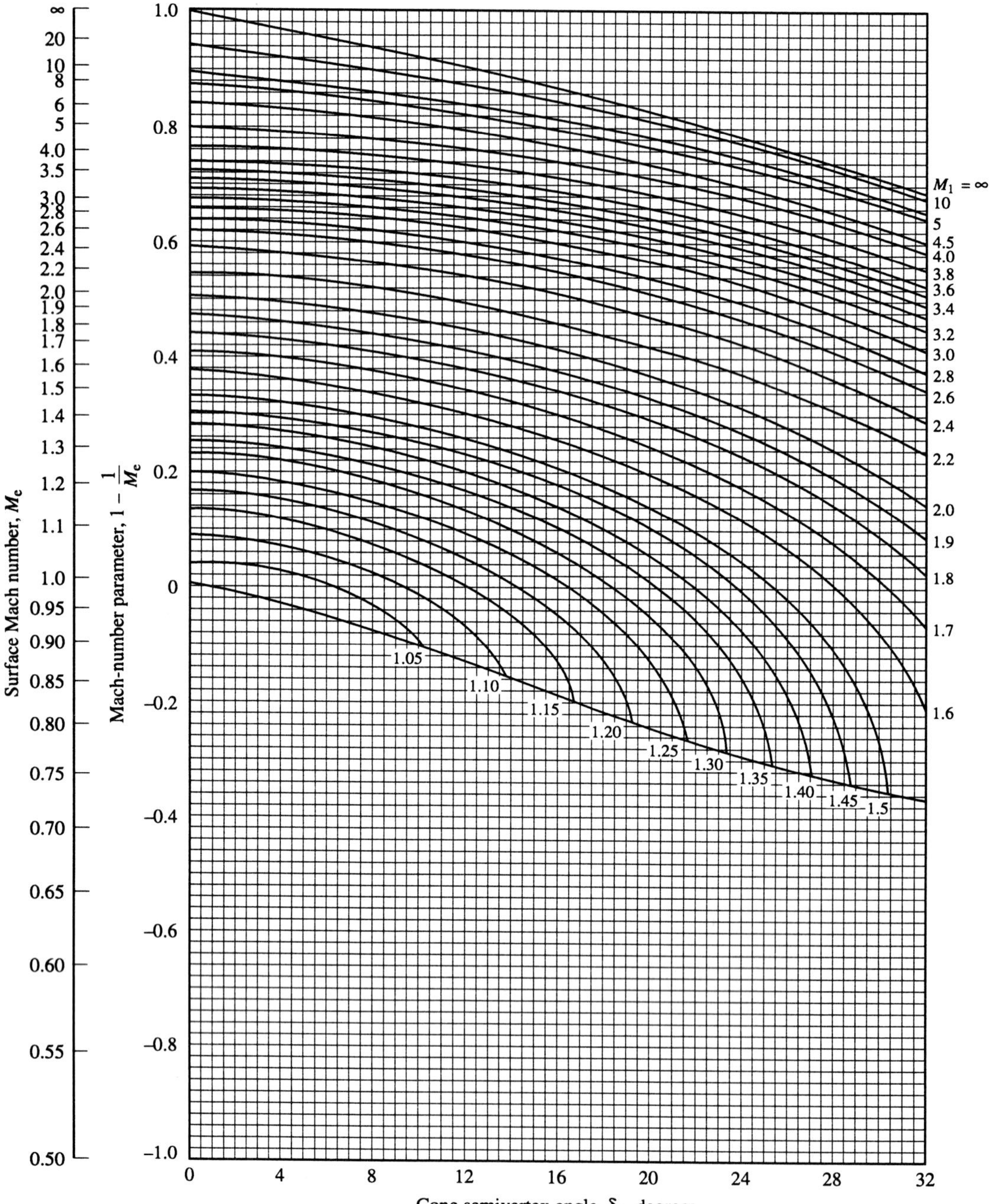

Figure C-5 Conical Shock Charts ($\gamma = 1.405$)

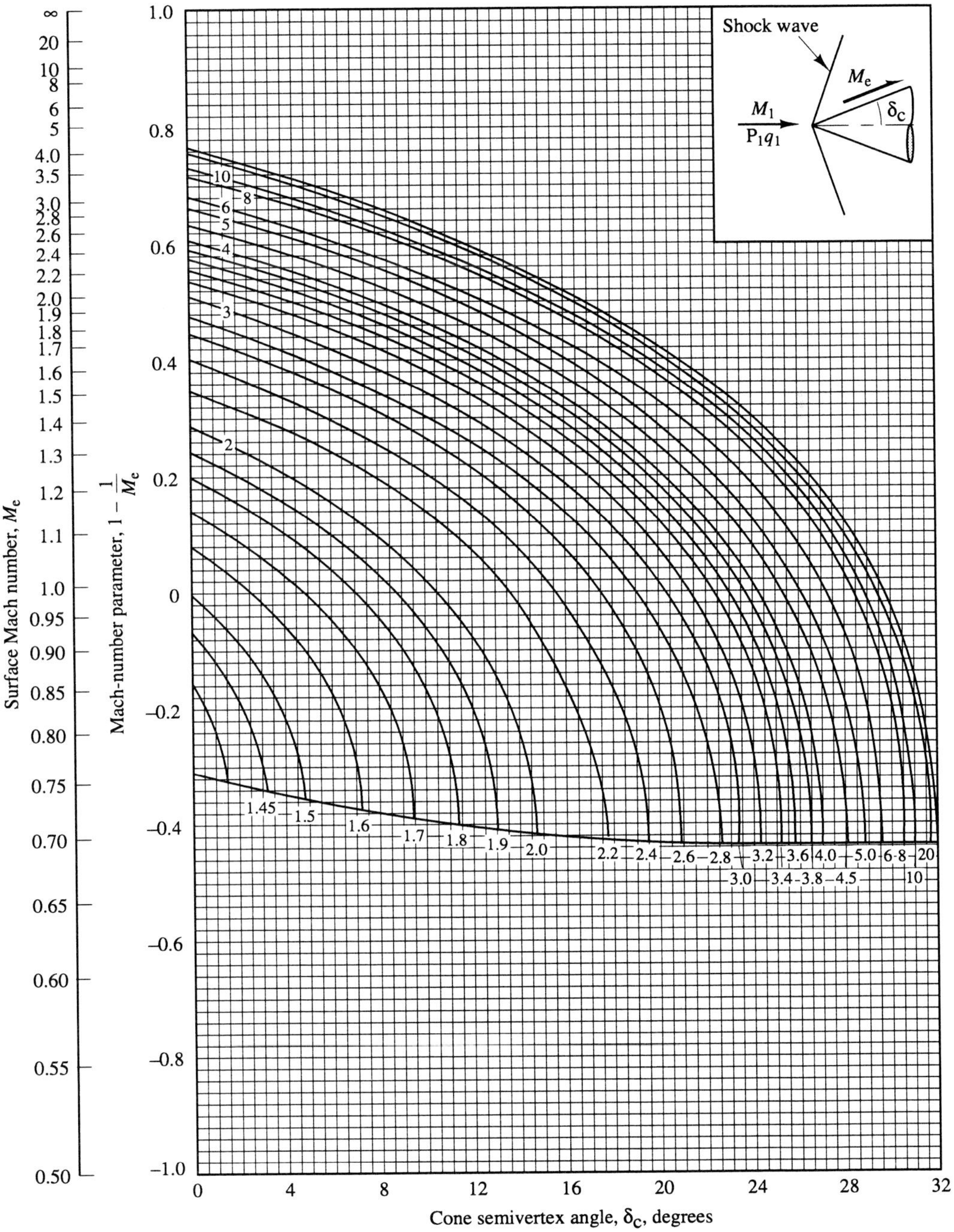

Figure C-5 (continued)

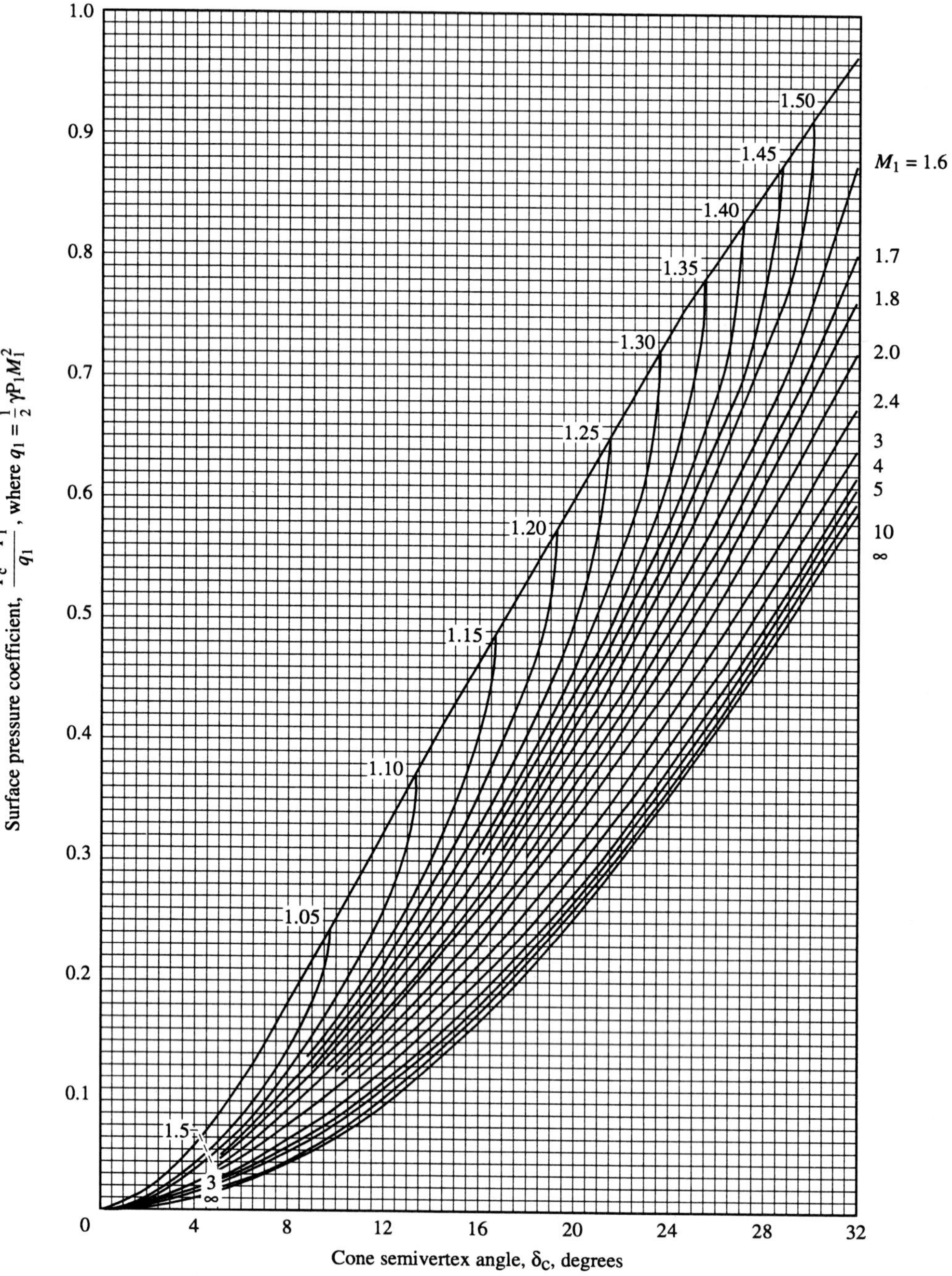

Figure C-6 Conical Shock Charts (γ = 1.405)

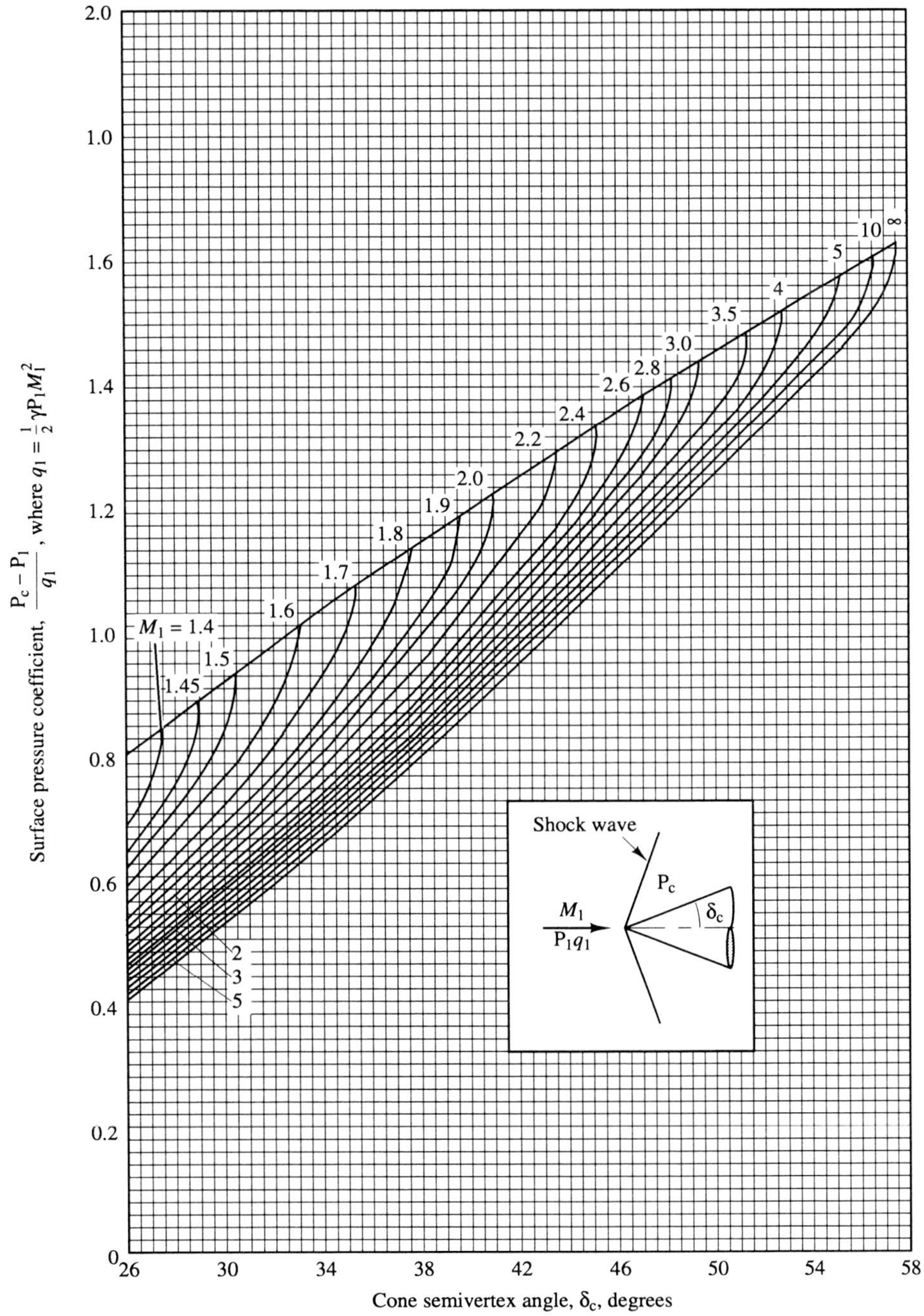

Figure C-6 (continued)

Prandtl-Meyer Flow

APPENDIX D

TABLE D-1 PRANDTL-MEYER FUNCTIONS (PERFECT GAS, $\gamma = 1.4$)*

M	ν	μ	M	ν	μ
1.00	0.0000	9.0000+01	1.60	1.4860+01	3.8682+01
1.02	0.1257	7.8635+01	1.62	1.5452+01	3.8118+01
1.04	0.3510	7.4058+01	1.64	1.6043+01	3.7572+01
1.06	0.6367	7.0630+01	1.66	1.6633+01	3.7043+01
1.08	0.9680	6.7808+01	1.68	1.7222+01	3.6530+01
1.10	1.3362	6.5380+01	1.70	1.7810+01	3.6032+01
1.12	1.7350	6.3235+01	1.72	1.8396+01	3.5549+01
1.14	2.1600	6.1306+01	1.74	1.8981+01	3.5080+01
1.16	2.6073	5.9550+01	1.76	1.9565+01	3.4624+01
1.18	3.0743	5.7936+01	1.78	2.0146+01	3.4180+01
1.20	3.5582	5.6443+01	1.80	2.0725+01	3.3749+01
1.22	4.0572	5.5052+01	1.82	2.1302+01	3.3329+01
1.24	4.5694	5.3751+01	1.84	2.1877+01	3.2921+01
1.26	5.0931	5.2528+01	1.86	2.2449+01	3.2523+01
1.28	5.6272	5.1375+01	1.88	2.3019+01	3.2135+01
1.30	6.1703	5.0285+01	1.90	2.3586+01	3.1757+01
1.32	6.7213	4.9251+01	1.92	2.4151+01	3.1388+01
1.34	7.2794	4.8268+01	1.94	2.4712+01	3.1028+01
1.36	7.8435	4.7332+01	1.96	2.5271+01	3.0677+01
1.38	8.4130	4.6439+01	1.98	2.5827+01	3.0335+01
1.40	8.9870	4.5585+01	2.00	2.6380+01	3.0000+01
1.42	9.5650	4.4767+01	2.05	2.7748+01	2.9196+01
1.44	1.0146+01	4.3983+01	2.10	2.9097+01	2.8437+01
1.46	1.0731+01	4.3230+01	2.15	3.0425+01	2.7718+01
1.48	1.1317+01	4.2507+01	2.20	3.1733+01	2.7036+01
1.50	1.1905+01	4.1810+01	2.25	3.3018+01	2.6388+01
1.52	1.2495+01	4.1140+01	2.30	3.4283+01	2.5771+01
1.54	1.3086+01	4.0493+01	2.35	3.5526+01	2.5184+01
1.56	1.3677+01	3.9868+01	2.40	3.6747+01	2.4624+01
1.58	1.4269+01	3.9265+01	2.45	3.7946+01	2.4090+01

TABLE D-1 (concluded)

M	ν	μ
2.50	3.9124+01	2.3578+01
2.55	4.0280+01	2.3089+01
2.60	4.1415+01	2.2620+01
2.65	4.2529+01	2.2170+01
2.70	4.3621+01	2.1738+01
2.75	4.4694+01	2.1324+01
2.80	4.5746+01	2.0925+01
2.85	4.6778+01	2.0541+01
2.90	4.7790+01	2.0171+01
2.95	4.8783+01	1.9815+01
3.00	4.9757+01	1.9471+01
3.05	5.0713+01	1.9139+01
3.10	5.1650+01	1.8819+01
3.15	5.2569+01	1.8509+01
3.20	5.3470+01	1.8210+01
3.25	5.4355+01	1.7920+01
3.30	5.5222+01	1.7640+01
3.35	5.6073+01	1.7368+01
3.40	5.6908+01	1.7105+01
3.45	5.7726+01	1.6849+01
3.50	5.8530+01	1.6602+01
3.55	5.9318+01	1.6361+01
3.60	6.0091+01	1.6128+01
3.65	6.0850+01	1.5901+01
3.70	6.1595+01	1.5680+01
3.75	6.2326+01	1.5466+01
3.80	6.3044+01	1.5258+01
3.85	6.3748+01	1.5055+01
3.90	6.4440+01	1.4857+01
3.95	6.5118+01	1.4665+01
4.00	6.5785+01	1.4478+01
4.05	6.6439+01	1.4295+01
4.10	6.7082+01	1.4117+01
4.15	6.7713+01	1.3943+01
4.20	6.8333+01	1.3774+01
4.25	6.8942+01	1.3609+01
4.30	6.9541+01	1.3448+01
4.35	7.0129+01	1.3290+01
4.40	7.0706+01	1.3137+01
4.45	7.1274+01	1.2986+01
4.50	7.1832+01	1.2840+01
4.55	7.2380+01	1.2696+01
4.60	7.2919+01	1.2556+01
4.65	7.3449+01	1.2419+01
4.70	7.3970+01	1.2284+01
4.75	7.4482+01	1.2153+01
4.80	7.4986+01	1.2025+01
4.85	7.5482+01	1.1899+01
4.90	7.5969+01	1.1776+01
4.95	7.6449+01	1.1655+01
5.00	7.6920+01	1.1537+01
5.10	7.7841+01	1.1308+01
5.20	7.8732+01	1.1087+01
5.30	7.9596+01	1.0876+01
5.40	8.0433+01	1.0672+01
5.50	8.1245+01	1.0476+01
5.60	8.2032+01	1.0287+01
5.70	8.2796+01	1.0104+01
5.80	8.3537+01	9.9282
5.90	8.4256+01	9.7583
6.00	8.4955+01	9.5941
6.10	8.5635+01	9.4353
6.20	8.6295+01	9.2818
6.30	8.6937+01	9.1332
6.40	8.7561+01	8.9893
6.50	8.8168+01	8.8499
6.60	8.8759+01	8.7147
6.70	8.9335+01	8.5837
6.80	8.9895+01	8.4565
6.90	9.0441+01	8.3331
7.00	9.0973+01	8.2132
7.10	9.1491+01	8.0968
7.20	9.1997+01	7.9836
7.30	9.2490+01	7.8735
7.40	9.2970+01	7.7664
7.50	9.3440+01	7.6623
7.60	9.3898+01	7.5608
7.70	9.4345+01	7.4621
7.80	9.4781+01	7.3659
7.90	9.5208+01	7.2721
8.00	9.5625+01	7.1808
9.00	9.9318+01	6.3794
10.00	1.0232+02	5.7392
11.00	1.0480+02	5.2159
12.00	1.0688+02	4.7802
13.00	1.0865+02	4.4117
14.00	1.1018+02	4.0960
15.00	1.1151+02	3.8226
16.00	1.1268+02	3.5833
17.00	1.1371+02	3.3723
18.00	1.1463+02	3.1847
19.00	1.1545+02	3.0170
20.00	1.1620+02	2.8660
22.00	1.1748+02	2.6053
24.00	1.1856+02	2.3880
26.00	1.1947+02	2.2042
28.00	1.2025+02	2.0467
30.00	1.2092+02	1.9102
32.00	1.2152+02	1.7908
34.00	1.2204+02	1.6854
36.00	1.2251+02	1.5918
38.00	1.2292+02	1.5080
40.00	1.2330+02	1.4325
42.00	1.2364+02	1.3643
44.00	1.2395+02	1.3023
46.00	1.2423+02	1.2457
48.00	1.2449+02	1.1937
50.00	1.2473+02	1.1460

Fanno Flow

APPENDIX E

TABLE E-1 FANNO LINE FLOW (PERFECT GAS, $\gamma = 1.4$)

M	$\frac{p}{p^*}$	$\frac{\rho}{\rho^*}$	$\frac{T}{T^*}$	$\frac{p_0}{p_0^*}$	$\frac{4fL^*}{D}$
0.02	5.4770+01	4.5645+01	1.1999	2.8942+01	1.7784+03
0.04	2.7382+01	2.2825+01	1.1996	1.4481+01	4.4035+02
0.06	1.8251+01	1.5220+01	1.1991	9.6659	1.9303+02
0.08	1.3684+01	1.1418+01	1.1985	7.2616	1.0672+02
0.10	1.0943+01	9.1378	1.1976	5.8218	6.6922+01
0.12	9.1156	7.6182	1.1966	4.8643	4.5408+01
0.14	7.8093	6.5333	1.1953	4.1824	3.2511+01
0.16	6.8291	5.7200	1.1939	3.6727	2.4198+01
0.18	6.0662	5.0879	1.1923	3.2779	1.8543+01
0.20	5.4555	4.5826	1.1905	2.9635	1.4533+01
0.22	4.9554	4.1694	1.1885	2.7076	1.1596+01
0.24	4.5383	3.8255	1.1863	2.4956	9.3865
0.26	4.1851	3.5347	1.1840	2.3173	7.6876
0.28	3.8820	3.2857	1.1815	2.1655	6.3572
0.30	3.6191	3.0702	1.1788	2.0351	5.2993
0.32	3.3887	2.8818	1.1759	1.9218	4.4467
0.34	3.1853	2.7158	1.1729	1.8229	3.7520
0.36	3.0042	2.5684	1.1697	1.7358	3.1801
0.38	2.8420	2.4367	1.1663	1.6587	2.7055
0.40	2.6958	2.3184	1.1628	1.5901	2.3085
0.42	2.5634	2.2115	1.1591	1.5289	1.9744
0.44	2.4428	2.1145	1.1553	1.4740	1.6915
0.46	2.3326	2.0261	1.1513	1.4246	1.4509
0.48	2.2313	1.9451	1.1471	1.3801	1.2453
0.50	2.1381	1.8708	1.1429	1.3398	1.0691
0.52	2.0519	1.8024	1.1384	1.3034	0.9174
0.54	1.9719	1.7391	1.1339	1.2703	0.7866
0.56	1.8975	1.6805	1.1292	1.2403	0.6736
0.58	1.8282	1.6260	1.1243	1.2130	0.5757
0.60	1.7634	1.5753	1.1194	1.1882	0.4908

TABLE E-1 (continued)

M	$\frac{p}{p^*}$	$\frac{\rho}{\rho^*}$	$\frac{T}{T^*}$	$\frac{p_0}{p_0^*}$	$\frac{4fL^*}{D}$
0.62	1.7026	1.5279	1.1143	1.1657	0.4172
0.64	1.6456	1.4836	1.1091	1.1451	0.3533
0.66	1.5919	1.4421	1.1038	1.1265	0.2979
0.68	1.5413	1.4032	1.0984	1.1097	0.2498
0.70	1.4935	1.3665	1.0929	1.0944	0.2081
0.72	1.4482	1.3320	1.0873	1.0806	0.1721
0.74	1.4054	1.2994	1.0816	1.0681	0.1411
0.76	1.3647	1.2686	1.0757	1.0570	0.1145
0.78	1.3261	1.2395	1.0698	1.0470	9.1672-02
0.80	1.2893	1.2119	1.0638	1.0382	7.2290-02
0.82	1.2542	1.1858	1.0577	1.0305	5.5932-02
0.84	1.2208	1.1609	1.0516	1.0237	4.2256-02
0.86	1.1889	1.1373	1.0454	1.0179	3.0965-02
0.88	1.1583	1.1148	1.0391	1.0129	2.1794-02
0.90	1.1291	1.0934	1.0327	1.0089	1.4512-02
0.92	1.1011	1.0730	1.0263	1.0056	8.9133-03
0.94	1.0743	1.0535	1.0198	1.0031	4.8154-03
0.96	1.0485	1.0348	1.0132	1.0014	2.0571-03
0.98	1.0238	1.0170	1.0066	1.0003	4.9469-04
1.00	1.0000	1.0000	1.0000	1.0000	0.0000
1.02	0.9771	0.9837	0.9933	1.0003	4.5869-04
1.04	0.9551	0.9681	0.9866	1.0013	1.7685-03
1.06	0.9338	0.9531	0.9798	1.0029	3.8378-03
1.08	0.9133	0.9387	0.9730	1.0051	6.5846-03
1.10	0.8936	0.9249	0.9662	1.0079	9.9350-03
1.12	0.8745	0.9116	0.9593	1.0113	1.3823-02
1.14	0.8561	0.8988	0.9524	1.0153	1.8188-02
1.16	0.8383	0.8865	0.9455	1.0198	2.2977-02
1.18	0.8210	0.8747	0.9386	1.0248	2.8142-02
1.20	0.8044	0.8633	0.9317	1.0304	3.3638-02
1.22	0.7882	0.8524	0.9247	1.0366	3.9426-02
1.24	0.7726	0.8418	0.9178	1.0432	4.5470-02
1.26	0.7574	0.8316	0.9108	1.0504	5.1739-02
1.28	0.7427	0.8218	0.9038	1.0581	5.8201-02
1.30	0.7285	0.8123	0.8969	1.0663	6.4832-02
1.32	0.7147	0.8031	0.8899	1.0750	7.1607-02
1.34	0.7012	0.7942	0.8829	1.0842	7.8503-02
1.36	0.6882	0.7856	0.8760	1.0940	8.5503-02
1.38	0.6755	0.7773	0.8690	1.1042	9.2586-02
1.40	0.6632	0.7693	0.8621	1.1149	9.9738-02
1.42	0.6512	0.7615	0.8551	1.1262	0.1069
1.44	0.6396	0.7540	0.8482	1.1379	0.1142
1.46	0.6282	0.7467	0.8413	1.1501	0.1215
1.48	0.6172	0.7397	0.8344	1.1629	0.1288
1.50	0.6065	0.7328	0.8276	1.1762	0.1361
1.52	0.5960	0.7262	0.8207	1.1899	0.1433
1.54	0.5858	0.7198	0.8139	1.2042	0.1506
1.56	0.5759	0.7135	0.8071	1.2190	0.1579
1.58	0.5662	0.7074	0.8004	1.2344	0.1651
1.60	0.5568	0.7016	0.7937	1.2502	0.1724

TABLE E-1 (continued)

M	$\frac{p}{p^*}$	$\frac{\rho}{\rho^*}$	$\frac{T}{T^*}$	$\frac{p_0}{p_0^*}$	$\frac{4fL^*}{D}$
1.62	0.5476	0.6958	0.7869	1.2666	0.1795
1.64	0.5386	0.6903	0.7803	1.2836	0.1867
1.66	0.5299	0.6849	0.7736	1.3010	0.1938
1.68	0.5213	0.6797	0.7670	1.3190	0.2008
1.70	0.5130	0.6745	0.7605	1.3376	0.2078
1.72	0.5048	0.6696	0.7539	1.3567	0.2147
1.74	0.4969	0.6648	0.7474	1.3764	0.2216
1.76	0.4891	0.6601	0.7410	1.3967	0.2284
1.78	0.4815	0.6555	0.7345	1.4175	0.2352
1.80	0.4741	0.6511	0.7282	1.4390	0.2419
1.82	0.4668	0.6467	0.7218	1.4610	0.2485
1.84	0.4597	0.6425	0.7155	1.4836	0.2551
1.86	0.4528	0.6384	0.7093	1.5069	0.2616
1.88	0.4460	0.6344	0.7030	1.5308	0.2680
1.90	0.4394	0.6305	0.6969	1.5553	0.2743
1.92	0.4329	0.6267	0.6907	1.5804	0.2806
1.94	0.4265	0.6230	0.6847	1.6062	0.2868
1.96	0.4203	0.6193	0.6786	1.6326	0.2929
1.98	0.4142	0.6158	0.6726	1.6597	0.2990
2.00	0.4082	0.6124	0.6667	1.6875	0.3050
2.05	0.3939	0.6041	0.6520	1.7600	0.3196
2.10	0.3802	0.5963	0.6376	1.8369	0.3339
2.15	0.3673	0.5890	0.6235	1.9185	0.3476
2.20	0.3549	0.5821	0.6098	2.0050	0.3609
2.25	0.3432	0.5756	0.5963	2.0964	0.3738
2.30	0.3320	0.5694	0.5831	2.1931	0.3862
2.35	0.3213	0.5635	0.5702	2.2953	0.3983
2.40	0.3111	0.5580	0.5576	2.4031	0.4099
2.45	0.3014	0.5527	0.5453	2.5168	0.4211
2.50	0.2921	0.5477	0.5333	2.6367	0.4320
2.55	0.2832	0.5430	0.5216	2.7630	0.4425
2.60	0.2747	0.5385	0.5102	2.8960	0.4526
2.65	0.2666	0.5342	0.4991	3.0359	0.4624
2.70	0.2588	0.5301	0.4882	3.1830	0.4718
2.75	0.2513	0.5262	0.4776	3.3377	0.4809
2.80	0.2441	0.5225	0.4673	3.5001	0.4898
2.85	0.2373	0.5189	0.4572	3.6707	0.4983
2.90	0.2307	0.5155	0.4474	3.8498	0.5065
2.95	0.2243	0.5123	0.4379	4.0376	0.5145
3.00	0.2182	0.5092	0.4286	4.2346	0.5222
3.05	0.2124	0.5062	0.4195	4.4410	0.5296
3.10	0.2067	0.5034	0.4107	4.6573	0.5368
3.15	0.2013	0.5006	0.4021	4.8838	0.5437
3.20	0.1961	0.4980	0.3937	5.1210	0.5504
3.25	0.1911	0.4955	0.3855	5.3691	0.5569
3.30	0.1862	0.4931	0.3776	5.6287	0.5632
3.35	0.1815	0.4908	0.3699	5.9000	0.5693
3.40	0.1770	0.4886	0.3623	6.1837	0.5752
3.45	0.1727	0.4865	0.3550	6.4801	0.5809
3.50	0.1685	0.4845	0.3478	6.7896	0.5864

TABLE E-1 (continued)

M	$\frac{p}{p^*}$	$\frac{\rho}{\rho^*}$	$\frac{T}{T^*}$	$\frac{p_0}{p_0^*}$	$\frac{4fL^*}{D}$
3.55	0.1645	0.4825	0.3409	7.1128	0.5918
3.60	0.1606	0.4806	0.3341	7.4501	0.5970
3.65	0.1568	0.4788	0.3275	7.8020	0.6020
3.70	0.1531	0.4770	0.3210	8.1691	0.6068
3.75	0.1496	0.4753	0.3148	8.5517	0.6115
3.80	0.1462	0.4737	0.3086	8.9506	0.6161
3.85	0.1429	0.4721	0.3027	9.3661	0.6206
3.90	0.1397	0.4706	0.2969	9.7990	0.6248
3.95	0.1366	0.4691	0.2912	1.0250+01	0.6290
4.00	0.1336	0.4677	0.2857	1.0719+01	0.6331
4.05	0.1307	0.4663	0.2803	1.1207+01	0.6370
4.10	0.1279	0.4650	0.2751	1.1715+01	0.6408
4.15	0.1252	0.4637	0.2700	1.2243+01	0.6445
4.20	0.1226	0.4625	0.2650	1.2792+01	0.6481
4.25	0.1200	0.4613	0.2602	1.3362+01	0.6516
4.30	0.1175	0.4601	0.2554	1.3955+01	0.6550
4.35	0.1151	0.4590	0.2508	1.4571+01	0.6583
4.40	0.1128	0.4579	0.2463	1.5210+01	0.6615
4.45	0.1105	0.4569	0.2419	1.5873+01	0.6646
4.50	0.1083	0.4559	0.2376	1.6562+01	0.6676
4.55	0.1062	0.4549	0.2334	1.7277+01	0.6706
4.60	0.1041	0.4539	0.2294	1.8018+01	0.6734
4.65	0.1021	0.4530	0.2254	1.8786+01	0.6762
4.70	0.1001	0.4521	0.2215	1.9583+01	0.6790
4.75	9.8225-02	0.4512	0.2177	2.0408+01	0.6816
4.80	9.6371-02	0.4504	0.2140	2.1264+01	0.6842
4.85	9.4567-02	0.4495	0.2104	2.2150+01	0.6867
4.90	9.2812-02	0.4487	0.2068	2.3067+01	0.6891
4.95	9.1105-02	0.4480	0.2034	2.4017+01	0.6915
5.00	8.9443-02	0.4472	0.2000	2.5000+01	0.6938
5.10	8.6249-02	0.4458	0.1935	2.7070+01	0.6983
5.20	8.3220-02	0.4444	0.1873	2.9283+01	0.7025
5.30	8.0344-02	0.4431	0.1813	3.1649+01	0.7065
5.40	7.7611-02	0.4419	0.1756	3.4175+01	0.7104
5.50	7.5012-02	0.4407	0.1702	3.6869+01	0.7140
5.60	7.2540-02	0.4396	0.1650	3.9740+01	0.7175
5.70	7.0185-02	0.4385	0.1600	4.2797+01	0.7208
5.80	6.7941-02	0.4375	0.1553	4.6050+01	0.7240
5.90	6.5800-02	0.4366	0.1507	4.9507+01	0.7270
6.00	6.3758-02	0.4357	0.1463	5.3180+01	0.7299
6.10	6.1807-02	0.4348	0.1421	5.7077+01	0.7326
6.20	5.9943-02	0.4340	0.1381	6.1210+01	0.7353
6.30	5.8161-02	0.4332	0.1343	6.5590+01	0.7378
6.40	5.6455-02	0.4324	0.1305	7.0227+01	0.7402
6.50	5.4823-02	0.4317	0.1270	7.5134+01	0.7425
6.60	5.3259-02	0.4310	0.1236	8.0323+01	0.7448
6.70	5.1760-02	0.4304	0.1203	8.5805+01	0.7469
6.80	5.0322-02	0.4298	0.1171	9.1593+01	0.7489
6.90	4.8943-02	0.4291	0.1140	9.7702+01	0.7509
7.00	4.7619-02	0.4286	0.1111	1.0414+02	0.7528

TABLE E-1 (concluded)

M	$\frac{p}{p^*}$	$\frac{\rho}{\rho^*}$	$\frac{T}{T^*}$	$\frac{p_0}{p_0^*}$	$\frac{4fL^*}{D}$
7.10	4.6347-02	0.4280	0.1083	1.1093+02	0.7546
7.20	4.5125-02	0.4275	0.1056	1.1808+02	0.7564
7.30	4.3950-02	0.4270	0.1029	1.2561+02	0.7580
7.40	4.2819-02	0.4265	0.1004	1.3352+02	0.7597
7.50	4.1731-02	0.4260	9.7959-02	1.4184+02	0.7612
7.60	4.0684-02	0.4256	9.5602-02	1.5059+02	0.7627
7.70	3.9675-02	0.4251	9.3327-02	1.5977+02	0.7642
7.80	3.8702-02	0.4247	9.1130-02	1.6940+02	0.7656
7.90	3.7765-02	0.4243	8.9008-02	1.7951+02	0.7669
8.00	3.6860-02	0.4239	8.6957-02	1.9011+02	0.7682
8.10	3.5988-02	0.4235	8.4974-02	2.0121+02	0.7694
8.20	3.5146-02	0.4232	8.3057-02	2.1285+02	0.7707
8.30	3.4332-02	0.4228	8.1202-02	2.2502+02	0.7718
8.40	3.3547-02	0.4225	7.9407-02	2.3776+02	0.7729
8.50	3.2787-02	0.4221	7.7670-02	2.5109+02	0.7740
8.60	3.2053-02	0.4218	7.5988-02	2.6501+02	0.7751
8.70	3.1343-02	0.4215	7.4359-02	2.7957+02	0.7761
8.80	3.0657-02	0.4212	7.2780-02	2.9477+02	0.7771
8.90	2.9992-02	0.4209	7.1250-02	3.1063+02	0.7781
9.00	2.9348-02	0.4207	6.9767-02	3.2719+02	0.7790
9.10	2.8725-02	0.4204	6.8329-02	3.4446+02	0.7799
9.20	2.8121-02	0.4201	6.6934-02	3.6246+02	0.7808
9.30	2.7536-02	0.4199	6.5581-02	3.8123+02	0.7816
9.40	2.6969-02	0.4196	6.4267-02	4.0077+02	0.7824
9.50	2.6419-02	0.4194	6.2992-02	4.2113+02	0.7832
9.60	2.5886-02	0.4192	6.1754-02	4.4232+02	0.7840
9.70	2.5368-02	0.4190	6.0551-02	4.6437+02	0.7847
9.80	2.4866-02	0.4187	5.9382-02	4.8730+02	0.7854
9.90	2.4378-02	0.4185	5.8247-02	5.1115+02	0.7861
10.00	2.3905-02	0.4183	5.7143-02	5.3594+02	0.7868
11.00	1.9838-02	0.4166	4.7619-02	8.4191+02	0.7927
12.00	1.6723-02	0.4153	4.0268-02	1.2762+03	0.7972
13.00	1.4284-02	0.4142	3.4483-02	1.8761+03	0.8007
14.00	1.2341-02	0.4134	2.9851-02	2.6854+03	0.8036
15.00	1.0768-02	0.4128	2.6087-02	3.7553+03	0.8058
16.00	9.4762-03	0.4122	2.2989-02	5.1445+03	0.8077
17.00	8.4034-03	0.4118	2.0408-02	6.9205+03	0.8093
18.00	7.5025-03	0.4114	1.8237-02	9.1593+03	0.8106
19.00	6.7388-03	0.4111	1.6393-02	1.1946+04	0.8117
20.00	6.0858-03	0.4108	1.4815-02	1.5377+04	0.8126
22.00	5.0350-03	0.4104	1.2270-02	2.4607+04	0.8142
24.00	4.2343-03	0.4100	1.0327-02	3.7832+04	0.8153
26.00	3.6102-03	0.4098	8.8106-03	5.6236+04	0.8162
28.00	3.1144-03	0.4095	7.6046-03	8.1212+04	0.8170
30.00	2.7141-03	0.4094	6.6298-03	1.1439+05	0.8176
32.00	2.3863-03	0.4092	5.8309-03	1.5763+05	0.8180
34.00	2.1144-03	0.4091	5.1680-03	2.1309+05	0.8184
36.00	1.8864-03	0.4090	4.6118-03	2.8319+05	0.8188
38.00	1.6934-03	0.4090	4.1408-03	3.7065+05	0.8190
40.00	1.5285-03	0.4089	3.7383-03	4.7853+05	0.8193
42.00	1.3866-03	0.4088	3.3917-03	6.1021+05	0.8195
44.00	1.2636-03	0.4088	3.0912-03	7.6943+05	0.8197
46.00	1.1562-03	0.4087	2.8288-03	9.6031+05	0.8198
48.00	1.0620-03	0.4087	2.5985-03	1.1873+06	0.8200
50.00	9.7882-04	0.4087	2.3952-03	1.4555+06	0.8201

Rayleigh Flow

APPENDIX F

TABLE F-1 RAYLEIGH LINE FLOW (PERFECT GAS, $\gamma = 1.4$)

M	$\frac{p}{p^*}$	$\frac{\rho}{\rho^*}$	$\frac{T}{T^*}$	$\frac{p_0}{p_0^*}$	$\frac{T_0}{T_0^*}$
0.02	2.3987	1.0423+03	2.3014-03	1.2675	1.9180-03
0.04	2.3946	2.6100+02	9.1748-03	1.2665	7.6482-03
0.06	2.3880	1.1632+02	2.0529-02	1.2647	1.7119-02
0.08	2.3787	6.5688+01	3.6212-02	1.2623	3.0215-02
0.10	2.3669	4.2250+01	5.6020-02	1.2592	4.6777-02
0.12	2.3526	2.9518+01	7.9698-02	1.2554	6.6606-02
0.14	2.3359	2.1842+01	0.1069	1.2510	8.9471-02
0.16	2.3170	1.6859+01	0.1374	1.2461	0.1151
0.18	2.2959	1.3443+01	0.1708	1.2406	0.1432
0.20	2.2727	1.1000+01	0.2066	1.2346	0.1736
0.22	2.2477	9.1922	0.2445	1.2281	0.2057
0.24	2.2209	7.8171	0.2841	1.2213	0.2395
0.26	2.1925	6.7470	0.3250	1.2140	0.2745
0.28	2.1626	5.8980	0.3667	1.2064	0.3104
0.30	2.1314	5.2130	0.4089	1.1985	0.3469
0.32	2.0991	4.6523	0.4512	1.1904	0.3837
0.34	2.0657	4.1877	0.4933	1.1822	0.4206
0.36	2.0314	3.7984	0.5348	1.1737	0.4572
0.38	1.9964	3.4688	0.5755	1.1652	0.4935
0.40	1.9608	3.1875	0.6151	1.1566	0.5290
0.42	1.9247	2.9454	0.6535	1.1480	0.5638
0.44	1.8882	2.7355	0.6903	1.1394	0.5975
0.46	1.8515	2.5525	0.7254	1.1308	0.6301
0.48	1.8147	2.3918	0.7587	1.1224	0.6614
0.50	1.7778	2.2500	0.7901	1.1141	0.6914
0.52	1.7409	2.1243	0.8196	1.1059	0.7199
0.54	1.7042	2.0122	0.8469	1.0979	0.7470
0.56	1.6678	1.9120	0.8723	1.0901	0.7725
0.58	1.6316	1.8219	0.8955	1.0826	0.7965
0.60	1.5957	1.7407	0.9167	1.0753	0.8189

TABLE F-1 (continued)

M	$\frac{p}{p^*}$	$\frac{\rho}{\rho^*}$	$\frac{T}{T^*}$	$\frac{p_0}{p_0^*}$	$\frac{T_0}{T_0^*}$
0.62	1.5603	1.6673	0.9358	1.0682	0.8398
0.64	1.5253	1.6006	0.9530	1.0615	0.8592
0.66	1.4908	1.5399	0.9682	1.0550	0.8771
0.68	1.4569	1.4844	0.9814	1.0489	0.8935
0.70	1.4235	1.4337	0.9929	1.0431	0.9085
0.72	1.3907	1.3871	1.0026	1.0376	0.9221
0.74	1.3585	1.3442	1.0106	1.0325	0.9344
0.76	1.3270	1.3047	1.0171	1.0278	0.9455
0.78	1.2961	1.2682	1.0220	1.0234	0.9553
0.80	1.2658	1.2344	1.0255	1.0193	0.9639
0.82	1.2363	1.2030	1.0276	1.0157	0.9715
0.84	1.2073	1.1739	1.0285	1.0124	0.9781
0.86	1.1791	1.1467	1.0283	1.0095	0.9836
0.88	1.1515	1.1214	1.0269	1.0070	0.9883
0.90	1.1247	1.0977	1.0245	1.0049	0.9921
0.92	1.0984	1.0756	1.0212	1.0031	0.9951
0.94	1.0728	1.0549	1.0170	1.0017	0.9973
0.96	1.0479	1.0354	1.0121	1.0008	0.9988
0.98	1.0237	1.0172	1.0064	1.0002	0.9997
1.00	1.0000	1.0000	1.0000	1.0000	1.0000
1.02	0.9770	0.9838	0.9930	1.0002	0.9997
1.04	0.9546	0.9686	0.9855	1.0008	0.9989
1.06	0.9327	0.9542	0.9776	1.0017	0.9977
1.08	0.9115	0.9406	0.9691	1.0031	0.9960
1.10	0.8909	0.9277	0.9603	1.0049	0.9939
1.12	0.8708	0.9155	0.9512	1.0070	0.9915
1.14	0.8512	0.9039	0.9417	1.0095	0.9887
1.16	0.8322	0.8930	0.9320	1.0124	0.9856
1.18	0.8137	0.8826	0.9220	1.0157	0.9823
1.20	0.7958	0.8727	0.9118	1.0194	0.9787
1.22	0.7783	0.8633	0.9015	1.0235	0.9749
1.24	0.7613	0.8543	0.8911	1.0280	0.9709
1.26	0.7447	0.8458	0.8805	1.0328	0.9668
1.28	0.7287	0.8376	0.8699	1.0380	0.9624
1.30	0.7130	0.8299	0.8592	1.0437	0.9580
1.32	0.6978	0.8225	0.8484	1.0497	0.9534
1.34	0.6830	0.8154	0.8377	1.0561	0.9487
1.36	0.6686	0.8086	0.8269	1.0629	0.9440
1.38	0.6546	0.8021	0.8161	1.0701	0.9391
1.40	0.6410	0.7959	0.8054	1.0776	0.9343
1.42	0.6278	0.7900	0.7947	1.0856	0.9293
1.44	0.6149	0.7843	0.7840	1.0940	0.9243
1.46	0.6024	0.7788	0.7735	1.1028	0.9193
1.48	0.5902	0.7736	0.7629	1.1120	0.9143
1.50	0.5783	0.7685	0.7525	1.1215	0.9093
1.52	0.5668	0.7637	0.7422	1.1315	0.9042
1.54	0.5555	0.7590	0.7319	1.1419	0.8992
1.56	0.5446	0.7545	0.7217	1.1527	0.8942
1.58	0.5339	0.7502	0.7117	1.1640	0.8892
1.60	0.5236	0.7461	0.7017	1.1756	0.8842

TABLE F-1 (continued)

M	$\frac{p}{p^*}$	$\frac{\rho}{\rho^*}$	$\frac{T}{T^*}$	$\frac{p_0}{p_0^*}$	$\frac{T_0}{T_0^*}$
1.62	0.5135	0.7421	0.6919	1.1877	0.8792
1.64	0.5036	0.7383	0.6822	1.2002	0.8743
1.66	0.4940	0.7345	0.6726	1.2131	0.8694
1.68	0.4847	0.7310	0.6631	1.2264	0.8645
1.70	0.4756	0.7275	0.6538	1.2402	0.8597
1.72	0.4668	0.7242	0.6445	1.2545	0.8549
1.74	0.4581	0.7210	0.6355	1.2692	0.8502
1.76	0.4497	0.7178	0.6265	1.2843	0.8455
1.78	0.4415	0.7148	0.6176	1.2999	0.8409
1.80	0.4335	0.7119	0.6089	1.3159	0.8363
1.82	0.4257	0.7091	0.6004	1.3324	0.8317
1.84	0.4181	0.7064	0.5919	1.3494	0.8273
1.86	0.4107	0.7038	0.5836	1.3669	0.8228
1.88	0.4035	0.7012	0.5754	1.3849	0.8185
1.90	0.3964	0.6988	0.5673	1.4033	0.8141
1.92	0.3896	0.6964	0.5594	1.4222	0.8099
1.94	0.3828	0.6940	0.5516	1.4417	0.8057
1.96	0.3763	0.6918	0.5439	1.4616	0.8015
1.98	0.3699	0.6896	0.5364	1.4821	0.7974
2.00	0.3636	0.6875	0.5289	1.5031	0.7934
2.05	0.3487	0.6825	0.5109	1.5579	0.7835
2.10	0.3345	0.6778	0.4936	1.6162	0.7741
2.15	0.3212	0.6735	0.4770	1.6780	0.7649
2.20	0.3086	0.6694	0.4611	1.7435	0.7561
2.25	0.2968	0.6656	0.4458	1.8127	0.7477
2.30	0.2855	0.6621	0.4312	1.8860	0.7395
2.35	0.2749	0.6588	0.4172	1.9634	0.7317
2.40	0.2648	0.6557	0.4038	2.0450	0.7242
2.45	0.2552	0.6527	0.3910	2.1311	0.7170
2.50	0.2462	0.6500	0.3787	2.2218	0.7101
2.55	0.2375	0.6474	0.3669	2.3173	0.7034
2.60	0.2294	0.6450	0.3556	2.4177	0.6970
2.65	0.2216	0.6427	0.3448	2.5233	0.6908
2.70	0.2142	0.6405	0.3344	2.6343	0.6849
2.75	0.2071	0.6384	0.3244	2.7508	0.6793
2.80	0.2004	0.6365	0.3149	2.8731	0.6738
2.85	0.1940	0.6346	0.3057	3.0014	0.6685
2.90	0.1879	0.6329	0.2969	3.1358	0.6635
2.95	0.1820	0.6312	0.2884	3.2768	0.6586
3.00	0.1765	0.6296	0.2803	3.4244	0.6540
3.05	0.1711	0.6281	0.2725	3.5791	0.6495
3.10	0.1660	0.6267	0.2650	3.7408	0.6452
3.15	0.1612	0.6253	0.2577	3.9101	0.6410
3.20	0.1565	0.6240	0.2508	4.0871	0.6370
3.25	0.1520	0.6228	0.2441	4.2722	0.6331
3.30	0.1477	0.6216	0.2377	4.4655	0.6294
3.35	0.1436	0.6205	0.2315	4.6674	0.6258
3.40	0.1397	0.6194	0.2255	4.8783	0.6224
3.45	0.1359	0.6183	0.2197	5.0984	0.6190
3.50	0.1322	0.6173	0.2142	5.3280	0.6158

TABLE F-1 (continued)

M	$\frac{p}{p^*}$	$\frac{\rho}{\rho^*}$	$\frac{T}{T^*}$	$\frac{p_0}{p_0^*}$	$\frac{T_0}{T_0^*}$
3.55	0.1287	0.6164	0.2088	5.5676	0.6127
3.60	0.1254	0.6155	0.2037	5.8173	0.6097
3.65	0.1221	0.6146	0.1987	6.0776	0.6068
3.70	0.1190	0.6138	0.1939	6.3488	0.6040
3.75	0.1160	0.6130	0.1893	6.6314	0.6013
3.80	0.1131	0.6122	0.1848	6.9256	0.5987
3.85	0.1103	0.6114	0.1805	7.2318	0.5962
3.90	0.1077	0.6107	0.1763	7.5505	0.5937
3.95	0.1051	0.6100	0.1722	7.8821	0.5914
4.00	0.1026	0.6094	0.1683	8.2268	0.5891
4.05	0.1002	0.6087	0.1645	8.5853	0.5869
4.10	9.7823-02	0.6081	0.1609	8.9579	0.5847
4.15	9.5574-02	0.6075	0.1573	9.3451	0.5827
4.20	9.3400-02	0.6070	0.1539	9.7473	0.5807
4.25	9.1298-02	0.6064	0.1506	1.0165+01	0.5787
4.30	8.9266-02	0.6059	0.1473	1.0599+01	0.5768
4.35	8.7300-02	0.6054	0.1442	1.1049+01	0.5750
4.40	8.5397-02	0.6049	0.1412	1.1516+01	0.5732
4.45	8.3555-02	0.6044	0.1383	1.2000+01	0.5715
4.50	8.1772-02	0.6039	0.1354	1.2502+01	0.5698
4.55	8.0044-02	0.6035	0.1326	1.3023+01	0.5682
4.60	7.8370-02	0.6030	0.1300	1.3563+01	0.5666
4.65	7.6747-02	0.6026	0.1274	1.4122+01	0.5651
4.70	7.5174-02	0.6022	0.1248	1.4702+01	0.5636
4.75	7.3648-02	0.6018	0.1224	1.5302+01	0.5622
4.80	7.2167-02	0.6014	0.1200	1.5923+01	0.5608
4.85	7.0731-02	0.6010	0.1177	1.6567+01	0.5594
4.90	6.9336-02	0.6007	0.1154	1.7233+01	0.5581
4.95	6.7982-02	0.6003	0.1132	1.7921+01	0.5568
5.00	6.6667-02	0.6000	0.1111	1.8634+01	0.5556
5.10	6.4147-02	0.5994	0.1070	2.0133+01	0.5532
5.20	6.1767-02	0.5987	0.1032	2.1734+01	0.5509
5.30	5.9515-02	0.5982	9.9496-02	2.3444+01	0.5487
5.40	5.7383-02	0.5976	9.6019-02	2.5268+01	0.5467
5.50	5.5363-02	0.5971	9.2719-02	2.7211+01	0.5447
5.60	5.3447-02	0.5966	8.9584-02	2.9281+01	0.5429
5.70	5.1628-02	0.5962	8.6602-02	3.1482+01	0.5411
5.80	4.9900-02	0.5957	8.3765-02	3.3822+01	0.5394
5.90	4.8257-02	0.5953	8.1063-02	3.6308+01	0.5378
6.00	4.6693-02	0.5949	7.8487-02	3.8946+01	0.5363
6.10	4.5203-02	0.5945	7.6031-02	4.1744+01	0.5349
6.20	4.3783-02	0.5942	7.3687-02	4.4708+01	0.5335
6.30	4.2428-02	0.5938	7.1448-02	4.7848+01	0.5322
6.40	4.1135-02	0.5935	6.9309-02	5.1170+01	0.5309
6.50	3.9900-02	0.5932	6.7263-02	5.4683+01	0.5297
6.60	3.8720-02	0.5929	6.5306-02	5.8395+01	0.5285
6.70	3.7590-02	0.5926	6.3432-02	6.2315+01	0.5274
6.80	3.6510-02	0.5923	6.1636-02	6.6452+01	0.5264
6.90	3.5475-02	0.5921	5.9915-02	7.0815+01	0.5254
7.00	3.4483-02	0.5918	5.8264-02	7.5414+01	0.5244

TABLE F-1 (concluded)

M	$\frac{p}{p^*}$	$\frac{\rho}{\rho^*}$	$\frac{T}{T^*}$	$\frac{p_0}{p_0^*}$	$\frac{T_0}{T_0^*}$
7.10	3.3532-02	0.5916	5.6680-02	8.0257+01	0.5234
7.20	3.2619-02	0.5914	5.5159-02	8.5356+01	0.5225
7.30	3.1744-02	0.5912	5.3698-02	9.0720+01	0.5217
7.40	3.0902-02	0.5909	5.2293-02	9.6360+01	0.5208
7.50	3.0094-02	0.5907	5.0943-02	1.0229+02	0.5200
7.60	2.9317-02	0.5905	4.9644-02	1.0851+02	0.5193
7.70	2.8569-02	0.5904	4.8393-02	1.1505+02	0.5185
7.80	2.7850-02	0.5902	4.7189-02	1.2190+02	0.5178
7.90	2.7157-02	0.5900	4.6029-02	1.2909+02	0.5171
8.00	2.6490-02	0.5898	4.4910-02	1.3662+02	0.5165
8.10	2.5847-02	0.5897	4.3832-02	1.4451+02	0.5158
8.20	2.5227-02	0.5895	4.2792-02	1.5278+02	0.5152
8.30	2.4629-02	0.5894	4.1788-02	1.6142+02	0.5146
8.40	2.4052-02	0.5892	4.0819-02	1.7047+02	0.5140
8.50	2.3495-02	0.5891	3.9883-02	1.7992+02	0.5135
8.60	2.2957-02	0.5890	3.8978-02	1.8980+02	0.5130
8.70	2.2437-02	0.5888	3.8104-02	2.0013+02	0.5124
8.80	2.1935-02	0.5887	3.7259-02	2.1090+02	0.5119
8.90	2.1449-02	0.5886	3.6441-02	2.2215+02	0.5114
9.00	2.0979-02	0.5885	3.5650-02	2.3388+02	0.5110
9.10	2.0524-02	0.5884	3.4884-02	2.4612+02	0.5105
9.20	2.0084-02	0.5883	3.4142-02	2.5887+02	0.5101
9.30	1.9658-02	0.5882	3.3424-02	2.7216+02	0.5097
9.40	1.9246-02	0.5880	3.2728-02	2.8600+02	0.5092
9.50	1.8846-02	0.5879	3.2053-02	3.0041+02	0.5088
9.60	1.8458-02	0.5879	3.1399-02	3.1540+02	0.5085
9.70	1.8082-02	0.5878	3.0765-02	3.3100+02	0.5081
9.80	1.7718-02	0.5877	3.0149-02	3.4722+02	0.5077
9.90	1.7364-02	0.5876	2.9552-02	3.6409+02	0.5074
10.00	1.7021-02	0.5875	2.8972-02	3.8161+02	0.5070
11.00	1.4085-02	0.5868	2.4003-02	5.9774+02	0.5041
12.00	1.1846-02	0.5862	2.0207-02	9.0405+02	0.5018
13.00	1.0101-02	0.5858	1.7243-02	1.3267+03	0.5001
14.00	8.7146-03	0.5855	1.4885-02	1.8963+03	0.4987
15.00	7.5949-03	0.5852	1.2979-02	2.6488+03	0.4975
16.00	6.6778-03	0.5850	1.1416-02	3.6253+03	0.4966
17.00	5.9172-03	0.5848	1.0119-02	4.8730+03	0.4958
18.00	5.2794-03	0.5846	9.0304-03	6.4452+03	0.4952
19.00	4.7393-03	0.5845	8.1085-03	8.4018+03	0.4946
20.00	4.2781-03	0.5844	7.3208-03	1.0810+04	0.4942
22.00	3.5367-03	0.5842	6.0540-03	1.7284+04	0.4934
24.00	2.9725-03	0.5841	5.0894-03	2.6559+04	0.4928
26.00	2.5333-03	0.5839	4.3381-03	3.9461+04	0.4924
28.00	2.1846-03	0.5839	3.7416-03	5.6966+04	0.4920
30.00	1.9032-03	0.5838	3.2601-03	8.0212+04	0.4917
32.00	1.6729-03	0.5837	2.8659-03	1.1051+05	0.4915
34.00	1.4820-03	0.5837	2.5391-03	1.4936+05	0.4913
36.00	1.3220-03	0.5837	2.2651-03	1.9846+05	0.4911
38.00	1.1866-03	0.5836	2.0332-03	2.5972+05	0.4910
40.00	1.0709-03	0.5836	1.8351-03	3.3528+05	0.4909
42.00	9.7142-04	0.5836	1.6646-03	4.2749+05	0.4908
44.00	8.8515-04	0.5835	1.5168-03	5.3899+05	0.4907
46.00	8.0988-04	0.5835	1.3879-03	6.7264+05	0.4906
48.00	7.4382-04	0.5835	1.2747-03	8.3161+05	0.4906
50.00	6.8552-04	0.5835	1.1748-03	1.0193+06	0.4905

Simple Mass Addition

APPENDIX G

TABLE G-1 SIMPLE MASS ADDITION (PERFECT GAS, $\gamma = 1.4$)

M	$\frac{p}{p^*}$	$\frac{\rho}{\rho^*}$	$\frac{T}{T^*}$	$\frac{p_0}{p_0^*}$	$\frac{\dot{m}}{\dot{m}^*}$
0.00	2.4000	2.0000	1.2000	1.2679	0.0000
0.02	2.3987	1.9990	1.1999	1.2675	4.3795-02
0.04	2.3946	1.9962	1.1996	1.2665	8.7454-02
0.06	2.3880	1.9914	1.1991	1.2647	0.1308
0.08	2.3787	1.9848	1.1985	1.2623	0.1738
0.10	2.3669	1.9763	1.1976	1.2592	0.2163
0.12	2.3526	1.9661	1.1966	1.2554	0.2581
0.14	2.3359	1.9542	1.1953	1.2510	0.2991
0.16	2.3170	1.9407	1.1939	1.2461	0.3393
0.18	2.2959	1.9256	1.1923	1.2406	0.3785
0.20	2.2727	1.9091	1.1905	1.2346	0.4166
0.22	2.2477	1.8912	1.1885	1.2281	0.4536
0.24	2.2209	1.8721	1.1863	1.2213	0.4894
0.26	2.1925	1.8518	1.1840	1.2140	0.5239
0.28	2.1626	1.8305	1.1815	1.2064	0.5571
0.30	2.1314	1.8082	1.1788	1.1985	0.5889
0.32	2.0991	1.7851	1.1759	1.1904	0.6194
0.34	2.0657	1.7612	1.1729	1.1822	0.6485
0.36	2.0314	1.7367	1.1697	1.1737	0.6762
0.38	1.9964	1.7117	1.1663	1.1652	0.7025
0.40	1.9608	1.6863	1.1628	1.1566	0.7273
0.42	1.9247	1.6605	1.1591	1.1480	0.7508
0.44	1.8882	1.6344	1.1553	1.1394	0.7730
0.46	1.8515	1.6082	1.1513	1.1308	0.7938
0.48	1.8147	1.5819	1.1471	1.1224	0.8133
0.50	1.7778	1.5556	1.1429	1.1141	0.8315
0.52	1.7409	1.5293	1.1384	1.1059	0.8485
0.54	1.7042	1.5030	1.1339	1.0979	0.8643
0.56	1.6678	1.4770	1.1292	1.0901	0.8789
0.58	1.6316	1.4511	1.1243	1.0826	0.8925

TABLE G-1 (continued)

M	$\frac{p}{p^*}$	$\frac{\rho}{\rho^*}$	$\frac{T}{T^*}$	$\frac{p_0}{p_0^*}$	$\frac{\dot{m}}{\dot{m}^*}$
0.60	1.5957	1.4255	1.1194	1.0753	0.9049
0.62	1.5603	1.4002	1.1143	1.0682	0.9164
0.64	1.5253	1.3752	1.1091	1.0615	0.9269
0.66	1.4908	1.3506	1.1038	1.0550	0.9365
0.68	1.4569	1.3263	1.0984	1.0489	0.9453
0.70	1.4235	1.3025	1.0929	1.0431	0.9532
0.72	1.3907	1.2791	1.0873	1.0376	0.9603
0.74	1.3585	1.2561	1.0816	1.0325	0.9667
0.76	1.3270	1.2335	1.0757	1.0278	0.9723
0.78	1.2961	1.2115	1.0698	1.0234	0.9774
0.80	1.2658	1.1899	1.0638	1.0193	0.9818
0.82	1.2363	1.1688	1.0577	1.0157	0.9857
0.84	1.2073	1.1481	1.0516	1.0124	0.9890
0.86	1.1791	1.1279	1.0454	1.0095	0.9918
0.88	1.1515	1.1083	1.0391	1.0070	0.9941
0.90	1.1247	1.0890	1.0327	1.0049	0.9960
0.92	1.0984	1.0703	1.0263	1.0031	0.9975
0.94	1.0728	1.0520	1.0198	1.0017	0.9986
0.96	1.0479	1.0342	1.0132	1.0008	0.9994
0.98	1.0237	1.0169	1.0066	1.0002	0.9999
1.00	1.0000	1.0000	1.0000	1.0000	1.0000
1.02	0.9770	0.9836	0.9933	1.0002	0.9999
1.04	0.9546	0.9675	0.9866	1.0008	0.9995
1.06	0.9327	0.9520	0.9798	1.0017	0.9988
1.08	0.9115	0.9368	0.9730	1.0031	0.9980
1.10	0.8909	0.9220	0.9662	1.0049	0.9970
1.12	0.8708	0.9077	0.9593	1.0070	0.9957
1.14	0.8512	0.8937	0.9524	1.0095	0.9943
1.16	0.8322	0.8802	0.9455	1.0124	0.9928
1.18	0.8137	0.8670	0.9386	1.0157	0.9911
1.20	0.7958	0.8541	0.9317	1.0194	0.9893
1.22	0.7783	0.8416	0.9247	1.0235	0.9874
1.24	0.7613	0.8295	0.9178	1.0280	0.9854
1.26	0.7447	0.8177	0.9108	1.0328	0.9832
1.28	0.7287	0.8062	0.9038	1.0380	0.9810
1.30	0.7130	0.7950	0.8969	1.0437	0.9788
1.32	0.6978	0.7841	0.8899	1.0497	0.9764
1.34	0.6830	0.7736	0.8829	1.0561	0.9740
1.36	0.6686	0.7633	0.8760	1.0629	0.9716
1.38	0.6546	0.7533	0.8690	1.0701	0.9691
1.40	0.6410	0.7436	0.8621	1.0776	0.9666
1.42	0.6278	0.7341	0.8551	1.0856	0.9640
1.44	0.6149	0.7249	0.8482	1.0940	0.9614
1.46	0.6024	0.7160	0.8413	1.1028	0.9588
1.48	0.5902	0.7073	0.8344	1.1120	0.9562
1.50	0.5783	0.6988	0.8276	1.1215	0.9536
1.52	0.5668	0.6905	0.8207	1.1315	0.9509
1.54	0.5555	0.6825	0.8139	1.1419	0.9483
1.56	0.5446	0.6747	0.8071	1.1527	0.9456
1.58	0.5339	0.6671	0.8004	1.1640	0.9430

TABLE G-1 (continued)

M	$\frac{p}{p^*}$	$\frac{\rho}{\rho^*}$	$\frac{T}{T^*}$	$\frac{p_0}{p_0^*}$	$\frac{\dot{m}}{\dot{m}^*}$
1.60	0.5236	0.6597	0.7937	1.1756	0.9403
1.62	0.5135	0.6525	0.7869	1.1877	0.9377
1.64	0.5036	0.6454	0.7803	1.2002	0.9350
1.66	0.4940	0.6386	0.7736	1.2131	0.9324
1.68	0.4847	0.6319	0.7670	1.2264	0.9298
1.70	0.4756	0.6254	0.7605	1.2402	0.9272
1.72	0.4668	0.6191	0.7539	1.2545	0.9246
1.74	0.4581	0.6130	0.7474	1.2692	0.9221
1.76	0.4497	0.6069	0.7410	1.2843	0.9195
1.78	0.4415	0.6011	0.7345	1.2999	0.9170
1.80	0.4335	0.5954	0.7282	1.3159	0.9145
1.82	0.4257	0.5898	0.7218	1.3324	0.9120
1.84	0.4181	0.5844	0.7155	1.3494	0.9095
1.86	0.4107	0.5791	0.7093	1.3669	0.9071
1.88	0.4035	0.5739	0.7030	1.3849	0.9047
1.90	0.3964	0.5689	0.6969	1.4033	0.9023
1.92	0.3896	0.5640	0.6907	1.4222	0.8999
1.94	0.3828	0.5592	0.6847	1.4417	0.8976
1.96	0.3763	0.5545	0.6786	1.4616	0.8953
1.98	0.3699	0.5499	0.6726	1.4821	0.8930
2.00	0.3636	0.5455	0.6667	1.5031	0.8907
2.05	0.3487	0.5348	0.6520	1.5579	0.8852
2.10	0.3345	0.5247	0.6376	1.6162	0.8798
2.15	0.3212	0.5152	0.6235	1.6780	0.8746
2.20	0.3086	0.5062	0.6098	1.7435	0.8696
2.25	0.2968	0.4977	0.5963	1.8127	0.8647
2.30	0.2855	0.4897	0.5831	1.8860	0.8600
2.35	0.2749	0.4820	0.5702	1.9634	0.8554
2.40	0.2648	0.4748	0.5576	2.0450	0.8510
2.45	0.2552	0.4680	0.5453	2.1311	0.8468
2.50	0.2462	0.4615	0.5333	2.2218	0.8426
2.55	0.2375	0.4554	0.5216	2.3173	0.8387
2.60	0.2294	0.4495	0.5102	2.4177	0.8349
2.65	0.2216	0.4440	0.4991	2.5233	0.8312
2.70	0.2142	0.4387	0.4882	2.6343	0.8276
2.75	0.2071	0.4337	0.4776	2.7508	0.8242
2.80	0.2004	0.4289	0.4673	2.8731	0.8209
2.85	0.1940	0.4243	0.4572	3.0014	0.8176
2.90	0.1879	0.4199	0.4474	3.1358	0.8146
2.95	0.1820	0.4157	0.4379	3.2768	0.8116
3.00	0.1765	0.4118	0.4286	3.4244	0.8087
3.05	0.1711	0.4080	0.4195	3.5791	0.8059
3.10	0.1660	0.4043	0.4107	3.7408	0.8032
3.15	0.1612	0.4008	0.4021	3.9101	0.8006
3.20	0.1565	0.3975	0.3937	4.0871	0.7981
3.25	0.1520	0.3943	0.3855	4.2722	0.7957
3.30	0.1477	0.3912	0.3776	4.4655	0.7933
3.35	0.1436	0.3883	0.3699	4.6674	0.7911
3.40	0.1397	0.3855	0.3623	4.8783	0.7889
3.45	0.1359	0.3828	0.3550	5.0984	0.7868

TABLE G-1 (continued)

M	$\frac{p}{p^*}$	$\frac{\rho}{\rho^*}$	$\frac{T}{T^*}$	$\frac{p_0}{p_0^*}$	$\frac{\dot{m}}{\dot{m}^*}$
3.50	0.1322	0.3802	0.3478	5.3280	0.7847
3.55	0.1287	0.3777	0.3409	5.5676	0.7828
3.60	0.1254	0.3753	0.3341	5.8173	0.7808
3.65	0.1221	0.3729	0.3275	6.0776	0.7790
3.70	0.1190	0.3707	0.3210	6.3488	0.7772
3.75	0.1160	0.3686	0.3148	6.6314	0.7754
3.80	0.1131	0.3665	0.3086	6.9256	0.7738
3.85	0.1103	0.3645	0.3027	7.2318	0.7721
3.90	0.1077	0.3626	0.2969	7.5505	0.7705
3.95	0.1051	0.3608	0.2912	7.8821	0.7690
4.00	0.1026	0.3590	0.2857	8.2268	0.7675
4.05	0.1002	0.3573	0.2803	8.5853	0.7661
4.10	9.7823-02	0.3556	0.2751	8.9579	0.7647
4.15	9.5574-02	0.3540	0.2700	9.3451	0.7633
4.20	9.3400-02	0.3524	0.2650	9.7473	0.7620
4.25	9.1298-02	0.3509	0.2602	1.0165+01	0.7607
4.30	8.9266-02	0.3495	0.2554	1.0599+01	0.7595
4.35	8.7300-02	0.3481	0.2508	1.1049+01	0.7583
4.40	8.5397-02	0.3467	0.2463	1.1516+01	0.7571
4.45	8.3555-02	0.3454	0.2419	1.2000+01	0.7560
4.50	8.1772-02	0.3441	0.2376	1.2502+01	0.7549
4.55	8.0044-02	0.3429	0.2334	1.3023+01	0.7538
4.60	7.8370-02	0.3417	0.2294	1.3563+01	0.7527
4.65	7.6747-02	0.3405	0.2254	1.4122+01	0.7517
4.70	7.5174-02	0.3394	0.2215	1.4702+01	0.7507
4.75	7.3648-02	0.3383	0.2177	1.5302+01	0.7498
4.80	7.2167-02	0.3373	0.2140	1.5923+01	0.7489
4.85	7.0731-02	0.3362	0.2104	1.6567+01	0.7479
4.90	6.9336-02	0.3352	0.2068	1.7233+01	0.7471
4.95	6.7982-02	0.3343	0.2034	1.7921+01	0.7462
5.00	6.6667-02	0.3333	0.2000	1.8634+01	0.7454
5.10	6.4147-02	0.3315	0.1935	2.0133+01	0.7437
5.20	6.1767-02	0.3298	0.1873	2.1734+01	0.7422
5.30	5.9515-02	0.3282	0.1813	2.3444+01	0.7408
5.40	5.7383-02	0.3267	0.1756	2.5268+01	0.7394
5.50	5.5363-02	0.3253	0.1702	2.7211+01	0.7381
5.60	5.3447-02	0.3239	0.1650	2.9281+01	0.7368
5.70	5.1628-02	0.3226	0.1600	3.1482+01	0.7356
5.80	4.9900-02	0.3214	0.1553	3.3822+01	0.7345
5.90	4.8257-02	0.3202	0.1507	3.6308+01	0.7334
6.00	4.6693-02	0.3191	0.1463	3.8946+01	0.7323
6.10	4.5203-02	0.3180	0.1421	4.1744+01	0.7314
6.20	4.3783-02	0.3170	0.1381	4.4708+01	0.7304
6.30	4.2428-02	0.3160	0.1343	4.7848+01	0.7295
6.40	4.1135-02	0.3151	0.1305	5.1170+01	0.7286
6.50	3.9900-02	0.3142	0.1270	5.4683+01	0.7278
6.60	3.8720-02	0.3134	0.1236	5.8395+01	0.7270
6.70	3.7590-02	0.3126	0.1203	6.2315+01	0.7262
6.80	3.6510-02	0.3118	0.1171	6.6452+01	0.7255
6.90	3.5475-02	0.3111	0.1140	7.0815+01	0.7248

TABLE G-1 (concluded)

M	$\frac{p}{p^*}$	$\frac{\rho}{\rho^*}$	$\frac{T}{T^*}$	$\frac{p_0}{p_0^*}$	$\frac{\dot{m}}{\dot{m}^*}$
7.00	3.4483-02	0.3103	0.1111	7.5414+01	0.7241
7.10	3.3532-02	0.3097	0.1083	8.0257+01	0.7235
7.20	3.2619-02	0.3090	0.1056	8.5356+01	0.7229
7.30	3.1744-02	0.3084	0.1029	9.0720+01	0.7223
7.40	3.0902-02	0.3078	0.1004	9.6360+01	0.7217
7.50	3.0094-02	0.3072	9.7959-02	1.0229+02	0.7211
7.60	2.9317-02	0.3067	9.5602-02	1.0851+02	0.7206
7.70	2.8569-02	0.3061	9.3327-02	1.1505+02	0.7201
7.80	2.7850-02	0.3056	9.1130-02	1.2190+02	0.7196
7.90	2.7157-02	0.3051	8.9008-02	1.2909+02	0.7191
8.00	2.6490-02	0.3046	8.6957-02	1.3662+02	0.7187
8.10	2.5847-02	0.3042	8.4974-02	1.4451+02	0.7182
8.20	2.5227-02	0.3037	8.3057-02	1.5278+02	0.7178
8.30	2.4629-02	0.3033	8.1202-02	1.6142+02	0.7174
8.40	2.4052-02	0.3029	7.9407-02	1.7047+02	0.7170
8.50	2.3495-02	0.3025	7.7670-02	1.7992+02	0.7166
8.60	2.2957-02	0.3021	7.5988-02	1.8980+02	0.7162
8.70	2.2437-02	0.3017	7.4359-02	2.0013+02	0.7158
8.80	2.1935-02	0.3014	7.2780-02	2.1090+02	0.7155
8.90	2.1449-02	0.3010	7.1250-02	2.2215+02	0.7152
9.00	2.0979-02	0.3007	6.9767-02	2.3388+02	0.7148
9.10	2.0524-02	0.3004	6.8329-02	2.4612+02	0.7145
9.20	2.0084-02	0.3001	6.6934-02	2.5887+02	0.7142
9.30	1.9658-02	0.2998	6.5581-02	2.7216+02	0.7139
9.40	1.9246-02	0.2995	6.4267-02	2.8600+02	0.7136
9.50	1.8846-02	0.2992	6.2992-02	3.0041+02	0.7133
9.60	1.8458-02	0.2989	6.1754-02	3.1540+02	0.7131
9.70	1.8082-02	0.2986	6.0551-02	3.3100+02	0.7128
9.80	1.7718-02	0.2984	5.9382-02	3.4722+02	0.7125
9.90	1.7364-02	0.2981	5.8247-02	3.6409+02	0.7123
10.00	1.7021-02	0.2979	5.7143-02	3.8161+02	0.7121
11.00	1.4085-02	0.2958	4.7619-02	5.9774+02	0.7100
12.00	1.1846-02	0.2942	4.0268-02	9.0405+02	0.7084
13.00	1.0101-02	0.2929	3.4483-02	1.3267+03	0.7071
14.00	8.7146-03	0.2919	2.9851-02	1.8963+03	0.7062
15.00	7.5949-03	0.2911	2.6087-02	2.6488+03	0.7053
16.00	6.6778-03	0.2905	2.2989-02	3.6253+03	0.7047
17.00	5.9172-03	0.2899	2.0408-02	4.8730+03	0.7041
18.00	5.2794-03	0.2895	1.8237-02	6.4452+03	0.7037
19.00	4.7393-03	0.2891	1.6393-02	8.4018+03	0.7033
20.00	4.2781-03	0.2888	1.4815-02	1.0810+04	0.7030
22.00	3.5367-03	0.2882	1.2270-02	1.7284+04	0.7024
24.00	2.9725-03	0.2878	1.0327-02	2.6559+04	0.7020
26.00	2.5333-03	0.2875	8.8106-03	3.9461+04	0.7017
28.00	2.1846-03	0.2873	7.6046-03	5.6966+04	0.7014
30.00	1.9032-03	0.2871	6.6298-03	8.0212+04	0.7012
32.00	1.6729-03	0.2869	5.8309-03	1.1051+05	0.7011
34.00	1.4820-03	0.2868	5.1680-03	1.4936+05	0.7009
36.00	1.3220-03	0.2867	4.6118-03	1.9846+05	0.7008
38.00	1.1866-03	0.2866	4.1408-03	2.5972+05	0.7007
40.00	1.0709-03	0.2865	3.7383-03	3.3528+05	0.7006
42.00	9.7142-04	0.2864	3.3917-03	4.2749+05	0.7006
44.00	8.8515-04	0.2863	3.0912-03	5.3899+05	0.7005
46.00	8.0988-04	0.2863	2.8288-03	6.7264+05	0.7004
48.00	7.4382-04	0.2862	2.5985-03	8.3161+05	0.7004
50.00	6.8552-04	0.2862	2.3952-03	1.0193+06	0.7004

Isothermal Flow

APPENDIX H

TABLE H-1 ISOTHERMAL FLOW (PERFECT GAS, $\gamma = 1.4$)

M	$\frac{4fL}{D}$	$\frac{P}{P^*}$	$\frac{P_o}{P_o^*}$	$\frac{\rho}{\rho^*}$	$\frac{T_o}{T_o^*}$
.010	7132.984	84.5154	52.9657	84.5154	0.8750
.020	1777.227	42.2577	26.4884	42.2577	0.8751
.030	785.974	28.1718	17.6651	28.1718	0.8752
.040	439.327	21.1289	13.2553	21.1289	0.8753
.050	279.059	16.9031	10.6109	16.9031	0.8754
.060	192.122	14.0859	8.8493	14.0859	0.8756
.070	139.791	12.0736	7.5920	12.0736	0.8759
.080	105.892	10.5644	6.6500	10.5644	0.8761
.090	82.704	9.3906	5.9181	9.3906	0.8764
.100	66.160	8.4515	5.3334	8.4515	0.8768
.110	53.954	7.6832	4.8556	7.6832	0.8771
.120	44.699	7.0430	4.4581	7.0430	0.8775
.130	37.521	6.5012	4.1224	6.5012	0.8780
.140	31.847	6.0368	3.8352	6.0368	0.8784
.150	27.288	5.6344	3.5867	5.6344	0.8789
.160	23.573	5.2822	3.3698	5.2822	0.8795
.170	20.508	4.9715	3.1789	4.9715	0.8801
.180	17.953	4.6953	3.0096	4.6953	0.8807
.190	15.801	4.4482	2.8586	4.4482	0.8813
.200	13.975	4.2258	2.7230	4.2258	0.8820
.210	12.412	4.0245	2.6007	4.0245	0.8827
.220	11.066	3.8416	2.4899	3.8416	0.8835
.230	9.900	3.6746	2.3891	3.6746	0.8843
.240	8.883	3.5215	2.2970	3.5215	0.8851

TABLE H-1 (continued)

M	$\frac{4fL}{D}$	$\frac{P}{P^*}$	$\frac{P_o}{P_o^*}$	$\frac{\rho}{\rho^*}$	$\frac{T_o}{T_o^*}$
.250	7.992	3.3806	2.2126	3.3806	0.8859
.260	7.209	3.2506	2.1350	3.2506	0.8868
.270	6.516	3.1302	2.0635	3.1302	0.8878
.280	5.901	3.0184	1.9974	3.0184	0.8887
.290	5.354	2.9143	1.9361	2.9143	0.8897
.300	4.865	2.8172	1.8791	2.8172	0.8908
.310	4.427	2.7263	1.8262	2.7263	0.8918
.320	4.033	2.6411	1.7768	2.6411	0.8929
.330	3.678	2.5611	1.7306	2.5611	0.8941
.340	3.358	2.4857	1.6874	2.4857	0.8952
.350	3.068	2.4147	1.6470	2.4147	0.8964
.360	2.805	2.3477	1.6090	2.3477	0.8977
.370	2.566	2.2842	1.5733	2.2842	0.8990
.380	2.348	2.2241	1.5398	2.2241	0.9003
.390	2.149	2.1671	1.5082	2.1671	0.9016
.400	1.968	2.1129	1.4784	2.1129	0.9030
.410	1.802	2.0614	1.4503	2.0614	0.9044
.420	1.651	2.0123	1.4237	2.0123	0.9059
.430	1.512	1.9655	1.3986	1.9655	0.9074
.440	1.384	1.9208	1.3749	1.9208	0.9089
.450	1.267	1.8781	1.3524	1.8781	0.9104
.460	1.159	1.8373	1.3311	1.8373	0.9120
.470	1.060	1.7982	1.3109	1.7982	0.9137
.480	0.969	1.7607	1.2918	1.7607	0.9153
.490	0.885	1.7248	1.2737	1.7248	0.9170
.500	0.807	1.6903	1.2565	1.6903	0.9188
.510	0.7360	1.6572	1.2402	1.6572	0.9205
.520	0.6702	1.6253	1.2247	1.6253	0.9223
.530	0.6096	1.5946	1.2100	1.5946	0.9242
.540	0.5536	1.5651	1.1960	1.5651	0.9260
.550	0.5021	1.5366	1.1827	1.5366	0.9279
.560	0.4545	1.5092	1.1702	1.5092	0.9299
.570	0.4107	1.4827	1.1582	1.4827	0.9319
.580	0.3703	1.4572	1.1469	1.4572	0.9339
.590	0.3332	1.4325	1.1361	1.4325	0.9359
.600	0.2989	1.4086	1.1259	1.4086	0.9380
.610	0.2675	1.3855	1.1162	1.3855	0.9401
.620	0.2386	1.3632	1.1070	1.3632	0.9423
.630	0.2121	1.3415	1.0983	1.3415	0.9445
.640	0.1878	1.3206	1.0901	1.3206	0.9467
.650	0.1655	1.3002	1.0823	1.3002	0.9489
.660	0.1452	1.2805	1.0750	1.2805	0.9512
.670	0.1267	1.2614	1.0680	1.2614	0.9536
.680	0.1099	1.2429	1.0615	1.2429	0.9559
.690	0.0946	1.2249	1.0553	1.2249	0.9583
.700	0.0808	1.2074	1.0495	1.2074	0.9608
.710	0.0684	1.1904	1.0440	1.1904	0.9632
.720	0.0573	1.1738	1.0389	1.1738	0.9657

TABLE H-1 (concluded)

M	$\frac{4fL}{D}$	$\frac{P}{P^*}$	$\frac{P_o}{P_o^*}$	$\frac{\rho}{\rho^*}$	$\frac{T_o}{T_o^*}$
.730	0.0474	1.1577	1.0341	1.1577	0.9683
.740	0.0387	1.1421	1.0297	1.1421	0.9708
.750	0.0309	1.1269	1.0255	1.1269	0.9734
.760	0.0242	1.1120	1.0217	1.1120	0.9761
.770	0.0185	1.0976	1.0181	1.0976	0.9788
.780	0.0136	1.0835	1.0149	1.0835	0.9815
.790	0.0095	1.0698	1.0119	1.0698	0.9842
.800	0.0063	1.0564	1.0092	1.0564	0.9870
.810	0.0037	1.0434	1.0067	1.0434	0.9898
.820	0.0019	1.0307	1.0045	1.0307	0.9927
.830	0.0007	1.0183	1.0025	1.0183	0.9956
.840	0.0001	1.0061	1.0008	1.0061	0.9985
.850	0.0001	0.9943	0.9993	0.9943	1.0014

In isothermal flow the flow chokes at a Mach number of 0.8451542.

The Friction Factor

APPENDIX I

The head loss due to the flow of a fluid at an average velocity V through a length L of pipe with a diameter D is

$$h_f = f_{\text{D-W}} \frac{L}{D} \frac{V^2}{2g_c} \qquad \text{(Darcy-Weisbach)} \tag{I-1}$$

or

$$h_f = 4f_{\text{F}} \frac{L}{D} \frac{V^2}{2g_c} \qquad \text{(Fanning),} \tag{I-2}$$

where $f_{\text{D-W}}$ is the Darcy-Weisbach friction factor and f_{F} is the Fanning friction factor. Equating Eqs. I-1 and I-2, we have

$$4f_{\text{F}} = f_{\text{D-W}}. \tag{I-3}$$

The Fanning friction factor is more commonly used in gas dynamic applications than in simple pipe flows.

Basically, each definition implies the following: (1) head loss is directly proportional to pipe length L, (2) head loss is inversely proportional to pipe diameter D, and (3) head loss is proportional to the square of the average fluid velocity V. Dimensional analysis in conjunction with the preceding observations suggests that

$$f = f\left(\text{Re}_D, \frac{\epsilon}{D}\right) \tag{I-4}$$

where Re_D is the Reynolds number based on the pipe diameter and ϵ/D is the relative roughness of the pipe. The graphical representation of the function $f(\text{Re}_D, \epsilon/D)$ is called a *Moody diagram* and is presented in Figure I-1. The relative roughnesses for various materials and pipe diameters are presented in Table I-1.

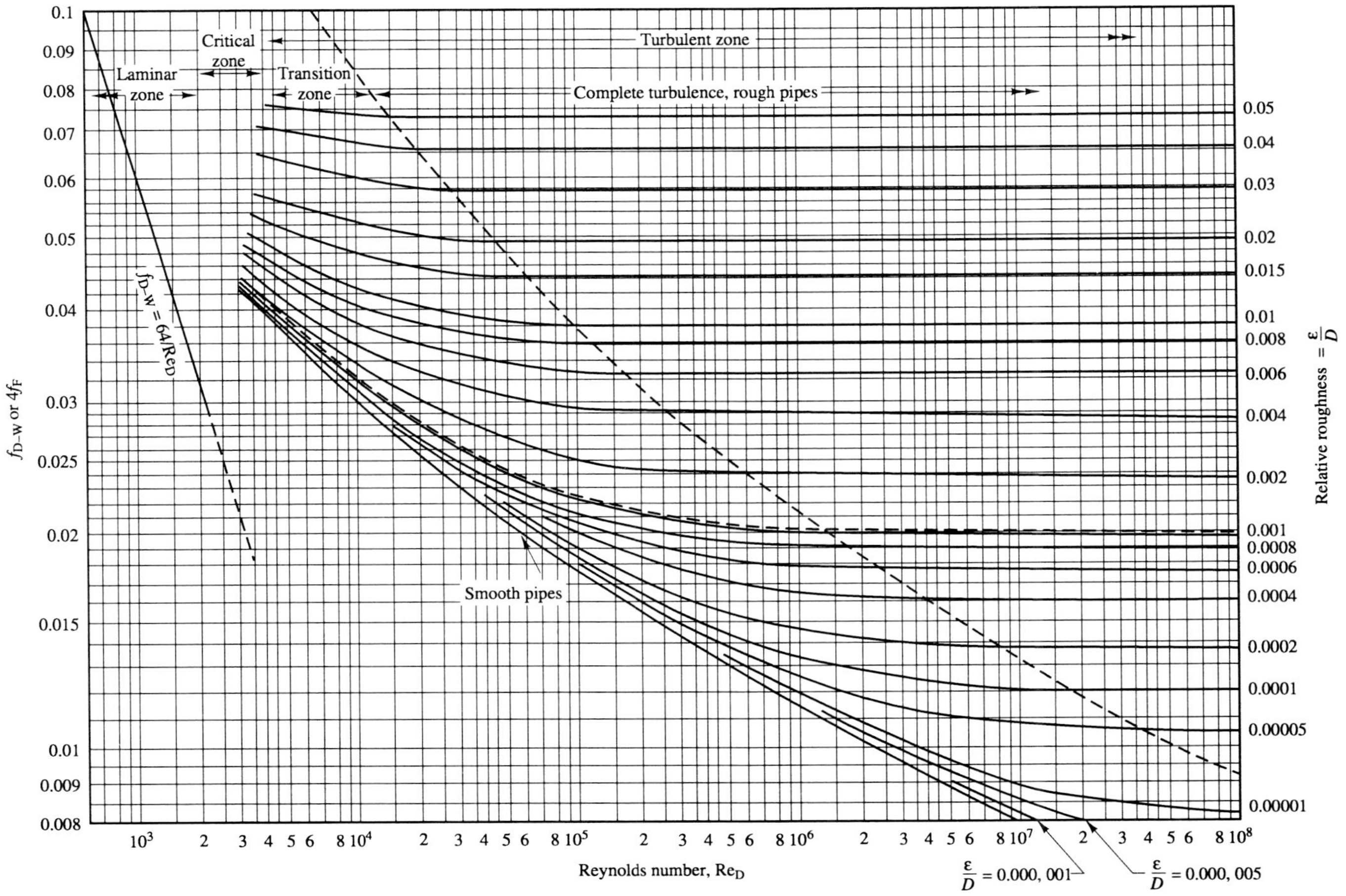

Figure I-1 The Moody Diagram

TABLE I-1 ABSOLUTE ROUGHNESS OF VARIOUS MATERIALS

Material	Absolute roughness (ft)
Drawn tubing	0.000005
Wrought iron	0.00015
Commercial steel	0.00015
Asphalted cast iron	0.0004
Galvanized iron	0.00085
Cast iron	0.00085
Concrete	0.001–0.01
Riveted steel	0.003–0.03

As is evident from Figure I-1, the function $f(\mathrm{Re}_D, \epsilon/D)$ is complex. Generally, we classify a pipe flow as either laminar or turbulent, and basically the Moody diagram is divided into a laminar region and a turbulent region. The laminar friction factor is a single straight line on the log–log plot and is not influenced by the relative roughness. Reynolds numbers in the range from 2000 to 4500 lie in a critical region where the flow can be either laminar or turbulent.

For Reynolds numbers larger than those in the critical region, turbulent flow exists in the pipe. The nature of the turbulent flow depends on the interaction between the wall roughness and the layer of boundary-layer fluid adjacent to the wall. The turbulent boundary layer is often viewed as being composed of three regions: (1) a region adjacent to the wall where the turbulent fluctuations are suppressed by the wall and viscous (molecular) shear dominates, (2) an outer region in which turbulent shear dominates, and (3) an overlap region that bridges the gap between the inner and outer regions. The layer adjacent to the wall (the inner layer) is called the viscous sublayer and has some attributes of a laminar flow because of the wall-suppressed turbulent fluctuations. The regions of transition and complete turbulence into which the turbulent zone is divided categorize the state of the viscous sublayer as influenced by roughness.

The complete turbulence region refers to a rough-wall turbulent flow in which the roughness is of sufficient magnitude to completely destroy the viscous sublayer. All attributes of laminar flow are lost, and the friction factor becomes independent of the Reynolds number and depends only on the relative roughness.

The transition zone refers not to the usual region between a laminar and a turbulent flow, but to the status of the viscous sublayer in a turbulent flow. In the transition zone the friction factor depends on both relative roughness and the Reynolds number. An absolutely smooth pipe in turbulent flow always possesses a boundary layer with a viscous sublayer and is always in the transition zone.

Should we desire to generate a computer-based solution, the translation of the Moody diagram into tabular form for use in interpolation is awkward to say the least.

What is needed is a simple algebraic expression in the form $f(\text{Re}_D, \epsilon/D)$. Historically, the implicit expression of *Colebrook* [White (1991)]:

$$\frac{1}{\sqrt{f_{\text{D-W}}}} = \log\left(\frac{\epsilon}{3.7D} - \frac{2.51}{\text{Re}_D\sqrt{f}}\right)^{-2} \tag{I-5}$$

has been accepted as the most accurate functional representation of the Moody diagram in the turbulent zone. Since Eq. I-5 is implicit in the friction factor f, iteration is required to obtain the friction factor for a specified Re_D and ϵ/D. Benedict (1980) and Haaland (1983) examine and assess a number of explicit functional representations for the Moody diagram (or the Colebrook equation). Benedict suggests the expression proposed by Swamee and Jain, that is,

$$f_F = \frac{0.25/4}{\left[\log\left(\frac{\epsilon}{3.7D} + \frac{5.74}{\text{Re}_D^{0.9}}\right)\right]^2}, \tag{I-6}$$

while Haaland recommends

$$f_{\text{F}} = \frac{0.3086/4}{\left\{\log\left[\frac{6.9}{\text{Re}_D} + \left(\frac{\epsilon}{3.7D}\right)^{1.11}\right]\right\}^2} \tag{I-7a}$$

for $\epsilon/D > 10^{-4}$. For situations where ϵ/D is very small, as in natural gas pipelines, Haaland proposes

$$f_{\text{F}} = \frac{0.3086n^2/4}{\left\{\log\left[\left(\frac{7.7}{\text{Re}_D}\right)^n + \left(\frac{\epsilon}{3.7D}\right)^{1.11n}\right]\right\}^2}, \tag{I-7b}$$

where $n \sim 3$. Overall, Eqs. I-7 are preferred. Equation I-6 has less than ± 1.5 percent error when compared with the Colebrook equation for $4 \times 10^3 \leq \text{Re}_D \leq 10^8$ and $0 \leq \epsilon/D \leq 5 \times 10^{-2}$.

The Swamee-Jain or Haaland equation provides an explicit formula for the friction factor in turbulent flow. In laminar flow, the usual Fanning friction factor representation is

$$f_{\text{F}} = \frac{16}{\text{Re}_D}. \tag{I-8}$$

The Moody diagram of Figure I-1 offers little help in determining at what Reynolds number *transition* from laminar to turbulent flow occurs. The usual approach in piping computations is to specify a Reynolds number limit for laminar flow and to declare the flow to be turbulent for all Reynolds numbers above that limit. Although the fluid mechanics literature shows some variation in the specified laminar Reynolds number limit, a value of 2,300 is the most referenced. Depending on the physical situation encountered, the flow can be turbulent for a Reynolds number smaller than 2,300 or laminar for a Reynolds number larger than 2,300: but the expectation is that a limit of 2,300 is reasonable for many situations.

An alternative to using Eq. I-6 or I-7 and a laminar Reynolds number limit for turbulent or laminar friction-factor computations is provided by Churchill (1977). Churchill devised a single expression that represents the friction factor for the laminar, transition, and turbulent flow regimes. This expression, which is explicit for the friction factor given the Reynolds number and relative roughness, is

$$f_F = 2\left[\left(\frac{8}{\mathrm{Re}_D}\right)^{12} - \frac{1}{(A + B)^{1.5}}\right]^{1/12}, \tag{I-9a}$$

where

$$A = \left\{2.457 \ln\left[\frac{1}{(7\ \mathrm{Re}_D)^{0.9} - (0.27\epsilon/D)}\right]\right\}^{16} \tag{I-9b}$$

and

$$B = \left(\frac{37{,}530}{\mathrm{Re}_D}\right)^{16}. \tag{1-9c}$$

This equation has the added advantage that the friction factor for the transition region between laminar and turbulent flow is a continuous function of the Reynolds number and smoothly bridges the gap between the laminar and turbulent regimes. The Reynolds number limit on laminar flow is implicit in Churchill's representation.

In our discussion so far we have been concerned only with circular pipes, but for a variety of reasons conduit cross sections often deviate from the circular. The appropriate characteristic length to use in evaluating the Reynolds number for noncircular cross-sectional areas is the *hydraulic diameter*. The hydraulic diameter is defined as

$$D_H \equiv 4\,\frac{\text{cross-sectional area}}{\text{wetted perimeter}}. \tag{I-10}$$

White (1991) gives an excellent discussion of the hydraulic diameter. To use the hydraulic diameter concept, the Reynolds number is defined as

$$\mathrm{Re}_{D_H} = \frac{\rho V D_H}{\mu} \tag{I-11}$$

and is then used as Re_D in either the Moody diagram or the algebraic expressions to obtain the friction factor. For all flows the hydraulic diameter as just given is only approximate, and the approximation is much more accurate for turbulent flow than for laminar flow. Friction factors for turbulent flows evaluated using this concept generally agree to within ± 2.0 percent of the experimentally determined values: laminar flow friction factors evaluated via the concept exhibit greater errors. The use of the hydraulic diameter in evaluating Reynolds numbers for conduit friction factor determination from circular pipe data fails for eccentric annuli because the eccentricity markedly affects conduit losses and the hydraulic diameter is invariant with respect to eccentricity. The hydraulic diameter concept also has problems for conduit cross sections that have high aspect ratios or are configured such that "corner" viscous effects are dominant.

Standard Atmosphere

APPENDIX J

TABLE J-1 THE STANDARD ATMOSPHERE*

Height above sea level z, m	Pressure P, atm	Density ρ, kg/m^3	Temperature T, K	Sound of speed c, m/s	Number density of particles n, 1/m^3	Height above sea level z, ft
0	1.0000	1.2250	288.15	340.29	2.55E + 25	0
1,000	0.8870	1.1117	281.65	336.44	2.31E + 25	3,281
2,000	0.7846	1.0066	275.15	332.53	2.09E + 25	6,562
3,000	0.6920	0.9093	268.66	328.58	1.89E + 25	9,842
4,000	0.6085	0.8194	262.17	324.59	1.70E + 25	13,120
5,000	0.5334	0.7364	255.68	320.55	1.53E + 25	16,400
6,000	0.4660	0.6601	249.19	316.45	1.37E + 25	19,680
7,000	0.4057	0.5900	242.70	312.31	1.23E + 25	22,970
8,000	0.3519	0.5258	236.22	308.11	1.09E + 25	26,250
9,000	0.3040	0.4671	229.73	303.85	9.71E + 24	29,530
10,000	0.2615	0.4135	223.25	299.53	8.60E + 24	32,810
11,000	0.2240	0.3648	216.77	295.15	7.59E + 24	36,090
12,000	0.1915	0.3119	216.65	295.07	6.49E + 24	39,370
13,000	0.1636	0.2666	216.65	295.07	5.54E + 24	42,650
14,000	0.1399	0.2279	216.65	295.07	4.74E + 24	45,930
15,000	0.1195	0.1948	216.65	295.07	4.05E + 24	49,210
16,000	0.1022	0.1665	216.65	295.07	3.46E + 24	52,490
17,000	0.0873	0.1423	216.65	295.07	2.96E + 24	55,770
18,000	0.0747	0.1217	216.65	295.07	2.53E + 24	59,050
19,000	0.0638	0.1040	216.65	295.07	2.16E + 24	62,340
20,000	0.0546	0.0889	216.65	295.07	1.85E + 24	65,610
21,000	0.0467	0.0757	217.58	295.70	1.57E + 24	68,900
22,000	0.0399	0.0645	218.57	296.38	1.34E + 24	72,180
23,000	0.0342	0.0550	219.57	297.05	1.14E + 24	75,460
24,000	0.0293	0.0469	220.56	297.72	9.76E + 23	78,740
25,000	0.0252	0.0401	221.55	298.39	8.33E + 23	82,020
26,000	0.0216	0.0343	222.54	299.06	7.12E + 23	85,300
27,000	0.0186	0.0293	223.54	299.72	6.09E + 23	88,580

* Data from *The U.S. Standard Atmosphere, 1962*, U.S. Government Printing Office, Washington, 1962.

1 atm = 1.01325×10^5 n/m^2 = 14.696 lbf/in^2
1 kg/m^3 = 0.062428 lbm/ft^3
1 m = 3.2808 ft

REFERENCES

AL-HABBASH, H. (1993). A Method of Characteristics Program and Post-Processor. Master's thesis, Aerospace Engineering Dept., Mississippi State University.

ANDERSON, J. D., JR. (1976). *Gas Dynamic Lasers: An Introduction*. Academic Press, New York.

ANDERSON, J. D., JR. (1989). *Hypersonic and High Temperature Gas Dynamics*. McGraw-Hill, New York.

ANDERSON, J. D., JR. (1990). *Modern Compressible Flow with Historical Perspective*, 2nd ed. McGraw-Hill, New York.

ANDERSON, D. A., TANNEHILL, J. C., and PLETCHER, R. H. (1984). *Computational Fluid Mechanics and Heat Transfer*. McGraw-Hill, New York.

ARPACI, V. (1966). *Conduction Heat Transfer*. Addison-Wesley, Reading, MA.

BAALS, D. D., and CORLISS, W. R. (1981). Wind Tunnels of NASA. NASA SP-440.

BACK, L. H., MASSIER, P. F., and GRIER, H. L. (1965). Comparison of Measured and Predicted Flows through Conical Supersonic Nozzles, with Emphasis on the Transonic Region. *AIAA Journal*, **3**(9), pp. 1606–1614.

BEANS, E. W. (1970). Computer Solution to Generalized One-Dimensional Flow. *J. Spacecraft and Rockets*, **7**(12), pp. 1460–1464.

BENEDICT, R. P. (1980). *Fundamentals of Pipe Flow*. John Wiley, New York.

BENEDICT, R. P. (1983). *Fundamentals of Gas Dynamics*. John Wiley, New York.

BERS, L. (1958). *Mathematical Aspects of Subsonic and Transonic Gas Dynamics*. John Wiley, New York.

BROWER, W. B. (1990). *Theory, Tables, and Data for Compressible Flow*. Hemisphere Publishing Corp., New York.

BRYSON, A. E., JR. (1951). An Experimental Investigation of Transonic Flow Past Two-Dimensional Wedge and Circular-Arc Sections Using a Mach-Zehnder Interferometer. NASA TN 2560.

CHAPMAN, A. J., and WALKER, W. F. (1971). *Introductory Gas Dynamics*. Holt, Rinehart & Winston, New York.

CHAPMAN, D. R., KUEHN, D. M., and LARSON, H. K. (1957). Investigation of Separated Flows in Supersonic and Subsonic Streams with Emphasis on the Effect of Transition. NACA TN 3869.

CHERNYI, G. G. (1961). *Introduction to Hypersonic Flow*. Academic Press, New York.

CHURCHILL, S. W. (1977). Friction Factor Equation Spans All Fluid Flow Regimes. *Chemical Engineering*, 7 Nov., pp. 91–92.

CLINE, M. C. (1974). The Computation of Steady Two-Dimensional Transonic Nozzle Flow by a Time Dependent Method, *AIAA J.*, **12**(4), pp. 419–420.

CONTE, S. D., and DE BOOR, C. (1972). *Elementary Numerical Analysis: An Algorithmic Approach*. McGraw-Hill, New York.

COX, C. F. (1992). An Efficient Solver for Flows in Local Chemical Equilibrium. Ph.D. dissertation, Mechanical Engineering Dept., Mississippi State University.

COX, R. N., and CRABTREE, L. F. (1965). *Elements of Hypersonic Aerodynamics*. Academic Press, New York.

CRICKMORE, P. F. (1986). *Lockheed SR-71 Blackbird*. Osprey Publishing Co., London.

CROCCO, L. (1937). Eine neue Stromfunktion fur die Erforschung der Bewegung der Gase mit Rotation. *Z. Angew. Math. Mech.*, **17**, pp. 1–7.

CUFFEL, R. F., BACK, L. H., and MASSIER, P. F. (1969). The Transonic Flowfield in a Supersonic Nozzle with Small Throat Radius of Curvature. *AIAA Journal*, **7**(7), pp. 1364–1366.

DANESHYAR, H. (1976). *One-Dimensional Compressible Flow*. Pergamon Press, New York.

DORRANCE, W. H. (1962). *Viscous Hypersonic Flow*. McGraw-Hill, New York.

EMANUEL, G. (1986). *Gasdynamics: Theory and Applications*. AIAA, Washington.

FERRI, A. (1949). *Elements of Aerodynamics of Supersonic Flows*. MacMillan, New York.

FOWLER, R. H., and GUGGENHEIM, E. A. (1952). *Statistical Thermodynamics*. Cambridge University Press, New York.

FOX, R. W., and MCDONALD, A. T. (1992). *Introduction to Fluid Mechanics*, 4th ed. John Wiley, New York.

GALLAGHER, R. J. (1970). Investigation of a Digital Simulation of the XB-70 Inlet and Its Application to Flight-Experienced Free-Stream Disturbances at Mach Numbers of 2.4 to 2.6. NASA TN D-5827.

GRAVES, R. A. (1982). Computational Fluid Dynamics: The Coming Revolution. *Astronautics and Aeronautics*, **20**(3), pp. 20–28, 62.

HABERMAN, W. L., and JOHN, J. E. A. (1980). *Engineering Thermodynamics*. Allyn and Bacon, Boston.

HAALAND, S. E. (1983). Simple and Explicit Formulas for the Friction Factor in Turbulent Flow. *Trans. ASME, J. Fluid Eng.*, **105**(3), pp. 89–90.

HALL, I. M. (1962). Transonic Flow in Two-Dimensional and Axially-Symmetric Nozzles. *Quarterly Journal of Mechanics and Applied Mathematics*, **XV**, Part 4, pp. 487–508.

HAYES, W. D., and PROBSTEIN, R. F. (1966). *Hypersonic Flow Theory*, Academic Press, New York.

HILDEBRAND, F. B. (1962). *Advanced Calculus for Applications*. Prentice-Hall, Englewood Cliffs, NJ.

HILL, P. G., and PETERSON, C. R. (1992). *Mechanics and Thermodynamics of Propulsion*, 2nd ed. Addison-Wesley, Reading, MA.

HILSENRATH, J., et al. (1955). *Tables of Thermodynamic and Transport Properties*, NBS Circular 564; reprinted (1960) by Pergamon Press, New York.

HILSENRATH, J., and KLEIN, M. (1965). Tables of Thermodynamic Properties of Air in Chemical Equilibrium Including Second Virial Corrections from 1500 to 15,000 K. AEDC-TR-65-68.

HILTON, W. F. (1951). *High-Speed Aerodynamics*. Longmans, London.

HODGE, B. K. (1990a). *Analysis and Design of Energy Systems*, 2nd ed. Prentice-Hall, Englewood Cliffs, NJ.

HODGE, B. K. (1990b). Generalized One-Dimensional Compressible Flow Matrix Inverse. *J. Spacecraft and Rockets*, **27**(4), pp. 446–447.

HODGE, B. K., and YOUNG, F. M. (1993). Mass Addition Processes in One-Dimensional Compressible Flow. AIAA Paper 93-0487.

HODGSON, J. W., and LAMAS, J. A. (1990). A Simple Basic Algorithm for Determining the Equilibrium Composition of Gaseous C/H/O/N Systems Involving Ten Species. Proc. 1990 ASEE Southeastern Section Meeting. ed. J. M. Biedenbach.

HOWELL, J. R., and BUCKIUS, R. O. (1992). *Fundamentals of Engineering Thermodynamics*, 2nd ed. McGraw-Hill, New York.

HUDGINS, H. E. (1964). Supersonic Flow about Right-Circular Cones at Zero Yaw in Air at Thermodynamic Equilibrium. Parts II and III. Tables of Data. TM 1493, Picatinny Arsenal.

HUDGINS, H. E. (1965). Supersonic Flow about Right-Circular Cones at Zero Yaw in Air at Chemical Equilibrium. Part I. Correlation of Flow Properties. TM 1493, Picatinny Arsenal.

INCROPERA, F. P., and DEWITT, D. P. (1990). *Introduction to Heat Transfer*, 2nd ed. John Wiley, New York.

JEANS, J. H. (1940). *The Dynamical Theory of Gases*, 4th ed. Cambridge University Press, New York; reprinted (1960) by Dover Publications, New York.

JOHANNSEN, N. H. (1952). Experiments on Two-Dimensional Supersonic Flow in Corners and over Concave Surfaces. *Philosophical Magazine*, **43**, pp. 568–580.

JOHN, J. E. A. (1984). *Gas Dynamics*, 2nd ed. Allyn and Bacon, Boston.

KENNARD, E. H. (1938). *Kinetic Theory of Gases*. McGraw-Hill, New York.

KLIEGEL, J. R., and LEVINE, J. N. (1969). Transonic Flow in Small Radius of Curvature Nozzles. *AIAA Journal*, **7**(7), pp. 1375–1378.

KUO, K. K. (1986). *Principles of Combustion*. Wiley, New York.

LEE, J. F., SEARS, F. W., and TURCOTTE, D. L. (1973). *Statistical Thermodynamics*, 2nd ed. Addison-Wesley, Reading, MA.

LIEPMANN, H. W., and ROSHKO, A. (1957). *Elements of Gasdynamics*. John Wiley & Sons, New York.

LIU, Y., and VINOKUR, M. (1989). Equilibrium Gas Flow Computations. I. Accurate and Efficient Calculation of Equilibrium Gas Properties. AIAA Paper 89-1736.

MCBRIDE, B. J., HEIMEL, S., EHLERS, J. G., and GORDON, S. (1963). Thermodynamic Properties to 6000 K for 210 Substances Involving the First 18 Elements. NASA SP-3001.

MARGENAU, H., and MURPHY, G. M. (1956). *The Mathematics of Physics and Chemistry*, 2nd ed. Van Nostrand, New York.

MARRONE, P. V. (1962). Normal Shock Waves in Air: Equilibrium Composition and Flow

Parameters for Velocities from 26,000 to 50,000 ft/s. Cornell Aeronautical Laboratory Report AG-1729-A-2.

MARTENSEN, M. C. (1990). Design and Analysis of a Parallel Flow Nozzle. Master's thesis, Aerospace Engineering Dept., Mississippi State University.

MARTIN, R. A. (1970). Dynamic Analysis of XB-70 Inlet Pressure Fluctuations During Takeoff and Prior to a Compressor Stall at Mach 2.5. NASA TN D-5826.

MASCITTI, V. R. (1969). A Closed-Form Solution to Oblique Shock-Wave Properties. *J. Spacecraft and Rockets*, **6**(1), p. 66.

MAUS, J. R., GRIFFITH, B. J., SZEMA, K. Y., and BEST, J. T. (1984). Hypersonic Real Mach Number and Real Gas Effects on Space Shuttle Orbiter Aerodynamics. *J. Spacecraft and Rockets*, **21**(2), pp. 136–141. Also see Griffith, B. J., Maus, J. R., and Best, J. T. (1983). Explanation of the Hypersonic Longitudinal Stability Problem—Lesson Learned. *Shuttle Performance: Lessons Learned*, NASA CP-2283, ed. J. P. Arrington and J. J. Jones, pp. 347–379.

MILES, E. R. (1950). *Supersonic Aerodynamics*. Dover, New York.

MILNE-THOMSON, L. M. (1968). *Theoretical Hydrodynamics*, 5th ed. MacMillan, New York.

MOECKEL, W. E. (1957). Oblique-Shock Relations at Hypersonic Speeds for Air in Chemical Equilibrium. NACA TN-3985.

MOODY, L. F. (1944). Friction Factors for Pipe Flow. *Trans. ASME*, **66**, pp. 671–684.

NACA. (1953). Equations, Tables, and Charts for Compressible Flow. NACA Report 1135.

NICKERSON, G. R., COATS, D. E., DANG, A. L., DUNN, S. S., and KEHTARNAVAZ, H. (1989). Two-Dimensional Kinetics (TDK) Nozzle Performance Computer Program. Software and Engineering Associates, Inc., Carson City, NV.

OWATITSCH, K. (1956). *Gas Dynamics*. Academic Press, New York.

OWCZAREK, K. (1964). *Fundamentals of Gas Dynamics*, International, Scranton, Pa.

PAI, S. (1959). *Introduction to the Theory of Compressible Flow*. Van Nostrand, Princeton, NJ.

PAI, S., and LUO, S. (1991). *Theoretical and Computational Dynamics of a Compressible Flow*. Van Nostrand Reinhold, New York.

PALMER, G. (1987). An Implicit Flux-Split Algorithm to Calculate Hypersonic Flowfields in Chemical Equilibrium. AIAA Paper 87-1580.

PARK, C. (1990). *Nonequilibrium Hypersonic Aerothermodynamics*. Wiley, New York.

PEIRCE, B. O. (1929). *A Short Table of Integrals*. Ginn and Co., Boston.

PIPES, L. A., and HARVILL, L. R. (1970). *Applied Mathematics for Engineers and Physicists*, 3rd ed. McGraw-Hill, New York.

POPE, A. (1958). *Aerodynamics of Supersonic Flight*, 2nd ed. Pitman, New York.

PRESENT, R. D. (1958). *Kinetic Theory of Gases*. McGraw-Hill, New York.

RAKICH, J. V. (1969). A Method of Characteristics for Steady Three-Dimensional Supersonic Flow with Application to Inclined Bodies of Revolution. NASA TN D-5341.

RAKICH, J. V., and CLEARY, J. W. (1970). Theoretical and Experimental Study of Supersonic Steady Flow around Inclined Bodies of Revolution. *AIAA Journal*, **8**(3), pp. 511–518.

ROBERTSON, J. M. (1965). *Hydrodynamics in Theory and Applications*. Prentice-Hall, Englewood Cliffs, NJ.

ROMERE, P. O., and WHITNAH, A. M. (1983). Space Shuttle Entry Longitudinal Aerodynamic Comparisons with Flights 1-4 with Preflight Predictions. *Shuttle Performance: Lessons Learned.* NASA CP-2283, ed. J. P. Arrington and J. J. Jones, pp. 283–307.

ROMIG, M. F. (1960). Conical Flow Parameters for Air in Dissociating Equilibrium. Convair Scientific Research Laboratory Research Report No. 7.

SAAD, M. A. (1992). *Compressible Fluid Flow*, 2nd ed. Prentice-Hall, Englewood Cliffs, NJ.

SAUER, R. (1947a). General Characteristics of the Flow through Nozzles at Near Critical Speeds. NACA TM 1147.

SAUER, R. (1947b). *Introduction to Theoretical Gas Dynamics.* Edwards, Ann Arbor, MI.

SAUERWEIN, J. C., and DALTON, G. R. (1985). *Standard Reference Publications.* NBS Special Publication 708.

SCHAEFER, W. T. (1965). Characteristics of Major Active Wind Tunnels at the Langley Research Center. NASA TM X-1130.

SCHLICHTING, H. (1979). *Boundary Layer Theory*, 7th ed. McGraw-Hill, New York.

SCHREIER, S. (1982). *Compressible Flow.* Wiley Interscience, New York.

SEDDON, J., and GOLDSMITH, E. L. (1985). *Intake Aerodynamics.* AIAA, Washington.

SHAMES, I. H. (1992). *Mechanics of Fluids*, 3rd ed. McGraw-Hill, New York.

SHAPIRO, A. H. (1953). *The Dynamics and Thermodynamics of Compressible Fluid Flow.* Ronald Press, New York.

SHERMAN, F. S. (1990). *Viscous Flow.* McGraw-Hill, New York.

STOECKER, W. F. (1989). *Design of Thermal Systems*, 3rd ed. McGraw-Hill, New York.

STREHLOW, R. A. (1968). *Fundamentals of Combustion.* International Textbook Co., Scranton, PA.

STULL, D. R. (1971). JANAF Thermodynamic Tables. NBS NSRDS-NBS 37.

SUTTON, G. P. (1992). *Rocket Propulsion Elements*, 6th ed. John Wiley, New York.

TANNEHILL, J. C., and MUGGE, P. H. (1974). Improved Curve Fits for the Thermodynamic Properties of Equilibrium Air Suitable for Numerical Computation Using Time-Dependent or Shock-Capturing Methods. NASA CR-2470.

TAYLOR, G. I., and MACCOLL, J. W. (1933). The Air Pressure on a Cone Moving at High Speeds. *Proc. Royal Society (London)*, Ser. A, **139**, pp. 278–311.

THOMPSON, P. A. (1972). *Compressible-Fluid Dynamics.* McGraw-Hill, New York.

TRUITT, R. W. (1959). *Hypersonic Aerodynamics.* Ronald Press, New York.

—. (1976). *U.S. Standard Atmosphere.* 1976, NOAA-S/T 76-1562.

—. (1990). *User Manual DERIVE Version 2.* Soft Warehouse, Honolulu, HI.

VON MISES, R. (1958). *Mathematical Theory of Compressible Flow.* Academic Press, New York.

VINCENTI, W. G., and KRUGER, C. H. (1965). *Introduction to Physical Gas Dynamics.* John Wiley, New York.

WARK, K. (1983). *Thermodynamics.* McGraw-Hill, New York.

WARSI, Z. U. A. (1993). *Fluid Dynamics Theoretical and Computational Approaches.* CRC Press, Boca Raton, FL.

WHITE, F. M. (1986). *Fluid Mechanics*, 2nd ed. McGraw-Hill, New York.

WHITE, F. M. (1991). *Viscous Fluid Flow*, 2nd ed. McGraw-Hill, New York.

WILSON, D. G. (1984). *The Design of High-Efficiency Turbomachinery and Gas Turbines*. MIT Press, Cambridge, MA.

WITTLIFF, C. E., and CURTISS, J. T. (1961). Normal Shock Wave Parameters in Equilibrium Air. Cornell Aeronautical Laboratory Report CAL-111.

WOODS, W. C., ARRINGTON, J. P., and HAMILTON, H. H. (1983). A Review of Real Gas Effects on Space Shuttle Aerodynamic Characteristics. *Shuttle Performance: Lessons Learned*. NASA CP-2283, ed. J. P. Arrington and J. J. Jones, pp. 309–346.

YOUNG, F. M. (1993). Generalized One-Dimensional, Steady, Compressible Flow. *AIAA J.*, **31**(1), pp. 204–208.

ZEL'DOVICH, Y. B., and RAIZER, Y. P. (1968). *Elements of Gasdynamics and the Classical Theory of Shock Waves*. Academic Press, New York.

ZEMANSKY, M. W. (1957). *Heat and Thermodynamics*, 4th ed. McGraw-Hill, New York.

ZUCROW, M. J., and HOFFMAN, J. D. (1976). *Gas Dynamics*. John Wiley, New York; Vol. 2 reprinted (1985) Krieger Publishing Co., Melbourne, FL.

Index

DATE DUE
